THERAPEUTIC MEDICINAL PLANTS
From Lab to the Market

THERAPEUTIC MEDICINAL PLANTS

From Lab to the Market

Editors

Marta Cristina Teixeira Duarte
CPQBA/UNICAMP – State University of Campinas
Paulinía, São Paulo
Brazil

Mahendra Rai
Department of Biotechnology
S.G.B. Amravati University
Amravati, Maharashtra
India

CRC Press
Taylor & Francis Group
Boca Raton London New York

CRC Press is an imprint of the
Taylor & Francis Group, an **informa** business

A SCIENCE PUBLISHERS BOOK

Cover illustrations reproduced by kind courtesy of Rodney Alexandre Ferreira Rodrigues and Alexandre Nunes Ponezi.

CRC Press
Taylor & Francis Group
6000 Broken Sound Parkway NW, Suite 300
Boca Raton, FL 33487-2742

First issued in paperback 2020

© 2016 by Taylor & Francis Group, LLC
CRC Press is an imprint of Taylor & Francis Group, an Informa business

No claim to original U.S. Government works

ISBN-13: 978-1-4822-5403-7 (hbk)
ISBN-13: 978-0-367-73761-0 (pbk)

Library of Congress Cataloging-in-Publication Data

Therapeutic medicinal plants from lab to the market / editors: Marta Cristina Teixeira Duarte, Mahendra Rai.
   pages cm
  Includes bibliographical references and index.
   ISBN 978-1-4822-5403-7 (hardcover : alk. paper) 1. Materia medica, Vegetable. 2. Medicinal plants. 3. Botanical drug industry. I. Duarte, Marta Cristina Teixeira, editor. II. Rai, Mahendra, editor.

RS164.T52 2016
615.3'21--dc23

   2015028938

Visit the Taylor & Francis Web site at
http://www.taylorandfrancis.com

and the CRC Press Web site at
http://www.crcpress.com

Foreword

II

Traditional or ethno-medicine represents a series of empirical practices included in the knowledge of a social group, frequently transmitted orally from one generation to another, with the intention of solving health problems. It is often strongly connected to religious beliefs and practices of the respective culture. The knowledge of medicinal or herbal plants is an important component of traditional medicine. Traditional medicine is used globally and has rapidly growing economic importance. In developing countries, traditional medicine is often the only accessible and affordable treatment available. In many Asian countries traditional medicine is widely used, even though western medicine is often readily available. In our time of constantly expanding pharmacological research it is easy to forget that medicinal plants still continue to play a leading role in primary health care for 80% of the world's population living in developing countries. Natural products and medicinal products derived from them also constitute an essential element in the system of health care for the remaining 20% of the world population.

In Japan, 60–70% of allopathic doctors prescribe traditional medicines for their patients. In the US the number of visits to providers of Complementary Alternative Medicine (CAM) now exceeds by far the number of visits to all primary care physicians. The expense for the use of Traditional and Complementary Alternative Medicine is exponentially growing in many parts of the world. Traditional knowledge has proven to be an important source for therapeutic drugs.

The global inventory of plant diversity consists currently of about 350,000 species, and most current estimates expect about 420,000 plant species to exist. This tremendous diversity accounts for a wide range of phytochemicals, and a high variation of compound composition even within one single species, depending on growth conditions (soil, climate, nutrient status, etc.), and harvest practices and timing, not even taking intraspecific variation into account. While traditional plant use and medicine preparation normally take these details into account, they are often seen as of marginal importance in the herbal trade. In the United States, botanical supplements are supposed to be labeled, with the requirement to include the correct scientific name. However, in practice this does not prevent accidental or deliberate adulterations, or can contain heavy-metal contaminations. The most problematic occurrence in herbal medicine trade is, however, linked to the purchase and use, either in medication or research, of botanicals that are either accidentally or purposefully wrongly identified, or are simply collected under a vernacular name without any subsequent taxonomic treatment, and often without having any vouchered material that could later be used for the verification of plant identity. A much more frequent occurrence is, however, the often deliberate adulteration of botanicals with more common and cheaper species, which, although generally not toxic, might completely lack efficacy. Bulk herbs are readily available unprocessed, which allows for the retention of material for a botanical voucher. In contrast, raw botanicals are also often provided in ground or powdered form, which makes morphological identification very difficult or virtually impossible. In addition, little research has been done on the efficacy of traditionally made medicinal preparations, because most efforts have focused on the elucidation of lead compounds and subsequent clinical trials, with little regard to the correct harvesting or cultivation, and botanical identification of the source material. One example of the problems of plant collections and markets was illustrated by recent studies in the markets of La Paz, Bolivia, that focused on plants used for the treatment of urinary infections. One of the most frequently mentioned herbs was 'cola de caballo', horsetail, which normally signifies species of *Equisetum*. However, every single vendor in La Paz instead sold it as a species of *Ephedra*. Not only does *Ephedra* not have any properties related to treating urinary infections or inflammations, but also its main compounds can lead to serious side effects. Without the collection of botanical vouchers, this serious health risk would not have been discovered. This clearly illustrates the great need of more studies that in fact follow therapeutic medicinal plants from the source to the laboratory.

The present volume is indeed a very worthy effort to outline and address these issues on a global scale. M.C. Texeira and M. Rai have done great job in bringing a broad field of accomplished contributors together, who in 19 chapters, illuminate all aspects of medicinal plant use from local collectors and markets to clinical trials in modern drug development.

Dominguez et al. provide an up to date view on traditional medicine in Mexico, and its translation into a source for the medicines of tomorrow, while Noun Jihad et al. address the possibilities and limitations of such an approach by using Lebanon as case study. In their contribution on Cuba, Escalona et al. highlight the great efforts of this country to create an independent, high-class medicinal system that offers patients the best of both traditional therapies and their applications in an allopathic setting. The chapter by Bussmann and Sharon follows a similar trajectory in illustrating the long way from traditional plant collectors to using medicinal plant extracts in bioassays.

A second set of contributions provides direct insights in applied research. Muñoz-Acevedo et al. review species from Latin American that could be promising for the cosmetics industry, while Ilhan et al. look at new remedies for hyperlipidemia, and Rojas and Buitrago as well as Teixeira Duarte and Teixeira Duarte address the possibilities of finding new antibacterial agents derived from essential oils. This approach culminates in a plethora of chapters focusing on individual species and the long way from traditional use to lead compounds in drug discovery: Vinet et al. (*Vitis vinifera* polyphenols), Al-Nahain et al. (*Centella asiatica*), Ríos and Andújar (*Crocus sativus*), Santana et al. (*Euphorbia hirta* and *E. hyssopifolia*), Patil and Lade (*Tribulus terrestris*), Rai et al. (*Vitex negundo*) and Ortega Hernández-Agero (*Melissa officinalis*).

In a last group of contribution the focus lies on the problems of production, quality control, toxicity and efficacy testing of herbal medicines. Melillo de Magalhães addresses the challenges in plant cultivation as first step to provide standardized source material for drug development, while Araujo focuses on the issue of toxicity in traditionally used medicinal plants. Rodrigues et al. take up the discussion on how to actually improve the properties and quality of plant extracts, and Mootoosamy and Mahomoodally finally review the current status of clinical trials of medicinal herbs, and highlight chances and challenges of future development.

It is to hope that 'Therapeutic Medicinal Plants: From lab to the market' will be widely read, and will become a standard reference for researchers in the whole chain of traditional medicines, from documentation of traditional medicinal practices, to plant harvest, production and markets, extraction, clinical studies and finally the elaboration of standardized herbal medicines for a global environment.

Professor Rainer W. Bussmann
William L. Brown Center
Missouri Botanical Garden, P.O. Box 299
St. Louis, MO 63166-0299, USA
phone: (314) 577-9503, facsimile: (314) 577-0800
Email: rainer.bussmann@mobot.org

Preface

||||||||||||||||||||||||||||||||||||||||||||||||||||||||||||||||||||||||||||||||||||||||||||||||||||||||||||||||||||||||||||||||||||||||||||||||||||||||||||||||||||||||||||||||

Medicinal plants are known as a natural resource for the cure of different diseases since the dawn of civilization. They have been used in the prevention, diagnosis and elimination of diseases, and solely based on practical experience of thousands of years.

There are various reports of medicinal/ethnomedicinal plants by researchers all over the world. Unfortunately, the research being carried out in laboratories are still restricted to the 'four walls' of the laboratories. There is a greater need to initiate and transform these researches in fruitful formulations leading to the development of newer products for the cure of diseases with special reference to new and emerging diseases like AIDS, cancer, hepatitis and also for coping with multidrug resistance problems.

In 21st century, there is a greater need to validate the available knowledge of medicinal plants. World Health Organization (WHO) emphasized on use of herbal medicines after its validation. The next steps are formulation and finally the development of medicinal products.

The purpose behind editing a new book, is to gather recent developments in medicinal plant research for different diseases, formulation of products and market strategy.

The book would be immensely useful for botanists, medicos, ayurvedic experts, traditional healers, pharmacologists and common people who are interested in curative properties of medicinal plants.

MKR wishes to thank Dr. D.P. Rathod, and Dr. Shubhangi Ingole for their help in editing this work.

Marta Cristina Teixeira Duarte
Mahendra Rai

Contents

Mexican Traditional Medicine: Traditions of Yesterday and Phytomedicines of Tomorrow

Fabiola Domínguez,[1] Angel Josabad Alonso-Castro,[2]
Maricruz Anaya,[1] Ma. Eva González-Trujano,[3]
Hermelinda Salgado-Ceballos[4] and Sandra Orozco-Suárez[4,*]

Introduction

The arrival of the Spaniards radically modified the native medicine practices of the Aztecs and the use of medicinal plants. Diverse colonial documents, such as those of Martín de la Cruz, Juan Badiano, Bernardino de Sahagún and Francisco Hernandez, provide examples of the use of medicinal plants from the viewpoint of the Aztecs in works such as *Libellus de Medicinalibus Indorum Herbis* (Little Book of the Medicinal Herbs of the Indians). Additional works describe the actions of Mexican medicinal plants and suggest their usefulness, such as *Historia de las cosas dela Nueva España* (General History of the Things of the New Spain) by Fray Bernardino de Sahagún (Viesca 1992, Viesca 1996).

Libellus de Medicinalibus is a manuscript completed in Mexico in 1552 which refers to native medicinal plants, the curative effects of which were indicated in Latin texts. This medicinal '*herbarium*' was written with the intention of showing the King of Spain the rich variety of medicinal plants that grew in the New World. An elderly Indian physician and a native of Tlatelolco, Martín de la Cruz, described the use of each of plant and provided drawings of them, and Juan Badiano (a young Indian) translated the material from Nahuatl into Latin. This book was housed in the archives of the Vatican Library in Rome and was rediscovered in the 20th century by U.S. historians, who confused it with a pre-Hispanic Aztec codex and renamed it the *Cruz-Badiano Codex* or the *Badiano Codex* (Garibay 1964, De la Cruz 1964). This codex is currently found at the National Museum of Anthropology in Mexico City.

[1] Centro de Investigación Biomédica de Oriente, IMSS.
[2] Departamento de Farmacia, División de Ciencias Naturales y Exactas, Universidad de Guanajuato.
[3] Dirección de Investigaciones en Neurociencias, Instituto Nacional de Psiquiatría Ramón de la Fuente Muñiz.
[4] Unidad de Investigación Médica en Enfermedades Neurológicas, Hospital de Especialidades Centro Médico Nacional Siglo XXI, IMSS. Av. Cuauhtémoc 330 Col. Doctores, 06720, México DF. México.
* Corresponding author: sorozco5@hotmail.com

In modern times, formal medical research of the codex and other texts began during the Porfirian era, in the last two decades of the 19th century. A major progression in formal medical research occurred in 1888, when the National Medical Institute in Mexico was created by the order of President Porfirio Díaz. The objective of the Institute was to conduct studies of the Mexican medicinal flora with the goal of incorporating medicinal plants into therapeutics at the national level. By 1915, the herbarium possessed 14,000 classified species and approximately 1,000 chemical compounds that were obtained from plants (García 1981). However, the modern era of interest in the chemistry of natural products surged in Mexico from 1940–1960 during the boom of steroidal sapogenins from inedible Mexican yams, which were used as a source of progesterone (Gereffi 1978). Indeed, this became the cornerstone of the Syntex Company that was founded in Mexico. Syntex initiated a true worldwide revolution in the organic synthesis of steroidal hormones; it was the first to achieve the synthesis of progesterone and cortisone. Additionally, the Syntex Company provided the basis for the first contraceptive, which was derived from the chemical and ethnobotanical studies of Russell Earl Marker concerning the chemical diosgenin that was obtained from the 'cabeza de negro' (black head) plant (*Dioscorea mexicana*) and later from the 'barbasco' (*Dioscorea composita*) plant. The barbasco plant is an endemic species of Mexico (Soto 2005). In 1975, the Mexican Institute for the Study of Medicinal Plants (IMEPLAM) was created. This institute was founded for the multidisciplinary study of the plants most widely used in Mexico to treat common illnesses. The Institute included historians, agronomical engineers, botanists, physicians, physiologists, chemists, and pharmacists that were under the direction of Dr. Xavier Lozoya. During the existence of IMEPLAM (1975–1980), numerous publications were produced, thus establishing the Institute as an icon in the research of medicinal plants in Latin America and reactivating this type of research (Lozoya 1976). At the same time, the Institute initiated the formation of the Medicinal Herbolarium, currently known as the Medicinal Herbolarium of the Mexican Institute of Social Security (IMSS), which is located at the Twenty-First Century National Medical Center in Mexico City. Its legacy comprises >120,000 specimens (Zolla 1980, Montes and Montes 2005). In 1980, the IMEPLAM became part of the IMSS Medical Research System.

The IMEPLAM embodied the tendencies of the 1960s with respect to the increased research interest in natural plant drugs, including: a) the *Indian Rauwolfia* drugs and their derivatives, which were demonstrated to successfully treat various mental disorders and other diseases; and b) the plant-derived drugs that induce psychotic symptoms akin to symptoms of mental illnesses. Indeed, interest in such psychosis-inducing drugs, which were traditionally used by healers and medicine men of 'primitive' cultures, resulted in an extensive search for substances possessing hallucinogenic properties (Viesca et al. 2000).

At the end of the 20th century, a second stage of studies on medicinal plants and their therapeutic potential ushered in a new era of phytopharmaceuticals. Currently, new mechanism actions are being proposed to explain the pharmacological effects of these plant extracts on several activities and functions.

In recent years, there has been a rapid increase in global technological and economic potential that has resulted in an increased ability to overcome problems related to poverty and poor health. However, many developing countries have an impaired health status due to the resurgence of infectious diseases and an increasing burden of noncommunicable diseases.

There are numerous methods to improve results in the healthcare sector including prevention, healing with an existing treatment, and research into better methods of prevention, diagnosis or treatment. Research results obtained for the prevention, diagnosis and treatment of diseases are reported to apply the gained knowledge in solving health problems. Generally, research results are disclosed in articles published in biomedical journals and/or theses.

However, commercially sponsored research results for drug, diagnostic and treatment techniques must be intellectually protected to obtain a health registry for proper marketing. One or more patents are used for the intellectual protection of research results. A patent has two important functions: protection, which allows the patent owner to exclude others from commercially exploiting the invention for a period of 20 years; and disclosure, because the patent provides information that can stimulate technological innovation.

In Mexico, the Mexican Institute of Industrial Property grants government patents. This type of intellectual protection is regulated by the Industrial Property Act and its rules. There have been no objections to granting phytomedicine patents; however, the invention must meet three basic criteria for patentability: novelty, inventive step, and industrial application.

The novelty criterion specifies that the invention is not derived from technical knowledge available to the public worldwide prior to the first filing date of the relevant patent application. The inventive criterion specifies that the invention would not have been obvious to a person skilled within the applicable field of technology. The industrial application criterion specifies that the invention must be capable of being used for industrial or business purposes.

In Mexico, there are few phytomedicine patents. As shown (Table 1.1), the first patent application in this area was in 2005 and to date there have been a total of 27 patent applications. It should be noted that 12 applications belong to the Mexican Social Security Institute. Medical applications of these patent applications include substances that have anti-inflammatory, anti-microbial, anti-neoplastic, and anti-hypertensive activity.

After scientists have identified specifically targeted new entities for disease diagnosis, treatment or pharmaceutical purposes, preclinical studies are undertaken comprising *in vitro* studies, animal testing and pharmacodynamic responses.

Table 1.1. Mexican patent applications of phytomedicines.

Phytomedicines	Medical application	Patent
Agave marmorata	Treating chronic degenerative diseases	MX2012010587
Tournefortia densiflora	Treating microbial infections in skin	MX2012007255
Buddleia cordata	Treating the discomforts generated by any type of gastritis	MX2012006934
Salvia elegans	Treating the comorbidity of high blood pressure disorders with anxiety	MX2012006426
Ageratina pichinchensis	Treating chronic venous ulcers	MX2012002783
Sechium chinantlense & Sechium compositum	Preventing and/or treating neoplasms	MX2012002675
Cydonia oblonga	Anti-inflammatory	MX2012002190
Bougainvillea xbuttiana	Anti-inflammatory	MX2011013522
Matricaria recutita & Calendula officinalis	Promoting the integrity of the corneal epithelium	MX2011013407
Stevia rebaudiana	Treating ocular diseases	MX2011013620
Taxus globosa	Treating anxiety	MX2011010853
Capsicum annum	Treating gastrointestinal problems	MX2011009963
Loselia mexicana	Treating anxiety	MX2011009446
Hibiscus sabdariffa	Anti-hypertensive	MX2011006660
Ageratina pichinchensis	Treating wound healing, and tinea	MX2011005607 MX2007013011
Petiveria alliacea	Treating of rheumatic diseases	MX2011003459
Amphipterygium adstringens	Treating skin lesions	MX2010013512
Heteropterys brachiata	Treating anxiety	MX2010007355
Cladocolea loniceroides	Treating breast cancer	MX2010006779
Psittacanthus calyculatus	Anti-hypertensive	MX2010004628
Galphimia glauca	Treating anxiety	MX2009007792
Fluorensia cernua	Anti-microbial agent against periodontopathogenic bacteria	MX2009007244
Psidium sartorianum	Anti-parasitic	MX2009004174
Allium spp.	Treating diabetic foot	MX2007016417
Echinacea angustifolia	Treating gingivitis	MX2005013173
Psidium guajava	Treating gastrointestinal problems	MX2005002081

It is necessary to have the proper authorization to market phytomedicines. In Mexico, the Federal Commission against Health Risk (COFEPRIS, for its acronym in Spanish) is responsible for granting the medical authorization for proper marketing. The COFEPRIS approval process begins when a manufacturer requests permission, by submitting an investigational new drug application, to begin human testing. The application must provide high quality preclinical data to justify the testing of the drug in humans.

The next stage is clinical trials, which use human subjects. Clinical trials include biopharmaceutical, pharmacokinetic, pharmacodynamic, efficacy, safety and studies designed to demonstrate the proposed therapeutic application. After application approval, the innovating company is allowed to distribute and market the drug.

However, despite the existence of phytomedicine patent applications, there have not been any clinical trials conducted that led to marketing authorization. There are various clinical trials of Mexican phytomedicines (Table 1.2). However, these clinical trials were not registered with the COFEPRIS.

For example, a double-blinded randomized clinical trial was conducted in 197 women with primary dysmenorrhea. Four intervention groups were defined: two extract doses of *Psidium guajava* (3 and 6 mg/day), ibuprofen (1200 mg/day) and placebo (3 mg/day). Participants were individually followed up for four months. The main outcome variable was abdominal pain intensity measured according to a visual analogue scale. The standardized phytomedicine reduced menstrual pain significantly compared with conventional treatment and placebo (Doubova et al. 2007). Currently, the patent for this phytomedicine has been granted and the Mexican Social Security Institute (IMSS) has granted an operating license to the company, Genomma Lab®. This phytomedicine is marketed under the name QG5®.

Table 1.2. Mexican clinical trials of phytomedicines.

Phytomedicines	Clinical trial	Reference
Solanum chrysotrichum	Safety and effectiveness for the treatment of *Tinea pedis*	Herrera-Arellano et al. 2003
Hibiscus sabdariffa	Anti-hypertensive effectiveness and tolerability	Herrera-Arellano et al. 2004
Galphimia glauca	Efficacy and tolerability on generalized anxiety disorder	Herrera-Arellano 2007
Ageratina pichinchensis	Effectiveness and tolerability on patients with mild to moderate onychomycosis	Romero-Cerecero et al. 2008

Data on patent applications and registration of herbal medicines demonstrate that Mexico lacks an adequate structure for phytomedicine development. This weakness is not due to Mexican scientific research, because research on natural products in Mexico has flourished since 1960 and has resulted in extensive literature concerning secondary metabolites in Mexican plants (Huerta-Reyes and Aguilar-Rojas 2009) including anti-cancer (Alonso-Castro et al. 2011) not given in reference section and anti-diabetic agents (Mata et al. 2013).

Despite the high scientific, technical level and experience of Mexican researchers, the low numbers of phytomedicine patent applications may be attributed to several factors, such as: scientific evaluation criteria, which are mainly based on recognition by the scientific community from publication within indexed periodicals; lack of information about the patent database (both free access and private); and low knowledge about patentability requirements, particularly the novelty criterion. Similarly, a strong policy of disclosure in the drug registration system is required because regulatory constraints lead to high costs for companies and there is a lack of information about the key steps required for medical authorization. Finally, a policy allowing cohesion between university-business-government to foster phytomedicine research, health registration and extensive marketing is necessary.

Herbalists Drugs (Phytomedicines)

Serenoa repens (W. Bartram) Small (Arecaeae) (001P2001, Tegrata; 044P2003, prostasan; oleomed p, 042P2003), a shrub-like species native to Mexico, the southeastern USA and West Indies, is used for the treatment of prostatic hyperplasia (Capasso et al. 2003). This plant contains fatty acids and their glycerides (oleic, caprylic, myristic, etc.), sterols (e.g., β–sitosterol, campesterol, and cycloartenol) and sitosterol derivatives (Capasso et al. 2003). The therapeutic effects of this plant have been associated with the down-regulation of inflammation-related genes and the activation of the nuclear factor-kappa B pathway in prostate tissue (Silvestri et al. 2013). A clinical study indicated that this preparation improved physical symptoms caused by prostatic hypertrophy (Coulson et al. 2013).

Hypericum perforatum L. (Hypericaceae) (005P2001, Conexit; 024P2003, Motivare), a plant native to Europe, is used for the treatment of wounds, eczema, burns, trauma, rheumatism, neuralgia, gastroenteritis, ulcers, hysteria, bedwetting and depression (Ghasemi-Pirbalouti et al. 2014). The chemical constituents reported in this plant are α and β pinene, hypericin and hyperforin (Crupi et al. 2013). It has been proposed that the pharmacological effects of this plant are involved in the regulation of genes that control hypothalamic-pituitary-adrenal axis function and partially influence stress-induced effects on neuroplasticity and neurogenesis (Crupi et al. 2013). A review indicated that *Hypericum perforatum* significantly decreased depression when compared with placebo in 25 trials involving a total of 2129 patients (van der Watt et al. 2008).

Valeriana officinalis L. (Caprifoliaceae) (006P2001, Tegrarina; 011P2001, Insocaps; 020P2003, Lerisor; 010P2005, Ansisom), native to Europe and Asia and commonly known as valerian, has been used for the treatment of dysmenorrhea, anxiety, insomnia, seizures and migraine. Felgentreff et al. (2012) reported the presence of valerenic acid and acetoxy valerenic acid in this plant. Fernández-San-Martín et al. (2010) demonstrated that valerian extract increased subjective sleep quality compared with a placebo. However, other reports indicate that valerian decreases fatigue in patients, but its efficacy to improve sleep needs to be clarified (Barton et al. 2011). Furthermore, patients with generalized anxiety disorder who received valerian during a four-week period had a significant improvement in the Hamilton Rating Scale for Anxiety, but not in total anxiety scores (Andreatini et al. 2002). However, *Valeriana officinalis* is reported to induce hepatotoxicity (Cohen and Del Toro 2008, Vassiliadis et al. 2009).

Piper methysticum G. Forst (Piperaceae), commonly known as kava, is distributed throughout the South Pacific and is used for the treatment of anxiety and stress. Traditionally, kava extracts are prepared from masticated rhizome roots that are combined with coconut milk or water (La Porte et al. 2011). The major active constituents responsible for the pharmacological effects of kava are known as kavalactones or kavapyrones (Bilia et al. 2002). The anxiolytic activity of kava is controversial. There are reports that kava induces moderate anxiolytic effects (Sarris et al. 2012), whereas other reports indicate that kava lacks an anxiolytic effect (Sarris et al. 2009, Sarris et al. 2012). Nevertheless, prolonged treatment with kava has been demonstrated to induce hepatotoxicity (Teschke 2011).

Valeriana officinalis L. (Caprifoliaceae) and *Melissa officinalis* L. (Lamiaceae) (008P2001, Pokan; 003P2005, Isoren). *Melissa officinalis* L. (Lamiaceae), a perennial herb commonly known as lemon balm, is native to South Central Europe and the Mediterranean region. Its potentially active components include monoterpenoids and sesquiterpenes including geranial, neral, 6-methyl-5-hepten-2-one, citronellal, geranyl-acetate, b-caryophyllene and b-caryophyllene-oxide, and 1, 8 cineole (Tittel et al. 1982). Clinical trials have demonstrated that *Melissa officinalis* exerts an anxiolytic-like modulation of mood (Kennedy et al. 2002, Kennedy et al. 2004). The combination of *Melissa officinalis* and *Valeriana officinalis* reduced the levels of sleep disorders in menopausal women compared with a placebo (Taavoni et al. 2013).

Gingko biloba L. (Gingkoaceae) (009P2001, Tegramen; 001P2003, Nemoril, 003P2003, Kolob; 022P2004, Maxibiloba; Oleomed cer 014P2004; 013P2004, G-Kroll; 006P2004, Fylgoba) is used for failing memory, age-related dementias and poor cerebral and ocular blood flow. However, this species is under threat of extinction (IUCN 2012).

Tanacetum parthenium (L.) Sch. Bip., and *Matricaria chamomilla* L. (Asteraceae) (010P2001, Plusan). *Tanacetum parthenium* is an aromatic herb, commonly known as feverfew, native to the Balkan Peninsula. *Tanacetum parthenium* has been used as a folk remedy for fever, rheumatoid arthritis and migraines.

Parthenolide, its active compound, exerts anti-inflammatory activities in a dose-dependent manner by inhibition of thromboxane B2 and leukotriene B4. This compound also inhibits the release of pro-inflammatory mediators such as nitric oxide and TNF-α in macrophages (Sumner et al. 1992). In clinical trials, feverfew relieved symptoms associated with migraine (Ernst and Pittler 2000). The adverse effects associated with this plant are nervousness, tension headache, constipation, diarrhea and others (Ernst and Pittler 2000). Nevertheless, it was reported that feverfew did not affect the frequency of chromosomal aberrations in lymphocytes (Anderson et al. 1988). *Matricaria chamomilla* is native to Europe and Asia. This plant has been used for the treatment of flatulence, colic, hysteria and fever (Singh et al. 2011). Chamomile extract contains coumarins, herniarin and umbelliferone, and the phenolic compounds, herniarin and umbelliferone (coumarin), chlorogenic acid and caffeic acid (phenylpropanoids), apigenin, apigenin-7-O-glucoside, luteolin and luteolin-7-O-glucoside (flavones), quercetin and rutin (flavonols), and naringenin (flavanone) (Singh et al. 2011). In clinical trials, *Matricaria chamomilla* produces moderate anxiolytic effects (Amsterdam et al. 2009, Amsterdam et al. 2012).

Hedera helix L. (Araliaceae) (012P2001, Panot-s) leaf extract has been clinically studied as a cough treatment (Holzinger and Chenot 2011, Schmidt et al. 2012a).

Eucalyptus globulus Labill (Myrtaceae) (103P2001, Broncorub) in combination with menthol and turpentine. The chemical components reported in Eucalyptus globulus are 8-cineole, α-pinene, d-limonene and linalool acetate (Kumar et al. 2012). The essential oil from *Eucalyptus globulus* has minimum inhibitory concentrations (MIC) of 1.25 µL/mL against *Haemophilus influenzae*, *Haemophilus parainfluenzae* and *Stenotrophomonas maltophilia*. *Eucalyptus globulus* has also demonstrated anti-inflammatory activity *in vitro* in J774A.1 macrophages by decreasing nitric oxide production (Vigo et al. 2004).

Panax ginseng C.A. Mey (Araliaceae), *Gingko biloba* L. (Gingkoaceae), vitamins and minerals (015P2001, Biometrix; 018P2001, Centrum; 009P2003, Onesource; 019P2003, Pharseng; 030P2003, M Force; 011P2004, Wilvit; 002P2006, Pharmaton). *Panax ginseng*, commonly known as ginseng, is native to Asia. Ginseng is one of the most popular and best-selling herbal medicines worldwide.

The main chemical constituents of ginseng are glycosides, also called ginsenosides or ginseng saponin. Each ginsenoside has a common hydrophobic four ring steroid-like structure with attached carbohydrate moieties (Nah 2014). Multiple clinical studies have been performed to characterize ginseng's therapeutic properties including improving physical performance (Kulaputana et al. 2007), diabetes (Kim et al. 2011), hypertension (Rhee et al. 2011) and other diseases.

Passiflora incarnata L. (Passifloraceae), *Salix alba* L. (Salicaceae), *Valeriana officinalis* L. (Caprifoliaceae) and *Crataegus monogyna* Jacq. (016P2001, Passiflorine RN). *Passiflora edulis*, used to treat nervous anxiety and insomnia, contains the c-glycosyl flavonoids orientin, isoorientin, vitexin, isovitexin and vicenin 2 (Li et al. 2011a). *Passiflora edulis* demonstrated anxiolytic and sedative effects in mice (Li et al. 2011a). *Crataegus monogyna,* native to Europe, Northwest Africa and western Asia, is used to treat cardiac insufficiency. *Crataegus monogyna* demonstrated antioxidant effects *in vitro* and its chemical components are quinic acid, catechin, gallic acid, epigallocatechin gallate (Simirgiotis 2013). *Salix alba*, native to Europe and Asia, is used as an analgesic and anti-inflammatory agent, its primary component is salicylic acid. *Salix alba* and salicylic acid decreased the pro-inflammatory mediators TNF-α, IL-1β and IL-6 in THP1 macrophages. In addition, salicylic acid exerted antioxidant effects (Drummond et al. 2013).

Glycine max L. (Fabaceae), vitamin D and calcium carbonate (019P2001, Caltrate 600 + S; 015P2004, Prevefem; 002P2005, Prevefemcomplex; 007P2005, Advancebonde; 001P2006, Caflovan). *Glycine max*, native to Asia, is used to treat menopausal disorders such as bone loss (Potter et al. 1998). Its major and active component is the phytoestrogen, genistein. In a six month study, 90 mg/day of soy isoflavones protected postmenopausal women against lumbar spine bone loss (Potter et al. 1998). Supplementation with 120 mg/day soy isoflavone in healthy postmenopausal women over two years protected against general body bone loss, but did not specifically protect against lumbar spine, femoral neck, or total hip bone loss (Wong et al. 2009). The synergistic effect of vitamins, calcium and phytoestrogens remain to be elucidated.

Panax quinquefolium L. (Araliaceae) (004P2003, gingold American). This plant, native to United States and Canada, is considered 'vulnerable' according to its conservation status (IUCN 2012). Its active components are called ginsenosides of which there are two types: protopanaxadiol ginsenosides and protopanaxatriol ginsenosides. A protein fraction of American ginseng has demonstrated anti-fatigue and anti-oxidant effects in mice (Qi et al. 2014). In addition, *Panax quinquefolium* had cardioprotective effects *in vitro* and *in vivo* (Xu et al. 2013), anti-hyperglycemic effects (Sen et al. 2012) and anti-psychotic activity (Chatterjee et al. 2012).

Peumus boldus Molina (Monimiaceae) (006P2003, GESTISOR). This plant, native to South America, is used for the treatment of digestive and liver diseases (Falé et al. 2012). Its main active components are procyanidin B2, reticuline, norglaucin, quercitrin, kaempferitrin, boldine and others (Falé et al. 2012). *Peumus boldus* inhibits acetylcholinesterase activity and has antiproliferative effects against cancer cells. Furthermore, this extract is not modified during gastrointestinal digestion (Falé et al. 2012). However, there have been reports of neurotoxic effects, attributed to the presence of the alkaloid, boldine (Mejía-Dolores et al. 2014).

Arctium lappa L. (Asteraceae) (007P2003, aresor), commonly known as burdock, has been widely consumed as a vegetable in East Asia for centuries. *Arctium lappa* is used as a diuretic, depurative, digestive stimulant and anti-inflammatory (de Almeida et al. 2013). The chemical compounds isolated from this plant include arctigenin, arctiin, tannin, beta-eudesmol, caffeic acid, chlorogenic acid, lappaol, and diarctigenin (Chan et al. 2011). *Arctium lappa* has demonstrated potent antioxidant effects *in vitro* and *in vivo* (Liu et al. 2014) and intestinal anti-inflammatory effects in an acute experimental colitis model (de Almeida et al. 2013). In addition, this plant significantly improves metabolism in the dermal extracellular matrix and leads to visible wrinkle reduction *in vivo* (Knott et al. 2008).

Melissa officinalis L. (Lamiaceae), *Rosmarinus officinalis* L. (Lamiaceae), *Salvia officinalis* L. (Lamiaceae), *Tilia platyphyllos* Scop (Malvaceae) and *Thymus vulgaris* L. (Lamiaceae) (010P2003, Tisal-Sor 2). *Melissa officinalis*, *Rosmarinus officinalis*, *Salvia officinalis* and *Thymus vulgaris* inhibit acetylcholinesterase activity and have antioxidant properties. These plants have high contents of rosmarinic acid (>19 mg/g extract) (Vladimir-Knežević et al. 2014). *Thymus vulgaris*, native to Europe, is used as broncholytic and spasmolytic agent (Engelbertz et al. 2012). The chemical constituents of this plant are luteolin, apigenin, thymonin, 8-methoxycirsilineol and cirsilineol (Engelbertz et al. 2012). *Thymus vulgaris* has spasmolytic activity (Engelbertz et al. 2012) and antimicrobial effects against multidrug resistant pathogens (Sienkiewicz et al. 2012). *Tilia platyphyllos*, native to Europe, is used as a diuretic, anti-neuralgic and sedative (Yayalacı et al. 2014). The main constituents found in this plant are uercetin glycosides (rutin, quercitrin, and isoquercitrin), kaempferol glycosides and phenolic acids (caffeic, p-coumaric, and chlorogenic acids) (Yayalacı et al. 2014). However, there have been no conclusive pharmacological reports using this plant.

Ribes nigrum L. (*Grossulariaceae*), *Passiflora incarnata* L. (Passifloraceae), *Equisetum arvense* L. (Equisetaceae), *Fumaria officinalis* L. (Papaveraceae), *Viola tricolor* L. (Violaceae) and *Hyssopus officinalis* L. (Lamiaceae) (011P2003, Ribe-sor 23). *Ribes nigrum* is a woody shrub native to Europe and commonly known as blackcurrant. Its major chemical components are delphinidin-3-O-glucoside, delphinidin-3-O-rutinoside, cyanidin-3-O-glucoside, and cyanidin-3-O-rutinoside (Kapasakalidis et al. 2006). This plant is used for the treatment of arthritis, spasmodic cough, diarrhea, as a diuretic and sore throat treatment. *Ribes nigrum* has anti-influenza activity *in vitro* and *in vivo* without toxic effects or an induction of viral resistance (Ehrhardt et al. 2013). *Passiflora incarnata* is commonly used in clinical practice for the treatment of anxiety and sleep disorders (Miroddi et al. 2013). The chemical components of this plant are the flavonoids, vitexin, isovitexin, orientin, isoorientin, apigenin, kaempferol, vicenin, lucenin and saponarin and the indole alkaloids, harman, harmin, harmalin, harmol, and harmalol (Miroddi et al. 2013). The results from various clinical trials support its ethnomedical use by demonstrating potential effects for the treatment of generalized anxiety disorder, pre-surgery anxiety, insomnia, attention-deficit hyperactivity disorder, opiate withdrawal symptoms, and control of menopausal symptoms (Miroddi et al. 2013). *Equisetum arvense*, which is used as an anti-hemorrhagic agent, contains the chemicals apigenin

5-O-glucoside and kaempferol 3-O-glycoside (Mimica-Dukic et al. 2008). *Equisetum arvense* produced a diuretic effect comparable with hydrochlorothiazide (25 mg) with no alterations in liver or kidney function (Carneiro et al. 2014).

Desmodium adscendens D.C. (*012P2003, Ribe-sor 23*), native to Africa and South America, is used to treat asthma and other diseases associated with smooth muscle contraction (N'gouemo et al. 1996). The chemical characterization of this plant revealed the presence of vitexin and isovitexin, and soyasaponins such as soyasaponin I (Magielse et al. 2013). This plant has analgesic and anti-spasmodic activities (Addy and Dzandu 1986, N'gouemo et al. 1996). *Desmodium adscendens* demonstrated hepatoprotective effects, with similar potency to sylimarin, in rats with D-galactosamin induced acute liver damage (Magielse et al. 2013).

Olea europea L. (Oleaceae) (013P2003, Oleomed-pa) is used to treat respiratory infections.

Eleutherococcus senticosus (Rupr. & Maxim.) Maxim (014P2003, Men´s gin). This plant is a shrub native to northeastern Asia. This plant's chemical composition includes caffeic acid, isofraxidin, sesamin, sitosterol, β sitosterol and others (Davydov and Krikorian 2000). *Eleutherococcus senticosus* is used to promote physiological homeostasis (Davydov and Krikorian 2000). This plant increases sexual performance by inhibiting nitric oxide via the inhibition of cyclic GMP signal transduction (Goldstein et al. 1998).

Sambucus nigra L. (Adoxaceae), *Tilia platyphyllos* Scop (Malvaceae), *Echinacea angustifolia* D.C. (Asteraceae), *Plantago lanceolata* L. (Plantaginaceae), *Origanum vulgare* L. (016P2003, sambusor 29). *Sambucus nigra*, native to Africa, Europe and Asia, is used to treat constipation, increase diuresis, as a diaphoretic in upper respiratory tract infections, alleviation of low back and/or neuropathic pain, headache and toothache (Vlachojannis et al. 2010). The *Sambucus nigra* chemical components include hyperoside, isoquercitrin, rutoside, sambucin (cyanidin-3-O-rhamnoglucoside), sambucyanin (cyanidin-3-O-xyloglucoside) and others (Vlachojannis et al. 2010). Consuming large amounts of *Sambucus nigra* may cause adverse effects such as nausea and emesis. This plant has demonstrated anti-oxidant, anti-microbial and anti-proliferative effects against cancer cells *in vitro* and *in vivo* (Vlachojannis et al. 2010). However, controlled clinical trials have not been performed with this plant. *Plantago lanceolata*, native to America, Europe and Asia, is used in traditional medicine for the treatment of bronchitis, asthma and other respiratory diseases (Vigo et al. 2005). The chemical compounds isolated from this plant are luteolin, acteoside, plantamajoside, catalpol peracetate, catalpol, isoacteoside, lavandulifolioside and aucubin (Fleer and Verspohl 2007). *Plantago lanceolata* inhibited NO production and iNOS mRNA expression in LPS/IFN gamma stimulated J774A.1 murine macrophages (Vigo et al. 2005) and exerted antispasmodic activity on isolated guinea-pig ileum and tracheas (Fleer and Verspohl 2007).

Salix alba L. (Salicaceae), *Tilia platyphyllos* Scop (Malvaceae), *Melissa officinalis* L. (Lamiaceae), *Chamaemelum nobile* (L.) All. (Asteraceae) and *Citrus aurantium* L. (017P2003 Jake sor 22). *Chamaemelum nobile*, native to southern Europe, is used to treat dyspepsia, nausea, rheumatic pain, eczema, wounds, hemorrhoids and neuralgia (Zhao et al. 2014). The compounds isolated from this plant are derivatives of octulosonic acid (Zhao et al. 2014). *Chamaemelum nobile* and its chemical components have demonstrated *in vitro* anti-oxidant and anti-inflammatory effects (Zhao et al. 2014).

Grindelia hirsutula Hook. & Arn (Asteraceae) (synonym Grindelia robusta Nutt) (018P2003, Grine sor). *Grindelia hirsutula*, native to North America, is used as an anti-tussive and anti-asthmatic agent (La et al. 2010). This plant is primarily composed of the chemicals diosmetin-7-O-glucuronide-3'-O-pentoside+apigenin-7-O-glucuronide-4'-O-pentoside, apigenin-7-O-glucuronide+diosmetin-7-O-glucuronide and others (Ferreres et al. 2014). *Grindelia hirsutula* exhibited antioxidant and protective effects against oxidative stress (Ferreres et al. 2014), in addition to anti-inflammatory activities *in vitro* (La et al. 2010).

Ruscus aculeatus L. (Asparagaceae) (021P2003, Nikzon), native to Europe and Asia, has been used to relieve constipation, water retention and improve circulation (Aguilar-Peralta et al. 2007). The compounds teroidal saponins ruscogenin and neoruscogenin have been isolated from this plant (Mimaki et al. 1999).

Ruscus aculeatus has demonstrated anti-inflammatory and astringent properties (MacKay 2001). In clinical trials, *Ruscus aculeatus* decreased symptoms such as heavy lower legs, sensation of tension, and tingling sensations, in chronic venous insufficiency patients (Vanscheidt et al. 2002, Aguilar-Peralta et al. 2007).

Humulus lupulus L. (Cannabaceae) (022P2003, Asepxia). This plant, native to Europe, western Asia and North America, is used for the treatment of anxiety (Yamaguchi et al. 2009). The active compounds of this plant are humulones, lupulones, isohumulones and xanthohumol (Yamaguchi et al. 2009). *Humulus lupulus* and its active compounds have anti-microbial activities against the pathogens, *Propionibacterium acnes*, *Staphylococcus epidermidis*, *Staphylococcus aureus*, *Kocuria rhizophila* and *Staphylococcus pyogenes*, which produce acne. This plant also has anti-oxidant effects (Yamaguchi et al. 2009).

Aesculus hippocastanum L. (Sapindaceae), *Vitis vinifera* L. (Vitaceae), *Centella asiatica* (L.) Urban (Apiaceae) (023P2003, Goicotabs). These three plant species are used for the treatment of inflammatory diseases and varicose veins. *Aesculus hippocastanum* is native to southeastern Europe. Its main active compounds are aescin, quercetin and kaempferol (Bombardelli and Morazzoni 1996). *Aesculus hippocastanum* is used to decrease blue skin discoloration, edema, leg heaviness and pain in patients with venous insufficiency (Suter et al. 2006). *Vitis vinifera*, commonly known as grape vine, is native to Europe and Asia. This plant decreases heaviness and itching in patients with venous insufficiency (Costantini et al. 1999). The main compounds of *Vitis vinifera* are resveratrol and ε-viniferin. *Centella asiatica*, commonly known as centella, is native to Asia. Its active constituents are asiaticosides, madecassoside and madasiatic acid (Gohil et al. 2010). *Centella asiatica* effectively treats hypertensive microangiopathy and venous insufficiency by improving microcirculatory parameters (Veerendra Kumar et al. 2002).

Glycyrrhiza glabra L. (Fabaceae) (025P2003, Ulebex). This plant is an herbaceous perennial native to southern Europe. *Glycyrrhiza glabra* is widely used as a flavoring agent in candies, tobacco products, drug tablets, and soft drinks. Glycyrrhizin (glycyrrhizic acid), a triterpene saponin, is the primary water soluble constituent of *Glycyrrhiza glabra* and the main source of its sweet taste (Jiang et al. 2013). However, this plant species has been reported to induce nephrotoxicity (Allard et al. 2013).

Cassia senna L. (Fabaceae) (027P2003, Purifiq; 004P2005, Purifiq) is used as a laxative.

Garcinia cambogia (L.) N. Robson (Clusiaceae), *Cyamopsis tetragonoloba* L. (Fabaceae) and *Cassia angustifolia* Mill (Fabaceae) (031P2003, metaboltonics). The primary compound present in *Garcinia cambogia* is hydroxycitric acid, which inhibits the biosynthesis of lipids, promoting a hypotriglyceridemic effect independently of changes in leptin and insulin serum levels (Vasques et al. in press). *Cyamopsis tetragonoloba* helps to maintain nearly normal levels of anti-oxidant enzymes in the gastric and intestinal mucosa during ethanol-induced oxidative stress (Pande and Srinivasan 2013).

Prunus africana (Hook F) Kalkman (Rosaceae) (033P2003, PYG 500). This plant, native to sub-Saharan Africa, is used to improve self-rated symptoms, nocturia and post-void residual urine volume (Russo et al. 2013). Nevertheless *Prunus africana* and *Garcinia cambogia* are considered 'vulnerable' according to their conservation status (IUCN 2012).

Camellia sinensis L. Kuntze (Theaceae) (034P2003, ulcaps). This plant is native to Asia and its major flavonoids are catechins including (−)-epicatechin (EC), (−)-epicatechin-3-gallate (ECG), (−)-(EGC) and (−)-epigallo-catechin-3-gallate (EGCG), (+)-catechin (C), (+)-gallocatechin (GC), (+)-catechin gallate (CG), and (+)-gallocatechin gallate (GCG) (Bhardwaj and Khanna 2013). *Camellia sinensis* exerts anti-oxidant, anti-inflammatory, anti-platelet, anti-thrombotic effects and maintains vascular tone (Bhardwaj and Khanna 2013).

Amorphophallus konjac K. Koch (Araceae) (035P2003, Natural fit). This plant is used in tropical and subtropical Asia as a food source and as a traditional Chinese medicine. *Amorphophallus konjac* has demonstrated anti-obesity, anti-inflammatory, anti-hyperglycemic and hypercholesterolemia effects, in addition to laxative and prebiotic activities (Chua et al. 2010).

Artemisia vulgaris L. (Asteraceae) *Calendula officinalis* L. (Asteraceae), *Salvia officinalis* L. (Lamiaceae), *Cupressus sempervirens* L. (Cupressaceae), *Achillea millefolium* L. (Asteraceae), *Thymus vulgaris* L. (Lamiaceae) (046P2003, aurisor 15). *Artemisia vulgaris* is used to treat menstrual conditions such as amenorrhea, dysmenorrhea and oligomenorrhea, pregnancy disorders, and severe pain during labor and leucorrhea. It is also used as an emmenagogue, uterine sedative and postpartum tonic (de Boer and Cotingting 2014). The flavonoids, eriodictyol and apigenin, isolated from *Artemisia vulgaris* may have estrogen-like effects. The presence of these estrogen-like compounds may account for the folkloric use of *Artemisia vulgaris* for the treatment of menstrual disorders (Lee et al. 1998). *Calendula officinalis* has been traditionally used for the treatment of inflammation of internal organs, gastrointestinal ulcers and dysmenorrhea (Arora et al. 2013). The chemical constituents isolated from this plant are calendulaglycoside A, calenduladiol, rutin, kaempferol, heliantriol C, astragalin and others (Arora et al. 2013). Preparations of *C. officinalis* are generally applied as infusions, tinctures and ointments as a wound healing remedy for inflammation of the skin, mucous membranes, for poorly healing wounds, bruises, boils and rashes (Arora et al. 2013). *Achillea millefolium*, native to Europe and Asia, is prescribed for hemorrhoids, headaches and bleeding disorders (Akram 2013). The chemical constituents of *Achillea millefolium* are camphor, eucalyptol, β-pinene, α-terpineol, artemetin, dihydrodehydrodiconiferyl alcohol 9-O-β-D-glucopyranoside, apigenin and others (Akram 2013). This plant has exhibited estrogenic activity, as well as anti-hypertensive and anti-microbial effects (Akram 2013).

Psidium guajava Linn (Family: Myrtaceae), *Xalxócotl,*or guava (Register No. 003P2007, QG5) is an important food crop and a medicinal plant used in traditional and folk medicine worldwide, particularly in tropical and subtropical countries. *Psidium guajava* is native to Mexico and extends throughout South America, Europe, Africa and Asia (Gutierrez et al. 2008, Juárez et al. 2013).

Psidium guajava has long been used as a therapeutic agent for the treatment of numerous conditions such as arthritis, wounds, ulcers, toothaches, coughs, sore throats, inflamed gums, cancer, malaria, gastroenteritis, vomiting, diarrhea, dysentery, hypertension, obesity and type II diabetes mellitus. It also protects the kidney against diabetic nephropathy progression (Lutterodt 1992, Morales et al. 1994, Jaiarj et al. 1999, Karawya et al. 1999, Abdelrahim et al. 2002, Begum et al. 2004, Sunagawa et al. 2004, Ojewole 2006, SEA 2006, Chen et al. 2009, Lin and Yin 2012) via its anti-inflammatory and analgesic properties as well as its anti-oxidative, hypoglycemic, anti-glycative and anti-pathogenic microorganism effects (Rattanachaikunsopon and Phumkhachorn 2007, Ojewole et al. 2008, Shen et al. 2008, Soman et al. 2010, Livingston et al. 2012).

Phytochemical studies have demonstrated that the leaf, stem-bark and roots of *Psidium guajava* contain numerous tannins, polyphenolic compounds, flavonoids, ellagic acid, pentacyclic triterpenoids, meroterpenoids and triterpenes, guiajaverin, quercetin, reynoutrin, hyperoside and other chemical compounds (Begum et al. 2004, Ojewole 2006, Shu et al. 2010, Shu et al. 2012, Shao et al. 2012, Zhu et al. 2013).

Flavonol glycosides from *Psidium guajava* exert therapeutic effects against type II diabetes mellitus by inhibiting dipeptidyl peptidase activity (Eidenberger et al. 2013) and adipogenesis via the down-regulation of PPARγ and C/EBP alpha expression (Yang et al. 2012). While quercetin has anti-oxidant properties, quercetin-3-*O*-β-D-arabinopyranoside and quercetin-3-*O*-α-L-arabinofuranoside possess anti-bacterial and anti-fungal activities (Metwally et al. 2010). Hyperoside exhibits anti-inflammatory, anti-cancer and anti-oxidant activities (Liu et al. 2012). Reynoutrin inhibits a-glucosidase activity (Schmidt et al. 2012a, Schmidt et al. 2012b), Finally, 2,4,6-trihydroxy-3,5-dimethylbenzophenone 4-*O*-(6"-*O*-galloyl)-β-D-glucopyranoside has been demonstrated to significantly inhibit histamine release (Matsuzaki et al. 2010). *Psidium guajava* leaf extract significantly inhibited lipopolysaccharide (LPS)-induced production of nitric oxide and prostaglandin E_2 in a dose-dependent manner, suppressed the expression and activity of inducible nitric oxide synthase and cyclooxygenase-2 partially through the down-regulation of ERK1/2 activation in RAW264.7 macrophages. It also exhibits significant anti-inflammatory activity (Jang et al. 2014). Using a streptozotocin (STZ)-induced diabetes mellitus model, acute oral administrations of *Psidium guajava* extract reduced glycemia in normoglycemic and STZ-treated diabetic rats. In hypertensive Dahl salt-sensitive rats, acute intravenous administrations of the *Psidium guajava* extract reduced systemic arterial blood pressure and heart rates. Although the exact mechanisms of action of the *Psidium guajava* extract still remain a topic of speculation, it is possible that the *Psidium guajava* extract causes hypotension in Dahl salt-sensitive rats via cholinergic mechanisms, because its cardiodepressant effects are resistant to atropine pretreatment (Ojewole 2005).

It is known that prostaglandin endoperoxide H synthase (PGHS) is a key enzyme for the synthesis of prostaglandins (PGs), which play important roles in inflammation and carcinogenesis. *Psidium guajava* leaf extract inhibits the cyclooxygenase reaction of recombinant human PGHS-1 and PGHS-2, as well as the PG hydroperoxidase activity of PGHS-1. Quercetin, one of the major components of *Psidium guajava*, inhibits the cyclooxygenase activity of both isoforms and also partially inhibits PG hydroperoxidase activity. These activities explain the anti-inflammatory, analgesic and anti-proliferative properties of *Psidium guajava* extract (Kawakami et al. 2009). In addition, methanolic and ethanolic extracts of *Psidium guajava* have inhibitory activities against gram-positive bacteria (*Staphylococcus aureus* and *Bacillus cereus*). Therefore, it may be a good candidate for a natural antimicrobial agent (Biswas et al. 2013). All these biochemical properties of *Psidium guajava* lend pharmacological credence to the ethnomedical and folkloric uses of the plant previously described.

Psyllium husks, Plantaginis ovatae, Plantago ovata, ispaghula (Register number 001P2007, METAMUCIL). Psyllium is a natural polysaccharide that is derived from biodegradable and biocompatible polymers.

The polysaccharides extracted from the husk of *Plantago ovata* contain a high proportion of polysaccharides such as cellulose and hemicellulose and 20–30% mucilage (a highly branched acidic arabinoxylan comprised of a xylan backbone chain with xylose andarabinose forming the side chains), as well as oil, proteins and steroids (Kennedy et al. 1979, Vanaclocha and Cañigueral 2003). Psyllium has historically been used as a dietary supplement to regulate bowel movements, as a laxative and a fecal bulking agent. Although the bulking effect mechanism of psyllium is still unclear, one possible mechanism of how dietary fibers increase stool weight is through the physical presence of indigestible fiber in the colon. The fiber's water holding capacity, stimulation of microbial growth and production of gas may increase stool volume and cause colonic propulsion. Other potential mechanisms include stimulation of colonic motility, by the mechanical action of the fiber on the colon, and by an increase in certain colokinetic end products of fiber fermentation (Prynne and Southgate 1979, Vanaclocha and Cañigueral 2003).

It has been proposed that daily dietary fiber intake helps prevent many nutritional disorders, i.e., gut related problems, cardiovascular diseases, certain types of cancer, obesity and type II diabetes (Verma and Banerjee 2010). In type II diabetes, plant fibers, particularly water-soluble fibers such as psyllium, can moderate postprandial glucose and insulin concentrations in non-insulin dependent diabetics if administered with meals (Pastors et al. 1991, Brigenti et al. 1995). In 1998, the U.S. Food and Drug Administration authorized the use of health claims on food labels and food labeling detailing the association between soluble fiber from psyllium seed husks and a reduced risk of coronary heart disease by lowering cholesterol levels (Romero et al. 2002).

Although Rosendaal et al. demonstrated in 2004 that psyllium administration had no effect on cholesterol-lowering or lipid parameters, in 2008, Uehleke et al. demonstrated that psyllium husk preparations may be therapeutic in patients with mild to moderately elevated cholesterol levels. They studied 54 patients who had significantly decreased total cholesterol and low density lipoprotein (LDL)-cholesterol after three weeks of treatment. Additionally, gastrointestinal symptoms were rated lower at the end of the study compared with the beginning of the study. However, triglycerides and high density lipoprotein (HDL) were unchanged.

Plantago psyllium (Register No. 001P2008, ALEXCIBRAN). Although true psyllium comes from the plant *Plantago psyllium*, the husk and seed of *Plantago ovata* (family Plantaginaceae) is commonly referred to as *psyllium*. The genus Plantago includes approximately 200 species (Rahn 1996), some with medicinal value. Species of Plantago are small herbs, mostly growing as weeds.

Psyllium husk is obtained by milling the seed of *Plantago ovata* to remove the hulls. In some studies, the seed has been used instead of the husk and is also commercially available. Psyllium husk contains hemicellulose (xylan backbone linked with arabinose, rhamnose, and galacturonic acid units), while the seed consists of soluble (35%) and insoluble (65%) polysaccharides (cellulose, hemicellulose, and lignin).

While its center of diversity is believed to be in central Asia, some species are now widely dispersed, with maximum concentrations in temperate regions. Psyllium is widely used as a fiber supplement for the treatment of constipation and has been used as an indigenous Ayurvedic and Unani medicine for a wide range of bowel problems including chronic constipation, amoebic dysentery and diarrhea. In addition to being used as a laxative, psyllium is also used in ice creams, chocolates, cosmetics and printing and finishing. It is also used to lower blood cholesterol levels (Dhar et al. 2005, Dhar et al. 2011). Dietary fibers from psyllium are used extensively as pharmacological supplements, food ingredients, in processed food to aid weight control, for glucose regulation in diabetic patients and to reduce serum lipid levels in hyperlipidemics (Singh 2007). Constipation is a health problem that negatively affects quality of life and increases colon cancer risk (Watanabe et al. 2004). Therefore, increased dietary fiber intake has been recommended to treat constipation (Marlett et al. 2002). It should be noted that soluble fiber absorbs water and becomes a gelatinous, viscous substance that is fermented by bacteria in the digestive tract, while insoluble fiber has a bulking action (Anderson et al. 2009). Psyllium is classified as a mucilaginous fiber due to its powerful ability to form a gel in water. Psyllium is effective in approximately 20% of constipation patients dependent on the cause of constipation, while it is effective in 37% of patients with rectocele, internal prolapse, anismus, and rectal hyposensitivity (Voderholzer et al. 1997). Psyllium also decreases by 50% the occurrence of incontinent stools in individuals with fecal incontinence due to liquid stools or diarrhea (Bliss et al. 2001). Anaerobic fermentation of the soluble non-starch polysaccharides in psyllium seeds result in the production of the short-chain fatty acids, acetate, propionate, and butyrate in the intestine (Mortensen and Nordgaard-Andersen 1993). Butyric acid exhibits antineoplastic activity against colorectal cancer (Nordgaard et al. 1996), is the preferred oxidative substrate for colonocytes, and may be helpful for the treatment of ulcerative colitis (Fernandez-Banares et al. 1999). Psyllium also reduces bleeding and congested hemorrhoidal cushions in hemorrhoid patients (Webster et al. 1978). Psyllium reduces cholesterol levels through a mechanism that is not fully understood. In animal studies, psyllium increases more than twice cholesterol 7 alpha-hydroxylase or cytochrome 7A (CYP7A) activity, the rate-limiting enzyme in bile acid synthesis, compared with cellulose or oat bran, but less than cholestyramine (Matheson et al. 1995). Moreover, in animals fed a high-fat diet, psyllium increased the activity of cholesterol 7 alpha-hydroxylase and HMG-CoA reductase (Vergara-Jimenez et al. 1998) and reduced Apo B secretion and LDL catabolic rates. In humans, psyllium lowered LDL cholesterol (via stimulation of bile acid synthesis by stimulating cholesterol 7 alphahydroxylase), decreased cholesterol absorption and increased the fractional turnover of chenodeoxycholic and cholic acids (Everson et al. 1992). In addition, psyllium reduces the postprandial rise of glucose (Anderson et al. 1999) and regulates appetite, as psyllium appears to reduce fat intake (Turnbull and Thomas 1995). Recently, psyllium consumption has contributed to a 30% decline in the coronary artery disease death rate (Petchetti et al. 2007).

However, psyllium must be used with caution, as several cases detailing individual allergic and anaphylactic reactions to psyllium have been published (James et al. 1991, Freeman 1994).

Soybean [*Glycine max* (L.) Merrill] is a major crop worldwide and is a source of protein, oil, sugars and minerals. Soybean seed oil composition and content are important for agronomic, nutritional and industrial applications and new energy uses (Clemente and Cahoon 2009). Soybean is a major source of high quality protein and oil, whose quality is often determined by seed nutritional and anti-nutritional parameters. The free fatty acid and triglyceride content ranges from 31–71 mg 100 $g^{(-1)}$ oil and 90.1–93.9 g 100 $g^{(-1)}$ oil, respectively. The anti-nutritional components include: mg $g^{(-1)}$, TIA (41.5–85.0), phytate (2.3–5.6), total phenols (1.0–1.5), flavonols (0.20–0.34) and ortho-dihydroxy phenols (0.10–0.21) (Sharma et al. 2014). The oil content in soybean seeds ranges from 13 to 22% in various soybean cultivars. The oleic acid content of seed oil varies between 21.4 (ATAEM7) and 26.6% (Türksoy). The proportion of linoleic acid in soybean oil ranges from 49.0 (Türksoy) to 53.5% (ATAEM7), while palmitic acid varies between 9.2 (Adasoy) and 11.2% (Noya). The major tocopherols are ¥-tocopherol, α-tocopherol and δ-tocopherol (Matthaus and Ozcan 2014). Soybean oil contains high levels of polyunsaturated fatty acids (PUFA) (60.8%), with a PUFA: saturated fat ratio of 4.0. Soybean contains 54% C18:2n-6 and 7.2% C18:3n-3 with a C18:2n-6:C18:3n-3 ratio of 7:1. Recent dietary guidelines suggest decreasing consumption of total and saturated fat and cholesterol, an objective that can be achieved by substituting soybean for animal fats. Such changes have consistently resulted in decreased total and low-density-lipoprotein cholesterol, which is thought to decrease the risk of cardiovascular disease (Meydani et al. 1991).

Use of vegetable oils such soybean increases C18:2n-6, decreases C20:4n-6, and slightly elevates C20:5n-3 and C22:6n-3 in platelets, which partly inhibit platelet thromboxane generation and *ex vivo* aggregation. Whether chronic use of this oil effectively blocks thrombosis at sites of vascular injury, inhibits pathologic platelet vascular interactions associated with atherosclerosis or reduces the incidence of acute vascular occlusion in coronary or cerebral circulation is uncertain (Meydani et al. 1991). Linoleic acid is required for normal immune response and Essential Fatty Acid (EFA) deficiency impairs B and T cell-mediated responses. Therefore, soybeans can provide adequate linoleic acid for the maintenance of the immune response. However, excess linoleic acid supports tumor growth in animals, an effect that has not been verified by data from diverse human studies concerning the risk, incidence, or progression of breast and colon cancers (Meydani et al. 1991). Nevertheless, soybean oil may protect against breast and prostate cancer and also may exert a beneficial influence when used in combination with other oils (Li et al. 2014).

Soybean oligosaccharides (SBOS) are able to reduce oxidative stress and alleviate insulin resistance in pregnant women with gestational diabetes mellitus, which indicates that SBOS may play an important role to control gestational diabetes mellitus complications (Fei et al. 2014).

Soybean oil and olive oil (Register No. 002P2008, SMOFLIPID). Olive oil is obtained from olives (*Olea europaea*; family Oleaceae). Olive is a traditional tree crop in the Mediterranean Basin. There are many different olive varieties, each one with different applications such as human consumption, domestic cooking or catering, animal feed, cosmetic, pharmaceutical and industrial uses, fuel for traditional oil lamps or other engineering applications. Olive oil was common in ancient Greek and Roman cuisine and is now used worldwide, predominantly in Mediterranean countries, particularly Portugal, Spain, Italy and Greece. Spain produces more than 40% of the world's production of olive oil, Italy more than 20%, Greece approximately 12%, Syria 6% and Portugal 5%. Australia also produces a substantial amount of olive oil.

Olive oil is primarily composed of the mixed triglyceride esters of oleic acid and palmitic acid and of other saturated and unsaturated fatty acids such as linoleic acid. Olive oil also has traces of squalene and sterols. Olive oil is a source of at least 30 phenolic compounds such as esters of tyrosol and hydroxytyrosol (oleocanthal and oleuropein) aldehydic secoiridoids, flavonoids and lignans as acetoxypinoresinol and pinoresinol (Tuck and Hayball 2002, Tripoli et al. 2005). It is well known that the Mediterranean people, great consumers of olive oil, are generally less affected with atheromatosis than the Anglo-Saxon people (Brozek et al. 1957). Moreover, when the diet is supplemented with olive oil, the plasma lipid profile is favorably altered and the susceptibility of LDL cholesterol to lipid peroxidation may be decreased in hypercholesterolemic subjects (Chan et al. 2007). Clinical studies have provided evidence that consumption of olive oil may lower the risk of heart disease by decreasing risk factors such as blood cholesterol levels and LDL cholesterol oxidation (Bagigo 2013). Olive oil may also influence inflammatory, thrombotic, hypertensive and vasodilatory mechanisms (Keys et al. 1986, Covas 2007, Mayo Clinic 2007), most likely due to the content of oleic acid, vitamin E and oleuropein, a chemical that may affect the oxidation of LDL particles (Coni et al. 2000).

Oleocanthal derived from olive oil is a non-selective inhibitor of cyclooxygenase (COX), similar to classical NSAIDs, e.g., ibuprofen. It has been suggested that long-term consumption of small quantities of oleocanthal may be partially responsible for the low incidence of heart disease associated with the Mediterranean diet (Lucas et al. 2011). Furthermore, hydroxytyrosol (2-(3,4-Di-hydroxyphenyl)-ethanol or DHPE) is a phenolic component of extra-virgin olive oil, which inhibits platelet aggregation and eicosanoid (thromboxane B2) formation *in vitro* (Petroni et al. 1995). These effects may further decrease the incidence of heart disease. Preliminary studies indicate that olive oil may be a chemopreventive agent for peptic ulcers or gastric cancer (Romero et al. 2007) because it reduces oxidative damage to DNA and RNA, which are carcinogenic factors (Machowetz et al. 2007). Moreover, consumption of olive oil may prevent the onset of Alzheimer's disease, possibly through a mechanism related to oleocanthal, by inhibiting the fibrillization of tau proteins (Monti et al. 2011).

Matricaria recutita [L.] Rauschert, Asteraceae (Register No. 002P2009, KAMILEN OCEAN) *Chamomilla recutita* L., *Matricaria chamomilla*, *Chamomille* is a well-known medicinal plant species widely used in herbal remedies in ancient Egypt, Greece, and Rome (Issac 1989). It is also used in folk and traditional medicine, where it is known by an array of names such as Baboonig, Babuna, Babuna camornile, Babunj, German chamomile, Hungarian chamomile, Roman chamomile, English chamomile, Camomilla, Flos chamomile, single chamomile, sweet false chamomile, pinheads, and scented mayweed (Leung and Foster 1996, Franke 2005). Chamomile is an annual plant with thin spindle-shaped roots that only flatly penetrate the soil. The branched stem is erect, heavily ramified and grows to a height of 10–80 cm. The long and narrow leaves are bi- to tri-pinnate. The flower heads are placed separately, pedunculate and heterogamous with a diameter of 10–30 mm. The golden yellow tubular florets have five teeth that are 1.5–2.5 mm long and always end in a glandulous tube. The 11–27 white plant flowers are 6–11 mm long, 3.5 mm wide and arranged concentrically. The receptacle is 6–8 mm wide, flat in the beginning and conical later, hollow and without paleae. The fruit is a yellowish brown achene. Although this plant is native to southern and eastern Europe, Germany, Hungary, France, Russia, Yugoslavia and Brazil (Ivens 1979), it is currently cultivated worldwide and can be found in India, North Africa, Asia, North and South America, Australia and New Zealand (Singh et al. 2011).

The chamomile drug is included in the pharmacopoeia of many countries (Pamukov and Achtardziev 1986), including Mexico, and is an ingredient in several traditional, Unani, and homeopathic medicinal preparations due to its multi-therapeutic, cosmetic, and nutritional value (Lawrence 1987, Das et al. 1998, Kumar et al. 2001, Mann and Staba 2002, Singh et al. 2011). A large group of therapeutically active compounds such as sesquiterpenes, flavonoids, coumarins, and polyacetylenes have been identified in chamomile. The coumarins are represented in *M. chamomilla* by herniarin, umbelliferone, and other minor ones (Redaelli et al. 1981, Kotov et al. 1991). The glucoside precursors of herniarin, (Z)- and (E)-2-β-d-glucopyranosyloxy-4-methoxycinnamic acid (GMCA), have also been described in chamomile (Ohe et al. 1995). Eleven bioactive phenolic compounds (Gupta et al. 2010) such as herniarin and umbelliferone (coumarin), chlorogenic acid and caffeic acid (phenylpropanoids), apigenin, apigenin-7-O-glucoside, luteolin and luteolin-7-O-glucoside (flavones), quercetin and rutin (flavonols) and naringenin (flavanone) are also found in chamomile extract. More than 120 chemical constituents have been identified in chamomile flowers as secondary metabolites (Pino et al. 2002, Pirzad et al. 2006) including 28 terpenoids, 36 flavonoids (Kunde and Isaac 1980) and 52 additional compounds with potential pharmacologic activity (Mann and Staba 2002).

The primary pharmaceutically active components of chamomile flower oil are chamazulene and bisabolols. The α-bisabolol and cyclic ethers are anti-microbial (Isaac 1980, Manday et al. 1999), whereas chamazulene and α-bisabolol are antiseptic (Duke 1985). Chamazulene is a degradation product spontaneously formed during steam distillation from the sesquiterpene lactone, matricine, several bisabolol-type sesquiterpenes ((−)-α-bisabolol, bisabolol oxides), flavonoids and two en-in-dicycloethers. In addition to its flowers, chamomile roots and shoots are also rich in essential oils. Chamomile also contains sesquiterpene hydrocarbons and alcohols such as (E)-β-farnesene and spathulenol, respectively (Repcak et al. 1980, Reichling et al. 1984, Kumar et al. 2001, Schilcher et al. 2005). Additionally, chamomile was determined to have the most effective anti-leishmanial activity (Shnitzler et al. 1996).

In folk and traditional medicine, chamomile is used mainly as an anti-inflammatory, antiseptic, anti-spasmodic and mild sudorific (Mericli 1990). Chamomile is also used for disturbances of the stomach associated with pain or colic, sluggish digestion, diarrhea, nausea, flatulence, intermittent fever, inflammation of the urinary tract, painful menstruation and hysteria. In powder form, chamomile may be applied to wounds, skin eruptions and infections such as shingles and boils, hemorrhoids and inflammation of the mouth, throat, and eyes (Fluck 1988). During the last century it was demonstrated that a component of chamomile, umbelliferone, is fungistatic (Duke 1985). Additionally, using microbioassay *in vitro* studies, it was determined that chamomile flower essential oil may be a potential candidate to design effective anti-fungal formulations suitable for the treatment of medically important dermatophytosis, opportunistic saprophytes and other fungal infections (Jamalian et al. 2012). The anti-inflammatory effect of chamomile infusions at the gastric level was tested on phorbol 12-myristate 13-acetate-stimulated AGS cells and human neutrophil elastase. Chamomile infusion inhibited neutrophil elastase and gastric metalloproteinase-9 activity and secretion by inhibiting NF-kB driven transcription. This effect was due to flavonoid-7-glycosides, one of the major constituents of chamomile flowers (Bulgari et al. 2012).

In addition to its beneficial effects *in vitro*, recent chamomile extract studies in rats demonstrated potent anti-diarrheal and anti-oxidant properties, confirming their use in traditional medicine (Sebai et al. 2014). Similarly, the anti-hyperalgesic effects of bisabolol-oxide-rich matricaria oil has been examined in a rat inflammation model induced by carrageenan using a modified 'paw-pressure' test, while its anti-edematous effects have been examined in a rat inflammation model induced by carrageenan, dextran and histamine using plethysmometry. In both experiments, bisabolol-oxide-rich matricaria oil was effective against the pain and edema present in different inflammatory conditions, which supports matricaria's traditional use as anti-inflammatory and analgesic (Tomić et al. 2014). Moreover, a randomized double blind clinical trial was performed in 90 students comparing the effects of chamomile extract and mefenamic acid on the intensity of premenstrual syndrome symptoms. Chamomile was determined to be more effective than mefenamic acid for relieving the intensity of premenstrual syndrome and its associated symptomatic psychological pain (Sharifi et al. 2014).

Platelet-rich plasma obtained from healthy donors was treated with polyphenolic-polysaccharide conjugates from chamomile, and this treatment resulted in a dose-dependent decrease of platelet aggregation induced by agonists such as ADP, collagen and arachidonic acid. Moreover, chamomile also reduced platelet aggregation in platelet-rich plasma obtained from patients with cardiovascular disorders. Therefore, compounds obtained from chamomile could lead to the development of a new anti-platelet agent that may be an alternative to currently used anti-platelet drugs for preventing cardiovascular diseases (Bijak et al. 2013). Chamomile flower extract also potently prevented fatty liver disease without the adverse side effects of classical peroxisome proliferator-activated receptor (PPAR) agonists. PPARs are a family of nuclear receptors that play a central role in cellular differentiation, glucose and lipid homeostasis, and can suppress inflammatory processes. Pharmacologic modulation of PPARs is a common strategy to treat insulin resistance and dyslipidemia (Berger and Moller 2002). Three different subtypes of PPAR exist: PPARα, PPARβ/δ and PPARγ. PPARα, predominantly expressed in the liver, controls fatty acid oxidation and lipoprotein metabolism and is also involved in gluconeogenesis and ketone body biosynthesis. PPARβ/δ is ubiquitously expressed and has a central role in fatty acid oxidation and adaptive thermogenesis. PPARγ plays a key role in adipose tissue differentiation and maintenance by regulating energy storage and balance. PPARγ is highly expressed in adipocytes but also controls differentiation and metabolic processes in the liver, macrophages, bone cells and skeletal muscle. Because PPARγ is involved in glucose metabolism by regulating insulin sensitivity (Lehrke and Lazar 2005) and chamomile flower extracts can activate PPARγ, chamomile flower extracts have therapeutic effects in type 2 diabetes and dyslipidemia in insulin-resistant, high-fat diet-fed mice (Weidner et al. 2013). It has been reported that chamomile may also have clinically meaningful anti-depressant activities in addition to its anxiolytic activity (Amsterdam et al. 2012).

Moreover, *Chamomilla* is a promising nephroprotective compound, reducing cisplatin nephrotoxicity, most likely through its anti-oxidant activities and by inhibiting gamma glutamyl transferase activity (Salama 2012). In experiments that tested chamomile in human blood platelets, mouse fibroblast cultures L929 and human lung cells A549, there were no observed cytotoxic effects (Bijak et al. 2013). However, chamomile pollen present in teas for eye washing can induce allergic conjunctivitis (Subiza et al. 1990).

Hedera helix L., English ivy, Common ivy (Register No. 001P2013, FLUIR) is a well-known native and ornamental plant in Europe. It is an evergreen, dioecious, woody liana renowned for its ability to adhere to vertical surfaces and is one of the 15 species of the genus *Hedera*, Araliaceae family. The adventitious roots of *Hedera helix* are responsible for the production of an adhesive compound composed of polysaccharides and spherical nanoparticles (Xia et al. 2011, Lenaghan and Zhang 2012) with optical absorption, light scattering properties. These nanoparticles' increased safety over commonly used metal oxide nanoparticles make them attractive candidates for sunscreen protection agents or fillers (Li et al. 2010, Ligin et al. 2010, Xia et al. 2010). Although in folkloric medicine it is used to cure benign warts, the dry extract of *Hedera helix* is currently known to act as an anti-inflammatory (Suleyman et al. 2003, Gepdiremen et al. 2005), anti-bacterial, mucolytic and spasmolytic agent with a bronchodilatatory effect on cell cultures (Trute et al. 1997, Sieben et al. 2009). Most of these effects are attributable to *Hedera helix*'s triterpene saponin content (Bedir et al. 2000, Trute et al. 1997). Because pharmaceutical manufacturers have demonstrated the efficacy of *Hedera helix* for the treatment of cough symptoms during acute and chronic bronchitis, among non-antibiotic cough remedies, herbal preparations containing extracts from *Hedera helix* leaves are popular in many European countries (Guo et al. 2006, Coca and Nink 2008, Glaeske et al. 2008). In 2007, more than 80% of herbal expectorants prescribed in Germany comprised *Hedera helix* extract, amounting to nearly two million prescriptions nationwide (Coca and Nink 2008). The effect of dry extracts on the respiratory function of children with chronic bronchial asthma has been confirmed (Hofmann et al. 2003). Additionally, a post-marketing study in 9657 patients (5181 children) with bronchitis (acute or chronic bronchial inflammatory disease) where dried *Hedera helix* leaf extract was given over seven days resulted in the improvement or healing of symptoms in 95% of patients with an adverse event incidence of 2.1% (Fazio et al. 2009). However, Holzinger and Chenot (2011) maintain that evidence of the effectiveness of *Hedera helix* leaf extract for the treatment of acute upper respiratory tract infections in randomized controlled trials, nonrandomized controlled clinical trials and observational studies has not been convincing. For example, the therapeutic effect of orally administered *Hedera helix* on lung histopathology in a murine model of chronic asthma was not superior to dexamethasone treatment (Hocaoglu et al. 2012).

It has been proposed that the potent anti-inflammatory property of *Hedera* extract is similar to the effect of diclofenac. However, experimental animals could not tolerate peritoneal injections of more than 75 μl of ethanol *Hedera* extract. Therefore, additional animal studies need to be performed in this area (Rai 2013). Although *Hedera helix* is a potent herb for the treatment of arthritis in animals, the exact component possessing anti-inflammatory, analgesic and anti-arthritic properties needs to be isolated and tested. This could result in the use of *Hedera helix* as a cost-effective and potent herbal medicine for the treatment of inflammation and arthritis (Rai 2013).

Additionally, the antiviral activity of hederasaponin B, derived from *Hedera helix*, against EV71, which causes hand, foot and mouth disease, was evaluated in vero cells. Hederasaponin B demonstrated significant antiviral activity against the EV71 subgenotypes C3 and C4a by inhibiting viral VP2 protein expression, most likely due to the inhibition of viral capsid protein synthesis. Therefore, *Hedera helix* extract may be a novel drug candidate with broad-spectrum antiviral activity against various subgenotypes of EV71 (Song et al. 2014). Moreover, *Hedera helix* extract has been reported to have anti-oxidant properties (Gülçin et al. 2004) anti-allergic effects (Jones et al. 2009) and antitumor activities (Elias et al. 1990). The ripe fruits of *Hedera helix* crude extracts have a potential anthelmintic benefit (Eguale et al. 2007).

Astragalus membranaceus (astragalus) (Register No. 003RH2005, Commercial name, Astranaceus). *Astragalus* has historically been used in Chinese medicine, but is only now receiving attention in the U.S. and Europe. *Astragalus* is a perennial plant, approximately 16–36 inches tall, that is native to the northern and eastern parts of China, as well as Mongolia and Korea. It has hairy stems with leaves made up of 12–18 pairs of leaflets. The root is the medicinal part of the plant and is usually harvested from four-year-old plants. The Chinese name of the herb, huang qi, means 'yellow leader', and the herb was given this name because its root is yellow and is one of the most important herbs in Chinese medicine. It is often combined with other herbs to strengthen the body against disease. *Astragalus* is an adaptogen, meaning it helps protect the body against various stresses including physical, mental, or emotional stress. *Astragalus* may protect the body from diseases such as cancer and diabetes. It contains anti-oxidants, which protect cells against damage, and is also used to protect and support the immune system, preventing colds and upper respiratory tract infections, lowering blood pressure, treating diabetes, and protecting the liver. *Astragalus* has anti-bacterial and anti-inflammatory properties. It is occasionally used on the skin for wound care. In addition, studies have demonstrated that *Astragalus* has antiviral properties and stimulates the immune system, suggesting that it may prevent colds (Sinclair 1998).

Astragalus contains the plant pigments: formononetin, astraisoflavan, astrapterocarpan, 2'-3'-dihydroxy-7,4'-dimethooxyisoflavone, and isoliquiritigenin. Other major constituents include D-ß-asparagine, calycosin, cycloastragenol, astragalosides IVII, choline, betaine, kumatakenin, sucrose, glucuronic acid, ß-sitosterol 1, and soyasaponin I. Astragalan, a polysaccharide fraction with a molecular weight between 20,000 and 25,000, has been extracted and researched in China for its ability to enhance the *in vitro* secretion of tumor necrosis factor (Zhao and Kong 1993).

Astragalus has also been used in immunotherapy. The use of recombinant interleukin-2 (rIL-2) in immunotherapy is limited by the toxicity associated with higher doses. *Astragalus* was administered with 100 u/ml of rIL-2 versus 1,000 u/ml of rIL-2 alone in an *in vitro* study in murine renal carcinoma cells. The astragalus rIL-2 group had a tumor cell lysis rate of 88 versus 86% in the group with 1000 u/ml rIL-2 alone. This suggests a 10-fold potentiation in the *in vitro* antitumor activity of rIL-2 generated lymphokine-activated killer (LAK) cells (Wang et al. 1992). In the United States, researchers have examined astragalus as a possible treatment for patients whose immune systems have been compromised by chemotherapy or radiation. In these studies, astragalus supplements appear to help people recover faster and live longer. Research using astragalus in patients with AIDS has produced mixed results. Recent research in China suggests that because *Astragalus* is an anti-oxidant, it may benefit patients with severe forms of heart disease; relieving symptoms, lowering cholesterol levels, and improving heart function. At low-to-moderate doses, astragalus has few side effects. However, it does interact with a number of other herbs and prescription medications. Astragalus may also be a mild diuretic, meaning it helps rid the body of excess fluid. It has also demonstrated *in vitro* anti-bacterial activity against *Shigella dysenteriae, Streptococcus hemolyticus, Diplococcus pneumoniae*, and *Staphylococcus aureus.* The saponins contained in *Astragalus* had a positive effect on the function of the heart by inhibiting lipid peroxide formation in the myocardium and decreasing blood coagulation (Purmova et al. 1998, Cheng et al. 2011).

Capsicum annuum (chile, Mex) (Register No. 004RH2005, commercial name Green Marvel). In Mexico pepper fruits are important ingredients in a balanced diet; peppers are a vital source of compounds that offer health benefits and enrich the anti-oxidant pool of food products including vitamins C and E, provitamin A, carotenoids and phenolic compounds. Pepper belongs to the genus *Capsicum*, which is comprised of more than 200 varieties, with *Capsicum annuum, Capsicum baccatum, Capsicum chinense, Capsicum frutescens*, and *Capsicum pubescens* being the main five species (Zimmer et al. 2012). Peppers are consumed worldwide and their importance has gradually increased to place them among the most consumed spice crops in the world (Bown 2012). They are usually consumed as food and used as additives in the food industry. They also have a significant role in traditional medicine. In fact, in Indian, Native American, and Chinese traditional medicine, *Capsicum* species have been used for the treatment of arthritis, rheumatism, stomach aches, skin rashes, dog/snake bites, and flesh wounds. These therapeutic applications are related to the capsaicinoid, phenolic compounds, and carotenoid content of peppers (Zimmer et al. 2012). Carotenoids are the pigments responsible for the yellow, orange, and red color of many types of peppers; however, they are more than mere pigments and also play an important role as anti-oxidants. In their capacity as anti-oxidants, carotenoids protect cells and tissues from harmful Radical Oxygen Species (ROS), acting as scavengers of singlet molecular oxygen, peroxyl radicals, and Reactive Nitrogen Species (RNS) (Stahl and Siess 2003, Hernández-Ortega et al. 2013).

Capsicum annuum L. is reported to be an excellent source of polyphenols, particularly flavonoids such as quercetin and luteolin (Koo and Mohamed 2001). The principal pungent ingredient present in red peppers (*Capsicum annuum* L.) is the phenolic substance named capsaicin (8-methyl-N-vanillyl-trans-6-nonenamide). This compound has attracted considerable attention over the past two decades because of its chemoprotective properties against certain diseases. The presence of high concentrations of chlorophylls and carotenoids in a single food matrix, as occur in peppers at IRS, may be important because they can exert different protective effects and can protect from the same disease via different mechanisms. The consumption of carotenoids and chlorophylls has been associated with protective effects against atherosclerosis, some forms of cancer, osteoporosis, cataracts, neurodegenerative diseases, mutagenesis, and oxidative stress (Elliott 2005, Ferruzzi and Blakeslee 2007). The protective effects of carotenoids are mediated by their oxidant, anti-oxidant, redox sensitive cell signaling, induction of gene expression, and provitamin A properties (Elliott 2005). The contribution of chlorophylls to the anti-oxidant activity of fresh and processed Capsicum gene has been clearly demonstrated (Alvarez-Parrilla et al. 2011). Fox et al. (2005) observed that fruits of Robusta bell peppers at IRS presented a higher (29%) anti-oxidant activity than fruits at other ripening stages.

Capsicum annuum also exhibited significant peripheral analgesic activity at 5, 20, and 80 mg/kg and induced central analgesia at 80 mg/kg as well as indomethacin (7 mg/kg). Interestingly, guajillo pepper (dry pepper) carotenoid extract had a more prolonged effect than indomethacin, increasing the latency of response time even after five hours post treatment. The results suggest that the carotenoids in dried guajillo peppers have significant analgesic and anti-inflammatory benefits and may be useful for pain and inflammation relief (Hernández-Ortega et al. 2012). Due to abundant phytochemicals and their culinary use, red peppers have become an important source of chemopreventive agents in the Orient. Agricultural wastes of plant origin have attracted considerable attention as potential sources of bioactive phytochemicals that can be used for various purposes in the pharmaceutical, cosmetic and food industries.

Malva parviflora L. (Malvaceae) (Register No. 005RH2005, Commercial name AZUL). *Malva parviflora* L. Ahala, Malba, country mallow, Malva de castilla, Malva of cheeses; State of Mexico: du-Jan (Mazahua); Oaxaca: baldag malv (*zapoteco*), belongs to the family Malvaceae that includes trees, shrubs and herbs and is widely distributed throughout Africa. Plants from this family are noted for their economic, horticultural and medicinal importance. Traditional healers and herbalists in Lesotho use dried powder or an infusion made from the leaves and roots of *Malva parviflora* to clean wounds and sores. A hot poultice made from leaves is also used to treat wounds and swelling and is incorporated into a lotion to treat bruised and broken limbs (Shale et al. 1999). An ethnobotanical survey observed that the leaves and stems of this plant, with or without the addition of heated brown sugar, is applied as a hot poultice to wounds and boils by the Xhosa people of South Africa. Shale et al. (1999) also reported the use of its lotion to treat bruises and broken limbs and the dried powder or infusion of the leaves and roots to clean wounds and sores by herbalists in Lesotho. The methanol extract of *Malva parviflora* also possessed appreciable activity against Gram-negative and Gram-positive bacteria, as well as anti-inflammatory activity against COX-1 (Shale 1999). In 1999, it was possible to isolate a new compound present in *Malva parviflora*, 5α-estigmast-9 (11)-en-3-one; however, their secondary metabolites have been under-researched, because over the last 55 years, research has been directed towards the metabolism of the primary (protein). Some studies have demonstrated the free radical scavenging activity of the methanolic extract of Malva. In a study by Afolayan et al. (2008), the plant demonstrated the ability to quench radicals, inhibiting 94.3% of radical cations. The plant possesses higher flavonoid content compared with phenolics and proanthocyanidins and a positive linear correlation was established between the polyphenols and free radical scavenging activity. Additionally, the hexane extract of *M. parviflora* leaves can efficiently inhibit insulin resistance, lipid abnormalities and oxidative stress, indicating that its therapeutic properties may be due to the interaction plant components soluble in the hexane extract with any of the multiple targets involved in diabetes pathogenesis (Pérez-Gutierrez 2012).

Eucalyptus (commercial name AGRIFEN). Labillardiere, common names Eucalipto (Méx, Ecu., Perú y Ven); Ocalito; Eucalipto macho (Bol.); Blue gum, Eucalipto bouton, Gommier bleu (USA), belongs to the family Malvaceae and is native to Australia. The genus *Eucalyptus* includes approximately 600 species. *Eucalyptus globulus* is most widely cultivated in subtropical and Mediterranean regions. Essential oils from *Eucalyptus* species are used in folk medicine and also widely used in modern cosmetics, food, and pharmaceutical industries (Gray and Flatt 1998).

The medicinal use of this plant is indicated for respiratory disorders, primarily coughs. For this purpose, a decoction of the leaves is ingested before bedtime, in addition to inhaling the stems. For severe coughs, it is prepared with camellia flowers or purple bougainvillea (*Bougainvillea* sp.) and mullein (*Gnaphalium attenuatum*) or cinnamon (*Cinnamomum zeylanicum*). This remedy is also used in other lung conditions and is consumed warm when necessary. The essential oils exert antibiotic activity against *Staphylococcus aureus, Pseudomonas aureginosa* and other *Pseudomonas* species, *Escherichia coli, Bacillus subtilis, Proteus mirabilis, P. morganii, P. rettgeri, Salmonella typhi, S. Wien, Haemophilus influenzae, Mycobacterium tuberculosis, Klebsiella* species, *Streptococcus, Enterobacter* and the fungus, *Candida albicans*. These compounds also exhibit antiviral activity against the influenza A2 virus, smallpox, and herpes type 2. The ethanol extract of the branches is active against *Plasmodium falsiparum* FMN-13, and slightly active against other types of *Plasmodium falsiparum*. The ether extract of the leaves has anthelmintic activity against *Stronggyloides stercoralis* and *antianquilostoma*, specifically, *A. duodenale* and *Ancylostoma caninum*. The molluscicidal action exerted by the extract has also been demonstrated (Hammer et al. 1999, Cimanga et al. 2002).

Other effects have been demonstrated experimentally including the hypoglycemic action of the aqueous extract of the leaves when administered in the diet of mice with streptozotocin-induced hyperglycemia and administered by gastric intubation and subcutaneously to hyperglycemic mice induced by alloxan.

Leaves and essential oil expectorant were administered to rats as a diuretic. In rabbits and cats, it was administered orally in doses of 150 and 100 mg/kg. The anthelmintic activity of *Eucalyptus globulus* leaf, flower and fruit extracts *in vitro* against *Fasciola hepatica* were lethal to parasites at concentrations of 2.5 mg plant/ml and 5.0 mg plant/ml. The leaves contain essential oils: monoterpenes camphene, Cineol, para-cymene; euglobal IB, IC and II A, alpha and beta-phellandrene, geraniol and acetate, iso-fenchone, limonene, myrcene, alpha and beta-pinene, trans-pineocarvol, terpineol, alpha-isomer and its acetate, valeraldehyde, aromandreno sesquiterpenes, allo-aromandreno, caryophyllene, euglobal III, IV A and IV B, globulol, epiglobulol, ledol and viridiflorol. *Eucalyptus* leaves contain: crisin flavonoids eucaliptín, hyperoside, galloyl procianidín B-2, B-2 prodelfinidín galloyl, prodelfinidín B-5 and its digaloil, quercetin, isoquercetin, rutin, and 8-desmethyl sewderoxilín sideroxilín. The essential oil of the fruit contains the monoterpenes 1-8 cineol, linalool oxide, beta-pinene, piperitone, last-4-ol, alpha, beta and gamma-terpinene and its alpha-isomer, gamma-cadinene, eremofileno, and alpha-gurguneno globulol. Cineol, an essential oil, has significant antibiotic activity against bacteria, fungi and viruses and is an expectorant. Additionally, phytochemical analysis of *Eucalyptus* revealed that it has an anti-cariogenic substance, alpha-farnesene, which is a sesquiterpene. The hexanoic and ethyl acetate extracts of *E. globulus* plant leaves have inhibitory potential against *Lactobacillus acidophilus* and a panel of cariogenic bacteria (Kalpesh et al. 2013). The alcoholic extract of *Eucalyptus globulus* administered orally to diabetic rats for 21 days at 0.05, 0.10, 0.20 and 0.40 g/kg significantly decreased serum glucose levels. Further, it increased serum insulin levels in a dose-dependent manner, suggesting it has anti-diabetic activity.

Crataegus sp. (Register No. 007RH2005, commercial name Strauds drops, native name Tejocote (Mex.)), is the *Crataegus* genus (Rosaceae) and comprises approximately 280 species and is found in northern temperate regions of East Asia, Europe, and eastern North America. The common name for the *Crataegus* species is hawthorn and is known as Tejocote in Mexico. Texocotl (Nahuatl), 'stone sour'; Chiapas: kanal chishte, chamomile, manzanita, bighorn haws; Federal District: texocotl (Nahuatl); State of Mexico: npeni (Otomi); Michoacán: karhasi (purhépecha). *Crataegus* sp. contains a number of chemical compounds: acids, triterpene acids, organic acids, sterols and trace amounts of cardioactive amines. Several biological activities for this genus have been reported, such as antispasmodic, diuretic, and digestive activities, among others. These reports are from Europe using numerous preparations and in combination with other herbal extracts. In addition, the *Crataegus* sp. has been used in Mexican traditional medicine, as well as in other countries, for the treatment of asthma (Digital Library, UNAM). Recently, the leaves of *C. mexicana* were reported to have a tracheal relaxant effect in a bioassay-guided study employing guinea-pig isolated tracheal rings as an experimental model. Assays by HPLC-MS reveal that at least 14 compounds may exist in the hexane extract. In addition, the results suggest that the relaxant effect of the effective fraction was partially related to the activity of β-adrenergic receptors and not K+ ATP channels. This study provides preliminary scientific support for the popular practice employing *Crataegus mexicana* for the treatment of respiratory diseases (Arrieta et al. 2010).

A 30% syrup derived from the fruit has pulmonary effects, which is useful for the treatment of airway diseases when the dominant symptoms are cough, bronchial congestion and lung inflammation. The leaves have a diuretic effect and promote healthy renal stimulation; traditionally a patient ingests two or three cups a day of a 10% infusion before meals when there is noticeable irritation in the urinary tract or kidney pain. The root has upnentel diuretic and only leaves have an effect, and it is useful for the treatment of inflammation of the kidneys and bladder (nephritis, pyelitis, pielinefritis, cystitis), particularly when there is anasarca (fluid retention in the tissues especially leg), or localized or general edema (infiltration skin tissue). It is also an advantageous treatment for kidney dysfunction and cardiac or vascular deficiency, having a similar effect as digitalis, caffeine or kola nut without their adverse effects. Therefore, it is effective as a cardiac tonic and sedative and for slight nervous hypotension (Long et al. 2006). It has also been demonstrated to reduce the amount of glucose in the urine and blood and may be useful in diabetes, resulting in the control of diabetes in some patients.

Cedronella mexicana, Cassia senna (Register No. 015RH2001, NATROSOLVE). *Cedronella mexicana,* currently named *Agastache mexicana* (H.B:K.) Lint & Epling, is used to alleviate anxiety or abdominal pain in folk medicine (the medical equivalent of sedatives and antispasmodics). In Mexico, it is commonly called 'toronjil'. Due to large demand for this plant, it is cultivated in various regions such as Mexico City and the states of Hidalgo, Mexico, Morelos, Puebla and Veracruz. The blooming season is August and the seeds ripen in September. *A. mexicana* is used to treat the cultural disease known as 'empacho' (indigestion), 'mal de ojo' (evil eye) and for 'spiritual cleansings' (Argueta et al. 1994). As a medicinal remedy, all of the aerial parts and only the flowers of *A. mexicana* are usually prepared, fresh or dried, in boiling water as an infusion or decoction or as a maceration in ethanol which are used to treat anxiety, insomnia, cardiovascular disorders (Linares et al. 1988, Argueta et al. 1994), rheumatism, stomach pain, and gastrointestinal affections (Hernández 1942, Linares et al. 1988, Argueta et al. 1994, Linares et al. 1995). Inflorescences are preferred to alleviate pain and aerial parts produce a sedative effect (Madaleno 2007). *A. mexicana* extracts have recently been studied for pharmacological activity such as anxiogenic (Molina-Hernández et al. 2000), vasoactive and anti-oxidant effects (Ibarra-Alvarado et al. 2010). Its medicinal properties on nociception have been tested by systemic administration in different experimental models of nociception to identify what types of nociception may be alleviated by using this plant. These studies have provided experimental support for its use in traditional medicine in the treatment of abdominal, inflammatory and gouty arthritis pain (González-Ramírez et al. 2012). This vegetal species is often combined with *Cassia senna*. There are over 400 known species of Cassia. The leaves and seedpods (fruit) have laxative activity, due to the presence of anthraquinone compounds. Anthranoid laxatives are a group of substances generally described as herbal laxatives because of their natural origin (Laitinen et al. 2007). Sennosides, the most well-known members of the anthranoid family, are obtained from senna plant dried leaflets and pods. Anthranoid laxatives are commonly used in clinical practice as self-medication for chronic constipation. Although the short-term use of these laxatives is generally safe, results from *in vitro* and animal studies suggest that they are potentially tumorigenic (National Toxicology Program 2012). However, translation of animal studies to humans is problematic, as these results were obtained in an experimental setting with relatively high and lengthy exposures for the lifespan of the animals (van Gorkom et al. 1999). Currently available evidence does not support a genotoxic risk for patients who consume senna-based laxatives (Morales et al. 2009). However, several human studies have suggested possible carcinogenic effects after long-term administration (van Gorkom et al. 1999). Therefore, these substances should be used with caution and should not be chronically applied.

Damiana (Register No. 005RH2001, DAMIN). *Turnera aphrodisiaca* Ward (synonym *Turnera diffusa* Willd. family Turneraceae) is commonly known as '*Damiana*'. It is a small shrub with an aromatic leaf found on dry, sunny, rocky hillsides in South Texas, southern California, Mexico, and Central America. The leaf has been used as an aphrodisiac and to boost sexual potency by the native peoples of Mexico including the Mayan Indians. The two species used in herbal medicine, both which are referred to as damiana, are *Turnera aphrodisiaca* and *T. diffusa*. *Turnera diffusa* is registered alone (Register No. 017RH001, DEBORDER) and combined with *Tecoma stans* and *Medicago sativa* (Register No. 014RH2001, AZOTH). Historically, damiana has been used to relieve anxiety, nervousness, and mild depression, particularly when these symptoms have a sexual component. The herb is used as a general tonic to improve wellness and has also been used traditionally to improve digestion and to treat constipation. It is also used as a diuretic, cough treatment, and in large doses is thought to have a mild laxative effect. Studies of preparations of *T. aphrodisiaca* as tinctures have provided evidence of significant anxiolytic activity. These tinctures have similar classes of phytoconstituents such as flavonoids, alkaloids or steroids, which may be responsible for the activity of damiana (Kumar and Sharma 2005). In addition, it is reported that *T. diffusa* significantly reduced the post-ejaculatory interval, supporting an aphrodisiac effect, and may be effective against sexual dysfunction (Estrada-Reyes et al. 2009).

Echinacea angustifolia alone (Register No. 006RH2001, EQUINOL) and combined with *Marrubium vulgaris* and *Glycyrrhiza glabra* (Register No. 007RH2001, K-NUT). *E. angustifolia* plant preparations (family Asteraceae) are widely used in Europe and North America for common colds and the flu. Most consumers and physicians are not aware that products available under the term *Echinacea* differ appreciably in their composition. This variability is mainly due to the use of variable plant material, extraction methods and the addition of other components (Karsch-Völk et al. 2014). Despite its worldwide acceptance, only limited data are available on its prophylactic efficacy. Prophylactic treatment with *Echinacea* for over four months appeared to be beneficial, suggesting that *Echinacea* has an advantageous safety profile; there was not a greater incidence of adverse effects observed with *Echinacea* use compared with placebo treatment. Overall, the risk/benefit results from this clinical study suggested that long-term treatment with *E. purpurea* over four months can be recommended (Jawad et al. 2012). This study is reinforced by a study examining an early intervention with a standardized *Echinacea* formulation that resulted in reduced symptom severity in subjects with naturally acquired upper respiratory tract infections (Goel et al. 2004). Evidence from preclinical studies supports some of the traditional and modern uses for *Echinacea*, particularly its reputed immunostimulatory (or immunomodulatory) properties (Barnes et al. 2005).

Ganoderma lucidum (Lingzhi or Reishi, Register No. 001RH2001). These mushrooms are being developed as nutraceuticals to obtain the essence of mushrooms and ease consumption. Scientific validation of traditional knowledge has confirmed the benefit of consuming mushrooms, fresh or processed, on human health (Sabaratnam et al. 2013). The effectiveness of *Ganoderma lucidum*, commercially named 'Reishi mushroom' or 'Medicine of kings', has been extensively studied and resulted in data from laboratory and clinical studies for a variety of diseases and conditions. *Ganoderma lucidum* is known as a bitter mushroom and has remarkable health benefits. The active constituents present in mushrooms include polysaccharides, dietary fibers, oligosaccharides, triterpenoids, peptides, proteins, alcohols, phenols, mineral elements (such as zinc, copper, iodine, selenium, and iron), vitamins, and amino acids. The bioactive components in the *G. lucidum* mushroom have numerous health properties for the treatment of diseases such as hepatopathy, chronic hepatitis, nephritis, hypertension, hyperlipemia, arthritis, neurasthenia, insomnia, bronchitis, asthma, gastric ulcers, atherosclerosis, leukopenia, diabetes, anorexia, and cancer. Despite the voluminous literature available, *G. lucidum* is used mostly as an immune enhancer and a health supplement, not therapeutically (Batra et al. 2013).

Ginkgo biloba (Register No. 004RH2001, TALIESIN). *Ginkgo biloba* is a dioecious tree that has been historically used in traditional Chinese medicine. Although the seeds are most commonly employed in traditional Chinese medicine, recently standardized leaf extracts have been widely sold as a phytomedicine in Europe and a dietary supplement in the United States and Mexico. The primary active constituents of the leaves include flavonoid glycosides and unique diterpenes known as ginkgolides; the latter are potent inhibitors of platelet activating factors (Smith et al. 1996). Ginkgo is recommended for inflammation and asthma treatment (Mahmoud et al. 2000). Clinical studies have demonstrated that ginkgo extracts exhibit therapeutic activity in a variety of central nervous system disorders including Alzheimer's disease (Maclennan et al. 2002), failing memory (Sakatani et al. 2014), age-related dementias (Ahlemeyer and Krieglstein 2003), poor cerebral and ocular blood flow, and congestive symptoms of premenstrual syndrome (Ozgoli et al. 2009). Due in part to its potent anti-oxidant properties and ability to enhance peripheral and cerebral circulation, ginkgo's primary application lies in the treatment of cerebrovascular dysfunctions and peripheral vascular disorders (McKenna et al. 2001).

Heterotheca inuloides (Register No. 009RH2001, SPLENDID). *Heterotheca inuloides* Cass. (Asteraceae) grows abundantly in the cooler, temperate regions of Mexico. The flowers of this plant are widely used in Mexican traditional medicine. Internally, they are used for the treatment of inflammatory diseases, fever and other disorders; externally, to treat contusions and wounds. Products containing Mexican arnica can be ointments and lotions for external or topical application and teas, tablets, or homeopathic tinctures for internal application. Components of *H. inuloides*, such as sesquiterpenoids, have been identified as anti-microbial agents (Kubo et al. 1994). The aqueous extract obtained from *H. inuloides* flowers has been assessed for analgesic and anti-inflammatory activity in several experimental models (Gené et al. 1998), in which several chemical constituents have also been tested (Delgado et al. 2001).

Jacobina spicigera 'Muicle' (Register No. 018RH2001, RIKLY). *Jacobina spicigera* Schlecht., also named *Justicia spicigera* Schlecht. (Acanthaceae), is a plant used as an immunostimulant and for the empirical treatment of cervical cancer in Mexican traditional medicine. There is evidence that one of the active constituents is kaempheritrin, which exerts immunostimulatory effects mediated by splenocytes, macrophages, human peripheral blood mononuclear and NK cells (Del Carmen et al. 2013) and cytotoxic, anti-tumoral effects against HeLa cells (Alonso-Castro et al. 2013). It is also considered a significant plant because of the effect it may have in patients with hematopoietic disorders. However, experiments in different hematopoietic cells, human leukaemic cell lines, umbilical cord blood cells, and mouse bone marrow cells indicated that *Justicia spicigera* infusions do not produce any hematopoietic activity. However, it does induce apoptosis by inhibiting bcl-2 and is linked to cell proliferation, demonstrating cytotoxic activity (Cáceres et al. 2001).

Matricaria recutita combined with *Salvia officinalis* (Register No. 016RH2001, CARDON) and with *Lycopersicum esculentum* Mill and *Citrus aurantiifolia* (Register No. 001RH2010, EXHER). *Matricaria recutita* L. (Asteraceae) commonly known as chamomile is one of the most widely used and well-documented medicinal plants in the world. It is included in the pharmacopoeia of 26 countries (Salamon 1992a). The use of chamomile as a medicinal plant dates back to ancient Greece and Rome. The name 'chamomile' comes from two Greek words meaning 'ground apple' for its apple-like smell. The ancient Egyptians considered the herb a sacred gift from the sun god and used it to alleviate fever and sun stroke. In the sixth century, it was used to treat insomnia, back pain, neuralgia, rheumatism, skin conditions, indigestion, flatulence, headaches and gout. In Europe, it is considered a 'cure all', and in Germany, it is referred to as 'alles zutraut', meaning capable of anything (Berry 1995). Chamomile is widely used throughout the world. Its primary uses are as a sedative, anxiolytic and anti-spasmodic and as a treatment for mild skin irritation and inflammation. Chamomile's main active constituents are chamazulene, apigenin and bisabolol. Despite its widespread use as a home remedy, relatively few trials have evaluated chamomile's many purported benefits. Randomized controlled studies have produced conflicting results for the treatment of dermatologic and mucosal irritations including eczema and mucositis. Animal trials suggest efficacy as a sedative, anxiolytic and anti-spasmodic, but clinical studies in humans are still required (Salamon 1992b). An alcoholic extract of chamomile inhibited acetylcholine- and histamine-induced spasms. Essential oil of chamomile was comparable to papaverine in reducing isolated guinea pig ileum spasms. Apigenin and bisabolol have dose-dependent spasmolytic effects on isolated guinea pig ileum (Achterrath-Tuckermann et al. 1980). Extracts of chamomile flowers have an inhibitory effect on gastric acid secretion (Tamasdan et al. 1981). The anti-inflammatory effects of chamomile are well documented in animals. Bisabolol reduced inflammation, fever and adjuvant arthritis in animal studies. Bisabolol was also an anti-pyretic in yeast-induced fever in rats. Apigenin has demonstrated anti-inflammatory properties in animal studies. It demonstrated potent anti-inflammatory activity in carrageenan-induced rat paw edema and delayed type hypersensitivity in mice (Jakovlev et al. 1979, Ammon et al. 1996). Chamomile extracts significantly reduced locomotor activity in rats (Avallone et al. 1996). Chamomile has also been widely used for the treatment of digestive system disorders. A chamomile extract has demonstrated potent anti-diarrheal and anti-oxidant properties in rats, confirming their use in traditional medicine (Sebai et al. 2014). *Salvia officinalis* has been used in herbal medicine for many centuries. It has been suggested to, on the basis of traditional medicine, modulate mood and improve cognitive performance in humans. Therefore, *Salvia officinalis* may provide a novel natural treatment for Alzheimer's disease (Akhondzadeh et al. 2003, Russo et al. 2013). In the form of an infusion, the principal and most valued application, it is used as a wash to cure diseases of the mouth and as a gargle for inflamed sore throats, being excellent for the relaxation of the throat and tonsils and for ulcerated throats. Activity against inflammatory processes has been demonstrated after stimulation by chemical agents such as acetic-acid, formalin, glutamate, capsaicin and cinnamaldehyde (Rodrigues et al. 2012). This has been reinforced by observing that *Salvia officinalis* tinctures significantly reduced total leukocyte, monocyte percentages and activated circulating phagocytes (Oniga et al. 2007). Tomato (*Lycopersicum esculentum* or *Solanum lycopersicum* L.) is an important fruit crop in the Americas, southern Europe, Middle East and India with increasing production in China, Japan, and Southeast Asia. It is amenable to producing pharmaceuticals, particularly for oral delivery, for many of the same reasons that make it a popular vegetable. Its fruit is nontoxic and is palatable uncooked. It is easily processed and the plants can be propagated by seed or clonally from tip or shoot cuttings. Tomato plants have high fruit yields, there is a reasonable biomass and protein content and they are easily grown under containment (Van Eck et al. 2006). In preclinical studies, it has been demonstrated that guanosine from *S. lycopersicum* possesses anti-platelet (secretion, spreading, adhesion and aggregation) activity *in vitro* and inhibits the platelet inflammatory mediator of atherosclerosis (sCD40L) (Fuentes et al. 2013). The anti-oxidant effects of tomatoes have also been examined and there was an association with dietary intake and a lowered risk of cancer, neurodegenerative, and cardiovascular diseases (Li et al. 2011b, Dubois et al. 2013, Aydin et al. 2013).

Mentha piperita, Tila platyphyllos and *Crataegus oxyacantha* (Register No. 012RH2001, BIOCALM). *Mentha piperita* 'Peppermint' has a wide variety of health and medicinal uses. It is used to help treat the common cold, calm inflammation and soothe digestive problems. The ancient Egyptians, one of the most medically advanced ancient cultures, cultivated and used peppermint leaves for indigestion. The ancient Romans and Greeks also used peppermint to soothe their stomachs. The plant was used by Europeans in the 18th century, especially western Europe. Peppermint oil has the most uses and data, the oil is considered relevant to leaf extract formulations as well. Topical preparations of peppermint oil have been used to calm pruritus and relieve irritation and inflammation (Herro and Jacob 2010). Many of peppermint's health and medicinal uses have been verified by scientific trials. Its anti-nociceptive and anti-inflammatory activity has been tested in several experimental models (Atta and Alkofahi 1998). *Crataegus* spp. 'hawthorn' (a genus comprising approximately 300 species) has been utilized by many cultures for a variety of therapeutic purposes for many centuries. Cardiovascular disease has become one of the most significant causes of premature death and recent research into the therapeutic benefits of hawthorn preparations has focused primarily on its cardiovascular effects (Tassell et al. 2010).

Tilia mexicana, Olea europaea, and *Casimiroa edulis* (Register No. 013RH2001, ESPIGOL). *Tilia americana* var. *mexicana* (Tiliaceae) is distributed in 14 states of Mexico, from the northern states of Chihuahua and Coahuila to the southern states of Guerrero and Oaxaca. Although this plant has a relatively large geographical distribution, the populations of this species are confined to lower mountainous forests, which cover less than 1% of Mexican territory (Flores et al. 1971). Tilia is a tree used in traditional medicine primarily as a non-narcotic sedative for sleep disorders or anxiety. Flower infusions (teas) of this species have generally been regarded as non-toxic; diluted teas are commonly given to overanxious children as a mild sedative. The inflorescence of Tilia are sold year-around in popular town markets located in several regions of Mexico (Pérez-Ortega et al. 2008). It has been observed that during the flowering months (April–June), there is an increase in the marketing of inflorescences of this species, because it is believed that the medicinal effect is greater when the infusion is produced during this period (Pavón and Rico-Gray 2000). Moreover, Tilia inflorescences are sold in markets where it is stored for almost a year; these samples may be adulterated by mixing with other species such as *Ternstroemia pringlee* (Theaceae), which is also known as Tilia. Tilia inflorescences acquired from local markets or freshly collected in different states of Mexico induce similar sedative and anti-anxiety effects in experimental models, supporting its use in folk medicine (Aguirre-Hernández et al. 2007a, 2007b). The presence of flavonoids such as quercetin and kaempferol derivatives (Herrera-Ruiz et al. 2008, Aguirre-Hernández et al. 2007a, 2010) and beta-sitosterol are thought to be responsible for this activity (Aguirre-Hernández et al. 2007b).

Conclusions

Although Mexican traditional medicine is one of the most diverse and complete medicine around the world, there are no Mexican phytomedicine production facilities. Trying to understand this situation is hard because the use of medicinal plants by the Aztec Indians in México had been described in different colonial documents since 1552. Such documents provide examples of the use, actions, and usefulness of different medicinal plants. Moreover, the formal medicine research in this area was started in 1888 when the National Medical Institute in Mexico was created by the order of President Porfirio Díaz. By 1915, the Institute's herbarium possessed 14,000 classified species and approximately 1,000 chemical compounds obtained from plants. In 1975, the Mexican Institute for the Study of Medicinal Plants (IMEPLAM) was founded with the aim to study the most widely utilized plants in Mexico to treat common illnesses. This Medicinal Herbarium is now known as the Medicinal Herbarium of the Mexican Institute of Social Security (IMSS) which has a legacy of more than 120,000 specimens. The Institute was considered an icon in medicinal plant research in Latin America during the late 90s, and at the end of the 20th century. Actually, mechanisms of action, activity, and pharmacological effects of several medicinal plant extracts are under study, but commercially sponsored research results for drug, diagnostic, and treatment techniques must be intellectually protected to obtain a health registry, and the patents for proper marketing. Unfortunately, in Mexico, there are few phytomedicine patents registered by the Mexican Institute of Industrial Property. Moreover, the first patent application in this area was in 2005, and to date there have been just 27 patent applications including plants such as Agave marmorata, Tournefortia densiflora, Buddleia cordata, Salvia elegans, Ageratina pichinchensis, Sechium chinantlense & Sechium compositum, Cydonia oblonga, Bougainvillea xbuttiana, Matricaria recutita & Calendula officinalis, Stevia rebaudiana, Taxus globosa, Capsicum annuum, Loselia mexicana, Hibiscus sabdariffa, Galphimia glauca, and Psidium guajava.

Once the patent is obtained, it is necessary to have the proper authorization to market phytomedicines and the Federal Commission against Health Risk (COFEPRIS, for its acronym in Spanish) who grants the medical authorization for proper marketing in Mexico, if and when the application provides high quality preclinical data to justify the testing of the drug in humans. Because of the beneficial effects of medicinal plants, a large and growing global interest has emerged, both in developed and developing countries. However, it is important to consider that the fact that something which is natural does not necessarily make it safe or effective. Many pharmacological studies have validated their ethnomedicinal use. Nevertheless, clinical and toxicological studies should be conducted with medicinal plants in order to guarantee their safety and effectiveness. The isolation and identification of their active compounds is also highly desirable. Another consideration of importance is to perform pharmacokinetic studies with active compounds from medicinal plants. Taking this into consideration, the integration of phytomedicine in the health system should be developed with scientific evidence of effective therapeutic properties. Clinical trials should then be conducted in humans to test and validate biopharmaceutical, pharmacokinetic, pharmacodynamic, efficacy, safety, and therapeutic applications.

If the application is approved, the innovating company is allowed to distribute and market the drug. As of now, there have not been any clinical trials conducted in Mexico that have led to marketing authorization of a drug. This could be due to the rigidity of scientific evaluation criteria, lack of information about the patent database, low knowledge about patentability requirements, or a lack of information about the key steps required for medical authorization. In addition, updating pharmacopeias of Mexico with new information about the use of medicinal plants is necessary.

Quality control for efficacy and safety of herbal products is of great importance. The assurance of the safety of an herbal drug requires monitoring of the quality of the finished product as well as the quality of the consumer information on the herbal remedy. All of these points require appropriate attention and correction for producing phytomedicines. Another relevant point that needs attention is the creation of a policy that allows cohesion between university-business-government agencies to foster phytomedicine research, health registration, and extensive marketing.

References

Abdelrahim, S.I., Almagboul, A.Z., Omer, M.E.A. and Elegami, A. 2002. Antimicrobial activity of *Psidium guajava* L. Fitoterapia 73(7): 713–715.

Achterrath-Tuckermann, U., Kunde, R., Flaskamp, E., Isaac, O. and Thiemer, K. 1980. Pharmacological investigations with compounds of chamomile. V. Investigations on the spasmolytic effect of compounds of chamomile and Kamillosan on the isolated guinea pig ileum. Planta. Med. 39(1): 38–50.

Addy, M.E. and Dzandu, W.K. 1986. Dose-response effects of *Desmodium adscendens* aqueous extract on histamine response, content and anaphylactic reactions in the guinea pig. J. Ethnopharmacol. 18(1): 13–20.

Afolayan, A.J., Aboyade, O.M. and Sofidiya, M.O. 2008. Total phenolic content and free radical scavenging activity of *Malva parviflora* L. (Malvaceae). J. Biol. Sci. 8(5): 945–949.

Aguilar, A., Camacho, J.R., Chino, S., Jácquez, P. and López, M.E. 1994. Herbario medicinal del Instituto Mexicano del Seguro Social. IMSS, México, DF.

Aguilar, P.G., Arévalo, G.J., Llamas, M.F., Navarro, C.V., Mendoza, C.S. and Martínez, M.C. 2007. Clinical and capillaroscopic evaluation in the treatment of chronic venous insufficiency with *Ruscus aculeatus*, hesperidin methylchalcone and ascorbic acid in venous insufficiency treatment of ambulatory patients. International angiology: J. Int. Union. Angiol. 26(4): 378–384.

Aguirre-Hernández, E., Martínez, A.L., González-Trujano, M.E., Moreno, J., Vibrans, H. and Soto-Hernández, M. 2007a. Pharmacological evaluation of the anxiolytic and sedative effects of *Tilia americana* L. var. *mexicana* in mice. J. Ethnopharmacol. 109(1): 140–145.

Aguirre-Hernández, E., Rosas-Acevedo, H., Soto-Hernández, M., Martínez, A.L., Moreno, J. and González-Trujano, M.E. 2007b. Bioactivity-guided isolation of beta-itosterol and some fatty acids as active compounds in the anxiolytic and sedative effects of *Tilia americana* var. *mexicana*. Planta. Med. 73(11): 1148–55.

Aguirre-Hernández, E., González-Trujano, M.E., Martínez, A.L., Moreno, J., Kite, G., Terrazas, T. and Soto-Hernández, M. 2010. HPLC/MS analysis and anxiolytic-like effect of quercetin and kaempferol flavonoids from *Tilia americana* var. *mexicana*. J. Ethnopharmacol. 127(1): 91–97.

Ahlemeyer, B. and Krieglstein, J. 2003. Pharmacological studies supporting the therapeutic use of *Ginkgo biloba* extract for Alzheimer's disease. Pharmacopsych. 36(S 1): 8–14.

Akhondzadeh, S., Noroozian, M., Mohammadi, M., Ohadinia, S., Jamshidi, A.H. and Khani, M. 2003. *Salvia officinalis* extract in the treatment of patients with mild to moderate Alzheimer's disease: a double blind, randomized and placebo-controlled trial. J. Clin. Pharm. Therap. (28): 53–59.

Akhtar, N. and Haqqi, T.M. 2012. Current nutraceuticals in the management of osteoarthritis: a review. Therap. Advan. Musculosk. Dis. 4(3): 181–207.

Akram, M. 2013. Minireview on *Achillea millefolium* Linn. J. Mem. Biol. 246(9): 661–663.

Allard, T., Wenner, T., Greten, H.J. and Efferth, T. 2013. Mechanisms of herb-induced Nephrotoxicity. Curr. Med. Chem. 20(22): 2812–2819.

Allegra, M., Ianaro, A., Tersigni, M., Panza, E., Tesoriere, L. and Livrea, M.A. 2014. Indicaxanthin from cactus pear fruit exerts anti-inflammatory effects in carrageenin-induced rat pleurisy. J. Nut. 144(2): 185–192.

Alonso-Castro, A.J., Villarreal, M.L., Salazar-Olivo, L.A., Gomez-Sanchez, M., Dominguez, F. and Garcia-Carranca, A. 2011. Mexican medicinal plants used for cancer treatment: pharmacological, phytochemical and ethnobotanical studies. J. Ethnopharmacol. 133(3): 945–972.

Alonso-Castro, A.J., Ortiz-Sánchez, E., García-Regalado, A., Ruiz, G., Núñez-Martínez, J.M., González-Sánchez, I. and García-Carrancá, A. 2013. Kaempferitrin induces apoptosis via intrinsic pathway in HeLa cells and exerts antitumor effects. J. Ethnopharmacol. 145(2): 476–489.

Alvarez-Parrilla, E., de la Rosa, L.A., Amarowicz, R. and Shahidi, F. 2010. Antioxidant activity of fresh and processed Jalapeno and Serrano peppers. J. Agricul. Food Chem. 59(1): 163–173.

Ammon, H.P., Sabieraj, J. and Kaul, R. 1996. Chamomile: mechanisms of anti-inflammatory activity of chamomile extracts and components. Deutsche. Apoth. Zeit. 136: 17–18.

Amsterdam, J.D., Li, Y., Soeller, I., Rockwell, K., Mao, J.J. and Shults, J. 2009. A randomized, double-blind, placebo-controlled trial of oral *Matricaria recutita* (chamomile) extract therapy of generalized anxiety disorder. J. Clin. Psychopharmacol. 29(4): 378.

Amsterdam, J.D., Shults, J., Soeller, I., Mao, J.J., Rockwell, K. and Newberg, A.B. 2012. Chamomile (*Matricaria recutita*) may have antidepressant activity in anxious depressed humans-an exploratory study. Alt. Therap. Health. Med. 18(5): 44.

Anderson, D., Jenkinson, P.C., Dewdney, R.S., Blowers, S.D., Johnson, E.S. and Kadam, N.P. 1988. Chromosomal aberrations and sister chromatid exchanges in lymphocytes and urine mutagenicity of migraine patients: a comparison of chronic feverfew users and matched non-users. Hum. Exp. Toxicol. 7(2): 145–152.

Anderson, J.W., Allgood, L.D., Turner, J., Oeltgen, P.R. and Daggy, B.P. 1999. Effects of psyllium on glucose and serum lipid responses in men with type 2 diabetes and hypercholesterolemia. Amer. J. Clin. Nut. 70(4): 466–473.

Anderson, J.W., Baird, P., Davis, Jr., R.H., Ferreri, S., Knudtson, M., Koraym, A. and Williams, C.L. 2009. Health benefits of dietary fiber. Nut. Rev. 67(4): 188–205.

Andreatini, R., Sartori, V.A., Seabra, M.L. and Leite, J.R. 2002. Effect of valepotriates (valerian extract) in generalized anxiety disorder: a randomized placebo-controlled pilot study. Phytother. Res. 16(7): 650–654.

Angeles-López, G.E., González-Trujano, M.E., Déciga-Campos, M. and Ventura-Martínez, R. 2013. Neuroprotective evaluation of *Tilia americana* and *Annona diversifolia* in the neuronal damage induced by intestinal ischemia. Neurochem. Res. 38(8): 1632–1640.

Argueta, V.A., Cano, L. and Rodarte, M. 1994. Atlas of plants from Mexican traditional medicine, III. Indigenous National Institute, Mexico City. pp. 1355–1356.

Arora, D., Rani, A. and Sharma, A. 2013. A review on phytochemistry and ethnopharmacological aspects of genus Calendula. Pharmacog. Rev. 7(14): 179.

Arrieta, J., Siles-Barrios, D., García-Sánchez, J., Reyes-Trejo, B. and Sánchez-Mendoza, M.E. 2010. Relaxant effect of the extracts of *Crataegus mexicana* on guinea pig tracheal smooth muscle. Pharmacog. J. 2(17): 40–46.

Arteche, A. and Vanaclocha, B. 1998. Fitoterapia: Vademécum de Prescripción, 3ª edición. Masson, Barcelona, pp. 434–437.

Atta, A.H. and Alkofahi, A. 1998. Anti-nociceptive and anti-inflammatory effects of some Jordanian medicinal plant extracts. J. Ethnopharmacol. 60(2): 117–124.

Avallone, R., Zanoli, P., Corsi, L., Cannazza, G. and Baraldi, M. 1996. Benzodiazepine-like compounds and GABA in flower heads of *Matricaria chamomilla*. Phytother. Res. 10: S177–S179.

Aydin, S.S., Büyük, I. and Aras, S. 2013. Relationships among lipid peroxidation, SOD enzyme activity, and SOD gene expression profile in *Lycopersicum esculentum* L. exposed to cold stress. Gen. Mol. Res. 12(3): 3220–3229.

Baggio, G., Pagnan, A., Muraca, M., Martini, S., Opportuno, A., Bonanome, A. and Piccolo, D. 1988. Olive-oil-enriched diet: effect on serum lipoprotein levels and biliary cholesterol saturation. Am. J. Clin. Nut. 47(6): 960–964.

Bañuelos, G.S., Fakra, S.C., Walse, S.S., Marcus, M.A., Yang, S.I., Pickering, I.J. and Freeman, J.L. 2011. Selenium accumulation, distribution, and speciation in spineless prickly pear cactus: a drought-and salt-tolerant, selenium-enriched nutraceutical fruit crop for biofortified foods. Plant Physiol. 155(1): 315–327.

Barnes, J., Anderson, L.A., Gibbons, S. and Phillipson, J.D. 2005. Echinacea species (Echinacea angustifolia (DC.) Hell., *Echinacea pallida* (Nutt.) Nutt., *Echinacea purpurea* (L.) Moench): a review of their chemistry, pharmacology and clinical properties. J. Pharm. Pharmacol. 57(8): 929–954.

Barton, D.L., Atherton, P.J., Bauer, B.A., Moore, Jr., D.F., Mattar, B.I., LaVasseur, B.I. and Loprinzi, C.L. 2011. The use of *Valeriana officinalis* (valerian) in improving sleep in patients who are undergoing treatment for cancer: a phase III randomized, placebo-controlled, double-blind study: NCCTG Trial, N01C5. J. Supp. Oncol. 9(1): 24.

Batra, P., Sharma, A.K. and Khajuria, R. 2013. Probing Lingzhi or Reishi medicinal mushroom *Ganoderma lucidum* (Higher Basidiomycetes): a bitter mushroom with amazing health benefits. Int. J. Med. Mush. 15(2): 127–143.

Bedir, E., Kırmızıpekmez, H., Sticher, O. and Çalış, İ. 2000. Triterpene saponins from the fruits of *Hedera helix*. Phytochem. 53(8): 905–909.

Berger, J. and Moller, D.E. 2002. The mechanisms of action of PPARs. Ann. Rev. Med. 53(1): 409–435.

Begum, S., Hassan, S.I., Ali, S.N. and Siddiqui, B.S. 2004. Chemical constituents from the leaves of *Psidium guajava*. Nat. Prod. Res. 18(2): 135–140.

Berry, M. 1995. Herbal products: Part 6. Chamomiles. Pharmaceu. J. 254: 191–193.

Bhardwaj, P. and Khanna, D. 2013. Green tea catechins: defensive role in cardiovascular disorders. Chin. J. Nat. Med. 11(4): 345–353.

Biblioteca digital de la medicina tradicional mexicana [homepage on the Internet]. UNAM, [updated 2009 October 10. Available from: http://www.medicinatradicionalmexicana.unam.mx.

Bijak, M., Saluk, J., Tsirigotis-Maniecka, M., Komorowska, H., Wachowicz, B., Zaczyńska, E. and Pawlaczyk, I. 2013. The influence of conjugates isolated from *Matricaria chamomilla* L. on platelets activity and cytotoxicity. Int. J. Biol. Macromol. 61: 218–229.

Bilia, A.R., Gallori, S. and Vincieri, F.F. 2002. Kava-kava and anxiety: growing knowledge about the efficacy and safety. Life Sci. 70(22): 2581–2597.

Biswas, B., Rogers, K., McLaughlin, F., Daniels, D. and Yadav, A. 2013. Antimicrobial activities of leaf extracts of guava (*Psidium guajava* L.) on two gram-negative and gram-positive bacteria. Int. J. Micro. doi: 10.1155/2013/746165.

Bliss, D.Z., Jung, H.J., Savik, K., Lowry, A., LeMoine, M., Jensen, L. and Schaffer, K. 2001. Supplementation with dietary fiber improves fecal incontinence. Nur. Res. 50(4): 203–213.

Bombardelli, E., Morazzoni, P. and Griffini, A. 1996. *Aesculus hippocastanum* L. Fitoter. 67(6): 483–511.

Bown. 2001. Encyclopedia of Herbs and Their Uses. Kindersley Dorling, London, Herb Society of America, London, UK.

Brahmi, F., Chehab, H., Flamini, G., Dhibi, M., Issaoui, M., Mastouri, M. and Hammami, M. 2013. Effects of irrigation regimes on fatty acid composition, antioxidant and antifungal properties of volatiles from fruits of Koroneiki cultivar grown under Tunisian conditions.Pakis. J. Biol. Sci. PJBS 16(22): 1469–1478.

Brighenti, F., Pellegrini, N., Casiraghi, M.C. and Testolin, G. 1995. *In vitro* studies to predict physiological effects of dietary fibre. Eur. J. Clin. Nut. 49: S81–88.

Brozek, J., Buzina, R., Mikic, F., Horvat, A., Zebec, M. and Rao, M.M. 1957. Population studies on serum cholesterol and dietary fat in Yugoslavia. Am. J. Clin. Nut. 5(3): 279–285.

Bulgari, M., Sangiovanni, E., Colombo, E., Maschi, O., Caruso, D., Bosisio, E. and Dell'Agli, M. 2012. Inhibition of neutrophil elastase and metalloprotease-9 of human adenocarcinoma gastric cells by Chamomile (*Matricaria recutita* L.) Infusion. Phytoter. Res. 26(12): 1817–1822.

Burton, R. and Manninen, V. 1982. Influence of a *Psyllium*-based fibre preparation on faecal and serum parameters. Act. Med. Scan. 212(S668): 91–94.

CA, R. 1999. Pharmacology of rosemary (*Rosmarinus officinalis* Linn.) and its therapeutic potentials. Ind. J. Exp. Biol. 37: 124–131.

Cáceres-Cortés, J.R., Cantú-Garza, F.A., Mendoza-Mata, M.T., Chavez-González, M.A., Ramos-Mandujano, G. and Zambrano-Ramírez, I.R. 2001. Cytotoxic activity of *Justicia spicigera* is inhibited by bcl-2 proto-oncogene and induces apoptosis in a cell cycle dependent fashion. Phytother. Res. 15(8): 691–697.

Capasso, F., Gaginella, T.S., Grandolini G. and Izzo, A.A. (eds.). (2003). Phytotherapy: A Quick Reference to Herbal Medicine. Springer, Berlin, Heidelberg, New York.

Cárdeno, A., Sánchez-Hidalgo, M., Aparicio-Soto, M., Sánchez-Fidalgo, S. and Alarcón-de-la-Lastra, C. 2014. Extra virgin olive oil polyphenolic extracts downregulate inflammatory responses in LPS-activated murine peritoneal macrophages suppressing NFκB and MAPK signalling pathways. Food Funct. 5(6): 1270–1277.

Carneiro, D.M., Freire, R.C., Honório, T.C.D.D., Zoghaib, I., Cardoso, F.F.D.S., Tresvenzol, L.M.F. and Cunha, L.C.D. 2014. Randomized, double-blind clinical trial to assess the acute diuretic effect of equisetum arvense (Field Horsetail) in Healthy Volunteers. Evid. Based Complement. Alt. Med. Article ID 760683, 8 pages.

Chan, Y.M., Demonty, I., Pelled, D. and Jones, P.J. 2007. Olive oil containing olive oil fatty acid esters of plant sterols and dietary diacylglycerol reduces low-density lipoprotein cholesterol and decreases the tendency for peroxidation in hypercholesterolaemic subjects. Brit. J. Nut. 98(03): 563–570.

Chan, Y.S., Cheng, L.N., Wu, J.H., Chan, E., Kwan, Y.W., Lee, S.M.Y. and Chan, S.W. 2011. A review of the pharmacological effects of *Arctium lappa* (burdock). Inflammopharmacol. 19(5): 245–254.

Chatterjee, M., Singh, S., Kumari, R., Verma, A.K. and Palit, G. 2012. Evaluation of the antipsychotic potential of *Panax quinquefolium* in ketamine induced experimental psychosis model in mice. Neurochem. Res. 37(4): 759–770.

Chavez-Santoscoy, R.A., Gutierrez-Uribe, J.A. and Serna-Saldívar, S.O. 2009. Phenolic composition, antioxidant capacity and *in vitro* cancer cell cytotoxicity of nine prickly pear (*Opuntia* spp.) juices. Plant Foods Hum. Nutr. 64(2): 146–152.

Chen, K.C., Hsieh, C.L., Huang, K.D., Ker, Y.B., Chyau, C.C. and Peng, R.Y. 2009. Anticancer activity of rhamnoallosan against DU-145 cells is kinetically complementary to coexisting polyphenolics in *Psidium guajava* budding leaves. J. Agr. Food Chem. 57(14): 6114–6122.

Cheng, Y., Tang, K., Wu, S., Liu, L., Qiang, C., Lin, X. and Liu, B. 2011. *Astragalus* polysaccharides lowers plasma cholesterol through mechanisms distinct from statins. PloS One 6(11): e27437.

Chu, X., Ci, X., He, J., Wei, M., Yang, X., Cao, Q. and Deng, X. 2011. A novel anti-inflammatory role for ginkgolide B in asthma via inhibition of the ERK/MAPK signaling pathway. Molecule 16(9): 7634–7648.

Chua, M., Baldwin, T.C., Hocking, T.J. and Chan, K. 2010. Traditional uses and potential health benefits of *Amorphophallus konjac* K. Koch ex NE Br. J. Ethnopharmacol. 128(2): 268–278.

Cimanga, K., Kambu, K., Tona, L., Apers, S., De Bruyne, T., Hermans, N. and Vlietinck, A.J. 2002. Correlation between chemical composition and antibacterial activity of essential oils of some aromatic medicinal plants growing in the Democratic Republic of Congo. J. Ethnopharmacol. 79(2): 213–220.

Clemente, T.E. and Cahoon, E.B. 2009. Soybean oil: genetic approaches for modification of functionality and total content. Plant. Physiol. 151(3): 1030–1040.

Coca, V. and Nink, K. 2008. Supplementary statistical overview. pp. 963–1071. *In*: U. Schwabe and D. Paffrath (eds.). Pharmaceutical Prescription Report. Springer, Heidelberg, Germany.

Cohen, D.L. and Del Toro, Y. 2008. A case of valerian-associated hepatotoxicity. J. Clin. Gastroenterol. 42(8): 961–962.

Coni, E., Di Benedetto, R., Di Pasquale, M., Masella, R., Modesti, D., Mattei, R. and Carlini, E.A. 2000. Protective effect of oleuropein, an olive oil biophenol, on low density lipoprotein oxidizability in rabbits. Lipids 35(1): 45–54.

Costantini, A., De Bernardi, T. and Gotti, A. 1998. Clinical and capillaroscopic evaluation of chronic uncomplicated venous insufficiency with procyanidins extracted from *Vitis vinifera*. Minerva Cardioangiol. 47(1-2): 39–46.

Coulson, S., Rao, A., Beck, S.L., Steels, E., Gramotnev, H. and Vitetta, L. 2013. A phase II randomised double-blind placebo-controlled clinical trial investigating the efficacy and safety of ProstateEZE Max: A herbal medicine preparation for the management of symptoms of benign prostatic hypertrophy. Complem. Ther. Med. 21(3): 172–179.

Covas, M.I. 2007. Olive oil and the cardiovascular system. Pharmacol. Res. 55(3): 175–186.

Covas, M.I. and Koebnick, C. 2007. Effect of olive oils on biomarkers of oxidative DNA stress in Northern and Southern Europeans. FASEB. J. 21(1): 45–52.

Crupi, R., Abusamra, Y.A., Spina, E. and Calapai, G. 2013. Preclinical data supporting/refuting the use of *Hypericum perforatum* in the treatment of depression. Current Drug Targets-CNS. Neurol. Disord. 2(4): 474–486.

Cwientzek, U., Ottillinger, B. and Arenberger, P. 2011. Acute bronchitis therapy with ivy leaves extracts in a two-arm study. A double-blind, randomised study vs. and other extract. Phytomed. 18(13): 1105–1109.

Das, M., Mallavarapu, G.P. and Kumar, S. 1998. Chamomile (*Chamomilla recutita*). Economic botany, biology, chemistry, domestication and cultivation. J. Med. Aromatic. Plant Sci. 20: 1074–1109.

Davydov, M. and Krikorian, A.D. 2000. *Eleutherococcus senticosus* (Rupr. and Maxim.) Maxim. (Araliaceae) as an adaptogen: a closer look. J. Ethnopharmacol. 72(3): 345–393.

De Almeida, A.B., Sánchez-Hidalgo, M., Martín, A.R., Luiz-Ferreira, A., Trigo, J.R., Vilegas, W., dos Santos, L.C., Souza-Brito, A.R. and de la Lastra, C.A. 2013. Anti-inflammatory intestinal activity of *Arctium lappa* L. (Asteraceae) in TNBS colitis model. J. Ethnopharmacol. 146: 300–310.

De Boer, H.J. and Cotingting, C. 2014. Medicinal plants for women's healthcare in southeast Asia: A meta-analysis of their traditional use, chemical constituents, and pharmacology. J. Ethnopharmacol. 151(2): 747–767.

De la Cruz, M. 1964. Libellus de Medicinalibus Indorum Herbis, IMSS [Publisher]. México. 62.

Del Carmen Juárez-Vázquez, M., Josabad Alonso-Castro, A. and García-Carrancá, A. 2013. Kaempferitrin induces immunostimulatory effects *in vitro*. J. Ethnopharmacol. 148(1): 337–340.

Delgado, G., del Socorro Olivares, M., Chávez, M.I., Ramírez-Apan, T., Linares, E., Bye, R. and Espinosa-García, F.J. 2001. Antiinflammatory constituents from *Heterotheca inuloides*. J. Nat. Prod. 64(7): 861–864.

Dhar, M.K., Kaul, S., Sareen, S. and Koul, A.K. 2005. *Plantago ovata*: Cultivation, genetic diversity, chemistry and utilization. Plant Genetic Resources: Cultivation and Utilization. 3: 252–263.

Dhar, M.K., Kaul, S., Sharma, P. and Gupta, M. 2011. Plantago ovata: cultivation, genomics, chemistry, and therapeutic applications. *In*: Ram J. Singh (ed.). Genetic Resources, Chromosome Engineering, and Crop Improvement: Medicinal Plants. CRC Press, New York, USA. 6: 763–792.

Doubova, S.V., Morales, H.R., Hernández, S.F., Martínez-García, M.D.C., Ortiz, M.G.D.C., Soto, M.A.C. and Lozoya, X. 2007. Effect of a *Psidium guajava* folium extract in the treatment of primary dysmenorrhea: A randomized clinical trial. J. Ethnopharmacol. 110(2): 305–310.

Drummond, E.M., Harbourne, N., Marete, E., Martyn, D., Jacquier, J.C., O'Riordan, D. and Gibney, E.R. 2013. Inhibition of proinflammatory biomarkers in THP1 macrophages by polyphenols derived from chamomile, meadowsweet and willow bark. Phytother. Res. 27(4): 588–594.

Dubois, A.F., Leite, G.O. and Rocha, J.B.T. 2013. Irrigation of *Solanum lycopersicum* L. with magnetically treated water increases antioxidant properties of its tomato fruits. Electromagn. Biol. Med. 32(3): 355–362.

Duke, J.A. 1990. CRC Handbook of medicinal herbs. International Clinical Psychopharmacology 5(1): 74.

Eguale, T., Tilahun, G., Debella, A., Feleke, A. and Makonnen, E. 2007. Haemonchus contortus: *In vitro* and *in vivo* anthelmintic activity of aqueous and hydro-alcoholic extracts of *Hedera helix*. Exp. Parasitol. 116(4): 340–345.

Ehrhardt, C., Dudek, S.E., Holzberg, M., Urban, S., Hrincius, E.R., Haasbach, E. and Ludwig, S. 2013. A plant extract of Ribes nigrum folium possesses anti-influenza virus activity *in vitro* and *in vivo* by preventing virus entry to host cells. PloS One 8(5): e63657.

Eidenberger, T., Selg, M. and Krennhuber, K. 2013. Inhibition of dipeptidyl peptidase activity by flavonol glycosides of guava (*Psidium guajava* L.): A key to the beneficial effects of guava in type II Diabetes mellitus. Fitoterapia 89: 74–79.

Elias, R., De Meo, M., Vidal-Ollivier, E., Laget, M., Balansard, G. and Dumenil, G. 1990. Antimutagenic activity of some saponins isolated from *Calendula officinalis* L., *C. arvensis* L. and *Hedera helix* L. Mutagenesis. 5(4): 327–332.

Elliott, R. 2005. Mechanisms of genomic and non-genomic actions of carotenoids. Bioch. Bioph. Acta (BBA)-Mol. Basis. Dis. 1740(2): 147–154.

Engelbertz, J., Lechtenberg, M., Studt, L., Hensel, A. and Verspohl, E.J. 2012. Bioassay-guided fractionation of a thymol-deprived hydrophilic thyme extract and its antispasmodic effect. J. Ethnopharmacol. 141(3): 848–853.

Erci, B. 2012. Medical herbalism and frequency of use. pp. 195–206. *In*: Arup Bhattacharya (ed.). A Compendium of Essays on Alternative Therapy. Publisher in Tech Europe, Rijeka Croatia.

Ernst, E. and Pittler, M.H. 2000. The efficacy and safety of feverfew (*Tanacetum parthenium* L.): an update of a systematic review. Public Health Nut. 3(4a): 509–514.

Estrada-Reyes, R., Ortiz-López, P., Gutiérrez-Ortíz, J. and Martínez-Mota, L. 2009. *Turnera diffusa* Wild (Turneraceae) recovers sexual behavior in sexually exhausted males. J. Ethnopharmacol. 123(3): 423–429.

Everson, G.T., Daggy, B.P., McKinley, C. and Story, J.A. 1992. Effects of *psyllium* hydrophilic mucilloid on LDL-cholesterol and bile acid synthesis in hypercholesterolemic men. J. Lipid. Res. 33(8): 1183–1192.

Falé, P.L., Amaral, F., Amorim Madeira, P.J., Sousa Silva, M., Florêncio, M.H., Frazão, F.N. and Serralheiro, M.L.M. 2012. Acetylcholinesterase inhibition, antioxidant activity and toxicity of *Peumus boldus* water extracts on HeLa and Caco-2 cell lines. Food Chem. Tox. 50(8): 2656–2662.

Fazio, S., Pouso, J., Dolinsky, D., Fernandez, A., Hernandez, M., Clavier, G. and Hecker, M. 2009. Tolerance, safety and efficacy of *Hedera helix* extract in inflammatory bronchial diseases under clinical practice conditions: A prospective, open, multicentre postmarketing study in 9657 patients. Phytomedicine 16(1): 17–24.

Fei, B.B., Ling, L., Hua, C. and Ren, S.Y. 2014. Effects of soybean oligosaccharides on antioxidant enzyme activities and insulin resistance in pregnant women with gestational *Diabetes mellitus*. Food Chem. 158: 429–432.

Felgentreff, F., Becker, A., Meier, B. and Brattström, A. 2012. Valerian extract characterized by high valerenic acid and low acetoxy valerenic acid contents demonstrates anxiolytic activity. Phytomedicine 19(13): 1216–1222.

Fernandez-Banares, F., Hinojosa, J., Sanchez-Lombrana, J.L., Navarro, E., Martinez-Salmeron, J.F., Garcia-Puges, A. and Gassull, M.A. 1999. Randomized clinical trial of *Plantago ovata* seeds (dietary fiber) as compared with mesalamine in maintaining remission in ulcerative colitis. Amer. J. Gastroenterol. 94(2): 427–433.

Fernández-San-Martín, M.I., Masa-Font, R., Palacios-Soler, L., Sancho-Gómez, P., Calbó-Caldentey, C. and Flores-Mateo, G. 2010. Effectiveness of Valerian on insomnia: a meta-analysis of randomized placebo-controlled trials. Sleep Med. 11(6): 505–511.

Ferreres, F., Grosso, C., Gil-Izquierdo, A., Valentão, P., Azevedo, C. and Andrade, P.B. 2014. HPLC-DAD-ESI/MS n analysis of phenolic compounds for quality control of *Grindelia robusta* Nutt. and bioactivities. J. Pharmaceu. Biomed. Anal. 94: 163–172.

Ferruzzi, M.G. and Blakeslee, J. 2007. Digestion, absorption, and cancer preventative activity of dietary chlorophyll derivatives. Nutr. Res. 27(1): 1–12.

Feugang, J.M., Konarski, P., Zou, D., Stintzing, F.C. and Zou, C. 2006. Nutritional and medicinal use of cactus pear (*Opuntia* spp.) cladodes and fruits. Front. Bios. (1): 2574–2589.

Fleer, H. and Verspohl, E.J. 2007. Antispasmodic activity of an extract from *Plantago lanceolata* L. and some isolated compounds. Phytomedicine 14(6): 409–415.

Fluck, H. 1988. Medicinal Plants and Authentic Guide to Natural Remedies. W. Foulsham and Co. Ltd., London.

Fox, A.J., Del Pozo-Insfran, D., Lee, J.H., Sargent, S.A. and Talcott, S.T. 2005. Ripening-induced chemical and antioxidant changes in bell peppers as affected by harvest maturity and postharvest ethylene exposure. HortScience 40(3): 732–736.

Franke, R. and Schilcher, H. (eds.). 2005. Chamomile: Industrial Profiles. CRC Press, Boca Ratón, 279 pp.

Freeman, G.L. 1994. *Psyllium* hypersensitivity. Ann. Allergy 73(6): 490–492.

Fuentes, E., AlarcÛn, M., Astudillo, L., Valenzuela, C., Gutiérrez, M. and Palomo, I. 2013. Protective mechanisms of guanosine from *Solanum lycopersicum* on agonist-induced platelet activation: role of sCD40L. Molecules 18(7): 8120–8135.

Galati, E.M., Tripodo, M.M., Trovato, A., Miceli, N. and Monforte, M.T. 2002. Biological effect of *Opuntia ficus indica* (L.) Mill. (Cactaceae) waste matter. Note I: diuretic activity. J. Ethnopharmacol. 79(1): 17–21.

Ganzera, M., Muhammad, I., Khan, R.A. and Khan, I.A. 2001. Improved method for the determination of oxindole alkaloids in *Uncaria tomentosa* by high performance liquid chromatography. Planta. Med. 67(5): 447–450.

García, J.C. 1981. Historia de las instituciones de investigación en salud en América Latina: 1880–1930. Educ. Méd. Salud. 15: 71–88.

Garibay, A.M. 1964. Libellus de Medicinalibus Indorum Herbis. En: De la Cruz M. Instituto Mexicano del Seguro Social [eds.]. México, pp. 3–8.

Garzón-De la Mora, P., García-López, P.M., García-Estrada, J., Navarro-Ruíz, A., Villanueva-

Gené, R.M., Segura, L., Adzet, T., Marin, E. and Iglesias, J. 1998. *Heterotheca inuloides*: anti-inflammatory and analgesic effect. J. Ethnopharmacol. 60(2): 157–162.

Gepdiremen, A., Mshvildadze, V., Suleyman, H. and Elias, R. 2005. Acute anti-inflammatory activity of four saponins isolated from ivy: alpha-hederin, hederasaponin-C, hederacolchiside-E and hederacolchiside-F in carrageenan-induced rat paw edema. Phytomedicine 12(6-7): 440–444.

Gereffi, G. 1978. Drug firms and dependency in Mexico: the case of the steroid hormone industry. Int. Organ. 32(1): 237–286.

Ghasemi Pirbalouti, A., Fatahi-Vanani, M. Craker, L. and Shirmardi, H. 2014. Chemical composition and bioactivity of essential oils of *Hypericum helianthemoides*, *Hypericum perforatum* and *Hypericum scabrum*. Pharm. Biol. 52(2): 175–181.

Goel, V., Lovlin, R., Barton, R., Lyon, M.R., Bauer, R., Lee, T.D. and Basu, T.K. 2004. Efficacy of a standardized *echinacea* preparation (Echinilin) for the treatment of the common cold: a randomized, double-blind, placebo-controlled trial. J. Clin. Pharm. Ther. 29(1): 75–83.

Gohil, K.J., Patel, J.A. and Gajjar, A.K. 2010. Pharmacological review on *Centella asiatica*: A potential herbal cure-all. Indian. J. Pharm. Sci. 72(5): 546–556.

Goldstein, I., Lue, T.F., Padma-Nathan, H., Rosen, R.C., Steers, W.D. and Wicker, P.A. 1998. Oral sildenafil in the treatment of erectile dysfunction. New. Engl. J. Med. 338(20): 1397–1404.

González-Ramírez, A., González-Trujano, M.E., Pellicer, F. and López-Muñoz, F.J. 2012. Anti-nociceptive and anti-inflammatory activities of the *Agastache mexicana* extracts by using several experimental models in rodents. J. Ethnopharmacol. 142(3): 700–705.

González-Trujano, M.E., Peña, E.I., Martínez, A.L., Moreno, J., Guevara-Fefer, P., Déciga-Campos, M. and López-Muñoz, F.J. 2007. Evaluation of the antinociceptive effect of *Rosmarinus officinalis* L. using three different experimental models in rodents J. Ethnoparmacol. 111(3): 476–482.

Gray, A.M. and Flatt, P.R. 1998. Anti-hyperglycemic actions of *Eucalyptus globulus* (eucalyptus) are associated with pancreatic and extrapancreatic effects in mice. J. Nut. 128(12): 2319–23.

Gülçin, I., Mshvildadze, V., Gepdiremen, A. and Elias, R. 2004. Antioxidant activity of saponins isolated from ivy: alpha-hederin, hederasaponin-C, hederacolchiside-E and hederacolchiside-F. Planta Med. 70(6): 561–563.

Guo, R., Pittler, M.H. and Ernst, E. 2006. Herbal medicines for the treatmet of COPD: a systematic review. Eur. Resp. J. 28(2): 330–338.

Gupta, V., Mittal, P., Bansal, P., Khokra, S.L. and Kaushik, D. 2010. Pharmacological potential of *Matricaria recutita*-A review. Int. J. Pharm. Sci. Drug. Res. 2: 12–6.

Gutierrez, R.M., Mitchell, S. and Solis, R.V. 2008. *Psidium guajava*: A review of its traditional uses, phytochemistry and pharmacology. J. Ethnopharmacol. 117(1): 1–27.

Hamme, K.A., Carson, C.F. and Riley, T.V. 1999. Antimicrobial activity of essential oils and other plant extracts. J. Appl. Microbiol. 86(6): 985–990.

Haloui, M., Louedec, L., Michel, J.B. and Lyoussi, B. 2000. Experimental diuretic effects of *Rosmarinus officinalis* and *Centaurium erythraea*. J. Ethnopharmacol. 71(3): 465–472.

Hardin, S.R. 2007. Cat's claw: an Amazonian vine decreases inflammation in osteoarthritis. Complement. Ther. Clin. Practice 13(1): 25–28.

Hernández, F. 1942. History of the Plants of New Spain. University Press, Mexico City, 227 pp.

Herrera-Arellano, A., Rodríguez-Soberanes, A., de los Angeles Martínez-Rivera, M., Martínez-Cruz, E., Zamilpa, A., Alvarez, L. and Tortoriello, J. 2003. Effectiveness and tolerability of a standardized phytodrug derived from *Solanum chrysotrichum* on *Tinea pedis*: a controlled and randomized clinical trial. Planta. Med. 69(5): 390–539.

Herrera-Arellano, A., Flores-Romero, S., Chávez-Soto, M.A. and Tortoriello, J. 2004. Effectiveness and tolerability of a standardized extract from *Hibiscus sabdariffa* in patients with mild to moderate hypertension: a controlled and randomized clinical trial. Phytomedicine 11(5): 375–382.

Herrera-Arellano, A., Jiménez-Ferrer, E., Zamilpa, A., Morales-Valdéz, M., García-Valencia, C.E. and Tortoriello, J. 2007. Efficacy and tolerability of a standardized herbal product from *Galphimia glauca* on generalized anxiety disorder. A randomized, double-blind clinical trial controlled with lorazepam. Planta Med. 73(8): 713–717.

Herrera-Ruiz, M., Román-Ramos, R., Zamilpa, A., Tortoriello, J. and Jiménez-Ferrer, J.E. 2008. Flavonoids from *Tilia americana* with anxiolytic activity in plus-maze test. J. Ethnopharmacol. 118(2): 312–317.

Herro, E. and Jacob, S.E. 2010. Europe, and gained popularity for stomach ailments and menstrual disorders. *Mentha piperita* (peppermint). Dermatitis 21(6): 327–329.

Hernández-Ortega, M., Ortiz-Moreno, A., Hernández-Navarro, M., Chamorro-Cevallos, G., Dorantes-Alvarez, L. and Necoechea-Mondragón, H. 2012. Antioxidant, antinociceptive, and anti-inflammatory effects of carotenoids extracted from dried pepper (*Capsicum annuum* L.). J. Biom. Biotechnol. 524019.

Hocaoglu, A.B., Karaman, O., Erge, D.O., Erbil, G., Yilmaz, O., Kivcak, B., Bagriyanik, H.A. and Uzuner, N. 2012. Effect of *Hedera helix* on lung histopathology in chronic asthma. Iran. J. Allergy. Asthm. Immunol. 11(4): 316–323.

Hofmann, D., Hecker, M. and Völp, A. 2003. Efficacy of dry extract of ivy leaves in children with bronchial asthma: A review of randomized controlled trials. Phytomedicine 10(2-3): 213–220.

Holzinger, F. and Chenot, J.F. 2011. Systematic review of clinical trials assessing the effectiveness of ivy leaf (*Hedera helix*) for acute upper respiratory tract infections. Evidence Based Complem. Alter. Med. article id 382789.

Huerta-Reyes, M. and Aguilar-Rojas, A. 2009. Protection of inventions derived from plant research in a megadiverse country: The case of Mexico. Bol. Latinoam. Caribe 8(4): 239–244.

Ibarra-Alvarado, C., Rojas, A., Mendoza, S., Bah, M., Gutiérrez, D.M., Hernández-Sandoval, L. and Martínez, M. 2010. Vasoactive and antioxidant activities of plants used in Mexican traditional medicine for the treatment of cardiovascular diseases. Pharmaceut. Biol. 48(7): 732–739.

International Union for Conservation of Nature (IUCN). 2012. 2 ⟨http://www.iucnredlist.org⟩ Accessed on April 26, 2014.

Isaac, O. 1979. Pharmacological investigations with compounds of *chamomile*, I. On the pharmacology of (–)-alpha-bisabolol and bisabolol oxides. Planta. Med. 35(2): 118–124.

Isaac, O. 1980. Therapy with chamomile-experience and verification. Deut. Apoth. Zeit. 120: 567–70.

Issac, O. 1989. Recent Progress in Chamomile Research-Medicines of Plant Origin in Modern Therapy. Prague Press, Czecho-Slovakia.

Ivens, G.M. 1979. Stinking mayweed. New. Zeal. J. Agr. 138(3): 21–23.

Jaiarj, P., Khoohaswan, P. Wongkrajang, Y., Peungvicha, P., Suriyawong, P., Saraya, M.L. and Ruangsomboon, O. 1999. Anticough and antimicrobial activities of *Psidium guajava* Linn.leaf extract. J. Ethnopharmacol. 67(2): 203–212.

Jakovlev, V., Isaac, O., Thiemer, K. and Kunde, R. 1979. Pharmacological investigations with compounds of chamomile ii. New investigations on the antiphlogistic effects of (–)-α-bisabolol and bisabolol oxides. Planta Med. 35(2): 125–140.

Jamalian, A., Shams-Ghahfarokhi, M., Jaimand, K., Pashootan, N., Amani, A. and Razzaghi-Abyaneh, M. 2012. Chemical composition and antifungal activity of *Matricaria recutita* flower essential oil against medically important dermatophytes and soil-borne pathogens. J. Mycol. Méd. 22(4): 308–315.

James, J.M., Cooke, S.K., Barnett, A. and Sampson, H.A. 1991. Anaphylactic reactions to a *psyllium* containing cereal. J. Aller. Clin. Immunol. 88(3): 402–408.

Jang, M., Jeong, S.W., Cho, S.K., Ahn, K.S., Lee, J.H., Yang, D.C. and Kim, J.C. 2014. Anti-inflammatory effects of an ethanolic extract of guava (*Psidium guajava* L.) leaves *in vitro* and *in vivo*. J. Med. Food. 17(6): 678–685.

Jawad, M., Schoop, R., Suter, A., Klein, P. and Eccles, R. 2012. Safety and efficacy profile of *Echinacea purpurea* to prevent common cold episodes: a randomized, double-blind, placebo-controlled trial. Evid. Base. Compl. Altern. Med. article id 841315.

Jiang, J., Zhang, X., True, A.D., Zhou, L. and Xiong, Y.L. 2013. Inhibition of lipid oxidation and rancidity in precooked pork patties by radical-scavenging licorice (*Glycyrrhiza glabra*) extract. J. Food Sci. 78(11): 1686–1694.

Jones, J.M. and White, I.R., White, J.M. and McFadden, J.P. 2009. Allergic contact dermatitis to English ivy (*Hedera helix*): A case series. Contact. Dermatitis. 60: 179–80.

Juarez-Vazquez, M. del C., Carranza-Alvarez, C., Alonso-Castro, A.J. Gonzalez-Alcaraz, V.F. Bravo-Acevedo, E., Chamarro-Tinajero, F.J. and Solano, E. 2013. Ethnobotany of medicinal plants used in Xalpatlahuac, Guerrero, Mexico. J. Ethnopharmacol. 148(2): 21–527.

Kalpesh, B., Ishnava, Jenabhai, B., Chauhan., Mahesh, B. and Barad., 2013. Anticariogenic and phytochemical evaluation of *Eucalyptus globulus* Labill. Saudi. J. Biol. Scis. 20(1): 69–74.

Kapasakalidis, P.G., Rastall, R.A. and Gordon, M.H. 2006. Extraction of polyphenols from processed black currant (*Ribes nigrum* L.) residues. J. Agricul. Food Chem. 54(11): 4016–4021.

Karawya, M.S., Wahab, S.M.A., Hifnawy, M.S., Azzam, S.M. and Gohary, H.M.E. 1999. Essential oil of *Egyptian guajava* leaves. Egypt. J. Biomed. Sci. 40(2): 209–216.

Kawakami, Y., Nakamura, T., Hosokawa, T., Suzuki-Yamamoto, T., Yamashita, H., Kimoto, M., Tsuji, H., Yoshida, H., Hada, T. and Takahashi, Y. 2009. Antiproliferative activity of *guava* leaf extract via inhibition of prostaglandin endoperoxide H synthase isoforms. Prostag. Leukotr. Ess. 80(5-6): 239–45.

Kennedy, D.O., Scholey, A.B., Tildesley, N.T., Perry, E.K. and Wesnes, K.A. 2002. Modulation of mood and cognitive performance following acute administration of *Melissa officinalis* (lemon balm). Pharmacol. Biochem. Behavior. 72(4): 953–964.

Kennedy, D.O., Little, W. and Scholey, A.B. 2004. Attenuation of laboratory-induced stress in humans after acute administration of *Melissa officinalis* (Lemon Balm). Psychos. Med. 66(4): 607–613.

Kennedy, J.F., Sandhu, J.S. and Southgate, D.A. 1979. Structural data for the carbohydrate of ispaghula husk ex *Plantago ovata* forsk. Carbohyd. Res. 75: 265–274.

Keys, A., Menotti, A., Karvonen, M.J., Aravanis, C., Blackburn, H., Buzina, R., Djordjevic, B.S., Dontas, A.S., Fidanza, F. and Keys, M.H. 1986. The diet and 15-year death rate in Bagigo the seven countries study. Amer. J. Epidemiol. 124(6): 903–915.

Kim, S., Shin, B.C., Lee, M.S., Lee, H. and Ernst, E. 2011. Red ginseng for type 2 *Diabetes mellitus:* a systematic review of randomized controlled trials. Chin. J. Integ. Med. 17(12): 937–944.

Knott, A., Reuschlein, K., Mielke, H., Wensorra, U., Mummert, C., Koop, U. and Gallinat, S. 2008. Natural *Arctium lappa* fruit extract improves the clinical signs of aging skin. J. Cosmetic. Dermatol. 7(4): 281–289.

Koo, H.M. and Mohamed, S. 2001. Flavonoid (myricetin, quercetin, kaempferol, luteolin, and apigenin) content of edible tropical plants. J. Agricult. Food Chem. 49(6): 3106–3112.

Kotov, A.G., Khvorost, P.P. and Komissarenko, N.F. 1991. Coumarins of *Matricaria recutita*. Chem. Nat. Comp. 27(6): 753–753.

Krinsky, N.I. and Yeum, K.J. 2003. Carotenoid-radical interactions. Biochem. Bioph. Res. Comm. 305(3): 754–760.

Kubo, I., Muroi, H., Kubo, A., Chaudhuri, S.K., Sanchez, Y. and Ogura, T. 1994. Antimicrobial agents from *Heterotheca inuloides*. Planta. Med. 60(3): 218–221.

Kumar, P., Mishra, S., Malik, A. and Satya, S. 2012. Compositional analysis and insecticidal activity of *Eucalyptus globulus* (family: Myrtaceae) essential oil against housefly (*Musca domestica*). Acta Trop. 122(2): 212–218.

Kumar, S. and Sharma, A. 2005. Anti-anxiety activity studies on homoeopathic formulations of *Turnera aphrodisiaca* Ward. Evid. Compl. Alt. Med. 2(1): 117–119.

Kumar, S., Das, M., Singh, A., Ram, G., Mallavarapu, G.R. and Ramesh, S. 2001. Composition of the essential oils of the flowers, shoots and roots of two cultivars of *Chamomilla recutita*. J. Med. Aromat. Plant Sci. 23(4): 617–623.

Kunde, R. and Isaac, O. 1980. On the flavones of chamomile (*Matricaria chamomilla* L.) and a new acetylated apigenin-7-glucoside. Planta Med. 37: 124–30.

La Porte, E., Sarris, J., Stough, C. and Scholey, A. 2011. Neurocognitive effects of kava (*Piper methysticum*): a systematic review. Human Psychopharmacol. 26(2): 102–111.

La, V.D., Lazzarin, F., Ricci, D., Fraternale, D., Genovese, S., Epifano, F. and Grenier, D. 2010. Active principles of *Grindelia robusta* exert antiinflammatory properties in a macrophage model. Phytother. Res. 24(11): 1687–1692.

Laitinen, L., Takala, E., Vuorela, H., Vuorela, P., Kaukonen, A.M. and Marvola, M. 2007. Anthranoid laxatives influence the absorption of poorly permeable drugs in human intestinal cell culture model (Caco-2). Eur. J. Pharm. Biopharm. 66(1): 135–145.

Lawrence, B.M. 1987. Progress in essential oils. Perfume Flavorist. 12: 35–52.

Lee, S.J., Chung, H.Y., Maier, C.G.A., Wood, A.R., Dixon, R. and Mabry, T.J. 1998. Estrogenic flavonoids from *Artemisia vulgaris* L. J. Agricult. Food Chem. 46(8): 3325–3329.

Lehrke, M. and Lazar, M.A. 2005. The many faces of PPAR gamma. Cell. 123(6): 993–999.

Lenaghan, S. and Zhang, M. 2012. Real-time observation of the secretion of a nanocomposite adhesive from English ivy (*Hedera helix*). Plant Sci. 183: 206–211.

Li, H., Zhou, P., Yang, Q., Shen, Y., Deng, J., Li, L. and Zhao, D. 2011a. Comparative studies on anxiolytic activities and flavonoid compositions of *Passiflora edulis* 'edulis' and *Passiflora edulis* 'flavicarpa'. J. Ethnopharmacol. 133(3): 1085–1090.

Li, H., Deng, Z., Liu, R., Young, J.C., Zhu, H., Loewen, S. and Tsao, R. 2011b. Characterization of phytochemicals and antioxidant activities of a purple tomato (*Solanum lycopersicum* L.). J. Agricult. Food Chem. 59(21): 11803–11811.

Li, L., Tsao, R., Liu, Z., Liu, S., Yang, R., Young, J.C., Zhu, H., Deng, Z., Xie, M. and Fu, Z. 2005. Isolation and purification of acteoside and isoacteoside from *Plantago psyllium* L. by high-speed counter-current chromatography. J. Chromat. A. 1063(1-2): 161–169.

Li, Q., Xia, L., Zhang, Z. and Zhang, M. 2010. Ultraviolet extinction and visible transparency by ivy nanoparticles. Nanoscale Res. Lett. 5(9): 1487–1491.

Li, Y., Ma, W.J., Qi, B.K., Rokayya, S., Li, D., Wang, J., Feng, H.X., Sui, X.N. and Jiang, L.Z. 2014. Blending of soybean oil with selected vegetable oils: impact on oxidative stability and radical scavenging activity. Asian. Pac. J. Cancer. P. 15(6): 2583–2589.

Lin, C.Y. and Yin, M.C. 2012. Renal protective effects of extracts from guava fruit (*Psidium guajava* L.) in diabetic mice. Plant Food Hum. Nut. 67(3): 303–308.

Linares, E., Flores, B. and Bye, R. 1988. Selection of Medicinal Plants of Mexico. Limusa, Mexico City, 125 pp.

Linares, E., Flores, B. and Bye, R. 1995. Medicinal Plants of Mexico: Uses and Traditional Remedies, second ed. Electronic and Computer Technology Center and Biology Institute at the National Autonomous University of Mexico, Mexico City, pp. 1–155.

Linde, K., Barrett, B., Wolkart, K., Bauer, R. and Melchart, D. 2006. *Echinacea* for preventing and treating the common cold. Cochrane Database System Review, 2: CD000530.

Liu, R.L., Xiong, Q.J., Shu, Q., Wu, W.N., Cheng, J., Fu, H., Wang, F., Chen, J.G. and Hu, Z.L. 2012. Hyperoside protects cortical neurons from oxygen-glucose deprivation-reperfusion induced injury via nitric oxide signal pathway. Brain Res. 1469: 164–173.

Liu, W., Wang, J., Zhang, Z., Xu, J., Xie, Z., Slavin, M. and Gao, X. 2014. *In vitro* and *in vivo* antioxidant activity of a fructan from the roots of *Arctium lappa* L. Int. J. Biol. Macromol. 65: 446–453.

Livingston Raja, N.R. and Sundar, K. 2012. *Psidium guajava* Linn confers gastro protective effects on rats. Eur. Rev. Med. Pharmacol. Sci. 16(2): 151–156.

Long, S.R., Carey, R.A., Crofoot, K.M., Proteau, P.J. and Filtz, T.M. 2006. Effect of hawthorn (*Crataegus oxycantha*) crude extract and chromatographic fractions on multiple activities in a cultured cardiomyocyte assay. Phytomedicine. 13(9-10): 643–50.

Lozoya, X. 1976. El Instituto Mexicano para el estudio de las plantas medicinales, A.C. (IMEPLAM). pp. 243–248. *In*: X. Lozoya (ed). Estado actual del conocimiento en plantas medicinales mexicanas, IMEPLAM, México.

Lozoya-Legorreta, X., Rodríguez-Reynaga, D., Ortega-Galván, J. and Enriquez-Habib, R. 1978. Isolation of a hypotensive substance from seeds of *Casimiroa edulis*. Arch. Inv. Méd. 9(4): 565–573.

Lucas, L., Russell, A. and Keast, R. 2011. Molecular mechanisms of inflammation. Anti-inflammatory benefits of virgin olive oil and the phenolic compound oleocanthal. Curr. Pharm. Design 17(8): 754–768.

Lutterodt, G.D. 1992. Inhibition of Microlax-induced experimental diarrhoea with narcotic-like extracts of *Psidium guajava* leaf in rats. J. Ethnopharmacol. 37(2): 151–157.

Manday, E., Szoke EMuskath, Z. and Lemberkovics, E. 1999. A study of the production of essential oils in chamomile hairy root cultures. Eur. J. Drug. Metab. Ph. 24(4): 303–308.

Mann, C. and Staba, E.J. 1986. The chemistry, pharmacology, and commercial formulations of chamomile. pp. 235–280. *In*: L.E. Craker and J.E. Simon (eds.). Herbs, Spices, and Medicinal Plants: Recent Advances in Botany, Horticulture, and Pharmacology. Haworth Press Inc, USA.

Mata, R., Cristians, S., Escandón-Rivera, S., Juárez-Reyes, K. and Rivero-Cruz, I.J. 2013. Mexican antidiabetic herbs: valuable sources of inhibitors of α-glucosidases. J. Nat. Prod. 76(3): 468–483.

Matheson, H.B., Colon, I.S. and Story, J.A. 1995. Cholesterol 7 alpha-hydroxylase activity is increased by dietary modification with *psyllium* hydrocolloid, pectin, cholesterol and cholestyramine in rats. J. Nut. 125(3): 454–458.

Maclennan, K.M., Darlington, C.L. and Smith, P.F. 2002. The CNS effects of *Ginkgo biloba* extracts and ginkgolide B. Prog. Neurobiol. 67(3): 235–257.

MacKay, D. 2001. Hemorrhoids and varicose veins: a review of treatment options. Alt. Med. Rev. 6(2): 126–140.

Machowetz, A., Poulsen, H.E., Gruendel, S., Weimann, A., Fitó, M., Marrugat, J., de la Torre, R., Salonen, J.T., Nyyssönen, K., Mursu, J., Nascetti, S., Gaddi, A., Kiesewetter, H., Bäumler, H., Selmi, H., Kaikkonen, J., Zunft, H.J. and Madaleno, I.M. 2007. Ethno-Pharmacology in Latin America, an alternative to the globalization of the practices of cure. Cuadernos Geograficos 41: 61–95.

Magielse, J., Arcoraci, T., Breynaert, A., van Dooren, I., Kanyanga, C., Fransen, E., Van Hoof, V., Vlietinck, A., Apers, S., Pieters, L. and Hermans, N. 2013. Antihepatotoxic activity of a quantified *Desmodium adscendens* decoction and D-pinitol against chemically-induced liver damage in rats. J. Ethnopharmacol. 146(1): 250–256.

Magos, G.A., Vidrio, H. and Enríquez, R. 1995. Pharmacology of *Casimiroa edulis*; III. Relaxant and contractile effects in rat aortic rings. J. Ethnopharmacol. 47(1): 1–8.

Mahmoud, F., Abul, H., Onadeko, B., Khadadah, M., Haines, D. and Morgan, G. 2000. *In vitro* effects of ginkgolide B on lymphocyte activation in atopic asthma: comparison with cyclosporin A. Jpn. J. Pharmacol. 83(3): 241–245.

Marlett, J.A., McBurney, M.I. and Slavin, J.L. 2002. Position of the American Dietetic Association: health implications of dietary fiber. J. Am. Diet. Ass. 102(7): 993–1000.

Martínez, A.L., González-Trujano, M.E., Aguirre-Hernández, E., Moreno, J., Soto-Hernández, M. and López-Muñoz, F.J. 2009. Antinociceptive activity of *Tilia americana* var. *mexicana* inflorescences and quercetin in the formalin test and in an arthritic pain model in rats. Neuropharmacol. 56(2): 564–571.

Martínez, A.L., González-Trujano, M.E., Chávez, M., Pellicer, F., Moreno, J. and López-Muñoz, F.J. 2011. Hesperidin produces antinociceptive response and synergistic interaction with ketorolac in an arthritic gout-type pain in rats. Pharmacol. Biochem. Beh. 97(4): 683–689.

Martínez, S.M.M., De la Paz, J.S., Corral, A.S. and Martínez, C.R. 2004. Actividad diurética y antipirética de un extracto fluido de *Rosmarinus officinalis* L. en ratas. Rev. Cub. Plantas Meds. 9(1): 1–7.

Matsuzaki, K., Ishii, R., Kobiyama, K. and Kitanaka, S. 2010. New benzophenone and quercetin galloyl glycosides from *Psidium guajava* L. J. Nat. Med. 64(3): 252–256.

Matthaus, B. and Ozcan, M.M. 2014. Fatty acid and tocopherol contents of several soybean oils. Nat. Prod. Res. 28(8): 589–92.

McKenna, D.J., Jones, K. and Hughes, K. 2001. Efficacy, safety, and use of *Ginkgo biloba* in clinical and preclinical applications. Alt. Therap. Health. M. 7(5): 70–86, 88–90.

Mejía-Dolores, J.W., Mendoza-Quispe, D.E., Moreno-Rumay, E.L., Gonzales-Medina, C.A., Remuzgo-Artezano, F., Morales-Ipanaqué, L.A. and Monje-Nolasco, R.C. 2014. Neurotoxic effect of aqueous extract of boldo (*Peumus boldus*) in an animal model]. Rev. Per. Med. Exp. Salud. Pub. 31(1): 62–68.

Mericli, A.H. 1990. The lipophilic compounds of a Turkish *Matricaria chamomilla* variety with no chamazuline in the volatile oil. Int. J. Crude Drug Research 28: 145–147.

Metwally, A.M., Omar, A.A., Harraz, F.M. and El Sohafy, S.M. 2010. Phytochemical investigation and antimicrobial activity of *Psidium guajava* L. leaves. Pharmacog. Mag. 6(23): 212.

Meydani, S.N., Lichtenstein, A.H., White, P.J., Goodnight, S.H., Elson, C.E., Woods, M., Gorbach, S.L. and Schaefer, E.J. 1991. Food use and health effects of soybean and sunflower oils. J. Amer. Coll. Nut. 10(5): 406–428.

Michel, T., Villarreal-de Puga, L.M. and Casillas-Ochoa, J. 1999. *Casimiroa edulis* seed properties in rats. J. Ethnopharmacol. 68(1-3): 275–282.

Mimaki, Y., Kuroda, M., Yokosuka, A. and Sashida, Y. 1999. A spirostanol saponin from the underground parts of *Ruscus aculeatus*. Phytochemistry 51(5): 689–692.

Mimica-Dukic, N., Simin, N., Cvejic, J., Jovin, E., Orcic, D. and Bozin, B. 2008. Phenolic compounds in field horsetail (*Equisetum arvense* L.) as natural antioxidants. Molecules 13(7): 1455–1464.

Miroddi, M., Calapai, G., Navarra, M., Minciullo, P.L. and Gangemi, S. 2013. *Passiflora incarnate* L.: Ethnopharmacology, clinical application, safety and evaluation of clinical trials. J. Ethnopharmacol. 150(3): 791–804.

Molina-Hernández, M., Téllez-Alcántara, P. and Martínez, E. 2000. *Agastache mexicana* may produce anxiogenic-like actions in the male rat. Phytomedicine 7(3): 199–203.

Molina-Hernández, M., Tellez-Alcántara, N.P., García, J.P., Lopez, J.I. and Jaramillo, M.T. 2004. Anxiolytic-like actions of leaves of *Casimiroa edulis* (Rutaceae) in male Wistar rats. J. Ethnopharmacology 93(1): 93–98.

Montes, D. and F. Montes. 2005. La medicina alternativa y complementaria, una opción institucional. Rev. San. Mil. Mex. 59: 385–388.

Monti, M.C., Margarucci, L., Tosco, A., Riccio, R. and Casapullo, A. 2011. New insights on the interaction mechanism between tau protein and oleocanthal, an extra-virgin olive-oil bioactive component. Food Funct. 2(7): 423–428.

Mora, S., Diaz-Veliz, G., Lungenstrass, H., García-González, M., Coto-Morales, T., Poletti, C., De Lima, T.C., Herrera-Ruiz, M. and Tortoriello, J. 2005. Central nervous system activity of the hydroalcoholic extract of *Casimiroa edulis* in rats and mice. J. Ethnopharmacol. 97(2): 191–197.

Morales, M.A., Tortoriello, J., Meckes, M., Paz, D. and Lozoya, X. 1994. Calcium-antagonist effect of quercetin and its relation with the spasmolytic properties of *Psidium guajava* L. Arch. Med. Res. 25(1): 17–21.

Morales, M.A., Hernández, D., Bustamante, S., Bachiller, I. and Rojas, A. 2009. Is senna laxative use associated to cathartic colon, genotoxicity, or carcinogenicity? J.Toxicol. article id 287247.

Mortensen, P.B. and Nordgaard-Andersen, I. 1993. The dependence of the *in vitro* fermentation of dietary fibre to short-chain fatty acids on the contents of soluble non-starch polysaccharides. Scand. J. Gastroenterol. 28(5): 418–422.

Moss, M., Cook, J., Wesnes, K. and Duckett, P. 2003. Aromas of rosemary and lavender essential oils differentially affects cognition and mood in healthy adults. Int. J. Neurosc. 113(1): 15–38.

Motamedi, H., Darabpour, E., Gholipour, M. and Nejad, S.M.S. 2010. Antibacterial effect of ethanolic and methanolic extracts of *Plantago ovata* and *Oliveria decumbens* endemic in Iran against some pathogenic bacteria. Int. J. Pharmacol. 6(2): 117–122.

Nah, S.Y. 2014. *Ginseng* ginsenoside pharmacology in the nervous system: involvement in the regulation of ion channels and receptors. Front. Physiol. 5, article 98.

National Toxicology Program. 2012. Toxicology study of senna (CAS No. 8013-11-4) in C57BL/6NTAC Mice and toxicology and carcinogenesis study of senna in genetically modified C3B6. 129F1/Tac-Trp53tm1Brd haploinsufficient mice (Feed Studies). Natl. Toxicol. Progr. Genet. Mod. Model. Report. (15): 1–114.

Navarro-Ruíz, A., Bastidas Ramírez, B.E., García Estrada, J., García López, P. and Garzón, P. 1995. Anticonvulsant activity of *Casimiroa edulis* in comparison to phenytoin and phenobarbital. J. Ethnopharmacol. 45(3): 199–206.

N'gouemo, P., Baldy-Moulinier, M. and Nguemby-Bina, C. 1996. Effects of an ethanolic extract of *Desmodium adscendens* on central nervous system in rodents. J. Ethnopharmacol. 52(2): 77–83.

Nordgaard, I., Hove, H., Clausen, M.R. and Mortensen, P.B. 1996. Colonic production of butyrate in patients with previous colonic cancer during long-term treatment with dietary fibre (*Plantago ovate* seeds). Scand. J. Gastroenterol. 31(10): 1011–1020.

Ohe, C., Miami, M., Hasegawa, C., Ashida, K., Sugino, M. and Kanamori, H. 1995. Seasonal variation in the production of the head and accumulation of glycosides in the head of *Matricaria chamomilla*. Act. Hortic. 390: 75–82.

Ojewole, J.A. 2005. Hypoglycemic and hypotensive effects of *Psidium guajava* Linn. (Myrtaceae) leaf aqueous extract. Method. Find. Exp. Clin. Pharmacol. 27(10): 689–95.

Ojewole, J.A. 2006. Antiinflammatory and analgesic effects of *Psidium guajava* Linn. (Myrtaceae) leaf aqueous extract in rats and mice.Method. Find. Exp. Clin. Pharmacol. 28(7): 441–446.

Ojewole, J.A., Awe, E.O. and Chiwororo, W.D. 2008. Antidiarrhoeal activity of *Psidium guajava* Linn. (Myrtaceae) leaf aqueous extract in rodents. J. Smooth. Mus. Res. 44(6): 195–207.

Onanong Kulaputana, M.D. 2007. Ginseng supplementation does not change lactate threshold and physical performances in physically active Thai men. J. Med. Assoc. Thai. 90(6): 1172–9.

Ozgoli, G., Selselei, E.A., Mojab, F. and Majd, H.A. 2009. A randomized, placebo-controlled trial of *Ginkgo biloba* L. in treatment of premenstrual syndrome. J. Alt. Compl. Med. 15(8): 845–851.

Pande, S. and Srinivasan, K. 2013. Protective effect of dietary tender cluster beans (*Cyamopsis tetragonoloba*) in the gastrointestinal tract of experimental rats. Appl. Physiol. Nutr. Me. 38(2): 169–176.

Park, E.H., Kahng, J.H., Lee, S.H. and Shin, K.H. 2001. An anti-inflammatory principle from cactus. Fitoterapia 72(3): 288–290.

Pastors, J.G., Blaisdell, P.W., Balm, T.K., Asplin, C.M. and Pohl, S.L. 1991. *Psyllium* fiber reduces risein postprandial glucose and insulin concentrations in patients with non-insulin-dependent diabetes. Am. J. Clin. Nut. 53(6): 1431–1435.

Pavón, N.P. and Rico-Gray, V. 2000. An endangered and potentially economic tree of Mexico: *Tilia mexicana* (Tiliaceae). Econ. Bot. 54(1): 113–114.

Perales, Y.J. and Leysa, M. 2012. Phytochemical screening and antibacterial acitivity of *Bougainvillea glabra* plant extract as potential sources of antibacterial and resistance-modifying Agents. Int. Proc. Chem. Biol. Env. Engin. 45(25): 121–125.

Pérez-Gutierrez, R. 2012. Evaluation of hypoglycemic activity of the leaves of *Malva parviflora* in streptozotocin-induced diabetic rats.Food. Funct. 3(4): 420–427.

Pérez-Ortega, G., Guevara-Fefer, P., Chávez, M., Herrera, J., Martínez, A., Martínez, A.L. and González-Trujano, M.E. 2008. Sedative and anxiolytic efficacy of *Tilia americana* var. *mexicana* inflorescences used traditionally by communities of State of Michoacan, Mexico. J. Ethnopharmacol. 116(3): 461–468.

Petchetti, L., Frishman, W.H., Petrillo, R. and Raju, K. 2007. Nutriceuticals in cardiovascular disease: *psyllium*. Cardiol. Rev. 15(3): 116–122.

Petroni, A., Blasevich, M., Salami, M., Papini, N., Montedoro, G.F. and Gallia, C. 1995. Inhibition of platelet aggregation and eicosanoid production by phenolic components of olive oil. Thromb. Res. 78(2): 151–160.

Pino, J.A., Bayat, F., Marbot, R. and Aguero, J. 2002. Essential oil of *Chamomilla recutita* (L.) Rausch from Iran. J. Essent. Oil. Res. 14(6): 407–408.

Pirzad, A., Alyari, H., Shakiba, M.R., Zehtab-Salmasi, S. and Moammadi, A. 2006. Essential oil content and composition of German chamomile (*Matricaria chamomilla* L.) at different irrigation regimes. J. Agron. 5(3): 451–455.

Pittler, M.H., Schmidt, K. and Ernst, E. 2005. Adverse events of herbal food supplements for body weight reduction: systematic review. Obes. Review 6(2): 93–111.

Potter, S.M., Baum, J.A., Teng, H., Stillman, R.J., Shay, N.F. and Erdman Jr., J.W. 1998. Soy protein and isoflavones: Their effects on blood lipids and bone density in postmenopausal women. Am. J. Clin. Nut. 68(6): 1375S–1379S.

Prynne, C.J. and Southgate, D.A. 1979. The effects of a supplement of dietary fibre on faecal excretion by human subjects. Brit. J. Nut. 41(3): 495–503.

Purmova, J. and Opletal, L. 1995. Phytotherapeutic aspects of diseases of the cardiovascular system. 5. Saponins and possibilities of their use in prevention and therapy. Ceska a Slovenska farmacie: casopis Cesk. Farm. Spol. Slov. Farm. Spol. 44(5): 246–251.

Qi, B., Liu, L., Zhang, H., Zhou, G.X., Wang, S., Duan, X.Z. and Zhao, D.Q. 2014. Anti-fatigue effects of proteins isolated from *Panax quinquefolium*. J. Ethnopharmacol. 153(2): 430–434.

Rahn, K. 1996. A phylogenetic study of the *Plantaginaceae*. Bot. J. Linn. Soc. 120(2): 145–198.

Rai, A. 2013. The antiinflammatory and antiarthritic properties of ethanol extract of *Hedera helix*. Indian J. Pharm. Sci. 75(1): 99–102.

Ramírez, P., Senorans, F.J., Ibáñez, E. and Reglero, G. 2004. Separation of rosemary antioxidant compounds by supercritical fluid chromatography on coated packed capillary columns. J. Chromatog. A. 1057(1-2): 241–245.

Rattanachaikunsopon, P. and Phumkhachorn, P. 2007. Bacteriostatic effect of flavonoids isolated from leaves of *Psidium guajava* on fish pathogens. Fitoterapia 78(6): 434–436.

Redaelli, C., Formentini, L. and Santaniello, E. 1981. HPLC determination of coumarins in *Matricaria chamomilla*. Planta. Med. 43(4): 412–413.

Reichling, J., Bisson, W. and Becker, H. 1984. Comparative-study on the production and accumulation of essential oil in the whole plant and in the callus-culture of *Matricaria chamomilla*. Planta. Med. 50(4): 334–337.

Repcak, M., Halasova, J., Honcariv, R. and Podhradsky, D. 1980. The content and composition of the essential oil in the course of anthodium development in wild camomile (*Matricaria chamomilla* L.). Biol. Plantarum. 22(3): 183–191.

Rhee, M.Y., Kim, Y.S., Bae, J.H., Nah, D.Y., Kim, Y.K., Lee, M.M. and Kim, H.Y. 2011. Effect of Koreanred ginseng on arterial stiffness in subjects with hypertension. J. Alt. Compl. Med. 17(1): 45–49.

Rodrigues, M.R., Kanazawa, L.K., das Neves, T.L., da Silva, C.F., Horst, H., Pizzolatti, M.G., Santos, A.R., Baggio, C.H. and Werner, M.F. 2012. Antinociceptive and anti-inflammatory potential of extract and isolated compounds from the leaves of *Salvia officinalis* in mice. J. Ethnopharmacol. 139(2): 519–526.

Rojas-Duran, R., González-Aspajo, G., Ruiz-Martel, C., Bourdy, G., Doroteo-Ortega, V.H., Alban-Castillo, J., Robert, G., Auberger, P. and Deharo, E. 2012. Anti-inflammatory activity of mitraphylline isolated from *Uncaria tomentosa* bark. J. Ethnopharmacol. 143(3): 801–804.

Romero, A.L., West, K.L., Zern, T. and Fernandez, M.L. 2002. The seeds from *Plantago ovate* lower plasma lipids by altering hepatic and bile acid metabolism in guinea pigs. J. Nut. 132(6): 1194–1198.

Romero, C., Medina, E., Vargas, J., Brenes, M. and De Castro, A. 2007. *In vitro* activity of olive oil polyphenols against *Helicobacter pylori*. J. Agric. Foo. Chem. 55(3): 680–686.

Romero-Cerecero, O., Zamilpa, A., Jiménez-Ferrer, J.E., Rojas-Bribiesca, G., Román-Ramos, R. and Tortoriello, J. 2008. Double-blind clinical trial for evaluating the effectiveness and tolerability of *Ageratina pichinchensis* extract on patients with mild to moderate onychomycosis. A comparative study with ciclopirox. Planta. Med. 74(12): 1430–1435.

Russo, G.I., Cimino, S., Salamone, C., Madonia, M., Favilla, V., Castelli, T. and Morgia, G. 2013. Potential efficacy of some African plants in benign prostatic hyperplasia and prostate cancer. Mini-Rev. Med. Chem. 13(11): 1564–1571.

Russo, P., Frustaci, A., Del Bufalo, A., Fini, M. and Cesario, A. 2013. From traditional European medicine to discovery of new drug candidates for the treatment of dementia and Alzheimer's disease: acetylcholinesterase inhibitors. Curr. Med. Chem. 20(8): 976–983.

Sabaratnam, V., Kah-Hui, W., Naidu, M. and Rosie, P.D. 2013. Neuronal health—can culinary and medicinal mushrooms help? J. Trad. Compl. Med. 3(1): 62–68.

Sahranavard, S., Kamalinejad, M. and Faizi, M. 2014. Evaluation of anti-Inflammatory and anti-nociceptive effects of defatted fruit extract of *Olea europaea*. Iran. J. Pharm. Res. 3(Suppl.): 119–123.

Sakatani, K., Tanida, M., Hirao, N. and Takemura, N. 2014. Ginkobiloba Extract Improves Working Memory Performance in Middle-Aged Women: Role of Asymmetry of Prefrontal Cortex Activity During a Working Memory Task. In Oxygen Transport to Tissue XXXVI (pp. 295–301): Springer New York.

Salama, R.H. 2012. *Matricaria chamomilla* attenuates cisplatin nephrotoxicity. Saudi. J. Kid. Dis. Transpl. 23(4): 765.

Salamon, I. 1992a. Chamomile: a medicinal plant. The Herb, spice and medicinal plant digest. 10: 1–4.

Salamon, I. 1992b. Production of chamomile, *Chamomilla recutita* (L.) Rauschert. Slovak. J. Herbs. Spices Med. Plant 1(1-2): 37–45.

Sarris, J., Kavanagh, D.J., Deed, G. and Bone, K.M. 2009. St. John's wort and Kava in treating major depressive disorder with comorbid anxiety: a randomised double-blind placebo-controlled pilot trial. Hum. Psychopharm. 24(1): 41–48.

Sarris, J., Scholey, A., Schweitzer, I., Bousman, C., Laporte, E., Ng, C., Murray, G. and Stough, C. 2012. The acute effects of kava and oxazepam on anxiety, mood, neurocognition; and genetic correlates: a randomized, placebo-controlled, double-blind study. Hum. Psychopharm. 27(3): 262–269.

Schilcher, H., Imming, P. and Goeters, S. 2005. Active chemical constituents of *Matricaria chamomilla* L. syn. *Chamomilla recutita* (L.) Rauschert. pp. 56–76. *In*: R. Franke and H. Schilcher (eds.). Chamomile: Industrial Profiles. CRC Press, Boca Raton.

Schmidt, M., Thomsen, M. and Schmidt, U. 2012a. Suitability of ivy extract for the treatment of paediatric cough. Phytot. Res. 26(12): 1942–1947.

Schmidt, J.S., Lauridsen, M.B., Dragsted, L.O., Nielsen, J. and Staerk, D. 2012b. Development of a bioassay-coupled HPLC-SPE-ttNMR platform for identification of alpha-glucosidase inhibitors in apple peel (*Malus domestica* Borkh.). Food Chem. 135(3): 1692–1699.

Sebai, H., Jabri, M.A., Souli, A., Rtibi, K., Selmi, S., Tebourbi, O., El-Benna, J. and Sakly, M. 2014. Antidiarrheal and antioxidant activities of chamomile (*Matricaria recutita* L.) decoction extract in rats. J. Ethnopharmacol. 152(2): 327–332.

Seifi, H., Masoum, S., Seifi, S. and Ebrahimabadi, E.H. 2014. Chemometric resolution approaches in characterisation of volatile constituents in *Plantago ovata* seeds using gaschromatography-mass spectrometry: methodology and performance assessment. Phytochem. Analysis 25(3): 273–281.

Sen, S., Chen, S., Feng, B., Wu, Y., Lui, E. and Chakrabarti, S. 2012. Preventive effects of North American ginseng (*Panax quinquefolium*) on diabetic nephropathy. Phytomedicine 19(6): 494–505.

Shale, T.L., Stirk, W. and Van Staden, J. 1999. Screening of medicinal plants used in Lesotho for antibacterial and anti inflamatory activity. J. Ethnopharmacol. 67(3): 347–354.

Sharifi, F., Simbar, M., Mojab, F. and Majd, H.A. 2014. Comparison of the effects of *Matricaria chamomilla* (Chamomile) extract and mefenamic acid on the intensity of premenstrual syndrome. Complement. Ther. Clin. Pract. 20(1): 81–88.

Sharma, S. Kaur, M., Goyal, R. and Gill, B.S. 2014. Physical characteristics and nutritional composition of some new soybean (*Glycine max* (L.) Merrill) genotypes. J. Food Sci. Technol. 51(3): 551–557.

Shao, M., Wang, Y., Jian, Y.Q., Huang, X.J., Zhang, D.M., Tang, Q.F., Jiang, R.W., Sun, X.G., Lv, Z.P., Zhang, X.Q. and Ye, W.C. 2012. Guadial A and psiguadials C and D, three unusual meroterpenoids from *Psidium guajava*. Org. Lett. 14(20): 5262–5265.

Shen, S.C., Cheng, F.C. and Wu, N.J. 2008. Effect of guava (*Psidium guajava* L.) leaf soluble solids on glucose metabolism in type 2 diabetic rats. Phytot. Res. 22(11): 1458–1464.

Sheng, Y., Akesson, C., Holmgren, K., Bryngelsson, C., Giamapa, V. and Pero, R.W. 2005. An active ingredient of Cat's Claw water extracts identification and efficacy of quinic acid. J. Ethnopharmacol. 96(3): 577–584.

Shnitzler, A.C., Nolan, L.L. and Labbe, R. 1996. Screening of medicinal plants for antileishmanial and antimicrobial activity. Act. Hort. 426: 235–242.

Shu, J.C., Chou, G.X. and Wang, Z.T. 2010. One new galloyl glycoside from fresh leaves of *Psidium guajava* L. Acta Pharm. Sin. 45(3): 334–337.

Shu, J.C., Chou, G.X. and Wang, Z.T. 2012. One new diphenylmethane glycoside from the leaves of *Psidium guajava* L. Nat. Prod. Res. 26(21): 1971–1975.

Sieben, A., Prenner, L., Sorkalla, T., Wolf, A., Jakobs, D., Runkel, F. and Häberlein, H. 2009. α-Hederin, but not hederacoside c and hederagenin from *Hedera helix*, affects the binding behavior, dynamics, and regulation of β2-adrenergic receptors. Biochem. 48(15): 3477–3482.

Sienkiewicz, M., Łysakowska, M., Denys, P. and Kowalczyk, E. 2012. The antimicrobial activity of thyme essential oil against multidrug resistant clinical bacterial strains. Microbial. Drug. Resis. 18(2): 137–148.

Silvestri, I., Cattarino, S., Aglianò, A., Nicolazzo, C., Scarpa, S., Salciccia, S., Frati, L., Gentile, V. and Sciarra, A. 2013. Effect of *Serenoa repens* (Permixon®) on the expression of inflammation-related genes: analysis in primary cell cultures of human prostate carcinoma. J. Inflamm. 10: 11–19.

Simirgiotis, M.J. 2013. Antioxidant capacity and HPLC-DAD-MS profiling of Chilean peumo (*Cryptocarya alba*) fruits and comparison with German peumo (*Crataegus monogyna*) from southern Chile. Molecules 18(2): 2061–2080.

Sinclair, S. 1998. Chinese herbs: a clinical review of *Astragalus, Ligusticum,* and *Schizandrae.* Alt. Med. Rev. 3(5): 338–344.

Singh, B. 2007. *Psyllium* as therapeutic and drug delivery agent. Int. J. Pharm. 334(1-2): 1–14.

Singh, O., Khanam, Z., Misra, N. and Srivastava, M.K. 2011. Chamomile (*Matricaria chamomilla* L.). An overview. Pharmacog. Rev. 5(9): 82–95.

Smith, O. 2009. Soy isoflavone supplementation and bone mineral density in menopausal women: a 2-y multicenter clinical trial. Am. J. Clin. Nut. 90(5): 1433–1439.

Smith, P.F., Maclennan, K. and Darlington, C.L. 1996. The neuroprotective properties of the *Ginkgo biloba* leaf: a review of the possible relationship to platelet-activating factor (PAF). J. Ethnopharmacol. 50(3): 131–139.

Sneddon, J. 1997. Encyclopedia of Common Natural Ingredients Used in Food, Drugs, and Cosmetics. John Wiley and Sons, New York.

Soman, S., Rauf, A.A., Indira, M. and Rajamanickam, C. 2010. Antioxidant and antiglycative potential of ethyl acetate fraction of *Psidium guajava* leaf extract in streptozotocin-induced diabetic rats. Plant. Food. Hum. Nut. 65(4): 386–391.

Song, J.H., Yeo, S.G., Hong, E.H., Lee., B.R., Kim, J., Kim, J., Jeong, H., Kwon, Y., Kim, H., Lee, S., Park, J.H. and Ko, H.J. 2014. Antiviral Activity of Hederasaponin B from *Hedera helix* against Enterovirus 71 Subgenotypes C3 and C4a. Biomol. Ther. 22(1): 41–46.

Sotelo-Felix, J.I., Martínez-Fong, D., Muriel, P., Santillán, R.L., Castillo, D. and Yahuaca, P. 2002. Evaluation of the effectiveness of *Rosmarinus officinalis* (Lamiaceae) in the alleviation of carbon tetrachloride-induced acute hepatotoxicity in the rat. J. Ethnopharmacol. 81(2): 145–154.

Soto, L.G. 2005. Studies in history and philosophy of science Part C: Studies in history and philosophy of biological and biomedical sciences. Drug. Traject. 36: 743–760.

Spelman, K., Burns, J., Nichols, D., Winters, N., Ottersberg, S. and Tenborg, M. 2006. Modulation of cytokine expression by traditional medicines: a review of herbal immunomodulators. Alt. Med. Rev. 11(2): 128–150.

Sreekanth, D., Arunasree, M.K., Roy, K.R., Chandramohan, Reddy, T., Reddy, G.V. and Reddanna, P. 2007. Betanin a betacyanin pigment purified from fruits of *Opuntia ficus-indica* induces apoptosis in human chronic myeloid leukemia cell line K562. Phytomedicine 14(11): 739–746.

Stahl, W. and Sies, H. 2003. Antioxidant activity of carotenoids. Molecular Aspects of Medicine 24(6): 345–351.

Subiza, J., Subiza, J.L., Alonso, M., Hinojosa, M., Garcia, R., Jerez, M. and Subiza, E. 1990. Allergic conjunctivitis to chamomile tea. Ann. Allergy 65(2): 127–132.

Sumner, H., Salan, U., Knight, D.W. and Hoult, J.R.S. 1992. Inhibition of 5-lipoxygenase and cyclo-oxygenase in leukocytes by feverfew: Involvement of sesquiterpene lactones and other components. Biochem. Pharmacol. 43(11): 2313–2320.

Sunagawa, M., Shimada, S., Zhang, Z., Oonishi, A., Nakamura, M. and Kosugi, T. 2004. Plasma insulin concentration was increased by long-term ingestion of guava juice in spontaneous non-insulin dependent *diabetes mellitus* (NIDDM) rats. J. Health. Sci. 50(6): 674–678.

Suter, A., Bommer, S. and Rechner, J. 2006. Treatment of patients with venous insufficiency with fresh plant horse chestnut seed extract: a review of 5 clinical studies. Adv. Ther. 23(1): 179–190.

Taavoni, S., Nazem Ekbatani, N. and Haghani, H. 2013. Valerian/lemon balm use for sleep disorders during menopause. Comp. Ther. Clin. Pract. 19(4): 193–196.

Tamasdan, S., Cristea, E. and Mihele, D. 1981. Action upon gastric secretion of Robiniae flores, *Chamomillae* flores and *Strobuli lupuli* extracts. Farmacia. 29: 71–75.

Tassell, M.C., Kingston, R., Gilroy, G., Lehane, M. and Furey, A. 2010. Hawthorn (*Crataegus* spp.) in the treatment of cardiovascular disease. Pharmacog. Rev. 4(7): 32–41.

Teschke, R., Sarris, J. and Lebot, V. 2011. Kava hepatotoxicity solution: A six-point plan for new kava standardization. Phytomedicine 18(2-3): 96–103.

Tittel, G., Wagner, H. and Bos, R. 1982. Chemical-composition of the essential oil from melissa. Planta. Med. 46(2): 91–98.

Tomić, M. and Popović, V. Petrović, S. Stepanović-Petrović, R., Micov, A., Pavlović-Drobac, M. and Couladis, M. 2014. Antihyperalgesic and antiedematous activities of bisabolol-oxides-richmatricaria oil in a rat model of inflammation. Phytot. Res. 28(5): 759–766.

Tripoli, E., Giammanco, M., Tabacchi, G., Di Majo, D., Giammanco, S. and La Guardia, M. 2005. The phenolic compounds of olive oil: structure, biological activity and beneficial effects on human health. Nut. Res. Rev. 18(1): 98–112.

Trute, A., Gross, J., Mutschler, E. and Nahrstedt, A. 1997. *In vitro* antispasmodic compounds of the dry extract obtained from *Hedera helix*. Planta. Med. 63(2): 125–129.

Tuck, K.L. and Hayball, P.J. 2002. Major phenolic compounds in olive oil: metabolism and health effects. J. Nut. Biochem. 13(11): 636–644.

Turnbull, W.H. and Thomas, H.G. 1995. The effect of *Plantago ovata* seed containing preparation on appetite variables, nutrient and energy intake. Int. J. Obesity. Rel. Metabol. Disor. 19(5): 338–342.

Uehleke, B., Ortiz, M. and Stange, R. 2008. Cholesterol reduction using *psyllium* husks-do gastrointestinal adverse effects limit compliance? Results of a specific observational study. Phytomedicine 15(3): 153–159.

Uylaşer, V. and Yildiz, G. 2014. The historical development and nutritional importance of olive and olive oil constituted an important part of the Mediterranean diet. Crit. Rev. Food Sci. Nut. 54(8): 1092–1101.

Vanaclocha, B.V. and Folcara, S.C. (eds.). 2003. Fitoterapia: Vademécum de prescripción (pp. 153–154). Barcelona: Masson.

Van der Watt, G., Laugharne, J. and Janca, A. 2008. Complementary and alternative medicine in the treatment of anxiety and depression. Curr. Opin. Psych. 21(1): 37–42.

Vanscheidt, W., Jost, V., Wolna, P., Lücker, P.W., Müller, A., Theurer, C., Patz, B. and Grützner, K.I. 2002. Efficacy and safety of a Butcher's broom preparation (*Ruscus aculeatus* L. extract) compared to placebo in patients suffering from chronic venous insufficiency. Arzneimittel-forschung 52(4): 243–250.

Van Eck, J., Kirk, D.D. and Walmsley, A.M. 2006. Tomato (*Lycopersicum esculentum*). Method. Mol. Biol. 343: 459–474.

Van Gorkom, B.A., de Vries, E.G., Karrenbeld, A. and Kleibeuker, J.H. 1999. Review article: anthranoid laxatives and their potential carcinogenic effects. Aliment. Pharmacol. Ther. 13(4): 443–452.

Vassiliadis, T., Anagnostis, P., Patsiaoura, K., Giouleme, O., Katsinelos, P., Mpoumponaris, A. and Eugenidis, N. 2009. Valeriana hepatotoxicity. Sleep. Med. 10(8): 935.

Vasques, C.A., Schneider, R., Klein-Júnior, L.C., Falavigna, A., Piazza, I. and Rossetto, S. 2014. Hypolipemic effect of *Garcinia cambogia* in obese women. Phytother. Res. 28(6): 887–891.

Veerendra Kumar, M.H. and Gupta, Y.K. 2002. Effect of different extracts of *Centella asiatica* on cognition and markers of oxidative stress in rats. J. Ethnopharmacol. 79(2): 253–260.

Vergara-Jimenez, M., Conde, K., Erickson, S.K. and Fernandez, M.L. 1998. Hypolipidemic mechanisms of pectin and *psyllium* in guinea pigs fed high fat-sucrose diets: alterations on hepatic cholesterol metabolism. J. Lipid. Res. 39(7): 1455–1465.

Verma, A. and Banerjee, R. 2010. Dietary fibre as functional ingredient in meat products: a novel approach for healthy living—a review. J. Food. Sci. Technol. 47(3): 247–257.

Viesca, C. 1992. El *libellus* y su contexto histórico. pp. 49–85. *In*: Kumate Jesús (ed.). Estudios Actuales sobre el *Libellus de medicinalibus indorum herbis*. Secretaría de Salud, México DF.

Viesca, C. and Aranda, A. 1996. Las Alteraciones del Sueño en el *Libellus de medicinalibus indorum herbis*. Estudios de Cultura Náhuatl, XXVI, México, p. 156.

Viesca, C., Ramos de Viesca, M. and Aranda, A. 2000. Los Tratamientos Medicamentosos de las Enfermedades Mentales en la Medicina Náhuatl Prehispánica. Aceves Patrana, Patricia. (Editora). Tradiciones e Intercambios Científicos: Materia Médica, Farmacia y Medicina. UAM, México: 27–44.

Vigo, E., Cepeda, A., Gualillo, O. and Perez-Fernandez, R. 2004. *In-vitro* anti-inflammatory effect of *Eucalyptus globulus* and *Thymus vulgaris*: nitric oxide inhibition in J774A.1 murine macrophages. J. Pharm. Pharmacol. 56(2): 257–263.

Vigo, E., Cepeda, A., Gualillo, O. and Perez-Fernandez, R. 2005. *In-vitro* anti-inflammatory activity of *Pinus sylvestris* and *Plantago lanceolata* extracts: effect on inducible NOS, COX-1, COX-2 and their products in J774A.1 murine macrophages. J. Pharm. Pharmacol. 57(3): 383–391.

Vlachojannis, J.E., Cameron, M. and Chrubasik, S. 2010. A systematic review on the sambucifructus effect and efficacy profiles. Phytot. Res. 24(1): 1–8.

Vladimir-Knežević, S., Blažeković, B., Kindl, M., Vladić, J., Lower-Nedza, A.D. and Brantner, A.H. 2014. Acetylcholinesterase inhibitory, antioxidant and phytochemical properties of selected medicinal plants of the Lamiaceae family. Molecules 19(1): 767–782.

Voderholzer, W.A., Schatke, W., Mühldorfer, B.E., Klauser, A.G., Birkner, B. and Müller-Lissner, S.A. 1997. Clinical response to dietary fiber treatment of chronic constipation. Am. J. Gastroenterol. 92(1): 95–98.

Wang, Y., Qian, X.J., Hadley, H.R. and Lau, B.H. 1992. Phytochemicals potentiate interleukin-2 generated lymphokine-activated killer cell cytotoxicity against murine renal cell carcinoma. Mol. Biother. 4(3): 143–146.

Watanabe, T., Nakaya, N., Kurashima, K., Kuriyama, S., Tsubono, Y. and Tsuji, I. 2004. Constipation, laxative use and risk of colorectal cancer: The Miyagi Cohort Study. Eur. J. Cancer 40(14): 2109–2115.

Webster, D.J., Pugh, D.C. and Craven, J.L. 1978. The use of bulk evacuant in patients with hemorrhoids. Brit. J. Surg. 65(4): 291–292.

Weidner, C., Wowro, S.J., Rousseau, M., Freiwald, A., Kodelja, V., Abdel-Aziz, H., Kelber, O. and Sauer, S. 2013. Antidiabetic effects of chamomile flowers extract in obese mice through transcriptional stimulation of nutrient sensors of the peroxisome proliferator-activated receptor (PPAR) Family. PLOS ONE 8(11): e80335.

Williams, J.E. 2001. Review of antiviral and immunomodulating properties of plants of the Peruvian rainforest with a particular emphasis on uña de Gato and Sangre de Grado. Alt. Med. Rev. 6(6): 567–579.

Wong, W.W., Lewis, R.D., Steinberg, F.M., Murray, M.J., Cramer, M.A., Amato, P., Young, R.L., Barnes, S., Ellis, K.J., Shypailo, R.J., Fraley, J.K., Konzelmann, K.L., Fischer, J.G., Xia, L., Lenaghan, S.C., Zhang, M., Zhang, Z. and Li, Q. 2010. Naturally occurring nanoparticles from English ivy: an alternative to metal-based nanoparticles for UV protection. J. Nanobiotechnol. 8: 12–20.

Xia, L. Lenaghan, S.C., Zhang, M., Wu, Y., Zhao, X., Burris, J.N. and Stewart, Jr., C.N. 2011. Characterization of English ivy (*Hedera helix*) adhesion force and imaging using atomic force microscopy. J. Nanopart. Res. 13(3): 1029–1037.

Xu, H., Yu, X., Qu, S., Chen, Y., Wang, Z. and Sui, D. 2013. *In vivo* and *in vitro* cardioprotective effects of *panax quinquefolium* 20(S)-protopanaxadiol saponins (PQDS), isolated from *Panax quinquefolium*. Pharmazie. 68(4): 287–292.

Yamaguchi, N., Satoh-Yamaguchi, K. and Ono, M. 2009. *In vitro* evaluation of antibacterial, anticollagenase, and antioxidant activities of hop components (*Humulus lupulus*) addressing acne vulgaris. Phytomedicine 16(4): 369–376.

Yang, J., Wang, H.P. Zhou, L. and Xu, C.F. 2012. Effect of dietary fiber on constipation: a meta analysis. World. J. Gastroenterol. 18(48): 7378–7383.

Yang, L., Li, X.F., Gao, L., Zhang, Y.O. and Cai, G.P. 2012. Suppressive effects of quercetin-3-O- (6"-Feruloyl)-beta-D-galactopyranoside on adipogenesis in 3T3-L1 preadipocytes through down-regulation of PPARgamma and C/EBPalpha expression. Phytother. Res. 26(3): 438–444.

Yayalacı, Y., Celik, I. and Batı, B. 2014. Hepatoprotective and antioxidant activity of linden (*Tilia platyphyllos* L.) infusion against ethanol-induced oxidative stress in rats. J. Membrane. Biol. 247(2): 181–188.

Zhao, K.W. and Kong, H.Y. 1993. Effect of *Astragalan* on secretion of tumor necrosis factor in human peripheral mononuclear cells. Chin. J. Integr. Med. (hongguo Zhong Xi Yi Jie He Za Zhi) 13(5): 263–265.

Zhao, J., Khan, S.I., Wang, M., Vasquez, Y., Yang, M.H., Avula, B., Wang, Y.H., Avonto, C., Smillie, T.J. and Khan, I.A. 2014. Octulosonic acid derivatives from Roman chamomile (*Chamaemelum nobile*) with activities against inflammation and metabolic disorder. J. Nat. Prod. 77(3): 509–515.

Zhu, Y., Liu, Y., Zhan, Y., Liu, L., Xu, Y., Xu, T. and Liu. T. 2013. Preparative isolation and purification of five flavonoid glycosides and one benzophenone galloyl glycoside from *Psidium guajava* by High-Speed Counter-Current Chromatography (HSCCC). Molecules 18(12): 15648–15661.

Zimmer, A.R., Leonardi, B., Miron, D., Schapoval, E., Oliveira, J.R. and Gosmann, G. 2012. Antioxidant and anti-inflammatory properties of *Capsicum baccatum*: from traditional use to scientific approach. J. Ethnopharmacol. 139(1): 228–233.

Zolla, C. 1980. Traditional medicine in Latin America, with particular reference to Mexico. J. Ethnopharmacol. 2(1): 37–41.

Centella asiatica: From Ethnomedicinal Uses to the Possibility of New Drug Discovery

Abdullah Al-Nahain,[1] Rownak Jahan,[2] Taufiq Rahman,[3]
Md Nurunnabi[4] and Mohammed Rahmatullah[1,*]

Introduction

Observation of indigenous medicinal practices of various communities throughout the world has always proved to be an excellent route to the discovery of many important modern drugs (Gilani and Rahman 2005, Gohil et al. 2010). The importance of traditional medicinal practices has recently gained attention of the scientific community. A number of factors have contributed to this phenomenon. First, all ancient and quite a number of emerging modern diseases cannot be cured with allopathic medicine. Second, many modern (allopathic) medicines have developed drug-resistant vectors and microbes. Third, a number of allopathic medicines have adverse side-effects. Even common analgesic drugs like aspirin and paracetamol can develop gastric ulceration or hepatotoxicity from prolonged usage or over-dosage. Fourth, allopathic medicine is out of reach of many rural people because of high costs and lack of accessibility to modern doctors and clinics. Fifth, and finally, people still believe in traditional medicine from either habit or from finding such medicines beneficial (Jahan et al. 2012).

Among many reported medicinal plants, *Centella asiatica* has broadly contributed to treatment of different diseases (Ramaswamy et al. 1970, Brinkhaus et al. 2000, Höfling et al. 2010, Jahan et al. 2012). As a medicine it has been used in Ayurvedic tradition of India for thousands of years and listed in the historic '*Sushruta Samhita*', an ancient Indian medical text (Chopra et al. 1986, Diwan et al. 1991). This plant is also known as gotu kola and shady, moist, or marshy areas favor its growth. It is distributed widely in many parts of the world, including India, Sri Lanka, Madagascar, South Africa, Australia, China, and Japan (Zheng 2007, Satake et al. 2007).

---

[1] Department of Pharmacy, University of Development Alternative, Dhanmondi, Dhaka-1209, Bangladesh.
[2] Department of Biotechnology & Genetic Engineering, University of Development Alternative, Dhanmondi, Dhaka-1209, Bangladesh.
[3] Department of Pharmacology, University of Cambridge, Tennis Court Road, CB2 1PD, Cambridge, UK.
[4] Department of Chemical and Biological Engineering, Korea National University of Transportation, Chungju, Republic of Korea, 380-702.
* Corresponding author: rahamatm@hotmail.com

Centella asiatica (CA) is a small creeping herb with shovel shaped leaves emerging alternately in clusters at stem nodes. This is a prostate, sparingly hairy or nearly smooth herb. The stems root at the nodes. The leaves are rounded to reniform, 2 to 5 centimeters wide, horizontal, more or less cupped, rounded at the tip, and kidney-shaped or heart shaped at the base, the rounded lobes often overlapping. The petioles are erect and long. The peduncles occur in pairs of three, are less than 1 centimeter long, and usually bear three sessile flowers. The petals are dark-purple, ovate, and about 1 millimeter long. The fruit is minute, ovoid, white or green, and reticulate, each with nine sub similar longitudinal ridges. The runners lie along the ground and the inch long leaves with their scalloped edges rise above on long reddish petioles. The insignificant greenish- to pinkish-white flowers are borne in dense umbels on separate stems in the summer.

Since ancient times, it has been used to treat different diseases like wound healing, anti-inflammatory, memory enhancing, strength promoting, immune booster, anti-anxiety, anti-epilepsy and anti-stress substance (Han et al. 2003). *Centella asiatica* has been clinically used in mentally retarded children and also in treatment of anxiety neurosis. This plant has also been found to improve short-term memory and learning (Meena et al. 2012).

Although a large number of studies reported over the last few decades to search out biologically active components of the plant and their mechanisms of action, the outcome of these studies is still unsatisfactory. Although there have been several claims regarding the underlying mechanisms involved in the biological actions of this herb, more scientific data are needed to justify its ever increasing use. The potential therapeutic substances in *Centella asiatica* are saponin-containing triterpene acids and their sugar esters, of which asiatic acid, madecassic acid and asiaticosides are considered to be the most important (Brinkhaus et al. 2000).

The present chapter aims to list the detailed account of the plant, its activity regarding therapeutic uses, pharmacology, and mechanisms of action based on preclinical and clinical studies.

Ethnomedicinal uses

The major ethnomedicinal uses of the plant appears to be to alleviate gastrointestinal disorders like dysentery, constipation, stomach problems, indigestion and loss of appetite, and to enhance memory or to serve as nerve stimulant. Altogether 23 ethnomedicinal uses were collected from the available literature, of which six dealt with the plant's use in gastrointestinal disorders, and four uses were linked to memory or uses associated with brain functions like stimulation of nerve or for treatment of mental retardation. However, the uses of the plant were quite diverse overall, for the plant was used for treatment of headache, toothache, cuts and wounds, leucorrhea, skin disorders (like eczema, carbuncle), hemorrhoids, as an antidote to poison, urinary troubles and leucorrhea, pneumonia, syphilis, liver problems (like jaundice), sexual weakness in males, fever, sun stroke, rickets, cardiovascular disorders, leprosy, tuberculosis, asthma, and varicocele. Altogether, 12 ethnomedicinal reports were from India, three from Nepal, six from Bangladesh, and two from Africa. Thus the major ethnomedicinal uses of the plant seemed to be centered in the Indian sub-continent (Jahan et al. 2012).

Phytochemical constituents

It is reported that *Centella asiatica* contains mainly four active bio-active compounds. A list of summarized phytochemical constituents of *Centella asiatica* is presented in Table 2.1. Figure 2.1 shows the chemical structure of the relevant phytochemicals of *Centella asiatica*.

Pharmacological activity studies with CA and isolated constituents

Neuroprotective effect

In vivo studies in rats exhibited significant anti-oxidant properties by reducing brain regional lipid peroxidation (LPO) and protein carbonyl (PCO) levels after oral administration of *Centella asiatica* (300 mg/kg body weight/day) for 60 days (Subathra et al. 2005). In another study, significant reduction

Table 2.1. Phytochemical constituents of *Centella asiatica*.

Groups	Example	References
Triterpenes acid	Asiatic acid, madecassic acid, terminolic acid, centic acid, centoic acid, centelloic acid, indcentoic acid, brahmic acid, isobrahmic acid, betulic acid, madasiatic acid.	(Rumalla et al. 2010, Bhavna and Joyoti 2011)
Glycosides	Asiaticoside A, Asiaticoside B, madecassioside, centelloside.	(Bhavna and Joyoti 2011, Tabassum et al. 2013)
Alkaloids	hydrocotylin	(Bhavna and Joyoti 2011)
Flavonoids	Castiloferol, castelicetin	(Bhavna and Joyoti 2011)
Volatile and fatty acids	Oleic acid, linolenic acid, palmitic acid, stearic acid, lignoceric acid.	(Bhavna and Joyoti 2011)
Amino acids	Aspartic acid, glycine, glutamic acid, phenylalanine.	(Bhavna and Joyoti 2011)
Others	Sterols, tannins, sugars, inorganic salt.	(Bhavna and Joyoti 2011)

Asiatic acid   R1=R2=H
Madecassic acid   R1=OH, R2=H
Asiaticoside   R1=H, R2= Glu-Glu-Rha
Madecassoside   R1=OH, R2= Glu-Glu-Rha

Betulabuside A

Roseoside

chlorogenic acid

3,5-di-O-caffeoyl quinic acid (R2 = R4 = caffeoyl, R1 = R3 = H)

1,5-di-O-caffeoyl quinic acid (R1 = R4 = caffeoyl, R2 = R3 = H)

3,4-di-O-caffeoyl quinic acid (R2 = R3 = caffeoyl, R1 = R4 = H)

4,5-di-O-caffeoyl quinic acid (R3 = R4 = caffeoyl, R1 = R2 = H)

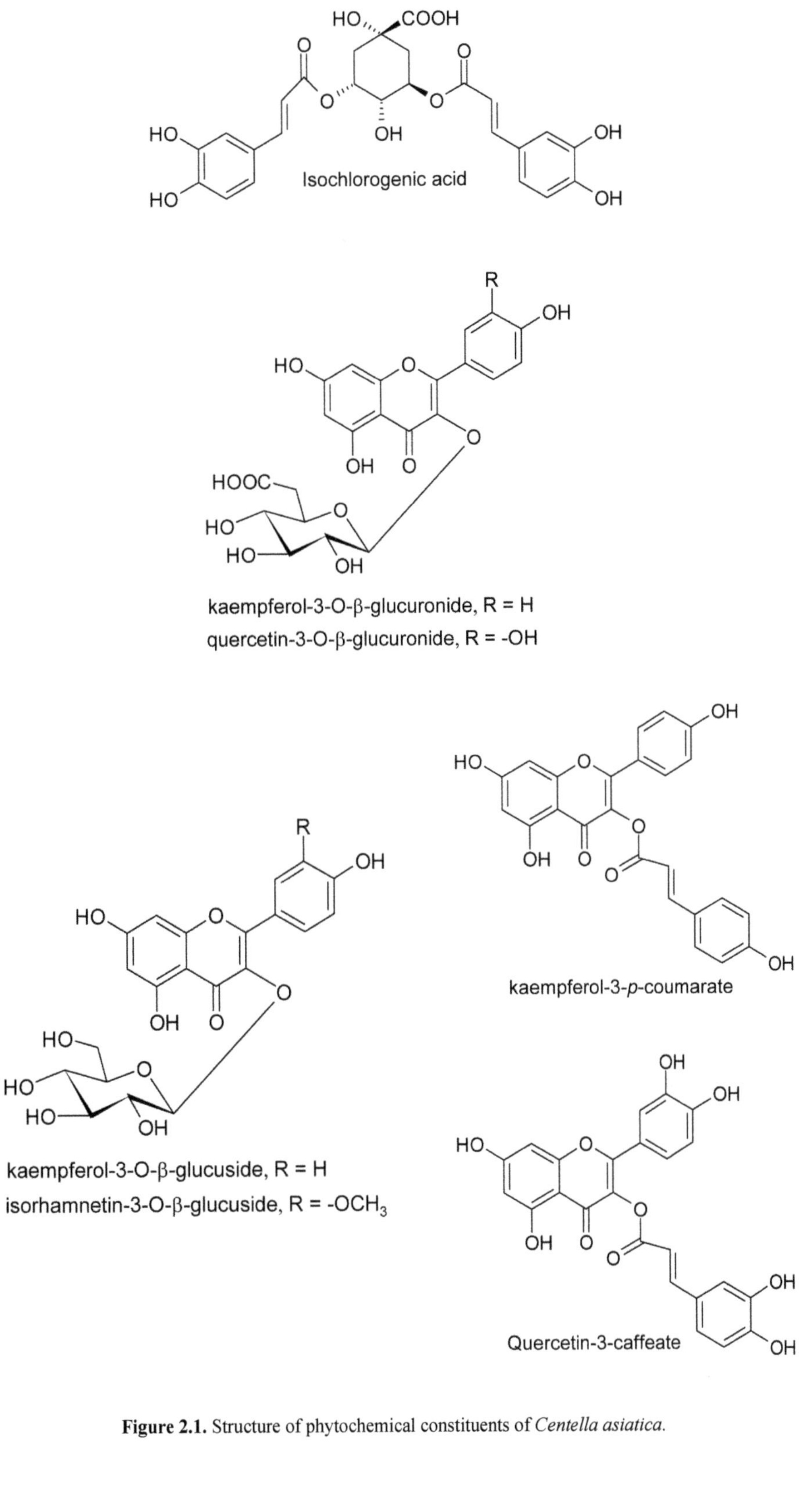

Isochlorogenic acid

kaempferol-3-O-β-glucuronide, R = H
quercetin-3-O-β-glucuronide, R = -OH

kaempferol-3-*p*-coumarate

kaempferol-3-O-β-glucuside, R = H
isorhamnetin-3-O-β-glucuside, R = -OCH₃

Quercetin-3-caffeate

Figure 2.1. Structure of phytochemical constituents of *Centella asiatica*.

of LPO and PCO by *Centella asiatica* aqueous extract was confirmed in 1-methyl-4-phenyl-1, 2, 3, 6-tetrahydropyridine (MPTP)-induced neurotoxicity in aged Sprague-Dawley rats (Haleagrahara et al. 2010). To investigate the effectiveness, the rats were treated with the extract at 300 mg/kg body weight for 21 days and effects on oxidative biomarker levels in corpus striatum and hippocampus homogenate was examined. Extract of *Centella asiatica* strongly inhibited neurotoxicity (Haleagrahara et al. 2010).

In addition to the whole plant extract, neuroprotective effect of isolated asiaticoside from *Centella asiatica* was carried out in MPTP induced rats model of Parkinsonism. On the 14th day, after administration of asiaticoside, rats were sacrificed and substantia nigra (SN) and striatum were dissected. Neuroprotective effect was quantitatively determined be estimating dopamine (DA) and its metabolites in striatum and malonyldialdehyde (MDA) contents, reduced glutathione (GSH) level and gene expression level in SN (Xu et al. 2012).

A study by Soumyanath et al. on male Sprague-Dawley rats with ethanolic and aqueous extracts of the plant showed that the ethanolic extract of the plant aided to accelerate functional recovery as well as axonal regeneration when the extract was provided in their drinking water (Soumyanath et al. 2005). Though this study failed to show the neuroprotective effect of aqueous extract of *Centella asiatica*, another study proved to have neuroprotective effect from aqueous extract against 3-nitropropionic-acid (3-NPA)-induced oxidative stress in the brain of prepubertal mice (Shinmol et al. 2008). The degree of oxidative stress in cytoplasm of brain regions of male mice (4 weeks-old) given CA prophylaxis (5 mg/kg bw) for 10 days followed by 3-NPA administration (i.p.75 mg/kg bw) on the last two days was determined. Significantly, *Centella asiatica* treated mice showed complete amelioration of 3-NPA-induced oxidative stress (Shinmol et al. 2008).

Ethanolic extract of *Centella asiatica* administration significantly improved neurobehavioral activity and diminished infarction volume along with the restored histological morphology of brain in MCAO rats (Tabassum et al. 2013).

Both methanolic and ethanolic extract as well as the phytochemical asiaticoside of *Centella asiatica* imparted anxiolytic activity (Wijeweera et al. 2006).

Fresh leaf extract of *Centella asiatica* was applied to find out its therapeutic benefits on adult rats on dendritic morphology of amygdaloid neurons, one of the regions concerned with learning and memory. Rats were fed with 2, 4 and 6 mL/(day per kg body weight) of fresh leaf extract for two, four and six weeks. *Centella asiatica* treated rats exhibited a significant increase in the dendritic length (intersections) and dendritic branching points in amygdaloid neurons of the rats depending on the dose and time (Rao et al. 2009). A similar effect following the same experimental procedure by this group was found in hippocampal CA3 neurons (Gadahad et al. 2008). Another study by Rao's group also evaluated the therapeutic benefits of leaf extracts of *Centella asiatica* during the rat growth spurt period on the dendritic morphology of hippocampal CA3 neurons. Both dendritic length (intersections) and dendritic branching points along the length of both apical and basal dendrites in rats were remarkably increased due to treatment with extract of *Centella asiatica* (Rao et al. 2006). Rao et al. showed that orally administered fresh leaf juice of *Centella asiatica* enhanced learning performance as well as memory retention (Rao et al. 2005).

Possible neuroprotective effects of *Centella asiatica* was explored against D-galactose induced cognitive impairment, oxidative and mitochondrial dysfunction in mice. The results showed significant improvement in behavioral alterations, mitochondrial enzyme complex activities as well as oxidative damage (Kumar et al. 2011). In their study, they found that extract of *Centella asiatica* significantly improved memories and cognitive deficit. Additionally, the extract was able to attenuate oxidative stress (Kumar et al. 2009).

The neuroprotective role of *Centella asiatica* was evaluated by examining Na, K and Ca/Mg ATPase in different regions of the brain during pentylenetetrazol (PTZ)-induced epilepsy and on antiepliptic treatment with n-hexane, ethyl acetate, n-butanol and aqueous extract of CA. Pentylenetetrazol (PTZ)-induced seizures rats were used to investigate anti-convulsant effect of different extracts of *Centella asiatica*. Considerable recovery of acetylcholine and acetylcholine esterase was observed in *Centella asiatica* pretreated rats (Visweswari et al. 2010).

Orally administered aqueous extract of *Centella asiatica* showed remarkable improvement in learning (Gupta et al. 2003).

Another study showed significant therapeutic benefits of *Centella asiatica* to treat conditions associated with increased phospholipase [PLA2] activity in the brain, such as epilepsy, stroke, multiple sclerosis and other neuropsychiatric disorders (Barbosa et al. 2008).

A clinical study in men was conducted to evaluate the beneficial effects of *Centella asiatica* against different neurological disorders. Seventy percent hydroethanolic extract of *Centella asiatica* was administered (500 mg/capsule, twice daily, after meals). On the basis of Hamilton's Brief Psychiatric Rating Scale (BPRS), it was observed that *Centella asiatica* not only attenuated anxiety related disorders but it also significantly ($p < 0.01$) reduced stress phenomenon and its correlated depression (Jana et al. 2010).

Methanolic, chloroform and aqueous extract of *Centella asiatica* was investigated to evaluate their effect on cognitive functions in rats. Effects on learning and memory were observed by using shuttle box, step through, step down and elevated plus maze paradigms. Among the extracts, only the aqueous extract showed significant cognitive improvement (Kumar and Gupta 2002).

ECa 233, a standardized extract of *Centella asiatica* containing triterpenoids not less than 80%, showed strong anxiolytic activity following a dark-light box and an open-field tests in rats (Wanasuntronwong et al. 2012).

Wound healing

Dose dependent wound healing activity was observed from asiaticoside, isolated from *Centella asiatica*, in guinea pig punch wounds. Tensile strength, epitheliazation, collagen content and hydroxypyroline amount were increased after topical application of 0.2% asiaticoside. A similar effect was also found in streptozocin diabetic rats, where healing was delayed, using 0.4% asiaticoside topically (Shukla et al. 1999).

In another study, cellular proliferation as well as synthesis of collagen was accelerated when different formulations like cream, ointment and gel containing *Centella asiatica* extract (CAE) was applied to open wounds in experimental rats thrice daily for 24 days (Sunilkumar et al. 1998). Epitheliazation was faster along with higher wound contraction. Effects of CAE on keratinization were also shown to enhance wound healing by thickening skin in areas of infection (Poizot et al. 1978).

In another study, ethanolic extract of the plant was applied on Wistar albino rats using incision, excision, and dead space wounds models. The extract exhibited significant wound healing activities by accelerating epitheliazation and increasing wound contraction (Shetty et al. 2006).

Supplementation of aqueous extract of *Centella asiatica* strongly inhibited proliferation of Rabbit Corneal Epithelial (RCE) cells in an *in vitro* wound healing model. The study also found that expression of differentiation markers and cell cycle were not altered due to application of the extract of *Centella asiatica* (Ruszymah et al. 2012).

Effectiveness of *Centella asiatica* was carried in 200 diabetic patients. Study groups were prescribed two capsules three times a day. Patients belonging to the study group experienced better wound contraction without any serious side effects (Paocharon et al. 2010). Water extract of *Centella asiatica* and asiaticoside (AC), one of the major phytochemical of its constituents, were observed in acetic acid induced gastric ulcers. Treatment with CE and AC showed significant wound healing depending on the dose (Cheng and Koo 2000).

Wound healing activities of sequential hexane, ethyl acetate, methanol and water extract of *Centella asiatica* was observed in incision and partial thickness burn wound models in Sprague-Dawley rats. All extracts showed significant wound healing activities by showing higher tensile strength. In addition, fully developed epitheliazation and keratinization was also observed (Somboonwong et al. 2012).

Topically applied ethanolic extract of *Centella asiatica* exhibited strong wound healing activities on rat dermal wound. The extract treated wounds enhanced epitheliazation and contraction of wounds compared to control wounds (Sugna et al. 1996).

Wound healing activities of *Centella asiatica* were accelerated when it was administered with collagen (Babu et al. 2011). *Centella asiatica* extract impregnated collagen and cross-linked collagen scaffolds were formulated and different concentrations were administered to wounds developed Wistar rats. It was found that 1.5% w/v of both formulations exhibited better wound-healing activities compared to the control (79.9 to 80.68%, respectively) (Babu et al. 2011).

Anti-cancer activity

Partially purified fractions of *Centella asiatica* were found to inhibit significant cell proliferation than crude extract of the plant in Ehrlich ascites tumor cells (EAC) and Dalton's lymphoma ascites tumor cells (DLA) while no toxic effects were observed when the extracts were applied to normal human lymphocytes (Babu et al. 1995). Partially purified fraction also considerably suppressed multiplication of mouse lung fibroblast cells at 8 μg/ml in long term culture. Both extracts retarded the development of solid and ascites tumors and increased the life span of these tumor-bearing mice after oral administration. *In vitro* experiment in HepG2 cell line showed that the juice of *Centella asiatica* exhibited significant cytotoxic effect on tumor cells while it was totally non-toxic to normal cells (Hussin et al. 2014).

Water extract of *Centella asiatica* was reported to have inhibitory effects on azomethane-induced aberrant crypt focus formation and carcinogenesis in the intestine of F344 rats (Bunpo et al. 2004).

Cytotoxic effects of asiatic acid (a constituent of CA) was found in U-87MG human glioblastoma cells depending on the dose (Cho et al. 2006). Topical application of asiatic acid showed anti-tumor promoting effect against 12-*O*-tetradecanoylphorbol 13-acetate (TPA)-mediated skin tumor genesis in 7, 12-dimethylbenz[*a*] anthracene (DMBA) initiated ICR mice (Park et al. 2007).

A recent study showed that asiatic acid from *Centella asiatica* significantly inhibited cell viability and induced apoptosis in human melanoma SK-MEL-2 cells depending on time and dose (Park et al. 2005). Anti-cancer effect by the same phytochemical was found in human adenocarcinoma cell line HT-29 (Bunpo et al. 2005).

Orally administered aqueous extract of *Centella asiatica* at 500 and 1000 mg/kg body weight for 30 days increased life span and decreased tumor volume remarkably (Rai et al. 2011). Significant cytotoxicities were found in brine shrimp lethality assays (Ullah et al. 2009). Asiatic acid markedly inhibited colon cancer cell proliferation and increased apoptosis (Tang et al. 2009).

Anti-tumor activity by asiatic acid isolated from *Centella asiatica* was observed by inhibiting angiogenesis (Kavitha et al. 2011).

Aqueous extract of *Centella asiatica* was investigated on the induction of spermatogenic cell apoptosis in male rats. Aqueous leaf extract was orally administered (10, 50, 80 and 100 mg/kg) daily for 60 days while the control group received just water. After 60 days, body and testis weight were measured and blood samples were taken from the heart. Apoptosis and histological changes were determined by collecting tissue samples obtained from rat testes following TUNEL assay and hematoxylin and eosin stain. Significant reduction of sperm count, motility, and viability and the number of spermatogenic cells in the seminiferous tubules were observed in *Centella asiatica* extract treated mice. Beside this number of apoptotic germ cells per seminiferous tubule cross-section was significantly increased in the experimental group (18.11 ± 3.5) compared with the control group (8.7 ± 0.81) (P < 0.05). Serum testosterone, follicle-stimulating hormone, and luteinizing hormone levels were reduced. In addition, weight of testis decreased remarkably (Heidari et al. 2012).

Anti-microbial activity

Different types of extract of *Centella asiatica* have been investigated to demonstrate anti-bacterial activity. In one study, it was found that petroleum ether, ethanol and chloroform extract exhibited significant anti-bacterial activity than water and n-hexane extract (Dash et al. 2011). On the other hand, there was no anti-bacterial activity by n-hexane extract against *Escherichia coli*.

Broad spectrum antimicrobial activity was observed against *E. coli* ATCC 25922, *Staphylococcus aureus* ATCC 25923, *Vibrio parahaemolyticus* ATCC 17802 and *Pseudomonas aeruginosa* ATCC 27853 and 45 fresh clinical isolates like *Vibrio cholera* 01, *V. cholerae* 0139, species of *Shigella*, *Salmonella typhimurium*, *Aeromonas hydrophila*, entero-aggregative *E. coli* and *Candida albicans* (Mamtha et al. 2004).

In another study, anti-microbial effect of petroleum ether, ethanol and water extract of *Centella asiatica* was demonstrated where only ethanolic extract showed better anti-microbial activity than the others (Jagtap et al. 2009).

Broad spectrum antibacterial activities against Gram-positive (*Bacillus subtilis, S. aureus*) and Gram-negative (*E. coli, Pseudomonas aeruginosa, Shigella sonnei*) organisms were found with the essential oil extract of *Centella asiatica* (Oyedeji et al. 2005).

Significant anti-bacterial and anti-fungal activities were demonstrated from calli and regenerated plant extracts *Centella asiatica* (Bibi et al. 2011).

Anti-microbial activity of aqueous extracts, aqueous-alcoholic extracts obtained with 30% ethanol, and aqueous-alcoholic extracts obtained with 60% ethanol were investigated by Kedzia et al. Among all extracts, aqueous-ethanol extracts (30% ethanol) exhibited significant bacteriostatic properties on Gram-positive and Gram-negative bacteria (Kedzia et al. 2007).

Depending on the concentration, ethanolic and aqueous extract of *Centella asiatica* showed strong inhibitory effects (determined by zone of inhibition), where ethanolic extract showed better activity than aqueous against *S. aureus* (Udoh et al. 2012). In another study, ethanolic extract of *Centella asiatica* showed significant anti-microbial activity against *S. aureus* from milk samples of dairy cows, while aqueous extract did not show any activity (Taemchuay et al. 2009). Strong anti-microbial activity by methanolic extract of *Centella asiatica* was also found by Rozarina et al. (2013).

Anti-microbial activity of aqueous extract of *Centella asiatica* was investigated in combination with four other medicinal plants against a wide range of Gram-positive and Gram-negative bacteria. The results showed that *Centella asiatica* exhibited the highest anti-microbial activity in terms of zone of inhibition (Das et al. 2013). This report also suggested potential anti-fungal activity by *Centella asiatica*.

Significant anti-microbial activity was observed against a wide range of Gram-positive, Gram-negative as well as fungi with hexane, carbon tetrachloride, and dichloromethane soluble fraction of the ethanolic extract and crude methanolic extract showing strong zones of inhibition (10 mm, 9 mm and 9 mm), respectively, at a concentration of 400 µg/disk (Rishikesh et al. 2012).

It is to be noted that while some reports have found anti-microbial effects from aqueous extract of *Centella asiatica*, other reports have not. There was only one finding that not only aqueous, but also ethanolic crude extract of this plant was not able to kill or inhibit the growth of *E. coli, K. pneumoniae* and *S. aureus* (Jacob and Shenbagaraman 2011).

Cardiovascular effects

Cardio-protective effect of madecassoside (MA), isolated from *Centella asiatica*, was found on isolated rat hearts and isolated cardiomyocytes against reperfusion injury (Bian et al. 2008). In their next study, the authors demonstrated that madecassoside inhibited myocardial cell apoptosis significantly on myocardial ischemia-reperfusion injury *in vivo* in rats. Function of left ventricle was monitored during the ischemia-reperfusion period by a multi-channel recorder. The levels of lactate dehydrogenase (LDH), creatinephosphokinase (CPK), malondialdehyde (MDA), super-oxide dismutase (SOD) and C-reactive protein (CRP) in serum were determined. Terminal-deoxynucleotidyl transferase mediated nick end labeling (TUNEL) staining was performed to determine cardiomyocytic apoptosis. Madecassoside pre-treated (50, 100 mg/kg body weight) attenuated myocardial damage showing decreased infarct size, and decreased LDH and CK release. In addition to these, activities of SOD were decreased, which consequently markedly reduced the level of MDA and the activity of CRP, and relieved myocardial cell apoptosis (Bian et al. 2008).

Aqueous extract of *Centella asiatica* strongly accelerated cardiprotective effect in adriamycin-induced cardiomyopathy in rats. Cardioprotective effect was evaluated by measuring various serum markers [LDH, CPK, glutamate oxaloacetic transaminase (GOT) and glutamate pyruvate transaminase (GPT) enzymes] and changes in the level of antioxidant enzymes [SOD, CAT (catalase), GPx (glutathione peroxidase), glutathione S-transferase (GST)]. *Centella asiatica* extract treated (200 mg/kg of body wt/oral) strongly prevented these alterations and restored the enzyme activities to near normal levels (Gnanapragasam et al. 2004).

Extract of *Centella asiatica* exhibited cardioprotective activity against myocardial injury in adriamycin-induced rats. Damaged mitochondria were observed from transmission electron microscopy from the control group, while restoration of mitochondria was found from *Centella asiatica* treated rats (Gnanapragasam et al. 2004).

Methanolic extract of *Centella asiatica* showed strong antithrombotic activity inhibition of shear-induced platelet activation and dynamic coagulation (Satake et al. 2007).

Effectiveness against dermatitis

UV irradiation leads to cell aging, senescence, apoptosis and cancer in human keratinocytes by inducing Reactive Oxygen Species (ROS), DNA damage and inflammatory and immunological reactions (An et al. 2012). Oral administration of extract of *Centella asiatica* significantly reduced radiation-induced DNA damage (Joy and Nair 2004).

Psoriasis is one of the common dermatological disorders. Aqueous extract of *Centella asiatica* showed remarkable inhibition of SVK-14 keratinocytes replication with IC_{50} value 209.9 ± 9.8 mg/ml. On the other hand, two phytochemical constituents of CA, namely, madecassoside and asiaticoside showed IC_{50} values of 8.6 ± 0.1 μM and 8.4 ± 0.6 μM, respectively (Sampson et al. 2001).

Analgesic and anti-inflammatory effects

Both *in vitro* and *in vivo* experiments have been carried out to demonstrate the anti-inflammatory activities with different extracts or individual constituent of *Centella asiatica*. Anti-inflammatory effect was found from methanolic extract of *Centella asiatica* (Chippada et al. 2011).

Madecassoside, isolated from *Centella asiatica* showed significant anti-inflammatory activity against ultraviolet (UV)-induced melanogenesis (Jung et al. 2013).

Anti-inflammatory and analgesic effects by isolated asiatic acid was investigated in acetic acid-induced writhing and formalin-induced pain in ICR mice. Asiatic acid treated mice showed inhibition in the number of writhing and formalin-induced pain significantly. Significant reduction was exhibited in carrageenan-induced paw edema (Huang et al. 2011). Both analgesic and anti-inflammatory activities were remarkably increased by water extract of *Centella asiatica* using acetic acid-induced writhing and hot-plate method in mice (Somchit et al. 2004). Significant anti-inflammatory effect from the extract was reported from clinical investigations (Maramaldi et al. 2014).

Chloroform and methanol extract of dried leaves reportedly showed anti-inflammatory and analgesic effects in carrageenan-induced paw edema model in Wistar albino rats (at doses of 100 and 200 mg/kg), and tail clip, tail flick, tail immersion and writhing assay tests in Swiss albino mice at doses of 50, 100 and 200 mg/kg (Saha et al. 2013).

Anti-arthritic effect

Few studies have been carried out to evaluate anti-arthritic effect by *Centella asiatica*. Anti-arthritic activity by the phytochemical madecassoside was investigated on collagen II (CII)-induced arthritis (CIA) in mice. Different doses of madecassoside (10, 20 and 40 mg/kg body weight) were orally administered. This study showed significant anti-arthritic activity (Liu et al. 2008).

In addition to madecassoside, methanolic extract of the plant exhibited strong protective effect against both inflammation and oxidative stress in collagen-induced arthritic animal model. Results showed that *C. asiatica* extract significantly inhibited the footpad swelling and arthritic symptoms in collagen-induced arthritic rats. Beside this, significant reduction in the production of pro-inflammatory cytokines was also found. Histopathological assessment showed reduced inflammatory cells infiltrate, tissue edema and bone erosion in joints of treated rats. Radiographic evidence of protection from bone resorption, soft tissue swelling and joint narrowing was apparent in the tibiotarsal (Sharma and Thakur 2011).

Anti-ulcerative effect

Fresh juice of *Centella asiatica*, exhibited strong anti-ulcerogenic activity against ethanol, aspirin, cold resistant stress and pyloric ligation-induced gastric ulcer when oral doses of 200 and 500 mg/kg body weight was given twice daily for five days (Sairam et al. 2001). Ethanol extract of *Centella asiatica* showed anti-ulcerogenic activity against ethanol-induced gastric mucosal injury in rats. Leaf extracts of the plant was orally administered (100, 200 and 400 mg/kg body weight) and exhibited significant mucosal protection by showing reduction or absence of edema and leucocytes infiltration of submucosal layer (Abdullah et al. 2010).

Significant inhibition of gastric ulceration induced by Cold and Restraint Stress (CRS) in Charles-Foster rats was observed from extract of *Centella asiatica* where the activity was compared with famotidine (H2-antagonist) and sodium valproate (anti-epileptic) (Chatterjee et al. 1992).

Different concentrations of aqueous extract of *Centella asiatica* (CE) and isolated asiaticoside (AC) were administered on acetic acid-induced gastric ulcers (kissing ulcers) in rats to evaluate the healing effect. The size of the ulcers was reduced at day 3 and 7 depending on the dose, with a concomitant attenuation of myeloperoxidase activity at the ulcer tissues. Besides this, epithelial cell proliferation and angiogenesis were accelerated (Cheng et al. 2004). In their previous study, they observed beneficial effects of CE on the prevention of ethanol-induced gastric lesions in rats. Oral administration of CE (0.05, 0.25, and 0.50 g/kg body weight) remarkably prevented gastric lesions formation (58 to 82% reduction) and decreased mucosal myeloperoxidase (MPO) activity in a dose-dependent manner (Cheng and Koo 2000).

Ethanol extract of roots of the plant also demonstrated anti-ulcer effect in eight hours restraint stress-induced ulcerization (Sarma et al. 2006).

Anti-diabetic effect

Both methanolic and ethanolic extracts of the leaves of *Centella asiatica* were found to have significant anti-diabetic activity by lowering blood glucose level compared with glibenclamide in alloxan-induced diabetic rats. The study revealed that ethanolic extract showed 48% reduction in blood glucose levels, where as it was 30% for methanolic extract (Chauhan et al. 2010).

Studies have found *Centella asiatica* possessing significant hypoglycemic activity in glucose tolerance test in rabbits. It was also found not to cause hypoglycemia in fasted rabbits compared to tolbutamide. The ethanolic extract of the plant showed an increased glycogen content in the liver, comparable to glibenclamide. Additionally, the extract showed lowered serum cholesterol and total lipid level. Methanolic extract was found to be more effective than ethanolic extract in lowering blood glucose (Kabir et al. 2014).

Venous insufficiency

Effect of different concentrations (60 and 30 mg thrice daily) of triterpenic fraction of *Centella asiatica* on Capillary Filtration Rate (CFR), Ankle Circumference (AC), and Ankle Edema (AE) was evaluated in patients with venous hypertension (ambulatory venous pressure > 42 mm Hg). Effectiveness was measured by determining capillary filtration rate. CFR, AC, and AET time in patients were reduced for TTFCA treated for four weeks patients (De Sancits et al. 2001).

Aqueous extract of *Centella asiatica* aided in improvement of signs and symptoms associated with chronic venous insufficiency (Chong and Aziz 2012).

Molecular mechanisms behind the observed pharmacological effects

Neuroprotective

Parkinson's disease is a progressive neurodegenerative disorder characterized by selective degeneration of nigrostriatal dopaminergic neurons in substantia nigra pars compacta region (Sampath and Janardhanam 2013). Decrease in anti-oxidant behaviors and increased level proxidant are considered to be responsible to developing Parkinson's disease. Asiaticoside inhibited symptoms of Parkinson's disease by increasing anti-oxidant properties and reducing proxidant levels (Sampath et al. 2012). On the other hand, activity of asiaticoside from aqueous extract of *Centella asiatica* exhibited its neuroprotective effect by inhibiting phospholipse A2 (cPLA2 and sPLA2) (Defillipo et al. 2012).

It has been hypothesized that total plant extract or individual component(s) exert neuroprotective activity by enhancing GSH (reduced glutathione), thiols and anti-oxidant pathways (Shinmol et al. 2008).

Aqueous extract of *Centella* showed effectiveness in preventing the cognitive deficits induced by malondialdehyde (MDA) through increase in glutathione (Kumar and Gupta 2003).

In depth mechanism behind the significant effect of *Centella asiatica* in alleviating impairment of memory and learning has been described. It was reported that dendritic branching accelerated learning and memory. Aqueous extract of *Centella asiatica* aided to enhance dendritic branching (Rao et al. 2006). During growth spurt period (neonatal), administration of fresh leaf juice of *Centella asiatica* facilitated and increased dendritic length and branches of neurons of amygdale and subsequently improved learning and memory observed in rat models. Besides this effect, the juice of the leaf not only enhanced dendritic length but also acted as an anti-anxiolytic agent (Rao et al. 2009). Another study showed that aqueous leaf extract of *Centella asiatica* stimulated neuronal dendritic growth (Gadahad et al. 2008).

New gene expression and protein synthesis (synaptic consolidation) were required to develop stable LTP (long-term potentiation) in response to high frequency stimulation. Several lines of evidence have implicated endogenous brain-derived neurotrophic factor-Tropomyosin related kinase B (BDNF-TrkB) signaling in synaptic consolidation. After LTP induction in the dentate gyrus, early gene Arc (activity-regulated cytoskeleton-associated protein) was strongly induced and transported to dendritic processes. BDNF activated Arc dependent synaptic consolidation. Increased dendritic arborization by *Centella asiatica* was hypothesized to be due to phytochemicals similar to BDNF and other neurotrophic factors, which would activate the expression of the early genes such as Arc, thereby leading to enhancement of the dendritic arborization as well as improved learning and memory (Rao et al. 2006). A recent study by this group strongly supported this mechanism (Rao et al. 2012).

Aqueous extract of *Centella asiatica* provided significant protection against δ-aminolevulinic acid dehydratase (ALAD), reduced glutathione GSH and thiobarbituric acid reactive substance (TBARS) levels, particularly at doses of 200 and 500 mg per kg body weight. *Centella asiatica* also provided significant recovery in the inhibited liver ALAD and G6PD (glucose 6-phosphate dehydrogenase) activities (Flora et al. 2007).

Improvement in neurobehavior by ethanolic extract of *Centella asiatica* was due to the ability to reduce the level of thiobarbituric acid reactive species, restoration of glutathione content and augmentation of the activities of anti-oxidant enzymes, namely, catalase, glutathione peroxidase, glutathione reductase, glutathione-S-transferase and superoxide dismutase. It was further reported that the bioactive triterpenes, asiatic acid, asiaticoside, madecassic acid and madecassoside were responsible for the observed effects (Tabassum et al. 2013).

It has also been shown that extract of *Centella asiatica* showed neuroprotective effect by altering the activity of Na^+, K^+-ATPase, Ca^{2+}-ATPase, and Mg^{2+}-ATPase (Visweswari et al. 2010).

Wound healing

Titrated extract from *Centella asiatica* (TECA) has been playing a pivotal role as a drug for many years in Europe for the treatment of wound healing. Among the phytochemicals of this plant, the three triterpines— asiatic acid, madecassic acid and asiaticoside acid wound healing. At a low dose, asiaticoside showed its effectiveness by stimulating collagen synthesis. In addition to collagen, the three components stimulated synthesis of glycosaminoglycan (Maquart et al. 1999).

Methanolic extract of *Centella asiatica* contains asiaticoside and madecassoside which induced type I collagen synthesis in human dermal fibroblast cells via activation of the TGFβ receptor I (TGF-βRI) kinase-independent Smad signaling pathway (Lee et al. 2006) and also elevated anti-oxidant levels thereby accelerating wound healing (Shukla et al. 1999).

Potential wound healing activity by asiaticoside has been found to occur through inhibiting type I and type III collagen protein and mRNA expressions. In addition, this phytochemical reduced the expression of transforming growth factor-β (TGF-β) receptor type I (TGF-βRI) and TGF-β receptor type II (TGF-βRII) at the transcriptional and translational level. On the other hand, it increased the expression of Smad7 protein and increased Smad7 mRNA levels [it is to be noted that Smad7 acts by recruiting the PPP1R15A-

PP1 complex to TGF-βR1, which promotes its dephosphorylation; dephosphorylation inhibits TGF-βR1 activity]. However, asiaticoside did not influence the expression of Smad2, Smad3, Smad4, phosphorylated Smad2, and phosphorylated Smad3 (Tang et al. 2011).

Asiaticoside was shown to induce the phosphorylation of both Smad2 and Smad3. In addition, it has been detected that asiaticoside induced binding of Smad3 and Smad4. The nuclear translocation of the Smad3 and Smad4 complex was induced via treatment with asiaticoside, pointing to the involvement of asiaticoside in Smad signaling. In addition, SB431542, an inhibitor of the TGFβ receptor I (TβRI) kinase, which is known to be an activator of the Smad pathway, was not found to inhibit both Smad2 phosphorylation and Type 1 collagen synthesis induced by asiaticoside. Therefore, the results show that asiaticoside can induce type I collagen synthesis via the activation of the TβRI kinase-independent Smad pathway (Lee et al. 2006).

Asiaticoside significantly increased the expression of inhibitory Smad7, but it had no effect on the expression of Smad2. A further study revealed that asiaticoside could induce Smad7 to enter the cytoplasm from the nucleus (Qi et al. 2008). Asiaticoside reportedly inhibited the hypertrophic scar fibroblasts by inhibiting expression of Smad7 mRNA (Pan et al. 2004).

Madecassoside-induced enhanced wound healing may occur through several mechanisms including anti-oxidative activity, collagen synthesis and angiogenesis (Liu et al. 2008).

Another study reported that extract of *Centella asiatica* increased cellular proliferation and accelerated collagen synthesis by increasing DNA, protein and collagen content of granulation tissues. Additionally, it also enhanced crosslinking of collagen which consequently increased tensile strength (Suguna et al. 1996). It has been hypothesized that flavonoids present in the aqueous extract of this plant possessed strong anti-oxidant properties which aided in the late stage of wound healing.

Cardiovascular

Generation of reactive oxygen species and mitochondrial dysfunction are considered to be responsible for myocardial injuries that ultimately develop into different cardiovascular diseases. Mitochondrial dysfunction is characterized by the accumulation of oxidized lipids, proteins and DNA, leading to disorganization of mitochondrial structure and systolic failure (Gnanapragasam et al. 2004). During cardiac damage, cardiac marker enzymes (lactate dehydrogenase, creatine phosphokinase, amino transferases), TCA cycle enzymes (isocitrate dehydrogenase, α-ketoglutarate dehydrogenase, malate dehydrogenase), respiratory marker enzymes (NADH-dehydrogenase, cytochrome-C-oxidase), mitochondrial antioxidant enzymes (GPx, GSH, SOD, CAT) are decreased and the level of lipid peroxidation is increased. Pre-co-treatment with aqueous extract of *Centella asiatica* (200 mg/kg body wt, oral) effectively counteracted the alterations in mitochondrial enzymes and the mitochondrial defense system. So, extract of *C. asiatica* not only possesses anti-oxidant properties but also reduced the extent of mitochondrial damage (Gnanapragasam et al. 2004). As such, cardioprotective effects shown by *C. asiatica* extract may involve both increase in anti-oxidants (with consequential cardio-beneficial effects as well as through reduction of mitochondrial damage (with consequential dysfunction and possible induction of apoptosis).

Anti-inflammatory

Prostaglandin (PG) E2 (PGE2) and prostaglandin F2α (PGF2α) are the main PGs synthesized extensively by keratinocytes in response to UV irradiation, and mediate post-inflammatory pigmentation by modulating melanin synthesis and melanocyte dendricity. It has been shown that by suppressing the production of these inflammatory mediators, madecassoside strongly inhibited inflammation. These results suggest that MA inhibits UVB-induced pigmentation by suppressing the production of PGE2 and PGF2α in keratinocytes. UVB irradiation induces expression of cyclooxygenase 2 and PAR2 (protease-activated receptor 2) pathways, which further accelerate production of PGE2. Madecossisode inhibited the expression of COX-2 and PAR2 which then inhibited inflammation. In addition, madacessoside also inhibited phagocytosis by blocking PAR2 (Jung et al. 2013).

Over expression of intrinsic nitric oxide (iNO), COX-2, interleukin-6, interleukin-1β (IL-1β), tumor necrosis factor-α accelerate development of inflammation. Asiatic acid reportedly down-regulated nuclear factor-κβ activation via suppression of IκB (interleukin κβ) kinase, and phosphorylation of mitogen-activated protein kinases (p38, ERK1/2, and JNK), and subsequently inhibited iNOS, COX-2, interleukin-6, IL-1β, and TNF-α expression (Huang et al. 2011).

Purified extract of *Centella asiatica* reduced inflammation by preventing DNA from ultraviolet light-induced damage and by decreasing the thymine photodimerization. Additionally, the extract also significantly suppressed expression of interleukin-1α, and inhibited carboxymethyl lysine (Maramaldi et al. 2013). It is to be noted that interleukin-1α can produce inflammation, and carboxymethyl lysine can induce endothelial cell injury, the latter being a consequence of UV light damage.

Anti-microbial activity

Exact molecular mechanism of *Centella asiatica* and its phytochemicals behind anti-microbial activity has not been clearly identified. There is no strong evidence to speculate which specific constituent of this plant is able to exhibit anti-microbial activities. Depending on the extraction method, *Centella asiatica* showed activity against various microorganisms. Investigation by Arumugam et al. showed that methanolic extract was mainly responsible for anti-microbial activity (Arumugam et al. 2011). One report suggested that tannins and cyanogenetic glycosides in the extract to be the responsible agents, as these have previously been reported to possess anti-microbial activities (Rishikesh et al. 2012). Jacob and Shenbagaraman reported that due to free radical scavenging by flavonoids and tannins of *Centella asiatica*, anti-microbial activity was found against some specific bacteria (Jacob and Shenbagaraman 2011).

Anti-cancer effect

The major phytochemicals, triterpines and rosmarinic acid are highly potent anti-cancer agents (Yoshida et al. 2005). The juice of *Centella asiatica* plays a pivotal role to alter the expression level of different genes like c-myc, c-fos, and c-erbB2 (Hussin et al. 2014). The juice showed anti-proliferative activity on hepatocellular G2 (HepGH2) cancer cell line depending on the dose. This was attributed to effects of anti-oxidants and apoptotic cell death (Hussin et al. 2014). On the other hand, asiatic acid extracted from *Centella asiatica*, induced apoptosis by increasing intracellular Ca^{2+} which in turn enhanced p53 expression in HepGH2 (Lee et al. 2002). Asiatic acid has also been claimed to have an anti-cancer effect in glioblastoma where it induced cell death via apoptosis and necrosis. Asiatic acid decreased mitochondrial membrane potential, activated the pro-apoptotic proteins—caspase-9 and caspase-3, and increased intracellular Ca^{++} (Cho et al. 2006). It has been reported that asiatic acid was able to prevent angiogenesis in melanoma cells (Rai et al. 2011). Asiatic acid accelerated production of reactive oxygen species (ROS), altered Bax/Bcl-2 ratio, and activated caspase-3. Asiatic acid enhanced cell apoptosis and arrested cell cycle in breast cancer cell by promoting the activity of extracellular signal regulated kinase pathways (Hsu et al. 2005, Chen et al. 2014). By disrupting endoplasmic reticulum and altering calcium homeostasis, asiatic acid also enhanced cytotoxicity.

Dermatological

By decreasing transforming growth factor-β1 (TGF-β1) expression and enhancing the expression of inhibitory Smad7, asiaticoside showed anti-scarring activity (Ju-Lin et al. 2008). This study reported that asiaticoside had no effect on the expression of Smad2.

By analyzing whole gene expression levels using microarrays, it was observed that *Centella asiatica* regulated Stress-Induced Premature Senescence (SIPS) by preventing repression of DNA replication and mitosis-related gene expression (Kim et al. 2011).

Anti-diabetic effect

The exact mechanism of anti-diabetic effect by *Centella asiatica* is not still clear. It is considered that *Centella asiatica* restores the function of pancreatic tissues (Rahman et al. 2012).

Anti-arthritic

Very few studies have been carried out to determine the mechanism for prevention and alleviation of arthritis by *Centella asiatica*. One study revealed that madecassoside alleviated infiltration of inflammatory cells and synovial hyperplasia as well as protected joint destruction through regulating the abnormal humoral and cellular immunity as well as protecting joint destruction (Liu et al. 2008).

Reduction of the symptoms associated with arthritis was observed due to decreased level of the endogenous biological defense system consisting of superoxide dismutase, reduced glutathione, catalase, glutathione peroxidase, glutathione reductase, glutathione transferase, lactate dehydrogenase, glyoxalase-I, lipid peroxide and protein carbonyl contents, which were decreased to a large extent in the spleen and joint cytosolic fraction by methanolic extract of *Centella asiatica* (Sharma and Thakur 2011).

By controlling the production of auto antigen and inhibiting denaturation of protein, membrane lysis and proteinase action, methanolic extract of the plant has been attributed to inhibit rheumatic disease (Chippada et al. 2011).

Anti-ulcer

Aqueous extract of *Centella asiatica* and asiaticoside showed anti-ulcerative activity by up-regulating expression of basic fibroblast growth factor, an important angiogenic factor in ulcer tissues (Cheng et al. 2004). In another study, they suggested that *Centella asiatica* inhibited ethanol induced gastric mucosal lesions by strengthening the mucosal barrier and reducing the damaging effects of free radicals (Cheng and Koo 2000).

Drawbacks of *Centella asiatica*

Though titrated extracts of *Centella asiatica* have been reported to have an important effect in various diseases, the low solubility of phytochemical constituents present in the plant is a great barrier to achieve a significant beneficial effect. However, a recent study has shown increased water solubility of constituents through using different surfactants (Hong et al. 2005).

Commercially available products of *Centella asiatica*

A list of products intended for therapeutic uses or use as a nutraceutical obtained from *Centella asiatica* is presented in Table 2.2 (European Medicines Agency 2012).

Table 2.2. Commercially available products of *Centella asiatica*.

Trade name	Extract	Pharmaceutical Form	Country
Centellase®	TECA (like French Madecassol®)	Tablets for oral use	Italy
Madecassol®	TECA	Tablets for oral use Cream for cutaneous use Cutaneous powder 2% Sterilised impregnated dressing (1 g of extract/100 g of mass; 2 g of mass/dm²)	France
Bilim	(like French Madecassol®)	Ointment 1% cutaneous powder 2%	Greece
Blastoestimulina	TECA	Cutaneous powder 2%	Spain

Besides the products mentioned above and manufactured in Europe, *Centella asiatica* individually or in combination with other plants is marketed in India, China and Bangladesh as 'Gotu kola' and 'Mandu kaparni' for enhancement of memory, as a digestive, and for other medicinal purposes.

Conclusion and future prospects

Centella asiatica and its various phytochemical constituents can play a major beneficial role in various disorders like loss of cognitive functions, heart disorders, wound healing, and arthritis. Some of these diseases cannot be treated wholly with allopathic medicine. As such, these phytochemicals can form a new generation of efficacious drugs for treatment of the above diseases. Triterpinoids are gaining increasing attention for their various pharmacological effects. *Centella asiatica* has a number of phytochemicals including triterpines, tannins, flavonoids and others. These phytochemicals need further scientific attention towards discovery of new and efficacious drugs.

References

Abdullah, M.A., AL-Bayaty, F.H., Younis, L.T. and Hassan, M.I.A. 2010. Anti-ulcer activity of *Centella asiatica* leaf extract against ethanol-induced gastric mucosal injury in rats. J. Med. Plants Res. 4: 1253–1259.

An, I.-S., An, S., Kang, S.-M., Choe, T.-B., Lee, S.N., Jang, H.H. and Bae, S. 2012. Titrated extract of *Centella asiatica* provides a UVB protective effect by altering microRNA expression profiles in human dermal fibroblasts. Int. J. Mol. Med. 30: 1194–1202.

Arumugam, T., Ayyanar, M., Pillai, Y.J.K. and Sekar, T. 2011. Phytochemical screening and antibacterial activity of leaf and callus extracts of *Centella asiatica*. Bangladesh J. Pharmacol. 6: 55–60.

Babu, M.K., Prasad, O.S. and Murthy, T.E.G.K. 2011. Comparison of the dermal wound healing of *Centella asiatica* extract impregnated collagen and crosslinked collagen scaffolds. J. Chem. Pharm. Res. 3: 353–362.

Babu, T.D., Kuttan, G. and Padikkala, J. 1995. Cytotoxic and anti-tumour properties of certain taxa of Umbelliferae with special reference to *Centella asiatica* (L.) Urban. J. Ethnopharmacol. 11: 53–57.

Babykutty, S., Padikkala, J., Sathiadevan, P.P., Vijayakurup, V., Azis, T.K.A., Srinivas, P. and Gopala, S. 2009. Apoptosis induction of *Centella asiatica* on human breast cancer cell. Afr. J. Trad. CAM 6: 9–16.

Barbosa, N.R., Pittella, F. and Gattaz, W.F. 2008. *Centella asiatica* water extract inhibits iPLA2 and cPLA2 activities in rat cerebellum. Phytomed. 15: 896–900.

Bian, X., Li, G.G., Yang, Y., Liu, R.T., Ren, J.P., Wen, L.Q., Guo, S.M. and Lu, Q. 2008. Madecassoside reduces ischemia-reperfusion injury on regional ischemia induced heart infarction in rat. Biolog Pharm. Bull. 31: 458–463.

Bibi, Y., Zia, M., Nisa, S., Habib, D., Waheed, A. and Chaudhary, F. 2011. Regeneration of *Centella asiatica* plants from non-embryogenic cell M. lines and evaluation of antibacterial and antifungal properties of regenerated calli and plants. J. Biolog Eng. 5: 13.

Brinkhaus, B., Lindner, M., Schuppan, D. and Hahn, E.G. 2000. Chemical, pharmacological and clinical profile of the East Asian medical plant *Centella asiatica*. Phytomed. 7: 427–448.

Bunpo, P., Kataoka, K., Arimochi, H., Nakayama, H., Kuwahara, T., Bando, Y., Lzumi, K., Vintketkumnuen, U. and Ohnishi, Y. 2004. Inhibitory effects of *Centella asiatica* on azoxymethane-induced aberrant crypt focus formation and carcinogenesis in the intestines of F344 rats. Food Chem. Toxicol. 42: 1987–1997.

Bunpo, P., Kataoka, K., Arimochi, H., Nakayama, H., Kuwahara, T., Vintketkumnuen, U. and Ohnishi, Y. 2005. Inhibitory effects of asiatic acid and CPT-11 on growth of HT-29 cells. J. Med. Invest. 52: 65–73.

Chatterjee, T.K., Chakraboty, A., Pathak, M. and Sengupta, G.C. 1992. Effects of plant extract *Centella asiatica* (Linn.) on cold restraint stress ulcer in rats. Ind. J. Exp. Biol. 30: 889–891.

Chauhan, P.K., Pnadey, I.P. and Dhatwalia, V.K. 2010. Evaluation of the anti-diabetic effect of ethanolic and methanolic extracts of *Centella asiatica* leaves extract on alloxan induced diabetic rats. Adv. Biol. Res. 4: 27–30.

Chen, J.Y., Chen, J.Y., Xu, Q.W., Xu, H. and Huang, Z.H. 2014. Asiatic Acid promotes p21WAF1/CIP1 protein stability through attenuation of NDR1/2 dependent phosphorylation of p21WAF1/CIP1 in HepG2 human hepatoma Cells. Asian Pac. J. Cancer Prev. 15: 963–967.

Cheng, C.L. and Koo, M.W. 2000. Effects of *Centella asiatica* on ethanol induced gastric mucosal lesions in rats. Life Sci. 67: 2647–2653.

Cheng, C.L., Guo, J.S. and Koo, M.W. 2004. The healing effects of *Centella* extract and asiaticoside on acetic acid induced gastric ulcers in rats. Life Sci. 74: 2237–2249.

Chippada, S.C. and Vangalapati, M. 2011. Antioxidant, an anti-inflammatory and anti-arthritic activity of *Centella asiatica* extracts. J. Chem. Bio. Phy. Sci. 1: 260–269.

Chippada, S.C., Volluri, S.S., Bammidi, S.R. and Vangalapati, M. 2011. *In vitro* anti inflammatory activity of methanolic extract of *Centella asiatica* by HRBC membrane stabilization. Rasyan J. Chem. 4: 457–460.

Cho, C.W., Choi, D.S., Cardone, M.H., Kim, C.W., Sinskey A.J. and Rha, C. 2006. Glioblastoma cell death induced by asiatic acid. Cell Biol. Toxicol. 22: 393–408.

Chong, N.J. and Aziz, Z. 2012. A systematic review of the efficacy of *Centella asiatica* for improvement of the signs and symptoms of chronic venous insufficiency. eCAM 2013: 10 pages.

Chopra, R.N., Nayar, S.L. and Chopra, I.C. 1986. Glossary of Indian Medicinal Plants (Including the Supplement). Council of Scientific and Industrial Research, New Delhi, 130 pp.

Das, M.P., Banerjee, A. and Anu. 2013. Comparative study on *in vitro* antibacterial and antifungal properties of five medicinal plants of west Bengal. Asian J. Plant Sci. Res. 3: 107–111.

Dash, B.K., Faruquee, H.M., Biswas, S.K., Alam, M.K., Sisir, S.M. and Prodhan, U.K. 2011. Antibacterial and antifungal activities of several extracts of *Centella asiatica* L. against some human pathogenic microbes. Life Sci. Med. Res. 2011: LSMR-35.

De sancits, M.T., Belcaro, G., Inacandela, L., Cesarone, M.R., Griffin, M., Ippolito, E. and Cacchio, M. 2001. Treatment of edema and increased capillary filtration in venous hypertension with total triterpenic fraction of *Centella asiatica*: a clinical, prospective, placebo-controlled, randomized, dose-ranging trial. Angiology 2: 55–59.

Defillipo, P.P., Raposo, A.H., Fedoce, A.G., Ferreira, A.S., Polonini, H.C., Gattaz, W.F. and Raposo, N.R. 2012. Inhibition of cPLA2 and sPLA2 activities in primary cultures of rat cortical neurons by *Centella asiatica* water extract. Nat. Prod. Commun. 7: 841–843.

Diwan, P.C., Karwande, I. and Singh, A.K. 1991. Anti-anxiety profile of mandukparni *Centella asiatica* Linn in animals. Fitoterapia 62: 255–257.

European Medicines Agency. 2012. Assessment report on *Centella asiatica* (L.) Urban, herba.

Flora, S.J. and Gupta, R. 2007. Beneficial effects of *Centella asiatica* aqueous extract against arsenic-induced oxidative stress and essential metal status in rats. Phytother. Res. 21: 980–988.

Gadahad, M.R., Rao, M. and Rao, G. 2008. Enhancement of hippocampal CA3 neuronal dendritic arborization by *Centella asiatica* (Linn) fresh leaf extract treatment in adult rats. J. Chin. Med. Assoc. 71: 6–13.

Gilani, A.H. and Rahman, A.U. 2005. Trends in ethnopharmacology. J. Ethnopharmacol. 100: 43–49.

Gnanapragasama, A., Ebenezara, K.K., Sathisha, V., Govindarajub, P. and Devakia, T. 2004. Protective effect of *Centella asiatica* on antioxidant tissue defense system against adriamycin induced cardiomyopathy in rats. Life Sci. 76: 585–597.

Gohil, K.J., Patel, J.A. and Gajjar, A.K. 2010. Pharmacological review on *Centella asiatica*: A potential herbal cure-all. Ind. J. Pharm. Sci. 72: 548–558.

Gupta, Y.K., Veerendra, K.M.H. and Srivastava, A.K. 2003. Effect of *Centella asiatica* on pentylenetetrazole-induced kindling, cognition and oxidative stress in rats. Pharmacol. Biochem. Behav. 74: 579–585.

Haleagrahara, N. and Ponnusamy, K. 2010. Neuroprotective effect of *Centella asiatica* extract (CAE) on experimentally induced parkinsonism in aged Sprague-Dawley rats. J. Toxicol. Sci. 35: 41–47.

Han, T., Qin, L., Rui, Y. and Zheng, H. 2003. Effect of total triterpenes from *Centella asiatica* on the depression behavior and concentration of amino acid in forced swimming mice. Zhong Yao Cai. 26: 870–873.

Heidari, M., Heidari-Vala, H., Sadeghi, M.R. and Akhondi, M.M. 2012. The inductive effects of *Centella asiatica* on rat spermatogenic cell apoptosis *in vivo*. J. Nat. Med. 66: 271–278.

Höfling, J.F., Anibal, P.C., Obando-Pereda, G.A., Peixoto, I.A., Furletti, V.F., Foglio, M.A. and Gonçalves, R.B. 2010. Antimicrobial potential of some plant extracts against Candida species. Braz. J. Bio. 70: 1065–1068.

Hong, S.S., Kim, J.H. and Shim, C.K. 2005. Advanced formulation and pharmacological activity of hydrogel of the titrated extract of *C. asiatica*. Arch. Pharm. Res. 28: 502–508.

Hsu, Y.L., Kuo, P.L., Lin, L.T. and Lin, C.C. 2005. Asiatic acid, a triterpene, induces apoptosis and cell cycle arrest through activation of extracellular signal-regulated kinase and p38 mitogen-activated protein kinase pathways in human breast cancer cells. J. Pharmacol. Exp. Ther. 313: 333–344.

Huang, S.S., Chiu, C.S., Chen, H.J., Hou, W.C., Sheu, M.J., Lin, Y.C., Shie, P.H. and Huang, G.J. 2011. Antinociceptive activities and the mechanisms of anti-inflammation of asiatic acid in mice. eCAM 2011: 10 pages.

Hussin, F., Ataollahi, S., Rahmat, A., Othman, F. and Akim, A. 2014. The *Centella asiatica* juice effects on DNA damage, apoptosis and gene expression in hepatocellular carcinoma (HCC). BMC Complemen. Altern. Med. 14: 32.

Jacob, S.J. and Shenbagaraman, S. 2011. Evaluation of antioxidant and antimicrobial activities of the selected green leafy vegetables. Int. J. Pharm. Tech. Res. 3: 148–152.

Jagtap, N.S., Khadabadi, S.S., Ghorpade, D.S., Banarase, N.B. and Naphade, S.S. 2009. Antimicrobial and antifungal activity of *Centella asiatica* (L.) Urban, Umbeliferae. research J. Pharm. Tech. 2: 328–330.

Jahan, R., Hossain, S., Seraj, S., Nasrin, D., Khatun, Z., Das, P.R., Islam, M.T., Ahmed, I. and Rahmatullah, M. 2012. *Centella asiatica* (L.) Urb.: Ethnomedicinal uses and their scientific validations. Am.-Eurasian J. Sustain. Agric. 6: 261–270.

Jana, U., Sur, T.K., Maity, L.N., Debnath, P.K. and Bhattacharyya, D. 2010. A clinical study on the management of generalized anxiety disorder with *Centella asiatica*. Nepal Med. Coll. J. 12: 8–11.

Joy, J. and Nair, C.K. 2004. Protection of DNA and membranes from gamma-radiation induced damages by *Centella asiatica*. J. Pharm. Pharmacol. 61: 941–947.

Ju-Lin, X., Shao-hai, Q., Tian-zeng, L., Bin, H., Jing-ming, Tang, Ying-bin, X., Xu-sheng, L., Bin, S., Hui-zhen, L. and Yong, H. 2008. Effect of asiaticoside on hypertrophic scar in the rabbit ear model. J. Cutan. Pathol. 36: 234–239.

Jung, E., Lee, J.-A., Shin, S., Roh, K.-B., Kim, J.-H. and Park, D. 2013. Madecassoside inhibits melanin synthesis by blocking ultraviolet-induced inflammation. Molecules 18: 15724–15736.

Kabir, A.U., Samad, Mehdi Bin, D'Costa Ninadh Malrina, Akhter Farjana, Ahmed Arif and Hannan, J.M.A. 2014. Anti-hyperglycemic activity of *Centella asiatica* is partly mediated by carbohydrase inhibition and glucose-fiber binding. BMC Complement and Alternat. Med. 14: 31.

Kavitha, C.V., Agarwal, C., Agarwal, R. and Deep, G. 2011. Asiatic acid inhibits pro-angiogenic effects of VEGF and human gliomas in endothelial cell culture models. PLos One 6: 227–245.

Kedzia, B., Bobkiewicz-Kozlowska, T., Furmanowa, M., Mikolaczak, P., Holderna Kedzia, E., Okulicz-Kozaryn, I., Wojcik, J., Guzewska, J., Buchwaldi, W., Mscisz, A. and Mrozikiewicz, P.M. 2007. Studies on the biological properties of extracts from *Centella asiatica* (L.) Urban herb. Herba Polonica 7: 34–44.

Kim, Y.J., Cha, H.J., Nam, K.H., Yoon, Y., Lee, H. and An, S. 2011. *Centella asiatica* extracts modulate hydrogen peroxide-induced senescence in human dermal fibroblasts. Exp. Dermatol. 20: 998–1003.

Kumar, A., Dogra, S. and Prakash, A. 2009. Neuroprotective effects of *Centella asiatica* against intracerebroventricular colchicine-induced cognitive impairment and oxidative stress. Int. J. Alzheimer's Dis. 2009: 8 pages.

Kumar, A., Prakash, A. and Dogra, S. 2011. *Centella asiatica* Attenuates D-Galactose-induced cognitive impairment, oxidative and mitochondrial dysfunction in mice. Int. J. Alzheimers Dis. 2009: 8 pages.

Kumar, M.H.V. and Gupta, Y.K. 2002. Effect of different extracts of *Centella asiatica* on cognition and markers of oxidative stress in rats. J. Ethnopharmacol. 79: 253–260.

Kumar, M.H.V. and Gupta, Y.K. 2003. Effect of *Centella asiatica* on cognition and oxidative stress in an intracerebroventricular streptozotocin model of Alzheimer's disease in rats. Clin. Exp. Pharmacol. Physiol. 30: 336–342.

Lee, J., Jung, E., Kim, Y., Park, J., Park, J., Hong, S., Kim, J., Hyun, C., Kim, Y.S. and Park. D. 2006. Asiaticoside induces human collagen I synthesis through TGFbeta receptor I kinase (TbetaRI kinase)-independent Smad signaling. Planta Med. 72: 324–328.

Lee, Y.S., Jin, D.Q., Kwon, E.J., Park, S.H., Lee, E.S., Jeong, T.C., Nam, D.H., Huh, K. and Kim, J.A. 2002. Asiatic acid, a triterpene, induces apoptosis through intracellular Ca2+ release and enhanced expression of p53 in HepG2 human hepatoma cells. Cancer Lett. 186: 83–91.

Liu, M., Dai, Y., Yao, X., Li, Y., Luo, Y., Xia, Y. and Gong, Z. 2008. Anti-rheumatoid arthritic effect of madecassoside on type II collagen-induced arthritis in mice. Int. Immunopharmacol. 8: 1561–1566.

Maquart, F.X., Chastang, F., Simeon, A., Birembaut, P., Gillery, P. and Wegrowski, Y. 1999. Triterpenes from *Centella asiatica* stimulate extracellular matrix accumulation in rat experimental wounds. Eur. J. Dermatol. 9: 289–296.

Mamtha, B., Kavitha, K., Srinivasan, K.K. and Shivanda, P.G. 2004. An *in vivo* study of the effect of *Centella asiatica* [Indian Pennywort] on enteric pathogen. Ind. J. Pharmacol. 36: 41–44.

Maramaldi, G., Tongi, S., Franceschi, F. and Lati, E. 2013. Anti-inflammaging and antiglycation activity of a novel botanical ingredient from African biodiversity (Centevita™). Clin. Cosmet. Investig. Dermatol. 7: 1–9.

Meena, H., Pandey, H.K., Pandey, P., Arya, M.C. and Ahmed, Z. 2012. Evaluation of antioxidant activity of two important memory enhancing medicinal plants Baccopa monnieri and *Centella asiatica*. Ind. J. Pharmacol. 44: 114–117.

Norzaharaini, M.G., Wan, N.W.S., Hasmah, A., Nor, I.N.J. and Rapeah, S.A. 2011. Preliminary study on the antimicrobial activities of asiaticoside and asiatic acid against selected gram positive and gram negative bacteria. Health Environment J. 2: 23–26.

Oyedeji, O.A. and Afolavan, A.J. 2005. Chemical composition and antibacterial activity of the essential oil of *Centella asiatica* growing in South Africa. Pharmaceutical Biol. 43: 249–252.

Pan, S., Li, T. and Li, Y. 2004. Effects of asiaticoside on cell proliferation and smad signal pathway of hypertrophic scar fibroblasts. Zhonggo Xiu Fu Chong Jian Wai Ke Za Zhi 18: 291–294.

Paocharon, V. 2010. The efficacy and side effects of oral *Centella asiatica* extract for wound healing promotion in diabetic wound patients. J. Med. Assoc. Thai 93: 166–170.

Park, B.C., Bosiro, K.O., Lee, E.S., Lee, Y.S. and Kim, J.A. 2005. Asiatic acid induces apoptosis in SK-MEL-2 human melanoma cells. Cancer Lett. 218: 81–90.

Park, B.C., Park, S.-H., Lee, Y.-S., Kim, S.-J., Lee, W.-S., Choi, H.G., Yong, C.S. and Kim, J.-A. 2007. Inhibitory effects of asiatic acid on 7, 12-Dimethylbenz[a]anthracene and 12-O-tetradecanoylphorbol 13-acetate-induced tumor promotion in mice. Biol. Pharm. Bull. 30: 176–179.

Poizot, A. and Dumez, D. 1978. Modification of the kinetics of healing after iterative exeresis in the rat. Action of a triterpenoid and its derivatives on the duration of healing. C.R. Acad. Sci. Hebd Seances Acad. Sci. D. 286: 789–792.

Qi, S.H., Xie, J.L., Pan, S., Xu, Y.B., Tang, J.M., Liu, X.S., Shu, B. and Liu, P. 2008. Effects of asiaticoside on the expression of Smad protein by normal skin fibroblasts and hypertrophic scar fibroblasts. Clin. Exp. Dermatol. 33: 171–175.

Rahman, Md. M., Sayeed, M.S.B., Haque, Md. A., Hassan, Md. M. and Islam, S.M.A. 2012. Phytochemical screening, antioxidant, anti-Alzheimer and anti-diabetic activities of *Centella asiatica*. J. Nat. Prod. Plant Resour. 2: 504–511.

Rai, N., Agrwal, R.C. and Khan, A. 2011. Chemopreventive potential of *Centella asiatica* on B6F10 melanoma cell lines in experimental mice. Pharmacology online 1: 748–758.

Ramaswamy, A.S., Pariyaswamy, S.M. and Basu, N. 1970. Pharmacological studies on *Centella asiatica* Linn. (Brahmamanduki) (N.O. Umbelliferae). J. Res. Indian Med. 4: 160.

Rao, K.G.M., Rao, S.M. and Rao, S.G. 2005. *Centella asiatica* (Linn) induced behavioral changes during growth spurt period in neonatal rats. Neuroanatomy 4: 18–23.

Rao, K.G.M., Rao, S.M. and Rao, S.G. 2006. *Centella asiatica* (L.) leaf extract treatment during the growth spurt period enhances hippocampal CA3 neuronal dendritic arborization in rats. Evid. Based Complement. Alternat. Med. 3: 349–357.

Rao, K.G.M., Rao, S.M. and Rao, S.G. 2009. Enhancement of amygdaloid neuronal dendritic arborization by fresh leaf juice of *Centella asiatica* (Linn) during growth spurt period in rats. Evid. Based Complement. Alternat. Med. 6: 203–210.

Rao, K.G.M., Rao, S.M. and Rao, S.G. 2012. Evaluation of amygdaloid neuronal dendritic arborization enhancing effect of *Centella asiatica* (Linn) fresh leaf extract in adult rats. Chin. J. Integr. Med. (Epub ahead of print).

Rishikesh, Md. Mofizur Rahman, Islam, S.M.S. and Rahman, Md. M. 2012. Phytochemical screening and antimicrobial investigation of the methanolic extract of *Centella asiatica* leaves. IJPSR 3: 3323–3330.

Rozarina, J.A. 2013. Antimicrobial potential of the methanolic extract of plants. IJSRR 2. Special Issue April–May 2013.

Ruszymah, B.H., Chowdhury, S.R., Mannan, N.A., Fong, O.S., Adenan, M.I. and Saim, A.B. 2012. Aqueous extract of *Centella asiatica* promotes corneal epithelium wound healing *in vitro*. J. Ethnopharmacol. 27: 333–338.

Saha, S., Guria, T., Singha, T. and Maity, T.K. 2013. Evaluation of analgesic and anti-inflammatory activity of chloroform and methanol extracts of *Centella asiatica* Linn. ISRN Pharmacol. 2013: 10 pages.

Sairam, K., Rao, C.V. and Goel, R.K. 2001. Effect of *Centella asiatica* Linn on physical and chemical factors induced gastric ulcereation and secretion in rats. Ind. J. Exp. Boil. 39: 137–142.

Sampath, U. and Janardhanam, V.A. 2013. Asiaticoside, a trisaccharide triterpene induces biochemical and molecular variations in brain of mice with parkinsonism. Transl. Neurodegener. 2: 23.

Sampath, U., Janardhanam, V.A. and Gopinath, G. 2012. Neuro-shielding effect of asiaticoside against MPTP induced variations in experimental mice. Int. J. Eng. Sci. Tech. 4: 960–965.

Sampson, J.H., Raman, A., Karlsen, G., Navasaria, H. and Leigh, I.M. 2001. *In vitro* keratinocyte antiproliferant effect of *Centella asiatica* extract and triterpenoid saponins. Phytomedicine. 8: 230–235.

Sarma, D.N.K., Khosa, R.L., Chansauria, J.P.N. and Saha, M. 2005. Antiulcer activity of *Tinospora cordifolia* Miers and *Centella asiatica* linn extracts. Phytother. Res. 9: 589–590.

Satake, T., Kamiya, K., An, Y., Oishi Nee Taka, T. and Yamamoto, J. 2007. The anti-thrombotic active constituents from *Centella asiatica*. Pharm Bull. 30: 5–940.

Sharma, S. and Thakur, S.C. 2011. *Centella asiatica* extract protects against both inflammation and oxidative stress in collagen-induced arthritis rat model. J. Nat. Sc. Biol. Med. 2: 117.

Shetty, B.S., Udupa, S.L., Udupa, A.L. and Somavaji, S.N. 2006. Effect of *Centella asiatica* L. (Umbelliferae) on normal and dexamethasone-suppressed wound healing in Wistar Albino rats. Int. J. Low Extrem. Wounds 5: 137–143.

Shinmol, G.K. and Muralidhara. 2008. Prophylactic neuroprotective property of *Centella asiatica* against 3-nitropropionic acid induced oxidative stress and mitochondrial dysfunctions in brain regions of prepubertal mice. Neurotoxicology 29: 48–57.

Shukla, A., Rasik, A.M., Jain, G.K., Shankard, R., Kulshrestha, D.K. and Dhawan, B.N. 1999. *In vitro* and *in vivo* wound healing activity of asiaticoside isolated from *Centella asiatica*. J. Ethnopharmacol. 65: 1–11.

Somboonwong, J., Kankaisre, M., Tantisira, B. and Tantisira, M.H. 2012. Wound healing activities of different extracts of *Centella asiatica* in incision and burn wound models: an experimental animal study. BMC Complemen. Altern. Med. 12: 103.

Somchit, M.N., Sulaiman, M.R., Zuraini, A., Samsuddin, L., Somchit, N., Israf, D.A. and Moin, S. 2004. Antinociceptive and antiinflammatory effects of *Centella asiatica*. Ind. J. Pharmacol. 36: 377–380.

Soumyanath, A., Zhong, Y.-P., Yu, X., Bourdette, D., Koop, D.R., Gold, S.A. and Gold, B.G. 2005. *Centella asiatica* accelerates nerve regeneration upon oral administration and contains multiple active fractions increasing neurite elongation *in-vitro*. J. Pharm. Pharmacol. 57: 1221–1229.

Subathra, M., Shila, S., Devi, M.A. and Panneerselvam, C. 2005. Emerging role of *Centella asiatica* in improving age-related neurological antioxidant status. Exp. Geronotol. 40: 707–715.

Suguna, L., Sivakumar, P. and Chandrakasan, G. 1996. Effects of *Centella asiatica* extract on dermal wound healing in rats. Ind. J. Exp. Biol. 34: 1208–1211.

Sunilkumar, Parameshwar, S. and Shivakumar, H.G. 1998. Evaluation of topical formulations of aqueous extract of *Centella asiatica* on open wounds in rats. Ind. J. Exp. Biol. 36: 569–572.

Tabassum, R., Vaibhav, K., Shrivastava, P., Khan, A., Ahmed, M.E., Javed, H., Islam, F., Ahmad, S., Siddiqui, M.S., Safhi, M.M. and Islam, F. 2013. *Centella asiatica* attenuates the neurobehavioral, neurochemical and histological changes in transient focal middle cerebral artery occlusion rats. Neurol. Sci. 34: 925–933.

Taemchuay, D., Rukkwamsuk, T., Sakpuaram, T. and Ruangwises, R. 2009. Antibacterial activity of crude extracts of *Centella asiatica* against *Staphylococcus aureus* in Bovine Mastitis. Kasetsart Veterinarians 19: 119–128.

Tang, B., Zhu, B., Liang, Y., Bi, L., Hu, Z., Chen, B., Zhang, K. and Zhu, J. 2011. Asiaticoside suppresses collagen expression and TGF-β/Smad signaling through inducing Smad7 and inhibiting TGF-βRI and TGF-βRII in keloid fibroblasts. Arch Deramtol. Res. 303: 563–572.

Tang, X.-U., Yang, X.-Y., Jung, H.-U., Kim, S.-Y., Jung, S.-Y., Choi, D.-Y., Park, W.-C. and Park, H. 2009. Asiatic acid induces colon cancer cell growth inhibition and apoptosis through mitochondrial death cascade. Biol. Pharm. Bull. 32(8): 1399–1405.

Udoh, D.I., Asamudo, N.U., Bala, B.D. and Enwongo, O. 2012. Inhibitory effect of varying concentrations of leaves' extracts of *Centella asiatica* (Gotu Kola) on some microorganisms of medical importance. Chem. Env. Pharma. Res. 3: 142–148.

Ullah, M.O., Sultana, S. and Haque, A. 2009. Antimicrobial, cytotoxic and antioxidant activity of *Centella asiatica*. Eur. J. Sci. Res. 30: 260–264.

Visweswari, G., Siva, P.K., Loknath, V. and Rajendra, W. 2010. The antiepileptic effect of *Centella asiatica* on the activities of Na+/K+, Mg2+ and Ca2+-ATPases in rat brain during pentylenetetrazol–induced epilepsy. Ind. J. Pharmacol. 42: 82–86.

Wanasuntronwong, A., Tantisira, M.H., Tantisira, B. and Watanabe, H. 2012. Anxiolytic effects of standardized extract of *Centella asiatica* (ECa 233) after chronic immobilization stress in mice. J. Ethnopharmacol. 143: 579–585.

Wijeweera, P., Amason, J.T., Koszycki, D. and Merali, Z. 2006. Evaluation of anxiolytic properties of Gotukola-(*Centella asiatica*) extracts and asiaticoside in rat behavioral models. Phytomedicine. 13: 668–676.

Xu, C.L., Wang, Q.Z., Sun, L.M., Li, X.M., Deng, J.M., Li, L.F., Zhang, J., Xu, R. and Ma, S.P. 2012. Asiaticoside: attenuation of neurotoxicity induced by MPTP in a rat model of Parkinsonism via maintaining redox balance and up-regulating the ratio of Bcl-2/Bax. Pharmacol. Biochem. Behav. 100: 413–418.

Yoshida, M., Fuchigami, M., Nagao, T., Okabe, H., Matsunaga, K., Takata, J., Karube, Y., Tsuchihashi, R., Kinjo, Mihashi, K. and Fujioka, T. 2005. Antiproliferative constituents from Umbelliferae Plants VII. Active triterpenes and rosmarinic acid from *Centella asiatica*. Biol. Pharm. Bull. 28: 173–175.

Zheng, C.J. and Qin, L.P. 2007. Chemical components of *Centella asiatica* and their bioactivities. Zhong Xi Yi Jie He Xue Bao. 3: 348–351.

Melissa officinalis L.: Overview of Pharmacological and Clinical Evidence of the Plant and Marketed Products in Spain

M.T. Ortega Hernández-Agero, O.M. Palomino Ruiz-Poveda,
M.P. Gómez-Serranillos Cuadrado and M.E. Carretero Accame*

Introduction

Melissa officinalis L., from the Lamiaceae family, is one of the vegetal species traditionally used in Europe for the treatment of nervousness, anxiety and sleep disorders, as well as for gastrointestinal complaints; it is usually known as melisa, toronjil or hierba limonera (Spanish), lemon balm (English), mélisse (French), melissa (Italian), melissenblätter or zitronenkraut (German), cidreira (Portuguese), citroenmelisse (Dutch) or φύλλο μελίσσης (Greek).

This species originated in southern Europe, but nowadays is naturalized around the world, from North America to New Zealand. Lemon balm occurs naturally in sandy and scrubby areas but has also been found on damp wasteland, at elevations ranging from sea level to the mountains, including most of the Iberian peninsula regions.

M. officinalis is an erect perennial aromatic herb which grows up to 70–150 cm tall with straight hairy stalks. The leaves grow in opposite pairs of toothed, smooth and ovate leaves growing on square, branching stems. They have a gentle lemon scent, close to mint, with 3–9 x 1.8–5 cm size and occasionally up to 15 x 8 cm. Flowers are two-lipped, grow in whorled clusters, and may be pale yellow, white or pinkish. The fruit is a tiny nutlet (1.6–2 x 0.8–1 mm), with a dark brown colour.

The herbal drug consists of the dried leaf of *Melissa officinalis* L. According to the European Pharmacopoeia monograph (2013), it contains a minimum of 1.0% of rosmarinic acid ($C_{18}H_{16}O_8$; M_r 360.3), calculated with reference to the dried drug. Other international organizations such as the European Scientific Cooperative on Phytotherapy (ESCOP), European Medicines Agency (EMA) and Commission E have also published a monograph for this species.

Department of Pharmacology, School of Pharmacy, Universidad Complutense de Madrid. Pza Ramón y Cajal s/n, 28040 Madrid, Spain.
Email: meca@farm.ucm.es
* Corresponding author

Melissa leaves chemical composition includes one essential oil which is mainly made up by sesquiterpenes (beta-caryophyllen, germacren-D) and especially monoterpenes (aldehydes: citral A and B, known as geranial and neral, respectively; citronellal; alcohols: linalol; esters: geranil acetate; hydrocarbon derivatives: beta-pinene); other groups have also been identified, including triterpenes (ursolic and oleanolic acids), phenolic acids derived from hydroxycinnamic acid (rosmarinic, caffeic, chlorogenic), flavonoids (quercitroside, apigenin derivatives, luteolin, quercetin and kaempferol) and tannins (Fig. 3.1).

The herbal drug has been used since ancient times. Greek and Roman civilizations used lemon balm preparations for oral and topical use, i.e., for wound healing and insect bites. Also the Persian medical doctor Avicenna recommended its topical use in wounds, ulcers and scabies. Later on, during the 15th century, Paracelsus used lemon balm for "all complaints supposed to proceed from a disordered state of the nervous system". Traditional use also assigned beneficial effects on the memory, as well as digestive, analgesic and sedative properties for the relief of insomnia, fever, common cold, etc. Ayurvedic traditional medicine used lemon balm for the treatment of dyspepsia associated with depression or anxiety.

Pharmacological and clinical studies have shown its antioxidant, sedative, anxiolytic, digestive, spasmolytic, hypoglycaemic, anti-inflammatory, hepatoprotective and antimicrobial (antibacterial, antifungal and antiviral) activities. Recent research studies point the possible protective role of lemon balm in Alzheimer's disease.

A summary of the scientific studies performed with *M. officinalis* L. leaves is shown in Tables 3.1 and 3.2 (pharmacological *in vitro* and *in vivo* studies, respectively).

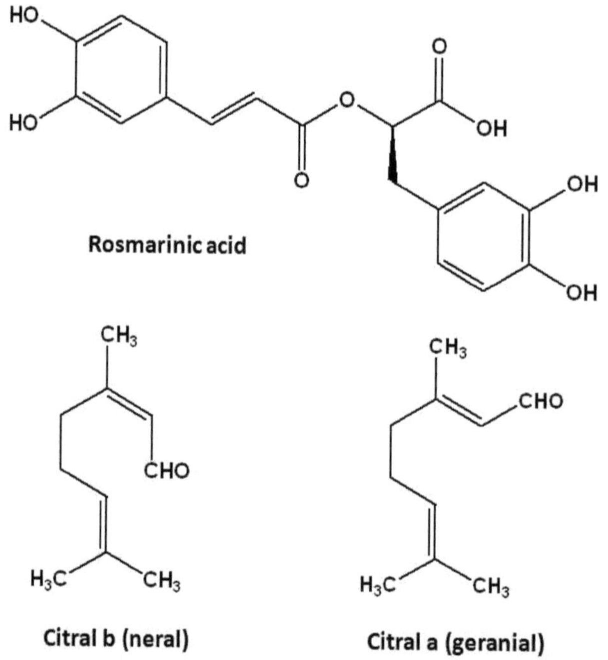

Figure 3.1. Main chemical components of *Melissa officinalis* L. leaf.

Central nervous system activity

Several *in vitro* and *in vivo* pharmacological studies on the activity of lemon balm on CNS have been performed. Some of them intended to elucidate its mechanism of action. This research line was completed with some clinical trials to prove its efficacy on human being.

Table 3.1. Summary of *in vitro* studies related to *Melissa officinalis* L. pharmacological activity.

	Extract/ Active principle	Assay	Model	Activity	Reference
Neuroprotection / Antioxidant	Hydroalcoholic extract	TBARS DPPH	-	Antioxidant	Pereira et al. 2009
	Aqueous extracts/ Poliphenols	ABTS	-	Antioxidant	Ivanova et al. 2005
	Aqueous extract Methanolic extract	MAO-A Inhibitory activity Acetylcholinesterase inhibitory activity ABTS Superoxide Radical Scavenging A Inhibition on Xanthine Oxidase	PC12 cells	Inhibition MAO-A No activity on acetylcholinesterase Antioxidant	López et al. 2009
	Hydroalcoholic extract	Inhibition of lipid peroxidation	Enzyme peroxidation system	Antioxidant	Hohmann et al. 1999
	Ethanolic extract (acidic/non acidic fraction)	Aβ-induced cytotoxicity Lipid peroxidation Glutation peroxidase activity	PC12 cells	Anticholinesterase activity Antioxidant	Sepand et al. 2013
	Ethanolic extract	CNS cholinergic receptor binding activity	Human cerebral cortical cell membranes	Displacement 3H-(N)-nicotine and 3H-(N)-scopolamine from muscarinic and nicotinic receptor	Kennedy et al. 2003
	Ethanol-water extract	TEAC	-	Antioxidant	Lahucky et al. 2010
	Ethanol extract	Cholinergic receptor binding activity		Displacement nicotinic and muscarinic receptors	Wake et al. 2000
	Hydroalcoholic extract	DPPH	-	Antioxidant	Lamaison et al. 1991
	Extracts containing rosmarinic acid	Inhibition amylase activity	Porcine pancreatic cells	Antioxidant	McCue and Shetty 2004
	Polar extract	DPPH/ TLC assay	-	Antioxidant	López et al. 2007
	Supercritical extract	EDTA-mediated oxidation of linoleic acid	-	Antioxidant	Marongiu et al. 2004
	Balm oil	Hypoxic-ischemic injury	Primary neuronal culture	LDH Caspase-3	Bayat et al. 2012
Antitumoral	Ethanolic extract Aqueous extract Rosmarinic acid	MTT/Neutral Red (NR) assays ABTS DPPH	Human colon cancer cell (HCT-116)	Cytotoxicity	Encalada et al. 2011
	Aqueous extract	Caspase 7 Apoptosis	Human cancer cells lines: MCF-7, MDA-MB-468, MDA-MB-231 cells		Saraydin et al. 2012
	Essential oil	DPPH/MTT assay	Human cancer cell lines: A549, MCF-7; Caco-2, HL-60; K562 Mouse cell line: B16F10	Antitumoral	De Sousa et al. 2004

Table 3.1. contd....

Table 3.1. contd.

System	Extract/Active principle	Assay	Model	Activity	Reference
Antimicrobial/Antiviral/Antifungal	Aqueous extract	Herpes simplex virus HSV-1; HSV-2 Plaque reduction assay	RC-37 cells	Antiviral	Nolkemper et al. 2006
	Aqueous Extract	HIV-1	MT-4 cells	Antiviral	Yamasaki et al. 1998
	Aqueous extract Caffeic, p-coumaric and rosmarinic acid	Plaque reduction assay Expression of viral protein ICP0	RC-37 cells Herpes simplex virus type 1(HSV-1)	Antiviral	Astani et al. 2012
	Aqueous, methanolic and ethyl acetate extracts	*Rhizopus* growth	*Rhizopus stolonifer*	Antifungal	Lopez et al. 2007
	Apigenine-7-O-glucoside from MO	Rhesus rotavirus-(RRV)	MA-104 cells	Antiviral activity	Knipping et al. 2012
	Lemon balm oil	*Candida albicans* growth	Gram-positive/ gram-negative strains	Anticandida	Hancianu et al. 2008
	Lemon balm oil	Herpes simplex virus type 1 (HSV-1) HSV-2	HSV-1 and HSV-2	Antiviral	Schnitzler et al. 2008
Central Nervous System	Aqueous extract	Inhibition of GABA-T activity	Rat brain homogenate	Anxiolytic	Awad et al. 2007
	Methanol extract	Inhibition of GABA transaminase	Rat brain homogenate	Anxiolytic	Awad et al. 2009
	Aqueous extract Methanolic extract	GABAA-receptor Assay	Cerebral cortex	No activity	López et al. 2009
	Essential oil	Inhibition of GABA receptor channel Inhibition of inhibitory and excitatory transmission	Radioligand binding Electrophysiological study	Sedative	Abuhamdah et al. 2008

Table 3.2. Summary of *in vivo* studies related to *Melissa officinalis* L. pharmacological activity.

Category	Extract/ Active principle	Assay	Model	Activity	Reference
Central Nervous System	Ethanolic/aqueous extract	Sleep induced by pentobarbital Staircase test Two compartment test	Mice	Sedative	Soulimani et al. 1991
	Ethanolic extract	EPM Forced swimming test Open field test	Rats(male and female)	Anxiolytic Antidepressant	Taiwo et al. 2012
	Ethanolic extract	EPM	Mice	Anxiolytic	Ibarra et al. 2010
	Ethanolic/aqueous extract	Hippocampal dentate gyrus functionality	Mice	Decrease corsticoesterone levels Increase GABA	Yoo et al. 2011
	Rosmaric acid	EPM	Rats	Anxiolytic	Pereira et al. 2005
Antioxidant	Essential oil	Trasient Hippocampal ischemia	Rats	Antioxidant	Bayat et al. 2012
	Aqueous extract	Mn Oxidative model	Mice	Antioxidant	Martins et al. 2012
	Ethanolic/aqueous extract	TBARS	Pigs	Lipidic peroxidation inhibition	Lahucky et al. 2010
Antiinflammatory/ analgesic	Ethanolic extract	Writhing test Formalin Test	Mice	Analgesic	Guginski et al. 2009
	Ethanolic/aqueous extract	Writhing test	Mice	Analgesic	Soulimani et al. 1991
	Essential oil	Paw oedema (carragenan) Traumatism	Rat	Antiinflammatory	Bounihi et al. 2013
	Rosmarinic acid	Paw oedema (CVF)	Rats	Antiinflammatory	Englberger et al. 1988
Antitumoral	Aqueous extract	Mammary tumors (DMBA)	Rats	Antitumoural	Saradyn et al. 2012
Others	Aqueous extract	Induced gastric ulcer	Rat	Cytoprotective	Khayyal et al. 2001
	Ethanolic extract	Fat rich diet Type 2 DM model	Obese mice	Glycemia and triglyceridemia reduction	Weidner et al. 2013
	Essential oil	Fat rich diet Type 2 DM model	Mice	Glycemia and triglyceridemia reduction	Chung et al. 2010
	Essential oil	*APOE2* (R158C) transgenic mice	Mice	Decrease TG	Jun et al. 2012
	Aqueous extract	Fat rich diet	Rats	Cholesterol, total lipids, ALT, AST, ALP reduction Hepatoprotective	Bolkent et al. 2005

EPM: elevated plus-maze

In vitro studies

Methanolic and aqueous extracts were tested in relation to their affinity to the GABA A-benzodiazepine Receptor, but no activity was detected (López et al. 2009).

Chemically-validated essential oil derived from *Melissa officinalis* elicited a significant dose-dependent reduction in both inhibitory and excitatory transmission, with a net depressant effect on neurotransmission, in contrast to the classical GABA-A antagonist picrotoxinin (Abuhamdah et al. 2008).

The aqueous extract of *Melissa officinalis* (lemon balm) exhibited the greatest inhibition of GABA transaminase (GABA-T) activity (IC_{50} = 0.35 mg/mL) *in vitro* rat brain homogenate assays (Awad et al. 2009). The same authors found that the methanol extract of lemon balm was a potent *in vitro* inhibitor of rat brain GABA-T, an enzyme target in the therapy of anxiety, epilepsy and related neurological disorders. Rosmarinic acid and the triterpenoids, ursolic acid and oleanolic acid, were the main compounds of the extract and thus were supposed to be responsible for the sedative effect. Phytochemical characterization of the crude extract determined that rosmarinic acid was the major compound responsible for activity (40% inhibition at 100 μg/mL) since it represented approximately 1.5% of the dry mass of the leaves. Synergistic effects may also play an important role (Awad et al. 2007).

In vivo studies

Sedative properties of the hydroalcoholic extract of *M. officinalis* leaf were proved in different animal models. In mice, intraperitoneal administration of low doses exerted a sedative and sleep inductive effect, as it improved the sleep induced by an infra-hypnotic dose of pentobarbital (Soulimani et al. 1991).

The anxiolytic and antidepressant effects have been tested in male and female rats. Ethanolic extract was capable of reducing anxiety in stressed rats, although the response was dependent on the treatment duration. The administration of 30, 100 or 300 mg/kg/day of the ethanolic extract for 10 days showed a percentage of open arm entries and open arm times for both male and female rats which was significantly higher than the control group alone, with similar values to diazepan group (1 mg/kg), when tested in the elevated plus-maze (EPM) test. With respect to the antidepressant activity which was tested by the forced swimming test, immobility duration was significantly lower in rats treated with the plant extract when compared to control-animals. Also in this assay, a significant influence of the animal genera and treatment duration in the efficacy was observed: a single administration was ineffective; the lowest extract dose was not active in male rats in subacute treatment. On the contrary, no effect on locomotor activity was observed in the open field test. Effects obtained with fluoxetine (10 mg/kg) were superior to lemon balm extract (Taiwo et al. 2012).

One marketed ethanolic extract (30%) of *M. officinalis* leaf (Cyracos®), standardized to a rosmarinic acid content superior to 7% and hydroxycinnamic acid derivatives superior to 15%, when administered orally to mice (120, 240 and 360 mg/kg) for 15 days significantly reduced anxiety-like reactivity dose-dependently in an elevated plus maze task. These doses did not alter those tests evaluating exploratory or circadian activities (Ibarra et al. 2010).

Administration of an infusion containing *Melissa* and *Passiflora caerulea* decreased immobilization—stress in mice. A significant decrease in plasma levels of corticosterone was observed in the treated group when compared to the control group (Feliú-Hemmelmann et al. 2013). The administration of a freeze-dried aqueous ethanolic (20%) extract of lemon balm leaf (50–200 mg/kg/day for three weeks) to 12 months old mice induced a dose-dependent increase in cellular proliferation, neuroblast differentiation and integration into granule cells in the hippocampal dentate gyrus; this structure may have a functional role in stress and depression. The observed activity was related to the decrease in corticosterone levels in serum and the GABA increase in that brain structure (Yoo et al. 2011).

Only a few studies have been performed with the isolated active principles in order to determine which of them are responsible for the pharmacological effects obtained with *M. officinalis* leaf. Rosmarinic acid exerts anxiolytic activity; intraperitoneal administration of low doses to rats increased exploratory conduct on elevated plus-maze test, without affecting locomotor activity and short or long term-memory (Pereira et al. 2005).

Interestingly, intraperitoneal administration of the essential oil did not induce a sedative effect, while oral administration (3.16 mg/kg or superior) exerted a sedative and narcotic effect (ESCOP 2003).

Clinical trials

Several clinical trials proved that lemon balm improves cognitive behaviour and mood, while decreasing induced stress and exerting anxiolytic properties after a single dose treatment. Doses of 600 mg/day of extract increased calm, decreased vigilance and improved the quality of attention as proved by a randomized, placebo-controlled, double-blind, balanced-crossover study including 20 healthy subjects treated with daily single doses of 300, 600 or 900 mg of a standardized, commercial extract of *M. officinalis* (30:70 methanol/ water) (Kennedy et al. 2002). The same research group studied the effect on mood and cognitive performance of the same herbal preparation at 300 and 600 mg on 18 healthy subjects. Participants underwent light psychological stress that was induced in the laboratory one hour after treatment. No adverse events were recorded and all the participants completed the study. Doses of 600 mg improved the subjective effects of laboratory-induced stress (Kennedy et al. 2004).

A prospective, open-label study lasting 15 days included 20 volunteers aged 18 to 70 years with mild to moderate anxiety and sleep disorders. A clear improvement in anxiety symptoms were observed after treatment with 600 mg/day of *M. officinalis* extract (standardized hydroalcoholic leaf extract at 30%, > 7% rosmarinic acid and > 15% hydroxycinnamic acid derivatives) (Cases et al. 2011).

As a general conclusion, lemon balm extract properties were attributed to the presence of some polyphenolic compounds, mainly rosmarinic acid, flavonoids, monoterpene glycosides and the triterpenoids ursolic and oleanoic acids. Again, the synergistic effect of the extract was superior to the addition of the activity of the isolated compounds.

M. officinalis essential oil seems to be more efficient in patients suffering from dementia; aromatherapy proved to be effective in 71 patients with severe dementia. For this double-blind, placebo-controlled study, patients received local administration of the essential oil combined in a lotion on the face and forearms. An improvement in the nervousness status and quality of life of patients was observed after four weeks' treatment when compared to the placebo group (less socially unassertive and more participative in different activities). Safety profile was positive, as no significant adverse events were recorded (Ballard et al. 2002).

Oral administration of lemon balm was also tested in Alzheimer's disease patients. One randomized, double blind, placebo-controlled trial evaluated the efficacy and safety of an alcoholic extract (45% V/V) standardized to 500 µg citral/ml for a four month period; a daily dose of 60 drops was administered to 42 patients of both sexes (21 received lemon balm extract; 21 received a placebo) aged between 65 and 80 years with light to moderate Alzheimer's disease. Thirty five patients completed the study and showed a significant beneficial effect on cognitive function. Related to the adverse events, no significant differences were observed between placebo and treated groups, except for the agitation frequency (not considered as an adverse event) which was superior in the placebo group. These results show a beneficial effect of *M. officinalis* treatment (Akhondzadeh et al. 2003).

The essential oil obtained by the leaves distillation showed inhibitory activity on cholinesterase. This activity may induce some improvement in symptoms from Alzheimer's disease. One double-blind parallel-group placebo-controlled randomized trial performed with 114 patients was designed to compare the administration of lemon balm essential oil twice daily (massage over the hands and upper arms) with a standard treatment (donepezil). Results showed that the essential oil was not better than the placebo or donepezil for the relief of agitation in patients suffering from Alzheimer's disease (Burns et al. 2011).

Antioxidant activity

In vitro studies

Current lifestyle is causing the overproduction of free radicals and Reactive Oxygen Species (ROS) and decreasing the physiological antioxidant capacity. Consistent evidence demonstrated that hydrogen peroxide (H_2O_2), the major ROS in the human body, is involved in molecular mechanisms by which oxidative stress causes cell death in several types of dementia including Alzheimer's disease (Tabner et al. 2005). Phytochemical compounds with antioxidants properties are considered as a promising preventive or therapeutic strategy for reducing H_2O_2-mediated oxidative stress (Speciale et al. 2011). Extracts from plants of the Lamiaceae family have been described for antioxidant and antibacterial effects which are linked to their polyphenolic composition.

Medicinal plants constitute the main source of natural antioxidants which protect the human body from free radicals, prevent oxidative stress and associated diseases. For these reasons they play a very important role in health care. Plants are a source of compounds with antioxidant activity such as phenolic acids, flavonoids, anthocyanins, tannins and carotenoids that may be used as pharmacologically active products.

Melissa officinalis exhibited high antioxidant potential, comparable to that of black and green tea (Ivanova et al. 2005). The aqueous-methanolic extract of lemon balm caused a considerable concentration-dependent inhibition of lipid peroxidation; the phenolics from the extract demonstrated antioxidant activity (Hohmann et al. 1999). Hydroalcoholic extracts from lemon balm have also shown significant antioxidant activities, by free radical scavenger effect on DPPH (N,N-Diphenyl-N-(2,4,6-trinitrophenyl)-hydrazyl), partly in relation to their rosmarinic acid content (Lamaison et al. 1991, McCue and Shetty 2004, Pereira et al. 2009).

Methanolic and aqueous extracts from *M. officinalis* were tested for protective effects on the PC12 cell line, free radical scavenging properties and neurological activities such as inhibition of monoaminooxidase A (MAO-A) and acetylcholinesterase enzymes, as well as the affinity to GABAA-benzodiazepine receptors. Results suggested a protective effect on hydrogen peroxide induced toxicity in PC12 cells (López et al. 2009).

Bayat et al. (2012) examined the effect of *Melissa officinalis* on a model of neuronal death induced by hypoxia in neuronal culture. Cytotoxicity assays showed a significant protection against hypoxia, which was confirmed by a conventional staining. *Melissa* treatment decreased caspase3 activity and also inhibited malondialdehyde (MDA) level, as a marker of lipid peroxidation.

In vivo studies

Related to the antioxidant and neuroprotective properties, the effect of lemon balm essential oil was evaluated by a transient hippocampal ischemia induced in male rats by carotid occlusion for 20 minutes. After reperfusion, a recovery of the antioxidant ability was observed in those animals treated with lemon balm; also a decrease in MDA levels was measured. These results indicate that lemon balm could attenuate oxidative damage induced by ischemic brain injury. In addition the antioxidant activity studies, *in vitro* tests correlated this activity with a possible inhibitory effect on HIF-1alfa and activation of caspase3; this mechanism minimized apoptosis induced by ischemia in neurons and astrocytes (Bayat et al. 2012).

Mice receiving an aqueous extract of *Melissa officinalis* for three months showed a decrease in oxidative cerebral damage induced by manganese, a metal which is capable of inducing neurodegenerative effects similar to Parkinson's disease after long exposure (Martins et al. 2012).

The antioxidant effect was also tested in pork, although results were mild. Animals received different doses of one hydroalcoholic extract of lemon balm (100 or 20 mL) alone or combined with vitamin E for 30 days. After this period, samples of the muscle *longissimus thoracis* were taken and the ability to protect lipid peroxidation induced by Fe^{2+}/ascorbate (TBARS) was measured. Results were significantly better for those animals supplemented with vitamin E (Lahucky et al. 2010).

Clinical trials

Infusion from the leaves of lemon balm (1.5 g/100 mL), when taken twice daily for 30 days, improved the oxidative stress and protected DNA from oxidative injury. These conclusions were obtained by means of a clinical study including 55 volunteers who were exposed to low radiation levels during the workday (Zeraatpishe et al. 2011).

Anti-inflammatory, analgesic activity

Essential oil from *M. officinalis* leaf is rich in nerol (30.44%), citral (27.03%) and isopulegol (22.02%). It exerts anti-inflammatory activity when tested in inflammation models in rat. Oral administration of 200 or 400 mg/kg induced a 61.76 and 70.58% decrease, respectively, in plantar oedema induced by carrageenan; these values were similar to those obtained with the reference standard (indomethacin, 10 mg/kg). When induced by traumatic injury, the anti-inflammatory activity got to values superior to 90%; also in this case, values were similar to oral administration of indomethacin. Results from this study indicate that the mechanism of action is related to an inhibition in the inflammatory mediators release, citral being mainly responsible due to its ability of inhibiting TNFalfa production (Bounihi et al. 2013).

Also isolated rosmarinic acid proved anti-inflammatory activity in different *in vivo* experimental models in which complement activation plays a role. Intramuscular administration of doses between 0.316–3.16 mg/kg reduced paw oedema induced by Cobra Venom Factor (CVF) in the rat. In mice, oral administration of doses between one and 100 mg/kg inhibited passive cutaneous anaphylaxis and provoked a decrease in the macrophagues activation induced by heat-killed *Corynebacterium parvum*, although it did not inhibit t-butyl hydroperoxide-induced paw oedema in the rat; these results indicate selectivity for complement-dependent processes (Englberger et al. 1988).

The ethanolic extract from *Melissa officinalis* leaves when administered orally to mice (3–1000 mg/kg) one hour prior to injury induction, decreased the visceral pain induced by acetic acid in a dose-dependent manner. It also decreased neurogenic and inflammatory pain when evaluated through the formalin-induced licking test. Both the extract and isolated rosmarinic acid (0.3–3 mg/kg), inhibited the pain induced by glutamate in a dose-dependent manner. The antinociceptive effect was not affected by intraperitoneal administration of naloxone. The authors proposed a mechanism of action in which the cholinergic system and L-arginine-nitric oxide pathway are implied; rosmarinic acid seems to be responsible of this activity (Guginski et al. 2009). High doses of the hydroalcoholic extract also exerted peripheral analgesic activity in mice (Soulimani et al. 1991), although studies conducted in mice discarded the inhibitory effect of hydroalcoholic extracts on acetylcholinesterase activity (Orhan and Aslan 2009).

Antimicrobial activity

In vitro studies

The results of studies on natural compounds are promising since several pure substances of plant origin have been shown to exhibit antiviral activity.

The antifungal activity of different extracts of lemon balm was reported (López et al. 2007) on *Rhizopus stolonifer* (Ehrenger: Fries) Lind. var. *stolonifer*; the dichloromethane extracts inhibited the microorganism growth over 20%.

Results from studies with different extracts and essential oil from *Melissa officinalis* showed antiviral activity against HSV-1 and HSV-2 (Yamasaki et al. 1998, Nolkemper et al. 2006, Schnitzler et al. 2008, Astani et al. 2012) and *Rhesus* rotavirus (Knipping et al. 2012). Rosmarinic acid was the main contributor to the antiviral activity of lemon balm extract (Astani et al. 2012).

Clinical trials

Several *in vitro* studies demonstrated the antiviral activity of *M. officinalis.* This effect could be useful in the treatment of *herpes labialis*. One clinical trial, double-blind, controlled versus placebo and lasting five days was conducted with 66 patients (34 verum, 32 placebo) with recurrent *herpes labialis* (at least four episodes per year). Results showed that topical administration of the cream containing a dry extract from lemon balm leaf was more efficient than a placebo: it decreased the symptom's intensity, with a shorter time for healing, a decrease in relapses and a better prevention of the infection propagation. Moreover, the cream avoided virus resistance. These results were in accordance with one previous pilot study conducted with the aim of calculating the number of patients to be included in order to obtain reliable results (Koytchev et al. 1999).

Melissa extract efficacy is higher when treatment starts during the first stages of the infection. This conclusion was obtained by means of two clinical trials: the first one was an open, multicentre study including 115 patients; the second one was a double-blind, controlled study versus placebo including 116 patients. Every patient received lemon balm extract diluted in one cream within the 72 hours after symptoms appeared, two-four times daily for five–10 days. Results were considered as 'very good' by 41% of the patients in the lemon balm group, compared to 19% in the placebo group (Wölbling and Leonhardt 1994).

Antitumoural activity

Cancer is one of the major human diseases which cause considerable suffering and economic loss worldwide. Approximately 60% of drugs currently used for cancer treatment have been isolated from natural products and the plant kingdom has been the most significant source (Gordaliza 2007).

In vitro studies

Some vegetal species, such as *Taxus baccata, Podophyllum peltatum, Camptotecha accuminata*, and *Vinca rosea* are known as strong antitumoural drugs in breast cancer, and several isolated compounds and their semisynthetic derivatives have been evaluated in clinical trials and marketed under this therapeutic indication (Mukherjee et al. 2001, Saraydin et al. 2012). The initial screening for plants used for cancer treatment are cell-based assays using established cell lines, in which the toxic effects of plant extracts or isolated compounds can be measured.

De Sousa et al. (2004) reported that *M. officinalis* essential oil inhibited the viability of several tumour cell lines in a concentration-dependent manner. It showed cytotoxic activity on three cancer cell lines (MCF-7, MDA-MB-468 and MDA-MB-231) with IC50 values 18 ± 2.0 µg/mL, 17 ± 1.4 µg/mL and 19 ± 1.8 µg/mL respectively; reduction in the cell viability was dose-dependent.

The research carried out by Encalada et al. (2011) showed a strong correlation between the phenolic content, antioxidant effect and antiproliferative activity in both the extract and fractions from *M. officinalis* on human colon cancer cell line (HCT-116). Rosmarinic acid could contribute to induce cytotoxic effect in this cell line. However, future research is needed to determine the chemical nature of other compounds present in lemon balm and the possible synergism of action with rosmarinic acid.

In vivo studies

Antitumoural activity of lemon balm leaf has been tested by *in vivo* assays. Aqueous extracts inhibited up to 40% the development of mammary tumours induced by 7,12-dimethylbenz(a)anthracene (DMBA) in rats, with respect to control animals. The antitumoural activity was related to caspase-7 activation, thereby promoting apoptosis of tumour cells (Saraydin et al. 2012).

Antiulcerogenic activity

Melissa extract exerted a dose dependent antiulcerogenic activity against induced gastric ulcers in rat. Antisecretory and cytoprotective activities were related to a decrease in acid release and leukotrienes levels, together with an increase in mucin secretion and prostaglandin E2 release. These effects may be related to the flavonoids content (Khayyal et al. 2001).

Results from *ex vivo* assays (isolated jejunum and aorta of rabbits; guinea pig ileum and tracheal muscle; duodenum and vas deferens of rat) showed the relaxant activity of lemon balm essential oil and several isolated compounds on smooth muscle. The essential oil and citral inhibited contractions of isolated rat ileum induced by KCl (80 mM), acetylcholine (320 nM) and 5-HT (1.28 µM) in a concentration-dependent manner (Sadraei et al. 2003).

Hypoglycaemic activity

The ethanolic extract from the leaf showed beneficial effects in the treatment of insulin resistance and dyslipidaemias in obese mice receiving a fat-rich diet; this experimental model resembles metabolic syndrome and type 2-diabetes in humans. Treatment with 200 mg/kg/day of extract for six weeks significantly reduced hyperglycaemia and insulin resistance; also a decrease in serum levels of triglycerides, non-esterified fatty acids and LDL/VLDL cholesterol was observed. The mechanism of action was related to the activation of the Peroxisome Proliferator-Activated Receptors (PPARs), which play key roles in the regulation of whole body glucose and lipid metabolism (Weidner et al. 2014).

Administration of a daily dose of 0.015 mg of lemon balm essential oil for six weeks to mice resembling type-2 diabetes, resulted in a glycaemia and triglyceridemia reduction (65% for the former), with an improvement in glucose tolerance and insulin levels when compared to control animals. These low concentrations of essential oil may induce an increase in glucose uptake and metabolism in the liver and adipose tissue; hepatic glyconeogenesis could also be inhibited (Chung et al. 2010).

Hypolipidemic activity

In male human *APOE2* transgenic mice (R158C), which develop extreme hypertriglyceridemia and accumulate TG-rich lipoproteins in plasma, oral administration of 12.5 μg/kg of lemon balm leaf essential oil for two weeks, induced a significant decrease in triglyceride serum levels when compared to the control group. The authors suggest that this effect may be related to the changes in the lipid metabolism, including inhibition in the synthesis of biliary acids and cholesterol and changes in fatty acids metabolism (Jun et al. 2012).

Another assay performed with hyperlipidemic rats showed a reduction in serum levels of cholesterol, total lipids, alanine transaminase (ALT), aspartate transaminase (AST) and alkaline phosphatase (ALP) after oral administration of a dry aqueous extract (2 g/kg/day for four weeks). Also lipid peroxidation levels in hepatic tissue were reduced, while glutathione levels increased. A protective hepatic effect was observed through a significant reduction in morphological changes induced by hyperlipidaemia such as vacuolization, picnotic nuclei, mononuclear cell infiltration and rupturing in the endothelium of the central veins in the hepatocytes (Bolkent et al. 2005).

Immunostimulant activity

The administration of one lemon balm leaf extract to animals by oral or subcutaneous route showed immunostimulant activity when compared to the reference compound levamisole. *M. officinalis* activity could imply both humoral and cellular response (Drozd and Anuszewska 2003).

Therapeutic indications

There exists a long traditional medicinal use of *M. officinalis* leaf in Europe, for the relief of mild symptoms of mental stress and to aid sleep, and for the symptomatic relief of mild gastrointestinal disorders including bloating and flatulence (ESCOP 2003). Also the topical use in *herpes labialis* patients is shown. The EMA monograph approves the following indications as a Traditional Herbal Medicinal Product (THMP): (1) relief of mild symptoms of mental stress and to aid sleep and (2) symptomatic treatment of mild gastrointestinal complaints including bloating and flatulence.

Posology

ESCOP (adults, oral use)

- Infusion: 2–3 g of the drug/2–3 times daily.
- Tincture: 1:5, in 45% ethanol V/V, 2–6 ml/1–3 times daily; or equivalent preparations.
- Topical use: cream containing 1% of a freeze-dried aqueous extract (70:1), 2–4 times daily.

EMA (adolescents over 12 years of age, adults and the elderly)

- Herbal tea: 1.5–4.5 g of the comminuted herbal substance in 150 ml of boiling water as an herbal infusion, 1–3 times daily.
- Powdered herbal substance: 0.19–0.55 g, 2–3 times daily.
- Liquid extract: 2–4 mL, 1–3 times daily.
- Tincture: 2–6 mL, 1–3 times daily.
- Dried water or ethanol (45–53% V/V) extracts in doses corresponding to the posology for tea, liquid extract and tincture above.

Safety and adverse events

M. officinalis leaf is considered as a safe herbal drug; no serious adverse events have been reported. Nonetheless, the use in children under 12 years of age is not recommended without medical supervision due to lack of adequate safety data. The same principle applies to the use during pregnancy and lactation. It may affect the ability to drive and operate machinery and also could increase the effect of other anxiolytic or sedative drugs.

Ethanolic extract of lemon balm leaf (250 or 500 mg/kg) do not induce mutagenic or genotoxic effects in mice. On the contrary, it exerts a protective effect as shown in mice blood cells treated with an alkilanting agent (methyl methanesulphonate).

Aqueous extract is also safe and is not genotoxic or mutagenic, although it does not exert the same protective properties than the ethanolic extract (de Carvalho et al. 2011).

Doses up to 2000 mg/kg of the essential oil did not induce toxicity or weight variations in rats (Bounihi et al. 2013). Isolated rosmarinic acid is not genotoxic for rat brain tissues at doses up to 8 mg/kg (Pereira et al. 2005).

Combinations of *Melissa officinalis* with other vegetal species

Several combinations including lemon balm leaves with other vegetal species are used for different indications (sedative, to aid sleep, digestive complaints, etc.). One multicentre clinical trial including 914 children (mean age 8.3 years) showed that the combination of lemon balm and valerian root was efficient for insomnia and anxiety treatment, and also to improve mood in children under 12 years of age with a very good tolerance (Müller and Klement 2006).

Another placebo-controlled, double-blind, parallel group, multicentre designed clinical trial involving 99 healthy adults showed a significant improvement in sleep quality when compared to a placebo. Treatment with 600 mg lemon balm/valerian lasted 30 days. Tolerance was also good, and no serious adverse events were recorded (Cerny and Schmid 1999). The same effect was obtained for patients with sleep disorders; results were similar to those obtained with 0.125 mg triazolam (Dressing et al. 1992). Another double-blind, randomized, placebo-controlled clinical trial including 24 healthy volunteers tested the efficacy of a standardized product containing lemon balm/valerian (600, 1200 or 1800 mg). Stress was induced under controlled conditions in the laboratory and results showed anxiolytic activity at 600 mg (Kennedy et al. 2006).

Patients diagnosed with irritable colon syndrome experienced an improvement in their symptoms after receiving a combination of *Melissa officinalis*, *Mentha spicata* and *Coriandrum sativum*. Despite the low number of subjects included in the study (n = 32), this was a well designed, double-blind, randomized, placebo-controlled, multicentre pilot study lasting eight weeks which demonstrated that the plant preparation decreases severity and frequency of pain and of bloating (Vejdani et al. 2006).

Previous studies performed with 24 patients diagnosed with chronic colitis did not show a remarkable improvement after treatment with a combination of *Taraxacum officinale*, *Hypericum perforatum*, *Calendula officinalis*, *Foeniculum vulgare* and *M. officinalis* (Chakürski et al. 1981).

Also one combination with *Matricaria recutita*, *Foeniculum vulgare* and *M. officinalis* was efficient to relieve colic in children. This randomized, double-blind, placebo-controlled study including 88 breastfed colicky infants for one week showed a decrease in the daily average crying time of the treated group when compared to the placebo (Savino et al. 2005).

One prospective, randomized, placebo-controlled clinical trial was conducted on 40 subjects with the aim of studying the anti-inflammatory activity of seven vegetal species, including *Melissa officinalis*. Melissa extract (2%) showed a moderate effect on the UV-induced erythema, with no significant difference. No adverse events including contact dermatitis were obtained. The authors conclude that treatment with lemon balm is a good choice and represents an alternative to topical treatment with corticoids (Beikert et al. 2013).

Marketed products containing *Melissa officinalis* in Spain

Herbal substances can be marketed as medicinal products or food. Herbal Medicinal Products in the Spanish market do need to comply with the European legal framework which is applied to any kind of medicinal product, including herbal drugs in any of the different marketing authorization type: Traditional Herbal Medicinal Product (THMP), Well Established Use (WEU) or a complete Medicinal product application.

Herbal medicinal products are defined as those medicines, for which the active principle corresponds to one or more herbal substances or preparations or their combinations. Nonetheless, scientific information supporting the medicinal use of herbals is quite variable: while some species have got several published studies demonstrating efficacy and safety, some others are lacking this information.

THMP include some therapeutic indications according to their special nature, as they are supposed to be used without medical supervision, a concrete posology, only by oral, external or inhalation route. Safety should be demonstrated by an expert assessment report. Apart from the singularities on efficacy and safety, one traditional herbal medicinal product has to fulfil the same quality requirements than the rest of medicines, and pharmaceutical assays (physic-chemical, biological or microbiological tests) have to demonstrate its pharmaceutical quality.

Medicinal products

According to the existing legislation, one herbal product or preparation is considered a medicinal product when it is presented to treat or prevent any disease in human beings or when endorsed with pharmacological, immunological or metabolic activity and so, it has to be registered or authorized by competent authorities.

According to the European legislation, all medicinal products, including herbal medicinal products, need a marketing authorization to be placed on the EU market. Due to their particular characteristics, notably their long tradition of use, a lighter, simpler and less costly registration procedure exists that provides the necessary guarantees of quality, safety and efficacy (Herbal Directive—Directive 2004/24/EC),[1] when compared to the requirements of a full marketing authorization. The Committee on Herbal Medicinal Products (HMPC) was then established in September 2004 with the aim of assisting the harmonization of procedures and provisions concerning herbal medicinal products laid down in EU Member States, and further integrating herbal medicinal products in the European regulatory framework. According to Article 16 h(3) of Directive 2001/83/EC,[2] as amended as regarding Traditional Herbal Medicinal Products (THMP), the HMPC has to establish Community herbal monographs for THMP which have to be taken in account by every Member State when examining an application.

The European market for herbal medicinal products has been increasing during the last years and largely exceeds prescription medicines. The main therapeutic indications for herbal medicines are respiratory tract diseases (cough, common cold), gastrointestinal diseases, anxiety and sleep disorders.

The structure of Community monographs follow the Summary of Product Characteristics (SPC) structure and sets out the agreed position on the medicinal product as distilled during the course of the assessment process. This information should, then, be the same in every member State.

---

[1] Directive 2004/24/EC on Traditional Herbal Medicinal Products http://www.emea.europa.eu/htms/human/hmpc/index.htm.
[2] Directive 2001/83/EC of the European Parliament and of the Council of 6 November 2001 on the Community code relating to medicinal products for human use.

Herbal Community monographs include two different columns: the left one with the WEU data and the right one with the THMP data. The WEU is adopted when scientific data are available on efficacy and safety for the proposed herbal product; traditional use is adopted when sufficient data on safety are available and efficacy is plausible due to a long standing use within the Community (at least 30 years). The importance of one Community Herbal Monograph derives from its usage as the reference material during the marketing authorization application by any pharmaceutical industry.

A Community herbal monograph on *Melissa officinalis* L., folium was first adopted in October, 2007 by the HMPC. The first systematic review took place five years after its approval and was finally adopted in May, 2013. This monograph includes all the necessary information for the use of a medicinal product: indication, posology (for adults, children, elderly...), duration of use and method of administration, together with safety information such as special warnings or interactions with other medicinal products and other forms of interaction.

Only the traditional use for Melissa leaf herbal preparations was included in the monograph: comminuted herbal substance, powdered herbal substance, liquid extract, (extraction solvent ethanol), tincture and dried water or ethanol extracts corresponding to the tea, liquid extract and tincture previously described. Pharmaceutical forms included comminuted herbal substance or herbal preparations in solid or liquid dosage forms for oral use.

The therapeutic indications accepted for melisa leaf at the European Union level arise from the long traditional medicinal use of *Melissa officinalis* L., folium in several European countries (Czech Republic, France, Germany, The Netherlands, Poland and Spain, between others), in the form of herbal tea, powdered herbal substance or aqueous/ethanolic extracts, for the relief of mild symptoms of mental stress and to aid sleep and for the symptomatic relief of mild gastrointestinal complaints including bloating and flatulence, and the reference in several handbooks (Madaus 1938, Steinegger and Hänsel 1972, British Herbal Pharmacopoeia 1983, ESCOP 2003).

Ten medicinal products containing Melissa are granted a marketing authorization in Spain, although three of them are not marketed nowadays. The oldest authorization comes from 1955 and corresponds to an oral solution containing eight vegetal components (*Arenaria rubra, Peumus boldus, Cynodon dactylon, Equisetum arvense, Melissa officinalis, Opuntia ficus-indica, Rosmarinus officinalis* and *Sideritis angustifolia*).

Most of the authorized medicinal products are included under one of the following categories: A16A (Other products for digestive system or metabolism) or N05C (Other hypnotic and sedative, alone or in combination) (Table 3.3).

One of the marketed medicines containing lemon balm is 'Agua del Carmen' (Carmen's water) which historical data indicate derives from a preparation used by Carmelite monks during the 16th century. According to the product SPC, its composition is a multiple combination of essential oils and fluid extracts obtained by hydroethanolic distillation of several medicinal plants. The highest extract percentage corresponds to lemon balm leaf, while the highest essentials oil concentrations are for orange (*Citrus aurantium* L.) and then, lemon balm. 'Agua del Carmen' is indicated for stress, symptoms related to stress such as gastrointestinal complaints and nervousness, and disturbances related to menstruation period.

Food products

Apart from its therapeutic use, *M. officinalis* leaf is used in cosmetic, perfumery and food industries, i.e., the aqueous freeze-dried extract is added as a natural antioxidant in several sauces receipts.

Herbal products may be classified and placed on the market as food provided that they do not fulfil the definition of medicinal products and that they do comply with the applicable food law. The Directive 2002/46/EC on food supplements and Regulation (EC) No. 1924/2006 on nutrition and health claims made on foods regulates the presence of herbal products in the form of food supplements.[3] Nonetheless, only a few member states do include herbals as general components in food supplements, such as Belgium

---

[3] Regulation (EC) No. 1925/2006 of the European Parliament and of the Council of 20 December 2006 on the addition of vitamins and minerals and of certain other substances to foods.

Table 3.3. Herbal medicinal products containing *Melissa officinalis* in Spain.

Composition	Marketing authorisation	Therapeutic indication/Posology
Ethanol, Citrus essence, Lemon essence, Hierba Luisa essence, Azahar essence, Bergamote essence, Vainillic essence, *Rosmarinus officinalis* essence, *Eucaliptus* essence, *Thymus* essence, Bitter almond essence, *Mentha piperita* essence, *Origanum* essence, *Melissa officinalis* essence, *Pimpinella anisum* essence, *Cinamomum* essence, Hierba Luisa, Tilia, *Cinamomum*, *Angelica archangelica* root, *Myristica fragrans*, *Hyssopus officinalis*, *Coriandrum*, Chamomilla, *Melissa officinalis* leaf	Since 1957	Stress-related symptoms such as gastrointestinal disorders and nervousness. 10–15 ml diluted in an infusion or sweet water
Melissa officinalis leaf	Since 2010	Stress-related symptoms and to aid sleep. Symptomatic treatment of mild gastrointestinal disorders including bloating and flatulence. *Adults and children over 12 years*: 1100 mg powdered drug/ day
Matricaria recutita L. Rauschert, *Glycyrrhiza glabra*, *Chelidonium major*, *Angelica archangelica* root, *Silybum marianum*, *Melissa officinalis* leaf, *Carum carvi*, *Mentha pulegium*, *Iberis amara*	Since 2012	Gastrointestinal disorders such as dyspepsia, gastritis, stomach ache, bloating, flatulence and stomach burning. *Adults and children over 12 years*: 20 drops, 3 times daily
Melissa officinalis dried extract, *Valerian* dried extract	Since 2003	Symptomatic treatment of temporary and mild nervousness and sleep disorders. *Adults and children over 12 years*: • *Nervousness*: 2–3 tablets (160–240 mg melissa dried extract), up to 3 times daily. • *Sleep disorders* 2–3 tablets (160–240 mg melissa dried extract) before bed time
Melissa officinalis dried extract, *Valerian* dried extract	Since 1999	Symptomatic treatment of temporary and mild nervousness and sleep disorders. *Adults and children over 12 years*: • *Nervousness*: 1 tablet (80 mg melissa dried extract), 3 times daily. • *Sleep disorders* 2 tablets (160–240 mg melissa dried extract) before bed time
Fumaria officinalis fluid extract, *Hamamelis* fluid extract, *Hydrastis canadensis* fluid extract, *Melissa officinalis* fluid extract, *Pyscidia erythrina* fluid extract, *Rheum palmatum* fluid extract, *Mentha piperita* essence, *Peumus boldus* fluid extract	Since 1958	Symptomatic treatment of mild gastrointestinal disorders including bloating and flatulence
Arenaria rubra, *Peumus boldus* leaf, *Cynodon dactylon*, *Equisetum arvense*, *Melissa officinalis*, *Opuntia ficus-indica*, *Rosmarinus officinalis*, *Sideritis angustifolia*	Since 1955	Antispasmodic (urological use)

(Moniteur Belge, 3/18/2009: C-2008/24355), as the inclusion of a botanical in any list of substances intended for use as food supplements cannot be interpreted as being classified as foodstuffs; each Member state could classify them according the relevant legal framework.

Melissa officinalis aerial part is included in the Annex B in the "Compendium of botanicals reported to contain naturally occurring substances as possible concern for human health when used in food and food supplement" (EFSA 2012). Botanicals included in this annex are those which appear on a negative list or are subjected to restricted use in at least one European member State but for which the Scientific Committee could not identify substances of concern.

The use of *M. officinalis* leaf in Spain as a substance to be added to food is allowed and so, several products can be found in the market in which the composition of this species is included because of its nutritional or physiological effect: yogurt or milk with *Tilia* and *Melissa*.

The most frequent category in the Spanish market under the food scheme for lemon balm products is covered by the national regulation (RD 3176/83, November 16). This regulation defines these species as those botanicals which are used in the human diet because of their physiological or organoleptic effects as infusions. An infusion is then defined as the liquid product obtained after pouring boiling water over the herbal product with the aim of extracting soluble components, whereas a soluble extract is defined as the water-soluble product which is obtained after partial or total evaporation of the corresponding infusion.

A list of vegetal species, together with the used part for each one, is included (Table 3.4). Aerial parts or leaves of lemon balm are included in the list and so several products can be found in the Spanish market that contain lemon balm alone or in combination with other species to be used as infusion.

According to the regulation in the field, no health or sanitary claims can be included in the labelling of these products. Only instructions for use or preservation, if necessary, can be included.

Table 3.4. Vegetal species intended for infusion to be used for Human Nutrition in Spain (Royal Decree 3176/1983).

Specie	Part used
Illicium verum	Fruit
Pimpinella anisum	Fruit
Citrus aurantium	Flower
Rosa canina	Fruit
Eucaliptus globulus	Leaf
Hibiscus sabdariffa	Flower
Lippia citriodora	Leaf
Foeniculum vulgare	Fruit
Malva sylvestris	Leaf, flower
Matricaria chamomilla	Aerial parts, flower
Anthemis nobilis	Flower
Santolina chamaeciparisus	Aerial parts
Origanum mejorana	Plant
Melissa officinalis	Plant, leaf
Mentha piperita	Plant, leaf
Mentha pulegium	Leaf
Rosmarinus officinalis	Leaf
Salvia officinalis	Leaf
Sambucus nigra	Flower
Tilia argentum/T. officinalis	Flower, bractea
Thymus vulgaris	Plant, leaf
Verbena officinalis	Plant, leaf
Smilax officinalis	Root

Conclusions

The European Framework for human medicines is also applicable to herbal medicinal products. The special characteristics shown by these kind of medicines led to the development of a simplified authorization procedure which results in a quicker, easier and cheaper procedure while keeping the required quality, safety and efficacy for a medicinal product for human use.

Melissa officinalis leaf and essential oil has traditionally been used due to its effects on the nervous system. Studies on neuroprotection and neurological activity have proven their protective, antioxidant, free radical scavenging properties and MAO-A inhibitory effects.

M. officinalis extracts has been analyzed for their antiviral and antifungal activity, showing an interesting antiherpetic effect *in vitro* which supports its traditional use in the topical treatment of recurrent herpes infection.

The cancer preventive or protective activities of *M. officinalis* are based in antiproliferative and immunohistochemistry experiments, results showing its potential as an antitumour agent against breast cancer. Future experiments are necessary to identify which of the components are responsible for these activities.

More well-designed clinical trials have to be performed in order to evaluate lemon balm leaf efficacy and thus prove a well established use. Most of the published studies in humans include very few subjects, while others are not well designed or give methanolic extracts different from those included in the EMA monograph for THMP.

References

Abuhamdah, S., Huang, L., Elliott, M.S., Howes, M.J., Ballard, C., Holmes, C., Burns, A., Perry, E.K., Francis, P.T., Lees, G. and Chazot, P.L. 2008. Pharmacological profile of an essential oil derived from *Melissa officinalis* with anti-agitation properties: focus on ligand-gated channels. J. Pharm. Pharmacol. 60: 377–384.

Akhondzadeh, S., Noroozian, M., Mohammadi, M., Ohadinia, S., Jamshidi, A.H. and Khani, M. 2003. *Melissa officinalis* extract in the treatment of patients with mild to moderate Alzheimer's disease: a double blind, randomized, placebo controlled trial. J. Neurol. Neurosurg. Psychiatry 74: 863–866.

Astani, A., Reichling, J. and Schnitzler, P. 2012. *Melissa officinalis* extract inhibits attachment of herpes simplex virus *in vitro*. Chemotherapy 58: 70–77.

Awad, R., Levac, D., Cybulska, P., Merali, Z., Trudeau, V.L. and Arnason, J.T. 2007. Effects of traditionally used anxiolytic botanicals on enzymes of the gamma-aminobutyric acid (GABA) system. Can. J. Physiol. Pharmacol. 85: 933–942.

Awad, R., Muhammad, A., Durst, T., Trudeau, V.L. and Arnason, J.T. 2009. Bioassay-guided fractionation of lemon balm (*Melissa officinalis* L.) using an *in vitro* measure of GABA transaminase activity. Phytother. Res. 23: 1075–1081.

Ballard, C.G., O'Brien, J.T., Reichelt, K. and Perry, E.K. 2002. Aromatherapy as a safe and effective treatment for the management of agitation in severe dementia: the results of a double-blind, placebo-controlled trial with *Melissa*. J. Clin. Psychiatry 63: 553–558.

Bayat, M., Azami Tameh, A., Hossein Ghahremani, M., Akbari, M., Mehr, S.E., Khanavi, M. and Hassanzadeh, G. 2012. Neuroprotective properties of *Melissa officinalis* after hypoxic-ischemic injury both *in vitro* and *in vivo*. Daru. 20: 42.

Beikert, F.C., Schönfeld, B.S., Frank, U. and Augustin, M. 2013. Antiinflammatorische wirksamkeit von 7 pflanzenextrakten im ultraviolett-erythemtest. Eine randomisierte, placebokontrollierte studie. Hautarzt. 64: 40–46.

Blumenthal, M., Goldberg, A. and Brinckmann, J. (eds.). 1998. The Complete German Commission E Monographs. American Botanical Council, Austin, TX.

Bolkent, S., Yanardag, R., Karabulut-Bulan, O. and Yesilyaprak, B. 2005. Protective role of *Melissa officinalis* L. extract on liver of hyperlipidemic rats: a morphological and biochemical study. J. Ethnopharmacol. 99: 391–398.

Bounihi, A., Hajjaj, G., Alnamer, R., Cherrah, Y. and Zellou, A. 2013. *In vivo* potential anti-inflammatory activity of *Melissa officinalis* L. essential oil. Adv. Pharmacol. Sci. 2013:101759. doi:10.1155/2013/101759.

British Herbal Pharmacopoeia. British Herbal Medicine Association. London, 1983.

Burns, A., Perry, E., Holmes, C., Francis, P., Morris, J., Howes, M.J., Chazot, P., Lees, G. and Ballard, C. 2011. A double-blind placebo-controlled randomized trial of *Melissa officinalis* oil and donepezil for the treatment of agitation in Alzheimer's disease. Dement. Geriatr. Cogn. Disord. 31: 158–164.

Cases, J., Ibarra, A., Feuillère, N., Roller, M. and Sukkar, S.G. 2011. Pilot trial of *Melissa officinalis* L. leaf extract in the treatment of volunteers suffering from mild-to-moderate anxiety disorders and sleep disturbances. Med. J. Nutrition Metab. 4: 211–218.

Cerny, A. and Schmid, K. 1999. Tolerability and efficacy of valerian/lemon balm in healthy volunteers (a double-blind, placebo-controlled, multicentre study). Fitoterapia 70: 221–228.

Chakŭrski, I., Matev, M., Koîchev, A., Angelova, I. and Stefanov, G. 1981. Treatment of chronic colitis with an herbal combination of *Taraxacum officinale*, *Hypericum perforatum*, *Melissa officinalis, Calendula officinalis* and *Foeniculum vulgare* l. Vutr. Boles. 20: 51–54.

Chung, M.J., Cho, S.Y., Bhuiyan, M.J., Kim, K.H. and Lee, S.J. 2010. Anti-diabetic effects of lemon balm (*Melissa officinalis*) essential oil on glucose- and lipid-regulating enzymes in type 2 diabetic mice. Br. J. Nutr. 104: 180–188.

de Carvalho, N.C., Corrêa-Angeloni, M.J., Leffa, D.D., Moreira, J., Nicolau, V., de Aguiar Amaral, P., Rossatto, A.E. and de Andrade, V.M. 2011. Evaluation of the genotoxic and antigenotoxic potential of *Melissa officinalis* in mice. Genet. Mol. Biol. 34: 290–297.

de Sousa, A.C., Alviano, D.S., Blank, A.F., Alves, P.B., Alviano, C.S. and Gattass, C.R. 2004. *Melissa officinalis* L. essential oil: antitumoral and antioxidant activities. J. Pharm. Pharmacol. 56: 677–681.

Dressing, H., Riemann, D., Low, H., Schredl, M., Reh, C., Laux, P. and Muller, W.E. 1992. Insomnia: are valerian/balm combination of equal value to benzodiazepine? Therapiewoche 42: 726–736.

Drozd, J. and Anuszewska, E. 2003. The effect of the *Melissa officinalis* extract on immune response in mice. Acta. Pol. Pharm. 60: 467–470.

Encalada, M.A., Hoyos, K.M., Rehecho, S., Berasategi, I., de Ciriano, M.G., Ansorena, D., Astiasarán, I., Navarro-Blasco, I., Cavero, R.Y. and Calvo, M.I. 2011. Anti-proliferative effect of *Melissa officinalis* on human colon cancer cell line. Plant Foods Hum. Nutr. 66: 328–334.

Englberger, W., Hadding, U., Etschenberg, E., Graf, E., Leyck, S., Winkelmann, J. and Parnham, M.J. 1988. Rosmarinic acid: a new inhibitor of complement C3-convertase with anti-inflammatory activity. Int. J. Immunopharmacol. 10: 729–737.

European Food Safety Authority (EFSA). 2012. Compendium of botanicals reported to contain naturally occurring substances as possible concern for human health when used in food and food supplement. EFSA J. 10: 2663–2723.

European Medicines Agency (EMA). 2013. Community herbal monograph on *Melissa officinalis* L., folium. 14 May 2013. EMA/HMPC/196745/2012.

European Pharmacopoeia. EDQM editors. 7th ed. 2013. 2: 2355–2356. Strasbourg, France.

European Scientific Cooperative on Phytotherapy (ESCOP) Monographs 2nd ed. Melissa leaf. Thieme, Stuttgart 2003, 324–328.

Feliú-Hemmelmann, K., Monsalve, F. and Rivera, C. 2013. *Melissa officinalis* and *Passiflora caerulea* infusion as physiological stress decreaser. Int. J. Clin. Exp. Med. 6: 444–451.

Gordaliza, M. 2007. Natural products as leads to anticancer drugs. Clin. Transl. Oncol. 9: 767–776.

Guginski, G., Luiz, A.P., Silva, M.D., Massaro, M., Martins, D.F., Chaves, J., Mattos, R.W., Silveira, D., Ferreira, V.M., Calixto, J.B. and Santos, A.R. 2009. Mechanisms involved in the antinociception caused by ethanolic extract obtained from the leaves of *Melissa officinalis* (lemon balm) in mice. Pharmacol. Biochem. Behav. 93: 10–16.

Hãncianu, M., Aprotosoaie, A.C., Gille, E., Poiatà, A., Tuchiluş, C., Spac, A. and Stãnescu, U. 2008. Chemical composition and *in vitro* antimicrobial activity of essential oil of *Melissa officinalis* L. from Romania. Rev. Med. Chir. Soc. Med. Nat. Iasi. 112: 843–847.

Hohmann, J., Zupkó, I., Rédei, D., Csányi, M., Falkay, G., Máthé, I. and Janicsák, G. 1999. Protective effects of the aerial parts of *Salvia officinalis, Melissa officinalis* and *Lavandula angustifolia* and their constituents against enzyme-dependent and enzyme-independent lipid peroxidation. Planta Med. 65: 576–578.

Ibarra, A., Feuillère, N., Roller, M., Lesburgere, E. and Beracochea, D. 2010. Effects of chronic administration of *Melissa officinalis* L. extract on anxiety-like reactivity and on circadian and exploratory activities in mice. Phytomedicine 17: 397–403.

Ivanova, D., Gerova, D., Chervenkov, T. and Yankova, T. 2005. Polyphenols and antioxidant capacity of Bulgarian medicinal plants. J. Ethnopharmacol. 96: 145–150.

Jun, H.J., Lee, J.H., Jia, Y., Hoang, M.H., Byun, H., Kim, K.H. and Lee, S.J. 2012. *Melissa officinalis* essential oil reduces plasma triglycerides in human apolipoprotein E2 transgenic mice by inhibiting sterol regulatory element-binding protein-1c-dependent fatty acid synthesis. J. Nutr. 142: 432–440.

Kennedy, D.O., Scholey, A.B., Tildesley, N.T.J., Perry, E.K. and Wesnes, K.A. 2002. Modulation of mood and cognitive performance following acute administration of *Melissa officinalis* (lemon balm). Pharmacol. Biochem. Behav. 72: 953–964.

Kennedy, D.O., Wake, G., Savelev, S., Tildesley, N.T., Perry, E.K., Wesnes, K.A. and Scholey, A.B. 2003. Modulation of mood and cognitive performance following acute administration of single doses of *Melissa officinalis* (Lemon balm) with human CNS nicotinic and muscarinic receptor-binding properties. Neuropsychopharmacology 28: 1871–1881.

Kennedy, D.O., Little, W. and Scholey, A.B. 2004. Attenuation of laboratory-induced stress in human after acute administration of *Melissa officinalis* (lemon balm). Psychosom. Med. 66: 607–613.

Kennedy, D.O., Little, W., Haskell, C.F. and Scholey, A.B. 2006. Anxiolytic effects of a combination of *Melissa officinalis* and *Valeriana officinalis* during laboratory induced stress. Phytother. Res. 20: 96–102.

Khayyal, M.T., el-Ghazaly, M.A., Kenawy, S.A., Seif-el-Nasr, M., Mahran, L.G., Kafafi, Y.A. and Okpanyi, S.N. 2001. Antiulcerogenic effect of some gastrointestinally acting plant extracts and their combination. Arzneimittelforschung. 51: 545–553.

Knipping, K., Garssen, J. and van't Land, B. 2012. An evaluation of the inhibitory effects against rotavirus infection of edible plant extracts. Virol. J. 9: 137.

Koytchev, R., Alken, R.G. and Dundarov, S. 1999. Balm mint extract (Lo-701) for topical treatment of recurring *Herpes labialis*. Phytomedicine 6: 225–230.

Lahucky, R., Nuernberg, K., Kovac, L., Bucko, O. and Nuernberg, G. 2010. Assessment of the antioxidant potential of selected plant extracts—*in vitro* and *in vivo* experiments on pork. Meat. Sci. 85: 779–784.

Lamaison, J.L., Petitjean-Freytet, C. and Carnat, A. 1991. Medicinal Lamiaceae with antioxidant properties, a potential source of rosmarinic acid. Pharm. Acta. Helv. 66: 185–188.

López, V., Akerreta, S., Casanova, E., García-Mina, J.M., Cavero, R.Y. and Calvo, M.I. 2007. *In vitro* antioxidant and anti-rhizopus activities of Lamiaceae herbal extracts. Plant Foods Hum. Nutr. 62: 151–155.

López, V., Martín, S., Gómez-Serranillos, M.P., Carretero, M.E., Jäger, A.K. and Calvo, M.I. 2009. Neuroprotective and neurological properties of *Melissa officinalis*. Neurochem. Res. 34: 1955–1961.

Madaus, G. 1938. Lehrbuch der biologischen Heilmittel. Georg Thieme Verlag, Leipzig.

Marongiu, B., Porcedda, S., Piras, A., Rosa, A., Deiana, M. and Dessi, M.A. 2004. Antioxidant activity of supercritical extract of *Melissa officinalis* subsp. *officinalis* and *Melissa officinalis* subsp. *inodora*. Phytother. Res. 18: 789–792.

Martins, E.N., Pessano, N.T., Leal, L., Roos, D.H., Folmer, V., Puntel, G.O., Rocha, J.B., Aschner, M., Ávila, D.S. and Puntel, R.L. 2012. Protective effect of *Melissa officinalis* aqueous extract against Mn-induced oxidative stress in chronically exposed mice. Brain Res. Bull. 87: 74–79.

McCue, P.P. and Shetty, K. 2004. Inhibitory effects of rosmarinic acid extracts on porcine pancreatic amylase *in vitro*. Asia Pac. J. Clin. Nutr. 13: 101–106.

Müller, S.F. and Klement, S. 2006. A combination of valerian and lemon balm is effective in the treatment of restlessness and dyssomnia in children. Phytomed. 13: 383–387.

Mukherjee, A.K., Basu, S., Sarkar, N. and Ghosh, A.C. 2001. Advances in cancer therapy with plant based natural products. Curr. Med. Chem. 8: 1467–1486.

Nolkemper, S., Reichling, J., Stintzing, F.C., Carle, R. and Schnitzler, P. 2006. Antiviral effect of aqueous extracts from species of the Lamiaceae family against Herpes simplex virus type 1 and type 2 *in vitro*. Planta Med. 72: 1378–1382.

Orhan, I. and Aslan, M. 2009. Appraisal of scopolamine-induced antiamnesic effect in mice and *in vitro* antiacetylcholinesterase and antioxidant activities of some traditionally used Lamiaceae plants. J. Ethnopharmacol. 122: 327–332. Erratum in: J. Ethnopharmacol. 2011; 133: 251.

Pereira, P., Tysca, D., Oliveira, P., da Silva Brum, L.F., Picada, J.N. and Ardenghi, P. 2005. Neurobehavioral and genotoxic aspects of rosmarinic acid. Pharmacol. Res. 52: 199–203.

Pereira, R.P., Fachinetto, R., de Souza Prestes, A., Puntel, R.L., Santos da Silva, G.N., Heinzmann, B.M., Boschetti, T.K., Athayde, M.L., Bürger, M.E., Morel, A.F., Morsch, V.M. and Rocha, J.B. 2009. Antioxidant effects of different extracts from *Melissa officinalis*, *Matricaria recutita* and *Cymbopogon citratus*. Neurochem. Res. 34: 973–983.

Sadraei, H., Ghannadi, A. and Malekshahi, K. 2003. Relaxant effect of essential oil of *Melissa officinalis* and citral on rat ileum contractions. Fitoterapia 74: 445–452.

Saraydin, S.U., Tuncer, E., Tepe, B., Karadayi, S., Özer, H., Şen, M., Karadayi, K., Inan, D., Elagöz, Ş., Polat, Z., Duman, M. and Turan, M. 2012. Antitumoral effects of *Melissa officinalis* on breast cancer *in vitro* and *in vivo*. Asian Pac. J. Cancer Prev. 13: 2765–2770.

Savino, F, Cresi, F., Castagno, E., Silvestro, L. and Oggero, R. 2005. A randomized double-blind placebo-controlled trial of a standardized extract of *Matricaria recutita*, *Foeniculum vulgare* and *Melissa officinalis* (ColiMil) in the treatment of breastfed colicky infants. Phytother. Res. 19: 335–340.

Schnitzler, P., Schuhmacher, A., Astani, A. and Reichling, J. 2008. *Melissa officinalis* oil affects infectivity of enveloped herpes viruses. Phytomedicine 15: 734–740.

Sepand, M.R., Soodi, M., Hajimehdipoor, H., Soleimani, M. and Sahraei, E. 2013. Comparison of neuroprotective effects of *Melissa officinalis* total extract and its acidic and non-acidic fractions against A β-induced toxicity. Iran J. Pharm. Res. 12: 415–423.

Soulimani, R., Fleurentin, J., Mortier, F., Misslin, R., Derrieu, G. and Pelt, J.M. 1991. Neurotropic action of the hydroalcoholic extract of *Melissa officinalis* in the mouse. Planta Med. 57: 105–109.

Speciale, A., Chirafisi, J., Saija, A. and Cimino, F. 2011. Nutritional antioxidants and adaptive cell responses: an update. Curr. Mol. Med. 11: 770–789.

Steinegger, E. and Hänsel, R. 1972. Pharmakognosie. 3rd ed. Springer Verlag, Berlin, Heidelberg.

Tabner, B.J., Turnbull, S., El-Agnaf, O.M. and Allsop, D. 2002. Formation of hydrogen peroxide and hydroxyl radicals from A(beta) and alpha-synuclein as a possible mechanism of cell death in Alzheimer's disease and Parkinson's disease. Free Radic. Biol. Med. 32: 1076–1083.

Taiwo, A.E., Leite, F.B., Lucena, G.M., Barros, M., Silveira, D., Silva, M.V. and Ferreira, V.M. 2012. Anxiolytic and antidepressant-like effects of *Melissa officinalis* (lemon balm) extract in rats: Influence of administration and gender. Indian J. Pharmacol. 44: 189–192.

Vejdani, R., Shalmani, H.R., Mir-Fattahi, M., Sajed-Nia, F., Abdollahi, M., Zali, M.R., Mohammad Alizadeh, A.H., Bahari, A. and Amin, G. 2006. The efficacy of an herbal medicine, Carmint, on the relief of abdominal pain and bloating in patients with irritable bowel syndrome: a pilot study. Dig. Dis. Sci. 51: 1501–1507.

Wake, G., Court, J., Pickering, A., Lewis, R., Wilkins, R. and Perry, E. 2000. CNS acetylcholine receptor activity in European medicinal plants traditionally used to improve failing memory. J. Ethnopharmacol. 69: 105–114.

Weidner, C., Wowro, S.J., Freiwald, A., Kodelja, V., Abdel-Aziz, H., Kelber, O. and Sauer, S. 2014. Lemon balm extract causes potent antihyperglycemic and antihyperlipidemic effects in insulin-resistant obese mice. Mol. Nutr. Food Res. 58: 903–907.

Wölbling, R.H. and Leonhardt, K. 1994. Local therapy of herpes simplex with dried extract from *Melissa officinalis*. Phytomedicine 1: 25–31.

Yamasaki, K., Nakano, M., Kawahata, T., Mori, H., Otake, T., Ueba, N., Oishi, I., Inami, R., Yamane, M., Nakamura, M., Murata, H. and Nakanishi, T. 1998. Anti-HIV-1 activity of herbs in Labiatae. Biol. Pharm. Bull. 21: 829–833.

Yoo, D.Y., Choi, J.H., Kim, W., Yoo, K.Y., Lee, C.H., Yoon, Y.S., Won, M.H. and Hwang, I.K. 2011. Effects of *Melissa officinalis* L. (lemon balm) extract on neurogenesis associated with serum corticosterone and GABA in the mouse dentate gyrus. Neurochem. Res. 36: 250–257.

Zeraatpishe, A., Oryan, S., Bagheri, M.H., Pilevarian, A.A., Malekirad, A.A., Baeeri, M. and Abdollahi, M. 2011. Effects of *Melissa officinalis* L. on oxidative status and DNA damage in subjects exposed to long-term low-dose ionizing radiation. Toxicol. Ind. Health. 27: 20.

Saffron crocus (*Crocus sativus*): From Kitchen to Clinic

José Luis Ríos[1,*] and Isabel Andújar[2]

Introduction

Crocus sativus L., a plant from the iris family (Iridaceae), is mainly grown in the Mediterranean and south-western Asia (Gresta et al. 2008). Commercial saffron consists of the dried red stigma of the flower with a small portion of the yellowish style attached (Fig. 4.1). While saffron is principally used in cooking and baking, it is also used to add flavor and color to both alcoholic and nonalcoholic beverages. Other less extended applications include its use as a dye in the textile industry and as an excipient in pharmaceutical preparations (Ríos et al. 1996). Saffron is recognized as the most expensive, and therefore the most interesting and attractive, spice in the world, prized for its coloring, bitterness, and aromatic power of its dried stigmas (Gresta et al. 2008).

Various authors have published interesting reviews on saffron, covering all aspects of this spice throughout history, including the use of the dried red stigmas and saffron's cultivation, processing, chemistry and standardization, with a final overview and outlook for the future (Sampathu et al. 1984, Ríos et al. 1996). More recently, Gresta et al. (2008) conducted an extensive review of various agricultural aspects of saffron such as its origins and distribution, genetic traits, botanical characteristics, culture adaptation, management techniques, and qualitative characteristics, including the latest instrumental methods for quality assessment. Other reviews include Kumar et al.'s (2009) compilation of the most recent agronomic advances in saffron's commercial flower and corm production: Carmona et al.'s (2007) review of saffron's aromatic characteristics, Winterhalter and Straubinger's (2000) paper on saffron's chemical composition and the most recent findings on saffron aroma formation, Maggi et al.'s (2010) analysis of changes in the spice's volatile profile depending on its storage time, and Carmona et al.'s (2005) study on the influence of different drying and aging conditions on the spice's constituents.

Commercial saffron for use in cooking, that is, the dried stigmas of the flower itself, should contain no more than 12% water and 7% mineral matter (Ríos et al. 1996). The main components responsible for its flavor and aroma are the essential oil, bitter principles, and dye material. The dye material consists

---

[1] Departament de Farmacologia, Universitat de Valencia, 46100 Burjassot (Spain).
  Email: riosjl@uv.es
[2] COMAV, Universidad Politécnica de Valencia, 46022 Valencia (Spain).
  Email: isanpe@upvnet.upv.es
* Corresponding author

Figure 4.1. Saffron flower and stigmas.

mainly of both water-soluble and fat-soluble carotenoids (8%), with the latter containing a remarkable amount of lycopene, α-carotene, β-carotene, and zeaxanthin. The relevant water soluble components are carotenoid glycosides such as crocin and the free aglycone *trans*-crocetin, although other minor water-soluble compounds are also present, such as flavonoids (kaempferol, quercetin, and naringenin derivatives), cianidins (delphinidin and petunidin glucosides), and anthraquinones (emodin derivatives), which are the principal phenolic compounds (Ríos et al. 1996, D'Auria et al. 2004, Gresta et al. 2008, Padmavati et al. 2011).

Among the bitter principles, picrocrocin is the most important. It can be hydrolyzed to glucose and the volatile aglycone, safranal (Fig. 4.2). The volatile fraction or essential oil (0.3–1.5%) is a colorless liquid, which gives the spice its characteristic, intense odor. It is highly unstable due to its high capacity for oxygen absorption and browning. It is slightly levorotatory, with a density between 0.9514 and 0.9998. The essential oil is made up of monoterpenes, mainly aldehydes (70%), with safranal as the most abundant component (47% of the volatile fraction). All the aldehydes have a similar structure, with 2,6,6-trimethyl-cyclohexene-1-carboxaldehyde being the most prevalent. For a complete chemical overview, see the reviews from Ríos et al. (1996), D'Auria et al. (2004), Gresta et al. (2008) and Padmavati et al. (2011).

Botanical aspects

Saffron is a stemless perennial geophyte herb, which flowers in autumn. It is a sterile triploid derived from saffron, a triploid species (x = 8; 2n = 3x = 24) that is self- and out-sterile and mostly male-sterile and therefore unable to produce seed, so it is propagated principally through corms. It is derived from a wild ancestor, probably through fertilization of either a diploid unreduced egg cell by a haploid sperm cell or a haploid egg by two haploid sperm cells (Gresta et al. 2008, Kumar et al. 2009). Some authors consider *C. cartwrightianus* Herb. cv. *albus* to be a possible ancestor of *C. sativus* since ultrastructural observations have revealed that the pollen of both taxa show numerous pollen germination anomalies which are different from those of other closely related species such as *C. thomasii* Ten. or *C. hadriaticus* Herb. (Siracusa et al. 2013). Moreover, flowering in *C. cartwrightianus* shares close similarities to that in *C. sativus* (Gresta et al. 2008). Morphologically, it has a tuberous-bulb formation referred to as a corm, which is covered

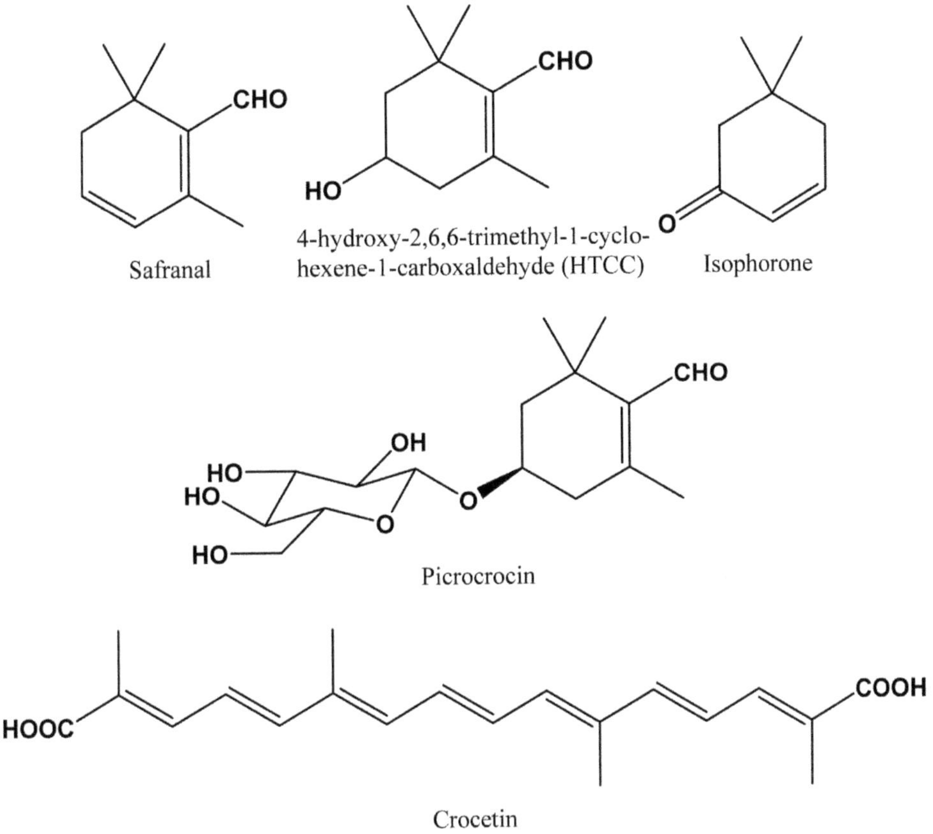

Safranal

4-hydroxy-2,6,6-trimethyl-1-cyclo-
hexene-1-carboxaldehyde (HTCC)

Isophorone

Picrocrocin

Crocetin

Figure 4.2. Chemical composition of saffron. Major components.

by several reticulated fibrous tunics. Leaves are erect, narrow, grass-like, and dark green in color. The flowers (one to 12) have six violet petals that are connate at the base in a long and narrow tube, with a pistil composed of an inferior ovary from which a slender style divides into three dark red branches, called stigmas (Gresta et al. 2008). The dry stigmas constitute the valuable spice.

Saffron grows best in friable, loose, low-density, well-watered, and well-drained clay calcareous soils (Kumar et al. 2009). The biological cycle of saffron starts with the first autumn rains with the emission of the aerial parts, leaves, and flowers, and finishes with the replacement the corms. The flowering season is from mid-October to the end of November, depending on climatic conditions (Gresta et al. 2008). Traditional cultivation methods generally respect this biological cycle. As for the soil, it should be completely cleared of weeds, plowed at a depth of 25–30 cm, and left to rest, either for a few weeks or for the entire winter. Disinfestation of soil before planting avoids fungal infection and sowing by hand is recommended. Climate and soil, planting time, seed/corm rate, planting depth, corm size/weight, crop density, nutrient management, weed management, growth regulators, harvest, and post-harvest management all influence saffron quality and yield (Kumar et al. 2009).

However, with the application of new mechanical techniques in its cultivation and harvest, some relevant modifications can be made to the biological cycle of saffron. For example, the induction of hysteranthy, or flowering prior to leaf appearance, may be of great interest as it facilitates the mechanization of the flower harvest without damaging the leaves. This physiological phenomenon can be induced by controlling the temperature during corm storage (Gresta et al. 2008).

Use and applications as dye, perfume and in food

While the essential oil of saffron is not directly used in food science, it has great potential as a condiment in food: both solid foods and beverages can be improved with the use of this special spice. Saffron not only improves the smell, flavor, and taste of meals, but it can also improve the health of those who consume it daily.

The use of saffron as a dye stems from the presence of water-soluble carotenoids, principally α-crocetin, which is mostly responsible for the yellow color saffron gives to food. It was extensively used in ancient cultures (Assyria, Egypt, Greece, and Rome) as a dye for wool, silk, and hair, among other things. However, its use has decreased with the introduction of cheaper synthetic dyes. Today, saffron is used as a coloring agent mostly in the food and beverage industry. Fresh saffron is odorless, but as it dries, its characteristic aroma appears. This odor is due to safranal, which is released after the enzymatic or thermal hydrolysis of its corresponding glycoside, picrocrocin, present in the fresh stigmas.

Saffron is an extensively used spice in many typical Mediterranean and Asian dishes and is highly appreciated for its bitter taste and the luminous yellow-orange color it gives to foods. There are many reasons for using saffron in cooking, including its colorant properties, pleasing flavor, delicate aroma, and bitter taste. It is used as a condiment in both meat and fish dishes, in soups or cakes, and in sauces or liquor. Saffron is also used in rice dishes, creams, cheese, chicken, mayonnaise, bouillabaisse, and other traditional foods, as well as an additive in alcoholic and nonalcoholic beverages worldwide (Sampathu et al. 1984). The recommended final concentrations in which saffron should be used to avoid an overwhelming taste or smell, which would change the final properties of the food and beverages containing it are given in Table 4.1.

Because the consumption of saffron has stabilized during the past few years, alternative applications have been created to open up new markets. The use of saffron for beverage production is one of these new alternatives. For example, in Spain saffron is used to flavor soft drinks and other beverages, whereas in Italy is added to liquors and in Greece is used in the preparation of alcoholic distillates. The concentration of saffron in non-alcoholic beverages should generally be around 1.3 ppm, whereas in alcoholic beverages the concentration is higher, about 200 ppm (Table 4.1). In the USA, tinctures of saffron are used for flavoring liquors, such as 'Boonekamp', a special bitter brandy, or 'Strega', in which saffron gives color, taste, and flavor to this highly alcoholic liquor, otherwise known as 'saffron gin'.

Aside from its organoleptic properties and its potential as a medicinal drug, saffron contains a series of highly interesting nutrients and minerals. For example, the principal elements present in saffron are carbohydrates, vitamin C, and manganese (Table 4.2). Of the total content in carbohydrates, about 20% are reducing sugars (glucose, fructose, gentiobiose, xylose, and rhamnose), but pentosans, gums, and dextrins are also present (10%). The lipid fraction is around 3–8%, with the presence of fatty acids such

Table 4.1. Final concentration of saffron in ppm in the final product.

	Saffron powder	Saffron extract
Nonalcoholic beverages	1.3	1.3–7.5
Alcoholic beverages	200	–
Baked goods	10	1.9–14.0
Meats	260	–
Ice cream, ices, …	–	1.3–9.0
Candy	–	6.3
Condiments	–	50

(Sampathu et al. 1984)

Table 4.2. Nutritional values per 100 g of saffron. Source United States Department of Agriculture, Agricultural (USDA) National Nutrient Database. Percentages are expressed as the recommended daily allowance (RDA).

Principle	Nutrient Value	Percentage of RDA
Energy	310 Kcal	15.5%
Carbohydrates	65.37 g	50%
Protein	11.43 g	21%
Total lipid (fat)	5.85 g	29%
Cholesterol	0 mg	0%
Dietary fiber	3.9 g	10%
Vitamins	**Nutrient Value**	**Percentage of RDA**
Folates	93 µg	23%
Niacin	1.46 mg	9%
Pyridoxin	1.01 mg	77%
Riboflavin	0.267 mg	20%
Vitamin A	530 UI	18%
Vitamin C	80.8 mg	135%
Electrolytes	**Nutrient Value**	**Percentage of RDA**
Sodium	148 mg	10%
Potassium	1724 mg	37%
Minerals	**Nutrient Value**	**Percentage of RDA**
Calcium	111 mg	11%
Copper	0.33 mg	37%
Iron	11.10 mg	139%
Magnesium	264 mg	66%
Manganese	28.41 mg	1235%
Phosphorus	252 mg	36%
Selenium	5.6 µg	10%
Zinc	1.09 mg	10%

as palmitic, stearic, oleic, linoleic, and linolenic acids, as well as the phytosterols sitosterol, campesterol, and stigmasterol. The mineral content is 1.0–1.5%, with a high proportion of magnesium and potassium salts. The presence of proteins, amino acids, and nitrogen compounds is over 11–13% (Ríos et al. 1996, Padmavati et al. 2011). The spice also contains vitamins, especially riboflavin and thiamine (Table 4.2).

Varieties and quality of saffron

There are different qualities of saffron and various denominations depending on its origin. According to both national and international specifications, saffron can be classified in different ways. The International Organization for Standardization (ISO) has created a classification system for saffron based on the minimal requirements for each quality level (ISO 3632). Thus, ISO 3632 establishes four categories (I–IV) of quality in which different points, such as color (due to crocin), flavor (due to picrocrocin), and aroma (due to safranal), are evaluated. In addition, other parameters such as floral waste and foreign matter are also taken into account (Table 4.3).

Category IV has the poorest quality, with a maximum floral waste mass fraction of 10, double that stipulated for category III. Moreover, each country establishes its own rules and categories. In Spain, for example, the standards and categories established by the government are Coupe, Mancha, Rio, Standard,

Table 4.3. Classification of saffron in categories by physical criteria in filament form (ISO/TS 3632-2: 2003) and crocin content (absorbance at 440 nm).

Characteristics	Categories			
	I	II	III	IV
Floral waste, mass fraction, % max	0.5	3.0	5.0	10
Foreign matter, mass fraction, % max	0.1	0.5	1.0	1.0
Crocin content (absorbance at 440 nm)	> 190	150–190	110–150	80–110

and Sierra, in which the coloring strengths, expressed as direct reading of the absorbance of crocin at about 440 nm, are 190, 180, 150, 145, and 110, respectively. In other countries with a different legal codex, there are relevant differences in quality. For example, the Indian Pharmaceutical Codex establishes the maximum values of water and volatile matter at 103°C to be 14 and 8 for saffron in filaments and in powder, respectively (Sampathu et al. 1984), whereas in the international codex (ISO 3632), these values are 12 and 10, respectively (Table 4.4).

Table 4.4. Quality requirements for saffron in function of its categories according to the ISO/TS 3632-1: 2003's specifications.

	Categories		
	I	II	III
Water and volatile matter at 103°C maximum (% w/w)			
Filament	12	12	12
Powder	10	10	10
Total ash on the dry basis maximum (%)	8	8	8
Ash insoluble in HCl on the dry basis (% w/w) max	1.0	1.0	1.5
Extract soluble in cold water on the dry basis (% w/w) min	70	55	40
Coloring strength, on dry basis, min			
257 nm, min (maximum absorbance of picrocrocine)	70	55	40
330 nm, min (maximum absorbance of safranal)	20	20	20
330 nm, max (maximum absorbance of safranal)	50	50	50
440 nm, min (maximum absorbance of crocines)	190	150	100
Artificial water-soluble acid colorants	0	0	0

Saffron is the most expensive spice in the world and for this reason, its adulteration is common. This is easier in the case of powder, but it also occurs when presented as filaments. There are different kinds of adulterants, most of them of plant origin. They usually include styles, stamen, and strips of the corolla of the saffron flower itself or stigmas from other *Crocus* species such as *C. vernus* or *C. speciosus*. Mixing new saffron with condensed or older saffron or addition of other parts of the saffron flower are also common adulteration techniques. Sometimes, saffron weight can be increased by the addition of substances such as water to increase the humidity percentage. Other techniques include soaking in syrup, honey, glycerin, or olive oil; adding mineral salts such as barium sulfate, calcium carbonate, potassium hydroxide, potassium nitrate, monopotassium tartrate, and sodium borate; or the addition of organic compounds such as lactose, starch, and glucose.

In other cases, the additives used come from other species, for example, florets of safflower (*Carthamus tinctorius*), maize (*Zea mays*), and calendula (*Calendula officinalis*), sometimes colored with methyl orange. Other adulterants are obtained from arnica (*Arnica montana*), pomegranate (*Punica granatum*), common golden thistle (*Scolymus hispanicus*), sliced poppy (*Papaver rhoeas*) flowers, Cape saffron (*Cassine peragua*), red sandalwood (*Santalum paniculatum*), madder (*Rubia tinctorum*), or the outer skin of onions (*Allium cepa*). In these cases, the uncolored material is cut into similarly sized pieces and dyed with eosin. Other natural colorants used as adulterants are carnation perianths (*Dianthus caryophyllus*), turmeric (*Curcuma longa*), annatto (*Bixa orellana*), ground red pepper (*Capsicum frutescens*), and small roots from allium porrum (*Allium ampeloprasum* var. *porrum*). Sometimes even salted and dried meat fibers or colored gelatin fibers are added.

Adulteration is easier in the case of saffron powder, in which different colorants are used. These include Martins yellow, tropeolin, fuchsine, picric acid, tartrazine, erythrosine, azorubine, cochineal red A, orange yellow, naphthol yellow, rocelline yellow, and methyl orange, among others. There are different methods for detecting the presence of contaminants or adulterants in saffron; these differ depending on the presentation (styles or powder). For example, the ISO/TS 3632, 2003 recommends the use of High-Performance Liquid Chromatography (HPLC) as it is the most sensitive method for detecting adulterants. However, for non-specialized laboratories or in the home, other methods exist.

An examination of the macro- and micromorphology of saffron stigmas could be a first step in the detection of adulteration and an analysis of general quality. For example, a stigma length over 30 mm with styles measuring 23 to 24 mm, hard brilliant color, and a strong aroma are all characteristic of the highest quality (high select) whereas poor quality saffron tends to have stigmas that are either broken or < 20 mm long (Table 4.5).

Table 4.5. Spanish quality of saffron. Morphologic and organoleptic characteristics.

	Stigma length	Style length	Organoleptic characteristics
Very select	30 mm	23–24 mm	Hard brilliant color, strong aroma
Select	30 mm	23 mm	Brilliant dark red color, thick thread
Superior	28 mm	22 mm	Dark red color, whole strong threads
Medium	25 mm	21 mm	Good odor, color and appearance
Ordinary	20–24 mm	20–24 mm	Pleasant odor
Slack	< 20 mm	23–24 mm	Broken stigma, dark color

Unfortunately, the quality grades for saffron differ from country to country. With this in mind, one good way to assess product purity is through microscopic examination to identify several characteristic traits. Presence of the epidermis of the style, a vascular bundle with a rounded section, upper epidermis, and parenchyma, together with the characteristic papilla and smooth grains of pollen, are all easily identifiable, thus facilitating evaluation. In addition, pollen should not contain more than three germinal pores (Fig. 4.3).

Another simple way to detect contamination or fraud is Thin-Layer Chromatography (TLC), which is cheaper and easier to carry out than HPLC. The standard TLC analysis uses silica gel as the stationary phase and different mobile phases, such as butanol/acetic acid/water 4:1:1 or ethyl acetate/isopropanol/water 65:25:10. In the latter system, crocin and crocetin appear as yellow spots (R_f 0.15–0.25) in daylight, but show fluorescence-quenching in UV_{254} (as does picrocrocine: R_f 0.55), becoming dark violet-blue when anisaldehyde-sulfuric acid is used as a reagent. The lack of these characteristic spots helps detect different contaminants or adulterants, such as curcuma, safflower, or calendula. The use of high-performance thin-layer chromatography (HPTLC) has also been proposed, with silica gel F_{254} plates and chloroform/methanol/acetic acid 10:1:0.13 as the mobile phase.

Yet another simple method to evaluate product quality is the spectrophotometric assay, taking into account the characteristics outlined in the ISO/TS 3632 standard. The aqueous extract of saffron presents three characteristic absorption maximums at 443, 330, and 257 nm, corresponding to crocine and its derivatives, safranal and picrocrocine, respectively. When the spice has been adulterated with foreign substances, alterations appear in the spectrum. For example, the minimum extinction values for water extracts at 440 nm (corresponding to the maximum absorbance of crocines) give the values of 190, 150, 100, and 80 for categories I, II, III and IV, respectively (see Table 4.4).

Analysis with HPLC is the best method for detecting all kinds of contaminants and adulterants. Different systems can be used for a good separation. For example, in reverse phase HPLC, the suggested

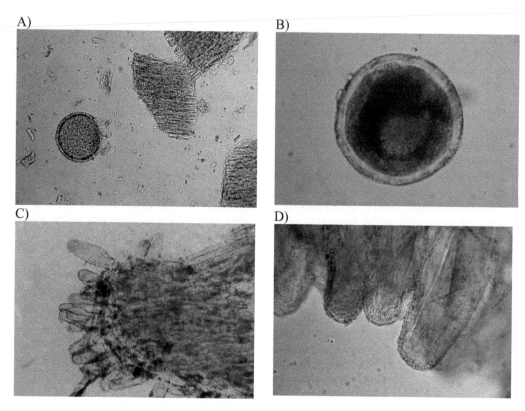

Figure 4.3. Typical microscopic elements from saffron. (A) Pollen grain and rest of parenchyma and epidermis ×100; (B) Pollen grain ×400; (C) Papilla ×100; (D) Papilla ×400.

mobile phases are either methanol/water (60:40) at a flow rate of 1 mL/minute and detection at 440 nm, or a gradient with methanol 5 until 95% (lineal 3.3% per minute) and detection at 254 nm for picrocrocin and 438 nm for crocin. Another alternative is the use of acetonitrile instead of methanol, albeit with the same gradient (over 30 minutes, 1.0 mL/minute) or water/acetonitrile gradient from water 10%, linear gradient to 40% (45 minutes), isocratic (5 minutes), and back to initial conditions in the last 12 minutes (Lechtenberg et al. 2008). Some more recent chromatographic techniques have been used to analyze saffron, including a non-destructive method based on a supercritical carbon dioxide extraction combined with HPLC and Gas Chromatography (GC) for determining the safranal content (Lozano et al. 2000); an LC-ESI-MS system for the determination of crocetin esters, picrocrocin and its related compounds (Carmona et al. 2006); and a micellar electrokinetic chromatographic (MEKC) method for quantification of the main metabolites (Gonda et al. 2012).

Pharmacological properties of saffron and its components

The medicinal uses and pharmacological properties of saffron have been extensively studied and reviewed. For example, Adbullaev (2002) and Fernández (2006) reviewed the anticancer and chemopreventive properties of saffron, while Ríos et al. (1996), Schmidt et al. (2007), Srivastava et al. (2010), Bathaie and Mousavi (2010), Mousavi and Bathaie (2011), and Hosseinzadeh and Nassiri-Asl (2013) published all the known information on the pharmacological properties of this spice and its components. In addition, Ulbricht et al. (2011) carried out an evidence-based systematic review, which included written and statistical analysis of scientific literature, expert opinion, and pharmacological and toxicological data.

Still, other researchers have studied the different activities of compounds isolated from saffron. Thus, Mehdizadeh et al. (2013) looked into the cardioprotective effects of safranal while Rezaee and

Hosseinzadeh (2013) examined its anticonvulsant, anti-anxiety, and hypnotic properties. Another isolate, crocetin, has not only been studied in cancer treatment and prevention (Gutheil et al. 2012), but has also been demonstrated to improve cerebral oxygenation, reduce damage in experimental atherosclerosis and arthritis, and suppress liver and bladder toxicity induced by different agents (Giaccio 2004).

Some negative effects have been also reported, such as the hemorrhagic complications after the intake of saffron. However, a study carried out by Ayatollahi et al. (2013) demonstrates that the administration of saffron tablets (200 or 400 mg/day, for seven days) does not induce any significant modification on coagulation or anticoagulation system in comparison with the placebo group. Therefore, the hemorrhagic complications reported could be related to other causes, such as high doses, long duration of consumption or idiosyncrasy effects.

Cardioprotective effects

Kamalipour and Akhondzadeh published a review in which they collected the different studies carried out up to 2011 on the protective cardiovascular effect of saffron. As they highlight, the consumption of saffron prevents different cardiac problems due to its content in anti-oxidants, such as crocetin, which protects from the development of atherosclerosis. In agreement with this report, Mehdizadeh et al. (2013) demonstrated that pretreatment of male Wistar rats with an aqueous extract of saffron (80 and 160 mg/kg) or safranal (0.050 and 0.075 mL/kg) daily for a period of nine days before the induction of a myocardial infarction, normalized the activity of two cardiac injury markers (creatine kinase muscle-brain and lactate dehydrogenase). Moreover, pretreatment with safranal (0.025, 0.050, 0.075 mL/kg) or the aqueous extract of saffron (20, 40, 80 and 160 mg/kg) could reduce the level of malondialdehyde content as a marker of lipid peroxidation and the histological examination also demonstrated a reduced cardiac tissue damage in the treated groups. Finally, the authors concluded that these cardioprotective effects observed were modulated, at least partly, through a reduction of the oxidative stress.

The cardioprotective activity of another component of saffron, crocin, has also been recently evaluated (Razavi et al. 2013a). These authors use a model of subchronic intoxication with an organophosphorous ester pesticide, diazinon, in male Wistar rats to demonstrate that the intraperitoneal administration of crocin (12.5, 25 or 50 mg/kg) three times per week restored the hypotensive effect of diazinon. In addition, the tachycardia caused by diazinon was restored to the level of the control group. Moreover, crocin protected from lipid peroxidation induced by diazinon in aortic tissue through scavenging of free radicals, as demonstrated previously by Mehdizadeh et al. (2013). These authors also demonstrate that crocin did not have effect on the inhibition of the cholinesterase activity induced by diazinon, which confirms that crocin modulates the alterations caused by diazinon through scavenging of free radicals and anti-oxidant activity. This oxidative stress is an important mechanism of diazinon toxicity especially in chronic exposure.

These same authors (Razavi et al. 2013b) later delved into the protective effect of crocin in this same rat model. Administration of diazinon induces morphological damages in the cardiac tissue and elevates the cardiac marker creatine kinase muscle-brain, which is associated with increased oxidative stress and apoptosis in cardiac tissues. Co-treatment with crocin (25 and 50 mg/kg) reduced the increase of the protein level as well as the mRNA expression of the Bax/Bcl-2 ratio, the cytochrome c level in the cytosolic fraction and inhibited the activation of caspase 3 induced by diazinon in heart tissue. Therefore, they conclude that crocin can maintain a redox balance in cardiac tissue and modulate the susceptibility of biological membranes to interact with Reactive Oxygen Species (ROS) and subsequently prevent tissue apoptosis. More recently, Razavi et al. (2014) studied the effects of diazinon in this same rat model on contractile and relaxant responses in rat aorta as well as the *ex vivo* anti-oxidant actions of crocin. Diazinon not only decreased the contractile responses to KCl and phenylephrine, but also attenuated the relaxant response to acetylcholine. However, treatment with crocin improves the toxic effects of diazinon via the reduction of lipid peroxidation and restoration of altered contractile and relaxant responses in rat aorta.

Chahine et al. (2013) studied this cardioprotective effect of saffron against acute myocardium damage in rabbits. In their experiment, they compared the effect of saffron perfusion against cardiac damage by anthracyclines and by free radicals. Their results demonstrate that administration of saffron during free-radical damage helps trap ROS, significantly improving cardiac function. However, it is less effective against anthracycline-induced damage, suggesting that other mechanisms, aside from oxidative stress, underlie anthracycline cardiotoxicity. Moreover, saffron is also effective against lethal cardiac arrhythmias induced by heart ischemia-reperfusion in rat (Joukar et al. 2013). In this sense, pretreatment with saffron (100 mg/kg/day), attenuates the susceptibility and incidence of fatal ventricular arrhythmia during the reperfusion period due to the reduction of electrical conductivity and prolongation of the action potential duration.

Antidepressant and anti-anxiety properties

The antidepressant effects have been reported *in vivo*, both in animal studies and in clinical pilot studies, making saffron a potential agent for including in a rational phytotherapy (Schmidt et al. 2007). The group of Akhondzadeh et al. (2004) studied the efficacy of the stigmas of saffron in the treatment of mild to moderate depression. They carried out different studies in a six-week pilot double-blind, randomized trial with groups of 30 to 40 adult outpatients who met the Diagnostic and Statistical Manual (DSM) of Mental Disorders for major depression with a baseline Hamilton Rating Scale for Depression (HAM-D) score. Different studies comparing saffron versus reference drugs or placebo were carried out. In this way, in a first assay, patients who received saffron (30 mg/day) showed similar effects to those who were treated with imipramine (100 mg/day) for mild to moderate depression. However, patients of the imipramine group showed anticholinergic effects, such as dry mouth and sedation (Akhondzadeh et al. 2004). In another trial with 40 adult outpatients, they compared a hydro-alcoholic extract of saffron stigma versus fluoxetine, and the saffron group (30 mg/day) gave similar effects to those of fluoxetine (20 mg/day) in the treatment of mild to moderate depression during the six-week study. In this case, there were no significant differences in the two groups in terms of observed side effects (Noorbala et al. 2005). In a complementary study with 40 patients, Akhondzadeh et al. (2005) observed that the saffron group (30 mg/day) produced a significantly better outcome on the HAM-D versus placebo, with no significant differences in the two groups in terms of the observed side effects. After these studies, the authors justified and proposed a large-scale trial with saffron.

Short-term therapy with saffron capsules (30 mg/day) showed the same antidepressant efficacy than fluoxetine (40 mg/day) in a randomized double-blind parallel-group study with 40 patients with a history of percutaneous coronary intervention who were suffering from depression (Shahmansouri et al. 2014). Hausenblas et al. (2013) conducted a meta-analysis of published randomized controlled trials examining the effects of saffron supplementation on symptoms of depression among participants, aged 18 and older, with major depressive disorders. A large effect size was found for saffron supplementation vs. placebo in treating depressive symptoms. A null effect size was evidenced between saffron supplementation and the antidepressant-reference group indicating that both treatments were similarly effective in reducing depression symptoms. The authors concluded that saffron supplementation could improve symptoms of depression in adults with major depressive disorders.

Satiereal is a patented extract obtained from saffron stigma (Inoreal, Plerin, France). Its consumption produces a reduction of snacking and creates a satiating effect that could contribute to body weight loss. A randomized controlled trial with 60 overweight women for eight weeks was carried out to study the potential of saffron for the reduction of snacking. After the administration of 176.5 mg/day of saffron, a significant reduction of snacking frequency and hunger and a trend towards lower body weight was observed, with a mean difference from placebo of 0.9 kg (Gout et al. 2010, Mattes 2012). The activity of Satiereal was due to the anxiolytic and antidepressant effects of this extract (Mattes 2012).

The saffron stigma is one of the world's most expensive spices. Therefore, the group of Akhondzadeh designed a study to demonstrate the effectiveness of saffron petals, since they are a sub-product of the saffron industry and have a reduced cost compared with the stigma. They studied the effect of petals in a similar trial, working with the same number and characteristics than the previous studies carried out with stigmas. Patients treated with petals (15 mg, morning and evening) during eight-weeks showed results that were similar to those of fluoxetine (10 mg, morning and evening). In both cases, the remission rate was about 25%, with no significant differences in the two groups on side effects (Basti et al. 2007). In a similar study, 40 adult outpatients received a capsule of saffron petals (30 mg/day) and were compared against a placebo. After six weeks, petals of *Crocus sativus* produced a significantly better outcome on HDR scale than a placebo, with no significant differences in the two groups in terms of observed side effects (Moshiri et al. 2006). In conclusion, it could be considered that the effects of saffron stigma and petal in mild-to-moderate depression compare favorably to results observed in other systematic trials, such as the ones carried out with St. John's wort extracts. The administration of both the stigma and petal resulted in reductions in HAM-D scores over a six-week period of about 54 and 56%, respectively (Dwyer et al. 2011).

Some studies aimed to clarify the potential active principles present in saffron, but no clinical trials were performed with isolated compounds. Wang et al. (2010) studied the lipophilic fractions from *C. sativus* corms, and obtained antidepressant-like effects in predictive behavioral models of antidepressant properties. There are different studies in which the activity is associated to crocin and/or safranal. The mechanism by which these compounds could be implicated are the increase of re-uptake of monoamines (dopamine, noradrenaline, and serotonin), antagonism with NMDA receptor, and agonist with GABA-A (Sarris et al. 2011). The effects of crocin were tested against CREB, BDNF and VGF levels and it was observed that, in a dose-dependent manner, crocin significantly increased the levels of CREB and BDNF, also increasing p-CREB and the transcript levels of BDNF. However, it did not modify CREB and VGF transcript levels. The authors concluded that the antidepressant-like effect of crocin is due to the increasing CREB, BDNF and VGF levels in hippocampus in rats (Vahdati Hassani et al. 2014).

These trials were double-blind trials, carried out using different dosing schedules ranging from 20 to 200 mg/day and for a duration period of 10 days to 16 weeks. However, no studies were carried out to compare the effects of saffron and crocin and to clearly specify dosage correlation of the two components. As saffron contains approximately 10% crocin, the proposed dose for this constituent in future clinical trials is around 10–20 mg.

Sexual dysfunction

Saffron has been used in the treatment of different sexual disorders, including sexual impairment in men (Modabbernia et al. 2012), male erectile dysfunction (Shamsa et al. 2009), sexual dysfunction in women (Kashani et al. 2013), and premenstrual syndrome (Agha-Hosseini et al. 2008).

Agha-Hosseini et al. (2008) carried out a double-blind and placebo-controlled trial to study the effects of saffron stigma on symptoms of premenstrual syndrome with women (20–45 years) with regular menstrual cycles and experience of premenstrual syndrome symptoms for at least six months. The saffron group received 30 mg/day (15 mg twice a day; morning and evening) of saffron and was compared versus a placebo group (twice a day) for two menstrual cycles (cycles 3 and 4). Saffron was found to be effective in relieving symptoms of premenstrual syndrome, with significant differences in efficacy in cycles 3 and 4 in the total premenstrual daily symptoms and HAM-D.

Shamsa et al. (2009) studied the effect of saffron on a group of 20 male patients with erectile dysfunction. The saffron group received 200 mg in the morning. After day 10, saffron treated-group had a statistically significant improvement in tip rigidity and tip tumescence as well as base rigidity and base tumescence. Modabbernia et al. (2012) studied the effects of saffron on 30 married male patients with major depressive disorders whose depressive symptoms had been stabilized on fluoxetine and had subjective complaints of sexual impairment. After four weeks, the saffron-treated group (15 mg twice per day) resulted in a significantly greater improvement of the erectile function, intercourse satisfaction domains and total scores than the placebo group. Moreover, the effect of saffron did not differ significantly from that of a placebo in orgasmic function, overall satisfaction, and sexual desire domains scores. The frequency of side effects were similar between the two groups.

In a similar study, but carried out with women, Kashani et al. (2013) studied the effects of saffron on selective serotonin reuptake inhibitor-induced sexual dysfunction in women. A randomized double-blind placebo-controlled study was carried out with 38 women with major depression who were stabilized on fluoxetine (40 mg/day for a minimum of six weeks) and experienced subjective feeling of sexual dysfunction. At the end of the fourth week, patients in the saffron group (30 mg/daily) had experienced a significant improvement in total Female Sexual Function Index (FSFI), arousal, lubrication, and pain domains of

FSFI but not in desire, satisfaction, and orgasm domains, versus placebo group. The frequency of side effects was similar between the two groups.

Chemoprevention and anticancer properties

Several studies highlight saffron's chemopreventive and anticancer properties, which most authors link to its free radical-scavenging and anti-oxidant properties. Bathaie et al. (2013a) demonstrated that the aqueous extract of saffron inhibits the progression of gastric cancer in rats in a dose-dependent manner. In this study, the authors proved that 20% of the rats treated with 175 mg/kg of saffron aqueous extract were completely normal at the end of the experiment and that no adenoma could be detected in treated animals. Moreover, apoptosis was increased, and serum LDH and plasma anti-oxidant activity was normalized. This pro-apoptotic activity of saffron aqueous extract has also been described *in vitro* in alveolar human lung cancer cells (Samarghandian et al. 2013). These authors confirmed that saffron has a good cytotoxicity (which is dose- and time-dependent) against the A549 cells but is less sensitive to normal human lung fibroblast cells (MRC-5), indicating a selective behavior. In addition, saffron-induced apoptosis is mediated by caspase-dependent pathways, as demonstrated by the elevation of these proteases in a time-dependent manner.

More recently, Manoharan et al. (2013) also tested the aqueous extract of saffron (100 mg/kg) in another *in vivo* model of cancer: 7,12-dimethylbenz[α]anthracene-induced hamster buccal pouch carcinogenesis. Oral administration of saffron completely prevented the tumor formation, although some pre-neoplastic lesions (hyperplasia, keratosis and dysplasia) could be noticed. Moreover, treatment with saffron restored the status of liver and buccal mucosa phase I and II detoxification enzymes favoring the excretion of the carcinogenic metabolites, and it restored the status of lipid peroxidation and the anti-oxidant capacity in plasma. However, the authors point out that severe hyperplasia and dysplasia could be detected after 14 weeks, therefore further studies should be done to confirm whether saffron delayed or inhibited the tumor formation.

Saffron contains, among other molecules, crocin, crocetin, safranal, as well as other anti-oxidant carotenoids, such as lycopene. These constituents have been demonstrated to have anticancer properties in different studies. In this sense, dietary crocin (100 and 200 ppm) effectively suppresses colitis and colitis-associated colorectal carcinoma in a model of Dextran Sulfate Sodium (DSS)-induced chronic ulcerative colitis (Kawabata et al. 2012) without showing clinical or histopathological toxicity. Furthermore, dietary crocin suppresses the proliferation activity in adenocarcinomas. These effects were associated with the anti-inflammatory activity of crocin: in this *in vivo* model, the dietary feeding of crocin significantly inhibited mRNA expression levels of COX-2, iNOS, TNF-α, IL-1β, and IL-6, and the NF-κB expression in the colorectal mucosa of the mice that received DSS. The authors also report an increase of Nrf2 expression after crocin treatment, which has a protective role against xenobiotic toxicity. Previously, Aung et al. (2007) demonstrated the inhibitory activity of saffron extract and crocin (1 mM) on the proliferation of three colorectal cancer cell lines (HCT-116, SW-480 and HT-29). Crocin significantly reduced cell proliferation, but did not affect normal cells; therefore, this compound could be mainly responsible for saffron's ability to inhibit colorectal cancer cell proliferation.

Interestingly, Hoshyar et al. (2013) demonstrate that crocin causes cytotoxicity in a dose- and time-dependent manner in gastric adenocarcinoma by raising sub-G_1 population, activating caspases and rising Bax/Bcl-2 ratio; however, the same factors remained unchanged in human normal fibroblast skin cells. In addition, this antiproliferative activity has been demonstrated in human prostate cancer cell lines (D'Alessandro et al. 2013). These authors investigated the antiproliferative effects of saffron extract and crocin on different malignant and non-malignant prostate cancer cell lines. Both saffron extract (IC_{50} 0.4–4 mg/mL) and crocin (IC_{50} 0.26–0.95 mM/mL) reduced cell proliferation in all malignant cell lines in a time- and concentration-dependent manner, without affecting non-malignant cells. Most cells were arrested at G_0/G_1 phase, with a significant presence of apoptotic cells and a dramatic down-regulation of Bcl-2 and up-regulation of Bax. Moreover, analysis of caspase activity indicated a caspase-dependent pathway with involvement of caspase-9 activation, suggesting an intrinsic pathway.

Along the same lines, other authors have also demonstrated crocin's anti-proliferative activity in leukemia cells: Rezaee et al. (2013) recently published that crocin (250 µM) increased apoptosis and reduced cell viability and growth on MOLT-4 human leukemia cells, and Sun et al. (2013) demonstrated this same effect in human leukemia HL-60 cells. In this sense, crocin (0.625–5 mg/mL) induced apoptosis in HL-60 cells in a dose-dependent manner inducing G_0/G_1 phase arrest, but necrosis occurred at higher doses (10 mg/mL), as reported by Rezaee et al. (2013) and D'Alessandro et al. (2013). Moreover, Sun et al. (2013) confirmed these *in vitro* results in a xenograft model and demonstrated that crocin (6.25 and 25 mg/kg) had strong inhibitory effect on HL-60 cell growth in nude mice, while the high dose (100 mg/kg) had no significant inhibitory effect, perhaps due to the toxic effects. In agreement with other authors, they demonstrate that the doses of 6.25 and 25 mg/kg induced apoptosis by increasing Bax expression and decreasing Bcl-2.

Crocin was proposed as an adjuvant second-line anticancer drug (Li et al. 2013). In combination with cisplatin, crocin (2 M) enhances the killing effect of osteosarcoma cells (MG63 and OS732), inhibiting cell growth in a dose-dependent manner and through a caspase-dependent mechanism, and it diminishes their invasive ability.

Crocetin has also demonstrated significant anticancer activity, especially in the lung, pancreatic, breast and leukemia cells. Gutheil et al. (2012) extensively reviewed this activity, and studies on crocetin's anticancer activity continue to be published. In this sense, Bathaie et al. (2013b) recently demonstrated that crocetin prevents tumor progression in a rat model of gastric cancer, normalizing altered biochemical parameters such as serum anti-oxidant activity and lactate dehydrogenase. Moreover, *in vitro* they demonstrate that crocetin induces apoptosis by suppression of Bcl-2 and up-regulation of Bax in human adenocarcinoma gastric cancer cells (AGS) without having effect on normal human fibroblasts (HFSF-P13).

Safranal, a third bioactive compound in *C. sativus*, has also been proved to possess cytotoxic activity toward human prostate cancer cells (Samarghandian and Shabestari 2013) and in HeLa and MCF7 (Malaekeh-Nikouei et al. 2013) in a dose- and time-dependent manner, but was less sensitive to normal human cells.

An important concern in the use of these carotenoids in therapeutics is their low solubility in water. To overcome this problem, several researchers have studied the encapsulation of crocin and safranal in nanoliposomes (Malaekeh-Nikouei et al. 2013, Rastgoo et al. 2013). In all cases, nanoliposome encapsulation increased the antitumor activity of both substances.

Anti-oxidant properties

Although it is mainly used for adding color and taste, saffron also confers other relevant properties to food (Kamalipour and Akhondzadeh 2011). For example, Pellegrini et al. (2006) analyzed and compared the anti-oxidant power of 11 herbs and spices using different *in vitro* experimental protocols, such as the Ferric Reducing-Anti-oxidant Power (FRAP), Total Radical-trapping Anti-oxidant Parameter (TRAP), and Trolox Equivalent Anti-oxidant Capacity (TEAC) assays. In all cases, saffron showed the highest anti-oxidant capacity, with values of 739 mmol F^{2+}/kg (FRAP), 374 mmol/Trolox/kg (TRAP), and 53 mmol Trolox/kg (TEAC). According to these tests, saffron was 3.6, 4.2, and 1.2 times more active, respectively, than the other herb/spice assayed (bay leaf). One year later, Kanakis et al. (2007) tested the anti-oxidant activity of the isolated components from saffron, crocetin, dimethylcrocetin, and safranal, by the DPPH· assay, and their IC_{50} values were compared to that of well-known anti-oxidants such as Trolox and Butylated Hydroxy Toluene (BHT). The IC_{50} values of crocetin and safranal were 17.8 and 95 µg/mL, respectively, and values of Trolox and BHT were 5.2 and 5.3 µg/mL, respectively. The inhibition of dimethylcrocetin reached a point of 39%, which corresponds to a concentration of 40 µg/mL, and then started to decrease. Recently, another study established the anti-oxidant properties of saffron aqueous extract by the DPPH test and obtained an IC_{50} value of 3.76 mg dry weight whilst FRAP assay indicated that one gram of dried sample presented the same anti-oxidant power of 2.53 mM $FeSO_4$ (Gismondi et al. 2012).

This anti-oxidant capacity could be of interest in food preparation and preservation, since it prevents food oxidation while offering an added nutraceutical value to foods that contain it. Indeed, the systematic use of saffron is part of the well-known Mediterranean diet. Widely used in Spain, Greece, and Italy, it may be, at least in part, responsible for the protective effects of the Mediterranean diet on hypercholesterolemia, atherosclerosis, and heart disease (Kamalipour and Akhondzadeh 2011). In a complementary study, Serrano-Díaz et al. (2012) studied the anti-oxidant potential of the separate parts of saffron' flowers by four *in vitro* assays, and all of them showed potentiality as anti-oxidant agents. All the saffron samples tested (flowers, tepals, stamens, styles, and floral bio-residues) showed LOO˙, OH˙, and ABTS˙⁻ radicals scavenging activity, while stigmas were active only as LOO˙, and ABTS˙⁻ scavengers. All samples studied improved the oxidative stability of sunflower oil in the Rancimat test.

The anti-oxidant properties of saffron may also be of interest for obtaining good oxidative stability in shell eggs and liquid yolks. In a recent experiment, hens were given diets enriched with 10 or 20 mg/kg saffron. After six weeks, eggs were collected and the rate of lipid oxidation was determined in shell eggs and yolks, adjusted to a pH of 6.2 or 4.2, respectively, and stored under refrigeration in the presence of light. Saffron demonstrated a dose-dependent anti-oxidant activity, with eggs from hens that received 20 mg/kg exhibiting greater protection against oxidation. It was also observed that lipid oxidation in shell eggs did not change with storage, and that pH influenced the oxidation of liquid yolks. Thus, the eggs showed a higher protection at pH 6.2, whereas at pH 4.2, although the pattern was similar, there was an increased rate of oxidation (Botsoglou et al. 2005).

Other relevant pharmacological properties of saffron and its constituents

The anticonvulsant activities of saffron constituents, safranal and crocin, were evaluated in mice using pentylenetetrazole-induced convulsions in mice. Safranal (0.15 and 0.35 mL/kg, i.p.) reduced the seizure duration, delayed the onset of tonic convulsions, and protected mice from death, whereas crocin (22 mg/kg) had no effect (Hosseinzadeh and Talebzadeh 2005).

Oral administration of saffron may be useful in the treatment of neurodegenerative disorders and related memory impairment. The experimental studies in mice and rats demonstrated that saffron, crocetin and crocin, improved memory and learning skills (Abe and Saito 2000, Tashakori-Sabzevar et al. 2013). Studies carried out by Ghahghaei et al. (2013) shed light on the possible mechanism for this protection against neurodegenerative diseases. As they demonstrate, crocin inhibits the formation of Aβ amyloid, the main constituent of the amyloid plaque found in the brains of patients with Alzheimer's disease, and disrupts already formed aggregates. Therefore, crocin could be of great importance in the search for new therapies that can inhibit and/or disrupt amyloid aggregation. Moreover, Purushothuman et al. (2013) demonstrated that saffron pre-treatment protects dopaminergic cells of the substantia nigra pars compacta and retina in a mouse model of Parkinson's disease. For future perspectives, see references: da Rocha et al. (2011), Howes and Houghton (2012), Poma et al. (2012), and Russo et al. (2013).

Saffron stigma exhibited antinociceptive effects in chemically induced pain test (acetic acid induced writhing) but not in the hot plate test. That is, the extracts have no analgesic effect acting on the central nervous system. The extracts have also anti-inflammatory activity on acute (xylene ear edema test) and chronic experimental models of inflammation (formalin test) (Hosseinzadeh and Younesi 2002). Some anti-inflammatory properties have been recently reviewed by Poma et al. (2012).

Crocin analogs isolated from saffron have positive effects on ocular blood flow and retinal function. Indeed, they significantly increased the blood flow in the retina and choroid and improved the retinal function recovery. The authors proposed their potential use to treat ischemic retinopathy as well as the age-related macular degeneration (Xuan 1999). This potentiality has been also studied by other authors, such as Maccarone et al. (2008), Shukurova and Babaev (2010), Falsini et al. (2010), Yamauchi et al. (2011), Piccardi et al. (2012), Fernández-Sánchez et al. (2012), Ishizuka et al. (2013), Qi et al. (2013).

Different studies demonstrated the relaxant properties of saffron aqueous and ethanol extracts and safranal against different agonists. The results suggested that they have a relatively potent stimulatory effect on β_2-adrenoreceptors, which is in part due to safranal. In addition, a possible inhibitory effect of the plant on histamine H_1-receptors was also suggested (Nemati et al. 2008).

Saffron can also improve the negative effects of psoriasis. Brown et al. (2004) carried out a study with five patients diagnosed with chronic plaque psoriasis (two men and three women; range age, 40–68 years), ranging from mild to severe. The five cases at the onset of the study, improved on all measured outcomes over a six-month period. These results suggest that a dietary regimen supplemented with saffron tea may be an effective medical nutrition therapy for the complementary treatment of psoriasis.

Potentiality as therapeutic agent and conclusions

After this study, different properties have been established for saffron and its constituents. The most extensively reviewed during the last decades are the anticancer and chemopreventive properties. These effects correlate with the demonstrated anti-oxidant activity of saffron. In the same way, other properties described, such as the protective effects on hypercholesterolemia, atherosclerosis, and heart disease, can be considered a consequence of this anti-oxidant capacity. With no intention of questioning the studies carried out to demonstrate these effects, it is true that the scientific community is rather reluctant to accept all of those effects only on the claim of being an anti-oxidant substance and more in depth studies on the mechanisms of action that are the ultimate responsible of those properties should be carried out.

Nevertheless, the antidepressant and anti-anxiety properties and its positive effects on sexual dysfunction seem to be the most promising studies in the last decade, as demonstrated by the encouraging results obtained in several clinical trials (Alavizadeh and Hosseinzadeh 2013). These studies give saffron a high potentiality for a future application as phytomedicine or as a therapeutic drug. However, in future studies the trials should be unified, using standardized extracts, duration of assays and same doses. For example, a dose of 10–20 mg of saffron, with a 10% concentration of crocin could be a good proposal for future trials. Double-blind, randomized trials with a relevant number of patients/voluntaries groups and comparing saffron versus reference drugs or placebo, are desirable. Saffron has positive effects on sexual disorders, including sexual impairment in men, male erectile dysfunction, sexual dysfunction in women, and premenstrual syndrome, but these effects are in part due to its antidepressant and anti-anxiety properties.

Saffron has been, for centuries, the most expensive, and sought after spice in the world, being used mostly as a condiment and as a dye. In the last 20 years, scientific research has uncovered an unknown therapeutic potential thus giving an added value to saffron. However, the potential medicinal use of saffron needs highly standardized extracts, instead of non-standard preparations. In the case of its potential as an antidepressant and anti-anxiety agent, the use of different sub-products, such as petals, could give *Crocus sativus* a further higher value and economic interest.

References

Abdullaev, F.I. 2002. Cancer chemopreventive and tumoricidal properties of saffron (*Crocus sativus* L.). Exp. Biol. Med. (Maywood) 227: 20–25.

Abe, K. and Saito, H. 2000. Effects of saffron and its constituent crocin on learning behavior and long-term potentiation. Phytother. Res. 14: 149–152.

Agha-Hosseini, M., Kashani, L., Aleyaseen, A., Ghoreishi, A., Rahmanpour, H., Zarrinara, A.R. and Akhondzadeh, S. 2008. *Crocus sativus* L. (saffron) in the treatment of premenstrual syndrome: a double-blind, randomised and placebo-controlled trial. BJOG. 115: 515–519.

Akhondzadeh, S., Fallah-Pour, H., Afkham, K., Jamshidi, A.H. and Khalighi-Cigaroudi, F. 2004. Comparison of *Crocus sativus* L. and imipramine in the treatment of mild to moderate depression: a pilot double-blind randomized trial [ISRCTN45683816]. BMC Complement. Altern. Med. 4: 12.

Akhondzadeh, S., Tahmacebi-Pour, N., Noorbala, A.A., Amini, H., Fallah-Pour, H., Jamshidi, A.H. and Khani, M. 2005. *Crocus sativus* L. in the treatment of mild to moderate depression: a double-blind, randomized and placebo-controlled trial. Phytother. Res. 19: 148–151.

Alavizadeh, S.H. and Hosseinzadeh, H. 2013. Bioactivity assessment and toxicity of crocin: a comprehensive review. Food Chem. Toxicol. 64: 65–80.

Aung, H.H., Wang, C.Z., Ni, M., Fishbein, A., Mehendale, S.R., Xie, J.T., Shoyama, C.Y. and Yuan, C.S. 2007. Crocin from *Crocus sativus* possesses significant anti-proliferation effects on human colorectal cancer cells. Exp. Oncol. 29: 175–180.

Ayatollahi, H., Javan, A.O., Khajedaluee, M., Shahroodian, M. and Hosseinzadeh, H. 2014. Effect of *Crocus sativus* L. (saffron) on coagulation and anticoagulation systems in healthy volunteers. Phytother. Res. 28: 539–543.

Basti, A.A., Moshiri, E., Noorbala, A.A., Jamshidi, A.H., Abbasi, S.H. and Akhondzadeh, S. 2007. Comparison of petal of *Crocus sativus* L. and fluoxetine in the treatment of depressed outpatients: a pilot double-blind randomized trial. Prog. Neuropsychopharmacol. Biol. Psychiatry 31: 439–442.

Bathaie, S.Z. and Mousavi, S.Z. 2010. New applications and mechanisms of action of saffron and its important ingredients. Crit. Rev. Food Sci. Nutr. 50: 761–786.

Bathaie, S.Z., Miri, H., Mohagheghi, M.A., Mokhtari-Dizaji, M., Shahbazfar, A.A. and Hasanzadeh, H. 2013a. Saffron aqueous extract inhibits the chemically-induced gastric cancer progression in the Wistar albino rat. Iran J. Basic Med. Sci. 16: 27–38.

Bathaie, S.Z., Hoshyar, R., Miri, H. and Sadeghizadeh, M. 2013b. Anticancer effects of crocetin in both human adenocarcinoma gastric cancer cells and rat model of gastric cancer. Biochem. Cell Biol. 91: 397–403.

Botsoglou, N.A., Florou-Paneri, P., Nikolakakis, I., Giannenas, I., Dotas, V., Botsoglou, E.N. and Aggelopoulos, S. 2005. Effect of dietary saffron (*Crocus sativus* L.) on the oxidative stability of egg yolk. Br. Poult. Sci. 46: 701–707.

Brown, A.C., Hairfield, M., Richards, D.G., McMillin, D.L., Mein, E.A. and Nelson, C.D. 2004. Medical nutrition therapy as a potential complementary treatment for psoriasis—five case reports. Alternat. Med. Rev. 9: 297–307.

Carmona, M., Zalacain, A., Pardo, J.E., López, E., Alvarruiz, A. and Alonso, G.L. 2005. Influence of different drying and aging conditions on saffron constituents. J. Agric. Food Chem. 53: 3974–3979.

Carmona, M., Zalacain, A., Sánchez, A.M., Novella, J.L. and Alonso, G.L. 2006. Crocetin esters, picrocrocin and its related compounds present in *Crocus sativus* stigmas and *Gardenia jasminoides* fruits. Tentative identification of seven new compounds by LC-ESI-MS. J. Agric. Food Chem. 54: 973–979.

Carmona, M., Zalacain, A., Salinas, M.R. and Alonso, G.L. 2007. A new approach to saffron aroma. Crit. Rev. Food Sci. Nutr. 47: 145–159.

Chahine, N., Hanna, J., Makhlouf, H., Duca, L., Martiny, L. and Chahine, R. 2013. Protective effect of saffron extract against doxorubicin cardiotoxicity in isolated rabbit heart. Pharm. Biol. 51: 1564–1571.

D'Alessandro, A.M., Mancini, A., Lizzi, A.R., De Simone, A., Marroccella, C.E., Gravina, G.L., Tatone, C. and Festuccia, C. 2013. *Crocus sativus* stigma extract and its major constituent crocin possess significant antiproliferative properties against human prostate cancer. Nutr. Cancer 65: 930–942.

D'Auria, M., Mauriello, G. and Rana, G.L. 2004. Volatile organic compounds from saffron. Flavour Fragr. J. 19: 17–23.

da Rocha, M.D., Viegas, F.P., Campos, H.C., Nicastro, P.C., Fossaluzza, P.C., Fraga, C.A., Barreiro, E.J. and Viegas, C., Jr. 2011. The role of natural products in the discovery of new drug candidates for the treatment of neurodegenerative disorders II: Alzheimer's disease. CNS Neurol. Disord. Drug Targets 10: 251–270.

Dwyer, A.V., Whitten, D.L. and Hawrelak, J.A. 2011. Herbal medicines, other than St. John's Wort, in the treatment of depression: a systematic review. Altern. Med. Rev. 16: 40–49.

Falsini, B., Piccardi, M., Minnella, A., Savastano, C., Capoluongo, E., Fadda, A., Balestrazzi, E., Maccarone, R. and Bisti, S. 2010. Influence of saffron supplementation on retinal flicker sensitivity in early age-related macular degeneration. Invest. Ophthalmol. Vis. Sci. 51: 6118–6124.

Fernández, J.A. 2006. Anticancer properties of saffron, *Crocus sativus* Linn. Adv. Phytomed. 2: 313–330.

Fernández-Sánchez, L., Lax, P., Esquiva, G., Martín-Nieto, J., Pinilla, I. and Cuenca, N. 2012. Safranal, a saffron constituent, attenuates retinal degeneration in P23H rats. PLoS One 7: e43074.

Ghahghaei, A., Bathaie, S.Z., Kheirkhah, H. and Bahraminejad, E. 2013. The protective effect of crocin on the amyloid fibril formation of $A\beta_{42}$ peptide *in vitro*. Cell Mol. Biol. Lett. 18: 328–339.

Giaccio, M. 2004. Crocetin from saffron: an active component of an ancient spice. Crit. Rev. Food Sci. Nutr. 44: 155–172.

Gismondi, A., Serio, M., Canuti, L. and Canini, A. 2012. Biochemical, antioxidant and antineoplastic properties of Italian saffron (*Crocus sativus* L.). Am. J. Plant Sci. 3: 1573–1580.

Gonda, S., Parizsa, P., Surányi, G., Gyémánt, G. and Vasas, G. 2012. Quantification of main bioactive metabolites from saffron (*Crocus sativus*) stigmas by a micellar electrokinetic chromatographic (MEKC) method. J. Pharm. Biomed. Anal. 66: 68–74.

Gout, B., Bourges, C. and Paineau-Dubreuil, S. 2010. Satiereal, a *Crocus sativus* L. extract, reduces snacking and increases satiety in a randomized placebo-controlled study of mildly overweight, healthy women. Nutr. Res. 30: 305–513.

Gresta, F., Lombardo, G.M., Siracusa, L. and Ruperto, G. 2008. Saffron, an alternative crop for sustainable agricultural systems. A review. Agron. Sustain. Dev. 28: 95–112.

Gutheil, W.G., Reed, G., Ray, A., Anant, S. and Dhar, A. 2012. Crocetin: an agent derived from saffron for prevention and therapy for cancer. Curr. Pharm. Biotechnol. 13: 173–179.

Hausenblas, H.A., Saha, D., Dubyak, P.J. and Anton, S.D. 2013. Saffron (*Crocus sativus* L.) and major depressive disorder: a meta-analysis of randomized clinical trials. J. Integr. Med. 11: 377–383.

Hoshyar, R., Bathaie, S.Z. and Sadeghizadeh, M. 2013. Crocin triggers the apoptosis through increasing the Bax/Bcl-2 ratio and caspase activation in human gastric adenocarcinoma, AGS, cells. DNA Cell Biol. 32: 50–57.

Hosseinzadeh, H. and Younesi, H. 2002. Petal and stigma extracts of *Crocus sativus* L. have antinociceptive and anti-inflammatory effects in mice. BMC Pharmacol. 2: 7.

Hosseinzadeh, H. and Talebzadeh, F. 2005. Anticonvulsant evaluation of safranal and crocin from *Crocus sativus* in mice. Fitoterapia 76: 722–724.

Hosseinzadeh, H. and Nassiri-Asl, M. 2013. Avicenna's (Ibn Sina) the Canon of Medicine and saffron (*Crocus sativus*): a review. Phytother. Res. 27: 475–483.

Howes, M.J. and Houghton, P.J. 2012. Ethnobotanical treatment strategies against Alzheimer's disease. Curr. Alzheimer Res. 9: 67–85.

Ishizuka, F., Shimazawa, M., Umigai, N., Ogishima, H., Nakamura, S., Tsuruma, K. and Hara, H. 2013. Crocetin, a carotenoid derivative, inhibits retinal ischemic damage in mice. Eur. J. Pharmacol. 703: 1–10.

Joukar, S., Ghasemipour-Afshar, E., Sheibani, M., Naghsh, N. and Bashiri, A. 2013. Protective effects of saffron (*Crocus sativus*) against lethal ventricular arrhythmias induced by heart reperfusion in rat: a potential anti-arrhythmic agent. Pharm. Biol. 51: 836–843.

Kamalipour, M. and Akhondzadeh, S. 2011. Cardiovascular effects of saffron: an evidence-based review. J. Tehran Heart Cent. 6: 59–61.

Kanakis, C.D., Tarantilis, P.A., Tajmir-Riahi, H.A. and Polissiou, M.G. 2007. Crocetin, dimethylcrocetin, and safranal bind human serum albumin: stability and anti-oxidative properties. J. Agric. Food Chem. 55: 970–977.

Kashani, L., Raisi, F., Saroukhani, S., Sohrabi, H., Modabbernia, A., Nasehi, A.A., Jamshidi, A., Ashrafi, M., Mansouri, P., Ghaeli, P. and Akhondzadeh, S. 2013. Saffron for treatment of fluoxetine-induced sexual dysfunction in women: randomized double-blind placebo-controlled study. Hum. Psychopharmacol. 28: 54–60.

Kawabata, K., Tung, N.H., Shoyama, Y., Sugie, S., Mori, T. and Tanaka, T. 2012. Dietary crocin inhibits colitis and colitis-associated colorectal carcinogenesis in male ICR mice. Evid. Based Complement. Alternat. Med. 2012: 820415.

Kumar, R., Singh, V., Devi, K., Sharma, M., Singh, M.K. and Ahuja, P.S. 2009. State of art of saffron (*Crocus sativus* L.) agronomy: a comprehensive review. Food Rev. Int. 25: 44–85.

Lechtenberg, M., Schepmann, D., Niehues, M., Hellenbrand, N., Wünsch, B. and Hensel, A. 2008. Quality and functionality of saffron: quality control, species assortment and affinity of extract and isolated saffron compounds to NMDA and σ_1 (sigma-1) receptors. Planta Med. 74: 764–772.

Li, X., Huang, T., Jiang, G., Gong, W., Qian, H. and Zou, C. 2013. Synergistic apoptotic effect of crocin and cisplatin on osteosarcoma cells via caspase induced apoptosis. Toxicol. Lett. 221: 197–204.

Lozano, P., Delgado, D., Gómez, D., Rubio, M. and Iborra, J.L. 2000. A non-destructive method to determine the safranal content of saffron (*Crocus sativus* L.) by supercritical carbon dioxide extraction combined with high-performance liquid chromatography and gas chromatography. J. Biochem. Biophys. Methods. 43: 367–378.

Maccarone, R., Di Marco, S. and Bisti, S. 2008. Saffron supplement maintains morphology and function after exposure to damaging light in mammalian retina. Invest. Ophthalmol. Vis. Sci. 49: 1254–1261.

Maggi, L., Carmona, M., Zalacain, A., Kanakis, C.D., Anastasaki, E., Tarantilis, P.A., Polissiou, M.G. and Alonso, G.L. 2010. Changes in saffron volatile profile according to its storage time. Food Res. Int. 43: 1329–1334.

Malaekeh-Nikouei, B., Mousavi, S.H., Shahsavand, S., Mehri, S., Nassirli, H. and Moallem, S.A. 2013. Assessment of cytotoxic properties of safranal and nanoliposomal safranal in various cancer cell lines. Phytother. Res. 27: 1868–1873.

Manoharan, S., Wani, S.A., Vasudevan, K., Manimaran, A., Prabhakar, M.M., Karthikeyan, S. and Rajasekaran, D. 2013. Saffron reduction of 7,12-dimethylbenz[a]anthracene-induced hamster buccal pouch carcinogenesis. Asian Pac. J. Cancer Prev. 14: 951–957.

Mattes, R.D. 2012. Spices and energy balance. Physiol. Behav. 107: 584–590.

Mehdizadeh, R., Parizadeh, M.R., Khooei, A.R., Mehri, S. and Hosseinzadeh, H. 2013. Cardioprotective effect of saffron extract and safranal in isoproterenol-induced myocardial infarction in Wistar rats. Iran. J. Basic Med. Sci. 16: 56–63.

Modabbernia, A., Sohrabi, H., Nasehi, A.A., Raisi, F., Saroukhani, S., Jamshidi, A., Tabrizi, M., Ashrafi, M. and Akhondzadeh, S. 2012. Effect of saffron on fluoxetine-induced sexual impairment in men: randomized double-blind placebo-controlled trial. Psychopharmacology 223: 381–388.

Moshiri, E., Basti, A.A., Noorbala, A.A., Jamshidi, A.H., Hesameddin Abbasi, S. and Akhondzadeh, S. 2006. *Crocus sativus* L. (petal) in the treatment of mild-to-moderate depression: a double-blind, randomized and placebo-controlled trial. Phytomedicine 13: 607–611.

Mousavi1, S.Z. and Bathaie, S.Z. 2011. Historical uses of saffron: identifying potential new avenues for modern research. Avicenna J. Phytomed. 1: 57–66.

Nemati, H., Boskabady, M.H. and Ahmadzadef Vortakolaei, H. 2008. Stimulatory effects of *Crocus sativus* (saffron) on β_2 adrenoreceptors of guinea pig tracheal chains. Phytomedicine 15: 1038–1045.

Noorbala, A.A., Akhondzadeh, S., Tahmacebi-Pour, N. and Jamshidi, A.H. 2005. Hydro-alcoholic extract of *Crocus sativus* L. versus fluoxetine in the treatment of mild to moderate depression: a double-blind, randomized pilot trial. J. Ethnopharmacol. 97: 281–284.

Padmavati, J., Kumar, Ch.P., Saraswathi, V.S., Saravanan, D., Lakshmi, I.A., Bindu, N.H.S. and Hemafaith, V. 2011. Pharmacological, pharmacognostic and phytochemical review of saffron. Int. J. Pharm. Technol. 3: 1214–1234.

Pellegrini, N., Serafini, M., Salvatore, S., Del Rio, D., Bianchi, M. and Brighenti, F. 2006. Total antioxidant capacity of spices, dried fruits, nuts, pulses, cereals and sweets consumed in Italy assessed by three different *in vitro* assays. Mol. Nutr. Food Res. 50: 1030–1038.

Piccardi, M., Marangoni, D., Minnella, A.M., Savastano, M.C., Valentini, P., Ambrosio, L., Capoluongo, E., Maccarone, R., Bisti, S. and Falsini, B. 2012. A longitudinal follow-up study of saffron supplementation in early age-related macular degeneration: sustained benefits to central retinal function. Evid. Based Complement. Alternat. Med. 2012: 429124.

Poma, A., Fontecchio, G., Carlucci, G. and Chichiriccò, G. 2012. Anti-inflammatory properties of drugs from saffron crocus. Antiinflamm. Antiallergy Agents Med. Chem. 11: 37–51.

Purushothuman, S., Nandasena, C., Peoples, C.L., El Massri, N., Johnstone, D.M., Mitrofanis, J. and Stone, J. 2013. Saffron pre-treatment offers neuroprotection to nigral and retinal dopaminergic cells of MPTP-treated mice. J. Parkinsons Dis. 3: 77–83.

Qi, Y., Chen, L., Zhang, L., Liu, W.B., Chen, X.Y. and Yang, X.G. 2013. Crocin prevents retinal ischaemia/reperfusion injury-induced apoptosis in retinal ganglion cells through the PI3K/AKT signalling pathway. Exp. Eye Res. 107: 44–51.

Rastgoo, M., Hosseinzadeh, H., Alavizadeh, H., Abbasi, A., Ayati, Z. and Jaafari, M.R. 2013. Antitumor activity of PEGylated nanoliposomes containing crocin in mice bearing C26 colon carcinoma. Planta Med. 79: 447–451.

Razavi, B.M., Hosseinzadeh, H., Movassaghi, A.R., Imenshahidi, M. and Abnous, K. 2013a. Protective effect of crocin on diazinon induced cardiotoxicity in rats in subchronic exposure. Chem. Biol. Interact. 203: 547–555.

Razavi, B.M., Hosseinzadeh, H., Abnous, K. and Imenshahidi, M. 2014. Protective effect of crocin on diazinon induced vascular toxicity in subchronic exposure in rat aorta *ex vivo*. Drug Chem. Toxicol. 37: 378–383.

Razavi, M., Hosseinzadeh, H., Abnous, K., Motamedshariaty, V.S. and Imenshahidi, M. 2013b. Crocin restores hypotensive effect of subchronic administration of diazinon in rats. Iran. J. Basic Med. Sci. 16: 64–72.

Rezaee, R. and Hosseinzadeh, H. 2013. Safranal: from an aromatic natural product to a rewarding pharmacological agent. Iran. J. Basic Med. Sci. 16: 12–26.

Rezaee, R., Mahmoudi, M., Abnous, K., Zamani Taghizadeh Rabe, S., Tabasi, N., Hashemzaei, M. and Karimi, G. 2013. Cytotoxic effects of crocin on MOLT-4 human leukemia cells. J. Complement. Integr. Med. 10: 105–112.

Ríos, J.L., Recio, M.C., Giner, R.M. and Máñez, S. 1996. An update review of saffron and its active constituents. Phytother. Res. 10: 189–193.

Russo, P., Frustaci, A., Del Bufalo, A., Fini, M. and Cesario, A. 2013. Multitarget drugs of plants origin acting on Alzheimer's disease. Curr. Med. Chem. 20: 1686–1693.

Samarghandian, S. and Shabestari, M.M. 2013. DNA fragmentation and apoptosis induced by safranal in human prostate cancer cell line. Indian J. Urol. 29: 177–183.

Samarghandian, S., Borji, A., Farahmand, S.K., Afshari, R. and Davoodi, S. 2013. *Crocus sativus* L. (saffron) stigma aqueous extract induces apoptosis in alveolar human lung cancer cells through caspase-dependent pathways activation. Biomed. Res. Int. 2013: 417928.

Sampathu, S.R., Shivashankar, S. and Lewis, Y.S. 1984. Saffron (*Crocus sativus* L.). Cultivation, processing, chemistry and standardization. CRC Crit. Rev. Food Sci. Nutr. 20: 123–157.

Sarris, J., Panossian, A., Schweitzer, I., Stough, C. and Scholey, A. 2011. Herbal medicine for depression, anxiety and insomnia: a review of psychopharmacology and clinical evidence. Eur. Neuropsychopharmacol. 21: 841–860.

Schmidt, M., Betti, G. and Hensel, A. 2007. Saffron in phytotherapy: pharmacology and clinical uses. Wien Med. Wochenschr. 157: 315–319.

Serrano-Díaz, J., Sánchez, A.M., Maggi, L., Martínez-Tomé, M., García-Diz, L., Murcia, M.A. and Alonso, G.L. 2012. Increasing the applications of *Crocus sativus* flowers as natural antioxidants. J. Food Sci. 77: 1162–1168.

Shahmansouri, N., Farokhnia, M., Abbasi, S.H., Kassaian, S.E., Noorbala Tafti, A.A., Gougol, A., Yekehtaz, H., Forghani, S., Mahmoodian, M., Saroukhani, S., Arjmandi-Beglar, A. and Akhondzadeh, S. 2014. A randomized, double-blind, clinical trial comparing the efficacy and safety of *Crocus sativus* L. with fluoxetine for improving mild to moderate depression in post percutaneous coronary intervention patients. J. Affect. Disord. 155: 216–222.

Shamsa, A., Hosseinzadeh, H., Molaei, M., Shakeri, M.T. and Rajabi, O. 2009. Evaluation of *Crocus sativus* L. (saffron) on male erectile dysfunction: a pilot study. Phytomedicine 6: 690–693.

Shukurova, P. and Babaev, R. 2010. A study into the effectiveness of the application of saffron extract in ocular pathologies in experiment. Georgian Med. News 182: 38–42.

Siracusa, L., Gresta, F., Avola, G., Albertini, E., Raggi, L., Marconi, G., Lombardo, G.M. and Ruberto, G. 2013. Agronomic, chemical and genetic variability of saffron (*Crocus sativus* L.) of different origin by LC-UV-vis-DAD and AFLP analyses. Genet. Resour. Crop Evol. 60: 711–721.

Srivastava, R., Ahmed, H., Dixit, R.K., Dharamveer and Saraf, S.A. 2010. *Crocus sativus* L.: A comprehensive review. Pharmacogn. Rev. 4: 200–208.

Sun, Y., Xu, H.J., Zhao, Y.X., Wang, L.Z., Sun, L.R., Wang, Z. and Sun, X.F. 2013. Crocin exhibits antitumor effects on human leukemia HL-60 cells *in vitro* and *in vivo*. Evid. Based Complement. Alternat. Med. 2013: 690164.

Tashakori-Sabzevar, F., Hosseinzadeh, H., Motamedshariaty, V.S., Movassaghi, A.R. and Mohajeri, S.A. 2013. Crocetin attenuates spatial learning dysfunction and hippocampal injury in a model of vascular dementia. Curr. Neurovasc. Res. 10: 325–334.

Ulbricht, C., Conquer, J., Costa, D., Hollands, W., Iannuzzi, C., Isaac, R., Jordan, J.K., Ledesma, N., Ostroff, C., Serrano, J.M., Shaffer, M.D. and Varghese, M. 2011. An evidence-based systematic review of saffron (*Crocus sativus*) by the Natural Standard Research Collaboration. J. Diet. Suppl. 8: 58–114.

Vahdati Hassani, F., Naseri, V., Razavi, B.M., Mehri, S., Abnous, K. and Hosseinzadeh, H. 2014. Antidepressant effects of crocin and its effects on transcript and protein levels of CREB, BDNF, and VGF in rat hippocampus. Daru 22: 16.

Wang, Y., Han, T., Zhu, Y., Zheng, C.J., Ming, Q.L., Rahman, K. and Qin, L.P. 2010. Antidepressant properties of bioactive fractions from the extract of *Crocus sativus* L. J. Nat. Med. 64: 24–30.

Winterhalter, P. and Straubinger, M. 2000. Saffron-renewed interest in an ancient spice. Food Rev. Int. 16: 39–59.

Xuan, B., Zhou, Y.H., Li, N., Min, Z.D. and Chiou, G.C. 1999. Effects of crocin analogs on ocular flow and retinal function. J. Ocul. Pharmacol. Ther. 15: 143–152.

Yamauchi, M., Tsuruma, K., Imai, S., Nakanishi, T., Umigai, N., Shimazawa, M. and Hara, H. 2011. Crocetin prevents retinal degeneration induced by oxidative and endoplasmic reticulum stresses via inhibition of caspase activity. Eur. J. Pharmacol. 650: 110–119.

Clinical Studies of Traditional Therapeutic Herbs before Marketing: Current Status and Future Challenges

Anushka Mootoosamy and M. Fawzi Mahomoodally*

|||||||||||||||||||||||||||||||||||||||||||||||||||||||||||||||||||||||||||||||||||||||||||||||||||||||||||||||||||||||||||||||||||||||||||||||||||||||||||||

Introduction

According to the World Health Organisation (WHO), traditional medicine is defined as the sum total of the knowledge, skill and practices based on the theories, beliefs, and experiences indigenous to diverse cultures used in the maintenance of health as well as in the prevention, diagnosis, improvement or treatment of physical and mental ailment (WHO 2013). Approximately, 4000 million people in developing countries have recourse to traditional medicine including phytotherapy (use of plants) and zootherapy (use of animals) on a regular basis (Alzweiri et al. 2011). One reason for this is the fact that despite accessibility of allopathic medicine in these countries, herbal remedies have retained their supremacy, owing to the belief of their long-standing use and less side effects (Alzweiri et al. 2011). Recently, the world has witnessed renewed interest in herbal remedies in both developed and developing countries for the management and/ or treatment of various diseases. It is generally argued that herbal remedies are relatively inexpensive when compared to prescription medications and they may be viewed as cost-effective alternatives to conventional pharmaceutical therapies. Moreover, since they are derived directly from nature they are assumed to be safe. They are also free from the negative stigma attached to many commonly prescribed conventional medicines (Ashar and Dobs 2004). There is a great possibility that medicinal plants may turn out to be much more significant sources of bioactive molecules that has ever been imagined for the development of novel pharmaceuticals (Bougel 2007). Currently, much research is being geared towards phytochemicals owing to their modulation ability towards non-communicable diseases like cancer, cardiovascular diseases, diabetes, arthritis, neurodegenerative diseases, and cataract as well as against other infectious diseases amongst others. The underlying reason that has contributed to the reemergence of interest in bioactive phytochemicals is scientific proof acquired from well-designed epidemiological,

---

Department of Health Sciences, Faculty of Science, University of Mauritius, Réduit, Mauritius.
* Corresponding author: f.mahomoodally@uom.ac.mu

toxicological and experimental studies conducted during the previous two decades. Investigations that are being carried out on plants are mostly focused towards their prophylactic approach such as antioxidant properties due to their rich flavonoids content, which is responsible for their free radical scavenging properties (Srivastava and Vankar 2012).

Furthermore, based on the current growth rate, it is anticipated that by the year 2020, the world population will reach 11.5 billion. The drastic increase in population, insufficient supply of quality drugs in certain regions of the globe, unaffordable cost of treatment for common diseases, serious adverse effects of numerous modern conventional drugs presently in use and above all the development of resistance to currently used drugs for infectious diseases have inevitably shifted our attention towards the use of phytotherapy against the management and/or treatment of a multitude of ailments (DMAPR 2012). Medicinal plants have greatly contributed to the management and/or treatment of diseases for instance HIV/AIDS, malaria, diabetes, sickle-cell anemia and mental disorders (Elujoba et al. 2005). Moreover, *Illicium verum* (Star anise) which serves as the industrial source of shikimic acid, a principal ingredient used to produce the antiviral drug, Tamiflu (oseltamivir phosphate) which is so far, the only existing drug that may diminish the severity of bird flu (Wang et al. 2011). Studies carried out have documented the therapeutic effect of many medicinal plants that are utilized worldwide for their therapeutic properties and hence it is undeniable that the knowledge of medicinal plants that the indigenous populations possess is a starting point for modern science to discover new drugs (Clark 2012).

Natural products play a fundamental role in pharmaceutical biology and drug development. According to Joseph et al. (2010) many drugs which are on the market either mimic naturally occurring bioactive molecules or have structures that are fully or partly derived from natural motifs. Medicinal plants continue to contribute significantly to modern prescription drugs by providing novel lead compounds upon which the synthesis of new drugs can be made. According to Newman et al. (2003), 60% of anticancer drugs and 75% of anti-infectious disease drugs approved from 1981–2002, were of natural origins. Moreover, 61% of all novel chemical entities introduced worldwide as drugs during the same period were inspired by natural products. The use of, and search for drugs and dietary supplements derived from plants have increased exponentially during the recent years. Pharmacologists, microbiologists, biochemists, botanists, and natural-products chemists all over the world are currently investigating medicinal herbs for phytochemicals and lead compounds that could be developed for treatment of various diseases (Acharya and Shrivastava 2008). After decades of serious obsession with the modern medicinal system, people are now looking at the ancient healing systems like Ayurveda, Siddha, Unani and Chinese traditional medicines with the view of treating diverse types of ailments. The main reason for this shift is the adverse effects associated with synthetic drugs and their narrow therapeutic window and index (Bigoniya et al. 2011).

According to the Dietary Supplements Health and Education Act (DSHEA) of 1994, premarket testing of herbal products to determine their safety and efficacy was no longer required. Supplements were assumed to be safe unless proven otherwise by the Food and Drug Administration (FDA). This has resulted in a plethora of herbal products flooding neighborhood supermarkets and drugstores. The lack of regulation of herbal supplements is problematic due to various reasons. Firstly, no universal standards are in place to guarantee homogeneity among different herbal products. Secondly, there is also no quality assurance that the active ingredient from the plant is present and variations exist between batches from the same manufacturer due to differences in plant composition, handling and preparation. Another issue is that in many cases the active ingredient(s) is unknown, thus making standardization and identification impossible (Ashar and Dobs 2004).

It is also a known fact that not all plants are the same or even members of the same species. While some are safe and effective for specific diseases and purposes, others are not. The general perception that herbal remedies are very safe and devoid from side effects is not true and should not be taken for granted. Plants can produce undesirable side effects which can be lethal. A particular plant part will possess an array of active constituents and some of them may be toxic (Wickramasinghe 2006). Moreover, the risks of herb-drug interactions cannot be overlooked or ignored. Hence, the pressing need of conducting clinical trials to allow safety and efficacy data to be amassed for herbal products are becoming a basic need for both the scientific community and consumers.

By definition, clinical trial is a systematic study of medicine in human subjects including patients or healthy volunteers, intended to validate or reveal their action, adverse reaction, and/or absorption, distribution, metabolism, and excretion, and to identify their efficacy and safety (ICH 1997, Yuan et al. 2011, www.ich.org). It is generally agreed that traditional medicinal products should be subjected to more clinical research as this would help to select products of interest for further investigations in ethnopharmacology and it could also provide immediate recommendations for the population using the assessed local treatments (Graz et al. 2007). Clinical trial is also argued to be more important than preclinical experimental investigations to evaluate efficacy and safety of both modern and traditional herbal medicines. This is because the bioactivity, pharmacodynamic and toxic results obtained from *in vitro* bioassays and *in vivo* animal studies might differ significantly from the efficacy and toxicity on human bodies, due to absorption, metabolism and species differences between animals and human beings (Yuan et al. 2011).

This chapter reviews the various ways in which clinical studies of traditional herbal medicine can be performed, summarizes data generated from key clinical trials conducted to determine safety and efficacy of herbal products and ends by highlighting key challenges of clinical studies on herbal medicine which need to be addressed.

Clinical trial design for traditional medicine

Clinical studies of traditional medicine, allows for screening plants with higher chances to yield novel drug candidates and may also result in information of immediate interest for patients and practitioners who make use of the traditional formulations, particularly in relation to their efficacy, toxicity and also dosage (Teixeira and Fuchs 2006). A clinical trial design is defined as a scientific, thoughtful and detailed plan which must be established prior to any stage(s) of clinical trial, so as to enable the trials to run smoothly and to obtain accurate and reliable results (ICH 1997, Yuan et al. 2011, www.ich.org). The main purposes of clinical trials for herbal medicine are to scientifically validate the therapeutic effects, safety and efficacy of herbal medicine even though many of them have been used since time immemorial and to elucidate its possible adverse effects and toxic reactions (Yuan et al. 2011). Diallo et al. (2006) and Graz et al. (1999) reported that clinical studies of traditionally used medicinal plants can be conducted at relatively low cost, when the study is conducted in a setting where the treatment of interest is commonly used and has a known safety profile. If the herbal product is a commonly used medicinal plant (e.g., medicinal food plant or spice) then the clinical trial should be set up based on data obtained from common users of the medicinal plant. In order to avoid variations in the outcomes of a clinical trial with an herbal product, a number of questions and challenges as elaborated by Yuan et al. (2011) should be taken into consideration as depicted in Table 5.1.

Toxicological studies prior to clinical studies

Clinical trials of traditional medicine are made easier than the usual clinical trials because a good documentation of traditional use alleviates the constitution of the toxicology/safety file, so long as the clinical study is conducted with the traditional medicine being prepared and applied according to the selected local customary recipe (WHO 2000). Information about known side effects of medicinal plants can pave the way for further research on its toxicity. According to Graz et al. (2007), proponents of toxicology studies argue that long-used products may have an unknown long-term toxicity. On the other hand, opponents refute that what we commonly eat are generally accepted as safe despite a lack of formal *in vitro* or *in vivo* toxicology studies hence common traditional treatments should be regarded in the same way. Nonetheless, before proposing a clinical study feasible, potential conflicts on the necessity of preliminary toxicology studies (*in vitro* or *in vivo*) must be addressed and resolved, with a special emphasis on long term use and reproductive toxicity. When conducting clinical trial of an herbal remedy, information on medicinal plant collection (taxonomy), preparation, dose, and frequency of administration are fundamental to reduce the occurrence of toxicity and future serious adverse effects.

Table 5.1. Major questions and challenges to be taken into consideration (Yuan et al. 2011).

Questions and challenges which need to be addressed
• Which species of plant or which taxonomic group does the plant material belong?
• In which region or season of the year should the plant be collected?
• Will the herbal product be administered (e.g., dosage, forms) the same way as it is used traditionally?
• Which population will be targeted as study subjects?
• How will the participants be allotted into the experimental or control group?
• Participants belonging to which age group will be selected?
• Which sample size will be best suited to be used in this study?
• Which allopathic drug (or will it be another known herbal product) will be used to make comparison or use of placebo?
• In case a placebo is being used, will it be able to mimic the strong aroma and taste of the herbal remedy?
• How will the effectiveness of herbal product be determined over time?
• Do you have a clear knowledge on the onset time and natural course of the ailment treated and/or managed with the herbal product?
• Any method or tools available to determine pharmacokinetic and pharmacodynamic data?

Randomized controlled trial

According to the FDA (2004), clinical trial for herbal remedy can directly start on phase III. Phase III clinical trial is designed to confirm the safety and therapeutic efficacy of an herbal medicine as it is used traditionally. Hence, a clinical trial of herbal medicine or product is mostly randomized, double-blind, and a placebo-controlled trial.

Randomized controlled trials are research studies that aim to evaluate the efficacy and toxicity of a drug treatment (Saturni et al. 2014). In randomized controlled trials, each subject is randomly allocated to receive the herbal product or the placebo. Randomly assigning a single subject into a treatment or control group will ensure that the populations are similar in both groups (Yuan et al. 2011). Moreover, random assignment to treatment groups minimizes bias (Jaillon 2007). There are various ways to achieve randomization, for instance lot drawing, checking random digital tables or random arrangement tables amongst others (Yuan et al. 2011). Blinding is utilized in clinical trial to reduce biases resulting from management, treatment, assessment, or interpretation of results from subjects or researchers knowing the assigned treatment, thus ensuring that the outcomes are not affected by knowledge of treatment assigned. In a single—blind trial, only the subjects are not aware whether they receive test treatment or a placebo whereas in double-blind trial, neither the subjects nor the researchers know which subject is given the treatment or a placebo. During phase III clinical trial, routine examination of the subjects should be carried out using a fixed schedule and time of administration of the herbal medicine should be well established and documented. Proper training should be given to personnel to ensure smooth running of the clinical trial as per international guidelines and to prevent mistakes (ICH 1997, Yuan et al. 2011, www.ich.org).

Clinical trial protocol

A clinical trial protocol is a document that describes the aim(s) and objective(s), scientific motivation, treatment procedure (dosage, course, and measurement), study population, control group, statistical analysis and organization of a clinical trial. It should also provide details on the composition of the research team (ICH 1997, Yuan et al. 2011, www.ich.org). Graz et al. (2007) have suggested that when planning a clinical study on a traditional medicine, it is very important to have an interdisciplinary research group, with the collaboration of traditional practitioner/healers, ethical investigators, skilled clinicians, botanists/pharmacists, epidemiologists/statisticians and patients. Sharing questions regarding the design of research protocols and the interpretation of the results can help save much time and may avoid significant mistakes and bias. It has also been proposed that it is fundamental to provide an exact template for trial implementations so that all investigators involved perform the clinical trial in exactly the same way (Yuan et al. 2011). As in any conventional clinical trial, the clinical trial protocol should be ready before the start of the clinical trial and should satisfy any administrative and legal criteria set by the health authorities in the country in question. The relevance of the research question, the rigor of the data collection and analysis, and the importance of the observed effects are key factors which make a clinical study valuable and valid (Graz et al. 2007).

Superiority and non-inferiority clinical trials

A clinical study of superiority design is intended to demonstrate that the herbal treatment is more effective than a placebo in a placebo-controlled trial. Nonetheless, sometimes the investigational product, in this case an herbal product or polyherbal formulation is compared to a reference treatment without the objective of showing superiority. These control trials are designed to demonstrate that the efficacy of the herbal product is not worse than that of the active comparative treatment (for instance a known conventional drug). These trials are termed as non-inferiority clinical trials (Karlberg and Speers 2010).

Sample size

Phase III clinical trial should be conducted with sufficient number of participants in order to obtain reliable results. Therefore, to ensure that neither too many nor too few subjects are included in the herbal intervention, the sample size should be calculated prior the start of the study. The sample size is defined as the total number of subjects participating in a clinical trial. A sample size which is too small will lead to an incorrect conclusion due to statistical bias. On the other hand a sample size which is too large is more expensive and time-consuming to conduct. According to Röhrig et al. (2010), the chances of success in a clinical trial and the quality of the research results depend to a large extent on sample size planning. Sample size planning should always be conducted in collaboration with a biometrician or an expert statistician. The unpaired t-test is often used to determine sample size. Provision should be made to recruit more subjects since drop outs are common in clinical trials and additional subjects need to be added in each group.

Recruitment of subjects

The appropriate subjects can be identified and recruited from a cabinet of physicians, public hospitals and/or private clinics or can also be recruited using newspaper and/or media advertising (www.ich.org). Participants are selected based on inclusion and exclusion criteria, which are established by investigators of the study. Inclusion criteria are defined as characteristics which make prospective subjects eligible to participate in the study while exclusion criteria are characteristics which prevent prospective participants to be part of the study. Inclusion and exclusion criteria include factors for instance age, participant's ability to give informed consent, type and stage of the disease, participant's treatment history, whether the participant is pregnant or breast feeding, participant's agreement to come for follow ups amongst others (ICH 1997, www.ich.org). The informed consent is a legal document which informs prospective participants the nature, aims and objectives, procedures, duration, possible risks and benefits of the study before they make their decision whether to participate or not. Participants who agree to participate in the clinical trial should sign the informed consent. They also have the right to withdraw from the clinical trial at any time during the course of the study (Li et al. 2009).

Controls

In many regions of the world, strong belief that herbal medicines will be beneficial and safe may introduce bias, which can be minimized by careful attention to study design including appropriate control groups (WHO 2005). The establishment of a control group is fundamental for comparison. Control studies are used to determine whether there is a significant statistical difference between the herbal product/formulation and the positive control in terms of the therapeutic effects. Therapeutic effects through clinical treatment may be achieved by the herbal product or some other factor, for instance psychological effect. Therefore, control studies are adopted to minimize the influence of other factors to the utmost degree (Yuan et al. 2011). Figure 5.1 summarizes three types of control groups which can be adopted when conducting clinical trial of traditional medicine.

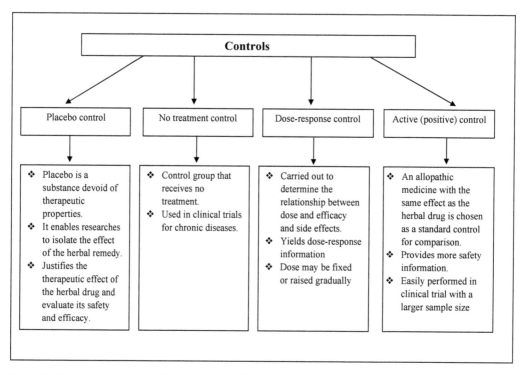

Figure 5.1. Types of control groups (Yuan et al. 2011, ICH 2001).

Cross-over trials

The design of randomized controlled trials consists of patients being randomly placed into two different, parallel treatment groups. However, not all randomized controlled trials are of this design. In contrast, to a parallel group trial, in cross-over trial each subject receives two or more treatments in a random order; that is the sequence of treatment which is randomized. For instance, a cross-over trial is conducted to investigate the effect of medicinal plant Y to reduce blood glucose level in diabetic patients. The subjects participating in the trial are assigned a regimen of one week of the medicinal plant Y and one week of a matching placebo in a random order. At the end of each week, blood glucose level is measured to determine efficacy of the medicinal plant Y. The particular advantage of cross-over trial is that the treatment is evaluated on the same patient, allowing comparison at the individual level rather than at the group level. Since each patient receives both interventions, cross-over trials require half the number of patients to produce the same precision as in parallel group trial. Moreover, variation in repeated responses within the same patient is usually less as compared to different patients. However, the major concern with cross-over trial is the risk of carry-over effect whereby the treatment given in the first period has an effect that persists into the second period (Elbourne et al. 2002).

Consolidated Standards of Reporting Trials (CONSORT)

The CONSORT statement was developed in order to aid investigators, reviewers, authors and editors on the information of interest to be included in reports of controlled clinical trials. The CONSORT statement is also relevant to any herbal medicinal products (Gagnier et al. 2006). The CONSORT statement consists of a checklist and flow diagram for reporting a randomized controlled trial. The checklist and diagram together are called CONSORT. They are primarily intended for use in writing, reviewing, or assessing reports of simple two-group parallel randomized controlled trial. Preliminary data indicate that the use of CONSORT helps to improve the quality of reports of randomized controlled trial (Moher et al. 2001). Figure 5.2 illustrates an example of subjects' disposition through the different phases of a randomized

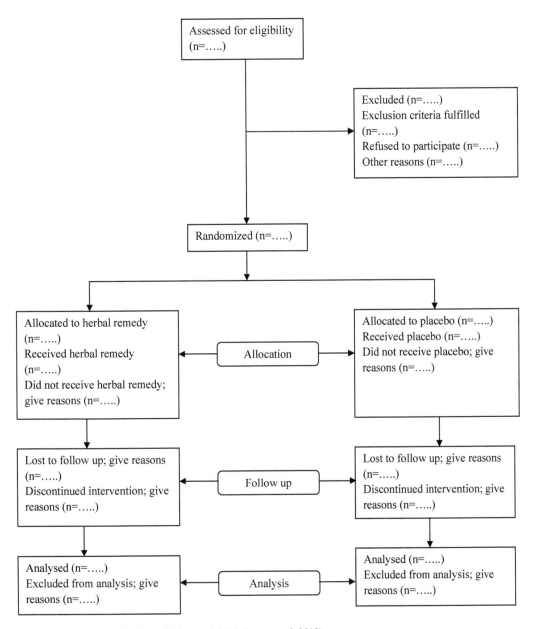

Figure 5.2. Disposition of subjects (Moher et al. 2001, Saxena et al. 2012).

control trial. However, according to Gagnier et al. (2006), controlled trials of herbal medicine do not report the information suggested in CONSORT adequately. Hence, reporting recommendations were developed in order to elaborate more on several items of the CONSORT. These recommendations if applied accordingly will make CONSORT more relevant and complete for randomized controlled trials of herbal medicine and will therefore lead to more complete and accurate reporting of clinical trials on herbal medicine. A summary of the recommendations is illustrated in Table 5.2.

The advantages and limitations of randomized controlled trial are summarized in Table 5.3.

Table 5.2. Recommendations to be applied to CONSORT when reporting trials on herbal medicine (Gagnier et al. 2006).

Recommendations for reporting trials on herbal medicine
1. It is fundamental to have a complete description of the control group and the way in which it is disguised to mimic the herbal medicine.
2. It is recommended to work with healthcare practitioners who are well-trained, experienced and who able to work in different environments.
3. A description of subjects participating in the trial enables the clinicians to evaluate how relevant the trial is to a specific patient.
4. Researches should clearly specify any concomitant use of allopathic medicine or medication or herbal product.
5. The researches should clearly state the dosage of the herbal medicine which resulted in their findings.
6. The researches should discuss where possible, how the medicinal plant used relates to what is currently used in self-care.
7. Researches must make sure to report any eligibility criteria and exclude subjects with previous use of a specific herb since the use of that plant prior to the start of the trial may potentiate the amount of adverse effects.
8. Precise details of the clinical trial must be included for instance: part of the plant used, whether the plant should be fresh or dry, how the plant was authenticated, the dosage, the duration of treatment amongst others.

Table 5.3. Advantages and limitations of randomized controlled trials (Saturni et al. 2014).

Randomized controlled trials	
Advantages	**Disadvantages**
Rigorous experimental design	Patients selection based on inclusion and exclusion criteria
Selection bias is avoided	Do not consider factors which influence outcomes in real life
Blinding increases reliability of the results	Higher frequency of visits and assessments leads to monitoring bias
Presence of control for comparison	Short duration of study
Rigorous analysis methods	Expensive

Observational clinical study

An observational study can be regarded as a strategy to provide information on effectiveness and toxic side-effect(s) of the traditional medicine being evaluated (Graz et al. 2007). Randomized controlled trials have long been used in clinical studies of herbal products, however, they are time-consuming, expensive, and difficult to conduct and in certain cases unnecessary, impossible, inappropriate, or inadequate (Black 1996). Observational studies are designed to monitor and describe real life management/treatment of clinical conditions. They can be both prospective and retrospective and consist of cohort, case-control and cross-sectional studies (Price et al. 2013). Observational studies are conducted in conditions closer to real practice and therefore provide a meaningful evaluation of traditional treatments (Elisabetsky and Setzer 1985). They have a longer duration than a randomized controlled trial therefore they enable identification of late side effects (Saturni et al. 2014). Observational studies also require informed consent from participants before they are included in the study and they are allowed to drop out from the study at any time. A medical team should also be available during patient follow-up (Graz et al. 2007). A follow up study can provide reliable answers to questions such as: "Does the traditional medicine lower blood glucose level similar to an allopathic hypoglycemic agent?" An observational study can answer question: "Does the blood glucose level of diabetic patients who use traditional medicine better controlled than those patients who do not use traditional medicine?" Phase IV clinical trial or post-marketing surveillance is fundamental for the detection of relatively rare adverse effects or events of special interest (WHO 2004).

Prospective cohort study

A prospective cohort study is an observational investigation and involves following a cohort of subjects over a long period of time. The outcome of interest should be common to all members of cohort in order for the study to be successful. For instance ethnopharmacological data supports that medicinal plant x cures cataract. A prospective cohort study can be conducted whereby all members of the cohort suffering from cataract and the preparation of traditional remedy is administered to each member of the cohort to determine whether all members are cured following administration of the traditional remedy over a certain period of time (Sedgwick 2013).

Retrospective treatment-outcome study

In order to determine the most effective traditional treatment for a given ailment, a retrospective treatment-outcome study can be used. This type of study analyses records of patient progress for a given ailment, and all the different traditional treatments used against that specific ailment. The retrospective treatment-outcome study makes it possible to perform correlation tests between the traditional treatment used and outcome in order to provide evidence of safety and efficacy of the traditional treatment. Moreover, a retrospective treatment-outcome study provides important information about precautions, contra-indications and also any drug-herb interactions (Graz et al. 2005, Diallo et al. 2006). The study is conducted using a questionnaire administered to a sample of patients suffering from the disease of interest. The sample size should be sufficient in order to perform statistical analysis to detect correlations between patient progress and different traditional treatments (Graz et al. 2007). If a given traditional treatment when used alone, is followed by a rapid and complete recovery, with no signs of side effects or adverse reactions, it can be concluded that it deserves further investments for scrutiny as it is usually considered to be a success for future drug development (Diallo et al. 2006).

Comparison of prognosis and outcome study

In order to evaluate the results of a traditional treatment, a comparison of prognosis and outcome study can be used. This type of study explores whether with the traditional treatment, patient progress is as good as or better than what can be expected with allopathic medicine. The general practitioner's (GP) prognosis is the result expected with allopathic medicine and the actual patient's progress is the outcome. The patient's progress (outcome) when treated with traditional medicine can also be compared with the traditional healer's prognosis. A four-point relative scale allows comparison between expected and observed patient progress (Graz et al. 2003).

Examples of clinical trial conducted on herbal products

Nahid et al. (2009) conducted a randomized, double-blind, placebo-controlled clinical trial among 180 female participants between the ages of 18–27 years who suffered from primary dysmenorrhea. The participants were randomly divided into three groups: herbal drug, mefenamic acid, and a placebo. The herbal drug consisted of a highly purified extract of saffron, celery seed and anise. The result showed significant reductions in pain scores and pain duration scores in the groups that took the herbal drug and mefenamic acid. The magnitude of the reduction was significantly greater in the herbal drug group than in the mefenamic acid and placebo groups.

A randomized, double-blinded, placebo-controlled trial was conducted with 55 subjects suffering from temporomandibular joint and masticatory muscle pain in order to compare the effectiveness of using Ping On ointment and using petroleum jelly in the treatment of temporomandibular joint and masticatory muscle pain. The main ingredients of Ping On ointment include peppermint oil, menthol, natural camphor, birch oil, sandalwood oil, eucalyptus oil, bee wax and aromatic oil. The result showed that Ping On ointment significantly reduced the symptoms of temporomandibular joint and masticatory muscle pain (Li et al. 2009).

Saxena et al. (2012) conducted a randomized, double-blind, placebo-controlled study to evaluate the efficacy of OciBest, an extract of *Ocimum tenuiflorum* Linn. in symptomatic control of general stress. The result showed an overall improvement in the OciBest group as compared to the placebo group.

Challenges of clinical trials with herbal medicine

A number of factors particularly affect the outcome of clinical trials with herbal products which include:

- Different dosages of herbal preparations are used in clinical trials, which may not yield uniform outcomes. It is fundamental that the dose is administered according to the traditional recipe.
- Absence of stringent quality control procedures of medicinal formulation used in clinical trials.
- Lack of appropriate standardization procedures for medicinal plants used in clinical trials. The problems are often accentuated when working with polyherbal formulations. Special attention must also be paid on standardization of the extracts using bioactive markers (e.g., phytochemicals present in such extracts) (Sharma et al. 2010).
- The number of subjects in most clinical trials is insufficient in order to perform statistical analysis. This can be overcome by conducting multicenter trials.
- Wide variations exist in the duration of treatments using herbal medicine (Sharma et al. 2010).
- According to Rothman and Michels (1994), clinical trials on traditional medicine are quite challenging because the herbal study drug may exhibit strong aroma, a specific distinguished taste and these cannot be exactly imitated while manufacturing a placebo.
- Thatte (1993) reported that storage conditions can alter the bioavailability of herbal remedies, which may entail loss of therapeutic activity, contamination by microorganisms resulting into batch to batch variation.
- Herbal materials have a number of inconsistencies in view of their geographical locations, climatic conditions, environmental threats, harvesting procedures, collection protocols, amongst others which makes it difficult to standardize the finished product for a reproducible quality (Bauer and Tittel 1996).
- As most herbal medicines used in traditional systems are not subjected to any toxicity studies, their embryotoxic, fetotoxic, and carcinogenic effects are likely to remain unrecognized in traditional practice (Sharma et al. 2010).
- Another important issue with the use of herbal remedies is that it is not always clear whether interactions exist between other medicinal plants and other allopathic medicines (Sharma et al. 2010).
- According to Kristofferson (1996), randomized clinical trials cannot be conducted with traditional medicine because of the holistic approach that is required and the fact that the traditional medicine is usually tailored specifically for each individual.

Conclusion

Clinical trials of traditionally used herbal medicine are becoming increasingly important due to the resurgence and popularity of medicinal plants for the treatment of various human ailments. Clinical trials are essential to establish the true efficacy and safety of medicinal plants as they are used in traditional practices and to evaluate the possible adverse effects or drug-herb interactions. Given the strong aroma and taste of certain medicinal plants, a placebo is often difficult to design and administer. Nonetheless, the strong smell can be masked in order to cover the aroma and taste of medicinal plants which will facilitate blinding methods in clinical trials. The indigenous community from whom the knowledge on the therapeutic properties of the medicinal plant originates should be consulted during the course of the clinical trials and the results and benefits of the study should be shared with them (WHO 2005). Successful clinical trials on herbal medicine may provide information of interest to patients, who are common users of these traditional formulations and inform them about the efficacy and possible toxicity of these traditional remedies. Moreover, outcomes of clinical trials will serve as a guide to patients and physicians toward the judicious use of herbal medicine (Ashar and Dobs 2004). As a conclusive note, clinical trial on herbal medicine should be conducted cautiously according to traditional recipes and based on thoughtful validated protocol since an inaccurate design of clinical trial of a herbal product may not only entail a wrong conclusion on the efficacy and safety of the herbal product but also lead to loss of time, money and manpower (Yuan et al. 2011).

References

Acharya, D. and Shrivastava, K. 2008. Indigenous Herbal Medicines: Tribal Formulations and Traditional Herbal Practices. Aavishkar Publishers Distributor, Jaipur, India. 440 pp.

Alzweiri, M., Al-Sarhan, A., Mansi, K., Hudaid, M. and Aburjai, T. 2011. Ethnopharmacological survey of medicinal herbs in Jordan, the Northern Badia region. J. Ethnopharmacol. 137(1): 27–35.

Ashar, B.H. and Dobs, A. 2004. Clinical trials for herbal extracts. pp. 53–72. *In*: L. Packer, N.C. Ong and B. Halliwell (eds.). Herbal and Traditional Medicine Molecular Aspects of Health. Marcel Dekker, New York.

Bauer, R. and Tittel, G. 1996. Quality assessment of herbal preparations as a precondition of pharmacological and clinical studies. Phytomedicine 2: 193–198.

Bigoniya, P., Singh, C.S. and Shukla, A. 2011. Pharmacognostical and physicochemical standardisation of ethnopharmacologically important seeds of *Lepidium sativum* Linn. and *Wrightia tinctoria* R. Br. Indian J. Nat. Prod. Res. 2(4): 464–471.

Black, N. 1996. Why we need observational studies to evaluate the effectiveness of health care? Br. Med. J. 312: 1215–1218.

Bougel, A. 2007. Laureate for her exploration and analysis of plants from Mauritius and their bio-medical application. Available at http://agora.forwomeninscience.com/index.php/2007/02/2007-laureate-for-africa-and-arab-states-meet-prof-ameenah-gurib-fakim/Accessed March 26, 2014.

Clark, J. 2012. Are plants used in modern medicine? Available at http://health.howstuffworks.com/medicine/modern-technology/plants-used-in-medicine.htm Accessed April 1st 2014.

Diallo, D., Graz, B., Falquet, J., Traore, A.K., Giani, S., Mounkoro, P.P., Berthé, A., Sacko, M. and Diakité, C. 2006. Malaria treatment in remote areas of Mali: use of modern and traditional medicines, patient outcome. Trans. R. Soc. Trop. Med. Hyg. 100: 515–520.

DMAPR. 2012. Available at http://www.dmapr.org.in/ Accessed March 25 2014.

Elbourne, D.R., Altman, D.G., Higgins, J.P.T., Curtin, F., Worthington, H.V. and Vauk, A. 2002. Meta-analyses involving cross-over trials: methodological issues. Int. J. Epidemiol. 31: 140–149.

Elisabetsky, E. and Setzer, R. 1985. Caboclo concepts of disease, diagnosis and therapy: implications for ethnopharmacology and health systems in Amazonia. pp. 243–278. *In*: E.P. Parker (ed.). The Amazon Caboclo: Historical and Contemporary Perspectives, Vol. 32. Studies on Third World Societies Publication Series, Williamsburgh.

Elujoba, A.A., Odeleye, O.M. and Ogunyemi, C.M. 2005. Traditional medical development for medical and dental primary health care delivery system in Africa. Afr. J. Tradit. Complement. Altern. Med. 2(1): 46–61.

Food and Drug Administration (FDA). 2004. Guidance for Industry Botanical Drug Products. Center for Drug Evaluation and Research.

Gagnier, J.J., Boon, H., Rochon, P., Moher, D., Barnes, J. and Bombardier, C. 2006. Recommendations for reporting randomized controlled trials of herbal interventions: explanation and elaboration. J. Clin. Epidemiol. 59: 1134–1149.

Graz, B., Xu, J.M., Yao, Z.S., Han, S.R. and Kok, A. 1999. Trachoma: can Trichiasis be treated with a sticking-plaster? a randomised controlled trial in China. Trop. Med. Int. Health. 4: 222–228.

Graz, B., Falquet, J. and Morency, P. 2003. Assessment of alternative medicine through a comparison of the expected and observed progress of patients: a feasibility study of the prognosis/follow-up method. J. Altern. Complement. Med. 9: 755–761.

Graz, B., Diallo, D., Falquet, J., Willcox, M. and Giani, S. 2005. Screening of traditional herbal medicine: first, do a retrospective study, with correlation between diverse treatments used and reported patient outcome. J. Ethnopharmacol. 101: 338–339.

Graz, B., Elisabetsky, E. and Falquet, J. 2007. Beyond the myth of expensive clinical study: assessment of traditional medicines. J. Ethnopharmacol. 113: 382–386.

ICH, 1997. ICH M3-Maintenance of the ICH Guideline on Non-Clinical Safety Studies for the Conduct of Human Clinical Trials for Pharmaceuticals.

ICH, 2001. ICH Topic E 10 Note for Guidance on Choice of Control Group in Clinical Trials. European Medicines Agency.

Jaillon, P. 2007. Controlled randomized clinical trials. Bulletin de l'Academie Nationale de Medecine 191(4-5): 739–756.

Joseph, B. and Raj, S.J. 2010. Pharmacognostic and phytochemical properties of *Aloe vera* linn—an overview. Int. J. Pharm. Sci. Rev. Res. 4(2): 106–110.

Karlberg, J.P.E. and Speers, M.A. 2010. Reviewing Clinical Trials: A Guide for the Ethics Committee. Hong Kong, PR China.

Kristofferson, S. 1996. Uptake of alternative medicine in Australia. Lancet 347: 569–573.

Li, L.C.F., Orth, M., Wong, R.W.K. and Rabie, A.B.M. 2009. Clinical effect of a topical herbal ointment on pain in temporomandibular disorders: a randomized placebo-controlled trial. J. Altern. Complement. Med. 15(12): 1311–1317.

Moher, D., Schulz, K.F. and Douglas, D.G. 2001. The CONSORT statement: revised recommendations for improving the quality of reports of parallel-group randomised trials. The Lancet 357: 1191–1194.

Nahid, K., Fariborz, M., Ataolah, G. and Solokian, S. 2009. The effect of an Iranian herbal drug on primary Dysmenorrhea: a clinical controlled trial. J. Midwifery Womens Health 54(5): 401–404.

Newman, D.J., Cragg, G.M. and Snader, K.M. 2003. Natural products as sources of new drugs over the period 1981–2002. J. Nat. Prod. 66: 1022–1032.

Price, D., Hillyer, E.V. and Van der Molen, T. 2013. Efficacy versus effectiveness trials: informing guidelines for asthma management. Curr. Opin. Allergy Clin. Immunol. 13(1): 50–57.

Röhrig, B., Du Prel, J.B., Wachtlin, D., Kwiecien, R. and Blettner, M. 2010. Sample size calculation in clinical trials. Dtsch. Arztebl. Int. 107(31–32): 552–556.

Rothman, K.J. and Michels, K.B. 1994. The continuing unethical use of placebo controls. N. Engl. J. Med. 331: 394–398.

Saturni, S., Bellini, F., Braido, F., Paggiaro, P., Sanduzzi, A., Scichilone, N., Santus, P.A., Morandi, L. and Papi, A. 2014. Randomized controlled trials and real life studies. Approaches and methodologies: a clinical point of view. Pulm. Pharmacol. Ther. 27: 129–138.

Saxena, R.C., Singh, R., Kumar, P., Negi, M.P.S., Saxena, V.S., Geetharani, P., Allan, J.J. and Venkateshwarlu, K. 2012. Efficacy of an extract of *Ocimum tenuiflorum* (OciBest) in the management of general stress: a double-blind, placebo-controlled study. J. Evid. Based Complementary Altern. Med. doi:10.1155/2012/894509.

Sharma, A.K., Kumar, R., Mishra, A. and Gupta, R. 2010. Problems associated with clinical trials of Ayurvedic medicines. Braz. J. Pharmacog. 20(2): 276–281.

Srivastava, J. and Vankar, P.S. 2012. Antioxidant profile in north eastern India: traditional herbs. J. Nutr. Food. Sci. 42(1): 26–33.

Teixeira, C.C. and Fuchs, F.D. 2006. The efficacy of herbal medicines in clinical models: the case of jambolan. J. Ethnopharmacol. 108: 16–19.

Thatte, U.M., Rege, N.N., Phatak, S. and Dahanukar, S.A. 1993. The flip side of Ayurveda. J. Postgrad. Med. 39: 179–182.

Wang, G., Hu, W., Huang, B. and Qin, L. 2011. *Illicium verum*: A review on its botany, traditional use, chemistry and Pharmacology. J. Ethnopharmacol. 136: 10–20.

Wickramasinghe, M.B. 2006. Quality Control, Screening, Toxicity, and Regulation of Herbal Drugs. Available at http://faculty.ksu.edu.sa/23494/Documents/Quality%20Control%20of%20Herbal%20Drugs.pdf Accessed March 17 2014.

World Health Organisation (WHO). 2000. General Guidelines for Methodologies on Research and Evaluation of Traditional Medicine. World Health Organisation, Geneva.

World Health Organisation (WHO). 2004. WHO Guidelines on Safety Monitoring of Herbal Medicines in Pharmacovigilance Systems. World Health Organisation, Geneva.

World Health Organisation (WHO). 2005. Operational Guidance: Information Needed to Support Clinical Trials of Herbal Products. World Health Organisation, Geneva.

World Health Organisation (WHO). 2013. Traditional Medicine Strategy 2014–2023. World Health Organisation, Geneva.

Yuan, H., Yang, G.-P. and Huang, Z.-J. 2011. Clinical study of traditional herbal medicine. pp. 343–377. *In*: W.J.H. Liu (ed.). Traditional Herbal Medicine Research Methods Identification, Analysis, Bioassay, and Pharmaceutical and Clinical Studies. John Wiley & Sons, New Jersey.

Opportunities and Limitations in Medicinal and Aromatic Plants' Markets and Research in Developing Countries: Lebanon as a Case Study

Karam Nisrine,[1,a,2,*] Noun Jihad,[3] Yazbeck Mariana,[4] Soubra Noura[5] and Talhouk Rabih[5,b,6,*]

Introduction

Lebanon, a country on the eastern coast of the Mediterranean, has a rich floral diversity. Relative to its area, Lebanon has more wild plant species density than any European country. With a surface area of only 10452 square kilometer, more than 2600 species of wild plants can be found. The flora of Lebanon has been of interest to several botanists, most popular being Post (1932) and Mouterde (1966, 1970, 1983) who have published monumental works documenting the plants of Lebanon and neighboring countries. Following their steps, Tohme and Tohme (2007, 2013) have compiled the most recent comprehensive flora of the country. Their data indicated that out of the 2597 plants identified, 109 are endemic to Lebanon, about 52% of Lebanese flora is not found in Europe and 1185 species are particularly known in the eastern Mediterranean region. This richness, added to a diversity of different cultural practices in the Lebanese social strata, has contributed to a variety of approaches across generations for utilization of Medicinal and

[1] Department of Biotechnology, Lebanese Agricultural Research Institute (LARI).
[a] Email: nisrine_karam@hotmail.com
[2] Departments of Plant Production, Faculty of Agriculture.
[3] Biology, Faculty of Sciences, Lebanese University.
[4] Genetic Resources Section, International Center for Agricultural Research in the Dry Areas (ICARDA).
[5] Department of Biology, Faculty of Arts and Sciences.
[b] Email: rtalhouk@aub.edu.lb
[6] Nature Conservation Center (NCC), American University of Beirut, Lebanon.
* Corresponding authors

Aromatic Plants (MAP) that is still, to a certain extent, in practice to-date. However, this richness of plant biodiversity and more particularly of MAP is at risk due to degradation of natural habitats as a result of uncontrolled growth of urbanization, habitat destruction, and overexploitation of the resources. This further contributes to the loss of the MAP genetic diversity (Noun 2007, MoA/UNEP/GEF 1996, UNDP 2013).

In Lebanon, a country historically rich in MAP diversity and in indigenous MAP-associated knowledge, several attempts have been made to document this heritage (Ruwaiha 1981, Jabr 1988, Akil 1996, Baalbaki 2000, Kadamah 2000, Hayek 1996–2001, Ismail 2001). Unfortunately, most of these attempts have focused on compiling existing knowledge without any significant effort to add to it since the golden age of the traditional medicine of Ibn Sina (1025), Daoud Al Antaki (1877) and others. Recently, at the national level, an exhaustive work on plants has been published by the Ministry of Agriculture (MOA), the United Nations Environmental Programme (UNEP) and Global Environmental Facility (GEF). In this milestone study (MOA/UNEP/GEF 1996), around 300 species were listed to have medicinal and aromatic properties. An update of their work combined with other more recent publications on the listing of MAP of Lebanon (Hayek 1996–2001, Abou Chaar et al. 2004, El Beyrouthy et al. 2008, Deeb et al. 2013) has led to a total of 385 MAP species distributed over 260 genera and 90 botanical families. These have been compiled and listed in Table 6.1; many of these plants are endemic to Lebanon and/or under threat due to their niche distribution while others are endemic to the Levant region. The main families rich in diversity of genera of MAP are *Lamiaceae* and Asteraceae with more than 40 MAP species each followed by *Fabaceae* and *Apiaceae* with more than 20 each.

Medicinal and aromatic plants in Lebanon

The degree of knowledge of these MAP species varies and depends to a large extent on a diversity of published literature starting from the local inherited knowledge that is communicated verbally across generations, to the documented traditional medicine (Known as popular or folk medicine) and to the well-documented publications in the scientific literature. Accordingly our knowledge of the Lebanese MAP may be categorized into:

A. MAP with medicinal value based on scientific and folk literature.
B. MAP commonly used in folk medicine without validated scientific literature to support their use.
C. MAP with claimed medicinal value typically communicated in spoken folk literature.
D. Plants that are erroneously used as MAP and are typically, confused with other similar species assumed to have medicinal value. Examples of these are: *Ecballium elaterium* (L.) A. Rich, *Ferula hermonis* Boiss., *Phlomis fruticosa* L., and *Cercis siliquastrum* L.

The use of MAP and alternative medicine practices is gaining considerable attention from the scientific community and from entrepreneurial enthusiasts in Lebanon. This trend was also noted in recent years in developed countries of the Western Hemisphere that have traditionally practiced 'modern medicine'. According to the World Health Organization (WHO), traditional medicines, particularly herbal medicines, are increasingly used worldwide over the past two decades (WHO 2003). However more importantly, in developing countries, MAP have played a key role in the lives of the populace of rural marginalized communities, who have been using these plants in a sustainable manner for centuries. For these marginalized communities, MAP have been viewed as a valuable commodity and from being commonly used as a potential source of income rather than a source for preparing traditional remedies for treating ailments. As a result, harvesting of some wild MAP, now conceived as cash crops that can contribute to economic growth, is growing at an alarming pace. This trend is increasing the risk of disrupting sustainable use of plants that has been in practice by local communities for centuries. In Lebanon and neighboring countries the local knowledge of MAP uses and consumption dates back to thousands of years (Malychef 1989, Deeb et al. 2013). However with the predominance of 'modern' medical practices the local knowhow of folk medicine and its uses became restricted to few individuals in rural village communities. This, across the generations, is leading to a gradual loss of folk knowledge and appreciation of the value of MAP and the need to sustain their natural habitats.

Table 6.1. Compiled list of reported MAP in Lebanon during the period of 1996–2013.

Family	Species Record	References
Alismataceae	*Alisma lanceolatum* With.	Hayek 1996–2001
Amaranthaceae	*Chenopodium vulvaria* L.	El Beyrouthy et al. 2008
Amaranthaceae	*Chenopodium album* L.	MoA/UNEP/GEF 1996
Amaranthaceae	*Dysphania ambrosioides* (L.) Mosyakin & Clemants	MoA/UNEP/GEF 1996
Amaranthaceae	*Dysphania botrys* (L.) Mosyakin & Clemants	Hayek 1996–2001
Amaryllidaceae	*Narcissus tazzetta* L.	Deeb et al. 2013
Anacardiaceae	*Pistacia lentiscus* L.	MoA/UNEP/GEF 1996
Anacardiaceae	*Pistacia palaestina* Boiss.	Hayek 1996–2001
Anacardiaceae	*Rhus coriaria* L.	Abou Chaar 2004, Deeb et al. 2013, MoA/UNEP/GEF 1996
Apiaceae	*Ammi majus* L.	Abou Chaar 2004, MoA/UNEP/GEF 1996
Apiaceae	*Ammi visnaga* (L.) Lam.	Abou Chaar 2004, Deeb et al. 2013, MoA/UNEP/GEF 1996
Apiaceae	*Anethum graveolens* L.	Deeb et al. 2013, MoA/UNEP/GEF 1996
Apiaceae	*Apium graveolens* L.	Abou Chaar 2004, Deeb et al. 2013, MoA/UNEP/GEF 1996
Apiaceae	*Apium nodiflorum* (L.) Lag.	El Beyrouthy et al. 2008
Apiaceae	*Bupleurum rotundifolium* L.	Hayek 1996–2001
Apiaceae	*Chaerophyllum libanoticum* Boiss. & Kotschy	Hayek 1996–2001
Apiaceae	*Conium maculatum* L.	MoA/UNEP/GEF 1996
Apiaceae	*Crithmurn maritimum* L.	Deeb et al. 2013, MoA/UNEP/GEF 1996
Apiaceae	*Daucus carota* L. ssp. *maxirnus* (Desf.) Thell	Deeb et al. 2013, MoA/UNEP/GEF 1996
Apiaceae	*Eryngium creticum* Lam.	Abou Chaar 2004, MoA/UNEP/GEF 1996
Apiaceae	*Eryngium falcatum* F.Delaroche	Hayek 1996–2001
Apiaceae	*Eryngium maritimum* L.	Abou Chaar 2004, Deeb et al. 2013, MoA/UNEP/GEF 1996
Apiaceae	*Ferula hermonis* Boiss.	Abou Chaar 2004, Deeb et al. 2013
Apiaceae	*Ferula tingitana* L.	Abou Chaar 2004
Apiaceae	*Ferulago syriaca* Boiss.	Abou Chaar 2004
Apiaceae	*Foeniculum vulgare* Mill.	Deeb et al. 2013
Apiaceae	*Prangos aspegula* Boiss.	UNDP 2013
Apocynaceae	*Nerium oleander* L.	Abou Chaar 2004, MoA/UNEP/GEF 1996
Apocynaceae	*Vinca herbacea* Waldst. & Kit.	Abou Chaar 2004, MoA/UNEP/GEF 1996

Table 6.1. contd....

Table 6.1. contd.

Family	Species Record	References
Araceae	*Arum maculatum* L.	El Beyrouthy et al. 2008
Araceae	*Arum palaestinum* Boiss.	Deeb et al. 2013
Araceae	*Fritillaria persica* L.	Hayek 1996–2001
Araliaceae	*Hedera helix* L.	MoA/UNEP/GEF 1996
Aristolochiaceae	*Aristolochia clematitis* L.	El Beyrouthy et al. 2008
Aristolochiaceae	*Aristolochia sempervirens* L.	El Beyrouthy et al. 2008
Aspleniaceae	*Aspleniumtrichomanes* L.	Hayek 1996–2001
Asteraceae	*Achillea falcata* L.	Hayek 1996–2001
Asteraceae	*Achillea fragrantissima* (Forssk) Schulz Bip.	Abou Chaar 2004, Deeb et al. 2013
Asteraceae	*Achillea odorata* L.	Hayek 1996–2001
Asteraceae	*Arctium lappa* L.	Abou Chaar 2004
Asteraceae	*Artemisia arborescens* (Vaill.) L.	El Beyrouthy et al. 2008
Asteraceae	*Artemisia herba-alba* Asso	Abou Chaar 2004, Deeb et al. 2013, MoA/UNEP/GEF 1996
Asteraceae	*Bellis perennis* L.	MoA/UNEP/GEF 1996
Asteraceae	*Bidens pilosa* L.	Hayek 1996–2001
Asteraceae	*Carduus argentatus* L.	Deeb et al. 2013
Asteraceae	*Carthamus tinctorius* L.	Hayek 1996–2001
Asteraceae	*Centaurea benedicta* (L.) L.	Abou Chaar 2004, MoA/UNEP/GEF 1996
Asteraceae	*Centaurea calcitrapa* L.	MoA/UNEP/GEF 1996
Asteraceae	*Centaurea eryngioides* Lam.	MoA/UNEP/GEF 1996
Asteraceae	*Centaurea iberica* Trevir. ex Spreng.	Senatore et al. 2005
Asteraceae	*Chrysanthemum coronarium* L.	Deeb et al. 2013
Asteraceae	*Cichorium intybus* L.	MoA/UNEP/GEF 1996
Asteraceae	*Cichorium pumilum* Jacq.	El Beyrouthy et al. 2008
Asteraceae	*Cota tinctoria* (L.) J.Gay	Hayek 1996–2001
Asteraceae	*Cyanus segetum* Hill.	MoA/UNEP/GEF 1996
Asteraceae	*Cynara scolymus* L.	Abou Chaar 2004
Asteraceae	*Dittrichia viscosa* (L.) Greuter	Abou Chaar 2004, Deeb et al. 2013, MoA/UNEP/GEF 1996
Asteraceae	*Erigeron canadensis* L.	Hayek 1996–2001, MoA/UNEP/GEF 1996
Asteraceae	*Eupatorium cannabinum* L., var. *indivisum* D.C.	MoA/UNEP/GEF 1996
Asteraceae	*Helichrysum pallassii* (Spreng.) Ledeb.	Formisano et al. 2009
Asteraceae	*Helichrysum plicatum* D.C.	MoA/UNEP/GEF 1996

Table 6.1. contd....

Table 6.1. contd.

Family	Species Record	References
Asteraceae	*Helichrysum stoechas* subsp. *barrelieri* (Ten.) Nyman	Abou Chaar 2004
Asteraceae	*Lactuca serriola* L.	Hayek 1996–2001
Asteraceae	*Lapsana communis* L.	Hayek 1996–2001
Asteraceae	*Matricaria aurea* (Loefl.) Sch. Bip.	El Beyrouthy et al. 2008
Asteraceae	*Matricaria chamomilla* L.	Abou Chaar 2004, Deeb et al. 2013, MoA/UNEP/GEF 1996
Asteraceae	*Onopordum cynarocephalum* Boiss. & Blanche	El-Najjar et al. 2007
Asteraceae	*Picnomon acarna* (L.) Cass.	Abou Chaar 2004
Asteraceae	*Ptilostemon chamaepeuce* (L.) Less	Abou Chaar 2004
Asteraceae	*Scolymus maculatus* L.	Abou Chaar 2004
Asteraceae	*Senecio vulgaris* L.	Deeb et al. 2013, MoA/UNEP/GEF 1996
Asteraceae	*Silybum marianum* (L.) Gaertner	Abou Chaar 2004, Deeb et al. 2013, MoA/UNEP/GEF 1996
Asteraceae	*Tanacetum cilicium* (Boiss.) Grierson	El Beyrouthy et al. 2008
Asteraceae	*Tanacetum parthenium* (L.) Sch. Bip.	El Beyrouthy et al. 2008
Asteraceae	*Taraxacum campylodes* G.E. Haglund	MoA/UNEP/GEF 1996
Asteraceae	*Taraxacum phaleratum* G.E. Haglund	Hayek 1996–2001
Asteraceae	*Tussilago farfara* L.	Deeb et al. 2013, MoA/UNEP/GEF 1996
Asteraceae	*Xanthium strumarium* L.	Hayek 1996–2001
Athyriaceae	*Athyrium filix-femina* (L.) Roth.	El Beyrouthy et al. 2008
Berberidaceae	*Berberis libanotica* Ehrenb. ex C.K. Schneid.	Abou Chaar 2004, Deeb et al. 2013, MoA/UNEP/GEF 1996
Berberidaceae	*Bongardia chrysogonum* (L.) Spach	Abou Chaar 2004
Berberidaceae	*Leontice leontopetalum* L.	Abou Chaar 2004, MoA/UNEP/GEF 1996
Boraginaceae	*Anchusa azurea* Mill.	MoA/UNEP/GEF 1996
Boraginaceae	*Borago officinalis* L.	Deeb et al. 2013, MoA/UNEP/GEF 1996
Boraginaceae	*Lithospermum officinale* L.	MoA/UNEP/GEF 1996
Brassicaceae	*Alliaria petiolata* (Bieb.) Cav. & Grande	MoA/UNEP/GEF 1996
Brassicaceae	*Brassica nigra* (L.) Koch.	MoA/UNEP/GEF 1996
Brassicaceae	*Capsella bursa-pastoris* (L.) Medik.	El Beyrouthy et al. 2008
Brassicaceae	*Descurainia sophia* (L.) Webb ex Prantl	MoA/UNEP/GEF 1996
Brassicaceae	*Diplotaxis erucoides* (L.) D.C.	MoA/UNEP/GEF 1996
Brassicaceae	*Fibigia clypeata* (L.) Medik.	Abou Chaar 2004, MoA/UNEP/GEF 1996
Brassicaceae	*Fibigia eriocarpa* (D.C.) Boiss.	Abou Chaar 2004

Table 6.1. contd....

Table 6.1. contd.

Family	Species Record	References
Brassicaceae	*Isatis tinctoria* L.	Hayek 1996–2001
Brassicaceae	*Lepidium latifolium* L.	Hayek 1996–2001
Brassicaceae	*Nasturtium officinale* R. Br.	MoA/UNEP/GEF 1996
Brassicaceae	*Raphanus raphanistrum* L.	MoA/UNEP/GEF 1996
Brassicaceae	*Sinapis alba* L.	El Beyrouthy et al. 2008
Brassicaceae	*Sinapis arvensis* L.	El Beyrouthy et al. 2008
Cactaceae	*Opuntia ficus-indica* (L.) Mill.	MoA/UNEP/GEF 1996
Campanulaceae	*Campanula rapunculus* L.	Hayek 1996–2001
Capparidaceae	*Capparis spinosa* L.	Abou Chaar 2004, Deeb et al. 2013, MoA/UNEP/GEF 1996
Caprifoliaceae	*Sambucus ebulus* L.	MoA/UNEP/GEF 1996
Caprifoliaceae	*Sambucus nigra* L.	MoA/UNEP/GEF 1996
Caryophyllaceae	*Paronychia argentea* Lam.	Abou Chaar 2004
Caryophyllaceae	*Silene vulgaris* (Moench) Garcke	Abou Chaar 2004
Caryophyllaceae	*Spergularia media* (L.) C. Presl	Abou Chaar 2004
Caryophyllaceae	*Stellaria cilicica* Boiss. & Bal.	MoA/UNEP/GEF 1996
Caryophyllaceae	*Stellaria media* (L.) Vill.	El Beyrouthy et al. 2008
Colchicaceae	*Colchicum luteum* Baker	El Beyrouthy et al. 2008
Convolvulaceae	*Convolvulus arvensis* L.	MoA/UNEP/GEF 1996
Convolvulaceae	*Cuscuta epithymum* (L.) L.	MoA/UNEP/GEF 1996
Convolvulaceae	*Cuscuta monogyna* Vahl.	Hayek 1996–2001
Convolvulaceae	*Ipomoea cairica* (L.) Sweet	Abou Chaar 2004
Crassulaceae	*Sedum sediforme* (Jacq.) Pau	El Beyrouthy et al. 2008
Crassulaceae	*Umbilicus intermedius* Boiss.	El Beyrouthy et al. 2008
Crassulaceae	*Umbilicus rupestris* (Salisb.) Dandy	MoA/UNEP/GEF 1996
Cucurbitaceae	*Bryonia syriaca* Boiss.	El Beyrouthy et al. 2008
Cucurbitaceae	*Citrullus colocynthis* (L.) Schrader	MoA/UNEP/GEF 1996
Cucurbitaceae	*Ecballium elaterium* (L.) A. Rich.	Abou Chaar 2004, Deeb et al. 2013, MoA/UNEP/GEF 1996
Cupressaceae	*Cupressus sempervirens* L.	Abou Chaar 2004
Cupressaceae	*Juniperus oxycedrus* L.	MoA/UNEP/GEF 1996
Cyperaceae	*Cyperus longus* L.	Hayek 1996–2001
Dioscoreaceae	*Dioscorea communis* (L.) Caddick & Wilkin	El Beyrouthy et al. 2008
Dioscoreaceae	*Dioscorea orientalis* (J. Thiébaut) Caddick & Wilkin	El Beyrouthy et al. 2008
Dipsacaceae	*Cephalaria syriaca* (L.) Schrad. ex Roem. & Schult.	Abou Chaar 2004

Table 6.1. contd....

Table 6.1. contd.

Family	Species Record	References
Dipsacaceae	*Dipsacus fullonum* L.	MoA/UNEP/GEF 1996
Droseraceae	*Drosera rotundifolia* L.	MoA/UNEP/GEF 1996
Dryopteridaceae	*Dryopteris libanotica* (Ros.) A. Christensen	El Beyrouthy et al. 2008
Eleagnaceae	*Elaeagnus angustifolia* L.	Abou Chaar 2004
Ephedraceae	*Ephedra foeminea* Forssk.	Abou Chaar 2004
Ephedraceae	*Ephedra fragilis* Desf.	Abou Chaar 2004
Equisetaceae	*Equisetum palustre* L.	Hayek 1996–2001
Equisetaceae	*Equisetum telmateia* Ehrh.	MoA/UNEP/GEF 1996
Ericaceae	*Arbutus andrachne* L.	Hayek 1996–2001
Ericaceae	*Arbutus unedo* L.	Deeb et al. 2013, MoA/UNEP/GEF 1996
Ericaceae	*Erica manipuliflora* Salisb.	Abou Chaar 2004
Ericaceae	*Erica sicula* Guss.	El Beyrouthy et al. 2008
Ericaceae	*Erica verticillata* P.J. Bergius	Abou Chaar 2004
Ericaceae	*Rhododendron ponticum* L.	El Beyrouthy et al. 2008
Euphorbiaceae	*Euphorbia peplus* L.	Hayek 1996–2001
Euphorbiaceae	*Mercurialis annua* L.	MoA/UNEP/GEF 1996
Euphorbiaceae	*Euphorbia hierosolymitana* Boiss.	El Beyrouthy et al. 2008
Euphorbiaceae	*Euphorbia parviflora* L.	El Beyrouthy et al. 2008
Euphorbiaceae	*Ricinus communis* L.	Abou Chaar 2004, Deeb et al. 2013
Fabaceae	*Acacia farnesiana* (L.) Willd	Abou Chaar 2004
Fabaceae	*Alhagi maurorum* Medik.	Abou Chaar 2004
Fabaceae	*Anagyris foetida* L.	MoA/UNEP/GEF 1996
Fabaceae	*Astragalus gummifer* Labill.	Abou Chaar 2004, MoA/UNEP/GEF 1996
Fabaceae	*Calycotome villosa* (Vahl) Link	Deeb et al. 2013
Fabaceae	*Ceratonia siliqua* L.	Abou Chaar 2004, Deeb et al. 2013, MoA/UNEP/GEF 1996
Fabaceae	*Colutea cilicica* Boiss. & Bal.	MoA/UNEP/GEF 1996
Fabaceae	*Glycyrrhiza glabra* L.	Abou Chaar 2004
Fabaceae	*Lotus corniculatus* L., var. *alpinus* (Scheich) Boiss.	MoA/UNEP/GEF 1996
Fabaceae	*Lupinus albus* L.	Abou Chaar 2004, Deeb et al. 2013, MoA/UNEP/GEF 1996
Fabaceae	*Medicago sativa* L.	Hayek 1996–2001
Fabaceae	*Melilotus albus* Medik.	Hayek 1996–2001
Fabaceae	*Melilotus officinalis* (L.) Lam.	MoA/UNEP/GEF 1996

Table 6.1. contd....

Table 6.1. contd.

Family	Species Record	References
Fabaceae	*Ononis natrix* L.	Hayek 1996–2001
Fabaceae	*Ononis spinosa* L., ssp. *leiosperma* (Boiss.) Sirj.	MoA/UNEP/GEF 1996
Fabaceae	*Retama raetam* (Forssk.) Webb.	MoA/UNEP/GEF 1996
Fabaceae	*Robinia pseudo-acacia* L.	MoA/UNEP/GEF 1996
Fabaceae	*Spartium junceum* L.	Abou Chaar 2004, MoA/UNEP/GEF 1996
Fabaceae	*Trifolium arvense* L.	MoA/UNEP/GEF 1996
Fabaceae	*Trifolium pratense* L.	El Beyrouthy et al. 2008
Fabaceae	*Trifolium repens* L.	El Beyrouthy et al. 2008
Fabaceae	*Trigonella berythea* Boiss. and Bl.	El Beyrouthy et al. 2008
Fabaceae	*Trigonella foenum-graecum* L.	Abou Chaar 2004, Deeb et al. 2013, MoA/UNEP/GEF 1996
Fagaceae	*Quercus coccifera* L.	Deeb et al. 2013
Gentianaceae	*Centaurium erythraea* Rafn	Abou Chaar 2004, MoA/UNEP/GEF 1996
Geraniaceae	*Erodium cicutarium* (L.) L'Hér.	Hayek 1996–2001
Geraniaceae	*Erodium moschatum* (L.) L'Hér.	Hayek 1996–2001
Geraniaceae	*Geranium maculatum* L.	Hayek 1996–2001
Geraniaceae	*Geranium robertianum* L.	Deeb et al. 2013, MoA/UNEP/GEF 1996
Geraniaceae	*Geranium rotundifolium* L.	Hayek 1996–2001
Grossulariaceae	*Ribes orientale* Desf.	MoA/UNEP/GEF 1996
Hypericaceae	*Hypericum elodeoides* Choisy	Abou Chaar 2004
Hypericaceae	*Hypericum hircinum* L.	Abou Chaar 2004
Hypericaceae	*Hypericum lanuginosum* Lam.	Abou Chaar 2004
Hypericaceae	*Hypericum libanoticum* N. Robson., sp. nov.	Abou Chaar 2004
Hypericaceae	*Hypericum lydium* Boiss.	Abou Chaar 2004
Hypericaceae	*Hypericum montbretii* Spach	Abou Chaar 2004
Hypericaceae	*Hypericum nanum* Poiret	Abou Chaar 2004
Hypericaceae	*Hypericum perforatum* L.	Abou Chaar 2004
Hypericaceae	*Hypericum saturejifolium* Jaub. & Spach	Abou Chaar 2004
Hypericaceae	*Hypericum scabrum* L.	Abou Chaar 2004
Hypericaceae	*Hypericum tetrapterum* Fries.	Abou Chaar 2004
Hypericaceae	*Hypericum thymifolium* Banks et Sol.	Abou Chaar 2004
Hypericaceae	*Hypericum triquetrifolium* Turra	Abou Chaar 2004
Hypericaceae	*Hypericum venustum* Fenzl	Abou Chaar 2004
Iridaceae	*Iris pseudacorus* L.	MoA/UNEP/GEF 1996

Table 6.1. contd....

Table 6.1. contd.

Family	Species Record	References
Juglandaceae	*Juglans regia* L.	Deeb et al. 2013, MoA/UNEP/GEF 1996
Lamiaceae	*Ballota nigra* L.	MoA/UNEP/GEF 1996
Lamiaceae	*Ballota saxatilis* Sieber ex C. Presl	Bruno et al. 2001
Lamiaceae	*Cistus creticus* L.	Hayek 1996–2001
Lamiaceae	*Cistus salviifolius* L.	El Beyrouthy et al. 2008
Lamiaceae	*Clinopodium libanoticum* (Boiss.) Kuntze	Abou Chaar 2004, Deeb et al. 2013
Lamiaceae	*Clinopodium serpyllifolium* subsp. *barbatum* (P.H. Davis) Bräuchler	UNDP 2013
Lamiaceae	*Cyclotrichium origanifolium* (Labill.) Manden. & Scheng.	Hayek 1996–2001
Lamiaceae	*Lavandula stoechas* L.	Abou Chaar 2004, Deeb et al. 2013
Lamiaceae	*Marrubium vulgare* L.	MoA/UNEP/GEF 1996
Lamiaceae	*Marrubium globosum* subsp. *libanoticum* (Boiss.) P.H. Davis	Rigano et al. 2006
Lamiaceae	*Melissa officinalis* L.	Abou Chaar 2004, MoA/UNEP/GEF 1996
Lamiaceae	*Mentha aquatica* L.	Abou Chaar 2004, MoA/UNEP/GEF 1996
Lamiaceae	*Mentha longifolia* (L.) L.	Abou Chaar 2004, Deeb et al. 2013, MoA/UNEP/GEF 1996
Lamiaceae	*Mentha pulegium* L.	MoA/UNEP/GEF 1996
Lamiaceae	*Mentha spicata* subsp. *condensata* (Briq.) Greuter & Burdet	Deeb et al. 2013
Lamiaceae	*Micromeria juliana* (L.) Benth. ex Rchb.	Abou Chaar 2004, Deeb et al. 2013
Lamiaceae	*Nepeta cataria* L.	MoA/UNEP/GEF 1996
Lamiaceae	*Nepeta curviflora* Boiss.	Mancini et al. 2009
Lamiaceae	*Nepeta nuda* subsp. *albiflora* (Boiss.) Gams	Mancini et al. 2009
Lamiaceae	*Origanum ehrenbergii* Boiss.	Abou Chaar 2004
Lamiaceae	*Origanum libanoticum* Boiss.	Abou Chaar 2004
Lamiaceae	*Origanum syriacum* L.	Abou Chaar 2004, Deeb et al. 2013, MoA/UNEP/GEF 1996
Lamiaceae	*Origanum* × *barbarae* Bornm.	Abou Chaar 2004
Lamiaceae	*Phlomoides laciniata* (L.) Kamelin & Makhm.	Abou Chaar 2004, Deeb et al. 2013, MoA/UNEP/GEF 1996
Lamiaceae	*Prunella orientalis* Bornm.	MoA/UNEP/GEF 1996
Lamiaceae	*Rosmarinus officinalis* L.	Abou Chaar 2004, Deeb et al. 2013, MoA/UNEP/GEF 1996
Lamiaceae	*Salvia bracteata* Banks & Sol.	Cardile et al. 2009

Table 6.1. contd....

Table 6.1. contd.

Family	Species Record	References
Lamiaceae	*Salvia fruticosa* Mill.	Abou Chaar 2004, Deeb et al. 2013, MoA/UNEP/GEF 1996
Lamiaceae	*Salvia hierosolymitana* Boiss.	Mancini et al. 2009
Lamiaceae	*Salvia multicaulis* Vahl.	Mancini et al. 2009
Lamiaceae	*Salvia palaestina* Benth.	Senatore et al. 2005
Lamiaceae	*Salvia rubifolia* Boiss.	Cardile et al. 2009
Lamiaceae	*Salvia sclarea* L.	Abou Chaar 2004, MoA/UNEP/GEF 1996
Lamiaceae	*Salvia viridis* L.	Hayek 1996–2001
Lamiaceae	*Satureja cuneifolia* Ten.	Hayek 1996–2001
Lamiaceae	*Satureja thymbra* L.	El Beyrouthy et al. 2008
Lamiaceae	*Sideritis pullulans* Vent.	Abou Chaar 2004
Lamiaceae	*Teucrium creticum* L.	Abou Chaar 2004
Lamiaceae	*Teucrium divaricatum* Sieber ex Heldr.	Formisano et al. 2009
Lamiaceae	*Teucrium montbretii* Benth.	Bruno et al. 2001
Lamiaceae	*Teucrium polium* L.	Abou Chaar 2004, Deeb et al. 2013
Lamiaceae	*Teucrium scordium* L.	Abou Chaar 2004, MoA/UNEP/GEF 1996
Lamiaceae	*Teucrium socinianum* Boiss.	Abou Chaar 2004
Lamiaceae	*Thymbra capitata* (L.) Cav.	Abou Chaar 2004
Lauraceae	*Laurus nobilis* L.	Abou Chaar 2004, Deeb et al. 2013, MoA/UNEP/GEF 1996
Liliaceae	*Allium sphaerocephalon* subsp. *arvense* (Guss.) Arcang.	Hayek 1996–2001
Liliaceae	*Asparagus acutifolius* L.	Deeb et al. 2013, MoA/UNEP/GEF 1996
Liliaceae	*Asparagus horridus* L.	El Beyrouthy et al. 2008
Liliaceae	*Asphodelus microcarpus* Salzm. et Viv.	MoA/UNEP/GEF 1996
Liliaceae	*Drimia maritima* (L.) Stearn	Abou Chaar 2004, Deeb et al. 2013, MoA/UNEP/GEF 1996
Liliaceae	*Leopoldia comosa* (L.) Parl.	MoA/UNEP/GEF 1996
Liliaceae	*Lilium candidum* L.	MoA/UNEP/GEF 1996
Liliaceae	*Ruscus aculeatus* L.	Abou Chaar 2004, Deeb et al. 2013, MoA/UNEP/GEF 1996
Linaceae	*Linum bienne* Mill.	MoA/UNEP/GEF 1996
Linaceae	*Linum usitatissimurn* L.	Abou Chaar 2004, Deeb et al. 2013
Loranthaceae	*Viscum album* L.	Abou Chaar 2004, MoA/UNEP/GEF 1996
Lythraceae	*Lythrum salicaria* L.	MoA/UNEP/GEF 1996

Table 6.1. contd....

Table 6.1. contd.

Family	Species Record	References
Malvaceae	*Alcea rosea* L.	Abou Chaar 2004
Malvaceae	*Alcea setosa* (Boiss.) Alef.	Deeb et al. 2013, MoA/UNEP/GEF 1996
Malvaceae	*Althaea officinalis* L.	MoA/UNEP/GEF 1996
Malvaceae	*Althaea damascena* Mouterde	UNDP 2013
Malvaceae	*Malva nicaeensis* All.	Abou Chaar 2004
Malvaceae	*Malva parviflora* L.	Abou Chaar 2004
Malvaceae	*Malva sylvestris* L.	Abou Chaar 2004, Deeb et al. 2013, MoA/UNEP/GEF 1996
Malvaceae	*Malva multiflora* (Cav.) Soldano, Banfi & Galasso	El Beyrouthy et al. 2008
Moraceae	*Ficus carica* L.	Deeb et al. 2013, MoA/UNEP/GEF 1996
Moraceae	*Morus alba* L.	Abou Chaar 2004, MoA/UNEP/GEF 1996
Moraceae	*Morus nigra* L.	MoA/UNEP/GEF 1996
Myrtaceae	*Myrtus communis* L.	Abou Chaar 2004, Deeb et al. 2013, MoA/UNEP/GEF 1996
Nympheaceae	*Nuphar lutea* (L.) Sm.	MoA/UNEP/GEF 1996
Oleaceae	*Fraxinus ornus* L.	Hayek 1996–2001
Oleaceae	*Fraxinus angustifolia* subsp. *syriaca* (Boiss.) Yalt.	El Beyrouthy et al. 2008
Oleaceae	*Olea eurapea* L.	Abou Chaar 2004, Deeb et al. 2013, MoA/UNEP/GEF 1996
Oleaceae	*Phillyrea latifolia* L.	El Beyrouthy et al. 2008
Onagraceae	*Epilobium angustifolium* L.	MoA/UNEP/GEF 1996
Orchidaceae	*Anacamptis morio* (L.) R.M. Bateman, Pridgeon & M.W. Chase	Hayek 1996–2001
Orobanchaceae	*Orobanche caryophyllacea* Sm.	Hayek 1996–2001
Osmundaceae	*Osmunda regalis* L.	MoA/UNEP/GEF 1996
Oxalidaceae	*Oxalis pes-caprae* L.	Deeb et al. 2013
Papaveraceae	*Fumaria macrorarpa* Parl.	MoA/UNEP/GEF 1996
Papaveraceae	*Fumaria officinalis* L.	El Beyrouthy et al. 2008
Papaveraceae	*Glaucium flavum* Crantz	MoA/UNEP/GEF 1996
Papaveraceae	*Papaver rhoeas* L.	MoA/UNEP/GEF 1996
Papaveraceae	*Papaver somniferum* L.	MoA/UNEP/GEF 1996
Periplocaceae	*Periploca graeca* L.	El Beyrouthy et al. 2008
Phytolaccaceae	*Phytolacca pruinosa* Fenzl.	Hayek 1996–2001
Pinaceae	*Abies cilicica* Ant. Et Ky.	El Beyrouthy et al. 2008
Pinaceae	*Cedrus libani* A. Rich.	Abou Chaar 2004

Table 6.1. contd....

Table 6.1. contd.

Family	Species Record	References
Pinaceae	*Pinus brutia* Ten.	El Beyrouthy et al. 2008
Pinaceae	*Pinus halepensis* Mill.	Abou Chaar 2004
Plantaginaceae	*Plantago major* L.	Deeb et al. 2013, MoA/UNEP/ GEF 1996
Plantaginaceae	*Plantago afra* L.	MoA/UNEP/GEF 1996
Plantaginaceae	*Plantago cretica* L.	Abou Chaar 2004
Plantaginaceae	*Plantago lanceolata* L.	Abou Chaar 2004, MoA/UNEP/ GEF 1996
Platanaceae	*Platanus orientalis* L.	MoA/UNEP/GEF 1996
Poaceae	*Arundo donax* L.	MoA/UNEP/GEF 1996
Poaceae	*Avena sterilis* L.	Deeb et al. 2013
Poaceae	*Cynodon dactylon* (L.) Pers.	Hayek 1996–2001
Poaceae	*Elymus repens* (L.) Gould	Deeb et al. 2013, MoA/UNEP/ GEF 1996
Poaceae	*Hyparrhenia hirta* (L.) Stapf.	El Beyrouthy et al. 2008
Poaceae	*Lolium perenne* L.	El Beyrouthy et al. 2008
Poaceae	*Lolium temulentum* L.	El Beyrouthy et al. 2008
Poaceae	*Phragmites australis* (Cav.) Trin. ex Steud.	El Beyrouthy et al. 2008
Poaceae	*Secale cereale* L.	MoA/UNEP/GEF 1996
Polygonaceae	*Atriplex halimus* L.	El Beyrouthy et al. 2008
Polygonaceae	*Polygonum arenastrum* Boreau	Hayek 1996–2001
Polygonaceae	*Polygonum aviculare* L.	Abou Chaar 2004, MoA/UNEP/ GEF 1996
Polygonaceae	*Rheum ribes* L.	Abou Chaar 2004, Deeb et al. 2013, MoA/UNEP/GEF 1996
Polygonaceae	*Rumex acetosella* L.	Hayek 1996–2001
Polygonaceae	*Rumex angustifolius* Campd.	Hayek 1996–2001
Polygonaceae	*Rumex bucephalophorus* L.	Hayek 1996–2001
Polygonaceae	*Rumex chalepensis* Mill.	Hayek 1996–2001
Polygonaceae	*Rumex occultans* Sam. Unresolved	Hayek 1996–2001
Polypodiaceae	*Polypodium australe* Fee	MoA/UNEP/GEF 1996
Portulacaceae	*Portulaca oleracea* L.	MoA/UNEP/GEF 1996
Primulaceae	*Anagallis arvensis* L.	MoA/UNEP/GEF 1996
Primulaceae	*Cyclamen libanoticum* Hildebr.	Hayek 1996–2001
Primulaceae	*Cyclamen persicum* Mill.	El Beyrouthy et al. 2008
Primulaceae	*Primula vulgaris* Huds	MoA/UNEP/GEF 1996
Pteridaceae	*Adiantum capillus-veneris* L.	Deeb et al. 2013, MoA/UNEP/ GEF 1996

Table 6.1. contd....

Table 6.1. contd.

Family	Species Record	References
Ranunculaceae	*Adonis aestivalis* L.	Abou Chaar 2004, MoA/UNEP/GEF 1996
Ranunculaceae	*Adonis annua* L.	Hayek 1996–2001
Ranunculaceae	*Anemone coronaria* L.	MoA/UNEP/GEF 1996
Ranunculaceae	*Clematis flammula* L.	El Beyrouthy et al. 2008
Ranunculaceae	*Clematis vitalba* L.	MoA/UNEP/GEF 1996
Ranunculaceae	*Delphinium staphisagria* L.	Abou Chaar 2004, MoA/UNEP/GEF 1996
Ranunculaceae	*Ficaria verna* subsp. *ficariiformis* (Rouy & Foucaud) B. Walln.	El Beyrouthy et al. 2008
Ranunculaceae	*Nigella sativa* L.	Hayek 1996–2001
Ranunculaceae	*Ranunculus arvensis* L.	El Beyrouthy et al. 2008
Ranunculaceae	*Ranunculus constantinopolitanus* (D.C.) d'Urv.	Fostok et al. 2009
Rhamnaceae	*Rhamnus alaternus* L.	Abou Chaar 2004
Rhamnaceae	*Rhamnus cathartica* L.	MoA/UNEP/GEF 1996
Rhamnaceae	*Ziziphus jujuba* Mill.	Abou Chaar 2004
Rosaceae	*Crataegus monogyna* Jacq.	Abou Chaar 2004, MoA/UNEP/GEF 1996
Rosaceae	*Crataegus rhipidophylla* Gand.	Abou Chaar 2004
Rosaceae	*Geum urbanum* L.	MoA/UNEP/GEF 1996
Rosaceae	*Neurada procumbens* L.	Abou Chaar 2004
Rosaceae	*Potentilla anserina* L.	MoA/UNEP/GEF 1996
Rosaceae	*Potentilla reptans* L.	MoA/UNEP/GEF 1996
Rosaceae	*Prunus avium* (L.) L.	Abou Chaar 2004
Rosaceae	*Rosa canina* L.	MoA/UNEP/GEF 1996
Rosaceae	*Rosa* × *damascena* Herrm.	El Beyrouthy et al. 2008
Rosaceae	*Rubus sanctus* Schneb.	Abou Chaar 2004, MoA/UNEP/GEF 1996
Rosaceae	*Rubus ulmifolius* Schott	Abou Chaar 2004
Rosaceae	*Rubus canescens* D.C.	Deeb et al. 2013
Rosaceae	*Rubus hedycarpus* Focke	Hayek 1996–2001
Rosaceae	*Rubus praecox* Bertol.	Abou Chaar 2004
Rosaceae	*Sanguisorba verrucosa* (Link ex G. Don) Ces.	Abou Chaar 2004, MoA/UNEP/GEF 1996
Rosaceae	*Sarcopoterium spinosum* (L.) Spach	Abou Chaar 2004, Deeb et al. 2013
Rubiaceae	*Galium aparine* L.	MoA/UNEP/GEF 1996
Rubiaceae	*Galium verum* L.	MoA/UNEP/GEF 1996

Table 6.1. contd....

Table 6.1. contd.

Family	Species Record	References
Rubiaceae	*Rubia tinctorum* L.	MoA/UNEP/GEF 1996
Rutaceae	*Citrus aurantium* L.	Abou Chaar 2004
Rutaceae	*Ruta chalepensisbracteosa* (D.C.) Boiss.	Abou Chaar 2004, Deeb et al. 2013
Rutaceae	*Ruta graveolens* L.	Abou Chaar 2004
Salicaceae	*Populus nigra* L.	MoA/UNEP/GEF 1996
Salicaceae	*Salix alba* L. subsp. *micans* (Andersson) Rech.f.	Abou Chaar 2004, Deeb et al. 2013, MoA/UNEP/GEF 1996
Salicaceae	*Salix fragilis* L.	Abou Chaar 2004
Salicaceae	*Salix pedicellaris* Pursh	Abou Chaar 2004
Salicaceae	*Salix pedicellata* Desf.	El Beyrouthy et al. 2008
Saxifragaceae	*Saxifraga tridactylites* L.	Hayek 1996–2001
Scrophulariaceae	*Antirrhinum majus* L., var. *angustifolium* Chav.	MoA/UNEP/GEF 1996
Scrophulariaceae	*Digitalis ferruginea* L.	Abou Chaar 2004, MoA/UNEP/GEF 1996
Scrophulariaceae	*Linaria cymbalaria* (L.) Mill.	MoA/UNEP/GEF 1996
Scrophulariaceae	*Verbascum gaillardotii* Boiss.	El Beyrouthy et al. 2008
Scrophulariaceae	*Verbascum tripolitanum* Boiss.	Hayek 1996–2001
Simarubiaceae	*Ailanthus altissimus* (Mill.) Swingle	MoA/UNEP/GEF 1996
Smilacaceae	*Smilax aspera* L.	MoA/UNEP/GEF 1996
Solanaceae	*Datura metel* L.	Abou Chaar 2004
Solanaceae	*Datura stramonium* L.	Abou Chaar 2004, Deeb et al. 2013, MoA/UNEP/GEF 1996
Solanaceae	*Hyoscyamus albus* L.	Abou Chaar 2004
Solanaceae	*Hyoscyamus aureus* L.	Abou Chaar 2004, MoA/UNEP/GEF 1996
Solanaceae	*Hyoscyamus niger* L.	Abou Chaar 2004
Solanaceae	*Hyoscyamus reticulatus* L.	Abou Chaar 2004
Solanaceae	*Lycium barbarum* L.	Deeb et al. 2013
Solanaceae	*Mandragora officinalis* Mill.	Hayek 1996–2001
Solanaceae	*Nicotiana tabacum* L.	MoA/UNEP/GEF 1996
Solanaceae	*Solanum dulcamara* L.	MoA/UNEP/GEF 1996
Solanaceae	*Solanum nigrum* L.	MoA/UNEP/GEF 1996
Solanaceae	*Withania somnifera* (L.) Dunal	El Beyrouthy et al. 2008
Styracaceae	*Styrax officinalis* L.	Hayek 1996–2001
Tamaricaceae	*Tamarix tetrandra* Pall. ex M. Bieb.	MoA/UNEP/GEF 1996
Thymelaeaceae	*Daphne libanotica* Mouterde	Hayek 1996–2001
Ulmaceae	*Celtis australis* L.	MoA/UNEP/GEF 1996

Table 6.1. contd....

Table 6.1. contd.

Family	Species Record	References
Ulmaceae	*Ulmus minor* Mill.	MoA/UNEP/GEF 1996
Urticaceae	*Parietaria judaica* L.	Abou Chaar 2004
Urticaceae	*Parietaria officinalis* L.	Abou Chaar 2004
Urticaceae	*Urtica dioica* L.	Deeb et al. 2013, MoA/UNEP/ GEF 1996
Urticaceae	*Urtica urens* L.	MoA/UNEP/GEF 1996
Verbenaceae	*Verbena officinalis* L.	Abou Chaar 2004, MoA/UNEP/ GEF 1996
Verbenaceae	*Vitex agnus-castus* L.	Abou Chaar 2004, Deeb et al. 2013
Violaceae	*Viola odorata* L.	MoA/UNEP/GEF 1996
Vitaceae	*Vitis vinifera* L.	Abou Chaar 2004
Zygophyllaceae	*Peganum harmala* L.	Abou Chaar 2004, Deeb et al. 2013, MoA/UNEP/GEF 1996

Scientific plant names of species and the authority of the name were according to:
The Plant List. 2013. Version 1.1. Published on the Internet; http://www.theplantlist.org/ (accessed May 2014).

Developing the sector of MAP in Lebanon is hindered by several limitations and needs. As it stands, most of the marketable MAP are collected from their ever-diminishing wild habitats. This exerts more selective pressure on the wild populations of the Lebanese MAP that have a documented medical value (i.e., MAP classified above under category 'A'). The unsustainable exploitation of these MAP from their natural resources increases the risk of degradation of their habitats and extinction in a small country like Lebanon. Therefore, there is urgent need to protect such habitats and adopt modes of use and cultivation of such plants in a sustainable manner. Alleviating pressure on wild populations of MAP has to be achieved through adopting legislations that prevent their collection beyond the levels required to maintain a sustainable resource of the respective MAP (Schippmann et al. 2006). However, this is proving to be increasingly difficult in Lebanon because of all the political turmoil that has plagued the country for so long disrupting many important legislative processes which leave MAP related legislations on the back-burner. As a result, many collectors interested in making a quick profit are depleting wild MAP resources at unprecedented rates. In contrast, older generations of the rural communities that depended on such resources for their wellbeing and who were closely associated with the environment throughout most of their lives clearly understood the consequences of over-harvesting and rarely did they abuse it.

Today in Lebanon only few MAP are managed through adequate regulations (see below), while cultivation efforts of some MAP is still limited despite a promising potential of revenue for the farmers.

Potentials and needs

The cultivation of MAP is needed mainly because it alleviates pressure on natural resources and reduces the demand on imports of MAP to meet market needs. Cultivation of MAP can provide a significant source of income particularly for marginalized farmers in fragile remote environments. In these degraded yet resilient habitats, MAP, like other native plants that are naturally adapted to poor soils and low rainfall, comprise a main component of conservation and low input agriculture and offer an opportunity to promote stability and health of agro-ecosystems. However, the economic return of MAP cultivation would be relatively low unless the cultivation is coupled to enhanced marketing strategies that would increase the economic return for local rural communities (ESCWA 2010). The cultivation of certain MAP species in Lebanon is an old tradition. The main MAP species that are cultivated at a commercial scale are *Rosa* x *damascena* Herrm. mainly in the Bekaa region, *Citrus aurantium* L. mainly in the coastal areas, *Rhus coriaria* L. in medium elevated mountain areas, *Coriandrum sativum* L. and *Satureja hortensis* L. in the coastal and interior plains. It is worth noting that such traditional cultivation practices are suffering from chronic neglect at the national level and lack government support as far as providing potential marketing strategies for the harvest and by-products. Furthermore, many MAP species are traditionally cultivated for household use (Malychef 1989) such as *Pelargonium odoratissimum* L., *Origanum majorana* L., *Alcea* sp., *Mentha* sp., *Petroselinum sativum*, *Foeniculum vulgare* Mill., *Ocimum basilicum* L., *Rosmarinus officinalis* L., *Lavandula* sp., *Salvia fruticosa* Mill., *Laurus nobilis* L., *Aloysia citriodora* Palau, *Celtis australis* L., *Sambucus nigra* L., *Cymbopogon citratus* (DC.) Stapf and *Matricaria chamomilla* L. among others. This traditional cultivation for household use still thrives in rural areas while in urban and peri-urban areas it is regressing.

To avoid depletion of natural resources of MAP that are not traditionally cultivated, domestication efforts are needed to certain select MAP. Selection of candidate MAP should identify threatened ones commonly used in folk medicine and at the same time that can be traded as a viable alternative crop to traditional agricultural crops. Domestication efforts have to generate adaptable agricultural practices that are species-specific that then need to be disseminated to farmers. In Lebanon, information on the domestication of medicinal plants is scattered and beyond the reach of farmers' community. Better efforts are needed to ensure effective communication between the researchers and the farming community. The domestication of *Origanum syriacum* L. (commonly known as Zaatar[1]) is a recent example of successful domestication of MAP and the subsequent communication of such a domestication protocol to farmers in rural communities (ESCWA 2010). Zaatar cultivation is mainly localized in South Lebanon in areas where limited water resources are found and where farmers' livelihood is highly dependent on agricultural revenues generated through tobacco mono-cropping. The cultivation and domestication efforts of zaatar started mainly after the year 2000. Joint efforts of development projects by local NGOs and international organizations allowed the establishment and dissemination of a successful method for the cultivation of zaatar in the area and in other remote rural areas in Lebanon (i.e., Akkar in the North Lebanon and the northern parts of the Bekaa Valley). Nurseries for the production of zaatar seedlings were established assuring the production of local seedlings mainly from seeds of wild or cultivated origin and cuttings. Despite the above, the total cultivated area is estimated not to exceed the 20 ha. It is worth noting that efforts to promote the cultivation for *Origanum syriacum* L., the bulk of the trade (up to 90%) nevertheless remains based on collection from the wild. In fact, wild MAP continue to be the most favored over cultivated MAP due to the widely-held perception that MAP grown in the wild have better medicinal value and taste better.

Other wild MAP species can be targeted with similar initiatives such as *Thymbra spicata* L. Another less promising crop is *Salvia fruticosa* Mill. that appeared to be sensitive to faulty irrigation practices which has led to damaged plants and loss of yield. This MAP in particular holds a promising market value and more efforts are needed to ensure its successful cultivation.

Policy, legislation and meeting market needs

In Lebanon, the Ministry of Agriculture is the sole government agency responsible for natural herbal resources. The organization chart of the Ministry of Agriculture issued in 1994 states that the Directorate of Rural Development and Natural Resources, the Department of Forestry and Natural Resources and the Department of Pastures and Parks are in charge to protect and develop medicinal plants by regulating and monitoring their exploitation from the wild. Despite what appears to be a successful organizational

---

[1] Zaatar designates a variety of aromatic perennial plants typically harvested from the wild, which generally belong to the Origanum, and other genera of the Lamiaceae (mint) family. The term zaatar is also used to designate the different herb mixes. Zaatar, whether fresh or dried, is a common item in Mediterranean diets and is witnessing burgeoning consumer demand (ESCWA 2010).

government set up for regulating MAP sector, the main legislative package regarding plant natural resources in Lebanon is the forest code issued in 1949. It organizes the forest exploitation for wood, pasture and other uses. However in as far as MAP are concerned, there is only one existing legislation that addresses the harvesting and exportation of few MAP such as *Origanum syriacum* L. and *Salvia fruticosa* Mill. (Ministerial Decision 340/1 issued in 1996 and updated in 2012 as per ministerial decision 179/1). More decrees are needed to govern a wider range of used and exploited MAP in Lebanon.

Harvesting of wild MAP in Lebanon is carried out for two different groups of users. Many MAP are collected for household use in rural areas where they are utilized as wild herbs in traditional dishes or consumed for folk medicinal purposes. Examples of this would be *Origanum* sp., *Thymus* sp., *Thymbra* sp., *Satureja* sp., *Cyclotrichium* sp., *Salvia* sp., *Micromeria* sp., *Matricaria* sp., *Asparagus* sp., *Fibigia* sp., *Centaurium* sp., *Hypericum* sp., *Althaea* sp., *Alcea* sp., *Crataegus* sp., and *Equisetum* sp. among others. Any marketing done at this level is restricted to the local populace trading within their own community or in niche markets. In contrast, for certain MAP species harvesting is done largely for commercial trading. The traders hire local hands mostly women and children from poor rural communities to do the harvesting. The main MAP traded in this fashion are *Origanum syriacum* L., *Salvia fruticosa* Mill., *Thymbra spicata* L., and *Satureja thymbra* L. whereby they are sold in local urban markets or transported to regional markets in neighboring countries. The existing competition between, and within the two groups, in seeking new sites of MAP for harvesting from the wild, is causing tremendous pressure on wild MAP populations. This is further compounded with the lack of sustainable harvesting regulations (i.e., avoiding early harvest before seed set, avoiding repeated harvests from same site and the harvest of the entire populations of MAP without taking into account regeneration potential for upcoming seasons).

It is worth noting that most collectors have direct access to the local market and often sell their harvest in major cities directly to the consumer. In other cases, collectors sell their harvest to wholesale markets or to middle sales persons while others supply traditional medicine practitioners and folk pharmacies upon request. The marketed MAP material include mainly packaged dried herbs, bottled distilled water extracts, alcohol extracts, essential oils, creams and ointments. The herbal tea (tisane, infusion) are the most demanded MAP end products. Moreover, the marketing of bottled essential oils extracts such as orange blossom water (*Citrus aurantium* L.) and rose water (*Rosa* × *damascena* Herrm.) are quite common and are produced in rural areas by local producers and by local industry as well. Most marketed MAP species in Lebanon are *Salvia fruticosa* Mill., *Origanum syriacum* L., *Thymus syriacus* Boiss., *Thymbra spicata* L., *Satureja thymbra* L., *Rhus coriaria* L., *Nigella sativa* L., *Alcea* sp., *Micromeria* sp., *Rosa* sp., *Coriandrum sativum* L., *Equisetum* sp., *Crataegus* sp., *Matricaria* sp., and *Mentha* sp. (Deeb et al. 2013).

Information and literature on the trade of Lebanese MAP is scarce and data is scattered at the national level. The estimated annual market value of medicinal and aromatic plants in Lebanon was US$18,600,000 (MoA/UNEP/GEF 1996). It increased up to US$29,600,000/year, based on 2004 figures (Ministry of Agriculture). The estimation of the annual production from the wild is 600–800 tons for *Salvia fruticosa* Mill. and 2000 tons for *Origanum syriacum* L. (UNDP 2013).

According to UNDP (2013), mills are most common processors of MAP and related spices numbering more than 140 in a country as small as Lebanon. Around 1/3 of the mills are considered as middle to large sized operations. Most of the large mills produce packaged and branded end products of herbs and MAP-related spices. Pharmaceuticals and cosmetics industries, health shops, Attareen, herbalists and retailers such as supermarkets and groceries constitute the market outlet of processed MAP. The main target species are *Origanum syriacum* L., *Salvia fruticosa* Mill., *Foeniculum vulgare* Mill., *Alcea* sp., *Althaea* sp., *Rheum ribes* L., *Micromeria* sp., *Malva* sp., *Ferula hermonis* Boiss., *Crataegus* sp., *Teucrium polium* L., *Lavandula* sp., *Rosa* sp. and *Matricaria chamomilla* L.

Research in the MAP sector

The research arm in Lebanon is spearheaded by few universities and research centers associated with the National Council for Scientific Research (CNRS-L). The Research and Development (R&D) arm that exists in the industries of developed countries is lacking in Lebanon. More so, and despite the continued efforts of CNRS-L, there are limited links and coordination between research centers and universities with interest in MAP-related fields on one hand and the needs of the industry and the production chain on the other hand. In Lebanon there exist more than 45 universities but not all are involved in related research. The main universities with noted research and graduate programs are the American University of Beirut (AUB), Beirut Arab University (BAU), Holy Spirit University of Kaslik (USEK), Lebanese American University (LAU), Lebanese University (LU), Saint Joseph University (USJ) and University of Balamand (UOB). National funded research is scarce and limited to what CNRS provides on competitive basis, and to research funds from the respective universities. Other sources of research funds are from EU and other international agencies, and these latter constitute the majority of research funding where Lebanon is one of several regional and international partners. Lebanon's role typically in such funded programs is limited to contributing as a partner to a research question that has regional or wider impact which at times fails to address the local needs of the industry, and in this case the industry of MAP. Although the CNRS-L issues a national agenda with priority areas of research that frequently includes health issues and other related domains that support MAP-related research, there is a lack of a comprehensive national agenda that addresses the issue of use of MAP, validation of folk medicinal practices, cultivation and sustainable use of wild habitats. The above and the lack of appreciable collaborations between industry, academia, and government have led to a deregulated growth of the MAP market. In other words, national priorities are not guided by market needs or industrial needs, and the industry itself is disconnected from academia and is not committed to developing R&D arms. All of this leaves researchers in academia and research centers developing their own research agendas that rarely lead to marketable products or drug discovery.

In reviewing recent research activities in Lebanon one notes substantial growth in MAP research. In fact close to 200 studies including technical extension booklets, thesis, book chapters, reviews and research manuscripts were done on MAP in Lebanon in the last 15 years. Over 85 MAP species were targeted in one or more of the above publications.

The targeted species are distributed over 22 botanical families out of the 90 botanical families of MAP in Lebanon and 56 genera out of the 260 genera assessed. However, only three species were addressed in more than 10 studies, namely *Nigella sativa* L., *Salvia fruticosa* Mill. and *Origanum syriacum* L.

The main research themes are those concerned in detection of bioactivities, chemical profiling and characterization of MAP. The majority of these studies are conducted on crude extracts of MAP. Amongst those bioactivities, anti-microbial and anti-tumor are predominant and constitute about 34% each of the total bioactivities screened for in MAP, while other bioactivities such as ethno-botany and anti-inflammatory constitute about 20 and 12% respectively. On the other hand chemical profiling of MAP, and mostly essential oils, are also a major undertaking by MAP researchers in Lebanon. Of the literature published in the past 15 years on chemical profiling of MAP, 40% of the studies addressed essential oil profiles (Senatore et al. 2005, Rigano et al. 2006, Mancini et al. 2009a,b, Najem et al. 2011, El Beyrouthy et al. 2013, Khoury et al. 2014) whereas tannins (20%), sesquiterpene lactones (14%), quinones (12%) and others including diterpeniod, flavonoids and fatty acids (13%) were the other constituents addressed in the profiling studies. Interestingly, and in recent years several studies isolated MAP-derived compounds and identified their mechanism of action in cell culture and animal models. While such studies hold promise for potential drug discovery and the chance to secure funds from regional or international pharmaceuticals, the cost and inability of research centers and academic institutes in Lebanon to hold pre-clinical animal Absorption, Distribution, Metabolism, and Excretion (ADME) studies renders the chance of funding from international pharmaceutical industry bleak, but possible. Later some of the recently identified pure compounds from Lebanese MAP and their mechanistic anti-tumor and anti-inflammatory activities are discussed.

Salograviolide A from *Centaurea ainetensis* Boiss. (Synonym of *Centaurea eryngioides* Lam.)

Centaurea ainetensis Boiss. is a Lebanese endemic plant belonging to the Asteraceae family. It grows in stony areas. This plant has been used in Lebanese folk medicine as an anti-fungal and anti-inflammatory MAP. Extensive *in vivo* and *in vitro* studies were done on crude extracts of this plant and demonstrated anti-inflammatory and anti-cancer properties (El-Najjar et al. 2007, Ghantous et al. 2008, Talhouk et al. 2008, Al-Saghir et al. 2009). These properties are mainly attributed to the secondary metabolites of the plant with the most common and highly investigated ones being sesquiterpene lactones (Ghantous et al. 2008, Saliba et al. 2009, Saikali et al. 2012). Salograviolide A (SA), a sesquiterpene lactones guaianolide is one of the main compounds in *C. ainetensis* Boiss. contributing to its medicinal value. The anti-inflammatory effect of the crude extract was proven to be mediated by Sal A which was shown to inhibit IL-6 in endotoxin treated mammary cells (Saliba et al. 2009). This compound also plays a major role in inhibiting tumor growth of different cell types including PAM212 skin cancer cells, neoplastic epidermal cells and HCT-116 and HT-29 human colon cancer cells. Sal A selectively inhibited growth of tumor cells, as well as the tumor promoter-induced proliferation and transformation of cancer cells, PAM212 in particular. The modulation of key genes involved in signaling pathways, at the transcriptional level, such as MMP-9, Bax, Bcl2, CDK1-p16 levels, can also be attributed to the activity of Salograviolide A. This activity was shown to take place during the reversible stage of tumor promotion and cell transformation, this compound is considered to be an attractive target for anti-cancer drugs (Saikali et al. 2012). Taken together, these studies suggest that *C. ainetensis* Boiss. in general and Salograviolide A, in specific, are potential candidates worth further investigation for possible uses in drug industry as anti-inflammatory and anti-tumor medicine.

Essential oils from *Salvia libanotica* (*S. fruticosa* Mill.)

Salvia libanotica, the Lebanese sage, is part of the mint family, *Lamiaceae*. It is a medicinal plant native to the Mediterranean region (Itani et al. 2008). Water extract of this plant has been widely used in folk medicine. Farhat et al. (2001) suggested that it is crucial to take caution while preparing extracts for treatment purposes since the plant extract might be toxic if collected during winter rather than the spring season depending on the essential oils composition, especially camphor, α- and β-thujone as well as camphene essential oils which differ between seasons. The growth inhibitory anti-tumor effect is the result of a synergistic effect of Linalyl acetate (Ly) and alpha-terpineol (Te). These compounds induced apoptosis and necrosis by suppressing NF-KB activity. They were shown to specifically down regulate the protein expression level of Bcl-2 and Bcl-x2, which are TNF-α-induced-NF-KB-dependent gene products. Ly and Te also inhibited TNF-α-induced-nuclear translocation of p65 by stabilizing the IKB/NF-KB complex (Deeb et al. 2011). In the absence of p53 the mechanism was caspase-independent, despite the presence of caspase-dependent features; DNA fragmentation and PARP cleavage is mainly induced by the translocation of AIF from the mitochondria to the nucleus, leading to cell death (Itani et al. 2008). These findings are of importance because they might help in improving the treatment of cancer patients. Ly and Te were shown to enhance apoptosis induced by chemotherapeutic agents, such as oxaliplatin and 5-FU, which means that combining them, at lower concentrations, with these agents helps in limiting their side effects (Deeb et al. 2011).

Thymoquinone from *Nigella sativa* L.

Nigella sativa L., known as the black seed, is an annual herb belonging to the family *Ranunculaceae*. It grows in the Middle East, western Asia and eastern Africa and is used in these regions as traditional medicine for the treatment of headaches, coughs, abdominal pain, diarrhea, asthma and many other symptoms (Gali-Muhtasib et al. 2004). Thymoquinone (TQ), a short-chain ubiquinone derivative, is known to be the major component of the plant responsible for its anti-oxidant, anti-inflammatory and anti-tumor effects (Gali-Muhtasib et al. 2008, El-Najjar et al. 2009). Thymoquinone was shown to have anti-tumor effects that were tested in many different cell types such as human pancreatic adenocarcinoma, uterine sarcoma, Ehrlich ascites carcinoma, Dalton's ascites lymphoma, human breast and ovarian adenocarcinoma, myeloblastic leukaemia, HL-60, squamous carcinoma, SCC VII, fibrosarcoma, FSSaR, laryngeal neoplastic cells-Hep-2, and prostate and pancreatic cell lines (Dergarabetian et al. 2013, Gali-Muhtasib et al. 2004). Many studies have been published on the effect of TQ on different colon cancer cell lines. Gali-Muhtasib et al. (2004, 2008) studied the anti-tumor activity of TQ on HCT-116 human colorectal cancer cells. The compound was shown to inhibit cell proliferation by inducing cell arrest at G1/S phase. Moreover, apoptosis was induced, both in cultured cells and xenografts introduced in mice, via a p-53 dependent mechanism. This mechanism was mediated through two possible pathways; down regulation of Bcl-2 protein levels and activation of the target gene P21[WAFI], the latter preventing the entry of cells into the S phase, or suppression of CHECK-1, DNA damage sensor, by interacting with its promoter thus enhancing apoptosis. Interestingly, El-Najjar et al. (2009) reported that the effect of TQ was shown to differ between different cell lines. By comparing two human colon colorectal adenocarcinoma cell lines, DLD-1 and HT-29, it was evident that although inhibition of proliferation took place in both, apoptosis induction was restricted to DLD-1 only. However, both of these effects were governed by the same cause: generation of ROS by TQ leading to oxidative stress. HT-29 cells were resistant to apoptosis since they possess DT-diaphorase reductase enzyme. The anti-tumor activity of thymoquinone was also shown to take place *in vivo*. Upon the administration of TQ to mice suffering from DMH-induced colorectal carcinoma, a reduction in the Aberrant Crypt Foci (ACF) and tumor multiplicity was recorded, as well as inhibition of tumor invasion (Gali-Muhtasib et al. 2008). Hence, thymoquinone extracted from *N. sativa* L. seems to be a promising compound to be investigated as a potential contributor to anti-tumor therapy.

Sesquiterpene lactones from *Achillea falcata* L.

Achillea falcata L., belonging to the *Astreraceae* family, is widely used in folk medicine, by people of its indigenous Middle eastern region (Tohme et al. 2013), for the treatment of various diseases and symptoms such as inflammation, infections, headaches and cancer (Saikali et al. 2012). Medicinal properties of this plant are attributed to specific compounds known as Sesquiterpene Lactones (SL); secondary metabolites found in other plants of the same family as well (Tohme et al. 2013). Sesquiterpene lactones can be purified from *A. falcata* L. to study their effects. Four compounds were purified by Ghantous et al. (2009): 3-β-methoxy-iso-seco-tanapartholide (β-tan), tanaphillin, isosecotanapartholide, and 8-hydroxy-3-methoxyisosecotanapartholide and their effect on HaCat cancer cells were studied. Results showed a correlation between the compound chemical structure and its potency. Molecules with higher lipophilicity, such as the first compound, are able to cross the plasma membrane more easily, thus having a greater effect in inhibiting growth.

Saikali et al. (2012) conducted a study to test for anti-tumor properties of a specific SL: β-tan. Using mouse epidermal cells, β-tan was shown to have a selective dose-dependent growth inhibition of PAM212 tumor cells. In addition, tumor promoter-induced proliferation and transformation were inhibited in TPA-induced JB6P+ cells. This inhibition was attributed to β-tan modulation of TPA-induced NF-KB activity by targeting key genes involved in its signaling pathway, thus affecting cell growth, migration and metastasis. The SL-induced decrease in MMP-9 levels, up regulation of p16, increase in Bax:Bcl-2 ratio and restore of IKBα levels, all together, contributed to a decrease in NF-KB transcriptional activity and thus inhibition of cell cycle progression and tumor promotion.

Another recent study by Tohme et al. (2013) showed that in addition to β-tan, the three other SL extracted from *A. falcata* L. showed to have anti-tumor activity on HCT-116 colorectal cancer cells. The purified molecules were shown to decrease the growth of HCT-116 cells.

Fatty acids from *Ranunculus constantinopolitanus* (D.C.) d'Urv.

Ranunculus constantinopolitanus (D.C.) d'Urv. is a medicinal plant belonging to the *Ranunculaceae* family. Due to its claimed medicinal value of anti-inflammatory, analgesic, anti-viral, anti-bacterial, anti-parasitic and anti-fungal activities, it has been used in folk medicine at the eastern Mediterranean region. These properties are suggested to be due to the presence of specific compounds including anemonin and some unusual Fatty Acids (FAs).

The anti-inflammatory effect of *Ranunculus constantinopolitanus* was studied by Fostok et al. (2009) on SCp-2 cells. The anti-inflammatory activity of Y_{2+3} purified compound(s) was suggested to be due to a Conjugated Linoleic Acid (CLA) and was evident by decreasing the levels of ET-induced IL-6. Interestingly, this effect was more significant when compared to most of the commercially available CLA and omega-3 fatty acids, claimed to have anti-inflammatory activities among other. In addition, this anti-inflammatory effect is not restricted to SCp-2 cells also but was also detected in Mode-K cells through a decrease in the IL-1 induced COX-2 levels.

Conclusion

The lack of national regulations and standards that govern the use of most of the harvested and traded MAP species and the lack of established protocols for Lebanese farmers to cultivate wild MAP with supporting guarantee of authenticity of active components for the purpose of safe and effective marketing have plagued the industry. In addition the absence of any notable collaboration between industry, academia and government suggests that growth of a regulated MAP sector may not be realized soon despite promising research outcomes concerned with production of cultivated MAP and with identifying bioactive components derived from indigenous plants. The economic potential of Lebanon's botanical diversity remains poorly investigated and does not contribute significantly to the agro-industrial and the pharmaceutical sector of the country even though extensive harvests of plant species with medicinal, culinary, and ornamental uses are ongoing.

Acknowledgements

The authors are grateful to Dana Bazzoun and Farah Nassar for critical reading of the manuscript and to Rana Nahhas for her help in preparation of this chapter. This work is supported by the Lebanese Council for Scientific Research (CNRS-L), the American University of Beirut-University Research Board (AUB-URB), and the Lebanese Agricultural Research Institute (LARI).

References

Abou Chaar, A.I. 2004. Medicinal plants of Lebanon. Archeology and History in Lebanon. 19: 70–85.

Akil, M. 1996. Diseases and remedies in ancient medicine (Arabic). Beirut, Lebanon: Dar al Mahajjat al Baydaa, 195 p.

Al Aantaki, D. 1877. Tazkirat Daoud Al-Antaki of the 16th century. Printed in Cairo.

Al-Saghir, J., Al-Ashi, R., Salloum, R., Saliba, N.A., Talhouk, R.S. and Homaidan, F.R. 2009. Anti-Inflammatory properties of Salograviolide A purified from Lebanese plant *Centaurea ainetensis*. BMC Complement. Altern. Med. 9: 36.

Baalbaki, I. 2000. Plants a remedy for all diseases (Arabic). Beirut, Lebanon: Dar al Sadah al Arabiah.

Bruno, M., Bondi, M.L., Piozzi, F., Arnold-Apostolides, N. and Simmonds, M.S. 2001. Occurrence of 18-hydroxyballonigrine in Ballota *saxatilis* ssp. *saxatilis* from Lebanon. Biochem. Syst. and Ecol. 29 (4): 429–431.

Cardile, V., Russo, A., Formisano, C., Rigano, D., Senatore, F., Arnold-Apostolides, N. and Piozzi, F. 2009. Essential oils of *Salvia bracteata* and *Salvia rubifolia* from Lebanon: chemical composition, antimicrobial activity and inhibitory effect on human melanoma cells. J. Ethnopharmacol. 126(2): 265–272.

Deeb, S.J., El-Baba, C.O., Hassan, S.B., Larsson, R.L. and Gali-Muhtasib, H.U. 2011. Sage components enhance cell death through nuclear factor kappa-B signaling. Frontiers in Bioscience 3: 410–420.

Deeb, T., Knio, K. and Shinwar, Z.K. 2013. Survey of medicinal plants currently used by herbalists in Lebanon. Pak. J. Bot. 45(2): 543–555.

Dergarabetian, E.M., Ghattass, K.I., El-Sitt, S.B., Al Mismar, R.M., El-Baba, C.O., Itani, W.S., Melhem, N.M., El-Hajj, H.A., Bazarbachi, A.A., Schneider-Stock, R. and Gali-Muhtasib, H.U. 2013. Thymoquinone induces apoptosis in malignant T-cells via generation of ROS. Frontiers in Bioscience (Elite Ed.) 5: 706–19.

El Beyrouthy, M., Arnold-Apostolides, N., Arnold, N., Annick Dusollier, A. and Frederic, D. 2008. Plants used as remedies antirheumatic and antineuralgic in the traditional medicine of Lebanon. Journal of Ethnopharmacology 120(3): 315–334.

El Beyrouthy, M., Arnold-Apostolides, N., Cazier, F., Najm, S., Abou Jaoudeh, C., Labaki, M., Dhifi, W. and Abou Kais, A. 2013. Chemical composition of the aerial parts of *Satureja thymbra* L. essential oil growing wild in Lebanon. Acta Horticulturae 997: 59–66.

El-Najjar, N., Saliba, N., Talhouk, S. and Gali-Muhtasib, H. 2007. *Onopordum cynarocephalum* induces apoptosis and protects against 1,2 dimethylhydrazine-induced colon cancer. Oncology Reports 17: 1517–1523.

El-Najjar, N., Chatila, M., Moukadem, H., Vuorela, H., Ocker, M., Gandesiri, M., Schneider-Stock, R. and Gali-Muhtasib, H.U. 2009. Reactive oxygen species mediate thymoquinone-induced apoptosis and activate ERK and JNK signaling. Apoptosis 15(2): 183–95.

ESCWA. 2010. Best Practices and Tools for Increasing Productivity and Competitiveness in the Production Sectors: Assessment of *Zaatar* Productivity and Competitiveness in Lebanon.

Farhat, G., Affara, N. and Gali-Muhtasib, H. 2001. Seasonal changes in the composition of the essential oil extract of East Mediterranean sage (*Salvia libanotica*) and its toxicity in mice. Toxicon 39: 1601–5.

Formisano, C., Mignola, E., Rigano, D., Senatore, F., Arnold-Apostolides, N., Bruno, M. and Rosselli, R. 2009. Constituents of leaves and flowers essential oils of *Helichrysum pallasii* (Spreng.) Ledeb. Growing wild in Lebanon. J. Med. Food. 12(1): 203–207.

Fostok, S.F., Ezzeddine, R.A., Homaidan, F.R., Al-Saghir, J.A., Salloum, R.G., Saliba, N.A. and Talhouk, R.S. 2009. Interleukin-6 and Cyclooxygenase-2 down regulation by fatty-acid fractions of *Ranunculus constantinopolitanus*. BMC Complement. Altern. Med. 9: 44.

Gali-Muhtasib, H.U., Diab-Assaf, M., Boltze, C., Al-Hmairai, J., Hartig, R., Roessner, A. and Schneider-Stock, R. 2004. Thymoquinone extracted from black seed triggers apoptotic cell death in human colorectal cancer cells via a p53-dependent mechanism. International Journal of Oncology 25: 857–866.

Gali-Muhtasib, H.U., Kuester, D., Mawrin, C., Bajbouj, K., Diestel, A., Ocker, M., Habold, C., Foltzer-Jourdainne, C., Schoenfeld, P., Peters, B., Diab-Assaf, M., Pommrich, U., Itani, W., Lippert, H., Roessner, A. and Schneider-Stock, R. 2008. Thymoquinone triggers inactivation of the stress response pathway sensor CHEK1 and contributes to apoptosis in colorectal cancer cells. Cancer Research 68(14): 5609–18.

Ghantous, A., Abou Tayyoun, A., Abou Lteif, G., Saliba, N.A., Gali-Muhtasib, H.U., El-Sabban, M. and Darwiche, N. 2008. Purified Salograviolide A isolated from *Centaurea ainetensis* causes growth inhibition and apoptosis in neoplastic epidermal cells. International Journal of Oncology 32: 841–849.

Ghantous, A., Nasser, N., Saab, I., Darwiche, N. and Saliba, N.A. 2009. Structure-activity relationship of seco-tanapartholides isolated from *Achillea falcata* for inhibition of HaCaT cell growth. European Journal of Medicinal Chemistry 44: 3794–3797.

Hayek, M. 1996–2001. Encyclopedia of Medicinal Plants (Arabic). Volumes I, II, III, IV and V. Library of Lebanon, Beirut.

Ibn Sina. 1025. Canon of Medicine. New York, NY: AMS Press, Inc.

Ismail, A. 2001. Easy Drug for the Cure of Diseases. Third edition, Beirut, Lebanon, 376 p.

Itani, W.S., El-Banna, S.H., Hassan, S.B., Larsson, R.L., Bazarbachi, A. and Gali-Muhtasib, H. 2008. Anti-colon cancer components from Lebanese sage (*Salvia libanotica*) essential oil. Cancer Biology & Therapy 7(11): 1765–1773.

Jabr, W. 1988. Healing with plants (Arabic). Beirut, Lebanon: Dar el Jil.

Kadamah, A. 2000. Dictionary of food and healing with plants (Arabic). Beirut, Lebanon: Dar el Nafaes.

Khoury, M., El Beyrouthy, M., Ouaini, N., Iriti, M., Eparvier, V. and Stien, D. 2014. Chemical composition and antimicrobial activity of the essential oil of *Juniperus excelsa* M.Bieb. Growing wild in Lebanon. Chemistry & Biodiversity 11: 825–830.

Malychef, P. 1989. Les plantes médicinales au Liban. The Lebanese Medicinal Journal 38: 59–60.

Mancini, E., Arnold-Apostolides, N., De Martino, L., De Feo, V., Formisano, C., Rigano, D. and Senatore, F. 2009a. Chemical Composition and Phytotoxic Effects of Essential Oils of *Salvia hierosolymitana* Boiss. and *Salvia multicaulis* Vahl. var. *simplicifolia* Boiss. Growing Wild in Lebanon. Molecules 14: 4725–4736.

Mancini, E., Arnold-Apostolides, N., De Feo, V., Formisano, C., Rigano, D., Piozzi, F. and Senatore, F. 2009b. Phytotoxic effects of essential oils of *Nepeta curviflora* Boiss., and *Nepeta nuda* L. subsp. albiflora growing wild in Lebanon. J. Plant Interact. 4(4): 253–259.

MoA/UNEP/GEF. 1996. Biodiversity country study. Coordinator Khouzami, M.

Mouterde, P. 1966. Nouvelle Flore du Liban et de la Syrie. Volume I. Dar El-Mashreq Editeurs, Beirut, Lebanon, pp. 565.

Mouterde, P. 1970. Nouvelle Flore du Liban et de la Syrie. Volume II. Dar El-Mashreq Editeurs, Beirut, Lebanon, pp. 727.

Mouterde, P. 1983. Nouvelle Flore du Liban et de la Syrie. Volume III. Dar El-Mashreq Editeurs, Beirut, Lebanon, pp. 578.

Najem, W., El Beyrouthy, M., Hanna, L., Neema, C. and Ouaini, N. 2011. Essential oil composition of *Rosa damascena* Mill. from different localities in Lebanon. Acta Botanica Gallica 158(3): 365–373.

Noun, J. 2007. The state of diversity. pp. 17–22. *In*: Lamis Chalak and Nada Sabra (eds.). Second Report on the State of Plant Genetic Resources for Food and Agriculture. FAO, Rome.

Post, G.E. 1932. Flora of Syria, Palestine and Sinai. Volumes I and II. American University of Beirut, Beirut, Lebanon.

Rigano, D., Arnold-Apostolides, N., Bruno, M., Formisano, C., Grassia, A., Piacente, S., Piozzi, F. and Senatore, F. 2006. Phenolic compounds of *Marrubium globosum* ssp. *libanoticum* from Lebanon. Biochemical Systematics and Ecology 34: 256–258.

Ruwaiha, A. 1981. Healing with plants (Arabic). Beirut, Lebanon: Dar AlKalam.

Saikali, M., Ghantous, A., Halawi, R., Talhouk, S.N., Saliba, N.A. and Darwiche, N. 2012. Sesquiterpene lactones isolated from indigenous Middle Eastern plants inhibit tumor promoter-induced transformation of JB6 cells. BMC Complement. Altern. Med. 12: 89.

Saliba, N.A., Dakdouki, S., Homaidan, F.R., Kogan, J., Bouhadir, K., Talhouk, S.N. and Talhouk, R.S. 2009. Bio-guided identification of an anti-inflammatory guaianolide from *Centaurea ainetensis*. Pharmaceutical Biology 47(8): 701–707.

Schippmann, U., Leaman, D. and Cunningham, A.B. 2006. A comparison of cultivation and wild collection of medicinal and aromatic plants under sustainability aspects. pp. 75–95. *In*: R.J. Bogers, L.E. Craker and D. Lange (eds.). Medicinal and Aromatic Plants, Springer, Dordrecht.

Senatore, F., Arnold-Apostolides, N. and Bruno, M. 2005. Volatile components of *Centaurea eryngioides* Lam. and *Centaurea iberica* Trev. var. *hermonis* Boiss. Lam., two Asteraceae growing wild in Lebanon. Natural Product Research 19(8): 749–754.

Talhouk, R.S., El-Jouni, W., Baalbaki, R., Gali-Muhtasib, H., Kogan, J. and Talhouk, R.S. 2008. Anti-inflammatory bio-activities in water extract of *Centaurea ainetensis*. Journal of Medicinal Plants Research 2, 2(2): 024–033.

Tohme, G. and Tohme, H. 2007. Illustrated Flora of Lebanon. CNRS Publication, Beirut, Lebanon.

Tohme, G. and Tohme, H. 2013. Illustrated Flora of Lebanon. Second edition. CNRS Publication, Beirut, Lebanon.

Tohme, R., Al Aaraj, L., Ghaddar, T., Gali-Muhtasib, H., Saliba, N.A. and Darwiche, N. 2013. Differential growth inhibitory effects of highly oxygenated guaianolides isolated from the Middle Eastern indigenous plant *Achillea falcata* in HCT-116 colorectal cancer cells. Molecules 18: 8275–8288.

UNDP. 2013. Final report of the project Mainstreaming Biodiversity Management into Medicinal an Aromatic Plants (MAPs) Production Processes in Lebanon.

World Health Organization (WHO). 2003. Guidelines on Good Agricultural and Collection Practices (GACP) for Medicinal Plants. Geneva.

Mutagenicity, Genotoxicity and Cytotoxicity Assays of Medicinal Plants: First Step for Drug Development

Silvany de S. Araújo, Thaís C.C. Fernandes,
Maria A. Marin-Morales, Ana C. Brasileiro-Vidal and
Ana M. Benko-Iseppon*

Introduction

Over the centuries, medicinal plants have been the basis for the treatment of different diseases, considered as the main therapeutic source for innumerous communities and ethnic groups (Borges et al. 2013). Furthermore, their great chemical diversity offers unlimited opportunities for drug discovery (Sasidharan et al. 2011).

The systematic search for new substances useful for therapeutic purposes can be accomplished through various processes. Most of them regard the synthesis of new molecules, molecular modification of natural or synthetic substances with defined pharmacological properties, as well as extraction, isolation and purification of novel compounds from medicinal plants. Research work on medicinal plants, as a rule, give rise to drugs in a shorter time, with lower costs, being therefore more accessible to the population (Brito and Brito 1993, Ugaz et al. 1994).

Despite the therapeutic benefits, it is known that medicinal plants and their derivatives may exhibit toxic properties, including carcinogenic and mutagenic nature or, still, can cause changes in the DNA. Such changes may affect vital processes as replication and gene transcription, sometimes also chromosomal alterations that lead to cancer and cell death processes (Belcavello et al. 2012, Ping et al. 2012). They may also cause chromatin modifications associated to epigenetic reprogramming, often leading to drastic consequences to the affected organism (Csokaa and Szyf 2009).

Due to increasing reports on adverse effects of such products, regulatory and supervisory agencies from many countries require prior fulfillment of safety tests for medicinal plants, including *in vitro* and

Universidade Federal de Pernambuco, Recife, 50.670-420, Recife, Brazil.
* Corresponding author: ana.iseppon@gmail.com

in vivo toxicology testing, regarding cytotoxicity, genotoxicity and mutagenicity (Nunes et al. 2012). These are the first tests for the evaluation of potential clinical application of a new material, due to its greater reproducibility, reduced experimental period and need of small amounts of target compounds.

Therefore, it is clearly important to evaluate the cytotoxic potential of medicinal plants and their compounds using *in vitro* and *in vivo* systems. Throughout this chapter the main tests and model organisms adopted for such evaluations including cytotoxic, mutagenic and genotoxic potential of medicinal plants for the development of drugs will be presented.

Test organisms: their importance and application in the investigation of toxic effects of medicinal plants

The test systems may be divided into groups based on the used biological system and in the genetic endpoint detected. Bioassays with prokaryotes allow the detection of agents that induce gene mutation and primary damage in the DNA. Moreover, analyses with eukaryotes allow the detection of a greater extent of damages, ranging from gene mutations to chromosomal damage and aneuploidy (Houk 1992). In this context, genotoxicity and toxicity tests using microorganisms, plant and mammalian cells have been used alone or in combination to obtain more reliable results for the pharmaceutical industry (Zegura et al. 2009).

Microorganisms

The development and standardization of toxicity tests based on prokaryotes (bacteria) and eukaryotes, such as protozoa, unicellular algae and yeast—instead of higher organisms—allowed a rapid and inexpensive screening of samples against toxic and genotoxic effects. The first generation of bioassays was based on diverse naturally sensitive microbes, whereas the second generation included genetically modified microorganisms to achieve better sensitivity and/or specificity. Most parameters measured by microbial toxicity trials are population growth, substrate consumption, respiration, ATP luminescence and bioluminescence inhibition (Logar and Vodovnik 2007).

Bacteria such as *Salmonella typhimurium*, *Escherichia coli* and *Bacillus subtilis* are used in the most widely applied methods for detection of gene mutations. The rapid cell division and the relative ease with which large amounts of data can be generated (approximately 10^8 bacteria per test pate) turned bacterial assays to be the most widely used in routine tests for mutagenicity of chemical compounds.

The mutagenicity assay using *S. typhimurium*—also known as *Salmonella*/microsome, or Ames test—is actually the most common screening method to detect genotoxic carcinogenic substances, being validated in scale by several laboratories (Umbuzeiro and Vargas 2003). The Ames test is characterized by the use of indicator strains of *S. typhimurium* sensitive to substances able to induce various kinds of mutations. In the presence of mutagenic agents, these strains reverse their auxotrophic character to synthesize histidine, forming colonies in a medium deprived of this amino acid. Thus, by counting colonies per plate, the mutagenic effect of a given compound may be established in association to sampled concentrations (Mortelmans and Zeiger 2000). However, bacteria are evolutionarily distant from the human model, lacking true nuclei and also enzymatic pathways to activate most promutagenic intermediates necessary to form mutagenic compounds (Teaf and Middendorf 2000).

Some carcinogenic chemicals such as aromatic amines and polycyclic aromatic hydrocarbons, for example, are biologically inactive unless they are metabolized to active forms. In humans and lower animals, the metabolic oxidation of cytochrome P450 is able to metabolize a large number of these chemicals to electrophilic forms that can react with DNA (Mortelmans and Zeiger 2000). Since bacteria do not have the cytochrome P450 oxidation system used by vertebrates in biotransformation of exogenous compounds, it is important to mimic this system in the tests, so that the responses are close to those found in eukaryotes. This procedure is performed by adding rat liver cells homogenate pre-treated with the enzyme-inducer Aroclor 1254 (S9 fraction). Thus, substances that exert their mutagenic activity after metabolism via cytochrome P450 can be detected by the addition of S9, whereas those—who do not need this oxidation system to exert their mutagenic effects—are identified in the absence of S9 (Jarvis et al. 1996).

Besides bacteria, also some fungi have been used in genotoxicity trials. Yeasts as *Saccharomyces*, *Schizosaccharomyces*, as well as *Neurospora* and *Aspergillus* have been used in mutation assays, which are similar in design to the testing of reverse histidine mutation in *Salmonella* (Teaf and Middendorf 2000).

Plants

Plants are also recognized as excellent indicators of genotoxic and cytotoxic effects, being applicable in the evaluation of mutagenesis generated by chemical factors. Different types of assays are available in plant systems, including locus-specific testing in maize (*Zea mays* L., Poaceae) and multilocus assay systems in *Arabidopsis*. Cytogenetic tests were developed for *Tradescantia* (Commelinaceae family), especially the micronucleus assay, whereas assays for chromosomal aberrations in root tips of onion (*Allium cepa* L., Amaryllidaceae) and beans (*Vicia faba* L., Fabaceae). Finally, analysis of DNA adducts is applicable to somatic and reproductive plant cell systems (Teaf and Middendorf 2000).

Among the higher plants used as test models, *A. cepa* has been considered an excellent *in vivo* model, in which the roots grow in direct contact with the substance of interest (effluent, toxin, chemical compound, etc.), allowing prediction of possible damage to eukaryotic DNA (Tesdesco and Laughinghouse 2012). This test is of great value, since it has a high sensitivity to detect changes in cellular cycle and in chromosomes, from point mutations up to chromosomal aberrations (Leme and Marin-Morales 2008).

The *A. cepa* test shows good correlation and high sensitivity when compared to other test systems, especially with tests using mammals. It presented a correlation of 82% when compared to the carcinogenicity test in rodents. Moreover, the *A. cepa* test system presents almost the same sensitivity as the test system of algae and human lymphocytes (Fiskesjo 1985, Rank and Nielsen 1994, Leme and Marin-Morales 2007).

Another advantage is that *A. cepa* exhibits eight pairs of relatively large chromosomes (2n = 16) that allows easy detection of chromosomal aberrations. In addition, the plant material is easy to obtain and is available throughout the year at low cost, presents easy handling, high sensitivity, possibility of using the tissue without pretreatment, reliability, aiding studies of prevention of damage to human health (Bagatini et al. 2007, Leme and Marin-Morales 2009, Abdelmigid 2013). Added to these factors, it presents rapidly growing roots, large numbers of dividing cells and is tolerant to different natural conditions (Matsumoto et al. 2006). That is why it has been considered a useful tool for research on genotoxicity and cytotoxicity of chemicals, complex substances, such as plant extracts (Cuchiara et al. 2012).

The main advantage of the '*Allium cepa* assay' concerns its simple and fast protocol, as described in the Table 7.1, demanding only a standard optical microscope with few methodological steps, as illustrated in Fig. 7.1.

Additionally the root tips are relatively large and present a tip with a large meristem that is easily spread in the slides, with large cell nuclei and chromosomes with easily recognizable aberrant and/or normal mitotic phases (Fig. 7.2).

Rodents

Rodents are commonly used in tests with animals, especially rats and mice, but also guinea pigs, hamsters, mice, desert rats and others. Two rodent species stand out as used models for carcinogenicity testing and by the pharmaceutical industry: the mouse and the rat. The derived strain Sprague-Dawley rat is more commonly used in American Pharmaceutical toxicology laboratories, followed by the Wistar and Fischer 344 strain, while strains Evans and CFE (Carworth) are used less frequently. Considering mice, strain CD-1 is the most frequently used in the pharmaceutical industry. Other strains are less frequently used, as B6C3F1, NMRI, C57B1, BALB/c and Swiss (Gad 2008). *In vivo* assays using rodents have the advantage of exhibiting greater similarity to the human system and, consequently, to important mutations in the etiology of degenerative diseases such as cancer and other genetic diseases (Lima and Ribeiro 2003).

The micronucleus assay (MN) in the bone marrow of rodents is widely used for the detection of clastogenic agents (that chromosomes break) and aneugenic (which interfere with the spindle fibers) and

Table 7.1. *Allium cepa* procedure (as described by Marin-Morales 2009).

1. Collect the germinating root tips from the Petri dish (as shown in Fig. 7.1) after exposition to the test substance and fix in ethanol: acetic acid (3:1).
2. Wash the root tips three times in distilled water for 5 minutes each and remove excess water from the roots touching them gently on filter paper.
3. Hydrolyze roots in vials containing 1N HCl at 60°C in a water bath for 10 minutes.
4. Wash the root tips again three times in distilled water, removing excess water on filter paper (taking great care because the meristem is even more fragile).
5. Transfer the root tips to vials (brown glass, due to their sensitivity to light) containing Schiff's reagent (Merk), where they should remain for 2 hours in the dark.
6. Wash the root tips in distilled water with the aid of a Pasteur pipette, until all excess stain is removed.
7. Using a clean and identified slide, place an intact root meristem (with the terminal portion stained purple, standing out from the rest of the root stained in pink) over the slide.
8. Remove excess of water with the aid of a filter paper.
9. Cut meristem with the aid of a razor blade and discard the rest of the root (stained in pink).
10. Add a drop of 2% acetic carmine to the meristem in the slide and cover with a cover slip.
11. Hold the cover slip with a filter paper and spread the material with slightly tapping, using a whisk with a blunt tip until the meristem turns into a small patch of scattered cells.
12. Remove excess acetic carmine, carefully pressing the slide and the coverslip involved in a filter paper, being careful not to drag the material.
13. Heat the slide in flaming lamp, monitoring its temperature to avoid burning of the cells.
14. Freeze the slides in liquid nitrogen for a time varying from 40 seconds to 3 minutes.
15. After immersion in liquid nitrogen, remove the coverslip quickly with the aid of a scalpel blade controlling if the material remained fixed to the slide.
16. After drying the slides, add a drop of mounting medium (Entellan) and cover the material with a dry, clean coverslip.
17. Dry slides overnight and store them in a dry place until the time of analysis.

Methodology of *Allium cepa* Assay

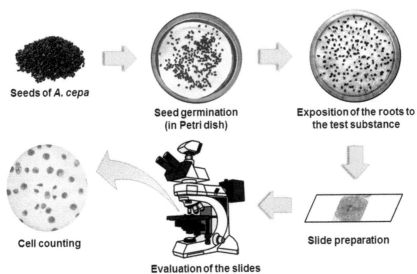

Seeds of *A. cepa*

Seed germination (in Petri dish)

Exposition of the roots to the test substance

Slide preparation

Evaluation of the slides

Cell counting

Figure 7.1. Methodological steps of the *Allium cepa* (onion) assay. For further details see Table 7.1.

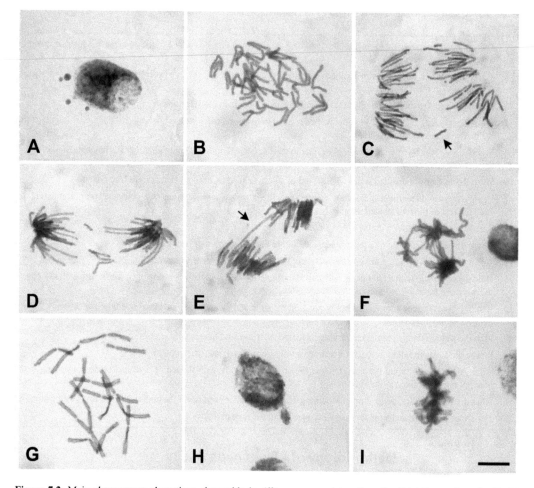

Figure 7.2. Main chromosome aberrations observable in *Allium cepa* meristematic cells. (A) Micronucleated cell. (B) Polyploid metaphase. (C) Anaphase with chromosomal breakage. (D) Metaphase with chromosomal loss. (E) Anaphase with chromosomal bridge. (F) Multipolar anaphase with chromosomal bridges. (G) Chromosomal adherence. (H) Nucleus with nuclear bud. (I) C-metaphase. Bar: 10 µm (for all pictures).

has been widely used to measure both *in vitro* and *in vivo* genotoxicity. The assay is characterized by observing the effect of the agent tested in immature enucleated erythrocytes, which have a short life, so that any MN observed represent recent chromosomal damage (Ribeiro 2003).

In vivo MN assay is especially relevant because it allows obtaining of additional information on experimental conditions, such as metabolism, pharmacokinetics and DNA repair processes. It is used first to assess the ability of the tested substance to induce structural or numerical chromosomal damage, both associated with the onset and/or progression of tumors (Krishna and Hayashi 2000). Because it is easy to identify and has a well-defined distribution, the positive results obtained with the *in vivo* MN test provide strong evidence of systemic genotoxicity of the evaluated chemical compound assessed under appropriate experimental conditions. On the other hand, the negative results support the conclusion that the tested substance is not genotoxic *in vivo* (Salvadori et al. 2003).

Because of its advantages, the *in vivo* MN assay in rodent bone marrow is widely accepted by international agencies and government institutions, as part of the battery of tests recommended to establish an evaluation and registration of new chemicals and pharmaceuticals entering annually on the world market (Ribeiro 2003, Tagliati et al. 2008), since the results are considered highly significant in the context of human health (Morita et al. 1997).

However, the *in vivo* toxicity tests are being mainly criticized due to the large number of animals required and suffering caused to them during some types of experiments. In this scenario, industries, government regulators and actors of quality control are under increasing pressure to replace *in vivo* testing by alternative methods which do not use living animals. Concern about the use of animals in experiments for toxicity tests has been a widely discussed topic (Bednarczuk et al. 2010). Therefore, *in vitro* studies have been increasingly used, since they limit the number of experimental variables, present simpler and faster implementation than the *in vivo* tests and can replace animals or, at least, serve as a previous screening prior to *in vivo* tests (Rogero et al. 2003).

Cell culture

In vitro cell culture method is the most widely used in pharmacology and toxicology for drug development and toxicological investigations, being considered the starting point for such studies (Spielmann et al. 2008). *In vitro* assays become more advantageous compared to *in vivo* testing since they provide better control over experimental conditions, allowing better reproducibility of testing conditions and associated results. They also present lower costs, higher speed, and simplicity with reduction of the number of animals used for overall assays.

Another significant advantage is the current availability of a variety of strains in cell banks, allowing increased access and use of *in vitro* models, becoming an alternative to toxicity assessments of many chemical substances. Thus, *in vitro* toxicity testing has been also proposed as substitutes for acute toxicity test (Bernauer et al. 2005).

Cell culture tests evaluate the potential of a substance to induce point mutations, clastogenesis and/or aneugenesis using mammalian cell lines or human primary cell cultures (Eastmond et al. 2009). With aid of these tests it is possible to evaluate the recombinogenic potential of any substance with ease, speed, security and good correlation to *in vivo* results (Pinhatti 2009). Results obtained in such *in vitro* systems direct *in vivo* testing, generating a rational basis for evaluating the real need for testing, the choice of the most suitable animal model as well as the type of test to be applied for the necessary prediction of the results considering the probable mechanism involved (Zucco et al. 2004).

Most cell lines correspond to cells that were initially derived from tumors. Because of their proliferative capacity, tumor cells remain in constant division, which permits their *in vitro* maintenance (Knasmuller et al. 1998). The use of tumor cell lines contributes to the discovery of new anticancer drugs with antineoplasic activity as well as for understanding the biological events involved in cancer. Most frequently used cell strains in cytotoxicity assays include L5178Y mouse lymphoma cells, CHO, CHO-AS52, Chinese hamster V79, human TK6 linphoblastonoids, as well as blood and bone marrow cells of different mammal species (Wang et al. 2009).

The human cell lines have a particular importance in research on the establishment of *in vitro* models of human diseases and also in investigating pharmacodynamic mechanisms of potentially active molecules (Gottfried et al. 2006). For example, tests with the cell lineage HepG2 (*human-derived hepatoma cell line*) derived from human hepatoma, have been considered important for assessing mutagenic and pro-mutagenic materials (Valentin-Severin et al. 2003, Knasmüller et al. 2004). These cells exhibit most of the morphological characteristics of liver cells and contain many enzymes responsible for the activation of various compounds (Valentin-Severin et al. 2003). They also express the activities of Phase I and Phase II metabolism of various xenobiotic enzymes that play key roles in the activation and/or detoxification of reactive carcinogens. They reflect the xenobiotic mechanisms in the human body in a most effective manner than other mammal cell lines as CHO, SHE or V79 (Hu et al. 2006). In this way, they have been considered an important tool for evaluating the genotoxicity because several genotoxic compounds are indirect mutagens (Valentin-Severin et al. 2003, Knasmüller et al. 2004). Therefore, HepG2 is useful to predict the metabolism, cytotoxicity and genotoxicity of chemicals in the human liver (Song et al. 2012).

Despite all above described advantages, cell culture methods also exhibit some disadvantages, such as maintenance of cultures in sterile conditions, gradual loss in the degree of differentiation of cells and the inability to reproduce the conditions of the organism, which turn responses only partially representative of the *in vivo* situation (O'Brien et al. 2000). Therefore, the use of *in vivo* and *in vitro* models—simultaneously or subsequently—as well as the use of different test organisms are necessary when the pharmacological action of a particular extract and/or active ingredient should be proved undoubtedly. Above all, they are important to elucidate the mechanism of action of whole extracts or of active ingredients isolated from plants aiming to manufacture of medicinal extracts.

Main assays used for the assessment of genotoxic and mutagenic effects of medicinal plants

Cells possess different protection mechanisms that allow them to withstand various environmental stresses to which they may be exposed. Among these mechanisms are those that prevent the entry of toxic compounds into the cell, until their inactivation or transformation into less harmful molecules to their biological systems. However, some compounds may directly or indirectly reach the DNA and promote changes in the genetic material, but—thanks to a system of extremely efficient DNA repair—most damages are repaired. As a last resort, the cell triggers a process of cell death in an attempt to prevent the spread of damage to the organism. Despite all these self-protective strategies to maintain cell viability, some substances are able to circumvent all defense mechanisms, penetrating into the cells and promoting non-repairable DNA damage, which, in turn, may lead to cell death and also initiate a process of immortality. Cells that carry mutations are more susceptible to further damage to the genetic material, that enhances the development of a carcinogenic process. The main steps that may (or not) reveal the cytotoxic, genotoxic, mutagenic and carcinogenic processes of a given compound are summarized in Fig. 7.3.

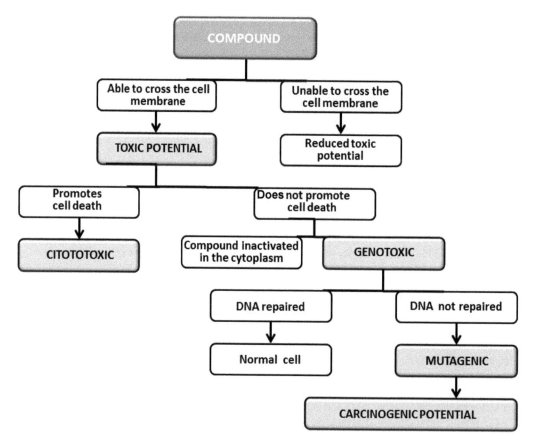

Figure 7.3. Flowchart illustrating the main toxic and genetic effects that a compound may present associated with a response (or its failure) in a given organism.

Some plants synthesize highly bioactive secondary compounds. Since many of them are considered medicinal plants, they have been used in folk medicine and also in the production of drugs for treatment of various diseases. Thus, researchers and health professionals are concerned about some abuse in the use of medicinal plants (*in natura*, or as manipulated or processed products), especially considering their indiscriminate use. Many medicinal products have already been studied, which apparently cause no harm to human health. They tend to be used improperly, in quantities and concentrations higher than those indicated due to the myth that "everything that is natural is healthy". Such a belief can trigger serious health problems, with immediate or long-term consequences.

Therefore, the use of tests that allow to investigate the mode of action and the effects of the active principles of medicinal plants on the cells and their genetic material, are of utmost importance because they will serve as possible alerts to users and health professionals about the potential risk that these compounds pose to the human body.

Cytotoxicity assays

Cytotoxicity assays are designed to evaluate possible changes in the cell structure and viability after exposition to a given compound or extract. Such tests use, in general, dyes or reagents that act as enzymatic markers of membrane integrity, i.e., are indicators of any change in markers of cellular homeostasis, allowing the identification of cellular changes caused by the process of induced cell death. They are commonly performed *in vitro*, due to the possibility of individual analysis of cell viability, being precisely quantified.

Besides *in vitro* assays, other tests have been used in plants, as those regarding counting of root meristem cells in order to monitor their division process under exposition of substances to be tested. However, a decrease in cell division not always means an existing cell death process, but may indicate a blockage of mitosis, which is also considered a cytotoxic effect. Such histological methods are also used to identify cell death processes, regarding exclusively qualitative methods.

Among the numerous methods for assessing *in vitro* cytotoxicity primarily related to the activity of plant extracts, some deserve mentioning, as the MTT assay (Mosmann 1983), the trypan blue exclusion test (Mischell and Shiingi 1980), the neutral red capture method (Hollert et al. 2000), the sulfo-rodamine B assay, MTS assay (Borefreund 1984) and the lactate dehydrogenase test (Henson 1971, Lucisano and Mantovani 1984).

MTT assay

The MTT assay is a colorimetric test widely used to determine the viability of isolated cells. The MTT [3-(4,5-dimetiltiazol-2yl)-2,5-difenil bromide tetrazoline], presents a molecular structure with a ring shape, being cleaved in the cell by a mitochondrial enzyme, the succinic dehydrogenase, yielding the formazan crystals, which are insoluble and present the violet color. Since viable cells present full activity of mitochondrial enzymes, they are stained in violet. In turn, dead cells are unable to form formazan crystals and, therefore, show no staining. The test reading is taken in a spectrophotometer which can assess by absorbance the number of cells that suffered cell death, since the greater the absorbance the greater will be the number of viable cells.

Tavares et al. (2010) used the MTT assay to investigate the cytotoxic activity of *Distichoselinum tenuifolium* (Lag.) García-Martín & Silvestre (Apiaceae family) a medicinal plant used for various therapeutic purposes of its essential oils (Khoshbakht et al. 2007, Edris 2007). According to Tavares et al. (2010), essential oils obtained from flowers and seeds of *D. tenuifolium* when investigated on its cytotoxic potential (concentration 0.64 U/ml and 1.25 U/ml) using the MTT assay for strains of mouse macrophage cells RAW 264.7 did not promote significant cell death.

Choedon et al. (2006) also investigated the cytotoxic potential of medicinal plant *Calotropis procera* (Aiton) W.T. Aiton (Asclepiadaceae) by means of the MTT assay. In traditional Indian medicine this plant has been used to treat a variety of diseases including leprosy, gastric ulcers, hemorrhoids and tumors. The authors prepared a methanolic extract using the latex of *C. procera* and evaluated its cytotoxic activity of different concentrations using two cell lines (human hepatoma—Huh-7 cells and African green monkey kidney—COS-1). Significant cell death was observed for both cell lines, even at low concentrations of the methanolic extract (0.1 mg/L). Naturally, more pronounced effects were observed for the highest concentrations evaluated (1 and 10 mg/L), showing that the use of this plant may be harmful and should be reevaluated.

Makhafola et al. (2014) appraised the cytotoxic potential of plant extracts of five species of the family Ochnaceae (*Ochna natalitia* Walp., *O. pretoriensis* E. Phillips, *O. pulchra* Hooke, *O. gamostigmata* du Toit and *O. serrulata* Walp.) by means of the MTT assay using culture cells of kidney monkey (Vero), human hepatocellular carcinoma (C3A) and bovine dermis. The species investigated in this study were characterized by Makhafola and Eloff (2012) as plants with antibacterial activity against *Staphylococcus aureus*, *Escherichia coli*, *Enterococcus faecalis*, and *Pseudomonas aeruginosa*. Despite the antibacterial effect, Makhafola et al. (2014) identified considerable cytotoxicity of all five tested plant extracts, for the three cell lines tested. The authors observed that the toxicity of the extracts was similar to the effects promoted against pathogenic bacteria, showing a non-selective toxicity for bacteria. Considering the results obtained, the authors concluded that the extracts can cause not only death of pathogenic bacteria, but also death of healthy cells of the patient. Makhafola et al. (2014) pointed out that the cytotoxicity test highlighted limitations in the use of plant extracts of Ochnaceae species in treatments against bacterial infections, emphasizing that further *in vivo* tests comprise important steps for better diagnostic of the action of such extracts on human cells.

Trypan Blue assay

The Trypan Blue exclusion method is based on a reaction to assess the integrity of the cell membrane. Cells exposed to a given compound are treated with Trypan Blue dye to check the selectivity potential of the membrane. Cytotoxicity is confirmed when a significant number of cells with purple coloring are counted, indicating that the compound affected their membranes, and thus allowed the entrance of the dye into its interior.

Besides the MTT assay, Sabini et al. (2013) used trypan blue exclusion test to evaluate the cytotoxicity of *Achyrocline saturejoides* Lam. DC (Asteraceae family). This species, apart from being largely consumed as a snack, has been used in infusions due to its important therapeutic properties. Among the properties attributed to this plant are anti-inflammatory, sedative, hepatoprotective, antioxidant, immunomodulatory, antimicrobial, antitumor, antiviral and photoprotective (Zanon et al. 1999, Kadarian et al. 2002, Ruffa et al. 2002, Arredondo et al. 2004, Bettega et al. 2004, Hnatyszyn et al. 2004, Polydoro et al. 2004, Morquio et al. 2005, Calvo et al. 2006, De Souza et al. 2007, Sabini et al. 2010).

Results obtained after testing aqueous extracts prepared from stems, leaves and flowers of *A. saturejoides* showed dose-dependent effects over VERO cells when evaluated by the trypan blue exclusion assay. However, this toxicity was not observed for the MTT assay with the same extracts, where all tests showed low toxicity on the cells studied. Thus, the authors concluded that the higher sensitivity of trypan blue assay was due to the fact that aqueous extract first attacks the cell membrane, and later interferes with the respiratory chain.

Kamdem et al. (2013) evaluated the cytotoxic potential of the ethanolic extract of *Melissa officinalis* L. (Lamiaceae) using the trypan blue exclusion assay in blood leukocytes of human volunteers aged 30 ± 12 years. *Melissa officinalis* is one of the oldest and most popular medicinal plants. Infusions of leaves of this species are used as sedatives and tranquilizers, as antispasmolytic, antibacterial, antiviral, anti-inflammatory, antioxidant and neuro-protective (Lamaison et al. 1991, Ulbricht et al. 2005, Kennedy et al. 2006, Dastmalchi et al. 2008, Pereira et al. 2009, Bayat et al. 2012). Earlier studies carried out by De Carvalho et al. (2011) showed that *M. officinalis* has antigenotoxic/antimutagenic properties in rats. However, according Kamdem et al. (2013) there were no previous reports about the cytotoxic and genotoxic potential over cell lineages using extracts of this species. Data obtained in their studies confirmed the lack of genotoxicity by the comet assay. Also no cytotoxic activity was observed by the exclusion method of trypan blue, even for high concentrations. Thus, based on their results and other published data, the authors claim that the popular use of extracts of *M. officinalis* will probably have no cytotoxic or genotoxic effects for organisms, but warns of the need of *in vivo* testing.

Neutral red assay

The neutral red assay is a test to evaluate the toxicity of substances by means of cell viability. Neutral red is a vital dye soluble in water, which crosses the cell membrane, concentrating in lysosomes and settling in anionic sites in the lysosomal matrix through hydrophobic electrostatic bonds. Many substances damage the membranes resulting in decreased capture and binding of neutral red. Therefore, it is possible to distinguish between living, dead or damaged cells by measuring the color intensity of the cell culture.

Edziri et al. (2011) investigated the cytotoxic potential of methanolic and ethyl acetate extracts of three medicinal plants from Tunisia [*Marrubium alysson* L. (Lamiaceae), *Retama raetam* (Forssk.) Webb (Papilionaceae) and *Peganum harmala* L. (Zygophillaceae)] using the neutral red assay, applied over a strain of human amniotic epithelial cells (FL). According to the authors, *M. alysson* is used in the treatment of hypertension, rheumatism and intestinal problems. In turn, *R. raetam* is applied in the treatment of hypertension and diabetes, while *P. harmala* has been used in cancer treatment, and also in emmenagogue and lactogogue fever. The authors found that 0.101 mg/mL of the ethyl acetate extract of *M. alysson*; 0.040 mg/mL of the ethyl acetate extract of and 1.171 mg/ml of methanolic extract of *R. raetam* were able to kill 50% of the exposed cells, whereas concentrations of the other extracts tested induced more than 50% mortality, showing that the extracts of these plants should be used with caution.

Extracts of nine medicinal plants used traditionally in Colombia for the treatment of a variety of diseases were tested *in vitro* for their antitumoral potential (cytotoxicity) by Betancur-Galvis et al. (1999). The cell lines used were MDBK cells (bovine kidney) and HEp-2 (laryngeal carcinoma cells) using the MTT assay and the neutral red. The results showed that none of the extracts examined show significant cytotoxic potential.

Ethanolic extracts of 20 plant species used to treat infectious diseases were investigated by Awadh Ali et al. (2001). According to the authors 14 of the 20 tested extracts showed varying degrees of antibacterial activity, while the extracts from *Calotropis procera* (Aiton) W.T. Aiton (Asclepidaceae), *Chenopodium murale* L. (Chenopodiaceae) *Pulicaria orientalis* Jaub. & Spach (Asteraceae), *Tribulus terrestris* L. (Zygophillaceae) and *Withania somnifera* L. Dunal (Solanaceae) displayed a remarkable cytotoxic activity, especially when evaluating the viability of FL cells by assaying neutral red assay after exposure during 72 hours to these extracts.

Sulforhodamine B assay

The SRB cytotoxicity evaluation method uses an aminoxanthine of bright pink color that displays two sulfonic acid groups as a dye, being able to bind to portions of protein components of cells fixed with trichloroacetic acid (Skehan et al. 1990). Unlike MTT, this method is independent of the metabolic activity of the cells. The test is a simple, sensitive, reproducible, and rapid method performed in microplates, whose readings are taken in a spectrophotometer after the test but not necessarily immediately, as is done in the MTT assay (Houghton et al. 2007).

Reyes-Chilpa et al. (2004) used the sulforhodamine B assay (SRB) to evaluate the cytotoxicity of *Calophyllum brasiliense* Cambess (Clusiaceae). The authors isolated different compounds from extracts of the leaves of this species and found the ability of these compounds and mixtures to inhibit the growth of tumor cell lines, through a possible cytotoxic potential. *C. brasiliense* is a tree species native in tropical forests of the American continent. From the hexane extract of the leaves the triterpenoids friedelin (1) and canophyllol (2) were isolated, as well as eight coumarins (3–10) which belong to the mammea type. The ketone extract showed the presence of several of the above-mentioned compounds (1, 2, 3, 4, 9 and 10), amentoflavone bioflavonoid (11) isomammeigin (12) and protocatechuic acid (13). Considering the methanolic extract, friedelin compounds (1) and shikimic acid (14) were obtained. All tested mammea coumarins exhibited cytotoxicity for the three strains of human tumor cells (PC3, K562 and U251). The most active were the mixtures of coumarin 3 plus 4, and 7 plus 8 that showed inhibition values of 88–100%. The mixture of compounds 9 and 10, as well as isomammeigin (12) inhibited only 38–69% of tumor cell growth. In turn the friedelin (1) inhibited 61.9% of PC3 cells, whereas U251 cells were less affected. As for the K562 cells the triterpene was harmless. The other triterpene [canophyllol (2)] and both shikimic and protocatechuic acids (14 and 13) showed no significant cytotoxicity for any of the strains tested. Structurally, friedelin (1) differs from canophyllol (2) only by the absence of a hydroxyl at C-28. Therefore, the presence of this functional group seems to decrease the cytotoxic activity of this tripertene.

Wesam et al. (2013) investigated the cytotoxicity of the water extract of *Erythroxylum cuneatum* Kurz (Erythroxylaceae) for two different cell lines (HepG2 hepatoma and human WRL-68 cells from normal human embryonic liver). For this study, two different methods that evaluate the cytotoxic potential of the MTS assay and the test of lactate dehydrogenase (LDH) were used.

MTS assay

MTS assay is based on the procedure originally developed by Borefreund (1984) for the screening of cytotoxic agents in general on a monolayer of cells. This test measures the amount of living cells after their exposure to a particular toxic agent. In this test the cells are incubated in a supravital dye of the tetrazolium compound (MTS) and an electron coupling agent. The tetrazolium is reduced in the cell into a soluble product in the culture medium and allows the quantification of incorporated dye by spectrophotometric analysis. The MTS amount incorporated by the cell population is directly proportional to the number of living cells in culture.

Lactate dehydrogenase test

Another test uses lactate dehydrogenase (LDH), considered an indicator of intact membrane. LDH is present throughout the cell cytoplasm and, when the membrane is damaged, there is leakage of the enzyme to the external environment. This enzyme can be detected indirectly by the catalytic reaction and conversion 2-(4-iodophenyl)-3-(4-nitrophenyl)-5-phenyl-2H-tetrazolium-chlorine (INT) into another dye called formazan. An increasing in the amount of formazan produced in the culture supernatant is directly correlated to an increase in cell lysis. Since the formazan dye is water soluble and can be detected in a spectrophotometer at 520 nm.

Wesam et al. (2013) performed an analysis using the MTS assay in HepG2 and WRL68 cells, concluding that the extract of *Erythroxylum cuneatum* (Miq.) Kurz (Erythroxylaceae) bears cytotoxic activity for both strains, with IC_{50} values of 125 ± 12 and 125 ± 14 mg/mL, respectively, after 72 hours exposure to the extract. In turn, the LHD test showed an IC_{50} of 251 ± 19 mg/mL for HepG2 and 199.5 ± 12 mg/mL for WRL68. The authors concluded from the cytotoxicity and genotoxicity tests, that the aqueous extract of *E. cuneatum* may have pharmacological potential.

As shown with different examples, investigations of the cytotoxic effects of medicinal plants by researchers bring different types of information for assessing their therapeutic potential. The identification of significant killing of tumor cell lines, as observed by Choedon et al. (2006) brings promising indication of an antineoplastic potential. However, if a broad-spectrum activity is detected for a given investigated agent, the exclusive application of this type of assay should be taken cautiously, as in the report of Makhafola and Eloff (2012).

Surveys also indicate that the combined use different tests for a given extract are advisable for safety recommendation. For example, Sabini et al. (2013) found a positive cytotoxicity using trypan blue assay, but not for the MTT assay, indicating that the use of similar methodological endpoints bring different markers, allowing a better identification of the possible toxicity, bringing evidence on the mechanism of action of the investigated compounds. Also extracts obtained by different isolation methods result in different compounds and therefore may have different levels of cytotoxicity, as found by Edziri et al. (2011). Negative results in relation to cytotoxicity may indicate that the plant has therapeutic properties that do not preclude the cells. However, viable cells are not always indicative of healthy cells, since the substance may not have triggered a process of death, but may promote damage on DNA detectable only after genotoxicity and mutagenicity tests.

Genotoxicity and mutagenicity tests

Natural compounds and their derivatives contribute to the production of about one third of the existing drugs on the market. Thus, there is interest from researchers in understanding the action and effects of plants used in folk medicine aiming to find potential new drugs, and also to alert on possible effects of these plants on human health (Zhou et al. 2010). The assays that investigate the genotoxicity and mutagenicity of chemical, biological and physical products are widely used to assess the genetic toxicology of these agents, being well accepted by the environmental regulatory and public health agencies. Among the tests used to verify the most toxic potential on the DNA molecule we highlight the Ames test (*Salmonella*/ Microsome), micronucleus assay, the comet assay and the chromosome aberration test.

Ames test (*Salmonella*/Microsome)

The *Salmonella*/Microsome test is widely used to detect mutagens in pure substances or complex mixtures and environmental samples. As explained before in this chapter, this assay is characterized by the use of strains of modified *Salmonella typhi* that in the presence of mutagens revert their auxotrophic character for histidine synthesis and start to form colonies in medium deprived of this amino acid (Varella et al. 2004). According to a review by these authors, many plant species have been evaluated for mutagenicity, exhibiting negative results after the *Salmonella*/Microsome test.

Zhang et al. (2004) evaluated the safety of aqueous extract of intermediate layers of bamboo culms [mainly species *Bambusa tuldoides* Munro (Poaceae), *Sinocalamus beecheyanus* (Munro) McClure (Poaceae) and *Phyllostachys nigra* (Lodd. ex Lindl.) Munro (Poaceae)] traditionally used in Chinese folk medicine (for stomach pains, nausea, diarrhea, inflammation of the chest, restlessness or excessive thirst) using the *Salmonella*/microsome test, with and without metabolic activation. According to the authors, there was an absence of revertant colonies for the four *Salmonella* strains tested (TA98, TA97, TA100 and TA102) indicating that the extract present no mutagenic activity.

Ozaki et al. (2002) also found no mutagenic activity after evaluation of the extract of gardenia fruits, by means of the Ames test, using *S. typhimurium* strains TA98 and TA100 with and without metabolic activation (S9). The fruits of *Gardenia jasminoides* (Rubiaceae) are widely used in Asian countries as a natural dye and in folk medicine. Mulaudzi et al. (2013) investigated 12 plant species used by the population of a region in South Africa. The species investigated are known for its uses in the treatment of gonorrhea, inflammation, headache, wounds and infertility. The extracts were prepared using different parts of each plant. For example, extracts of *Aloe chabaudii* Schönl. (Aloaceae) were prepared from the roots and in the case of *Adansonia digitata* L. (Bombacaceae) from the bark of the plant. The mutagenic potential was probed using different concentrations of all three extracts (50, 500, 5000 µg/mL) by means of the *Salmonella*/Microsome test using TA98 strain with and without metabolic activation (S9). The results indicated that the studied plants are safe to be used by the population, except for the extract from the bark of *Ekebergia capensis* Sparrm. (Meliaceae) which presented a weak mutagenic effect after metabolic activation, suggesting that further studies should be conducted for this species. However, the authors warn about the possibility of an antagonist action of compounds present in the extracts, which can mask positive results concerning mutagenic potential of these plants, and claim that despite extracts appearing to be safe for the treatment of inflammation associated with sexually transmitted diseases, other tests should be performed.

Lynch et al. (2013) performed an evaluation of the mutagenic potential of lyophilized aerial parts of *Hoodia parviflora* N.E.Br. (Asclepiadaceae) through reverse mutation in doses up to 5000 µg/plate using *S. typhimurium* (strains TA98, TA100, TA1535, TA1537), in the presence and absence of an exogenous source of metabolic activation (S9). During those trials, no mutagenic action for the chemical components of *H. parviflora* was detected, which led the authors to support their use in food or dietary supplements.

Although few studies with plant extracts have reported mutagenic activity when assessed using the *Salmonella*/Microsome test, many plant-derived products may have harmful effects. For example, *Ocotea duckei* Vatimmo-Gil (Lauraceae) is a plant widely used in Brazil as a condiment. Marques et al. (2003) evaluated the mutagenic potential of hydroalcoholic extract of its leaves and of a lignan called yangambina extracted from its leaves, also considered to have many pharmacological properties. According to the authors, yangambina is a selective antagonist of the Platelet Activating Factor (PAF) and, therefore, an effective pharmacological agent against cardiovascular collapse and death due to endotoxin shock, also presenting anti-allergenic effect. Although the authors did not observe mutagenic activity for yangambina, the hydroalcoholic extract of *O. duckei* was mutagenic to TA97, TA100 and TA102 strains in the presence and absence of metabolic activation, suggesting that the indiscriminate use of home preparations using this plant can be dangerous to human health.

Lopez et al. (2000) investigated basil (*Ocimun basilicum* L., Lamiaceae), a plant used as a culinary herb and also in folk medicine (in the form of infusion) as an antispasmodic, carminative, against colds and diarrhea. The authors tested a 70% ethanol extract and verified by means of the Ames test (*Salmonella/Microsome*), positive responses regarding the mutagenicity to TA98 strain with metabolic activation and TA1535 without metabolic activation.

Micronucleus assay

Micronuclei are portions of genetic material (whole chromosomes or their fragments) that are not properly forwarded to the pole of daughter cells in anaphase, due to changes in cell division. Non segregated chromosomes or their fragments remain in the cytoplasm as a small nucleus. The micronucleus assay is a test for evaluating the mutagenic activity of a xenobiont, which evaluates the capability of a particular agent to induce the formation of such structures, thus altering the process of cell division and hence causing a change in the genetic material of those cells. The mutagenic potential of the xenobiont is estimated by comparing the frequency of micronuclei of cell populations exposed to the evaluated agent as compared with negative controls.

The micronucleus assay, developed by Schimid et al. in 1971 was modified by Schimid (1975), being conducted primarily in bone marrow cells of mice (*in vivo* test), having its *in vitro* procedure introduced by Heddle (1976). Currently this test is applied in many cell types and is often used for the detection of both clastogenic agents (inducing chromosome breakage) as aneugenic agents (inducers of chromosome loss).

The micronucleus assay has also been traditionally used to assess mutagenic potential of extracts of medicinal plants. It was used for example in *Hoodia parviflora* N.E.Br. (Asclepiadaceae) a plant indicated due to its nutraceutical effect in weight loss treatments, being suggested as a surrogate species from *H. gordonii* Sweet (previously commercially available). Its effects have been attributed to the presence of glycosides, present in plant extracts from several species of the genus *Hoodia*. Lynch et al. (2013) evaluated the genotoxic potential of *H. parviflora* through the micronucleus test with human lymphocytes, observing negative results as to toxicogenetic activity of dry extract of *H. parviflora*, which allowed the researchers to indicate their nutraceutical use.

Cariddi et al. (2012) found that the aqueous extract of *Baccharis articulata* Pers. (Ba-CAE; Asteraceae family) induced death of mononuclear cells from human peripheral blood (PBMC, Peripheral Blood Mononuclear Cell) and exerted low mutagenic activity on mice six hours after administration, when evaluated by the micronucleus assay in bone marrow of these animals. Cariddi et al. (2012) reported the potential of Ba-CAE in the promotion of induced death in PBMCs and the ability of the extract to promote the formation of micronuclei in mice after a longer time of administration of the extract (24 and 48 hours). To investigate the mutagenic potential of BaCAE extract, concentrations of 1800, 900 and 450 mg/kg were injected in BALB/c mice with evaluation after 24 and 48 hours. The Ba-CAE exerted mutagenic effects for all concentrations tested due to an increase in the frequency of micronucleated polychromatic erythrocytes. The authors concluded that the time of exposure to the extract is a determinant factor for the risk of genotoxic damage (6 hours < 24 hours < 48 hours).

Byrsonima verbascifolia L.Dc. (Malpighiaceae) is used in folk medicine to treat diarrhea, intestinal infections, chronic wounds, Chagas disease, inflammations of the oral cavities and of the female genital tract (Panizza 1998). Using the micronucleus assay, Gonçalves et al. (2013) studied the effect of the hydroethanolic extract of *B. verbascifolia* (traditionally used in women during pregnancy) in pregnant mouse (*Mus musculus*) females. The micronucleus assay was conducted to investigate changes in peripheral blood cells of treated females. The results showed that there was no significant mutagenic activity for the treatments with hydro-alcoholic extract of *B. verbascifolia*, indicating no genetic toxicity for this plant.

Comet test 'Single Cell Gel Assay'

The comet assay or 'Single Cell Assay Gel' is a very sensitive test and may be performed with any type of eukaryotic cell that can be individualized. In this assay, cells are applied on an agarose gel placed on a microscope slide, where these cells are lysed and subjected to an electric field in alkaline buffer. The presence of single strand breaks, alkali-labile sites and crosslinks, resulting from the action of genotoxic compounds, alters the structure of DNA, which after suffering a relaxation, allows the identification of such damage by observing fragment migration in agarose gel toward the anode. Damaged DNA fragments migrate in the gel in similar way to a comet tail, which can be viewed upon application of specific dyes, visible in a fluorescence microscope. Damage is assessed in various ways, but basically by measuring the length of the tail of the comet and the amount of DNA present in the tail.

SMART test (Somatic Mutation and Recombination Test)

The SMART test (Recombination and Somatic Mutation Test) was developed by Becker (1986) and improved by Vogel (1987) aiming to detect loss of heterozygosity of gene-associated markers determining the phenotype of the *Drosophila melanogaster* wings. It is a method that uses somatic cells and can be performed in a single generation, i.e., in approximately 10 days. Furthermore, it can be used for a broad spectrum of genotoxic agents from different chemical classes as well as complex mixtures gaseous particles (Silva et al. 2003).

Also known as 'wing spot test', or SMART test, based on groups of cells (known as imaginal disks), which proliferate during development separately until they differentiate, during body metamorphosis, forming structures in the adult fly (eyes, wings, etc.). An advantage of this method is to enable the exposure of a large population of cells that are mitotically dividing in the imaginal disk of the larva. Therefore, when a genetic change occurs in a cell of the imaginal disk, it can be represented in all descendant cells and form a clone of mutant cells which can be detected as a spot on the surface of the fly's body (Frei and Würgler 1995). According to Silva et al. (2003) the spot size is related to the activation time of genotoxina throughout embryogenesis, indicating the time of induction of a given damage. Increases restricted to simple small spots may indicate increased aneugenesis or clastogenicity.

According Struwe (2002) and Scrirpsema et al. (2003), the Rubiaceae (coffee family) is distinguished by the presence of secondary metabolites—such as alkaloids from quinoline (quinine and cinchonine), isoquinolinico (emetine), purine bases, such as caffeine and indole (iombine) compounds. As an example, infusion of the roots from *Galianthe thalictroides* (Rubiaceae) has been used in Brazilian folk medicine for the treatment and prevention of cancer. Fernandes et al. (2013) evaluated the genotoxic potential of the aqueous and ethanolic extracts of the roots of this plant with two different methods, the comet assay and the SMART test. To perform the SMART test, the authors performed experimental crosses with three strains of *D. melanogaster* containing the recessive markers of the imaginal wing disk, located on chromosome 3. For the comet assay, human hepatoma cell line (HepG2) was used. The results demonstrated that aqueous extracts from the roots of *G. thalictroides* are not genotoxic. However, the ethanol extract was found to be genotoxic to the offspring of crossing High Bioactivation (HB) in SMART assay, while the comet assay failed to detect any significant effect on the DNA.

Aristolochia clematitis L. and *Asarum europaeum* L. (both of the Aristolochiaceae family) are medicinal plants used in Germany as homeopathic extracts. According to the compendium HAB (Homöopathisches Arzneibuch 2010) the extract of *A. clematitis* has been used for the treatment of respiratory, urinary and gastro-intestinal tract, as well as for gynecological indications. In turn, the homeopathic *A. europaeum* extract is applied in the treatment of diseases of the central nervous system, for respiratory and gastrointestinal infection by the influenza virus. Many herbs of the genera *Aristolochia* and *Asarum* contain Aristolochic Acid (AA) (Hashimoto et al. 1999). According to Arlt et al. (2002) AA has nephrotoxic and carcinogenic action in rodents. Renal toxicity of AA and the development of urothelial carcinoma are attributed mainly to AA type I (Schmeiser et al. 2009, Shibutani et al. 2007). Thus, Nitzsche et al. (2013) evaluated the genotoxicity of AAI, using metabolizer strains cells of human hepatoma (HepG2) with 32P-post-labeling assay, besides alkaline comet and micronucleus assays. The results showed that aristolochic acid is capable of forming DNA adducts with a significant frequency for the types dA-AAI (7-(deoxyadenosin-N6-yl) aristolactam I) and dG-AAI [7-(deoxyguanosin-N2-yl)aristolactam I]. In addition, the researchers also found significant induction of micronuclei at the highest concentrations tested. However, no significant damage on DNA as was detected by the comet assay.

According Phillips Arlt (2007), the 32P (post labeling) assay is an ultrasensitive method for detection and quantification of DNA adducts. The method consists of four main steps: (1) enzymatic digestion of DNA into nucleoside 3'-monophosphates; (2) enrichment of the fraction of the digested DNA adducts; (3) labeling of the 5' adducts by transfer of 32P-orthophosphate with gamma-32P-ATP-mediated polynucleotide kinase (PNK) and (4) separation by chromatography or by electrophoresis of adducts carrying labeled or modified nucleotides and quantify by measuring their radioactive decay. The assay is capable of detecting adducts with such low frequencies as one in 10 nucleotides, which makes it applicable for detection events resulting from environmental exposure or physiological experiments with different concentrations of inducing agents. The assay has a wide range of applications in *in vivo* and *in vitro*, being applicable to humans and animals and for the detection of adducts induced by a wide variety of classes of compounds or complex mixtures.

Chromosomal aberration test

One of the oldest and most used tools to undertake evaluation of genotoxicity and mutagenicity is the chromosomal aberration test. This test is based on classical cytogenetics and is one of the few direct methods used to measure changes in areas exposed to potential mutagens or carcinogens (Rank et al. 2002).

The analysis of Chromosomal Aberrations (CA) can be accomplished by chromosome in metaphase after colchicine pretreatment or during phases of cell division (see Fig. 7.3). The CA can be promoted by clastogenic and aneugenic agents. During cell division, DNA segments resulting from the replication change places symmetrically. These reciprocal translocation events occur in the same gene locus between the sister chromatids of a chromosome are called SCE (Sister Chromatid Exchange). Therefore SCE analysis comprises a sensitive cytogenetic method applicable to genotoxic studies. This exchange process is an indicator of the occurrence of a possible injury and repair of the DNA molecule in mammalian cells, since the greater the number of permutations, the more possibility of errors in assembly of the DNA strands will arise. The ratio between the induction of SCE and chromosomal mutation or aberration is different for each chemical agent tested. SCEs induction can be done both *in vitro* and *in vivo* in any organism where it is possible to obtain cells in culture. During cell culture markers are added that will stain the sister chromatids differently, highlighting eventual changes. Although sensitive, SCE assays are cumbersome and cannot be considered as mutagenesis tests but only as indicative of the action of a substance in promoting breaks in the genetic material, since dislocated sister chromatid exchanges do not necessarily reflect a mutation event (Silva et al. 2003).

Gadano et al. (2002) assessed the genetic damage induced by decoction and infusion of *Chenopodium ambrosioides* L. (Chenopodiaceae) an anthelmintic herb used in Latin American folk medicine. For this purpose different concentrations of the extracts were tested on human lymphocytes cultivated *in vitro* using Chromosomal Aberrations (CA) and Sister Chromatid Exchange (SCE) tests. The results revealed an increase in the percentage of cells with CA and the frequency of SCE in both preparations (decoction and infusion), suggesting a possible genotoxic effect of the studied species, probably due to different active ingredients released in each of the processes of preparation of the extracts.

Solanum lycocarpum A. St. Hil. (Solanaceae) is popularly known as wolf-fruit or wolf-apple being frequently used in folk medicine against diabetes and obesity. The fruits are also applied topically for snake bites, and when cooked and hot, are used in the treatment of tissue atrophy Dall Agnol et al. (2000). Munari et al. (2012) determined the possible genotoxic potential antigenotoxic and four concentrations of glycolalkaloid extracted from the fruit of *S. lycocarpum* in strains of fibroblasts from a Chinese hamster lung (V79 cells) by the comet assay and chromosome aberrations. Methyl methane sulfonate agent

(MMS, 22 µg/mL) was used to induce DNA damage combined with the tested extract or component aiming to evaluate a possible protective effect. The comet assays and chromosomal aberrations revealed that *S. lycocarpum* displayed no genotoxic activity. Moreover, the different concentrations studied showed a protective effect against the genomic and chromosomal damage induced by MMS.

De Pinho et al. (2010) analyzed the genotoxic effect *Baccharis trimera* Less. DC (Asteraceae) through the chromosomal aberration test. According to the authors, *B. trimera* (popularly known as carqueja) is often used as tea in southern Brazil for the treatment of kidney, intestinal and stomach diseases and mainly for weight loss. The authors used the *in vivo* and *in vitro* chromosomal aberration test using the *Allium cepa* test organism and culture of human lymphocytes, respectively to detect possible mutagenic action of teas prepared with this plant. The authors found no evidence of mutagenic effects of the tea on onion plant cells neither in human cells (lymphocytes) cultivated *in vitro*. It was also found that the effect is dose dependent and that, therefore, the carqueja tea should be consumed moderately.

The DNA molecule exhibits the same constitution in all living organisms, so it is possible to use different organisms or test systems to evaluate the genotoxic and mutagenic potential of various substances. However, specific physiological differences in each taxonomic group can give us more direct answers when the test organisms are evolutionarily closer to the target species in the study, while others give indirect answers, but serve in warning about possible adverse effects.

To assess possible DNA damage promoted by the activity of medicinal plants several organisms have been used, as can be seen in various studies previously mentioned, since bacteria, such as the *Salmonella/* Microsome (Ames) test; invertebrates in the SMART test (*Drosophila melanogaster*) to the use of human cells *in vivo* or *in vitro* that can be used in different trials as the micronucleus, comet, chromosomal aberrations, among many others.

Each of these tests measure different endpoints and give us complementary answers and all of them exhibit advantageous and disadvantageous aspects, as summarized in Table 7.2. The Ames test, for example, is a more specific test, once each strain used allows the verification of specific types of mutation. Nevertheless, it may be restrictive and not applicable as an indicative of the agent action in eukaryotes. Hence, the micronucleus assay is a more general test where it shows only if a substance promotes breakages or losses in the genetic material, but has as positive aspect the fact of being performed with human cells *in vivo* and/or *in vitro*. Still, there are those assays that allow us to assess DNA damage, as is the case with the comet assay; however, it remains unknown whether such damage will lead to a mutation or not.

Thus, to certificate if a plant can be safely used in folk medicine or not, many studies generally need to be made, especially regarding the damage that its compounds may cause to the human DNA, since many of the effects on the genetic material can endanger human health only in long term consuming, as observed in neoplastic processes.

Antimutagenic activity

Plant extracts can also contain a variety of substances with antimutagenic activity (Abdillahi et al. 2012). The antimutagenic agents are compounds that reduce the occurrence of spontaneous and/or induced mutations. These are classified into two types: desmutagenic and bioantimutagenic agents. Desmutagenic agents reduce the genotoxic effect of a xenobiotic agent, interacting directly with the mutagen or blocking its effects by inhibiting their metabolic activation and/or increasing its detoxification. In turn, bioantimutagenic agents act after the damage occurred, promoting DNA repair, increasing the fidelity of DNA replication errors and inhibiting prone replication of cells with damaged DNA (Mersch-Sundermann et al. 2004).

Plant extracts have great potential for use in folk medicine as substances capable of preventing damage to the human body, being easily accepted by the population (Cordell 1995). The sources investigated in search of antimutagenic compounds are highly diverse: fruits, leaves, stems, roots of plants, latex, etc.

In recent decades a wide range of evidence from epidemiological and laboratory studies have shown that consumption of certain medicinal plants or their active principles have protective effects against human carcinogenesis and mutagenesis. Despite the antimutagenic effect found in a given plant extract, it does not necessarily mean that it is anticarcinogenic, even though this is, anyway, a good indicator. So, medicinal

Table 7.2. Main assays used to evaluate the cytotoxic, genotoxic and mutagenic compounds from plants with medicinal properties.

Assay	Applicability	Advantages	Disadvantages	References
• Ames assay (*Salmonella/* Microsome) • *Escherichia coli*	• To detect mutations for shifting the reading frame or base pair substitutions in DNA	• Allows controlled changes in the genome • Experimental population superior to that used in animals or humans • Cost reduction and short experimental time • Validation possible on a large scale by several laboratories	• Are evolutionarily distant from the human model • Bacteria have no true nuclei and lack enzymatic pathways important in detoxification processes • Addition of secondary metabolic system is necessary	Mortelmans and Zeiger (2000)
Allium cepa assay	• Detects the cytotoxic, genotoxic and mutagenic potential using meristematic cells • Detects from point mutations to chromosomal aberrations	• High sensitivity and good correlation with other test organisms • Low cost, fast and easy handling • Identified chromosomal aberration helps in the understanding of the mechanism of action of each agent examined • Displays metabolic capacity to activate pro-mutagens without the prior addition of any exogenous metabolic system	• The oxidase enzyme systems of higher plants present at low concentration and limited specificity to several substrates when compared to cytochrome P450 from mammalian • Requires cell counting under a microscope, so it is laborious	Leme and Marin-Morales (2009) Rank and Nielsen (1997)
MTT assay	• Evaluates the cytotoxic potential by cell viability and cytostatic activity	• Easy implementation • Allows checking of the induced alterations in metabolic pathways or of the structural integrity related to cell death	• The rate of cell proliferation may interfere with the MTT reduction • The test conditions may influence the metabolic activity of cells	Moore et al. (2010) Mosmann (1983)
Trypan blue dye exclusion test	• Evaluates the cytotoxic potential through the cell membrane integrity	• Detects non-viable cells that have membrane damage • Preliminary test for the realization of the Comet Assay	• Cannot detect other types of damages such as those affecting cell adhesion or that may progress to cell death • Requires cell counting under a microscope, so it is laborious	• Ribeiro et al. (2003)
Neutral Red assay	• Evaluates the cytotoxic potential by quantification of membrane permeability and lysosomal activity	• Displays satisfactory sensitivity • Quick, practical and suitable test for handling a large number of cell cultures simultaneously	• In some cases this test is less sensitive • Its use is not recommended in studies of ion channels	• Fautz et al. (1991) • Weyermann et al. (2005)

Micronuclei assay	• Evaluates the mutagenic potential of a substance by the frequency of MNs • It is a biological indicator of clastogenic or aneugenic effect, revealing the genomic instability	• Can be applied both in animal or plant cells • Simple technique that can be used *in vivo* and/or *in vitro* • Can be used for monitoring populations exposed to mutagenic substances • Employed as a tool in the elucidation of the action mechanism of cytogenetic agents • Regards an early marker of carcinogenesis.	• The method uses cells that are multiplying constantly • Does not detect mitotic non disjunction • Does not detect aberrations associated to chromosomal rearrangements as translocations or inversions • Information on the extent of damage are not exact	• Fenech et al. (2011) • Ribeiro et al. (2003)
Single cell gel assay	• Evaluates the genotoxic potential of a xenobiont	• High sensitivity and fast performance • Can be applied to any cell type • Allows evaluation of DNA damage that are prone to repair processes • Uses a small number that not necessarily must be under division	• Demands a fluorescence microscope • Slides are not permanent and cannot be stored for further observation	• Wasson et al. (2008) • Azqueta et al. (2009)
SMART assay	Detects loss of heterozygosity of marker genes that determine phenotypes of the wings of *Drosophila melanogaster*	• Can be carried out in a single generation (approximately 10 days) • Can be used for a broad spectrum of genotoxic agents from different chemical classes as well as complex mixtures and gaseous particles • Enzymatic system similar to that found in mammalian S9 fraction	• Used less frequently for the evaluation of medicinal plants and their compounds	Graf et al. (1984) Graf and Singer (1992)

plants are candidates for cancer chemoprevention, because they may have chemopreventive agents with inhibitory effects on initiation, promotion and progression of carcinogenesis (Surh and Ferguson 2003).

The chemopreventive agents can be used not only to prevent cancers, but also in therapy, since many of them may be used in combination with chemotherapeutic agents to enhance the effect of the use of lower doses and thus minimize the toxicity induced by chemotherapy (Dorai and Aggarwal 2004). However, the use of protective agents is possible only on the basis of careful analysis of the reliable assessment of the risk-benefit after several experimental models in a variety of *in vitro*, *in vivo* and clinical tests (Steele and Kelloff 2005). In the future, most commonly applied mutagenicity assays can be modified and used to measure antimutagenicity.

Final considerations

The discovery of new chemicals with pharmacological activity is undoubtedly a multidisciplinary activity, where studies on efficacy, mechanisms of action, toxic and genotoxic potential depend on drug-toxicological bioassays *in vitro* and *in vivo* to confirm their activities. It is worth mentioning that besides the medicinal properties, many plants may also have antimutagenic and/or antineoplastic agents in its composition, so it is essential to apply antimutagenicity assays, aimed at formulating new, important chemopreventive drugs against diverse diseases, including cancer.

Due to the universal properties of the DNA structure in all living organisms, it is possible to use different organisms or test systems to evaluate the genotoxic and mutagenic potential of various substances. However, specific physiological differences in each taxonomic group remain a concern, whereas evolutionarily closely related targets to the study group can give us more straightforward answers, and other non-related organisms give indirect answers, but serve to alert us about possible problems.

To assess DNA damage promoted by the activity of medicinal plants several organisms are useful, as can be seen in various studies previously mentioned, including bacteria (Ames test, *Salmonella*/Microsome); invertebrates in the SMART test (*Drosophila melanogaster*) up to the use of human cells *in vivo* or *in vitro* that can be used in different trials as the micronucleus, comet and chromosomal aberrations assays, among many others. Each of the tests measure different endpoints and give us additional answers. The Ames test, for example, is a more specific test once each strain used allows the verification of specific types of mutation. However, it may be restrictive to be used as an indication of action of the agent in eukaryotes. In turn, the micronucleus test is a more general test, uncovering if the tested substance promotes breakages or losses in the genetic material, but may be performed with human cells *in vivo* and/or *in vitro*. Some assays allow the evaluation of DNA damage; however, it remains unknown whether these damages will lead to a mutation or not, as is the case of the comet assay.

Finally, for certification if a particular plant can be (or not) safely used in folk medicine, additional tests are required, especially regarding the damage that certain compounds can cause to the DNA of an organism, since many of the effects over the genetic material may endanger human health only in the long term, as it is observed in neoplastic processes.

References

Abdelmigid, H.M. 2013. New trends in genotoxicity testing of herbal medicinal plants. pp. 89–120. *In*: S. Gowder (ed.). New Insights into Toxicity and Drug Testing. InTech, Rijeka, Croatia.

Abdillahi, H.S., Verschaeve, L., Finnie, J.F. and Staden, J.V. 2012. Mutagenicity, antimutagenicity and cytotoxicity evaluation of South African *Podocarpus* species. J. Ethnopharmacol. 139: 728–738.

Arlt, V.M., Ferluga, D., Stiborova, M., Pfohl-Leszkowicz, A., Vukelic, M., Ceovic, S., Schmeiser, H.H. and Cosyns, J.P. 2002. Is aristolochic acid a risk factor for Balkan endemic nephropathy-associated urothelial cancer. Int. J. Cancer 101: 500–502.

Arredondo, M.F., Blasina, F., Echeverry, C., Morquio, A., Ferreira, M., Abin-Carriquiry, J.A., Lafon, L. and Dajas, F. 2004. Cytoprotection by *Achyrocline satureioides* (Lam.) DC and some of its main flavonoids against oxidative stress. J. Ethnopharmacol. 91: 13–20.

Awadh, Ali, N.A., Jülich, W.D., Kusnick, C. and Lindequist, U. 2001. Screening of Yemeni medicinal plants for antibacterial and cytotoxic activities. J. Ethnopharmacol. 74: 173–179.

Azqueta, A., Shaposhnikov, S. and Collins, A.R. 2009. DNA oxidation: investigating its key role in environmental mutagenesis with comet assay. Mutat. Res. 674: 101–108.

Bagatini, M.D., Silva, A.C.F. and Tedesco, S.B. 2007. Uso do sistema teste de *Allium cepa* como bioindicador de genotoxicidade de infusões de plantas medicinais. Rev. Bras. Farmacogn. 17: 444–447.

Bayat, M., Tameh, A.A., Ghahremani, M.H., Akbari, M., Mehr, S.E., Khanavi, M. and Hassanzadeh, G. 2012. Neuroprotective properties of *Melissa officinalis* after hypoxic-ischemic injury both *in vitro* and *in vivo*. DARU, 20–42.

Becker, H.J. 1986. Mitotic recombination. Gen. Biol. Drosophila, 1019–1087.

Bednarczuk, V.O., Verdam, M.C.S., Miguel, M.D. and Miguel, O.G. 2010. Testes *in vitro* e *in vivo* utilizados na triagem toxicológica de produtos naturais. V. Acadêmica. 11. 2.

Belcavello, L., Dutra, J.C.V., Freitas, J.V., Aranha, I.P. and Batitucci, M.C. 2012. Mutagenicity of ipriflavone *in vivo* and *in vitro*. Food Chem. Toxicol. 50: 996–1000.

Bernauer, U., Oberemm, A., Madle, S. and Gundert-Remy, U. 2005. The use of *in vitro* data in risk assessment. Basic Clin. Pharmacol. Toxicol. 96: 176–181.

Betancur-Galvis, L.A., Saez, J., Granados, H., Salazar, A. and Ossa, J.E. 1999. Antitumor and antiviral activity of Colombian medicinal plant extracts. Mem. Inst. Oswaldo Cruz. 94: 531–535.

Bettega, J.M.R., Teixeira, H., Bassani, V.L., Barardi, C.R.M. and Simoes, C.M.O. 2004. Evaluation of the antiherpetic activity of standardized extracts of *Achyrocline satureioides*. Phytother. Res. 18: 819–823.

Borefreund, P.A. 1984. Simple quantitative procedure using Monolayer cultures for Cytotoxicity Assay (HTD/NR-90). J. Tiss. Cult. Method. 9: 7–9.

Borges, C.C., Matos, T.F., Moreira, J., Rossato, A.E., Zanette, V.C. and Amaral, P.A. 2013. *Bidens pilosa* L. (Asteraceae): traditional use in a community of southern Brazil. Rev. Bras. Plantas Med. 15: 34–40.

Brito, A.R.M.S. and Brito, A.A.S. 1993. Forty years of Brazilian medicinal plant research. J. Ethnopharmacol. 39: 53–67.

Calvo, D., Cariddi, L.N., Grosso, M., Demo, M.S. and Maldonado, A.M. 2006. *Achyrocline satureioides* (Lam.) DC (Marcela): antimicrobial activity on *Staphylococcus* spp. and immunomodulating effects on human lymphocytes. Rev. Latinoam. Microbiol. 48: 247–255.

Cariddi, L., Escobar, F., Sabini, C., Torres, C., Reinoso, E., Cristofolini, A., Comini, L., Montoya, S.N. and Sabini. L. 2012. Apoptosis and mutagenicity induction by a characterized aqueous extract of *Baccharis articulata* (Lam.) Pers. (Asteraceae) on normal cells. Food Chem. Toxicol. 50: 155–161.

Choedon, T., Mathan, G., Arya, S.L., Kumar, V. and Kumar, V. 2006. Anticancer and cytotoxic properties of the latex of *Calotropis procera* in a transgenic mouse model of hepatocellular carcinoma. World J. Gastroentero. 12. 16: 2517–2522.

Cordell, J.L. 1995. A guide to developing clinical pathways. Med. Lab. Obs. 27. 4: 35–39.

Csokaa, A.B. and Szyf, M. 2009. Epigenetic side-effects of common pharmaceuticals: a potential new field in medicine and pharmacology. Med. Hypotheses 73: 770–780.

Cuchiara, C.C., Borges, C.S. and Bobrowski, V.L. 2012. Sistema teste de *Allium cepa* como bioindicador da citogenotoxicidade de cursos d'água. Tec. C. Agrop. 6(1): 33–38.

Dall'Agnol, R. and Von Poser, G.L. 2000. The use of complex polysaccharides in the management of metabolic diseases: the case of *Solanum lycocarpum* fruits. J. Ethnopharmacol. 71: 337–341.

Dastmalchi, K., Dorman, H.J.D., Oinonen, P.P., Darwis, Y., Laakso, I. and Hiltunen, R. 2008. Chemical composition and *in vitro* antioxidative activity of a lemon balm (*Melissa officinalis* L.) extract. Food Sci. Technol. 41: 391–400.

De Carvalho, N.C., Corrêa-Angeloni, M.J.F., Leffa, D.D., Moreira, J., Nicolau, V., Amaral, P.A., Rossatto, A.E. and De Andrade, V.M. 2011. Evaluation of the genotoxic and antigenotoxic potential of *Melissa officinalis* in mice. Genet. Mol. Biol. 34: 290–297.

De Pinho, D.S., Sturbelle, R.T., Martino-Roth, M.G. and Garcias, G.L. 2010. Avaliação da atividade mutagênica da infusão de *Baccharis trimera* (Less.) DC. em teste de *Allium cepa* e teste de aberrações cromossômicas em linfócitos humanos. Braz. J. Pharmacog. 20. 2: 165–170.

De Souza, K.C.B., Bassani, V.L. and Schapoval, E.E.S. 2007. Influence of excipients and technological process on anti-inflammatory activity of quercetin and *Achyrocline satureioides* (Lam.) D.C. extracts by oral route. Phytomedicine 14: 102–108.

Dorai, T. and Aggarwal, B.B. 2004. Role of chemopreventive agents in cancer therapy. Cancer Lett. 215: 129–140.

Eastmond, D.A., Hartwig, A., Anderson, D., Anwar, W.A., Cimino, M.C., Dobrev, I., Douglas, G.R., Nohmi, T., Phillips, D.H. and Vickers, C. 2009. Mutagenicity testing for chemical risk assessment: update of the WHO/IPCS Harmonized Scheme. Mutagenesis 24. 4: 341–349.

Ednarczuk, V.O., Verdam, M.C.S., Miguel, M.D. and Miguel, O.G. 2010. Testes *in vitro* e *in vivo* utilizados na triagem toxicológica de produtos naturais. V. Acadêmica 11: 2.

Edris, A.E. 2007. Pharmaceutical and therapeutic potentials of essential oils and their individual volatile constituents: a review. Phytother. Res. 21: 308–323.

Edziri, H., Mastouri, M., Mahjoub, A., Anthonissen, R., Mertens, B., Cammaerts, S., Gevaert, L. and Verschaeve, L. 2011. Toxic and mutagenic properties of extracts from Tunisian traditional medicinal plants investigated by the neutral red uptake, VITOTOX and alkaline comet assays. S. African J. Bot. 77: 703–710.

Fautz, R.B., Husein, B. and Hechenberger, C. 1991. Application of the Neutral Red assay to monolayer cultures of primary hepatocytes: rapid colorimetric viability determination for the unscheduled DNA synthesis test (UDS), Mutat. Res. 253: 173–179.

Fenech, M., Holland, N., Zeiger, E., Chang, W.P., Burgaz, S., Thomas, P., Bolognesi, C., Knasmueller, S., Kirsch-Volders, M. and Bonassi, S. 2011. The HUMN and HUMNxL international collaboration projects on human micronucleus assay in lymphocytes and buccal cells—past, present and future. Mutagenesis 26: 239–245.

Fernandes, L.M., Garcez, W.S., Mantovani, M.S., Figueiredo, P.O., Fernandes, C.A., Garcez, F.R. and Guterres, R. 2013. Assessment of the *in vitro* and *in vivo* genotoxicity of extracts and indole monoterpene alkaloid from the roots of *Galianthe thalictroides* (Rubiaceae). Food Chem. Toxicol. 59: 405–411.

Fiskesjo, G. 1985. The *Allium* test as a standard in environmental monitoring. Hereditas 102: 99–112.

Frei, H. and Würgler, F.E. 1995. Optimal experimental design and sample size for the statistical evaluation of data from somatic mutation and recombination tests (SMART) Drosophila. Mutat. Res. 334: 247–258.

Gad, S.C. 2008. Carcinogenicity studies. pp. 423–458. *In*: S.C. Gad (ed.). Preclinical Development Handbook: Toxicology. John Wiley Professional, Hoboken, USA.

Gadano, A., Gurni, A., López, P., Ferraro, G. and Carballo, M. 2002. *In vitro* genotoxic evaluation of the medicinal plant *Chenopodium ambrosioides* L. 81: 11–16.

Gonçalves, C.A., Siqueira, J.M., Carollo, C.A., Mauro, M.O., Davi, N., Cunha-Laura, A.L., Monreal, A.C.D., Castro, A.H., Fernandes, L., Chagas, R.R., Auharek, S.A. and Oliveira, R.J. 2013. Gestational exposure to *Byrsonima verbascifolia*: Teratogenicity, mutagenicity and immunomodulation evaluation in female Swiss mice. J. Ethnopharmacol. 150: 843–850.

Gottfried, E., Kunz-Schughart, L.A., Andreesen, R. and Kreutz, M. 2006. Brave little world: spheroids as an *in vitro* model to study tumor-immune-cell interactions. Cell Cycle 5: 691–695.

Graf, U. and Singer, D. 1992. Genotoxicity testing of promutagens in the wing somatic mutation and recombination test in *Drosophila melanogaster*. Rev. Int. Cont. Amb. 8: 15–27.

Graf, U., Wurgler, F.E., Katz, A.J., Frei, H., Juon, H., Hall, C.B. and Kale, P.G. 1984. Somatic mutation and recombination test in *Drosophila melanogaster*. Environ. Mol. Mutagen. 6: 347–377.

Hab, Homöopathisches Arzneibuch. 2010. Deutscher Apotheker, Stuttgart. Verlag. German.

Hashimoto, K., Higuchi, M., Makino, B., Sakakibara, I., Kubo, M., Komatsu, Y., Maruno, M. and Okada, M. 1999. Quantitative analysis of aristolochic acids, toxic compounds, contained in some medicinal plants. J. Ethnopharmacol. 64: 185–189.

Heddle, J.A. 1976. Measurement of chromosomal breakage in cultured cells by the micronucleus technique. pp. 191–200. *In*: H.J. Evans and D.C. Lloyd (eds.). Mutation-Induced Chromosome Damage to Man. Edinburgh University Press, Edinburgh, Scotland.

Henson, P.M. 1971. The immunologic release of constituents from neutrophil leukocytes. I. The role of antibody and complement on nonphagocytosable surfaces or phagocytosable particles. J. Immunol. 107. 6: 1535–1546.

Hnatyszyn, O., Moscatelli, V., Rondina, R., Costa, M., Arranz, C., Balaszczuk, A., Coussio, J. and Ferraro, G. 2004. Flavonoids from *Achyrocline satureioides* with relaxant effects on the smooth muscle of Guinea pig corpus cavernosum. Phytomedicine 11: 366–369.

Hollert, H.M., Durr, L., Erdinger, T. and *Braunbeck*. 2000. Cytotoxicity of settling particulate matter and sediments of the Neckar River (Germany) during a winter flood. Environ. Toxicol. Chem. 19. 3: 528–534.

Houk, V.S. 1992. The genotoxicity of industrial wastes and effluents—a review. Mutat. Res. 277: 91–138.

Hu, R., Xu, C., Shen, G., Jain, M.R., Khor, T.O., Gopalkirshnan, A., Lin, W., Reddy, B., Chang, J.Y. and Kong, A.N. 2006. Identification of Nrf-regulated genes induced by chemopreventive isothiocyanate PEITC by oligonucleotide microarray. Life Sci. 79: 1944–1955.

Jarvis, A.S., Hokeycett, M.E., McFarland, V.A., Bulich, A.A. and Bounds, H.C. 1996. A comparison of the Ames assay and Mutatox in assessing the mutagenic potential of contaminated dredged sediment. Ecotox. Environ. Safe. 1: 193–200.

Kadarian, C., Broussalis, A.M., Miño, J., Lopez, P., Gorzalczany, S., Ferraro, G. and Acevedo, C. 2002. Hepatoprotective activity of *Achyrocline satureioides* (Lam.) DC. Pharmacol. Res. 45: 57–61.

Kamdem, J.P., Adeniran, A.A., Boligon, A., Klimaczewski, C.V., Elekofehinti, O.O., Hassan, W., Ibrahim, M., Waczuk, E.P., Meinerz, D.F. and Athayde, M.L. 2013. Antioxidant activity, genotoxicity and cytotoxicity evaluation of lemon balm (*Melissa officinalis* L.) ethanolic extract: its potential role in neuroprotection. Ind. Crop. Prod. 51: 26–34.

Kennedy, D.O., Little, W., Haskell, C.F. and Scholey, A.B. 2006. Anxiolytic effects of a com-bination of *Melissa officinalis* and *Valeriana officinalis* during laboratory induced stress. Phytother. Res. 20: 96–102.

Khoshbakht, K., Hammer, K. and Pistrick, K. 2007. *Eryngium caucasicum* Trautv. Cultivated as a vegetable in the Elburz Mountains (Northern Iran). Genet. Resour. Crop. Ev. 54: 445–448.

Knasmuller, S., Parzefall, W., Sanyal, R., Ecker, S., Schwab, C., Uhl, M., Mersh-Sundermann, V., Williamson, G., Hietsch, G., Langer, L., Darroudi, F. and Natarajan, A.T. 1998. Use of metabolically competent human hepatoma cells for the detection of mutagens and antimutagens. Mutat. Res. 402: 185–202.

Knasmuller, S., Mersh-Sundermann, V., Kevekordes, S., Darroudi, F., Huber, W.W., Hoelzl, C., Bichler, J. and Majer, B.J. 2004. Use of human-derived liver cell lines for the detection of environmental and dietary genotoxicants; current state of knowledge. Toxicology 198: 315–328.

Krishna, G. and Hayashi, M. 2000. *In vivo* rodent micronucleus assay: protocol, conduct and data interpretation. Mutat. Res. 455: 155–166.

Lamaison, J.L., Petitjean-Freytet, C. and Carnat, A. 1991. Medicinal Lamiaceae with antioxidant properties, a potential source of rosmarinic acid. Pharm. Acta Helv. 66: 185–188.

Leme, D.M. and Marin-Morales, M.A. 2007. Avaliação da qualidade de águas impactadas por petróleo por meio de sistema-teste biológico (*Allium cepa*)—Um estudo de caso. 4º Cong. Bras. Pesq. Des. Petr. Gás Campinas: ABPG, pp. 1–10.

Leme, D.M. and Marin-Morales, M.A. 2008. Chromosome aberration and micronucleus frequencies in *Allium cepa* cells exposed to petroleum polluted water—A case study. Mutat. Res. 650: 80–86.

Leme, D.M. and Marin-Morales, M.A. 2009. *Allium cepa* test in environmental monitoring: a review on its application. Mutat. Res. 1: 71–81.

Lima, P.L.A. and Ribeiro, L.R. 2003. Teste de mutação genica em células de mamífero (*mouse lymphoma assay*). pp. 173–200. *In*: L.R. Ribeiro, D.M.F. Salvadori and E.K. Marques (eds.). Mutagênese ambiental. ULBRA, Canoas, Brazil.

Logar, R.M. and Vodovnik, M. 2007. The applications of microbes in environmental monitoring. pp. 329–339. *In*: A. Méndez-Vilas (eds.). Communicating Current Research and Educational Topics and Trends in Applied Microbiology. Formatex Research Center, Badajoz, Spain.

Lopez, A.G.L., Parra, A.V., Ruiz, A.R. and Piloto, J. 2000. Estudio toxicogenético de um extrato fluido de *Ocimun basilicum* I. (albahaca blanca). Rer. Cubana Plant. Méd. 3: 78–83.

Lucisano, Y.M. and Mantovani, B. 1984. Lysosomal enzyme release from polymorphonuclear leukocytes induced by immune complexes of IgM and of IgG. J. Immunol. 132: 2015–2020.

Lynch, B., Lau, A., Baldwin, N., Hofman-Hüther, H., Bauter, M.R. and Marone, P.A. 2013. Genotoxicity of dried *Hoodia parviflora* aerial parts. Food Chem. Toxicol. 55: 272–278.

Makhafola, T.J. and Eloff, J.N. 2012. Five Ochna species have high antibacterial activity and more than 10 antibacterial compounds. S. Afr. J. Sci. 108 1: 1–6.

Makhafola, T.J., McGaw, L.J. and Eloff, J.N. 2014. *In vitro* cytotoxicity and genotoxicity of five Ochna species (Ochnaceae) with excellent antibacterial activity. S. Afr. J. Bot. 91: 9–13.

Marin-Morales, M.A. 2009. A utilização de *Allium cepa* como organismo teste na detecção da genotoxicidade ambiental, Rio Claro, Brazil, pp. 1–43.

Marques, R.C.P., Medeiros, S.R.B., Dias, C.S., Barbosa-Filho, J.M. and Agnez-Lima, L.F. 2003. Evaluation of the mutagenic potential of yangabin and of the hydroalcoholic extract of *Ocotea duckei* by the Ames test. Mutat. Res. 536: 117–20.

Matsumoto, S.T., Mantovani, M.S., Malaguttii, M.I.A., Dias, A.L., Fonseca, I.C. and Marin-Morales, M.A. 2006. Genotoxicity and mutagenicity of water contaminate with tannery effluents, as evaluated by the micronucleus test and comet assay using the fish *Oreochromis niloticus* and chromosome aberrations in onion root-tips. Genet. Mol. Biol. 29: 148–158.

Mersch-Sundermann, V., Siegfried, K., Xim-Jiang, W.U., Darroudi, F. and Kassie, F. 2004. Use of a human-derived liver cell line for the detection of cytoprotective, antigenotoxic and cogenotoxic agents. Toxicology 198: 329–340.

Mischell, B.B. and Shiingi, S.M. 1980. Selected Methods in Cellular Immunology. W.H. Freeman Company, New York, USA.

Moore, P., Yedjou, C.G. and Tchounwou, P.B. 2010. Malation-induced oxidative stress, cytotoxicity and genotoxicity on human liver carcinoma (HepG2) cells. Environ. Toxicol. 25: 221–226.

Morita, T., Asano, N., Awogi, T., Sasaki, Y.F., Sato, S., Shimada, H., Sutou, S., Suzuki, T., Wakata, A., Sofuni, T. and Hayashi, M. 1997. Evaluation of the rodent micronucleus assay in the screening of IARC carcinogens. Mutat. Res. 389: 3–122.

Morquio, A., Rivera-Megret, F. and Dajas, F. 2005. Photoprotection by topical application of *Achyrocline satureioides* ('Marcela'). Phytother. Res. 19: 486–490.

Mortelmans, K. and Zeiger, E. 2000. The Ames *Salmonella* mutagenicity assay. Mutat. Res. 455: 29–60.

Mosmann, T. 1983. Rapid colorimetric assay for cellular growth and survival. J. Immunol. Methods 65: 55–63.

Mulaudzi, R.B., Ndhlala, A.R., Kulkarni, M.G., Finnie, J.F. and Vanstaden, J. 2013. Anti-inflammatory and mutagenic evaluation of medicinal plants used by Venda people against venereal and related diseases. J. Ethnopharmacol. 146: 73–179.

Munari, C.C., Oliveira, P.F., Lima, I.M.S., Martins, S.P.L., Da Costa, J.C., Bastos, J.K. and Tavares, D.C. 2012. Evaluation of cytotoxic, genotoxic and antigenotoxic potential of *Solanum lycocarpum* fruits glycoalkaloid extract in V79 cells. Food Chem. Toxicol. 50: 3696–3701.

Nitzsche, D., Melzig, M.F. and Arlt, V.M. 2013. Evaluation of the cytotoxicity and genotoxicity of aristolochic acid I—A component of Aristolochiaceae plant extracts used in homeopathy. Environ. Toxicol. Phar. 35: 325–334.

Nunes, L.G., Gontijo, D.C., Souza, C.J.A., Fietto, L.G., Carvalho, A.F. and Leite, J.P.V. 2012. The mutagenic, DNA-damaging and antioxidative properties of bark and leaf extracts from *Coutarea hexandra* (Jacq.) K. Schum. Environ. Toxicol. Phar. 33: 297–303.

O'brien, N.M., Woods, J.A., Aherne, S.A. and O'callaghan, Y.C. 2000. Cytotoxicity, genotoxicity and oxidative reactions in cell-culture models: modulatory effects of phytochemicals. Biochem. Soc. T. 28: 22–26.

Ozaki, A., Kitano, M., Furusawa, N., Yamaguchi, H., Kuroda, K. and Endo, G. 2002. Genotoxicity of gardenia yellow and its components. Food Chem. Toxicol. 40: 1603–10.

Panizza, S. 1998. Plantas que Curam: Cheiro de Mat. IBRASA, São Paulo, Brazil.

Pereira, R.P., Fachinetto, R., De Souza Prestes, A., Puntel, R.L., Santos Da Silva, G.N., Heinzmann, B.M., Boschetti, T.K., Athayde, M.L., Burger, M.E., Morel, A.F., Morsch, V.M. and Rocha, J.B. 2009. Antioxidant effects of different extracts from *Melissa officinalis*, *Matricaria recutita* and *Cymbopogon citratus*. Neurochem. Res. 34: 973–983.

Phillips, D.H. and Arlt, V.M. 2007. The 32P-postlabeling assay for DNA adducts. Nat. Protoc. 2: 2772–2781.

Ping, K.Y., Darah, I., Yusuf, U.K., Yeng, C. and Sasidharan, S. 2012. Genotoxicity of *Euphorbia hirta*: an *Allium cepa* assay. Molecules 17: 7782–7791.

Pinhatti, V.R. Avaliação das atividades biológicas e genotóxicas em dois derivados de guanilhidrazonas. 2009. 94 f. Dissertação (Mestrado em Biologia Celular e Molecular)—Universidade Federal do Rio Grande do Sul, Porto Alegre, Brazil.

Polydoro, M., De Souza, K.C.B., Andrades, M.E., Da Silva, E.G., Bonatto, F., Heydrich, J., Dal-Pizzol, F., Shapoval, E.E., Bassani, V.L. and Moreira, J.C. 2004. Antioxidant, a prooxidant and cytotoxic effects of *Achyrocline satureioides* extracts. Life Sci. 74: 2815–2826.

Rank, J. and Nielsen, M.H. 1994. Evaluation of the *Allium* anaphase–telophase test in relation to genotoxicity screening of industrial wastewater, Mutat. Res. 312: 17–24.

Rank, J., Lopez, L.C., Nielsen, M.H. and Moretton, J. 2002. Genotoxicity of maleic hydrazide, acridine and DEHP in *Allium cepa* root cells performed by two different laboratories. Hereditas 136: 13–18.

Reyes-Chilpa, R., Estrada-Muniz, E., Apan, T.R., Aumelas, B.A.A., Jankowski, C.K. and Vázquez-Torres, M. 2004. Cytotoxic effects of mammea type coumarins from *Calophyllum brasiliense*. Life Sci. 75: 1635–1647.

Ribeiro, L.R. 2003. Teste do micronúcleo em medula óssea de roedores *in vivo*. pp. 173–200. *In*: L.R. Ribeiro, D.M.F. Salvadori and E.K. Marques (eds.). Mutagênese ambiental. ULBRA, Canoas, Brazil.

Rogero, S.O., Lugão, A.B., Ikeda, T.I. and Cruz, A.S. 2003. Testes *in vitro* de citotoxicidade: estudo comparativo entre duas metodologias. Mater. Res. 6: 17–320.

Ruffa, M.J., Ferraro, G., Wagner, M.L., Calcagno, M.L., Campos, R.H. and Cavallaro, L. 2002. Cytotoxic effect of Argentine medicinal plant extracts on human hepatocellular carcinoma cell line. J. Ethnopharmacol. 79: 335–339.

Sabini, M.C., Escobar, F.M., Tonn, C.E., Zanon, S.M., Contigiani, S.M. and Sabini, L.I. 2010. Evaluation of antiviral activity of aqueous extracts from A. *satureioides* against Western Equine Encephalitis virus. Nat. Prod. Res. 26: 405–15.

Sabini, M.C., Cariddi, L.N., Escobar, F.M., Mañas, F., Comini, L., Reinoso, E., Sutil, S.B., Acosta, A.C., Núñez Montoya, S., Contigiani, M.S., Zanon, S.M. and Sabini, L.I. 2013. Evaluation of the cytotoxicity, genotoxicity and apoptotic induction of an aqueous extract of *Achyrocline satureioides* (Lam.) DC. Food Chem. Toxicol. 60: 463–470.

Salvadori, D.M.F., Ribeiro, L.R. and Fenech, M. 2003. Teste do micronúcleo em células humanas *in vitro*. pp. 201–224. *In*: L.R. Ribeiro, D.M.F. Salvadori and E.K. Marques (eds.). Mutagênese ambiental. ULBRA, Canoas, Brazil.

Sasidharan, S., Chen, Y., Saravanan, D., Sundram, K.M. and Yoga Latha, L. 2011. Extraction, isolation and characterization of bioactive compounds from plants' extracts. Afr. J. Tradit. Complement. Altern. Med. 8: 1–10.

Schimid, W., Arakaki, D.T., Breslau, N.A. and Culbertson, J.C. 1971. Chemical Mutagenesis. The Chinese hamster bone marrow as an *in vivo* test system. I Cytogenetic results on basic aspects of the methodology, obtained with alkylasing agents. Humangenetik 11: 103–118.

Schmeiser, H.H., Stiborova, M. and Arlt, V.M. 2009. Chemical and molecular basis of the carcinogenicity of Aristolochia plants. Curr. Opin. Drug Discovery Dev. 12: 141–148.

Schmid, W. 1975. The micronucleus test. Mutat. Res. 1: 9–15.

Scrirpsema, J., Dagnino, D. and Gosman, G. 2003. Alcaloides indólicos. pp. 679–705. *In*: C.M.O. Simões, E.P. Schenkel, G. Gosman, J.C.P. Mello, L.A. Mentz and P.R. Petrovick (eds.). Farmacognosia: da planta ao medicamento. 2. ed. Editora da Universidade UFRGS, Editora da UFSC, Florianópolis, Brazil.

Shibutani, S., Dong, H., Suzuki, N., Ueda, S., Miller, F. and Grollman, A.P. 2007. Selective toxicity of aristolochic acids I and II. Drug. Metab. Dispos. 35: 1217–1222.

Silva, J., Erdtmann, B. and Henriques, J.A.P. 2003. Genética Toxicologia. Alcance, Porto Alegre, Brazil.

Skehan, P., Storeng, R., Scudiero, D., Monks, A., McMahon, J., Vistica, D., Warren, J.T., Bokesch, H., Kenney, S. and Boyd, M.R. 1990. New colorimetric cytotoxicity assay for anticancer-drug screening. J. Natl. Cancer Inst. 82: 1107–1112.

Song, M.K., Kim, Y.J., Song, M., Choi, H.S., Park, Y.K. and Ryu, J.C. 2012. Formation of a 3,4-diol-1,2-epoxide metabolite of benz[a]anthracene with cytotoxicity and genotoxicity in a human *in vitro* hepatocyte culture system. Environ. Toxicol. Pharmacol. 33: 212–225.

Spielmann, H., Grune, B., Liebsch, M., Seiler, A. and Vogel, R. 2008. Successful validation of *in vitro* methods in toxicology by ZEBET, the National Centre for Alternatives in Germany at the BfR (Federal Institute for Risk Assessment). Exp. Toxicol. Pathol. 60: 225–233.

Steele, V.E. and Kelloff, G.J. 2005. Development of cancer chemopreventive drugs based on mechanistic approaches. Mutat. Res. 591: 16–23.

Struwe, L. 2002. Gentianales (Coffees, Dogbanes, Gentians and Milkweeds), Encyclopedia of Life Sciences. Macmillan Publishers, Nature Publishing Group, UK.

Surh, Y.J. and Ferguson, L.R. 2003. Dietary and medicinal antimutagens and anticarcinogens: molecular mechanisms and chemopreventive potential-highlights of a symposium. Mutat. Res. 523: 1–8.

Tagliati, C.A., Silva, R.P., Féres, C.A.O., Jorge, R.M., Rocha, O.A. and Braga, F.C. 2008. Acute and chronic toxicological studies of the Brazilian phytopharmaceutical product Ierobina®. Rev. Bras. Farmacogn. 18: 676–82.

Tavares, A.C., Gonçalves, M.J., Cruz, M.T., Cavaleiro, C., Lopes, M.C., Canhoto, J. and Salgueiro, L.R. 2010. Essential oils from *Distichoselinum tenuifolium*: Chemical composition, cytotoxicity, antifungal and anti-inflammatory properties. J. Ethnopharmacol. 130: 593–598.

Teaf, C.M. and Middendorf, P. 2000. Mutagenesis and genetic toxicology. pp. 239–264. *In*: P.L. Williams, R.C. James and S.M. Roberts (eds.). Principles and Practices of Toxicology: Industrial and Environmental Applications. John Wiley & Sons, Inc., New York, USA.

Tedesco, S.B. and Laughinghouse, IV H.D. 2012. Bioindicator of genotoxicity: the *Allium cepa* test. pp. 137–156. *In*: J.K. Srivastava (ed.). Environmental Contamination. InTech Publisher, Rijeka, Croácia.

Ugaz, O.L. 1994. Investigación Fitoquímica, 2 ed. Pontífica Universidade del Peru. Fondo editorial, Lima, Peru.

Ulbricht, C., Brendler, T., Gruenwald, J., Kligler, B., Keifer, D., Abrams, T.R., Woods, J., Boon, H., Kirkwood, C.D., Hackman, D.A., Basch, E. and Lafferty, H.J. 2005. Natural Standard Research Collaboration, Lemon balm (*Melissa officinalis* L.): an evidence-based systematic review by the Natural Standard Research Collaboration 5: 71–114.

Umbuzeiro, G.A. and Vargas, V.M.F. 2003. Teste de mutagenicidade com *Salmonella typhimurium* (Teste de Ames) como indicador de carcinogenicidade em potencial para mamíferos. pp. 81–112. *In*: L.R. Ribeiro, D.M.F. Salvadori and E.K. Marques (eds.). Mutagênese Ambiental. ULBRA, Canoas, Brazil.

Valentin-Severin, I., Hegarat, L.L., Lhuguenot, J.C., Bon, A.M.L. and Chagnon, M.C. 2003. Use of HepG2 cell line for direct or indirect mutagens screening: comparative investigation between comet and micronucleus assays. Mutat. Res. 536: 79–90.

Varella, S.D., Pozetti, G.L., Vilegas, W. and Varanda, E.A. 2004. Mutagenic activity of sweepings and pigments a household-wax factory assayed with *Salmonella typhimurium*. Food Chem. Toxicol. 42: 2029–35.

Vogel, E.W. and Szakmary, A. 1987. Evaluation of potential mamalian genotoxings using *Drosophila melanogaster*. Mutagenesis 3: 161–171.

Wang, J., Sawyer, J.R., Chen, L., Chen, T., Honma, M., Mei, N. and Moore, M.M. 2009. The mouse lymphoma assay detects recombination, deletion, and aneuploidy. Toxicol. Sci. 109: 96–105.

Wasson, G.R., McKelvey-Martin, J. and Downes, C.S. 2008. The use of the comet assay in the study of human nutrition and cancer. Mutagenesis 23: 153–162.

Wesam, R.K., Ghanya, A.N., Mizaton, H.H., Ilham, M. and Aisha, A. 2013. Assessment of genotoxicity and cytotoxicity of standardized aqueous extract from leaves of *Erythroxylum cuneatum* in human HepG2 and WRL68 cells line. Asian Pac. J. Trop. Med. 1: 811–816.

Weyermann, J., Lochmann, D. and Zimmer, A. 2005. A practical note on the use of cytotoxicity assays. Int. J. Pharm. 288: 369–376.

Zanon, S., Ceriatti, F., Rovera, M., Sabini, L. and Ramos, B. 1999. Search for antiviral activity of certain medicinal plants from Córdoba, Argentina. Rev. Latinoam. Microbiol. 41: 59–62.

Žegura, B., Heath, E., Černoša, A. and Filipič, M. 2009. Combination of *in vitro* bioassays for the determination of cytotoxic and genotoxic potential of wastewater, surface water and drinking water samples. Chemosphere 75: 1453–1460.

Zhang, Y., Wua, X., Ren, Y. and Fu, J. 2004. Safety evaluation of a triterpenoid-rich extract from bamboo shavings. Food Chem. Toxicol. 42: 1867–75.

Zhou, X., Li, Y. and Chen, X. 2010. Computational identification of bioactive natural products by structure activity relationship. J. Mol. Graphics Modell. 29: 38–45.

Zucco, F.D.E., Angelis, I., Testai, E. and Stammati, A. 2004. Toxicology investigations with cell culture systems: 20 years after. Toxicol. *In Vitro* 18: 153–163.

The Medicinal Species
Euphorbia hirta and *E. hyssopifolia*
(Euphorbiaceae): Uses, Phytochemical
Data and Genetic Diversity

Karla C.B. Santana,[1] Diego S.B. Pinangé,[1] Karina P. Randau,[1]
Marccus Alves[1] and Ana M. Benko-Iseppon[2,*]

Introduction

For many populations around the world, phytomedicines figure as the main form of treatment for a wide range of health problems, mostly being the first line of defense against potential health hazards (Rahman 2012). Although access to the modern medicine is available in most countries, the use of medicinal herbs has retained its popularity due to historical and cultural reasons (Agra et al. 2008). Based on the historical evolution of the use of medicinal plants, in 1978 the World Health Organization (WHO) started to recognize medicinal herbs as an alternative therapy for human diseases. Additionally, the use of natural compounds from biological sources, as well as ethnobotanical knowledge, has emerged as an important source for discovering new products (Albuquerque and Hanazaki 2006, Li and Vederas 2009, Desmachelier 2010, Newman and Cragg 2012). Furthermore, in recent years, bioinformatic approaches, associated to the advanced techniques of separation, structure elucidation, screening and combinatorial synthesis have led to the consolidation of the usage of plants as a source of new drugs (Saklani and Kutty 2008, Sharma and Sarkar 2013).

Brazil is recognized by high and well-documented biodiversity mainly based on abundant records of use of its flora for traditional uses and medicinal purposes (Desmachelier 2010, Nogueira et al. 2010, Brandão et al. 2012). Additionally, in the northeastern region of Brazil, natural products have been largely used for medicinal purposes. Further, in the poorest regions and even in big cities, medicinal plants are still commercialized in popular street markets, as well as cultivated in residential backyards (de Almeida 1993, Agra et al. 2008, Veiga Junior 2008, Pasa et al. 2011, Benko-Iseppon and Crovella 2010, Benko-Iseppon et al. 2012). In a recent survey conducted by Benko-Iseppon et al. (2012) regarding the potential use of

[1] Federal University of Pernambuco, Genetics Dept., Recife - PE, CEP 50.670-420 Brazil.
[2] Department of Genetics, Federal University of Pernambuco, Av. Prof. Moraes Rego s/n, CEP 50732-970 Recife, PE, Brazil
* Corresponding author: ana.benko.iseppon@pq.cnpq.br

medicinal plants in the Brazilian northeastern region, most species and even genera used for therapeutic and medicinal purposes were recognized as endemic. Thereafter, more multidisciplinary studies such as taxonomy, physiology, genetics and phytochemistry are mandatory in order to shed light on the identification and assignment of new compounds, their uses and applications. In this sense, population studies have proven to be essential, especially considering diverging genetic background together with environmental forces acting on plant phytochemical content.

Among several plant groups used in Brazilian folk medicine, Euphorbiaceae is one of the most recorded families for medicinal use (Agra et al. 2008, Benko-Iseppon and Crovella 2010, Forzza 2010, Benko-Iseppon et al. 2012). According to APG III (APG 2009) the so-called Euphorbiaceae *sensu stricto* consists of four subfamilies—Acalyphoideae, Crotoneideae, Euphorbioideae and Cheilosoideae. Within Euphorbioideae the genus *Euphorbia* L. figures as the largest Euphorbiaceae genus and one of the most diverse groups of the plant kingdom, with approximately 2,160 species (Webster 1994a, Steinmann and Porter 2002, Bruyns et al. 2006). *Euphorbia* encompasses a large morphological and adaptive diversity, reflected sometimes also in the types of metabolisms found as, for example, *E. tirucalli*, a species that displays the C3 or C4 photosynthetic systems depending on the plant part analyzed (Hans 1973, Van Damme 2001, Sage et al. 2011).

Regarding taxonomic data, *Euphorbia* has been largely discussed as a natural genus (Webster 1987, Webster 1994b, Park et al. 2000). Some studies such as those performed by Aqueveque et al. (1999) and Domsalla et al. (2010) have proposed a taxonomic delimitation based on chemotaxonomy of flavonoid and latex proteinase profiles. With respect to the current classification, some authors consider that the group support is weak and presents many inconsistencies (Steinmann and Porter 2002, Bruyns et al. 2006, Park and Jansen 2007, Tokuoka 2007). Nevertheless, Zimmermann et al. (2010) proposed the current most accepted division of *Euphorbia* in four subgenera: *Chamaesyce*, *Euphorbia*, *Esula* and *Rhizanthium*.

With regard to ecological and economic importance, *Euphorbia* species have often high invasive potential, with consequent damage to agriculture, including the reported *E. hirta* L. and *E. heterophylla* L. (Willard and Griffin 1993, Aarestrup et al. 2008). On the other hand, Bellini et al. (2008) reported the significance of some *Euphorbia* invasive species as pioneer plants and in the biological control of pest mites in plantations, hindering the spread of pathogens. Some *Euphorbia* taxa have commercial significance, as the crown-of-thorns (*E. milii*) and the Poinsettia (*E. pulcherrima*), a species largely used especially for Christmas decorations. Also the Baseball-plant (*E. obesa*) deserves mentioning as an ornamental plant (Lee 2000, Mwine and Van Damme 2011); besides, some species with abundant latex as *E. tirucalli* and *E. lathyris* are known for their potential for biodiesel production (Duke 1983, Van Damme 2001).

Biochemical diversity and medicinal use of *Euphorbia* species

Most *Euphorbia* species are useful in popular medicine, although often the dosages, efficacy and possible toxic effects remain unclear. Several species have been used in the treatment of various diseases, such as skin diseases, gonorrhea, migraines, intestinal parasites and warts (Singla and Pathak 1990, Appendino and Szallasi 1997, Shi et al. 2008). Some researchers have demonstrated that *Euphorbia* species also feature antitumor (de Melo et al. 2011, Zhang et al. 2011), antimicrobial (Sudhakar et al. 2006, Murugan et al. 2007, Akinrinmade and Oyeleye 2010, Benko-Iseppon and Crovella 2010) and antiviral activities (Hezareh 2005, Tian et al. 2010). Further effects as analgesic and antipyretic (Ma et al. 1997), anti-anaphylactic (Youssouf et al. 2007) and antioxidant (Barla et al. 2007) have been also reported. Records indicate activity in intestinal motility (Hore et al. 2006) and aid in the treatment of diabetes mellitus (Widharna et al. 2010). Such observations led to the development of several patented drugs by compounds from various *Euphorbia* species, as listed by Mwine and Van Damme (2011).

These medicinal properties are considered to be derived from the rich content of secondary metabolites in species of *Euphorbia*. A large number of studies have revealed the presence of alkaloids, diterpenes, glucosinolates, tannins, triterpenes, steroids and large amounts of phenolic compounds (Seigler 1994, Shi et al. 2008), even considering the still scarce amount of data available regarding the chemical diversity of these compounds among populations. In 1999, Aqueveque et al., in a chemotaxonomic study with 12 Chilean species of *Euphorbia* including *E. hirta*, isolated 22 flavonoids (13 flavonols, eight flavones and one flavanone) using thin layer chromatography. The flavonolic profile proved to be an excellent marker on *Euphorbia*, due to the high number of compounds observed, allowing an effective comparison among species, also uncovering a closer relationship among species of subgenus *Chamaesyce* when compared to the other *Euphorbia* members analyzed.

On the other hand, addressing the triterpenoid profiles from latex of 56 accessions of European leafy spurges, Holden and Mahlberg (1992) observed similar qualitative and quantitative profiles for the species *E. amygdaloides, E. agraria, E. cyparissias, E. lucida,* and *E. seguierana,* while 37 accessions of the *E. esula* complex were separated into 15 groups based on variability of the components, suggesting that different profiles exist within populations of Europe and North America. Further, the composition of triterpenes in *E. esula* presented a high level of qualitative and quantitative stability under diverse environmental and physiological conditions, indicating a genetic basis for triterpenoid synthesis (Mahlberg et al. 1987).

Genetic diversity in *Euphorbia* L.

Among several concepts adopted for biological diversity, perhaps the simplest one may be defined as the variation present in all species of plants and animals associated with their genetic material and the ecosystems in which they thrive. The relevance of biodiversity for humankind has been well recognized and reported in the recent decades and many consider that diversity is essential for allowing sustainable development of various human activities, in different forms and levels (Frankel and Bennet 1970, Shiva 1994, Swingland 2001, Rao and Hodgkin 2002). Thus, the importance of adopting a holistic view of biodiversity, including applied research, such as agricultural biodiversity, as well as linking conservation with sustainable use, has been the central theme in research activities conducted in the last decades (Dias and Kageyama 1991, Arora 1997, Frankham et al. 2002).

In the literature, the idea that different types of molecular markers have the ability to differentiate genomes of interest is already widespread, generating data on diversity levels and distribution, generally using genetic polymorphism as background. Information on genetic diversity, and therefore gene frequency, produce additional insights that can be combined with morphological, physiological and agronomic data, providing a complete analysis especially when combined with other research areas (Carvalho et al. 2000a, 2000b, Huang et al. 2002, Faleiro 2007), including ethnobotanical and phytochemical profiling, as in the present chapter.

Although *Euphorbia* is the largest genus within Euphorbiaceae, surprisingly, there are few studies involving analysis of genetic diversity via molecular markers in this group. Most of them evaluated *Euphorbia* species from North America and Europe (Nissen et al. 1992, Park et al. 1997, Rowe et al. 1997, Morden and Gregoritza 2005), often focusing on phenetic and phylogenetic data, as well as indexes of population polymorphisms and differentiation, using RAPD (Random Amplified Polymorphic DNA) markers (Rowe et al. 1997, Morden and Gregoritza 2005), Isoenzymes (Park et al. 1997), Alloenzymes (Park 2004), RFLPs (Restriction Fragment Length Polymorphism—Nissen et al. 1992, Rowe et al. 1997), sequencing of cpDNA (Chloroplast DNA—Nissen et al. 1992, Rowe et al. 1997) and ITS (Internal Transcribed Spacer—Morden and Gregoritza 2005), among others. Such evaluations allowed not only the identification of significant levels of polymorphism, but also some insights regarding the relationships between species and/or populations.

In Brazil, a single record is available regarding the analysis of diversity and structure of wild-poinsettia (*E. heterophylla*) populations performed by Frigo et al. (2009). In this evaluation, the esterase loci polymorphisms were analyzed revealing mostly high degree of differentiation in the sampled populations with subsequent deficit of heterozygosity probably caused by inbreeding.

Euphorbia hirta L. and *E. hyssopifolia* L.

Belonging to the subgenus *Chamaesyce, Section Anisophyllum* (Yang and Berry 2011), *E. hirta* and *E. hyssopifolia* (see Fig. 8.1) are widely used as medicinal herbs in folk medicine around the world. Both are sub-spontaneous and ruderal species, native to the New World, tolerant to high temperatures and drought (Stehmann et al. 2009, ZCZ 2013a, 2013b). They are widely distributed in tropical and subtropical

Figure 8.1. (A) *Euphorbia hyssopifolia* and (B) *E. hirta* (Credit to the authors).

regions, from the sea level up to 1500 m (Amorozo 2002, Schneider 2007, Euphorbia PBI 2013a, 2013b). In Brazil, they occur in all regions and biomes, where they inhabit degraded areas, roadsides, cultivated fields and gardens (Steinmann et al. 2013a, 2013b).

The species *E. hirta* (synonym *Chamaesyce hirta*, among others) is a semi-prostate or erect herb with simple, opposite and ovate-elliptic or lanceolate leaves, usually asymmetric, serrated margins. The inflorescence is a cyathium with tiny orbicular glands, without petaloid appendages and pubescent fruits. In English this species is called hairy-spurge, asthma plant, garden-spurge, pill-bearing-spurge, pill-pod broomspurge, pill-pod-sandmat, asthma weed, cat's-hair, spurge or milkweed (Johnson et al. 1999, ZCZ 2013b), whereas in Portuguese is known as '*erva-de-Santa-Luzia*', '*erva-de-cobra*', '*erva-de-sangue*', '*erva-andorinha*', '*burra-leiteira*', '*tranca-cu*' and '*quebra-pedra*' (Lorenzi 2000, Amorozo 2002, Guglieri-Caporal et al. 2011).

In turn *E. hyssopifolia* (*E. brasiliensis* and *Chamaesyce hyssopifolia*, among other synonyms) is an erect or semi-prostate herb with simple, opposite and oblong-elliptic to oblong-obovate leaves finely serrated margins. The inflorescence is a cyathium (terminal or pseudo-axillary) with oblong to reniform glands, petaloid appendages and glabrous fruits. It is recognized by the same common names in Portuguese—'*erva-de-andorinha*', '*erva-de-Santa-Luzia*', '*burra-leiteira*', '*quebra-pedra*', '*porca-parideira*', '*sete-sangrias*' and '*erva de cobra*' (Lorenzi 2000, Albuquerque et al. 2005, Agra et al. 2008, Pasa et al. 2011, Menezes et al. 2012) whereas in English it has been named hyssop-spurge, hyssop-leaf broomspurge, hyssop-leaf-sandmat and leafy-spurge (ZCZ 2013a).

In traditional medicine, several parts of the plant have been used: leaves, flowers, roots, latex, aerial parts and the whole plant. *E. hirta* has been studied more frequently, being widely used in different preparations for gastrointestinal disorders (gastritis, diarrhea, dysentery, colic, used as vermifuge, amoebicide, purgative, colagogue and to treat ulcers), bronchial and respiratory diseases (cough, coryza, hay asthma, asthma, bronchial affections), skin affections (acne, wart, rashes, furuncle, ringworm, measles, chickenpox, smallpox wounds, splinter removal), pinkeye, kidney stones, edema and hypertension, as galactagogue, and in snake bites (Watt and Breyer-Brandwijk 1962, Anjaria et al. 1997, Mhaskar et al. 2000, Agra et al. 2008, Kumar et al. 2010). In addition, it has been confirmed that a wide range of properties, such as antibacterial, antiamoebic, anthelmintic, antifungal, antiplasmodial, antiviral, antineoplastic, spasmolytic, antidiarrhoeic, decreased the gastrointestinal motility, sedative, antidepressant, anxiolytic, analgesic, antipyretic, anti-inflammatory, immunomodulatory, antiallergic, antihypertensive, as diuretic and as an adjunctive agent in diabetes mellitus treatment (Williams et al. 1997, Johnson et al. 1999, Tona et al. 2000, 2004, Hore et al. 2006, Ogbulie et al. 2007, Anuradha et al. 2008, Loh et al. 2009, Kumar et al. 2010, Ramesh and Padimavathi 2010, Shih et al. 2010, Widharna et al. 2010, Alisi and Abanobi 2012).

The second species, *E. hyssopifolia* is used to treat warts, corns, conjunctivitis and in external ulcers, as a purgative, also as a diuretic and in dysuria, as emmenagogue, as well as to expel placenta, against flu, coughs, fever, for eyes diseases and high blood pressure (Watt and Breyer-Brandwijk 1962, Arnason et al. 1980, Morton 1980, Agra et al. 2008, Brandão et al. 2012). It was also found that its metabolites have significant inhibitory effects on HIV-1 reverse transcriptase (Matsuse et al. 1999, Lim et al. 1997).

The problem: two species one common name

It is well known that *E. hirta* and *E. hyssopifolia* are used in folk medicine with recognized variety of pharmacological actions, sometimes shared by both. It reflects on the popular names they have in common: as *Erva de Santa Luzia* (related to the saint in charge of eye problems), *quebra-pedra* ('stone breaker' regarding its diuretic properties, a name also used for *Phylanthus niruri* L.), and *erva de cobra* ('snake-herb' related to its use after snake bites). However, most of the characteristic features of plant secondary metabolism regard its vast chemical diversity and intraspecific variation. Also, due to several reports regarding the toxicity related to species from this group, the quality of medicinal plants and their products should fulfill safety requirements, efficacy and stability in order to be used by pharmaceutical industries (WHO 2002, Mwine and Van Damme 2011, Kheyrodin and Ghazvinian 2012). Then, the use of homogeneous genotypes with well-known and stable phytochemical and biological properties is an important requirement for use considering pharmaceutical purposes.

Unfortunately, for many plant species data on metabolites stability or toxicity is not available, demanding a better knowledge of their chemical features, besides some variations that they may present due to their occurrence in contrasting environmental and ecological conditions. Especially in the case of non-cultivated medicinal plants, plant parts collected from the field may exhibit drastic differences in their compounds and therapeutic effects as well as in its toxicity. Therefore, it is critical to provide information about the phytochemical differences within species in different habitats as observed in *Arnica Montana*, considering different conditions of climate and altitude (Spitaler et al. 2006).

In the presented pilot study, two sources of polymorphisms were accessed: phytochemical (TLC—Thin Layer Chromatography) and genetic markers (ISSR—Inter Simple Sequence Region). For that we performed two experiments: (1) Chemical differentiation associated to color-associated characters (individuals of the same species with red versus green stem) and species distinction (leaves and stem of *E. hirta* versus the ones of *E. hyssopifolia*) and (2) Differences in secondary metabolism through methanolic screening, including flavonolic population diversity together with evaluation of population's genetic diversity. In the case of both species (*E. hirta* and *E. hyssopifolia*) the use of the same common name has induced herb sellers to provide parts of both species for different uses. Thus, the present study aimed to provide preliminary evidence on the diversity of the compounds contained in the species level and different areas of occurrence, and also an analysis of genetic diversity with ISSR.

Experimental design of an exemplary pilot test

Specimen collection

Specimens of *E. hirta* (EHIR) and *E. hyssopifolia* (EHYS) were collected in August 2009, in the city of Recife in urban Atlantic coastal rainforest fragments, as well as, in a locality in the semi-arid region in the inland of state of Pernambuco, Northeast of Brazil (Fig. 8.2). The populations from the city were ca. 85 km distance from the inland population and about 4 km from each other, including periurban and rural areas, with different grades of antropization (Table 8.1).

Search for polymorphisms

In the presented pilot study, two sources of polymorphisms were accessed: The phytochemical study was carried on into two experiments: the first regarding the differentiation of species and color character differences in secondary metabolism through methanolic screening (Experiment 1) while the second

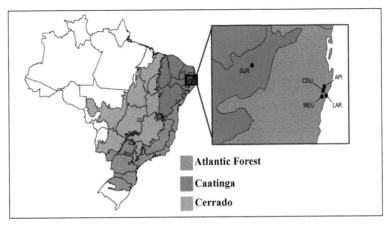

Figure 8.2. Collected populations from the Pernambuco state (Brazilian northeastern region), including four areas from Atlantic forest (API = Apipucos; CDU = Cidade Universitária; LAR = Lagoa do Araçá and MEU = Mata do Engenho Uchoa) and one (SUR = Surubim) from Caatinga (semi-arid) environment.

Table 8.1. Studied populations with their provenances and bio-geographical characteristics.

Samples plot		Cidade Universitária	Apipucos	Lagoa do Araçá	Mata do Engenho Uchoa	Surubim
Geographical coordinates	**Latitude**	8°3'4.09"S	8°1'10.34"S	8°5'32.41"S	8°5'53.91"S	7°50'35.82"S
	Longitude	34°56'47.38"O	34°56'0.56"O	34°54'51.04"O	34°57'37.08"O	35°42'7.95"O
Altitude		17 m	10 m	6 m	23 m	317 m
Sample abbreviation	*E. hirta*	EHIR-CDU	EHIR-API	EHIR-LAR	EHIR-MEU	EHIR-SUR
	E. hyssopifolia	EHYS-CDU	EHYS-API	EHYS-LAR	EHYS-MEU	EHYS-SUR
Plots characteristics	**Anthropization degree**	High	High	High	Moderate	Low
	Biome	Atlantic Forest	Atlantic Forest	Atlantic Forest	Atlantic Forest	Caatinga
	Environment	Urban	Urban	Urban	Ruderal	Rural

Climatic means in august 2009: (Source: INMET—Instituto Nacional de Meteorologia). Recife (CDU, API, LAR and SUR)—temperature 25°C, humidity 80%, radiation 1500 kjm, precipitation 12 mm/day; Surubim—temperature 23°C, humidity 70%, radiation 1700 kjm², precipitation 3 mm/day.

analyses focuses on population's flavonolic and genetic diversity using molecular markers (Experiment 2). The workflow presented here considering the mentioned approaches (Fig. 8.3), may be applied to any near-related species and their populations (as in the case of the present work) or even to a single species with different populations.

Experiment 1

The assembly of the experiment was based on the previous observation of color difference (reddish and greenish), existing on the stems (stem) of both species. Therefore, this experiment aimed to verify if the separation of species from its phytochemical profile, as well as assessing possible phytochemical variation associated with morphological differentiation of stem color.

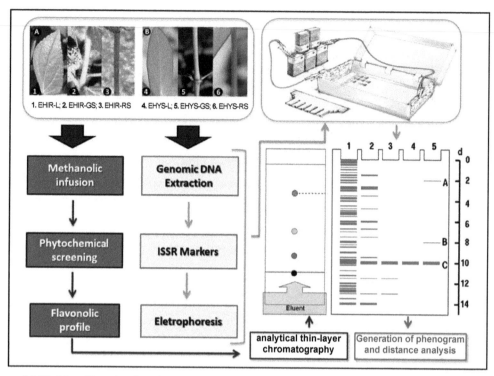

Figure 8.3. Workflow illustrating the main steps for phytochemical screening, chemical analysis and also genetic diversity with ISSR (Inter Simple Sequence Repeat) markers, also considering morphological and biogeographical peculiarities of the samples analyzed.

One specimen of each species was collected at the same time in one of the plots of the 'population' experiment—CDU (details in Table 8.1). Three samples were selected from each species: leaves—L; green stem—GS and red stem—RS, resulting six samples—EHIR-L, EHIR-GS, EHIR-RS, EHYS-L, EHYS-GS and EHYS-RS. Methanol extracts were obtained by using 10 ml of methanol per 5 g of sample. Secondary metabolites were separated thought thin-layer chromatography, according to the methodology described for each metabolite, and compared with the available patterns (Table 8.2). Finally, the plates were analyzed for presence/absence of metabolites and the number of compounds obtained in each population.

Experiment 2

The leaves collected for this experiment were split for genetic and phytochemical analysis. The environment chosen had different levels of degradation and human disturbance. Each population included 10 individuals, for both analyses. The individuals were grouped in a bulk with 5 g of stem/leaf mass per individual, in a total of 10 bulks. The characteristics of the locations and samples analyzed are shown in Table 8.1. For the phytochemical data, a phenolic (flavonoids and cinnamic derivatives) screening in chromatography was performed, where we used 10 ml of methanol per sample and the same method from Experiment 1.

In relation to the molecular analysis, the genomic DNA extraction from each bulk was performed according to the protocol described by Weising et al. (2004). The molecular marker employed was the dominant ISSR marker. The choice of ISSR marker was based on the premise that their target sequences are abundant throughout the eukaryotic genome, regard rapidly evolving regions and show high reproducibility (Fang and Roose 1997, Esselman et al. 1999, Pinangé 2009, Vasconcelos et al. 2012). Thus, the amplification reactions followed the protocol described by Bornet and Branchard (2001), with modifications performed

Table 8.2. Analyzed metabolites and employed chromatographic systems.

Metabolites	Mobile phase	Developer	Reference
Alkaloids	A-A-A-A	Dragendorff	Wagner and Bladt 1996
Mono- and sesquiterpenes	T-A(3%)	Vanillin sulphuric	Wagner and Bladt 1996
Triterpenes and steroids	T-A(10%)	Liebermann/Bouchardat	Sharma and Dawra 1991
Iridoids	A-A-A-A	Vanillin sulphuric	Wagner and Bladt 1996, Harborne 1998
Flavonoids	A-A-A-A	UV-NEU	Harborne 1998, Markham 1982, Neu 1956, Wagner and Bladt 1996
Cinnamic derivatives	A-A-A-A	UV-NEU	Wagner and Bladt 1996
Phenyl-propane-glycosides	A-A-A-A	UV-NEU	Wagner and Bladt 1996
Coumarins	T-E-A	UV-KOH 5%	Wagner and Bladt 1996
Condensed proanthocyanidin andleuco-anthocyanidin	A-A-A-A	Vanillin hydrochloric	Roberts et al. 1957
Hydrolysable tannins	T-E-A	UV-NEU	Xavier et al. 2002
Saponines	A-A-A-A	Anisaldehyde	Wagner and Bladt 1996

Chromatographic conditions employed: A-A-A-A = AcOEt – OHCOOH – AcOH – H_2O (100:11:11:26). T-E-A = Toluene-Et2O (1:1:Sat. AcOH 10%). T-A3% = Toluene-AcOEt (97:3). T-10% = Toluene-AcOEt (90:10). A-B-T = Me_2CO-n-BuOH-phosphate buffer pH 5,0 (5:4:1). W-H-M (Wagner-Harborne-Markham). W-H (Wagner-Harborne). NEU = sol. 1% of Diphenyl-boriloxi-ethilamine in MeOH. UV = ultraviolet 365 nm. Dragendorff = according to Munier and Macheboeuf.

by Amorim (2009). The amplified products were electrophoretically separated in ethidium-bromide agarose gel (1,8%), visualized and photographed under ultraviolet light with Sony Cyber W5 digital camera (3.0 megapixel optical zoom).

For data analysis, the criteria of presence (1) and absence (0) of bands (DNA fragments) were employed, in order to generate binary matrices for phenetic studies and identification of polymorphisms. A phenogram was generated from the Neighbor-Joining method (1000 bootstrap replicates), using the software MEGA 4.1. Concerning the analysis of polymorphisms, the rates of Genetic Divergence (GD%) was also calculated for each population. The genetic divergence was obtained from the ratio between the Total Number of Polymorphic Loci (TNPL) and the total number of loci (TNL)—total 'GD' as well as, the total number from each population ('Pop GD'), according to Nei (1972).

Phytochemical differentiation in *E. hirta* and *E. hyssopifolia* by methanolic screening

As already mentioned, this experiment aimed to analyze the chemical components of each species and also uncover distinctions between chemotypes in an intraspecific level (Fig. 8.4A). Thus, the methanolic screening (Fig. 8.4B–D) revealed several secondary metabolites in both species, as described in Table 8.2, with the observation of the following classes of metabolites: mono and sesquiterpenes, triterpenes and steroids, flavonoids and cynnamic derivatives. In addition, the generation of a phytochemical profile, especially phenolic compounds, permitted the separation of the studied species.

In *E. hirta*, the literature reports that tannins, flavonoids, phenolic acids, saponins and amino acids occur (Hore et al. 2006). A recent screening performed by Alisi and Abanobi (2012) showed that *E. hyssopifolia* and *E. hirta* exhibit positive reactions for alkaloids, saponins, tannins, flavonoids, cardiac glycosides and steroidal aglycones. In this study, the cyanogenic glycosides were present in *E. hyssopifolia*, but not detected in *E. hirta* and neither tannins nor saponins were found in detectable amount. One of the possible explanations is the different protocols used here and by Trease and Evans (1989). Nevertheless, since the study intended to observe polymorphisms, the compounds found were enough to distinguish chemotypes and/or species.

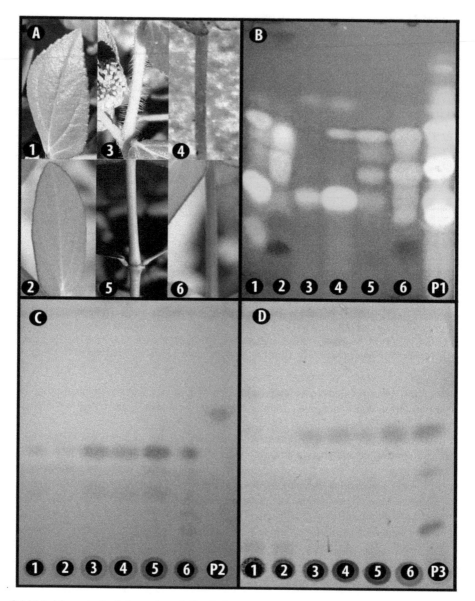

Figure 8.4. Material examined and observed metabolites. In (A), parts of plants used in the samples. The secondary metabolites are represented in (B) (flavonoids and phenyl-propane glycosides), (C) (mono and sesquiterpenes) and (D) (triterpenes and steroids). Samples: 1-EHIR-L; 2-EHYS-L; 3-EHIR-GS; 4-EHIR-RS; 5-EHYS-GS; 6-EHYS-RS. Standards: P_1-Artichoke (Rutin, luteolin-7-glicoside, cryptochlorogenic acid and cynarine); P_2-Camphor; P_3-β-Sitosterol, β-Amirin, ursolic acid.

Terpenes and steroids belong to a class of compounds widely spread for the genus (Haba et al. 2007, Ogunlesi et al. 2009). For example in *E. hirta,* some of the essential oils identified included perhydrofarnesyl acetone, hexadecanal, phytol and palmitic acid, including some minor constituents as butyl glycol, tetradecane, phthalic acid, malonic acid, oleic acid, heptadecynol, hexadecanol and phthalic acid (Ogunlesi et al. 2009). Two up to five mono/sesquiterpenes in *E. hirta* and *E. hyssopifolia*, were found, although, non-identified compounds in comparison to the standard. The foliar samples (EHIR-F, EHYS-F) exhibited traces of two compounds in both species. Those compounds were shared with stem samples of both species, with no concentration relation with the stem color. For *E. hyssopifolia*, besides these makers, we observed the occurrence of an additional band shared in EHYS-GS and EHYS-RS

(Fig. 8.4C). Still, the EHYS-RS sample showed two others metabolites not observed in EHYS-GS, indicating the distinction in two chemotypes the green and red stems samples in *E. hyssopifolia*.

In the search for triterpenes and steroids of *E. hirta* and *E. hyssopifolia*, all samples showed only the steroid β-sitosterol, already described for both species (Morton 1980, Johnson et al. 1999). β-sitosterol is a main dietary phytosterol found in plants, however, although they are structurally similar to cholesterol, they have been shown to exert significant unique biochemical effects in both animals and humans (Choi et al. 2010). In the present chapter, both species exhibited a slight increase in the concentration of this metabolite of red stems (EHIR-RS and EHYS-RS) when compared to the green ones (EHIR-GS and EHYS-GS), suggesting a possible relation to the morphological character. Additionally, the leaves showed a lower amount of this metabolite if compared to the stems samples. Furthermore, in the chemical characterization of *E. hirta* the compounds also included: α-amyrin, β-amyrin, campesterol, cycloartenol, friedelin, sitosterol, stigmasterol, taraxerol and taraxerone (Mors et al. 2000). Although the two first compounds (α-amyrin, β-amyrin) have an important role in the toxicity of the species (Kheyrodin and Ghazvinian 2012), it was not found in a detectable quantity by the methodology used here.

In relation to the phenolic compounds, the literature shows a high diversity of types in both species. In *E. hyssopifolia* quercetin, apigenin, corilagin, gallic acid, glucogallin, kaempferol and luteolin (and their glicosides) (Lim et al. 1997, Caballero-George and Gupta 2011) were reported. In turn, for *E. hirta* gallic acid, galloylquinic acid, glucogallin, myricitrin, quercetin and rhamnetin (and their glucosides) were described (Chen 1991, Mors et al. 2000). In the present study we observed a higher number and variety of compounds, among all the samples of the experiment, as it was expected. The leaf samples of both species showed a high amount of compounds, with the vast majority common to both species. In comparison to the standard, it was possible to distinguish the following compounds: rutin, luteolin-7-glucoside, cryptochlorogenic acid and cynarine. The first mentioned was observed in leaves of both species and in EHYS-GS. The presence of luteolin-7-glucoside was universal for all samples analyzed, in high quantity in leaves and red stems. Controversially, another flavonoid (not identified here) was common to all samples, but in higher levels in EHIR-L. Besides the inter-specifics similarities, the cryptochlorogenic acid was highly produced in EHIR, and exhibiting a higher concentration in EHIR-L and EHIR-RS, probably as a response to sun exposition and as an important UV-protector in the species, though it was found also for EHYS-L and EHYS-GS.

Cynarine was found exclusively in EHYS samples, highlighting the clear separation of species through this methodology, together with one cinnamic derivative present only in EHIR samples and another exclusive for EHIR-L. It is also interesting to note that while EHYS-GS exhibited rutin and cryptochlorogenic acid, EHYS-RS exhibited two others, not identified in comparison with the standard used. This is maybe due to a physiological balance of compounds necessary as an answer to environmental adversities. Thus, in a general comparison between species, *E. hirta* presented semi-quantitatively greater concentration of cinnamic derivatives compared to *E. hyssopifolia,* which excelled in the production of flavonoids (Fig. 8.4B). Again, semi-quantitative differences in metabolite concentration between samples of green and red stems in both species revealed more intensity for the red one.

Environmental conditions, including biotic and abiotic factors, modulate the occurrence rate of secondary metabolisms (Einhellig 1996). In *E. hyssopifolia*, one of the factors that may be responsible for the polymorphism is the soil composition. For a related species *E. thymifolia*, two ecotypes (red and green form) were recognized, based in its calcium uptake (Ramakrishnan 1965). Thus, it has been suggested that the existence of ecotypes for a variety of studies that registered the effects of soil composition in quali- and quantitative differences of flavonoid content (Chludil et al. 2008, Malusà et al. 2006, Ferraro et al. 2010). This suggestion may be the case for the species studied here, but further physiological analysis would be necessary to confirm.

In plants, phytosterols and triterpenes are the major secondary metabolites (Uchida et al. 2009). In the present experiment, those classes were responsible for the differences between the parts of the plants and the type of stems, by reveling a great number of compounds and its differential production. The chemotype presented by a given species can exhibit an increased regulation of particular metabolic pathway, leading to characteristic patterns, in which some chemical components are priority elements at the expense of the smaller ones. In such cases, the chemical cluster provides diversification data, which can result in the expression of several metabolic pathways from an entire population or system (Semmar et al. 2007).

So, it is important to emphasize the significance of noticing the same coloring on the use of extracts from the aerial parts, since for certain metabolites, concentration is largely increased in individuals with the red stem in relation to those with green stems (Fig. 8.4B–D). The chemotype distinction is even more important, considering that the differences observed in the pattern of presence/absence of mono/sesquiterpenes and polyphenols in *E. hyssopifolia* (Fig. 8.4C). Such polymorphisms may reflect both the toxicity and in obtaining the expected therapeutic effects on the use of these plants.

Phenolic and genetic diversification between populations of *E. hyssopifolia* L. and *E. hirta* L.

Phenolic diversity

A wide range of factors such as plant age, season of the year, microbial attack, radiation, temperature, competition, soil composition and nutritional status have proved an impact in the secondary metabolites profiles in Angiosperms (Harborne 1972). Differences on concentrations of phenolic compounds in plants may be the result of active adaptation to the environment or an effect of environmental constraints that regulate the tradeoff between plant growth and synthesis of secondary compounds (Stark et al. 2008). These differences in content were found in the present study. Through analysis of phenolic profile, cynarine was predominant in all samples of EHYS, while the same happened to cryptochlorogenic acid for EHIR. Luteolin-7-glucoside was present in all samples analyzed and also with components not identified in comparison with the standard marker: one second flavonoid and traces of a third one, as well as one cynnamic derivative.

Analyzing the chemical variation and heritability in *Betula pendula*, Laitinen et al. (2005), found that the environment had no significant effect on the accumulation of some compounds, whereas for others, a significant environmental effect and/or significant genotype by environment interaction was found, that suggested that its secondary chemistry is under strong genetic control and that the environmental effects depend on the studied chemical trait. In general, the semi-quantitative analyses of metabolites showed uniformity regarding the concentration of metabolites within populations of same species, perhaps for some neutralization of high and low levels of metabolites that may be presented in specimens that composed the population's bulk. The only reduction took place in *E. hirta* (EHIR-CDU and EHIR-API), which displayed a strong mark also found to match with trace marks in all samples of *E. hyssopifolia*, but apparently not shared with the other populations of *E. hirta* (Fig. 8.5A). The four flavonoids chemotypes found in *Astragalus caprinus* leaves by Semmar et al. (2007) allowed them to affirm that metabolic affinities or competitions between flavonoids would be responsible for the polymorphisms observed. Conversely, for the distinction of chemotypes found in *Cercis*, Isely (1975) proposed two alternatives: two distinct species or two varieties, based on the presence and absence of kaempferol. The first one was better accepted by Salatino et al. (2000), once the two chemical profiles supported environmental features: kaempferol bearers inhabit xerophytics environments while the non-bearers inhabit mesophytics. In the present study, there is no evidence of reproductive isolation within populations of *E. hyssopifolia*. Therefore, such observations can hypothesize the existence of varieties within this species with different genetic combinations for these characters or metabolites that can be a result from the influence of the environment on the secondary metabolisms of this plant.

A broad range of environmental factors changes with the elevation of the natural growing site, including precipitation, mean temperature, soil, wind speed, temperature extremes, duration of winter, length of the vegetation period, and radiation intensities (Körner 1999, Albert et al. 2009). The population from the Atlantic coast rainforest fragments (CDU, API, LAR and SUR) inhabit in coastal climate (about 10 m sea level, annual average of 25°C, 80% of humidity, 1500 kjm of radiation and precipitation of 12 mm/day), while the SUR population is from the semi-arid region, resulting in higher altitude, lower precipitation, but increase of radiation (317 m, 23°C, 70%, 1700 kjm², 3 mm/day, respectively). Although ultraviolet

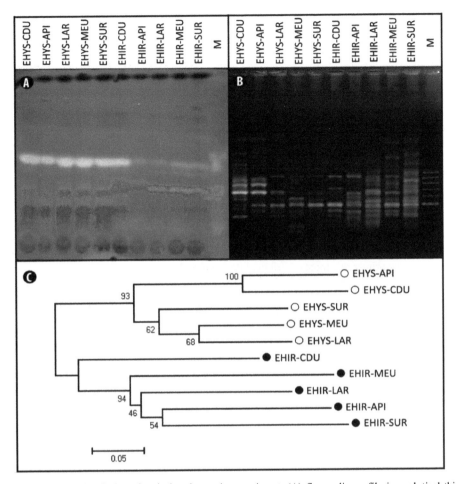

Figure 8.5. Interpopulational phytochemical and genetic experiment. (A) flavonolic profile in analytical thin-layer chromatography. (B) Fragments of DNA amplified via ISSR (primer 811). (C) Neighbor-Joining method Phenogram. M (Marker): In (A) Standard of Artichoke (Rutin, luteolin-7-glicoside, cryptochlorogenic acid and cynarin); in (B) DNA ladder (100 bp); In (C) numbers in the base of nodes regard bootstrap values, whereas bar regards genetic distance.

(UV)-B (280–315 nm) irradiation increases under clear sky conditions by approximately 18% per 1,000 m of altitude (Blumthaler et al. 1997), the variation of altitude from SUR to the coastal populations (about 300 m) was not enough to exhibit significant variation of phenolic contents (Table 8.2). Only recently, low temperatures were also linked with an increase in antioxidative secondary metabolites in plants (Bilger et al. 2007), but again the climatic distinction was not enough to impact in differential plant metabolism of SUR populations.

Regarding *E. hirta*, the literature relates the high adaptive pressure suffered by populations in CDU and API (Pinangé 2009, Santana et al. 2010). Such a scenario could influence the metabolite production pattern of some populations in a way that favor certain pathways to the detriment of others. Thus, the populations that exhibit this metabolite not shared with others are the closest geographically (ca. 4 km). The soil may have some influence in those plots (urban areas), which tend to be Ca$^+$ rich, the high adaptability of those plants in urban spaces is well documented (dos Reis et al. 2006). From an edaphic and climatic standpoint, the prevalence of stressful conditions might imply that the quali- and quantitative differences of polyphenols content exhibited by plants. Moreover, growing in these environmental conditions could be attributed to the existence of a different ecotype for these populations, a possible explanation of the successful spread and adaptation. Something similar was found for *E. thymifolia* by Ramakrishnan (1965) when ecotypes were recognized as facultative calciola and obligatory calcifuge. The author reinforced that the differences were genetically determined, once the characters were maintained under cultivation in a neutral substrate.

Genetic diversity assessment

Five ISSR primers allowed the amplification of 754 DNA fragments in 183 loci, from which 173 were polymorphic (Fig. 8.5B and Table 8.3). In an intraspecific analysis in *E. hyssopifolia* 127 loci were generated and 106 of which were polymorphic. For *E. hirta* 162 bands and 135 polymorphisms were obtained. Regarding the number of polymorphic loci accessed in populations of both species in both analytical spheres (inter- and intraspecific), the ISSR methodology was successful to infer levels of genetic variability in analyzed samples of both species, as evidenced by Pinangé (2009) and supported by the hypothesis of Telles et al. (2001) referring to the need of at least 60 loci for consistent analysis of genetic variation.

Chen et al. (2006) using five primers ISSR in genetic studies of *Caldesia grandis* populations, obtained a total of 60 fragments. On the other hand, Pinangé (2009) in the analysis of *Xylopia frutescens*, applying the same amount of primers obtained 126 polymorphic loci (97.6%) for a total of 129 loci with 997 fragments generated while the primer that most contributed to polymorphisms presented 34 loci.

With respect to the primers used, 827 was, with a large difference, the one which produced the largest number of loci (56 loci). However, 818 was the most informative with all loci polymorphic, resulting in the higher rate of divergence (100%), although together with 808 exhibits the smaller number of loci (29 loci). In the intraspecific level, the 818 still proved the most polymorphic, while the primer 811, which also showed the lowest overall Dg in interspecific analysis, with 82%, proved even more polymorphic, with Dgs 59 and 74% for *E. hyssopifolia* and *E. hirta*, respectively. Thus, together with the data provided by Pinangé (2009), Santana et al. (2010), and Pinangé et al. (2010), fingerprinting primers were indicated as promising access diversity in Euphorbiaceae genera.

Table 8.3. Primers using their features, polymorphims and their analysis.

Primer			808	811	818	827	843	Total	Mean
Primer sequence			$(AG)_8C$	$(GA)_8C$	$(CA)_8G$	$(AC)_8G$	$(CA)_8RT$	-	-
Healing temperature (°C)			52, 8	52,0	52, 8	52, 8	52, 8	-	-
Amplified DNA fragments			131	159	99	231	134	754	150,8
Interspecific analysis	Polimorphyc bands Global Dg	Loci number	29	33	29	56	36	183	36,6
			28	27	29	55	34	173	34,6
			97%	82%	100%	98%	94%	95%	94%
Intraspecific analysis	*E. hyssopifolia*	Loci number	24	22	26	37	18	127	25,4
		Polimorphyc bands	19	13	26	33	15	106	21,2
		Proper Dg	79%	59%	100%	89%	83%	83%	82%
	E. hirta	Loci number	27	31	25	45	34	162	32,4
		Polimorphyc bands	24	23	25	36	27	135	27
		Proper Dg	89%	74%	100%	80%	79%	83%	84%

In a close species (*Chamaesyce skottsbergii* =*Euphorbia skottsbergii*) also using a dominant marker (RAPD) besides ITS regions, Morden and Gregoritza (2005) found a high interpopulational polymorphism (> 95%), indicating differentiation among populations sampled, and further suggested that one of the three populations studied may concern a variety. Moreover, recent studies of genetic diversity in the species of *Euphorbia* studied here showed high intra-specific and intra-population diversity (Pinangé 2009, Santana et al. 2010). Similarly, nine populations of another medicinal species (*Ginkgo biloba* L., Ginkgoaceae family) from China were analyzed using RAPD uncovering 89% of variation within populations (Fan et al. 2004).

In such studies, variability depends mainly on the genetic composition of the sampled individuals. Thus, it can be noted that the degree of polymorphism depends on the degree of divergence of the accessions analyzed. The work undertaken by Pinangé (2009) and Santana et al. (2010) using a similar fingerprinting marker (DAF–DNA Amplification Fingerprinting) demonstrated both a phenetic separation at species level, refuting any evidence of breeding between *E. hyssopifolia* and *E. hirta,* supporting the well-substantiated morphological distinction. Additionally these data are in accordance with the structuring intrapopulational trend observed in both *Euphorbia* species analyzed in the present study (Fig. 8.5C).

Thus, in the phenetic analysis carried out here, there was a clear distinction at a specific level. At the population level, in *E. hyssopifolia*, there is a greater similarity between API and CDU, the geographically closest populations (ca. 4 km distant from each other) and from environments with similar characteristics (ecosystem, soil, temperature, etc.) (see Table 8.3). The work performed by Santana et al. (2010) indicated, from the phenetic data, a positive correlation between genetic and geographic distance, observing, therefore, that the most distant population (about 870 km from the others) behaved as the most dissimilar between sampled populations, positioning itself as an outgroup in phenogram. In the present work the same was not observed, although SUR has been the most geographically distant population (86 km away from the others, see Table 8.1), from the phenetic point of view it showed no clearcut differentiation. Moreover, unlike *E. hirta*, in *E. hyssopifolia* there was a positive correlation between increasing divergence through a greater distance between populations. This finding is probably due to fact that even furtherest, this distance was not significant enough to suggest a lack of genetic connectivity between the amplified loci.

On the other hand, in *E. hirta* no greater genetic correlation was observed between similar environments, antropization levels and/or geographical correlation, while the phenogram suggested that at least in the sampled population, there was no definite pattern or evidence of genetic structure between populations. Still, most bootstrap values obtained were below 50, while in the first species, all values were above 60. Also regarding *E. hirta* it is noteworthy that such data are corroborated both by Pinangé (2009) and Santana et al. (2010) that perceived greater phenotypic variability and wider scope and population density in the environments collected, which proved much more polymorphic inter- and intra-population towards *E. hyssopifolia* via DNA Amplification Fingerprinting. In the second study cited, the fragment data generated revealed nearly double polymorphisms both inter- and intrapopulation in *E. hirta* in relation to *E. hyssopifolia*. Additionally, according to Pinangé (2009) the high polymorphism found in relation to levels of divergence measured, especially in *E. hirta*, would indicate a possible retention of variability in their populations, as well as in those described by Morden and Gregoritza (2005) for *C. skottsbergii*. Yet in the work of Pinangé (2009) as well as Santana et al. (2010), the values of genetic connectivity Nm (which suggest evidence of gene flow between populations) indicated that in *E. hirta* such data were considerably higher when compared to those of *E. hyssopifolia*, so probably there is evidence of a greater success of pollinators of this species to maintain the gene pool shared between populations of such fragments being a possible explanation for its greater adaptive success, observed throughout these works.

Different genotypes may have different modes of chemical defense, and the magnitude of chemical responses of a genotype may partly depend on resource availability (Keinänen et al. 1999); but, despite genetically more distinct, the flavonolic pattern of *E. hyssopifolia* populations proved quite similar (Fig. 8.5). This fact, together with the lowest abundance of individuals in the visited plots and some restriction to certain environments could possibly indicate that, for the proper development of these species, conditions that favor certain pathways of these metabolites would be preferable. Eventually, or more likely, one may consider a low influence or pressure of environmental factors able to interfere with the secondary metabolism of these plants.

Moreover, in *E. hirta* the presence of a flavonoid (not identified by comparison with standard) was detected in the most anthropized and/or degraded sites (CDU and API), which does not appear in detectable quantity in other populations. This finding may indicate that, although arranged in more genetically distinct phenetic clades, environmental factors may be capable of generating a change in the production of given compounds.

Concluding remarks

Plant secondary chemistry is determined by both genetic and environmental factors. It is well known that medicinal plants collected at different times and from different localities may considerably vary in types and quantities of chemical components, therefore, resulting in different therapeutic efficacy. Although independent studies of either genetic or chemical profiles of medicinal plants have been widely conducted in the recent years, the combination of molecular markers and phytochemistry to reveal chemical diversity that is associated with genetic variation has received less attention. The combination of the two methods might shed valuable light for the understanding of secondary metabolite and genetic differentiation of the medicinal plant species.

From the phytochemical and genetic patterns obtained, it was possible to identify inter-specific and intra-specific polymorphisms, allowing the distinction of green and red chemotypes as well as an analysis of genetic variability with respect to secondary plant metabolism in the fragments studied. Although the TLC is a qualitative and semi-quantitative method, the analysis performed here proved to be an effective tool in disclosure of interspecific characteristics between *E. hirta* and *E. hyssopifolia*, especially in intra-specific level to the last one, in which the distinction of chemotype can be an important consideration for local use as medicinal plants. Phytochemical variability corroborated the existing genetic polymorphisms among populations of *E. hirta* towards *E. hyssopifolia*, while in the phenogram, the populations were distant, indicating that a common environmental factor may be responsible for increased expression of this metabolite pathway.

However, the phytochemical analysis performed here proved to be an effective tool to separate the chemotypes of the species observed here and, together with application of ISSR, in grouping and differentiation of populations. Therefore, the results observed in this study reinforce the importance of using molecular tools as aids in the characterization of populations, especially in studies involving ecological aspects of it, aided by interdisciplinary tools as verification of phytochemical standard including aspects susceptible to the environment. Such evaluations provide a solid basis for the combined use of chemical and genetic fingerprints in efficiently evaluating qualities and choosing favorable chemotypes with appropriate pharmacological properties of the *Euphorbia* analyzed here, in rapid and precise identification of the plant chemotypes helping to establishing good agricultural practices for medicinal plants.

The continuation of this research by increasing the number of individuals in the populations as well as other herbaceous species as bioindicators via deepening beyond analytical methodologies as High-Performance Liquid Chromatography or Gas Chromatography would enable the visualization of biological phenomena on a larger scale, although endorsing types of genetic markers used in the analysis of populations of importance in the construction of the ecological setting of these environments. Thus, the data presented can contribute to a better quality control of commercialized drug materials based on chemical compositions and genetic background, and facilitate the selection of superior genotypes with highly active components for utilization, cultivation, and conservation.

This study also sheds light on the question that even being significantly different in its components (including toxins) the two species evaluated have the same common name and are often commercialized to treat the same problems, being collected in nature in different regions and environments (since attempts at cultivation proved unsuccessful). Therefore, special care should be taken regarding the dosage and indication of these plants, aspects that our group intends to analyze in the near future.

Acknowledgements

The authors thank Prof. Dr. Suely Lins Galdino (in memorian) and Prof. Dr. Ana Christina Brasileiro-Vidal for relevant advice and interesting discussions. To Vanessa Cristina de Souza we are grateful for valuable technical help. This study received financial support from FACEPE (Fundação de Amparo à Ciência e Tecnologia do Estado de Pernambuco, Recife, PE, Brazil) and CNPq (Conselho Nacional de Desenvolvimento Científico e Tecnológico, Brasília, DF, Brazil).

References

Aarestrup, J.R., Karam, D. and Fernandes, G.W. 2008. Chromosome number and cytogenetics of *Euphorbia heterophylla* L. Genet. Mol. Res. 7(1): 217–222.

Agra, M.F., Silva, K.N., Basílio, I.J., Freitas, P.F. and Barbosa-Filho, J.M. 2008. Survey of medicinal plants used in the region Northeast of Brazil. Rev. Bras. Farmacogn. 18(3): 472–508.

Akinrinmade, J.F. and Oyeleye, O.A. 2010. Antimicrobial efficacy and tissue reaction of *Euphorbia hirta* ethanolic extract oncanine wounds. Afr. J. Biotechnol. 9(31): 5028–5031.

Albert, A., Sareedenchai, V., Heller, W., Seidlitz, H.K. and Zidorn, C. 2009. Temperature is the key to altitudinal variation of phenolics in *Arnica montana* L. cv. ARBO. Oecologia. 160(1): 1–8.

Albuquerque, U.P. and Hanazaki, N. 2006. Ethnodirected research in the discovery of new drugs of medical and pharmaceutical interest: flaws and perspectives. Rev. Bras. Farmacogn. 16(0): 678–689.

Albuquerque, U.P., Andrade, L.H.C. and Silva, A.C.O. 2005. Use of plant resources in a seasonal dry forest (Northeastern Brazil). Acta Bot. Bras. 19(1): 27–38.

Alisi, C.S. and Abanobi, S.E. 2012. Antimicrobial properties of *Euphorbia hyssopifolia* and *Euphorbia hirta* against pathogens complicit in wound, typhoid and urinary tract infections. IJTDH 2(2): 72–86.

Amorim, L.L.B. 2009. Construção de um mapa genético para feijão-caupi com marcadores moleculares ISSR, DAF e CAPS. Master's Thesis, Department of Genetics, Universidade Federal de Pernambuco, Recife, PE, 104 pp.

Amorozo, M.C.M. 2002. Uso e diversidade de plantas medicinais em Santo Antonio do Leverger, MT, Brasil. Acta Bot. Bras. 16(2): 189–203.

Anjaria, J., Parabia, M., Bhatt, G. and Khamar, R. 1997. Nature Heals: A Glossary of Selected Indigenous Medicinal Plants of India. Sristi Innovations, Ahmedabad, India, 20 pp.

Anuradha, H., Srikumar, B.N., Shankaranarayana, R.B.S. and Lakshmana, M. 2008. *Euphorbia hirta* reverses chronic stress-induced anxiety and mediates its action through the GABA(A) receptor benzodiazepine receptor-Cl(–) channel complex. J. Neural Transm. 115(1): 35–42.

APG—Angiosperm Phylogeny Group. 2009. An update of the angiosperm phylogeny group classification for the orders and families of flowering plants: APG III. Bot. J. Linn. Soc. 161: 105–121.

Appendino, G. and Szallasi, A. 1997. Euphorbium: modern research on its active principle, resiniferatoxin, revives an ancient medicine. Life Sci. 60(10): 681–696.

Aqueveque, P., Bittner, M., Ruiz, E. and Silva, M. 1999. Chemotaxonomy of Chilean species of the genus *Euphorbia* L. based on their flavonoids profile. Bol. Soc. Chil. Quim. 44: 61–65.

Arnason, T., Uck, F., Lambert, J. and Hebda, R. 1980. Maya medicinal plants of San Jose Succotz, Belize. J. Ethnopharmacol. 2(4): 345–364.

Arora, R.K. 1997. Biodiversity convention, global plan of action and the national programmes. pp. 26–29. *In*: M.G. Hossain, R.K. Arora and P.N. Mathur (eds.). Plant Genetic Resources—Bangladesh Perspective, Proceedings of a National Workshop on Plant Genetic Resources. Bangladesh Agricultural Research Council.

Barla, A., Oztürk, M., Kültür, S. and Oksüz, S. 2007. Screening of antioxidant activity of three *Euphorbia* species from Turkey. Fitoterapia. 78(6): 423–425.

Bellini, M.R., Feres, R.J.F. and Buosi, R. 2008. Ácaros (Acari) de seringueira (*Hevea brasiliensis* Muell. Arg., Euphorbiaceae) e de euforbiáceas espontâneas no interior dos cultivos. Neotrop. entomol. 37(4): 463–471.

Benko-Iseppon, A.M. and Crovella, S. 2010. Ethnobotanical bioprospection of candidates for potential antimicrobial drugs from Brazilian plants: state of art and perspectives. Curr. Protein Pept. Sci. 11(3): 189–194.

Benko-Iseppon, A.M., Pinangé, D.S.B., Chang, S.C. and Morawetz, W. 2012. Ethnobotanical uses of the native flora from Brazilian North-Eastern region. pp. 84–105. *In*: M. Rai, L. Rastrelli, M. Marinoff, J.L. Martinez and G. Cordell (eds.). Advances in the Study of Medicinal Plants. CRC Press, New Hampshire.

Bilger, W., Rolland, M. and Nybakken, L. 2007. UV screening in higher plants induced by low temperature in the absence of UV-B radiation. Photochem. Photobiol. Sci. 6(2): 190–195.

Blumthaler, M., Ambach, W. and Ellinger, R. 1997. Increase in solar UV radiation with altitude. J. Photoch. Photobio. B. 39(2): 130–134.

Bornet, B. and Branchard, M. 2001. Nonanchored Inter Simple Sequence Repeat (ISSR) markers: reproducible and specific tools for genome fingerprinting. Plant Mol. Biol. Rep. 19: 209–215.

Brandão, M.G., Pignal, M., Romaniuc, S., Grael, C.F. and Fagg, C.W. 2012. Useful Brazilian plants listed in the field books of the French naturalist Auguste de Saint-Hilaire (1779–1853). J. Ethnopharmacol. 143(2): 488–500.

Bruyns, P.V., Mapaya, R.J. and Hedderson, T. 2006. A new subgeneric classification for *Euphorbia* (Euphorbiaceae) in southern Africa based on its and PSBA-TRNH sequence data. Taxon 55(2): 397–420.

Caballero-George, C. and Gupta, M.P. 2011. A quarter century of pharmacognostic research on Panamanian flora: a review. Planta Med. 77(11): 1189–1202.

Carvalho, L.J.C.B., Schaal, B.A. and Fukuda, W.M.G. 2000a. Genetic diversity of cassava (*Manihot esculenta* Crantz) in a germplasm collection assessed by RAPD assay. pp. 80–92. *In*: International Scientific Meeting Cassava Biotechnology Network, 4. Proceedings, Brasília: Embrapa Recursos Genéticos e Biotecnologia.

Carvalho, L.J.C.B., Schaal, B.A. and Fukuda, W.M.G. 2000b. Morphological descriptors and random amplified polymorphic DNA (RAPD) marker used to assess the genetic diversity of cassava (*Manihot esculenta* Crantz). pp. 51–64. *In*: International Scientific Meeting Cassava Biotechnology Network, 4. Proceedings. Brasília: Embrapa Recursos Genéticos e Biotecnologia.

Chen, J.M., Gituru, W.R., Wang, Y.H. and Wang, Q.F. 2006. The extent of genetic diversity in the rare *Caldesia grandis* (Alismataceae): comparative results for RAPD and ISSR markers. Aquat. Bot. 84: 301–307.

Chen, L. 1991. Polyphenols from leaves of *Euphorbia hirta* L. Zhongguo Zhong Yao Za Zhi. 16(1): 38–39.

Chludil, H.D., Corbino, G.B. and Leicach, S.R. 2008. Soil quality effects on *Chenopodium album* flavonoid content and antioxidant potential. J. Agr. Food Chem. 56(13): 5050–5056.

Choi, J.S., Kang, N.S., Min, Y.K. and Kim, S.H. 2010. Euphorbiasteroid reverses P-glycoprotein-mediated multi-drug resistance in human sarcoma cell line MES-SA/Dx5. Phytother. Res. 24(7): 1042–1046.

de Almeida, E.R. 1993. Plantas medicinais brasileiras: conhecimentos populares e científicos. São Paulo: Hemus, 341 pp.

de Melo, J.G., Santos, A.G., de Amorim, E.L., do Nascimento, S.C. and de Albuquerque, U.P. 2011. Medicinal plants used as antitumor agents in Brazil: an ethnobotanical approach. Evid. Based Complement. Alternat. Med. Vol. 2011, 14 pp.

Dias, L.A.S. and Kageyama, P.Y. 1991. Variação genética em espécies arbóreas e conseqüências para melhoramento florestal. Agrotrópica 3(3): 119–127.

Domsalla, A., Görick, C. and Melzig, M.F. 2010. Proteolytic activity in latex of the genus *Euphorbia*-a chemotaxonomic marker? Pharmazie. 65(3): 227–230.

dos Reis, V.A., Lombardi, J.A. and de Figueiredo, R.A. 2006. Diversity of vascular plants growing on walls of a Brazilian city. Urban Ecosystems 9(1): 39–43.

Duke, J.A. 1983. Handbook of Energy Crops. Purdue University centre for new crops and plant products. Available at: http://www.hort.purdue.edu Accessed in September, 10th, 2011.

Einhellig, F.A. 1996. Interactions involving allelopathy in cropping system. Agron. J. 88: 886–893.

Esselman, E.J., Jianqiang, L., Crawford, D.J., Windus, J.L. and Wolfe, A.D. 1999. Clonal diversity in the rare *Calamagrostis porteri* ssp. insperata (Poaceae): comparative results for allozymes and random amplified polymorphic DNA (RAPD) and intersimple sequence repeat (ISSR) markers. Mol. Ecol. 8(3): 443–451.

Euphorbia PBI. 2013a. *Euphorbia* Planetary Biodiversity Inventory—*Euphorbia hyssopifolia*. Available at: http://app.tolkin.org/projects/72/taxa/92222? Accessed in December 20th 2013.

Euphorbia PBI. 2013b. *Euphorbia* Planetary Biodiversity Inventory—*Euphorbia hirta*. Available at: http://app.tolkin.org/projects/72/taxa/92671? Accessed in December 20th 2013.

Faleiro, F. 2007. Marcadores moleculares aplicados a programas de conservação e uso de recursos genéticos. Planaltina-DF: Embrapa Cerrados, 102 pp.

Fan, X.X., Shen, L., Zhang, X., Chen, X.Y. and Fu, C.X. 2004. Assessing genetic diversity of *Ginkgo biloba* L. (Ginkgoaceae) populations from China by RAPD markers. Biochem. Genet. 42(7-8): 269–278.

Fang, D.Q. and Roose, M.L. 1997. Identification of closely related citrus cultivars with inter-simple sequence repeat markers. Theor. Appl. Genet. 95(3): 408–417.

Ferraro, G., Rosana, F., del Pero, M.A., Basualdo, N., Mendoza, R. and García, I. 2010. Flavonoids of *Lotus tenuis* (Waldst. & Kit.) as markers of populations growing in soils of different saline and hydrologic conditions. J. Brazil Chem. Soc. 21(9): 1739–1745.

Forzza, R.C. 2010. Catálogo de Plantas e Fungos do Brasil. vol. 2. Rio de Janeiro: Jardim Botânico do Rio de Janeiro, 903 pp.

Frankel, O.H. and Bennett, E. 1970. Genetic Resources in Plants—Their Exploration and Conservation. IBP Handbook No. 11. Blackwell Scientific Publications, Oxford, 554 pp.

Frankham, R., Ballou, J.D. and Briscoe, D.A. 2002. Introduction to Conservation Genetics. Cambridge University Press, Cambridge, UK, 619 pp.

Frigo, M.J., Mangolin, C.A., Oliveira, R.S. and Machado, M.F.P.S. 2009. Esterase polymorphism for analysis of genetic diversity and structure of wild poinsettia (*Euphorbia heterophylla*) populations. Weed Sci. 57(1): 54–60.

Guglieri-Caporal, A., Caporal, F.J.M., Kufner, D.C.L. and Alves, F.M. 2011. Flora invasora de cultivos de aveia-preta, milho e sorgo em região de cerrado do Estado de Mato Grosso do Sul, Brasil. Bragantia 70(2): 247–254.

Haba, H., Lavaud, C., Harkat, H., Alabdul-Magid, A., Marcourt, L. and Benkhaled, M. 2007. Diterpenoids and triterpenoids from *Euphorbia guyoniana*. Phytochemistry 68(9): 1255–1260.

Hans, A.S. 1973. Chromosomal Conspectus of the Euphorbiaceae. Taxon 22(5/6): 591–636.

Harborne, J.B. 1972. Evolution and function of flavonoids in plants. Rescent Adv. Phytochem. 4: 107–141.

Harborne, J.B. 1998. Phytochemical Methods: A Guide to Modern Techniques of Plant Analysis. Chapman & Hall, New York, 288 pp.

Hezareh, M. 2005. Prostratin as a new therapeutic agent targeting HIV viral reservoirs. Drug News Perspect 18(8): 496–500.

Holden, A.N. and Mahlberg, P.G. 1992. Application of chemotaxonomy of leafy spurges (*Euphorbia* spp.) in biological control. Can. J. Bot. 70(8): 1529–1536.

Hore, S.K., Ahuja, V., Mehta, G., Kumar, P., Pandey, S.K. and Ahmad, A.H. 2006. Effect of aqueous *Euphorbia hirta* leaf extract on gastrointestinal motility. Fitoterapia 77(1): 35–38.

Huang, X., Börner, A., Röder, M. and Ganal, M. 2002. Assessing genetic diversity of wheat (*Triticum aestivum* L.) germplasm using microsatellite markers. Theor. Appl. Genet. 105: 699–707.

Isely, D. 1975. Leguminosae of the United States: II. Subfamily Caesalpinioideae. New York Botanical Garden 25(2): 1–228.

Johnson, P.B., Abdurahman, E.M., Tiam, E.A., Abdu-Aguye, I. and Hussaini, I.M. 1999. *Euphorbia hirta* leaf extracts increase urine output and electrolytes in rats. J. Ethnopharmacol. 65(1): 63–69.

Keinänen, M., Julkunen-Tiitto, R., Mutikainen, P., Walls, M., Ovaska, J. and Vapaavuori, E. 1999. Trade-offs in phenolic metabolism of silver birch: effects of fertilization, defoliation, and genotype. Ecology 80(6): 1970–1986.

Kheyrodin, H. and Ghazvinian, K. 2012. The toxicity material extraction from *Euphorbia* species. Iran Agric. Res. 31(2): 65–72.

Körner, C. 1999. Alpine Plant Life. Functional Plant Ecology of High Mountain Ecosystems. Springer, Berlin, 359 pp.

Kumar, S., Malhotra, R. and Kumar, D. 2010. *Euphorbia hirta*: its chemistry, traditional and medicinal uses, and pharmacological activities. Pharmacogn Rev. 4(7): 58–61.

Laitinen, M.L., Julkunen-Tiitto, R., Tahvanainen, J., Heinonen, J. and Rousi, M. 2005. Variation in birch (*Betula pendula*) shoot secondary chemistry due to genotype, environment, and ontogeny. J. Chem. Ecol. 31(4): 697–717.

Lee, I.M. 2000. Phytoplasma casts a magic spell that turns the fair poinsettia into a Christmas showpiece. Online. Plant Health Progress doi:10.1094/PHP-2000-0914-01-RV.

Li, J.W. and Vederas, J.C. 2009. Drug discovery and natural products: end of an era or an endless frontier? Science 325(5937): 161–165.

Lim, Y.A., Mei, M.C., Kusumoto, I.T., Miyashiro, H., Hattori, M., Gupta, M.P. and Correa, M. 1997. HIV-1 reverse transcriptase inhibitory principles from *Chamaesyce hyssopifolia*. Phytother. Res. 11(1): 22–27.

Loh, D.S., Er, H.M. and Chen, Y.S. 2009. Mutagenic and antimutagenic activities of aqueous and methanol extracts of *Euphorbia hirta*. J. Ethnopharmacol. 26(3): 406–414.

Ma, Q.G., Liu, W.Z., Wu, X.Y., Zhou, T.X. and Qin, G.W. 1997. Chemical studies of Lang-Du, a traditional Chinese medicine. Diterpenoids from *Euphorbia fischeriana*. Phytochemistry 44: 663–666.

Mahlberg, P.G., Davis, D.G., Galitz, D.S. and Manners, G.D. 1987. Laticifers and the classification of *Euphorbia*: the chemotaxonomy of *Euphorbia esula* L. Bot. J. Linn. Soc. 94: 165–180.

Malusà, E., Russo, M.A., Mozzetti, C. and Belligno, A. 2006. Modification of secondary metabolism and flavonoid biosynthesis under phosphate deficiency in bean roots. J. Plant Nutr. 29(2): 245–258.

Markham, K.R. 1982. Techniques of Flavonoid Identification, Vol. 31. Academic Press, London, 113 pp.

Matsuse, I.T., Lim, Y.A., Hattori, M., Correa, M. and Gupta, M.P. 1999. A search for anti-viral properties in Panamanian medicinal plants. The effects on HIV and its essential enzymes. J. Ethnopharmacol. 64(1): 15–22.

Menezes, C.M., Espinheira, M.J., Dias, F.J. and da Silva, V.I. 2012. Floristic composition and phytosociology of interval from the Coastal Vegetation North and South of Bahia State. Revista Biociências 18(1): 35–41.

Mhaskar, K.S., Blatter, E. and Caius, J.F. 2000. Kiritikar & Basu's Illustrated Indian Medicinal Plants, Vol. 9. Sri Satguru Publications, New Delhi, 366p.

Morden, C.W. and Gregoritza, M. 2005. Population variation and phylogeny in the endangered *Chamaesyce skottsbergii* (Euphorbiaceae) based on RAPD and ITS analyses. Cons. Gen. 6: 969–979.

Mors, W.B., Nascimento, M.C., Pereira, B.M. and Pereira, N.A. 2000. Plant natural products active against snake bite—the molecular approach. Phytochemistry 55(6): 627–642.

Morton, J.F. 1980. Caribbean and Latin American folk medicine and its influence in the United States. Pharm. Biol. 18(2): 57–75.

Murugan, S., Anand, R., Uma-Devi, P., Vidhya, N. and Rajesh, K.A. 2007. Efficacy of *Euphorbia milli* and *Euphorbia pulcherrima* on aflatoxin producing fungi *Aspergillus flavus* and *Aspergillus parasiticus*. Afr. J. Biotechnol. 6(6): 718–719.

Mwine, T.J. and Van Damme, P. 2011. Why do Euphorbiaceae tick as medicinal plants?—a review of Euphorbiaceae family and its medicinal features. J. Med. Plants Res. 5(5): 652–662.

Nei, M. 1972. Genetic distance between populations. Am. Nat. 106: 283–292.

Neu, R. 1956. A new reagent for differentiating and determining flavones on paper chromatograms. Vol. 43, Naturwissenchaften, 82 pp.

Newman, D.J. and Cragg, G.M. 2012. Natural products as sources of new drugs over the 30 years from 1981 to 2010. J. Nat. Prod. 75(3): 311–335.

Nissen, S.J., Masters, R.A., Lee, D.J. and Rowe, M.L. 1992. Comparison of restriction fragment length polymorphisms in chloroplast DNA of five leafy spurge (*Euphorbia* spp.) accessions. Weed Sci. 40(1): 63–67.

Nogueira, R.C., de Cerqueira, H.F. and Soares, M.B. 2010. Patenting bioactive molecules from biodiversity: the Brazilian experience. Expert Opin. Ther. Pat. 20(2): 145–157.

Ogbulie, J.N., Ogueke, C.C., Okoli, I.C. and Anyanwu, B.N. 2007. Antibacterial activities and toxicological potentials of crude ethanolic extracts of *Euphorbia hirta*. Afr. J. Biotechnol. 6(13): 1544–1548.

Ogunlesi, M., Okiei, W., Ofor, E. and Osibote, A.E. 2009. Analysis of the essential oil from the dried leaves of *Euphorbia hirta* Linn. (Euphorbiaceae), a potential medication for asthma. Afr. J. Biotechnol. 8(24): 7042–7050.

Park, K.R. 2004. Comparisons of allozyme variation of narrow endemic and widespread species of Far East *Euphorbia* (Euphorbiaceae). Bot. Bull. Acad. Sinica. 45: 221–228.

Park, K.R. and Elisens, W.J. 1997. Isozyme and morphological divergence within *Euphorbia* section *Tithymalopsis* (Euphorbiaceae). Int. J. Plant Sci. 158(4): 465–475.

Park, K.R. and Elisens, W.J. 2000. A phylogenetic study of tribe Euphorbieae (Euphorbiaceae). Int. J. Plant Sci. 161(3): 425–434.

Park, K.R. and Jansen, R.K. 2007. A Phylogeny of Euphorbieae subtribe Euphorbiinae (Euphorbiaceae) based on molecular data. J. Plant Biol. 50: 644–649.

Pasa, M.C. 2011. Saber local e medicina popular: a etnobotânica em Cuiabá, Mato Grosso, Brasil. Bol. Mus Para. Emílio Goeldi Ciênc Hum. 6(1): 179–196.

Pinangé, D.S.B. 2009. Análise da diversidade genética populacional nas espécies *Chamaesyce hirta* L., *Chamaesyce thymifolia* Millsp. (Euphorbiaceae) e *Xylopia frutescens* Aubl. (Annonaceae) através de *Fingerpriting* de DNA em fragmentos da Floresta Atlântica de Pernambuco. Master's Thesis, Department of Genetics, Universidade Federal de Pernambuco, Recife, PE, 104 pp.

Pinangé, D.S.B., Lira Neto, A.C., Santana, K.C.B., Carvalho, R., Alves, M. and Benko-Iseppon, A.M. 2010. Seleção de Primers para Avaliação de Diversidade Genética em dois Gêneros de Euphorbiaceae (*Croton* L. e *Chamaesyce* Gray) com Fingerprinting de DNA. *In*: XVIII Encontro de Genética do Nordeste—ENGENE, Jequié-BA. XVIII Encontro de Genética do Nordeste—ENGENE, 2010, 1 pp.

Rahman, A.U. 2012. Studies in Natural Products Chemistry. Vol. 36, Elsevier, 438 pp.

Ramakrishnan, P.S. 1965. Studies on edaphic ecotypes in *Euphorbia thymifolia* L.: II. Growth Performance, Mineral Uptake and Inter-Ecotypic Competition. J. Ecol. 53(3): 705–714.

Ramesh, K.V. and Padmavathi, K. 2010. Assessment of immunomodulatory activity of *Euphorbia hirta* L. Indian J. Pharm. Sci. 72(5): 621–625.

Rao, V.R. and Hodgkin, T. 2002. Genetic diversity and conservation and utilization of plant genetic resources. Plant Cell Tiss. Org. 68: 1–19.

Roberts, E.A.H., Cartwright, R.A. and Oldschool, M. 1957. Phenolic substances of manufactured tea. Fractionation and paper chromatography of water-soluble substances. J. Sci. Food Agr. 8: 72–80.

Rowe, M.L., Lee, D.J., Nissen, S.J., Bowditch, B.M. and Masters, R.A. 1997. Genetic variation in North American leafy spurge (*Euphorbia esula*) determined by DNA markers. Weed Sci. 45(3): 446–454.

Sage, T.L., Sage, R.F., Vogan, P.J., Rahman, B., Johnson, D.C., Oakley, J.C. and Heckel, M.A. 2011. The occurrence of C(2) photosynthesis in *Euphorbia* subgenus *Chamaesyce* (Euphorbiaceae). J. Exp. Bot. 62(9): 3183–3195.

Saklani, A. and Kutty, S.K. 2008. Plant-derived compounds in clinical trials. Drug Discov. Today 13(3-4): 161–171.

Salatino, A., Salatino, M.L. and Giannasi, D.E. 2000. Flavonoids and the taxonomy of *Cercis*. Biochem. Syst. Ecol. 28(6): 545–550.

Santana, K.C.B., Pinangé, D.S.B., Alves, M.V. and Benko-Iseppon, A.M. 2010. Similaridade e Estrutura Genética Populacional em *Euphorbia* spp. Através de Marcadores do Tipo DAF (Populacional Similarity and genetic structure in *Euphorbia* spp. by means of DAF markers). *In*: XIV Congreso Latinoamericano de Genética (ALAG), 1 pp.

Seigler, D.S. 1994. Phytochemistry and systematic of the Euphorbiaceae. Ann. Mo. Bot. Gard. 81: 380–401.

Semmar, N., Jay, M. and Nouira, S. 2007. A new approach to graphical and numerical analysis of links between plant chemotaxonomy and secondary metabolism from HPLC data smoothed by a simplex mixture design. Chemoecology 17(3): 139–155.

Sharma, O.P. and Dawra, R.K. 1991. Thin-layer chromatographic separations of lantandenes, the pentacyclic triterpenoids from (*Lantana camara*) plant. J. Chromatogr. 587: 351–354.

Sharma, V. and Sarkar, I.N. 2013. Bioinformatics opportunities for identification and study of medicinal plants. Brief Bioinform. 14(2): 238–250.

Shi, Q.W., Su, X.H. and Kiyota, H. 2008. Chemical and pharmacological research of the plants in genus *Euphorbia*. Chem. Rev. 108(10): 4295–4327.

Shih, M.F., Cheng, Y.D., Shen, C.R. and Cherng, J.Y. 2010. A molecular pharmacology study into the anti-inflammatory actions of *Euphorbia hirta* L. on the LPS-induced RAW 264.7 cells through selective iNOS protein inhibition. J. Nat. Med. 64(3): 330–335.

Shiva, V. 1994. Agriculture and food production. UNESCO/Environmental Education Dossiers 9: 2–3.

Singla, A.K. and Pathak, K. 1990. Phytoconstituents of Euphorbia species. Fitoterapia 61: 483–516.

Spitaler, R., Schlorhaufer, P.D., Ellmerer, E.P., Merfort, I., Bortenschlager, S., Stuppner, H. and Zidorn, C. 2006. Altitudinal variation of secondary metabolite profiles in flowering heads of *Arnica montana* cv. ARBO. Phytochemistry 67(4): 409–417.

Stark, S., Julkunen-Tiitto, R., Holappa, E., Mikkola, K. and Nikula, A. 2008. Concentrations of foliar quercetin in natural populations of white birch (*Betula pubescens*) increase with latitude. J. Chem. Ecol. 34(11): 1382–1391.

Stehmann, J.R. 2009. Plantas da floresta atlântica. Rio de Janeiro: Jardim Botânico do Rio de Janeiro. 1: 515.

Steinmann, V., Caruzo, M.B.R., Silva, O.L.M. and Riina, R. 2013a. *Euphorbia* in Lista de Espécies da Flora do Brasil. Jardim Botânico do Rio de Janeiro. Available at: <http://floradobrasil.jbrj.gov.br/jabot/floradobrasil/FB25518>. Accessed in 30 March 2014.

Steinmann, V., Caruzo, M.B.R., Silva, O.L.M. and Riina, R. 2013b. *Euphorbia* in Lista de Espécies da Flora do Brasil. Jardim Botânico do Rio de Janeiro. Available at: <http://floradobrasil.jbrj.gov.br/jabot/floradobrasil/FB22700>. Accessed in 30 March 2014.

Steinmann, V.W. and Porter, J.M. 2002. Phylogenetic relationships in Euphorbieae (Euphorbiaceae) based on its and NDHF sequence data. Ann. Mo. Bot. Gard. 89(4): 453–490.

Sudhakar, M., Rao, C.V., Rao, P.M., Raju, D.B. and Venkateswarlu, Y. 2006. Antimicrobial activity of *Caesalpinia pulcherrima*, *Euphorbia hirta* and *Asystasia gangeticum*. Fitoterapia. 77(5): 378–380.

Swingland, I.R. 2001. Definition of biodiversity. pp. 377–391. *In*: S.A. Levin (ed.). Encyclopedia of Biodiversity. Academic Press, San Diego.

Telles, M.P.C., Monteiro, M.S.R., Rodrigues, F.M., Soares, T.N., Resende, L.V., Amaral, A.G. and Marra, P.R. 2001. Marcadores RAPD na análise da divergência genética entre raças de bovinos e número de locos necessários para a estabilidade da divergência estimada. Ciênc Anim. Bras. 2(2): 87–95.

Tian, Y., Sun, L.M., Liu, X.Q., Li, B., Wang, Q. and Dong, J.X. 2010. Anti-HBV active flavone glucosides from *Euphorbia humifusa* Willd. Fitoterapia 81(7): 799–802.

Tokuoka, T. 2007. Molecular phylogenetic analysis of Euphorbiaceae *sensu stricto* based on plastid and nuclear DNA sequences and ovule and seed character evolution. J. Plant Res. 120(4): 511–522.

Tona, L., Kambu, K., Ngimbi, N., Mesia, K., Penge, O., Lusakibanza, M., Cimanga, K., De Bruyne, T., Apers, S., Totte, J., Pieters, L. and Vlietinck, A.J. 2000. Antiamoebic and spasmolytic activities of extracts from some antidiarrhoeal traditional preparations used in Kinshasa, Congo. Phytomedicine 7(1): 31–38.

Tona, L., Cimanga, R.K., Mesia, K., Musuamba, C.T., De Bruyne, T., Apers, S., Hernans, N., Van Miert, S., Pieters, L., Totté, J. and Vlietinck, A.J. 2004. *In vitro* antiplasmodial activity of extracts and fractions from seven medicinal plants used in the Democratic Republic of Congo. J. Ethnopharmacol. 93(1): 27–32.

Trease, G.E. and Evans, W.C. 1989. Trease and Evans' Pharmacognosy: A Physician's Guide to Herbal Medicine. 13th edition. Bacilliere Tinall Ltd., London. 616p.

Uchida, H., Yamashita, H., Kajikawa, M., Ohyama, K., Nakayachi, O., Sugiyama, R., Yamato, K.T., Muranaka, T., Fukuzawa, H., Takemura, M. and Ohyama, K. 2009. Cloning and characterization of a squalene synthase gene from a petroleum plant, *Euphorbia tirucalli* L. Planta. 229(6): 1243–1252.

Van Damme, P.L.J. 2001. Euphorbia tirucalli for high biomass production. pp. 169–187. *In*: A. Schlissel and D. Pasternak (eds.). Combating Desertification with Plants. Kluwer Academic Pub, New York.

Vasconcelos, S., Onofre, A.V.C., Milani, M., Benko-Iseppon, A.M. and Brasileiro-Vidal, A.C. 2012. Molecular markers to access genetic diversity of castor bean: current status and prospects for breeding purposes. pp. 201–222. *In*: I.Y. Abdurakhmonov (ed.). Plant Breeding, 1st ed. Intech, Rijeka.

Veiga Junior, V.F. 2008. Estudo do consumo de plantas medicinais na Região Centro-Norte do Estado do Rio de Janeiro: aceitação pelos profissionais de saúde e modo de uso pela população. Rev Bras Farmacogn. 18(2): 308–313.

Wagner, H. and Bladt, S. 1996. Plant Drug Analysis. 2ª ed. Springer, New York, 384 pp.

Watt, J.M. and Breyer-Brandwijk, M.G. 1962. The Medicinal and Poisonous Plants of Southern and Eastern Africa. E. & S. Livingstone, Edinburgh, 1457 pp.

Webster, G.L. 1987. The saga of the spurges: a review of classification and relationships in the Euphorbiales. Bot. J. Linn. Soc. 94: 3–46.

Webster, G.L. 1994a. Synopsis of the genera and suprageneric taxa of Euphorbiaceae. Ann. Mo. Bot. Gard. 81: 33–144.

Webster, G.L. 1994b. Classification of the Euphorbiaceae. Ann. Mo. Bot. Gard. 81: 3–32.

Weising, K., Nybom, H., Wolff, K. and Kahl, G. 2004. DNA fingerprinting in plants. CRC Press, Boca Raton, USA, 322 pp.

WHO. 2002. World Health Reports: Reducing Risks, Promoting Healthy Life. World Health Organization, Geneva, 248 pp.

Widharna, R.M., Soemardji, A.A., Wirasutisna, K.R. and Kardono, L.B.S. 2010. Antidiabetes mellitus activity *in vivo* of ethanolic extract and ethyl acetate fraction of *Euphorbia hirta* L. herb. Int. J. Pharmacol. 6(3): 231–240.

Willard, T.S. and Griffin, J.L. 1993. Soybean (*Glycine max*) yield and quality responses associated with poinsettia (*Euphorbia heterophylla*) control programs. Weed Technol. 7: 118–122.

Williams, L.A.D, Gossell-Williams, M., Sajabi, A., Barton, E.N. and Fleischhacker, R. 1997. Angiotensin converting enzyme inhibiting and anti-dipsogenic activities of *Euphorbia hirta* extracts. Phytother. Res. 11: 401–402.

Xavier, H.S., Sá Barreto, L.C.L., Randau, K.P.E. and Araújo, E.L. 2002. Caracterização de taninos hidrolizáveis. Resumos do 53° Congresso Nacional de Botânica, Recife, 20 pp.

Yang, Y. and Berry, P.E. 2011. Phylogenetics of the *Chamaesyce* clade (*Euphorbia*, Euphorbiaceae): reticulate evolution and long-distance dispersal in a prominent C4 lineage. Am. J. Bot. 98(9): 1486–1503.

Youssouf, M.S., Kaiser, P., Tahir, M., Singh, G.D., Singh, S., Sharma, V.K., Satti, N.K., Haque, S.E. and Johri, R.K. 2007. Anti-anaphylactic effect of *Euphorbia hirta*. Fitoterapia 78(7-8): 535–539.

ZCZ. 2013a. Zipcodezoo—*Chamaesyce hyssopifolia*. Available at: http://zipcodezoo.com/Plants/C/Chamaesyce_hyssopifolia/ Accessed in December 20th 2013.

ZCZ. 2013b. Zipcodezoo—*Chamaesyce hirta*. Available at: http://zipcodezoo.com/Plants/C/Chamaesyce_hirta/; Accessed in December 20th 2013.

Zhang, J.Y., Mi, Y.J., Chen, S.P., Wang, F., Liang, Y.J., Zheng, L.S., Shi, C.J., Tao, L.Y., Chen, L.M., Chen, H.B. and Fu, L.W. 2011. Euphorbia factor L1 reverses ABCB1-mediated multidrug resistance involving interaction with ABCB1 independent of ABCB1 downregulation. J. Cell Biochem. 112(4): 1076–1083.

Zimmermann, N.F.A., Christiane, M.R. and Hellwig, F.H. 2010. Further support for the phylogenetic relationships within *Euphorbia* L. (Euphorbiaceae) from nrITS and trnL-trnF IGS sequence data. Plant Syst. Evol. 286(1-2): 39–58.

Traditional Medicine in Cuba: Experience in Developing Products based on Medicinal Plants

Julio C. Escalona,[1] Renato Peres-Roses,[2] Jesús R. Rodriguez,[1] Claudio Laurido,[3] Raúl Vinet,[4] Ariadna Lafourcade,[1] Luisauris Jaimes[5] and José L. Martinez[6],*

Introduction

The use of medicinal plants is as old as the history of human beings. They set out to discover, use and convey knowledge that each of them had (Martinez and Marinoff 2008). The first text known of medicinal plants dates back to approximately 3000 BC, along with written evidence of civilizations such as China, India, North Africa and Greece, based mainly on the treatment of diseases, their description and categorization. For instance "The use of aloe leaves for burns" is one such specific pharmacological effect (Newall et al. 1996).

The exploitation of medicinal and aromatic plants by man occurred in ancient times, as recorded in the historical evidence belonging to different civilizations and cultures that have been evolving on Earth (Subhaktha et al. 2006, Martinez and Marinoff 2008). Plants with medicinal properties were first used empirically for the cure of diseases afflicting man; they also differentiated the ones that cured to those which killed, knowledge passed down orally from generation to generation, because of lack of writing in these early phases of human evolution. Subsequently, when writing developed and with the appearance of papyrus, men began collecting information, recording the heritage of a few in societies in which it has traversed humanity until today (Marinoff 2002). For example, the Bible describes approximately 200 medicinal plants and also its applications (CLIE 1985, Duke 2008).

[1] Facultad de Farmacia, Universidad de Oriente, Santiago de Cuba, Cuba.
[2] Unidad de Farmacología y Farmacognosia, Universidad de Barcelona, España.
[3] Facultad de Química y Biología, Universidad de Santiago de Chile, Santiago, Chile.
[4] Facultad de Farmacia, Universidad de Valparaíso & CREAS, Valparaíso, Chile.
[5] Facultad de Ciencias de la Educación, Universidad de Carabobo, Valencia, Venezuela.
[6] Vicerrectoría de Investigación, Desarrollo e Innovación, Universidad de Santiago de Chile, Santiago, Chile.
* Corresponding author: joseluis.martinez@usach.cl

The Ebers Papyrus, written nearly 3,500 years ago, describes diseases and directions to solve them by using plant species. During the 12th–13th centuries, the Arabian school—reknowned for its physicians-and that of Salerno in Italy, prescribed many herbal drugs, many of which are in use today. At that time, the famous Arab physician Ibn Wafid (born in Toledo in 1008 and passed away in 1074), author of 'The Pillow Book', a famous recipe book of the 11th century. His most original recipes, remedies included direct complaints or disturbances from the head to the toe nail. Numerous plants almost all of them from Spain and North Africa were used. In the 15th century, essences of bitter almonds, lavender, cinnamon, ginger, roses, sage and lavender among others, were recognized. A century later, more than 60 new scents were added to the list (Martinez and Marinoff 2008).

In 1511, in Barcelona, the first regional pharmacopoeia in the world was published, the 'Concordia Pharmacopolarum'. The famous Codex 'De la Cruz Badiano' written in 1552 by the Indian *Xochimilca Martin de la Cruz* and translated from *Nahuatl* into Latin by Juan Badiano in Mexico. It contains the herbal treasure of the ancient Mexicans. In the Middle Ages, the Arabs perfected the distillation of aromatic plants, hence encouraging the evolution of nascent and rudimentary pharmacy. In the 19(?) century, the first chemical analysis of perfumes and other active components of the plant, with the help of analytical chemistry and microscope were performed. Leading to pharmacochemistry. In the early nineteenth century, morphine was isolated from opium (Sosa 1998). It can be said therefore that phytotherapy consists of the usage of plant products for therapeutic uses, either to forestall, mitigate or cure a diseased state.

In societies, especially those that are not highly developed economically, home remedies based on medicinal plants, usually provided by the older people (grandparents) are a common practice (Martinez and Marinoff 2008). Phytotherapy is therefore the outcome of a slow evolution, backed by practical experience.

It is believed, in general, that the discovery includes all the necessary steps so that it can be guaranteed that the compound possesses a desirable activity profile; and comprises four principal stages: (1) Obtaining (from chemical synthesis, natural origin, or biotech variety), (2) Pharmaceutical formulation, (3) Preclinical pharmacology, and (4) Preclinical toxicology (studies in animal models). Thus, those compounds that carry through these stages are established as an accepted criterion for efficacy and safety for testing in man.

The development phase consists of two steps: (1) Clinical trials phases I, II, III and IV (human experimentation) and (2) The pharmacist registration of the new drug (Escalona-Arranz et al. 2008).

The use of medicinal plants based on scientific evidence and security—A social need?

The use of drugs based on medicinal plants has become an important worldwide development in recent years. Two different types of consumers exist: one that regards the ethnic heritage of peoples and is consumed locally according to well-defined populations, traditions-, and two, the one which is lawfully authorized to be made by doctors and sold by manufacturers and distributors. Both could 'illustrate' a society rooted in the use of natural products, but not necessarily the development of one of the two concurs with the parallel development of the other. It will be attempted to illustrate this statement with an example of each: a country with a long tradition in the use of medicinal plants may not sustain the resources necessary for the growth and commercialization of products derived from these plants, or alternatively, lack laws to regulate and control the products obtained based on these own medicinal plants, thereby preventing the medical indication and the selling of them. On the other hand, a state or government could 'facilitate' the importation of foreign natural products over those of domestic output, thus undermining feedback implying acceptance by the population of the national product obtained, and may even 'clear' the original ethnobotanical knowledge.

If these cases can be seen as actual probable events, it can be assumed that the formation of a regulatory instrument that protects and which in turn ensures the quality of the drug product marketed, will allow the progress in parallel to both consumer lines, increasing the levels of acceptance of this practice in the worldwide population. This will precede to a literal and comprehensive development of the role of medicinal plants.

The data available to the public on the role of medicinal plants (product of the accumulation of ancestral and cultural experiences), is used by universities and research centers to demonstrate the verity of that cognition. Grounded on this evidence, statute laws and regulations are promoted for disclosure, distribution and marketing, spreading that knowledge in society, which was the very root of the original knowledge.

This perspective of the development of natural products conforms exactly to what is technically divided into stages that defines the achievement of a new drug, and is taken over by the international scientific community; which supports two fundamental principles: the safe role of medicinal plants and established on scientific evidence.

From this viewpoint, the World Health Organization (WHO) in 1978 suggested incorporating traditional procedures (including the role of medicinal plants) to health systems in the context of each nation, in order to contribute to high levels of wellness. The same document states that: "Health systems can not dismiss the fact that medicinal plants and herbal medicines are a useful therapeutic resource, low cost and sustainable for all, if employed based on safety criteria, efficiency and quality" (WHO 1978, WHO 1998). The recognition by WHO of this need, emphasizes the importance that the organization presents to the involvement of governments, in society to establish comprehensive development policies and establish regulatory and legislative elements to sustain inquiry and evolution of products from medicinal plants. This appeal has impacted even first world nations, according to the current trend evidenced in a study conducted in 10 developed countries (Frass et al. 2012).

Other reasons why health organizations should prioritize the development of drugs from medicinal plants lie in the fact that the imbalance between the large transnational pharmaceutical development of new drug entities show by an overwhelming percentage, that the evolution of new drugs favors the elimination of health problems in industrialized nations with high income of its population. This further supports the fact that it is necessary that developing countries prioritize with endogenous resources (medicinal plants of traditional employment), studies related to pharmaceutical innovation. The existence of a scheme to ensure the protection of new products generated by the domestic industry, then emerges as essential to the survival of the same, so that innovative/production companies supported by their governments should take on IP systems (Trens 2000). This strategy will ensure a future, an increase in consumption by the population and an increase in the provision of therapeutic products associated with them.

Precisely these commercial aspects have been addressed very effectively in the study published by Alonso et al. (2008), when placed on the market of medicinal plants in the open market (depending on supply and demand) and the competitive market, where consumers can choose their treatment preferences according to their needs and their economic points. In essence, this view places all therapeutic products, food supplements and products based on medicinal plants as part of the free market (Martinez et al. 2009).

The final decades of the 20th century has seen a profound transformation of the pharmaceutical sector. The adoption of the TRIPS agreement (Trade Related Aspects of Intellectual Property Rights) with the recognition of intellectual property rights and widespread international patent system for medicinal products, as defined in the agreements and the GATT (General Agreement on Tariffs and Trade), and explicitly adopted in 1994 by most ministers of the 123 participating states, has clearly favored the growth of the sector, but also the imbalance in terms of meeting social needs. Given this reality, it becomes more important as postulated by WHO and, which can be summarized as: the accumulated knowledge of the people about the utility of medicinal plants should be exploited by the most economically disadvantaged countries for the sake of satisfying the social need it intends to cover, basing it on scientific evidence and security of the developed formulations.

Legislative and regulatory landscape in Latin America

Wellness, as a concept, is the state of perfect physical, mental and social wellbeing and not just the absence of disease or debility. The benefit of the highest attainable standard of health is one of the fundamental rights of every human being without distinction of race, religious belief, political opinion, economic or societal status.

Human rights refer mainly to the relationship between the nation and the individual; this generates state obligations and the individual's access to such rights. Thus, governments have the main obligation for public health, hence resulting from formulating national health policies and/or the acceptance of legal instruments that regulate it.

Undoubtedly, the progressive Latin American integration (started in the late 20s and strengthened at the outset of this century), has been one of the main tasks of governments and parties of every nation in the area. Mercosur and the Community of Latin American and Caribbean States are, to mention just two, some major pacts that dominate the economic and political landscape of the region. The fraternity among the peoples of America has also seen a major boost in the past two decades. The political will for integration and cooperation among the American masses is now more than a statement or an intention, a reality.

Although this emerging exchange until today covers the main exportable area as energy (principally of petroleum and its derivatives), the agro industry and other industrial profiles, other fields of collaboration are more focused on social issues concerning each country, which undoubtedly pass by health problems. The legislative and regulatory aspects of each Latin American country that will favor the pharmaceutical industry and natural products derived from, it can be part of these great economic operations with derivative social impact in small towns.

The conflicts between the laws of countries with large economic exchange for merchandise based on medicinal plants, has already generated more than one battle. Exemplars of this problem can be found in the common European market, especially prior to the European Directive 2004/24/EC which recognized that "these differences may hinder trade in traditional medicines in the residential area and lead to discrimination and distortions of competition between manufacturers of these wares. They can also impact on the protection of public health, as there always are currently offered the necessary guarantees of quality, safety and efficacy". This actually happened in Europe and can serve as an example in Latin America, to ensure that trade in medicinal plants and products thereof can enjoy a sound market based primarily on the quality and safety of the product sold (European Commission 2014).

The first real intention of harmonization among Latin American nations emerged during the First International Coordination Meeting of the Latin American Network of Phyto-pharmaceuticals (RIPROFITO), X. Sub Pharmaceutical Fine Chemicals Iberoamerican Program of Science and Technology for Development (CYTED) made in 1996, where it established one of its aims, the advancement of Latin American plant protection products legislation.

In 2004 (García et al. 2004), a work that assesses the status of laws, rules and/or ordinances referred to natural products and herbal medicines of the 16 countries in Latin America so far issued, were implemented (Argentina, Bolivia, Brazil, Colombia, Costa Rica, Cuba, Ecuador, Spain, Guatemala, Mexico, Nicaragua, Panama, Peru, Dominican Republic and Venezuela) (Cañigueral et al. 2003). The picture revealed is quite complex. There are multiple problems that undermine the foundation of a single law, and among them were:

- The existence of different terms in the aim of legislation.
- Only 12 of the 16 countries had a list of officially recognized medicinal plants.
- Only eight of the 16 countries had a list of toxic plants or restricted use.
- Only five nations had chosen to set up categories for enrollment and/or registration of the drug.
- Of the 16 countries, only 13 asked for product registration trial phytochemical identification of the active constituent or characteristic constituents.
- Only four of the 16 countries requested the test or experiment preclinical toxicity and only two do so with respect to clinical toxicity.
- Only eight of the 16 countries applied for clinical trial or efficacy.
- The genotoxic study is only requested by Cuba.
- The product stability tests are not demanded in two of the 16 countries (Costa Rica and Nicaragua).

They are also important differences corresponding to the implementation and impact of herbal productions among the countries studied. In Argentina, for instance, nearly half of the production of drugs comes from laboratories of national origin. Despite this panorama so encouraging, and being Argentina one traditionally agricultural country, pharmaceutical agribusiness is almost nonexistent. In Bolivia, although the practice of traditional is medicine is legally protected, there is no a national integration program for the practice to the general health and, therefore, the use of medicinal plants is not part of the therapeutic arsenal of resources of official medicine. Despite this situation, it is evident the importance of medicinal plants in the national culture, easily quantifiable when noticing the leading role present in shops and popular markets (Cañigueral et al. 2003).

The opposite side of this phenomenon was proposed by Brazil and Cuba, countries where there are legal mechanisms (for more than two decades ago) that give herbs a special position in the context of the conventional health system. In both areas, their own state structures encouraged research and the presentation of these solutions to daily therapy. This made it possible to use various preparations from medicinal plants, enshrined in the popular tradition and confirmed by scientific means for the main causes of action in the primary health network. It has also allowed them to offer an efficient treatment option to people at affordable prices.

Irrespective of the differences in legislation and regulatory orders of the Latin American regions, in regards to the organization of health concern, there is consensus on the need to see alternatives and treatment choices that take advantage of local technical and production infrastructure. The clear objective is to reduce system costs and realize efficiency and bargaining power. Rescue Practice of Phytotherapy, with its integration to therapeutic application systems, is a strategy that presents ample opportunities for consolidation, considering the great success that shows this practice for the general population.

A circle of natural processes to reach this end, especially those planned to surmount the troubles linked to technological dependence are definitely required. It is therefore an essential investment in the processes of research and development by state agencies, training of human resources and the junction with the productive sector. The possibility of use of phytotherapy in therapeutic practices, skill-based, multisectorial actions involving demands of primary production of medicinal plants to build quality control procedures of new textiles and medicines. These actions should be centered primarily on native plants present in the various ecosystems in the area, involving professionals from a broad scope of fields of knowledge, focusing on pharmaceutical and medical sciences, but also from chemistry, biology, agronomy, technology, among others.

The Cuban experience

Before addressing the development of herbal medicine in Cuban medicinal plants, it is necessary to define the characteristics of the Cuban health system, which is based on the principle that health is a right of the people and a state responsibility. Cuba has chosen to produce a drug policy based on the needs and demand, not supply and market mechanisms, while preserving its unique health system, free and worldwide coverage; but the fact remains that the economic needs of the rural area in the last years have been supported by the efficiency of this system of wellness maintenance.

In its strategy on drugs, WHO identified four central objectives: policy, access, quality and rational usage. It likewise said that an appropriate National Drug Policy is one that would ensure equitable availability and access to essential drugs for quality, safety and proven efficacy; and assure the rational exercise of them by promoting their proper therapeutic use. Without losing this perspective, the growth of natural medicine in Cuba found political support that has contributed to progress levels shown today.

Backdrop to the usage of medicinal plants in Cuba

Unlike other parts of Latin America, the natives who inhabited the Cuban archipelago left no written or pictorial evidence on the role of medicinal plants to treat their ailments. It's fast and violent extermination is another reason why it is considered that in Cuba, the introduction of herbal medicine began in the 15th century with the Spanish conquest. Apparently, it has been enriched over time with the arrival of migration from several regions of the globe, primarily from Africa and China, but also from other European nations such as France. In this operation of commingling of cultures accounted by the Cuban scholar Fernando Ortiz as Cuban acculturation, a civilization based on the use of herbal remedies has been well documented to this day. Without doubt, the greatest work collecting lore properties of medicinal plants used in Cuba is 'Medicinal Plants, Aromatic and Poisonous of Cuba', the first version in 1945 and written by Dr. Juan Tomás Roig and Mesa, investigating the mountains and fields all over the island over 15 years. It names a total of 595 species that were employed by the Cuban population for different therapeutic uses and clearly calls for the national scientific community to examine these plants in order to verify their safety and efficacy, with the resultant need to produce the domestic pharmaceutical industry (Roig 1945). While cultural differences between the communities of Cuba led to the same medicinal plant being recognized by various common names, which are often used even for more than one plant species, Dr. Juan Tomás Roig wrote the 'Botanical Dictionary of Vulgar Cuban Names', which in its second enlarged and revised edition in 1953, complemented the aforementioned valuable collection. During this time the maiden edition of the Flora of Cuba was also published, constituting undoubtedly the golden years of pre-revolutionary herbal medicine in Cuba, at least in regards to written evidence.

Nearly all the states, during the 60s, 70s and 80s of the last century formed the bulwark of synthetic medicine and left isolated attempts such as those noted below in the National Pharmaceutical Industry and in the NHS there was a marked predominance of so-called western medicine.

- In the 60s, scientific research on medicinal plants began to rise in the country in an ascending order, although in isolation by different researchers and mainly for academic purposes.
- In the 70s, the Experimental Station of Medicinal Plants 'Juan Tomás Roig' opened with the aim of studying medicinal plants with Cuba, but again with isolated efforts and poorly integrated into primary medical practice.
- In the late 80s the first Plan of Development of Traditional and Natural Medicine (PDMTN) by the Ministry of Public Health was founded. There were different effects of this program, particularly in the field of herbal medicine. Also in the last 80 years the National Center for Traditional Medicine that was responsible for setting up the methodological bases necessary to prepare, supervise and assess the implementation of PDMTN was created. In 1989 the State Center for Quality Control of Medicines and Drug Regulatory Authority of Cuba, charged with promoting and protecting public health through a regulatory system was formally made by the Ministry of Public Health capable of securing market access of medicines' quality, safety, efficacy and accurate data for intellectual use.

From purpose to practice: The explosion in the production of herbal medication

During the 90s a series of transformations began that helped the growth of the role of medicinal plants on the basis of scientific evidence. In 1991 the President of the Republic of Cuba, Fidel Castro Ruz ruled by starting an at home program that included the scientific role of known medicinal plants, processed by the nascent and thriving pharmaceutical industry, and taking the experience of most industrialized countries in the use of natural medicine with increasing force. The initiative also included the determination of the complexes containing herbal medicinal plants of popular use, its therapeutic effects, the essential clinical trials and the consequent generalization of the resulting experiences. These provisions allowed building a National Program for Development and Generalization of Traditional and Natural Medicine with participation by all units and institutions of the National Health Care System Health and other foundations of research and development.

A detailed survey of medicinal plants was published from 1899 to 1968 by Valero-González (1991).

As a result of this proposal, several edited copies known as serial FITOMED (I–V), which in its five parts divulged the medicinal attributes of many plants growing in the state and that in one way or another were part of the Cuban cultural wealth. The destiny of these booklets were doctors and associated health personnel. This was a first step for herbal medicines, which were now developing in an incipient manner, and were incorporated into the therapeutic arsenal of the Cuban National Health System. Among other matters, the FITOMED addressed the following issues:

- Common Names.
- Scientific name and synonyms.
- Useful part of the plant.
- Pharmacological properties scientifically tested under preclinical and clinical tests.
- Healing properties conferred by the population.
- Main components and chemical metabolites produced by the species.
- Major diseases in which they could be used.

In concert to these FITOMED, the 'Guía Terapéutica de Fitofármacos y Apifármacos' was released in 1992, which was meant for to community pharmacists, especially those operating in small clinics; who would be responsible for producing and selling these first products to the public. It issued the groundwork of some research in Cuba with Cuban medicinal plants and on reports from the scientific literature and international pharmacopoeia monographs. A total of 233 herbal medicine formulations were included for use in different therapeutic indications therein. While this was not called a national guide, it essentially served as such for many years (CECMED 1996). Parallel Branches Standards of the Ministry of Public Health, in which the quality standards to be satisfied by each plant drug used in the preparation of herbal medicines were imposed and the standards issued quality of the expressions obtained from them.

In 1993 the first national policy was set by the Minister of the Revolutionary Armed Forces (FAR) on the demand for comprehensive information on the role of medicinal plants as part of the therapeutic arsenal of the state for both wartime and peace. This policy was known as Directive 8/93. Analysis of the results obtained, mainly by FAR in the application and development of that program laid the foundation for a new National Directive (26/95) of the Minister of FAR and defined a generalized strategy for this type of medicine, which now includes not only the NHS but also those agencies and organizations associated with it (mainly the Ministry of Agriculture).

The Regulatory Bureau for the Protection of Public Health was established in 1996 as the highest state authority of health, the State Center for the Control and Quality of Medicines (CECMED) subordinated and others to organize the National Regulatory Agency for Health Protection. In the same year the requirements were set for applications for registration, renewal and alteration on the Registration of Pharmaceuticals for Human Use (CECMED 1996). In 2002, agreement No. 4282 by the Executive Committee of the Council of Ministers established the provisions for integration in the country's strategies of traditional medicine. Also in 2002, CECMED established the requirements for registration of herbal medicines (CECMED 2002) as part of regulation 16/2006 'Guidelines on Good Manufacturing Practices for Pharmaceutical Products' (Annex No. 3). The Registration of Pharmaceuticals for Human Use Natural Origin in the Republic of Cuba has a term of five years, and renewal which may be requested for equal and successive periods of 90 days before the expiration of the registry. In case of amendments, the Contractor Registry or manufacturer of the product needs to apply for approval of alterations in the duration of the register of the drug (Gonzalez-Ramirez et al. 2007). The structure of the requirements for registration of herbal medicines is:

a) *General information*: Includes general information about the products that will be inscribed and the category under which registration is sought (see Table 9.1).
b) *Terms and Definitions*: To standardize and facilitate the discernment of the subject, in this section the terms and definitions adopted by the CECMED in this area are included.
c) *Documentation of requests for paperwork*: It is composed of four parts that make up the documentation file for the drug:

PART I. Management Information. This includes general information about the applicant, manufacturer, and an overview of the finished product samples to include reference substances, packaging material and product information material that need to be promoted.

PART II. Information Biochemical Pharmaceutical and Biological. Whole prospects of product quality are submitted. It's chemical composition, filing form, dose, preparation and expiry date.

PART III. Preclinical information. This includes views of product toxicity and proven pharmacological properties.

PART IV. Clinical Information. This informs about the clinical aspects efficacy studies, clinical data significant for establishing biological safety and the intended dosage.

d) *Attachments*: Includes models and data that are tacked on as part of the substance of the rule.
e) *References*: Any consulted bibliographic information is gathered and composed for presentation and product registration.

Table 9.1. Categories of drugs of natural origin under Cuban law.

Category	Properties to meet
A	New Herbal Medicines and Traditional Herbal Medicines with a new route of administration and/or indication, supported by pharmacological, toxicological and clinical controlled studies.
B	Traditional Herbal Medicines backed by pharmacological and toxicological studies and some clinical studies (cohort studies, utilization studies, case control, case series and indexed publications).
C	Traditional Herbal Medicines supported by ethnoalimentary and ethnomedicines from the origin of the plant material, techno-scientific documentation and acute toxicity studies.

The implementation and generalization of the Regulation have had a great impact on the work of the regulatory authority in terms of regulation and control of drugs sold in the country; in order to have a national regulation which establishes the requirements for the registration and control of herbal medicines for human use, and also provides guidance for the evaluation and monitoring of such products in the research and development institutions, industry and pharmacies. The units of Secondary and Primary Health Care, have also taken on an significant part in the execution of this regulation, where more access to these drugs is higher and the conduct of clinical research back up the effectiveness of these while allowing their rational use (Gonzalez-Ramirez et al. 2007).

The existence of CECMED in Cuba ensures quality, safety and efficacy of medicinal plants and herbal medications. The growth of regulatory documents related to good agricultural practices, collection and preservation of medicinal plants on the quality, safety and efficacy of herbal medicines are also seen; and patterns provide the necessary information for the proper utilization of medicinal plants by the consumer.

In 2010, for the first time, the 'Formulario Nacional de Fitofármacos y Apifármacos' was published thus providing the 'Guía Terapéutica de Medicina Herbal y Apifármacos' of 1992, which had played a significant part in the development of Cuban Natural Medicine, but undoubtedly needs some adjustments, additions, review of pharmacological active principles of plants and exclusion of some species. The end outcome is a sum of 96 formulations based on 39 species of medicinal plants, ensuring adequate quality and consistency in the output.

Decidedly, since mid-1992, and only a few years into the implementation of legal and regulatory provisions for the usage of medicinal plants in the Cuban health system, many Cubans are of the opinion that the speedy success of herbal medicine in Cuba, was a consequence of the economic situation in the 90s, when the nation was entangled in a deep economic crisis known as the Special Period. At that time, the lack of synthetic products manufactured by the National Pharmaceutical Industry, or those imported medicines turned into a harsh reality for Cubans. Advocates of this hypothesis considered that shortages of drugs led to the 'acceptance' of this emerging, available and inexpensive source of wellness. While the denial of the above observation would be irrational, the authors think that denying the Cuban Health System, since each of its structures as well as the will and government support (building a regulatory and legislative apparatus thereon), decisively influenced the growth of natural medicine and especially of phytotherapy. If one of the items in the research above is analyzed (García-González et al. 2004) it can be observed how in 2004, the quality of the Cuban regulatory system stood as compared to most of the other 15 nations included in the survey.

Traditional and natural medicine national program

One of the most significant steps taken by the Cuban health system came in 1995 when the National Program for the Development and Generalization of Traditional Medicine was created (Programa Nacional de Medicina Tradicional y Natural 1999). In 1999, a new version of the National Program of Traditional and Natural Medicine was reported by the Ministry of Health (Ministry of Public Health), whose stated goal was: "To provide the technical basis for the development of traditional and natural medicine throughout the country as an element that contributes to improving the quality of Healthcare and public satisfaction with the health services they receive", involving many ministries through specific objectives that are emphasized among other things (CECMED 2002):

- Provide the health system of a forming tool with systemic, inclusive, dynamic and open access with a feedback mechanism for the evolution of a real Subsystem Health Care; whose functions are directed at achieving progressive introduction and spread upwards of traditional and natural medication across the country.
- Establish the foundations for accelerated training and training in Traditional and Natural Medicine (MTN) of the human resources, and the acquisition and development of technical resources and the essential means for the counterpane of specialization.
- Contribute to the Ministry of Agriculture (MINAGRI) developing a uniform system and planned growing, harvesting, profit, drying and quality control for the yield of the bleak material, to provide an adequate answer to the requirement of the Health System as a result of studies of territorial and national morbidity. Alongwith the basis of territorial self-sufficiency and compliance with delivery commitments to national production and for direct sale to the public and medical provisions for possible export plants. For this all frames of production should be utilized, including farms, the Basic Units of Cooperative Production (UBPCs), the Agricultural Production Cooperatives (CPA), and independent farmers and collectors.
- Develop in MINAGRI a systematic plan for the improvement and retraining of personnel working in the natural process, and maintaining close coordination with the Ministry of Education (MOE) for teachers and students at all layers of education and included in the training programs of Agricultural Engineering and Agronomy Polytechnics, under the study of medicinal plants.
- Contribute to the Ministry of Science, Technology and Environment (CITMA) in the development of a National Research Program that contributes to improving the quality of medical care. The program should include the fields of agricultural technology, and research on medicinal plants.
- Develop a Comprehensive Popular Outreach Program and to use all available means of bulk communication and to respond to the demands that arise in this society in the modalities and varieties of expression of this medicine that have been sanctioned by the Ministry of Public Health.

Several prospects are addressed in this plan. Those considered the most significant for the growth of herbal medicines are listed:

1. Actions linked to the production, distribution, prescription, and utilization of natural products.
2. Establishment and operation of services of Traditional and Natural Medicine in the network of the National Health System.
3. Activities to acquire the human resources training.
4. Actions to develop in the area of inquiry.
5. Outreach program.
6. Quality Control.
7. Guidelines for material-technical process planning.
8. Provision of mountain development (Plan Turquino-Manati).
9. Using natural formulations according to needs identified in each area of health.
10. Route reviewed for scientific research of herbal products.
11. Microbiological control of medicinal plants.
12. Equivalence of conventional generic medicines, natural medicines.
13. Medical conditions (syndromes) that at least can be treated with the proceedings of Traditional and Natural Medicine.
14. Listed Medicinal Plants that can be sold fresh to the population.
15. Basic Natural Medicines for sale in the community pharmacy that is necessary to ensure stability.
16. Utilization of plant-based drugs in the assorted nations of the globe.
17. Summary of technical data of Medicinal Plants.

While it is not possible to explain the claims and scope of each of these details; item 10 will be considered as having a vital impact on achieving the ultimate proposed goal.

Critical path for scientific research of herbal products

From the perspective of the authors, as mentioned above in point 10, it is demonstrated that different steps from the laboratory to the production and marketing of a medicine, was one of the reasons for writing this chapter. It is, therefore, undoubtedly a document that serves as a methodological guide for whoever proposes to make a new drug based on medicinal plants. This critical path consists of 13 stages, which are listed in a consistent order. The 13 stages for scientific research of herbal products are:

I. Choice of plants for investigation

For a start it is possible to increase the odds of success of research, both from the scientific point of view and its economic feasibility. There is no question about the relevant data concerning the traditional, national or international use is known or has been previously studied, the bibliographic information that it provides, not only of the activities tested or reported, but also the chemical composition. In many cases, plants are selected of which little or nothing about their chemical composition and pharmacological activity are known. If this were the case, it would not be appropriate to initiate studies of the same without first having to carry out a thorough inspection of the information useable for other species of the same genus. Many of the metabolites produced by medicinal plants are not unique to a single species, because they can be divided by other species taxonomically related to, and is undoubtedly the taxonomic position of the genus where there are most probable matches. However one cannot ignore that the very phylogenetic relationships of plants show that there are chemical characters which may even be common to well above the generic as the orders levels and subclasses, as dictated by the skill known as chemotaxonomy.

Some other prospects that may be important in selecting the species to be studied, are the outcomes of low local or specific investigations that have taken place principally in the setting in which the research group works. From the very beginning of the investigation it is important to find the availability of the species and the feasibility of growing the same. Many times, both from the academia and by the novelty and originality that any research result provides, plants are selected of endemic character and/or restricted distribution. However if the subsequent stages of the work involves a greater mass of the sample it will be stymied at the low accessibility of plant material. Additionally, it is important to keep in mind the fact that usually these endemic plants have a very limited and a low number of distributions, so the impact on the surroundings and especially the conservation of the species can be much more detrimental than the likely benefits of research to the development of new drug generated.

II. Botanical certification of the species to study

In many instances, especially when the selection of the plant to be studied is managed based on the ethnobotanical knowledge of the population, the preferred species are handled under different names. To achieve a particular nature of the material to be investigated, a competent botanist should identify, what are the species to be contemplated. To boot, it should be able to provide additional data on the material collected as: point of accumulation (if the GPS coordinates are possible), date and time of collection, state (within the life cycle of the plant) that is, and the part or parts of the plant to study. For research to be reproducible and traceable besides the collector's name and introduction, and institution, name of the specialist performing the botanical identification and herbarium number and details of the institution where the sample is deposited are needed. These aspects are vital, for any doubt on the nature of the species under study; one should always have a sample that will clarify any inconsistency or uncertainty generated.

III. Minimum phytochemical characterization of plant preparations

At this stage the investigations should be initiated correctly in the laboratory. The minimum phytochemical characterization of the species must be led by a comprehensive literature review. This is also applicable as indicated in stage I with respect to the information of other species of the genus and possibly a family. Depending on what is known from the literature and from what can be determined by qualitative chemical tests easy to perform (phytochemical screening), different types of extracts, each of which must be specified at least then prepared chemically by this phytochemical screening technique. These extracts should be prepared according to various criteria, including preparing as close as possible to the shape of traditional use, or using water and/or ethanol, as these solvents are better tolerated by humans. While water and ethanol are the solvents, most employees do not rule out the possibility of the extraction procedure used in other organic solvents, which must be entirely taken away (taking advantage of its low boiling point) after the extraction operation finishes. The resulting extracted material must be redissolved in hydro alcoholic mixtures, in order to move to the next stage, which almost always involves the use of animal models, which are also susceptible to the toxicity of organic solvents. As the extracts contain potentially toxic chemicals, it is necessary to discourage its traditional purpose and not continue the subject.

IV. Pharmacological study I (pre-clinical I)

This stage aims to confirm or identify a pharmacological action alleged in the plant. It is important to use suitable and valuable experimental models, both '*in vivo*' and '*in vitro*'. In all cases it is necessary to use positive and negative controls as well as evaluate doses that differ greatly between them, so as to define the ranges of the same. Additionally, it will ensure good laboratory practices to confirm the reproducibility and accuracy of the results. In particular careful selection of the animal model to be studied, as it should be as close as possible to the human system that it tries to reproduce in order to facilitate and further adjust the 'extrapolation' of information between animal species used and humans.

V. Toxicologists study I (acute pre-clinical toxicology)

Its aim is primarily to evaluate possible acute toxicities and genotoxicities of the plant extract with proven pharmacological activity and from studies of the previous stage. Similar issues should be considered to those brought up in the pre-clinical pharmacology in relation to the selected animal model. Additionally, strains of microorganisms must be employed for genotoxic evaluation of the extract or product to evaluate.

VI. Pharmaceutical (formulation)

It aims to develop the most appropriate pharmaceutical form, based on the same pharmacological indications and routes of administration more feasible and effective for the possible pathology to be treated. This stage takes important information from stages III and IV.

VII. Agrotechnic

The issues at this stage tend to be underestimated by the health personnel, but it is an extremely important part as these aspects of the availability of the species should have been addressed from the beginning of the investigation itself (Stage I), as already explained above. However, at this point of methodological route (because the product transcends the first pharmacological and toxicological tests), it is necessary to have a much higher intensity of the sample to the required time. Its aim is therefore to develop the necessary studies to ensure the plant material for the introduction of the result. Like the previous step, this does not lead to a culmination of the steps, but is a process that will continue to evolve even after the drug is introduced to the market.

These agro-technical studies involve a number of factors that can be defined as biotic and abiotic factors. The abiotic factors include climate, amount of water, sun, temperature and other aspects that could dramatically alter the lifespan of the plant and thence in the various chemical metabolites produced, immediately impacting the quality and strength of the new material to be applied. Another abiotic aspect is the nature of the soil, which will determine the use of various types of fertilizers. All these aspects will regulate the type of crop to be used, the distances between positions, etc. Biotic factors are also important such as the age of the plant to be collected, the state in the life cycle of the same (vegetative, flowering or fruiting), the subspecies or cultivar, optimal harvest time, among others. Once the basic planting, growing and harvesting are done, more complex studies must be performed which allow setting with greater precision the pharmacological localization of the stored metabolites.

VIII. Pharmacology II (pre-clinical II)

This phase complements preclinical pharmacological studies initiated in the previous pharmacological stage. Here aspects of pharmacodynamics and, if possible, pharmacokinetics (when the active substance is known) are measured. Additionally, aspects such as effective dose, power or relative phytodrug activity, therapeutic index or margin of safety, mechanism of action, duration of action (half life) should be included.

IX. Toxicology II (pre-clinical toxicology II)

This complements preclinical toxicology studies, initiated in Toxicology I including aspects such as: subchronic toxicity, genotoxicity, reproduction, chronic toxicity, carcinogens and all those tests to ensure medium and long term 'apparent' safety for humans.

X. Preparation of active ingredient

Refers to research to identify and/or obtain the active principles responsible for the pharmacological action. Significantly, it is not always possible to link the pharmacological action of a plant species to a single metabolite or a little group of them. At times, the final drug response is the result of small activities of multiple types of secondary metabolites acting on several mechanisms of action which in synergy exceed a threshold of sufficient activity to be considered efficient and effective in the treatment of a specific pathology. When this occurs as described above, usually a majority metabolite concentration and from the parameters of the quality of plant material to be used as raw material are established, as well as extracts derived from it are selected. Additionally, they may be used in aspects concerning the stability studies of the formulation itself.

After completion of these steps, sufficient experimental evidence should be registered to establish a level of animal models that is efficient as the pharmacological activity and suggests both secure and low toxicity to man. Therefore, concluding these 10 steps, it can be assumed that the drug will surpass the discovery phase and hence is ready to move to the second and final phase: development of the drug; these aspects were discussed at the beginning of this chapter.

XI. Clinical trial results

These studies should assess unequivocally the therapeutic activity and side effects in man. Although considered a single stage, in their clinical trials Phase I, II and III are the longest stages in time. During which the number of humans involved will progressively increase. One should also test the tolerance of the formulation (Phase I) by a small number of healthy volunteer patients (in case of intolerance, should be returned to stage VI), the efficiency and effectiveness of the drug in the pathology proposal (phase II, in ill patients), and a much broader measure of side effects and overall effects of the same study, also in ill patients (phase III).

XII. Registration standardization

Once the above 11 steps are completed, with a volume containing important information to proceed with the registration of the drug regulatory agency in the country. In the case of Cuba, and as already mentioned above, it is the State Center for Quality Control of Medicines (CECMED), which provides three categories (see Table 9.1).

XIII. Pharmacovigilance

These studies are also known as post-marketing studies or clinical trial phase IV, will be developed during the introduction of the drug to market. These subjects could include up to 15 years or more, and during this time all the observations related to their employment, potential drug interactions, potential contraindications, among others be collected. For the development of this important activity Cuba has established an institution known as the National Pharmacovigilance System, National Center for the Development of Pharmacoepidemiology (CDF). Pharmacoepidemiology is a branch of health whose primary aim is to collect data that contributes to the protection of the health of populations and improve the safety and efficacy of medications. Pharmacoepidemiology in Cuba has served to ease the access of drugs priority required as indicators of prevalence, mortality in the country, thus defined as a Basic Table of Drugs whose selection guarantees different stages of health concern. It has also regulated prescribing and dispensing of drugs in pharmaceutical establishments, ensuring the monitoring of their safety and strength through the effective pharmacovigilance system.

Conclusions

This chapter sought to address the Cuban experience in the context of publicity/production/marketing/consumption of medicinal plants, which is backed by two pillars: job security and evidence-based employment. The Yerbero (the source of traditional knowledge and dispensing of medicinal plants: this practice further promotes local communities on sustainable development that involves and benefits all of society as a cycle in a simplified way that could be described as presented) provides this knowledge to universities and research centers to show the veracity of that knowledge, API and job security (toxicity). Statute law is promoted from the authoritative information, and making regulations that knowledge is distributed in society. This is published, awareness campaigns among the medical and paramedical staff are prepared, training at several layers of human resources and regulatory data, safety wears, and are used by herbalists; the origin of knowledge and primary distribution, sometimes with a higher share of sales than their own pharmacies.

References

Alonso, M.J., Albarracín, G., Caminal, J. and Rodríguez, N. 2008. Práctica y productos terapéuticos en medicinas complementarias y alternativas, ¿mercado regulado o mercado libre? Atención Primaria 40: 571–575.

Cañigueral, S., Dellacassa, E. and Bandoni, A.L. 2003. Plantas Medicinales y Fitoterapia: ¿Indicadores de Dependencia o Factores de Desarrollo? Acta Farmacéutica Bonaerense 22: 265–278.

CECMED. 1996. Ministerio de Salud Pública. Centro para el Control Estatal de la Calidad de los Medicamentos. CECMED. Requisitos para las solicitudes de inscripción, renovación y modificación en el registro de medicamentos de uso humanos. Cuba.

CECMED. 2002. Requisitos para las solicitudes de trámites de inscripción, renovación y modificación de los medicamentos de origen natural de uso humano. CECMED Regulación N° 28, La Habana, Cuba.

CLIE. 1985. Nuevo Diccionario Bíblico Ilustrado. Impreso en los Talleres Gráficos de la M.C.E. Horeb, E.R. N° 265, S.G. Viladecavalls, Barcelona, España, 1232 pgs.

Duke, J.A. 2008. Duke's Handbook of Medicinal Plants of the Bible. CRC Press, Taylor and Francis, Group LLC, Boca Raton, Florida, USA, 552 pp.

Escalona-Arranz, J.C., Carrasco-Velar, R. and Padrón-García, J.A. 2008. Introducción al diseño racional de fármacos. Editorial Universitaria, Ciudad de La Habana, Cuba.

European Commission. 2014. http://ec.europa.eu/health/human.use/herbal-medicines/index_en.htm.

Frass, M., Strassl, R.P., Friehs, H., Müllner, M., Kundi, M. and Kaye, A.D. 2012. Use and acceptance of complementary and alternative medicine among the general population and medical personnel: a systematic review. Ochsner J. 12: 45–56.

García-González, M., Cañigueral, S. and Gupta, M. 2004. Legislación en Iberoamérica sobre fitofármacos y productos naturales medicinales. Revista de Fitoterapia 4: 53–62.

González-Ramírez, M., Remirez, D. and Jacobo, O.L. 2007. Antecedentes y situación reguladora de la medicina herbaria en Cuba. Bol. Latinoam. Caribe Plant. Med. Aromat. 6: 118–124.

Marinoff, M.A. 2002. Las plantas medicinales desde La Biblia a la actualidad. Revista Comunicaciones Científicas y Tecnológicas 53 (4 pp.).

Martinez, J.L. and Marinoff, M.A. 2008. Historia de las plantas medicinales: su evolución e implicancias en Chile. Capitulo 8 pp. 49–58. en: Comprender el mundo a través de la Historia de la Ciencias. Mario Quintanilla y Gerardo Saffer, compiladores. Una producción G.R.E.C.I.A.—Universidad Católica de Chile, Santiago de Chile.

Martinez, J.L., Barraza, F. and Samarotto, M. 2009. Comercio de Plantas Medicinales. Situación de Chile frente al mundo. pp. 157–170. Editado por Giovanna L. Reyes-Sánchez, Universidad Nacional de Colombia en Diálogo de saberes: plantas medicinales, salud y cosmovisiones, 205 pp.

Newall, C.A., Anderson, L.A. and Phillipson, J.E. 1996. Herbal Medicines: A Guide for Health Care Professionals. The Pharmaceutical Press, London, UK, 296 p.

Programa Nacional de Medicina Tradicional y Natural. 1999. Ministerio de Salud Pública, pp. 8–98.

Roig, J.T. 1945. Plantas medicinales, aromáticas o venenosas de Cuba, Editorial Científico-Técnica, La Habana, Cuba.

Sosa, R. 1998. El Poder Medicinal de las Plantas. Ed. ACES. Tomo I, 175 pgs., Tomo II, 240 pgs.

Subhaktha, P.K., Narayama, A., Sharma, B.K. and Rao, M.M. 2006. Diet, dietetics and flora of the holy bible. Bull Indian Inst. Hist. Med. Hyderabad 36: 21–42.

Trens, E. 2000. El desarrollo tecnológico y las políticas de salud. Rev. Fac. Med. UNAM 43. Enero-Febrero.

Valero-González, M. 1991. Studies of medicinal plants published by the Academy of Medical, Physical and Natural Science of Havana from 1899 to 1968. Asclepio 43: 89–100.

WHO. 1978. The Promotion and Development of Traditional Medicine. Geneva: WHO (Technical Report Series 622).

WHO. 2002. Guidelines for the appropriate use of herbal medicines. Manila (Technical Report Series No. 23).

Challenges for the Development of A Natural Antimicrobial from Essential Oils

Marta Cristina Teixeira Duarte* and Renata Maria Teixeira Duarte

Introduction

In the last two decades the search for new antimicrobial compounds from natural products has intensified around the world mainly due to the emergence of strains resistant to antibiotics. The well-known bacteria methicilin resistant *Staphylococcus aureus*—MRSA (Struelens 2009), vancomycin resistant *Enterococcus*—VRE (Willems et al. 2005), multiresistant *Mycobacterium tuberculosis*—MDR-TB (Pearson et al. 1992), *Clostridium difficile* and Enterobacteriaceae producing broad-spectrum β-lactamase—ESPL (Lee et al. 2003), besides the fungi *Candida* spp. and *Aspergillus* spp., represent one constant threat and has spread in hospitals and communities.

Antibiotic resistance is inevitable and irreversible. A natural consequence of adaptation of the bacterial cell exposure to antibiotics, mainly due to the indiscriminate use not only in human and animal medicine, but also in agriculture, cosmetics and as food preservatives (Duarte et al. 2012). The problem affects all countries of the world, developed or undeveloped, being one of the main public health problems. The currently proposed strategies to reduce resistance has been by minimizing the use of antimicrobials in different areas as well as the search for new antimicrobials able to reach different targets in the microorganisms cells from those in which conventional drugs attack (Ahmad and Beg 2001).

Concerning the use of antibiotics as a growth promoter in animal feed, therapeutic and prophylactic dosages can leave residues in edible products, contaminate the ecosystem and, contribute to development of resistant microbial strains. Although since 2006 the WHO (World Health Organization) announced a plan to ban the use of antimicrobial completely as a growth promoter, it is still in common use as chemical additives in feed.

In Brazil, the growth promoters have been used in prophylactic doses in animal feed since 1946, in order to control *Salmonella* infections in broiler chicks and by *Escherichia coli* in piglets. In 2004, the MAPA/Brazil (Ministry of Agriculture and Livestock Supply) banned the use of growth promoters such as tetracycline, penicillin, chloramphenicol, furazolidone, nitrofurazone and avoparcin with the goal of keeping the chicken exports to the European Union. In the mean while, other additives are still being used.

Research Center for Chemistry, Biology and Agriculture, CPQBA, State University of Campinas - UNICAMP, SP - Brasil.
* Corresponding author: mduarte@cpqba.unicamp.br

At the same time, the worldwide scientific community has focused on the search for natural alternatives. Among the alternatives for the replacement of antibiotics as growth promoters, are the use of appropriate feed (diet), control of the environmental conditions of creation, the use of prebiotics, probiotics and symbiotics, enzymes and finally the use of essential oils and spices (Silva 2004). In this context, extracts, essential oils, fractions and isolated compounds from medicinal and aromatic plants have been extensively studied. However, despite the efforts, little progress has been noted in the introduction of new drugs from natural products on the market.

The experience of our research group in the search for new antimicrobial agents from medicinal plants (Duarte et al. 2005, 2007, 2012, 2013) has shown that some work strategies, as the establishment of a data base containing information on the antimicrobial activity of plants able to control different panels of microorganisms, may direct the work and can help to get faster to an applicable result. This is also favored by the fact that some essential oils are considered GRAS (Generally Recognized as Safe) by the FDA (Food and Drug Administration). This was the case of a formulation containing essential oils developed for use as additives in feed for piglets in the nursery phase, aiming at replacing growth promoters (Duarte et al. 2013). This recent experience is described in this chapter.

Guided studies in the search for a natural antimicrobial from essential oils

The choice of the plant and microorganism

In the last decade numerous studies involving the assessment of plants for antimicrobial activity, in the form of crude extracts, essential oils, fractions and compounds isolated were performed in several continents and countries, including the flora of South Africa (Kamatou et al. 2008, Van Vuuren 2008), Lebanon (Formisano et al. 2010), Malaysia (Humeirah et al. 2010), India (Bhakshu and Raju 2009), Italy (De Martino et al. 2009) and Brazil (Duarte et al. 2005, Duarte et al. 2007, Silva et al. 2010), amongst others. From a recent survey, it is known that potential antimicrobial activity (Minimal Inhibitory Concentration—MIC until 1.000 µg.mL^{-1}) was observed only for 10% of the species studied (Duarte et al. 2012). These potentially active species belong to 30 botanical families, mainly to the families Apiaceae, Asteraceae, Boraginaceae, Lamiaceae, Lauraceae and Myrtaceae. The information available in the literature can now be used as a starting point for choosing plants for advanced studies. However, some aspects must be considered before choosing a group or species of plants for further investigation, such as the differences observed between different genotypes within the same plant species, oil yield as a function of harvest plant, in addition to differences observed as a result of culture, climatic and soil conditions, among others.

Since 1999, multidisciplinary studies have been developed by our research group focusing on exotic and native plant species in Brazil, potentially active against different panels of microorganisms, mostly belonging to the CPMA—Collection of Medicinal and Aromatic Plants from CPQBA—Research Center for Chemical, Biological and Agricultural, State University of Campinas, SãoPaulo, Brazil. The screening studies allowed the establishment of a database containing information about the activity of about 200 plant species besides information on the activity of their fractions, identification of its active compounds, oil yield and seasonal activity, among others. All this information is important to choose plants with potential to become a new antimicrobial. The information contained in the database has been used for conducting specific projects focusing on the control of microorganisms of importance in different aspects such as human and animal health, agriculture, food and cosmetics. Thus, guided studies for this purpose should be initially conducted for a number of plants. In order to have options to choose from, subsequently, those with desirable features not only for antimicrobial activity, but also according to essential oil yield and possibilities of production of the plant of interest on a commercial scale.

In this context, it is worth mentioning a recent study conducted firstly to determine the anti-*Escherichia coli* activity from the essential oil of 29 plant species commonly used in Brazil, where different serotypes of this bacterium were assessed (Duarte et al. 2007). The essential oils from three *Cymbopogon* spp., i.e., *C. citratus*, *C. martinii* and *C. Winterianus* exhibited a broad inhibition spectrum, presenting strong activity (MIC between 100 and 500 µg.mL^{-1}) against 10 out of 13 *E. coli* serotypes isolated from human and pigs with diarrhea, three being enterotoxigenic, two enterophatogenic, three enteroinvasive and two shiga-toxin serotype. The anti-*E. coli* activity was related to the presence of geraniol in the essential oil from *C. martinii*, the more active specie. *E. coli* is an important pathogen from the point of view of human and animal health. This information was important, subsequently, to the establishment of a project in partnership with an Animal Health Brazilian company, the Ouro Fino Animal Health, also supported by FAPESP-Foundation for Research Support of the State of SãoPaulo-SP, Brazil, within the program of technological innovation PITE/FAPESP, details of which are presented further ahead.

Multidisciplinary study: an ideal model

The importance of collaboration and performance of professionals from different fields, such as agronomists, biologists, chemists and pharmacists in research involving the search and implementation of new antimicrobials is currently the general consensus. The establishment of a team of researchers in these specific areas certainly ensures the success of the work, not only from the scientific point of view, but as a greater possibility of achieving a final product (Duarte et al. 2012).

Plant substances, especially essential oils are secondary metabolites representing a chemical interface between the plant and the environment, their biosynthesis often being affected by environmental conditions (Marques et al. 2012). Thus, the cultivation of the plant must be assisted by a professional with a wide knowledge of the different conditions required by the species of interest. According to Marques et al. (2012), a key point for essential oil production is the determination of agronomic practices that allow maximum exploitation of essential oils while presenting better quality and minimal losses. Other agronomic aspects still to be considered are the plant nutrition, planting density, seasonal photoperiod and variations and postharvest (Marques et al. 2012).

Chemical analysis and identification of the essential oils constituents involves several chromatographic and spectrometric techniques, especially Gas Chromatography (GC) coupled with Mass Spectrometer (GC-MS), besides nuclear magnetic resonance techniques, which are important tools for determining the structure of the different substances presents in the essential oils. As essential oils are mixtures containing 10s or even 100s of substances with diverse chemical composition, the use of these techniques, in turn, requires the expertise of professionals specialized in organic and analytical chemistry who should not only know how to perform the methods and operate the equipments, but also to interpret the mass spectra, in addition, must be able to perform the standardization of different samples or batches of essential oils, thereby ensuring the desired activity, fundamental for the commercial application.

The microbiologist plays a fundamental role in the search for new natural antimicrobials. The techniques involved in the analysis of antimicrobial activity of plant products, such as the determination of the minimal inhibitory concentration and its mode of action (bactericidal/fungicidal or bacteriostatic/fungistatic) are not complex. However, it is necessary to know in detail the nutritional and physicochemical requirements of the microorganisms of interest. While the details involved in the preparation and solubilization of plant samples, standardization of inocula, the correct reading of the inhibitory concentration through the use of specific dyes, and the possible interferences should be known.

Finally, the important role of the pharmacist in choosing the ideal formulation to the specific employment intended as well as the determination of the *in vitro* and *in vivo* toxicity of the crude essential oil, formulation and final product is worth mentioning.

Colibacillosis in pigs and the comproved action of essential oils

The colibacillosis caused by toxigenic strains of *Escherichia coli* (ETEC) is the enteric disease of major impact on the swine industry (Costa et al. 2006). The bacteria adhere to the intestinal mucosa and produce enterotoxins that cause diarrhea and dehydration, which may cause the death of piglets up to three days old in less than six hours. The use of antibiotics to control and treat these infections is widely spread. With the purpose of finding an additive which replaces antibiotics growth promoters in livestock, there is a demand for studying the impact of natural products and their therapeutic principles in the control of diseases that affect the breeding stocks.

In Brazil, it is estimated that around 15 to 20% of piglets die before weaning, due to colibacillosis (Baccaro et al. 2002). Furthermore, litters of piglets with diarrhea may weigh about 0.4 kg less at 30 days of age than litters of piglets without diarrhea (Ribeiro et al. 2005). This has accentuated the economic impact due to its negative effect on the production indices of the herd. In addition to colibacillosis, *E. coli* may cause colisepticemiain where bacteria invade the systemic circulation and internal organs (Smith and Huggins 1979). Moreover, the occurrence of shiga-toxin producing strains (STEC-similar to *Shigella* toxin) even in healthy animals but with the presence of bacteria in the feces has been responsible for outbreaks and isolated cases of infections in humans through the consumption of contaminated meat (Kaddu-Mulindwa et al. 2001).

Among the alternatives for the replacement of antibiotics as growth promoters, are the use of feed (diet), appropriate control of the environmental conditions of creation, the use of prebiotics, probiotics and symbiotics, enzymes and finally the use of essential oils and spices (Silva 2004). In this context, several studies have shown the favorable action of the essential oils on the intestinal microbiota *in vivo* (Manzanilla et al. 2004, Gong et al. 2008, Michiels et al. 2008, Wang et al. 2009, De Lange et al. 2010, Tiihonen et al. 2010, Michiels et al. 2011, Li et al. 2012). In a recent work, Li et al. (2012) suggested that diets treated with doses of microencapsulated essential oils (thymol and cinnamaldehyde) improved feed intake, growth rate and feed conversion ratio, and reduced the incidence of diarrhea in the first week post-weaning. Moreover, it had positive effects of the gut microbial populations measured in the feces, with a reduction of *E. coli* and an increase of *Lactobacillus* counts.

The encapsulation appears to contribute to the stability in feed and probably also in the animal body, and consequently the better bioefficacy in piglets. This improved feed intake, growth rate and feed conversion ratio, and reduction of the incidence of diarrhea was also observed in the case of a formulation containing a blend of essential oils from *Cymbopogon* spp. Developed by our research group for use as an additive in feed for piglets in the nursery phase.

Anti-*E. coli* activity of *Cymbopogon* spp.

During the evaluation of essential oils of medicinal and aromatic species belonging to CPMA (Collection of Medicinal and Aromatic Plants of CPQBA) to control different *E. coli* serotypes, three species of the genus *Cymbopogon* were able to inhibit 15 different serotypes of *E. coli* isolates from humans and 11 serotypes isolated from pigs with diarrhea (Duarte et al. 2007). These results showed that essential oils of the *Cymbopogon* species had potential application as natural antimicrobial in feed for pigs and, was then evaluated as a growth promoter in pigs in the nursery phase, in order to prove its *in vivo* effects.

Cymbopogon martinii (palmarosa) and *C. citratus* (lemongrass)—Poaceae, from CPMA—Medicinal and Aromatic Plants Collection at CPQBA, Research Center for Chemical, Biology and Agriculture, UNICAMP, State University of Campinas were collected during the morning after dew point, in the four seasons of the year 2009, namely, summer (February), autumn (May), winter (August), and spring (November). The essential oils were obtained from 100 g of aerial fresh plant parts by water distillation using a Clevenger-type system for 3 hours. The antimicrobial activity of essential oils from *Cymbopogon* species was evaluated against the standard strain *E. coli* ATCC 11775 and against six different strains of *E. coli* isolated from piglets with diarrhea on different Brazilian farms, kindly provided by a Cargill Company, NutronFoods, Campinas, SP—Brazil.

The *in vitro* antibacterial activity from *C. citratus* and *C. martinii* essential oils and from blend of oils (3:1; 1:3 and 1:1) against the different *E. coli* strains was evaluated through the microdilution method (CLSI 2003). In general, all combinations of oils provided better activities than each oil alone, especially when the oil was extracted from plants harvested in autumn and winter. The activity observed for the different strains of *E. coli* was strong to moderate (0.1 to 1.0 mg/mL), presenting good action spectrum. However, the blend of the oils from *C. citratus* and *C. martinii* 1:3 or 3:1 incurred the best activities. Whereas the essential oil of *C. citratus* is was found more easily and is cheaper than *C. martinii* oil, it is more apppealing in its use for animal feed. Thus, for the *in vivo* assays the microencapsulated mixture of oils from *C. citratus* and *C. martinii* in 3:1 proportion was used.

Active compounds in the essential oil of *Cymbopogon* spp.

The chemical analysis by GC-MS showed differences in the chemical composition of the essential oil obtained in the autumn, with a large number of compounds in the oil of *C. citratus* compared to *C. martinii*, and two common compounds between the two species, ocymene and geraniol. The concentration of geraniol found in *C. martinii* oil (79.65%) suggests that the best activity observed for this species is related to the presence of this compound. In fact, when geraniol was used singly against *E. coli* isolates, its activity was similar to chloramphenicol (Duarte et al. 2007).

Inclusion of microcapsules of essential oils in feed for piglets

Essential oils usually have the characteristic odor and flavor of the plant from which it was extracted. Some of them may improve the performance of animals not only by controlling enteric pathogens, but also by increasing palatability of diets (De Lange et al. 2010). However, some authors considered that their impact on growth performance of newly-weaned pigs was inconsistent, despite changes observed in the gut microbiota composition and improvement of gut health (Manzanilla et al. 2004, Gong et al. 2008, Michiels et al. 2011).

De Lange et al. (2010) suggested that the antimicrobial activity of essential oil compounds might be difficult to observe when they are tested *in vivo* because they are absorbed quickly by the animal. In piglets, essential oils were found to be absorbed nearly completely in the stomach and proximal small intestine within two hours after oral administration (Michiels et al. 2008). Thus, essential oils need protection until it reaches the target site within the pig's digestive tract and to exert their antimicrobial activity. Pigs fed Ca-alginate hydrogel microcapsules containing carvacrol effectively remained its high antimicrobial activity towards *E. coli* K88 in the small intestine (Wang et al. 2009). Also Tiihonen et al. (2010) evaluated the volatile compounds thymol and cinnamaldehyde microencapsulated in maltodextrin carrier to reduce evaporation and to increase shelf life of the product in feed. Studies revealed that this encapsulated product has a recovery rate over 90% for both thymol and cinnamaldehyde after six months storage. A recent work from Li et al. (2012) with weaner pigs treated with increasing doses of microencapsulated essential oils (thymol and cinnamaldehyde), suggest that the microencapsulation process contributes to the stability in feed and probably also in the animal body, and consequently the better bioefficacy in piglets.

Microencapsuled essential oils and its effects as growth promoter in piglets in the nursery phase

In the study reported in this chapter, the microcapsules were formulated as a mixture of essential oils from *C. citratus* and *C. martinii* (3:1), using maltodextrin as wall material, since this mixture showed the best anti-*E. coli* activities and action spectrum *in vitro*. A mini spray dryer Büchi model B-290 with inert loop B-295 (Flawil, Switzerland), was used for the microencapsulation process, under the following conditions: percentage of total solid was 30.0%; the percentage of wall material (maltodextrin) was 80.0% and the core material (essential oil) was 20.0%. The liquid flow rate during spray drying was 6.0 mL/min, the entering and the exiting air temperature in the spray dryer were 180.0 ± 5°C, respectively (Rodrigues 2004).

Microcapsules of the blend of essential oils from *C. citratus* and *C. martinii* and toxicity assays

The microcapsules containing the essential oils from *C. citratus* and *C. martinii* (3:1) were subsequently evaluated for toxicity *in vitro* and *in vivo*. The formulated material was tested *in vitro* on two normal cell lines—HaCat [human keratinocyte, donated by Professor Ricardo Della Coletta, Oral Diagnosis Department, Piracicaba Dental School, Unicamp SP Brazil] and VERO cells [epithelial cells from green monkey kidney, from Rio de Janeiro Cell Bank, RJ Brazil]. The acute oral toxicity was evaluated in rodents according to OECD (2001). The results from *in vitro* toxicity assays shows that the blend presented antiproliferative activity weak to moderate on VERO cells, showing TGI 84,2 µg/mL (Total Growth Inhibition), which suggests a small cytotoxic action on renal cells. However, the blend did not affect the growth of human keratinocyteline (HaCat)—TGI > 250 µg/mL. On the other hand, the *in vivo* test using the microencapulated showed no toxicity to animals.

Effects as growth promoter in piglets in the nursery phase

The effects of addition of the microencapsulated blend from *Cymbopogon* spp. On the diet of piglets in the nursery phase were assessed on the performance and diarrhea score. The study was conducted at the experimental farm sector swine at the Agro veterinary Sciences Center (CAV), State University of Santa Catarina (UDESC), Lages, Santa Catarina—Brazil, in the period from March 15 to April 20, 2012. A total of 72 piglets castrated males and females were used in equal numbers, in good health and nutrition, from Piglet Production System (UPL), a commercial strain AgPic (females Cambrou 22 x 420 male) from Orleans, Santa Catarina—Brazil, weaned at 28 days of age with an average initial weight of 7.5 kg ± 1.0 kg. The animals were divided into three experimental treatments with six replicates of four animals per experimental unit, two males and two females (Fig. 10.1). The experimental units were divided into groups according to a completely randomized design.

The microencapsulated blend of the essential oils, namely CB006-S, was added to treat groups at a dose of 2000 g/ton daily, once daily, during 28 to 63 days of age. On the other hand, the growth promoter standard MSD 08 was added at a dosage of 63 g/ton daily (5 ppm), once daily, during the same period. A negative control group (without additive) was included. The animals were evaluated for performance by getting the live weight, weight gain (DWG), feed intake (DFI) and feed conversion (FCR). The incidence of diarrhea (%) was performed by the same examiner, twice daily for the first 15 days of the experiment, through observation of feces from piglets to evaluate the presence of diarrhea, according to the procedure described by Morés et al. (1990) and Cristani (1997).

The results obtained for the performance (related to 100% of positive control) during the period of 28 to 63 days (Table 10.1), for the different groups show that the animals treated with the microencapsulated blend of *C. citratus* and *C. martinii* essential oils (1:3) presented weight gain 11.2% higher than the group treated with antibiotic (MSD 08). In the case of feed conversion, it was 9.3% better than the animals treated with the antibiotic.

Figure 10.1. Animals in collective boxes divided into blocks according to the initial weight.

Table 10.1. Results related to 100% of positive control: daily weight gain (DWG), daily feed intake (DFI) and feed conversion ratio (FCR) of piglets for the total period 28–63 days of age.

Group	GMDP	CMDR	FCR
Negative Control	93,9b	94,1b	101,1a
Positive Control MSD 08	100ab	100a	100a
Growth Promoter CB006-S	111,2a	101,1a	90,7a
VC%	9,80	4,44	9,49

Means followed by the same letter in the columns do not differ by Tukey test ($p > 0.05$).

The best results of Feed Conversion Ratio (FCR) directly reflected in the weight gain of the animals, showing that animals fed with the microcapsules of essential oils from *C. martinii* coupled with *C. citratus* (CB006-S) showed an improvement in nutrient digestibility of diets, emphasizing once more the action of these oils with regard to intestinal health and as improvement in the action of digestive enzymes. For the incidence of diarrhea (%) there was no difference among the treatments.

In conclusion, the results obtained from the present investigation corroborates those previously obtained in other recent works (Tiihonen et al. 2010, Li et al. 2012), showing that it is possible to use natural compounds as growth promoters in animal feed instead of synthetic antimicrobials used for the same purpose.

Recently, the data obtained were protected in a patent filed in Brazil and abroad (Duarte et al. 2013).

Conclusions

The development of new antimicrobial from natural products such as essential oils incurs several challenges. Despite the vast information available in the literature concerning the antimicrobial activity of essential oils of plant species little is known about the implementation of new antimicrobial drugs in the market. However, it has been seen that in some cases it is possible to reach a new product very soon. In order to achieve success, the research should be conducted by a multidisciplinary group.

Guided studies and the establishment of a database containing information about the activity of plant species, besides information on the activity of their fractions, identification of its active compounds, oil yield and seasonal activity, should be used for the planning of projects focusing on the control of specific microorganisms, with the goal of replacing antibiotics currently known by essential oils in different areas, such as human and animal health, agriculture, food, feed and cosmetics.

Among the alternatives currently proposed for the replacement of antibiotics as growth promoters in animal feed, the role of the essential oils, by providing them in microencapsulated form is already known. Animals fed diets containing added microencapsulated essential oils present improved feed conversion rate, which is directly reflected in weight gain, probably by improvement in intestinal health and action on digestive enzymes.

Acknowledgements

The authors are grateful to FAPESP—Foundation for Research Support of the State of SãoPaulo, SP—Brasil and Ouro Fino Animal Health for financial support.

References

Ahmad, I. and Beg, A.Z. 2001. Antimicrobial and phytochemical studies on 45 Indian plants against multi-drug resistant human pathogens. J. Ethnopharmacol. 74: 113–123.

Baccaro, M.R., Moreno, A.M., Corrêa, A., Ferreira, A.J.P. and Calderaro, F.F. 2002. Resistência antimicrobiana de amostras de *Escherichia coli* isoladas de fezes de leitões com diarreia. Arq. Inst. Biol. 69: 15–18.

Bhakshu, L.M. and Raju, R.R.V. 2009. Chemical composition and *in vitro* antimicrobial activity of essential oil of *Rhynchosia heynei*, an endemic medicinal plant from Eastern Ghats of India. Pharm. Biol. 47: 1067–1070.

CLSI. Clinical and Laboratory Standards Institute. 2003. Metodologia dos testes de sensibilidade a agentes antimicrobianos por diluição para bactéria de crescimento aeróbico. Clinical and Laboratorial Standards Institute, CLSI norma M7-A6, vol. 23 n°. 2, 6a. ed.

Costa, M.M., Silva, M.S., Spricigo, D.A., Witt, N.M., Marchioro, S.B., Kolling, L. and Vargas, A.P.C. 2006. Caracterização epidemiológica, molecular e perfil de resistência aos antimicrobianos de *Escherichia coli* isoladas de criatórios suínos no sul do Brasil. Pesq. Vet. Bras. 26: 5–8.

Cristani, J. 1997. Efeito do óxido de zinco (ZnO) no controle da diarréia pós-desmame em leitões experimentalmente desafiados com *Escherichia coli*. Pelotas, Universidade Federal de Pelotas, 74 p. (Master thesis).

De Lange, C.F.M., Pluske, J., Gong, J. and Nyachoti, C.M. 2010. Strategic use of feed ingredients and feed additives to stimulate gut health and development in young pigs. Livest. Sci. 134: 124–134.

De Martino, L., De Feo, V., Formisano, C., Mignola, E. and Senatore, F. 2009. Chemical composition and antimicrobial activity of the essential oils from three chemotypes of *Origanum vulgare* L. ssp. *hirtum* (Link) Ietswaart Growing Wild in Campania (Southern Italy). Molecules 14: 2735–2746.

Duarte, M.C.T., Figueira, G.M., Sartoratto, A., Rehder, V.L.G. and Delarmelina, C. 2005. Anti-*Candida* activity of Brazilian medicinal plants. J. Ethnopharmacol. 97: 305–311.

Duarte, M.C.T., Leme, E.E., Delarmelina, C., Soares, A.A., Figueira, G.M. and Sartoratto, A. 2007. Activity of essential oils from Brazilian medicinal plants on *Escherichia coli*. J. Ethnopharmacol. 111: 197–201.

Duarte, M.C.T., Duarte, R.M.T., Souza, D.P. and Bersan, S.M.F. 2012. 10. Antimicrobial activity and action mechanisms of essential oils. In: Medicinal Essential Oils: Chemical, Pharmacological and Therapeutic Aspects. Damião Pergentino de Souza (ed.) 1 ed. Nova York: Nova Science Publishers, Inc. 1: 173–200.

Duarte, M.C.T., Figueira, G.M., Foglio, M.A., Rodrigues, R.A.F., Coraucci Neto, D., Ruiz, A.L.T.G. and Carvalho, J.E. 2013. Micropartículas de óleos essenciais e seus usos para prevenção de doenças entéricas. BR10 2012 021975 1 Patent.

Formisano, C., Napolitano, F., Rigano, D., Arnold, N.A. and Piozzi, F.S.F. 2010. Essential oil composition of *Teucrium divaricatum* Sieb. ssp. *villosum* (Celak.) Rech. fil. growing wild in Lebanon. J. Med. Food 13: 1281–1285.

Gong, J., Yu, H., Liu, T., Li, M., Si, W., de Lange, C.F.M. and Dewey, C. 2008. Characterization of ileal bacterial microbiota in newly-weaned pigs in response to feeding lincomycin, organic acids or herbal extract. Livest. Sci. 116: 318–322.

Humeirah, A.G.S., Azah, M.A.N., Mastura, M., Mailina, J., Saiful, J.A., Muhajir, H. and Puad, A.M. 2010. Chemical constituents and antimicrobial activity of *Goniothalamus macrophyllus* (Annonaceae) from Pasoh Forest Reserve, Malaysia. Afr. J. Biotechnol. 9(34): 5511–5515.

Kaddu-Mulindwa, D.H., Aisu, T., Gleier, K., Zimmermann, S. and Beutin, L. 2001. Occurrence of Shiga toxin-producing *Escherichia coli* in fecal samples from children with diarrhea and from healthy zebu cattle in Uganda. Int. J. Food Microbiol. 66: 95–101.

Kamatou, G.P.P., Makunga, N.P., Ramogola, W.P.N. and Viljoen, A.M. 2008. South African *Salvia* species: a review of biological activities and phytochemistry. J. Ethnopharmacol. 119: 664–672.

Kirsti, T.K.H., Bento, M.H.L., Lahtinen, S., Ouwehand, A., Schulze, H. and Rautonen, N. 2010. The effect of feeding essential oils on broiler performance and gut microbiota. Br. Poult. Sci. 51: 381–392.

Lee, N., Yuen, K.-Y. and Kumana, C.R. 2003. Clinical role of β-lactam/β-lactamase inhibitor combinations. Drugs 63: 15: 1511–1524.

Li, S.Y., Ru, Y.J., Liu, M., Xu, B., Péron, A. and Sh, X.G. 2012. The effect of essential oils on performance, immunity and gut microbial population in weaner pigs. Livestock Science 145: 119–123.

Manzanilla, E.G., Perez, J.F., Martin, M., Kamel, C., Baucells, F. and Gasa, J. 2004. Effect of plant extracts and formic acid on the intestinal equilibrium of early-weaned pigs. J. Anim. Sci. 82: 3210–3218.

Marques, M.O.M., Facanali, R., Haber, L.L. and Vieira, M.A.R. 2012. 2. Essential oils: history, biosynthesis and agronomic aspects. In: Medicinal Essential Oils: Chemical, Pharmacological and Therapeutic Aspects. Damião Pergentino de Souza (ed.) 1 ed. Nova York: Nova Science Publishers, Inc. 1: 3–22.

Michiels, J., Missotten, J., Dierick, N., Fremaut, D., Maene, P. and De Smet, S. 2008. *In vitro* degradation and *in vivo* passage kinetics of carvacrol, thymol, eugenol and trans-cinnamaldehyde along the gastrointestinal tract of piglets. J. Sci. Food Agri. 88: 2371–2381.

Michiels, J., Missotten, J., van Hoorick, A., Ovyn, A., Fremaut, D., De Smet, S. and Dierick, N. 2011. Effects of dose and formulation of carvacrol and thymol on bacteria and some functional traits of the gut in piglets after weaning. Archives Anim. Nutr. 64: 136–154.

Morés, N., Marques, L.L.J., Sobestiansky, J., Oliveira, A. and Coelho, S.S.L. 1990. Influência do nível protéico e/ou da acidificação da dieta sobre a diarréia pós-desmame em leitões causada por *Escherichia coli*. Pesquisa Veterinária Brasileira 10(n.3/4): 85–88.

OECD Guideline for testing of chemicals—Acute Oral Toxicity: 420—adopted 17/12/2001. "Sistema Globalmente Harmonizado de Classificação e Rotulagem de Substâncias Químicas" (Globally Harmonized System of Classification and Labelling of Chemicals-GHS)—http://www.unece.org/trans/danger/publi/ghs/officialtext.html.

Pearson, M.L., Jereb, J.A., Frieden, T.R., Crawford, J.T., Davis, B.J., Dooley, S.W. and Jarvis, W.R. 1992. Nosocomial transmission of multidrug-resistant *Mycobacterium tuberculosis*: a risk to patients and health care workers. Ann. Intern. Med. 117(3): 191–196. doi:10.7326/0003-4819-117-3-191.

Ribeiro, A.M.L., Rudmik, L., Canal, C.W., Kratz, L.R. and Farias, C. 2005. Uso de gemas de ovos de aves hiperimunizadas contra *Escherichia coli* suína no controle da diarréia neonatal de leitões. R. Bras. Zootec. 34: 1234–1239.

Rodrigues, R.A.F. 2004. Preparo, caracterização e avaliação funcional de microcápsulas obtidas por spray drying, contendo extrato de café crioconcentrado. Tese de Doutorado, Faculdade de Engenharia de Alimentos, Universidade Estadual de Campinas, 98 pp.

Silva, C.J., Barbosa, L.C.A., Demuner, A.J., Montanari, R.M., Pinheiro, A.L., Dias, I. and Andrade, N.J. 2010. Chemical composition and antibacterial activities from the essential oils of myrtaceae species planted in Brazil. Quim. Nova 33: 104–108.

Silva, E.N. 2004. A polêmica da resistência a antibióticos em aves. III Simpósio Internacional de Inocuidade de Alimentos, São Paulo, Outubro.

Smith, H.W. and Huggins, M.B. 1979. Experimental infections of calves, piglets and lambs with mixtures of invasive and enteropathogenic strains of *Escherichia coli*. J. Med. Microbiol. 12: 507–510.

Struelens, M.J. 2009. Guidelines and indicators for methicillin-resistant *Staphylococcus aureus* control in hospitals: toward international agreement? Curr. Opin. Infect. Dis. 22(4): 337–8.

Van Vuuren, S.F. 2008. Antimicrobial activity of South African medicinal plants. J. Ethnopharmacol. 119: 462–472.

Wang, Q., Gong, J., Huang, X., Yu, H. and Xue, F. 2009. *In vitro* evaluation of the activity of microencapsulated carvacrol against *Escherichia coli* with K88 pili. J. Appl. Microbiol. 107: 1781–1788.

Willems, R.J., Top, J., van Santen, M., Robinson, D.A., Coque, T.M., Baquero, F., Grundmann, H. and Bonten, M.J. 2005. Global spread of vancomycin-resistant Enterococcus faecium from distinct nosocomial genetic complex. Emerg. Infect. Dis. 11: 821–828.

How to Improve Some Properties and Qualities of Plant Extracts and their Derivatives using Pharmacotechnical Technology Approach

Rodney A.F. Rodrigues,* Lais T. Yamane and Verônica S. de Freitas

Introduction

Currently, natural products exhibit huge potential for the discovery and development of novel strategies and opportunities for new alternative approaches in pharmaceutical, cosmetic, food, agrochemical and veterinary applications, among others.

The first medicines known to man were certainly made from locally grown wild plants. Knowledge regarding the actions of plants was compiled by trial and error and passed down orally from generation to generation (Holm and Hiltunen 2002). For many years, humans have relied on nature to supply their basic needs, among them are medicines that are used to treat and cure a wide spectrum of diseases. Specifically, herbs have formed the basis of traditional medicine in modern health care, with the earliest records documenting the use of approximately 1000 plant-derived substances since the earliest human civilizations in Mesopotamia (Cragg and Newman 2013). In 2010, among the list of approximately 250 drugs in the World Health Organization's essential medicine list, 11.0% are exclusively of plant origin, and nearly 80.0% of the African and Asian population depends on traditional medicines for their primary health care (Sahoo et al. 2010). Natural products are used as food and cosmetic excipients as well as pharmaceutically active compounds in varied applications such as natural colourants/antioxidants, flavouring agents and antimicrobials (Oliveira et al. 2010). As one seeks to develop a process for the manufacture of drug substances, that is, the active pharmaceutical agent itself, one must be aware of

Campinas State University, Chemical, Biological and Agricultural Research Center, Chemical of Natural Products Division, Campinas, Sao Paulo, Brazil.
* Corresponding author: rodney@cpqba.unicamp.br

certain requirements that are associated with the manufacturing process. These requirements are outlined in certain Food and Drug Administration guidelines (McChesney 1999). Standards for herbal drugs are being developed worldwide, but until now, there has been no common consensus regarding how these should be adopted (Sahoo et al. 2010). Several terms have been used for the classification of natural products, such as functional foods, medical foods, dietary supplement health foods, and nutraceuticals including botanicals and phytomedicines. Phytochemicals are secondary metabolites that are synthesized by plants including different groups of compounds, such as carotenoids, phenolics, alkaloids, terpenes, sterols, saponins, nitrogen-containing compounds, and organosulphurs. Currently, the herbal and nutraceutical processing sectors are moving increasingly toward the commercialization of standardized extracts or dried extracts that are manufactured to achieve a specified phytochemical dosage (Oliveira et al. 2010). Herbal medicines have distinctive characteristics that make them different from synthetic drugs. They contain more than one active compound, and the active principle is frequently unknown. Generally, these products are composed of a complex mixture of substances for which little information on physical and chemical properties is available (Sahoo et al. 2010, Oliveira et al. 2010). Thus, the production of standardized dried extracts represents an expanding field due to the current tendency of the pharmaceutical industry to substitute traditional fluid forms with dry plant extracts (Oliveira et al. 2010).

Because the discovery of drugs derived from new synthetic compounds requires several years and costs millions of dollars, and as this search provides a low rate of new therapeutic substances, plants have served as the starting point for countless drugs and crude active extracts on the market (McChesney 1999). Additionally, compounds are still isolated from plants if they cannot be easily synthesized; that is, the synthesis comprises too many steps or is economically unprofitable (Holm and Hiltunen 2002). Many natural products do not possess appropriate physical/chemical properties for direct development as pharmaceutical agents, and for this reason, their stability, solubility, and bioavailability may limit their application as a pharmaceutical agent (McChesney 1999).

Following this argument, the manufacturing process and pharmacotechnical factors that affect phytocompound liberation must be studied to achieve better results using a source such as natural products. The production process must match the stability of the active compound; that is, it must possess sufficient chemical/physical stability such that it can withstand processing and not change its composition during the transformation from an active substance to the final formulated pharmaceutical and such that the final formulated pharmaceutical is stable during different stages (McChesney 1999). Over the past several years, great advances have been achieved in the development of novel drug delivery systems for active plant compounds and their respective extracts. The variety of novel herbal formulations has been reported using bioactives and extracts. The novel formulations have been known to possess remarkable advantages over conventional formulations of active plant compounds and derivatives, including the enhancement of solubility and bioavailability, protection against toxicity, enhancement of pharmacological activity/ stability, improvement of tissue macrophage distribution, sustained delivery, and protection against physical/ chemical degradation (Saraf 2010). It is clear from this overview that significant efforts are needed to develop all of the necessary capabilities of the manufacturing process. Many studies have been published regarding the uneasiness to improve herbal product quality, which would be very effective to protect them during critical phases, especially transport, storage and shelf life, for a suitable period of time. In light of this approach, the aim of this chapter is to provide an overview of some of the techniques practiced in the majority of laboratories worldwide in recent decades, using an herbal source as a matrix for a formulation.

Many of the pharmacological properties of conventional drugs can be improved through the use of drug delivery systems, including particulate carriers composed primarily of lipids and/or polymers. Drug delivery systems are designed to alter the pharmacokinetics and biodistribution of their associated drugs, or to function as drug reservoirs (i.e., as sustained release systems), or both (Allen and Cullis 2004).

Herbal drugs present a number of advantages, including the enhancement of solubility, bioavailability, pharmacological activity, and stability, protection against toxicity, sustained delivery, and protection against physical and chemical degradation (Saraf 2010).

State-of-the art

Spray drying

Removal of water has been used for centuries to preserve foods and other products. Dehydration is a complex phenomenon that involves simultaneous mass and energy transport in a hygroscopic and shrinking system. The rate of dehydration is governed by the rate at which these processes occur. Many methods are used for drying, including hot air drying, vacuum drying, drum drying, spray drying, spray freeze drying, freeze drying, fluidized bed drying, spouted bed drying, superheated steam drying, microwave and radiofrequency drying. Dehydrated foods are preserved because their water activity is at a level in which no microbiological activity can occur and deteriorative chemical and biochemical reaction rates are reduced to a minimum (Araya-Farias and Ratti 2008).

Spray drying is the most common powder-generating method and represents an elegant one-step processing operation for turning a liquid feed into a dried particulate form by spraying the feed into a hot drying gas medium to produce biopharmaceutical formulations with unique particle characteristics. Spray drying consists of three steps of operation: atomization, dehydration, and powder collection (Ameri and Maa 2006). Although it reduces the product bulk weight and size, spray drying minimizes handling and preserves the product by reducing its water activity to a low level, which is required to stop bacterial growth. The production of dried particles from a liquid feed in a single processing step makes spray drying an exceptional and important method (Oliveira et al. 2010). Spray drying has been widely applied in the food and pharmaceutical industries for the preparation of fine powdered forms, such as microencapsulated active pharmaceutical ingredients, aroma and flavours, biopharmaceuticals and herbal medicines, and due to its associated ease of industrialization (Shu et al. 2006, Araújo et al. 2010).

Reluctance to use spray drying for formulation development may be because the process often subjects the therapeutic actives to temperatures in excess of 100°C utilizing a co-current spray dryer in single-stage mode, which is a concern for thermally labile compounds (Ameri and Maa 2006).

However, in contrast to common beliefs regarding high-temperature spray drying methods, the rapid evaporation associated with this process maintains a low droplet temperature that does not affect the composition of the final product. Thus, spray drying is appropriate for heat-sensitive products and for biological and pharmaceutical products (Rodrigues et al. 2011). Moreover, in favour of this tendency, the following characteristics should be highlighted: greater concentration, stability and ease of standardization of bioactive compounds; ease of transportation; reduced space required for product storage; and lower risk of microbial growth (Oliveira et al. 2010).

Microencapsulation is the science of packaging components denoted as core or active within a secondary material named an encapsulant or coat and delivering them in small particles (Sanguansri and Augustin 2006). For many years, this technique has been used in the pharmaceutical industry (Nori et al. 2011). In these microparticles, the active principle is protected from external conditions and can be released subsequently in a controlled manner. Additionally, microencapsulation can convert liquids into free-flowing powders, which may permit easy handling (Villacrez et al. 2014). There are various industrial applications of the microencapsulation process, such as carbonless paper, scratch and *sniff fragrance* sampling, *intelligent* textiles, controlled release of drugs, pesticides and cosmetics (Martins et al. 2008).

Microencapsulation is used as a means of isolating an ingredient from the reactions of the surrounding materials or environment. It may be used to stabilize a sensitive ingredient and for the controlled delivery of active components (Sanguansri and Augustin 2006). Different encapsulating agents are available for food applications, the following being the most commonly used agents: hydrolyzed and modified starches, Arabic gum and gelatine. Maltodextrin is hydrolyzed starches produced by the action of either acids or enzymes; it is commonly used as wall material due to its favourable properties in terms of emulsification, film formation, water solubility, low viscosity at high concentrations and biodegradability, as well as low cost (Villacrez et al. 2014). The choice of the polymer depends on the physical and chemical properties of the material to be encapsulated, the process used to form the microcapsules or microparticles, and the pharmacodynamic goals. Factors that can alter content retention during entrapment include the molecular mass, structural conformation, chemical function, physical state, and thickness and surface area of the microparticle. For microencapsulation by spray drying, gelatine is a good choice as a wall material because of its effective emulsification, film formation, water solubility, edibility and biodegradative properties, among others (Shu et al. 2006). Acacia gum, also known as Arabic gum, is the best known exudate gum in the world. Eighty percent of Arabic gum originates from *Acacia senegal* (*Fabaceae*), and 10–15% derives from other species, such as *Acacia seyal*, found in Chad, Mali, Mauritania, Nigeria, Senegal, and the Sudan. Acacia gum is a polysaccharide exudate from trees that is mainly composed of galactose (61.0%) and other sugars, including arabinose, rhamnose, glucose and glucuronic acid. It is odourless, tasteless, water soluble, slightly acidic, and nontoxic, and has little colour; it is also considered one of the food industry's best materials for manufacturing microcapsules or microparticles because of its capacity to promote emulsification, stabilization, and antioxidation and its very low viscosity compared with other gums. Its viscosity tends to increase at concentrations above 30.0%; however, it is still sufficiently low to permit the pumping solutions at concentrations of 55.0%, which is 10 times higher than the limit of other gums at approximately 5.0% (Rodrigues et al. 2011). There are some synonyms for Acacia gum in the *Handbook of Pharmaceutical Excipients*, such as Arabic gum, E-414, acacia gum, gummi africanum, Arabic gum, gummi arabicum, gummi mimosae and talha gum (Rowe et al. 2006). Problems associated with the use of Acacia gum in microencapsulation are the high cost and limited supply (Loksuwan 2007). Another special polymer is chitosan, a biopolymer that caught the attention of researchers because of its special properties and functionality. Chitosan is a linear polysaccharide that is insoluble in water and is formed from β-(1-4)-linked D-glucosamine and N-acetyl-D-glucosamine units obtained through the deacetylation of chitin. It is an atoxic, biocompatible/degradable material that has antibacterial properties. In acidic environments, the amino groups are protonated, and their positive charges can interact with polyanions such as alginate, carrageenan and hyaluronic acid, among others, forming polyelectrolyte complexes, which are increasingly used in the encapsulation of various biocomponents. Additionally, carrageenan is a very interesting polymer, defined as an anionic linear polysaccharide that is extracted from red seaweed (*Rhodophyceae*) which consists of alternating α-1,4 and β-1,3-linked anhydrogalactose residues that contain different numbers of sulphate groups. Depending on the number of sulphate groups in the galactose dimers, there are three major fractions: kappa (k)—carrageenan; iota (ι)—carrageenan and lambda (λ)—carrageenan (Dima et al. 2014).

Many oils in the food and flavour categories have properties such as a strong flavour and instability in an oxidative environment, thus requiring encapsulation in a core-shell material to reduced oxidative degradation to control the release rate or even to improve the shelf life of these materials. Microencapsulation techniques can be as diverse as coacervation, atomization, interfacial polymerization, spray drying and *in situ* polymerization (Martins et al. 2008).

According to Ameri and Maa (2006) and Rodrigues et al. (2011) the most important spray drying process parameters include the following: drying of the air inlet temperature, drying of the air outlet temperature, drying of the air flow rate, atomizing of the air flow rate, residence time, external humidity, and chamber geometry. All of these parameters can affect the properties of the final product, such as the particle shape and moisture content. Therefore, these variables must be studied and adjusted to achieve good productivity and reproducibility. In addition to the inherent properties of the process, the inherent properties of the product must also be taken into account. Such properties include the density, apparent volume, particle size distribution, hygroscopicity, agglomeration, and stability, among others (Rodrigues et al. 2011).

Spray drying has evolved into a mature technique for industrial-scale production up to a few tons per day; however, the production of high-valued powders is still limited to the laboratory bench-top. In conclusion, spray drying can not only serve as an effective research tool, but it also offers large-scale manufacturing capabilities (Ameri and Maa 2006).

Spray freeze drying

Spray freeze drying is a combination of two different drying processes, spray drying and freeze drying, and it combines the best features of both techniques to produce high quality and spherical porous particles in a wide range of sizes, especially for pharmaceutical and biochemical uses (Sadikoglu 2010).

Spray freeze drying can be performed at atmospheric pressure or under a vacuum into liquids, gases or over a fluidized bed. In liquid, a solution is atomized into a cryogenic medium to freeze the droplets, making it possible to obtain nanostructured particles with a large surface area. In gas, the solution is atomized into refrigerated air so that the droplets freeze instantly and are then freeze-dried in a classical freeze dryer or in an atmospheric freeze dryer (Kudra and Mujumdar 2009a).

This process generally results in yields greater than 90.0% because all atomized droplets can be collected and dried in a confined space. Furthermore, it allows control through the choice of excipient, atomization and drying conditions (Maa and Costantino 2004).

Among the challenges associated with spray freeze drying are the cost and logistics of handling and developing a scaled-up process for liquid nitrogen, which makes this technique more suitable for high value products (Filková et al. 2006, Abdul-Fattah and Truong 2010).

Freeze drying

Due to a significant increase in the commercialization of products from plants in recent years, the pharmaceutical industry has attempted to provide extracts that meet specific standards of quality and safety (Oliveira et al. 2010).

For this reason, concern regarding maintenance of the concentration and stability of bioactive compounds in these extracts has increased the search for technologies that efficiently preserve these compounds (Kudra and Mujumdar 2009b).

Much of the composition of fresh plant matter is water, which can interfere in different ways during processing by altering the biological and chemical characteristics of the material. Its removal, therefore, results in a longer shelf life due to an increase in the stability of the compounds and a decrease in the risk of microbiological contamination, facilitating transport and storage (Fissore and Velardi 2012).

Lyophilization or freeze drying is an interesting technology in this field because it permits the drying of thermolabile substances. The removal of water via this process is performed in an environment with a temperature that is usually below –10°C and a minimum absolute pressure of 2 mmHg (Bruttini and Liapis 2006, Adams 2007).

Under these conditions, the water present in the material undergoes sublimation and desorption. Sublimation is the transformation of ice directly into a gas without passing through a liquid phase (Nail 2005).

The lyophilization process can be divided into three distinct stages: freezing, primary drying and secondary drying. The freezing step is very important because it influences the distribution of ice crystals in the material and, consequently, its final form. Freezing the material slowly leads to the formation of larger pores, thereby reducing the drying time (Welti-Chanes et al. 2005, Kasper and Friess 2011).

Primary drying is responsible for the removal of approximately 90.0% of the water present in the plant through sublimation. In this step, there is a tendency toward a decrease in the temperature of the material because sublimation is an endothermic process; the presence of a heat source is necessary, but it must not be too hot to avoid melting the frozen material (Fissore and Velardi 2012).

Secondary drying is responsible for the desorption of water that did not crystallize and is strongly bound to the material. The removal of this water is usually slow and represents a large portion of the drying time (Nail 2005).

Knowledge of the complexity of this process is essential for successful lyophilization and will facilitate the development and control of all of the steps to achieve greater efficiency and stability. It is worthwhile to evaluate the effect of freeze drying on specific constituents of natural products because unexpected outcomes may occur (Abascal et al. 2005, Kasper and Friess 2011).

Extrusion

Pellets are spheres that range in diameter depending on the application and are found not only in the pharmaceutical industry but also in the agricultural market and in polymer development (Vervaet et al. 1995). The process went widely unnoticed in the pharmaceutical industry until 1970 (Erkoboni 2003).

This method as a drug delivery system offers not only therapeutic advantages, such as less irritation of the gastrointestinal tract, maximization of drug absorption, and a lowered risk of side effects due to dose dumping, but also technological advantages, for instance, better flowability, less friability, a narrow particle size distribution, ease of coating and uniform packing (Vervaet et al. 1995, Trivedi et al. 2007). The uniform dispersion of a drug into small dosage units reduces the risk of a high local drug concentration and their potentially irritating effects on the gastric mucosa (Dukić-Ott et al. 2009). Particles that are manufactured in the pharmaceutical area have a size between 500–1500 μm and are commonly delivered via hard gelatine capsules but can also be compressed into tablets (Vervaet et al. 1995, Trivedi et al. 2007). Some properties such as the flow, density, friability, porosity, and surface area are important properties for pellets or spherical granules that are intended for encapsulation or compression into tablets (Erkoboni 2003). They can be produced in different ways: spraying a solution or a suspension of a binder and a drug onto an inert core, building the pellet layer by layer, spraying melted fats/waxes from the top into a cold tower to form pellets as a consequence of the hardening of the molten droplets, spray drying a solution/suspension of drug-forming pellets due to evaporation of the fluid phase or spraying a binder solution into the whirling powder. The most popular method used to produce these particles is by the extrusion-spheronization of a wet mass, but pellets can also be achieved by drug layering on nonpareil sugar or microcrystalline cellulose beads, spray drying, spray congealing, rotogranulation, hot-melt extrusion, and spheronization of low melting materials (Vervaet et al. 1995, Trivedi et al. 2007). The extrusion-spheronization is also considered to be a granulation technique, and the process involves five important steps that consist of extrusion of a wet or heated mass into extrudates and spheronization of the resulting extrudates (Trivedi et al. 2007). The process is initiated with dry mixing, in which the dry powders achieve a uniform blend prior to wet granulation, preparation of the wet mass (or granulation), during which a wet mass with requisite plasticity or deformation characteristics is prepared, shaping of the wet mass into cylinders or extrusion, which is a continuous process that results in the formation of the wet mass into rod-shaped particles. The particles are forced through dyes, screens or moulds and shaped into small cylindrical particles with a uniform diameter, similar lengths and with their own unique weight; however, they must possess sufficient plasticity to deform but not so much that they adhere to other particles when collected or rolled in the equipment. Next, in the spheronization stage, the extrudate is broken apart, and the particles are rounded into spheres in a relatively simple machine, in which the working parts consist of a bowl with fixed sidewalls and a bottom plate that rapidly rotates ranges from 100 to 2000 rpm. The final step is performed to dry the particles, excluding optional steps such as coating and encapsulation or tableting. Because the process uses a large amount of water compared with the traditional granulation method and requires the uniform distribution of the water in the wetted mass, appropriate drying must be done, which may not be suitable for moisture and heat-sensitive compounds. There is an optional sixth step—the screening step. A method for the production of pharmaceutical pellets by means of the fluid-bed rotogranulator or by the centrifugal granulator is also used (Vervaet et al. 1995, Erkoboni 2003, Trivedi et al. 2007).

Extrusion-spheronization is a multi-step process that is capable of making uniformly sized particles that are commonly referred to as spheres or pellets. The spherical pellets can also be either coated with rate-limiting polymers or compressed into tablets to obtain slow-release, targeted-release, or controlled-release profiles. The major advantage of extrusion-spheronization over other methods of producing drug-loaded spheres is the ability to incorporate high levels of active components without producing an excessively large particle (Erkoboni 2003). This process is more efficient than other techniques, and the use of multiparticulate spherical pellets as a system for the sustained, controlled, or site-specific delivery of the drugs offers a broad number of options and capabilities. In the case of coated multi-particulates, every pellet acts as a single drug reservoir with its own release mechanism. Any coating imperfections would therefore only affect the release of a small portion of the drug, in contrast to complete dose dumping from a single-unit drug reservoir. In addition, there is the possibility to combine several active components, incompatible compounds or compounds with different release profiles in the same dosage unit (Erkoboni 2003, Trivedi et al. 2007, Dukić-Ott et al. 2009). The use of extrusion-spheronization to produce the pellets results in a robust product with unique properties compared to other pelletization techniques. A product that is produced using extrusion-spheronization is spherical and has a uniform size and shape (sometimes a slightly oval shape) and an acceptable aspect ratio/sphericity, but it can also result in barely shaped, irregular particles with physical properties similar to conventional granulation. Reasonably sized spherical particles can be prepared with high-drug loading ($\geq 90\%$) at a moderate cost using minimum excipients. In a multi-step process, the quality of the final product depends on the quality of the intermediate products obtained at the end of each step, which depend significantly on the types of excipients used. The characteristics can be modified by altering the composition of the particles, granulating fluid, or the processing conditions used that requires an experimental design to obtain better results and to understand not only the main effects but also the interactions that can have a large effect on the characteristics of the final particles. Additionally, these techniques are extremely useful during process development to understand the effect of each independent variable and to control the desired attributes. Experimental designs are useful for screening studies to determine the formulation and the process variables that have significant positive or negative effects (Erkoboni 2003, Trivedi et al. 2007).

Coacervation: simple and complex

Microencapsulation by complex coacervation is a phase separation process that is based on the simultaneous desolvation of oppositely charged polyelectrolytes induced by medium modifications. This technique is accomplished via the separation of one or many hydrocolloids from the initial solution and the subsequent deposition of the newly formed coacervate phase around the active ingredient, which is suspended or emulsified in the same reaction medium. This interaction is based on complexation resulting from the mixture of solutions of substances with opposite charges, forming coacervates that precipitate due to repulsion of the solvent. As a result, two phases are formed: one is rich in polymers containing the precipitate coacervate, and the other exhibits a low content of polymers containing the solution solvent (Nori et al. 2011). The first step consists of the formation of an oil-in-water emulsion (dispersion of the oil in an aqueous solution containing a surface-active hydrocolloid), the second comprises the formation of the coating (deposition of the polymer coating on the core material), and the final step is the stabilization of the coating (coating hardening, using thermal, cross-linking, or desolvation techniques, to form self-sustaining microcapsules) (Martins et al. 2008). Microcapsules produced by coacervation possess excellent controlled-release characteristics and heat-resistant properties (Xiao et al. 2011).

Coacervation in the aqueous phase can be classified as simple and complex according to the involved phase separation mechanism. In simple coacervation, the polymer is salted out by the action of electrolytes such as sodium sulphate, or it is desolvated by the addition of a water-miscible non-solvent such as ethanol, or by increasing/decreasing the temperature. These conditions promote macromolecule-macromolecule interactions rather than macromolecule-solvent interactions. However, complex coacervation is essentially driven by the attractive forces of oppositely charged polymers. The coexistence of a coacervate phase made of concentrated polyelectrolytes and a dilute equilibrium phase depends on the pH, ionic strength and polyion concentrations. Coacervation in the aqueous phase can only be used to encapsulate insoluble material in water, whereas coacervation in the organic phase allows the encapsulation of hydrosoluble material but requires the use of organic solvents (Martins et al. 2008).

Coacervation is potentially suitable for sensitive oils and hence could be an alternative for fish oil. The microencapsulation of fish oil by complex coacervation with Arabic gum and gelatine has been reported, but it is not easy to dry by spray drying due to the large amount of water in coacervates (Wu et al. 2005). Using the same approach to protect sweet orange oil by complex coacervation with soybean protein isolate/Arabic gum, Xiao et al. (2011) achieved good protection for core material.

To protect volatile components against the activity of physicochemical and technological agents, essential oils are encapsulated using different techniques such as the following: complex coacervation, ionotropic external/internal gelation, molecular inclusion, extrusion, freeze drying, spray drying, spray chilling and spray cooling. The encapsulation of flavour compounds through complex coacervation has a series of advantages considering that it utilizes a simple technology, has an encapsulation efficiency of more than 90% and does not use organic solvents. The microcapsules obtained through complex coacervation are stable at high temperature and enable the controlled release of components (Dima et al. 2014).

Fluid-bed coating

Every year, tons of food powders are required with some specific properties that are not offered by the natural product. The encapsulation of these products provides an alternative to fulfil this request. Fluid-bed coating is one of the various processes that can be employed for the encapsulation and coating of food ingredients or additives, such as extrusion, solvent extraction, coacervation, co-crystallization, spray drying, mixing and adhesion in rotating drums, among others. Its specificity allows it to completely coat shaped dry solid particles in powders; that is, the particles are engulfed by a coating material. This type of coating process leads to capsules called reservoir systems in which the particles are surrounded by a layer or multiple layers of coating materials (Teunou and Poncelet 2005).

The process in which polymers designated as coat or shell are applied to small solid particles, droplets of liquid, or gases for a variety of aesthetic and protective purposes, provides controlled or delayed release and reduces hygroscopicity. It also helps in changing the physical characteristics of the original material, such as the flowability and compression, dust reduction and density modification. The application of fluid-bed technology to coating, which is still a batch process, is an expensive and time-consuming process but is used in pharmaceutical and cosmetic industries that are able to compensate for the cost of the process by the high price of their final product. To reduce the cost of production, the continuous process appears to be an attractive alternative for powder coating (Teunou and Poncelet 2002, 2005).

Compared with a pharmaceutical technologist, a food technologist is more obliged to cut production costs and, therefore, should adopt a somewhat different approach to this rather expensive technology. The barrier characteristics of a top-spray coating is used in most food applications, but numerous variables are involved in film coating, and thorough insight regarding their effects is essential in establishing an appropriate thermodynamic operation and in countering the effect of weather as much as possible. The top-spray system has been successfully used to coat materials as small as 100 μm (Dewettinck and Huyghebaert 1998, 1999). The principle of a fluid bed is to maintain particles in suspension in a close area by blowing air through the powder bed. The spray consists of a coating material and a solvent in which the solute is dissolved or slurred; the liquid impinges and spreads on the particles. The fluidization air evaporates the solvent, leaving a layer of solute on the surface of the particle. Particle growth can occur by either inter-particle agglomeration or surface layering. Unwanted inter-particle agglomeration is the most serious problem that is encountered in the spray coating process, particularly for particles smaller than 100 μm, but agglomeration can occur when liquid bridges form between colliding particles that are strong enough to prevent strike back. The state of the fluid bed depends on the air velocity and the properties of the powder (Teunou and Poncelet 2002, Werner et al. 2007). The bed of particles then assumes the characteristics of a boiling liquid, which explains the term fluidization (Dewettinck and Huyghebaert 1999). The behaviour is like a liquid at the initiation of boiling; that is, the upper limited surface of the fluid bed remains horizontal if the bed is inclined; an object placed in the fluid bed will float depending on its density; particles will flow through any hole in the side wall of the fluid bed when a cylinder is immersed in the fluid bed; and there is an intense circulation of particles passing through the cylinder without any external supply of energy (Teunou and Poncelet 2002). Fluidized bed coating can be employed to enhance, time or tune the effect of functional ingredients and additives such as processing aids, preservatives, vitamins and minerals, and natural and synthetic flavour (Dewettinck and Huyghebaert 1999).

Air-suspension particle coating is a complex process that involves as many as 20 independent variables, in which thin coatings are applied to powdered particles (Werner et al. 2007). The fluid bed is still a very complex unit of operation, mostly because the trajectories of particles in the fluid bed are not predictable (Teunou and Poncelet 2002). This process differs from the spray drying encapsulation process, which produces particles consisting of a homogeneously blended matrix of the polymer entrapping the particle. Air-suspension particle coating, in contrast, layers a clearly defined film coating over a core. This process is typically performed by batch, in which powdered particles are recycled through the spray zone until the desired thickness is achieved. This feature makes it time-consuming and expensive, but most importantly, it permits a number of successive coatings of one or more materials that are useful for controlled release (Werner et al. 2007). Among the existing systems, top spray, bottom spray, and side spray with rotating disc, the *Wurster* fluid bed system is the most adequate system for particle coating. In most cases, the engineer has to optimize the process to enhance performance. However, in the case of fluid bed coating, the efficiency of the process also strongly depends on the preparation phase, a step by step procedure that is called the basis of successful coating (Teunou and Poncelet 2002). Advancement of the state of the art technique is necessary to speed product and process development. It is essential to develop guidelines for the selection of potential coating materials, equipment design and for determining the operating windows of the process. This requires a fundamental approach to develop microscopic coating quality models and linkages to the length scales of the process (Werner et al. 2007).

Several processes and types of equipment are available for the production of coated powders. The choice depends on the objectives of the coating process. In batch, they can all be used as granulators or coaters with more or less efficiency and a quality product. The use of more sophisticated tools has brought and will continue to bring understanding of the granulation and coating process. This will enable the generation of suitable models prior to the automation of these processes. The success of the continuous process depends on the degree of technical solutions that will be brought to the mentioned problems. The future challenge is to conduct research to develop a continuous fluid bed coater that is reasonably priced to avoid masking the advantages of the production cost (Teunou and Poncelet 2002).

Molecular inclusion

Cyclodextrins are a series of water-soluble cyclic oligomers that consist of six to eight D-glucose monomers linked by (1→4) glycosidic bonds, which form hydrophobic central cavities with hydrophilic external walls. Therefore, cyclodextrins are able to interact with a variety of hydrophobic compounds to form inclusion complexes (Shao et al. 2014). They have the capacity to alter the physicochemical and biological properties of hydrophobic substances by encapsulating molecules of a suitable size inside their ring (Nieddu et al. 2014). Molecular inclusion has been applied in the pharmaceutical, agrochemical, food and cosmetic fields (Mura 2014).

The high biocompatibility and negligible cytotoxicity of cyclodextrins have facilitated their use as drug excipients and other properties (Martina and Cravotto 2012), including an enhanced solubility of highly insoluble components, stabilization of labile components against the degradative effects of oxidation, visible or ultraviolet light, and heat, control of volatility and sublimation, physical isolation of incompatible compounds, chromatographic separations, taste modification by masking off flavours, unpleasant odours, and controlled and/or targeted release of drugs and flavours (Nieddu et al. 2014). These properties can be produced using ecologically friendly technologies; for example, semi-natural products are formed from starch via a simple enzymatic conversion and can be used as components of medicines, cosmetics and foods (Yaseen and Mo'ala 2014).

The most commonly used forms are α-, β-, and γ-cyclodextrin, which have, respectively, six, seven, and eight glucose units, differing only in their ring size and solubility (Rowe et al. 2006).

Among the cyclodextrins, β-cyclodextrin is useful in the pharmaceutical industry because of its high encapsulation efficiency, suitable cavity dimensions, and low cost (Shao et al. 2014), although it is the least soluble among the group (Rowe et al. 2006). The β-cyclodextrin derivatives are normally distributed based on their interactions with water molecules, i.e., hydrophilic, hydrophobic or ionisable derivatives (Pinho et al. 2014). Larger cyclodextrins have been identified: ξ-(zeta), η-(eta) and θ-(theta) cyclodextrin containing 11, 12 and 13 glucose units, respectively (Marques 2010).

Several types of forces are involved in inclusion complex formation, and their relative contribution depends on the guest and cyclodextrin type. These forces include hydrophobic interactions, the reduction of conformational strain, hydrogen bonding, dipole-dipole and electrostatic interactions, and van der Waals and dispersion forces (Mura 2014).

Cyclodextrins are produced from starch and related α-1,4-glucans via enzymatic conversion, degradation and cyclization using cyclodextrin glycosyltransferase and α-amylases. Corn starch and potato starch are the most commonly used, and maize starch and wheat starch are used but contain a higher percentage of amylose, which results in lower yields during cyclodextrin production (Marques 2010).

Various techniques are used for inclusion complex preparations as follows. In the sealed-heating method, the compound is produced when the container is heated to more than 70°C. Using these methods, inclusion complexes with drugs with a poor solubility in water can be obtained without using an organic solvent (Nieddu et al. 2014). In the kneading method, cyclodextrin is kneaded in a pestle with ethanol solution at ambient temperature until homogenization, and then the compound of interest is added to the paste. The resulting paste is dried at room temperature and then powdered (Marcolino et al. 2011). This method is particularly useful for guests with a poor solubility in water (Marques 2010). In the co-precipitation method, the compound of interest is dissolved in organic solvent, and cyclodextrin dissolved in water is added with agitation. When the cooling crystallization occurs, the crystals are washed with organic solvent and then dried to yield a powdered sample. This method is useful for substances that are not water-soluble (Marques 2010).

The growing interest in cyclodextrins, particularly in the pharmaceutical field, is directly related to the number of potential advantages. Furthermore, it is promising for the development of pharmaceutical products (Mura 2014).

Liposomes

Liposomes are formed by concentric lipid bilayers that are separated by an inner water volume. Having both hydrophilic and hydrophobic compartments in their structure, liposomes can be loaded with substances with different polarities, varying from water-soluble molecules, which are in the inner aqueous cavity, to hydrophobic molecules or amphiphilic substances (Munin and Edwards-Lévy 2011), which can be dissolved in the inner lipid bilayer. Because of their effective encapsulation capacity, liposomes have found numerous applications in various fields such as drug delivery vehicles (Martí et al. 2014).

Liposomes are often distinguished according to their number of lamellae, size and surface charge (anionic, cationic, or neutral) (Bonifácio et al. 2013). Small unilamellar vesicles, large unilamellar vesicles and large multilamellar vesicles or multivesicular vesicles are the main classes (Munin and Edwards-Lévy 2011).

Many processes are used to produce liposomes, and the most common are as follows. In the reverse-phase evaporation method, phospholipids (most commonly lecithin of soybean) are dissolved in an organic solvent and aqueous buffer and then sonicated to prepare the water/oil emulsion. The organic solvent is evaporated in a vacuum rotary evaporator until a thin film is formed on the walls (Gupta and Dixit 2011, Akhtar 2014). In the lipid film hydration method, the compound and phospholipid are added to a round bottom flask and dissolved in organic solvent. The solvent is then evaporated under a vacuum in a rotary evaporator to develop a thin film (Akhtar 2014, Kumar et al. 2014).

Liposomes are drug carriers that are loaded with a large variety of molecules such as small drug molecules, proteins, nucleotides, plasmids (Singh et al. 2014), intravenously administered, insoluble or toxic drugs, DNA, or contrast-enhancement agents for delivery to target sites (Lin et al. 2013).

Drug delivery systems such as liposomes possess particular advantages, including simple control of the composition, size, and stability, a relatively simple preparation and scale-up, good drug loading efficacy, the ability to specifically target small quantities of a targeting component (Piroyan et al. 2014), biocompatibility, biodegradability, a controllable drug release rate (Lin et al. 2013) and the absence of toxicity (Rashidinejad et al. 2014). Liposomes can be administered in various pharmaceutical forms: in suspension, in an aerosol, or in a semisolid form such as in a gel, cream or lotion, and delivered via ocular, pulmonary, nasal, oral, intramuscular, subcutaneous, topical and intravenous routes (Akhtar 2014).

The inclusion of stimulus-sensitive components in the liposomal formulation allows the vehicle to act in different physicochemical places, which can be unique to a pathological region or certain disease conditions (Piroyan et al. 2014). Modified liposomes such as ethosomes (ethanolic phospholipid) have also been utilized to impart deeper permeation compared to traditional liposomes (Elnaggar et al. 2014).

The incorporation of a high alcohol content in liposomes is capable of enhancing penetration into deep skin layers and the systemic circulation. Ethanol is an efficient penetration enhancer that increases membrane flexibility and decreases the density of the lipid multilayer of the cell membrane. Therefore, when incorporated into a vesicle membrane, it enhances the ability of the vesicle to penetrate the stratum corneum (Akhtar 2014, Park et al. 2014).

Because of their lipid bilayer, liposomes are structurally similar to cell membranes. Therefore, they are widely used to facilitate the permeability of compounds through cell membranes (Chen et al. 2012) and the activity of antioxidants on liposome peroxidation (Şerbetçi et al. 2012).

There are many factors that affect the encapsulation efficiency of a drug in liposomes and are derived from the properties of the liposomes and of the encapsulated drugs. The encapsulation efficiency is affected by the hydrophilic or lipophilic properties of the encapsulated drugs, which tend to interact with the lipid bilayer. However, the aqueous volume, surface area, membrane rigidity and preparation methods, are liposome properties that interfere with the encapsulation efficiency (Nii and Ishii 2005).

The liposomal system is passive, non-invasive is available for immediate commercialization and can be applied widely in pharmaceutical, veterinary and cosmetic fields (Akhtar 2014). It is also promising as carriers not only in topical delivery but also for transdermal, oral and intravenous drug delivery.

Colonic delivery

Conventional controlled-release formulations for oral administration normally lack any properties that would facilitate specific targeting in the gastrointestinal tract. However, any slow-release system with a drug-release time profile that extends beyond six to eight hours is likely to be present in the colon for the release of a high proportion of the compound (Mukherji and Wilson 2002b).

The targeted delivery of drugs to the colon is usually performed to achieve one or more objectives. The desired outcomes can be sustained delivery to reduce the dosing frequency, delayed delivery to the colon to achieve high local concentrations for the treatment of diseases of the distal gut, delayed delivery to a time that is appropriate to treat acute phases of disease, and delivery to a region that is less hostile metabolically (Wilson 2002). The mode of drug release from colon-targeted biopolymer systems can include one or more of the following four mechanisms: diffusion, polymer erosion, microbial degradation or enzymatic degradation (Mukherji and Wilson 2002b).

The most extensive application of a formulation strategy for colonic delivery has been the employment of enteric coatings on solid substrates. This is a natural development of conventional coating technologies to avoid gastric release, thus preventing problems such as degradation or pharmacological effects including gastric irritation and nausea. The underlying principle of this approach has been the employment of polymers that are able to withstand the lower pH values of the stomach, but disintegrate and release the drug as the pH increases in the small bowel. The principal difference is in the functional use of enteric polymers. Different strategies have been adopted in the market for colonic delivery, such as ethyl cellulose-coated pellets to provide lag time for drug release in the small intestine, tablets coated with methacrylic acid copolymer to result in the release of drug at pH 7, 5-amino salicylate linked to sulphapyridine through azobonds that are cleaved in the colonic microflora, and 5-amino salicylate dimer linked through azobonds and cleaved in the colonic microflora. Some enteric polymeric materials that are commercially available are phthalate-based enteric polymers, as well as methacrylic acid–based copolymers and miscellaneous enteric polymers such as a resin that is secreted by the female lac bug named Shellac, hydroxypropyl methylcellulose acetate succinate, poly-methyl vinyl ether monoethyl ester, poly-methyl vinyl maleic acid monoethyl ester, poly-methyl vinyl ether *n*-butyl ester, and poly-methyl vinyl maleic acid *n*-butyl ester. The physical form of the formulation strongly influences retention in the ascending colon because particulates are retained to a greater extent compared to monoliths. Thus, the variables of transit and pH combine to cause enteric systems to be somewhat unreliable (Mukherji and Wilson 2002a).

Herbal formulations: experimental conditions and effectiveness of technology

Villacrez et al. (2014) used the herbal formulation approach using spray drying technology to preserve the sensory characteristics (colour, odour and taste) of the fresh fruit of the Andes berry (*Rubus glaucus* Benth.), a tropical fruit with a high anthocyanin content and pleasant aroma. The Andes berry microencapsulates obtained with Hi-Cap® 100 and maltodextrin with dextrose equivalent 20 as wall material were chosen because of their sensory properties. The anthocyanin content did not change over 180 days at 18°C and a relative humidity of less than 60%. Krishnan et al. (2005) studied different polymers as wall materials for the protection of cardamom oleoresin by spray drying. They evaluated blends of Arabic gum, maltodextrin and a modified starch, demonstrated the excellent effects of the Acacia exudate. Shu et al. (2006) prepared lycopene microcapsules using a spray drying method with a wall system consisting of gelatine and sucrose. Lycopene is a type of carotenoid that is found in ripe tomato, watermelon and pink grapefruit, giving them their characteristic red pigmentation. A great number of publications suggest that supplementation with food rich in lycopene is associated with a decreased risk of cardiovascular disease and cancers, among others. The optimal conditions were determined in this publication are as follows: ratio of gelatine/sucrose of 3/7, ratio of core to wall material of 1/4, feed temperature of 55°C, inlet temperature of 190°C, homogenization pressure of 40 MPa, and lycopene purity of no less than 52.0%, at which the microencapsulated lycopene displayed some isomerization but good storage stability.

Wang et al. (2014) studied the effectiveness of hydroxypropyl methylcellulose as an anti-adherent during the spray drying of traditional Chinese herbal extracts, for example, *Crataegi fructus* decoction extract. A small amount of hydroxypropyl methylcellulose (≥ 8%) was sufficient to overcome problematic particle adhesion, whereas a large amount of maltodextrin (50.0%) was required for the same effect. According to Rowe et al. (2006), hydroxypropyl cellulose is compatible with a number of high-molecular-weight, high-boiling-temperature waxes and oils, and it can be used to modify certain properties of these materials. Examples of materials that are good solvents for hydroxypropyl cellulose at an elevated temperature are acetylated monoglycerides, glycerides, pine oil, polyethylene glycol, and polypropylene glycol.

We showed that many variables can affect the properties of the resulting products by the spray drying method. With the aim of studying the feasibility of a spray-dried sumac extract process together with the effects of the addition of maltodextrin and the effects of the inlet and outlet temperatures of drying air on the properties of the powdered product obtained from spray drying of the sumac extract, Caliskan and Dirim (2013) concluded that the interaction between temperature and maltodextrin are important for the analysis.

Similarly, Ozdikicierler et al. (2014) assessed *Saponaria officinalis* L. microencapsulated extract by spray drying and concluded that the productivity and drying rate showed a decreasing tendency as the air outlet temperature increased because high outlet temperatures required low feed flow rates.

Araújo et al. (2010) studied the preparation and characterization of ternary solid dispersions by direct spray drying of a liquid suspension containing curcumin to enhance the compound solubility and as a drying aid. The authors concluded that the results proved that spray drying acted as an attractive and promising method to obtain enhanced solubility of solid dispersions of ternary drugs.

Chen et al. (2014) employed a response surface methodology to obtain the optimal processing conditions for jujube pulp powder by spray drying. The inlet air temperature, weight ratio of maltodextrin and the dry matter weight of jujube pulp were chosen as independent variables, and the shift of the feed flow rate, moisture content, colour, vitamin C content and hygroscopicity were independent variables (responses). The results showed that the weight ratio of maltodextrin to the dry matter weight of the jujube pulp had an extremely significant effect on all of the responses.

Ezhilarasi et al. (2014) used *Garcinia cowa* fruit extract to obtain microencapsulated powder and studied three different wall materials by spray drying: whey protein isolate, maltodextrin and a combination of whey protein isolate and maltodextrin. These microencapsulated powders were further evaluated regarding their impact on bread quality and hydroxycitric acid retention. The authors concluded that the higher hydroxycitric acid contents of encapsulated incorporated breads were sufficient to claim the functionality of hydroxycitric acid in bread. Comparatively, maltodextrin contained efficiently encapsulated Garcinia fruit extract during spray drying and bread baking. Spray drying proved to be an excellent encapsulation technique for incorporation into the food system. With the same purpose, Pang et al. (2014) studied the effect of core material on the polyphenol content of *Orthosiphon stamineus* extracts and concluded that microencapsulation using a minimal amount of protein yielded better retention of rosmarinic acid (82.1%) compared with a higher protein concentration. In addition, 5.3% maltodextrin by weight provided the highest polyphenol retention, suggesting that rosmarinic acid is more susceptible to thermal degradation than other polyphenols during sample processing.

Souza et al. (2009) demonstrated the feasibility of obtaining granule products from spray drying the extract of *Phyllanthus niruri* L. Both dry and wet granulation techniques produced granules with appropriate technological characteristics and good operational performance. The presence of Eudragit® E 100 in proportions of 10% in granules obtained by wet granulation was efficient for the production of granules with a reduced sensitivity to moisture. The dry granules required greater compressive force and demonstrated plastic deformation, but with some degree of fragmentation, while the granules obtained by wet granulation presented an entirely plastic behaviour.

Materska (2014) focused on the effect of freeze drying on four different extracts of pepper fruits (*Capsicum annuum* L.): the varieties studied by the author were Capel hot, Red knight, Shanghai and Socrates. The total phenolic content (Folin-Ciocalteu method) revealed variations in phenolics in the ethanolic extracts of fresh and freeze-dried samples between the studied pepper varieties and anatomical parts of the fruit. The highest losses (40.0%) were noted in the placenta of the Capel hot variety, while the freeze-dried pericarp of the Shanghai variety increased by approximately 55.0% in comparison to the fresh fruits. The author suggested that this result could be explained by the higher sensitivity to drying exhibited by some varieties, which increases the porosity of the tissue and thus increases the extraction efficiency of phenolic compounds.

Azevedo et al. (2014) compared freeze drying and hot air drying techniques to dehydrate residues of camu-camu, a native Brazilian fruit with high levels of vitamin C that is mostly used to produce juice exported to Europe and Japan. The camu-camu residue was divided into four groups: fresh, hot air drying (temperatures of 50 and 80°C) and freeze drying. The authors observed that the ascorbic acid content decreased in all of the dried residues of camu-camu when compared with fresh material, yet the camu-camu residue powders still represented a relevant source of vitamin C. A significant reduction of anthocyanins, total phenolics and proanthocyanidins was also observed. The group that was subjected to hot air drying at 80°C presented the highest losses of these compounds (100, 63.9 and 46.7% for anthocyanins, total phenolics and proanthocyanidins, respectively), whereas the freeze-dried group presented the lowest losses (53.6, 42.2 and 15.8%, respectively).

Marigold (*Tagetes erecta* L.) has been used as a medicinal plant for a long time for a variety of ailments, and its flowers have commercial value as a carotenoid resource that provides yellow dye that is suitable as a food colourant (Williams 2013). Siriamornpun et al. (2012) evaluated the effects of different drying processes, such as freeze drying, hot air drying and combined far-infrared radiation with hot air convection, on carotenoids and phenolic compounds of marigold flowers. Both fresh and freeze-dried flowers maintained similar values in terms of the total flavonoid content, whereas hot air-dried flowers contained the lowest amount ($p < 0.05$). In the determination of the total phenolic content (Folin-Ciocalteu method), all of the treatments provided similar results. However, the authors observed different effects on carotenoid content because the hot air-dried flowers provided the highest content of β-carotene, while freeze drying, hot air drying and combined far-infrared radiation provided the highest levels of lutein and lycopene. These results suggest that the drying method should be chosen judiciously and in accordance with the desired bioactive compound.

Stoner et al. (2007) proposed the use of berries for the prevention of oesophageal and colon cancer. Blackberries, raspberries, strawberries and cranberries contain high concentrations of ellagic acid, a component that is active in oesophageal and colon tumours (Umesalma et al. 2014), and the quantity of this compound can be increased up to 10-fold by freeze drying. The authors freeze-dried black raspberries (*Rubus occidentalis*) and strawberries (*Fragaria ananasia*) and administered as 5 and 10% of the diet to rats with oesophageal-induced tumours. The components of freeze-dried black raspberries and strawberries remained stable for at least two years, and both berries were able to inhibit tumour initiation in the treated rats.

With the aim of developing a methodology to obtain alcohol-free powdered propolis with the potential for controlled release in foods and of characterizing the obtained material, Nori et al. (2011) studied complex coacervation as an alternative method. They concluded that it was possible to encapsulate propolis extract by complex coacervation with soy protein isolates and pectin and to obtain it in the form of a powder that is alcohol-free and stable and that has antioxidant properties, antimicrobial activity against *Staphylococcus aureus* and the potential for controlled release in foods.

Wu et al. (2005) applied microencapsulation of fish oil by simple coacervation followed by spray drying and investigated selected process parameters: the oxidative stability and microstructure of the microcapsules. The conclusion of this publication was that the oxidative stability of fish oil was improved by microencapsulation, and the best results were achieved when 40.0% of the maltodextrin was replaced with Acacia gum. Simple coacervation of hydroxypropyl methylcellulose could be observed only when the dextrose equivalent value of maltodextrin, the concentration of hydroxypropyl methylcellulose solution and the percentage of fish oil in the microcapsules was no more than 20.5 and 40.0%, respectively. Moreover, the microencapsulation efficiency was higher with a hydroxypropyl methylcellulose solution concentration of 4.0% and a fish oil percentage of less than 30.0%. The authors also concluded that the oxidative stability of fish oil was improved by microencapsulation via the simple coacervation of hydroxypropyl methylcellulose, and the best results were obtained by replacing 40.0% of the maltodextrin with Acacia.

Xiao et al. (2011) investigated coacervation between a soybean protein isolate and Arabic gum for sweet orange oil microencapsulation in terms of pH, ionic strength, the soybean protein isolate and Arabic gum ratio, the core material load and micromolecules. They concluded that the optimum pH for soybean protein isolate/Arabic gum coacervation was 4.0. A high ionic strength reduced the coacervation between the two biopolymers. The highest coacervate yield was achieved with a soybean protein isolate to Arabic gum ratio of 1:1, and the core material load for the highest microencapsulation efficiency and microencapsulation yield was 10.0%. The addition of sucrose to the sucrose/soybean protein isolate ratio of 1:1 increased the microencapsulation yield by 20.0%, reaching 78.0% compared to 65.0% in the control. The microcapsules were spherical, lacked holes on their surface by scanning electron microscopy observation and retained their flavour components in microcapsules according to the results of a gas chromatography–mass spectrometry analysis, indicating good protection of the core material.

For the encapsulation of the *Pimenta dioica* (L.) Merr., essential oil microspheres were investigated by Dima et al. (2014). The microspheres containing *Pimenta dioica* essential oil were prepared through the extrusion of the oil-in-water emulsions using chitosan and chitosan/k—carrageenan in different mass ratios as wall material; they proposed that microspheres can be used for the preparation of some food products, mainly for meat products.

Martins et al. (2008) selected coacervation to produce polylactide microcapsules using *Thymus vulgaris* L. (thyme oil), an antioxidant and antimicrobial agent, as the core material. According to the authors, the novelty of this approach consists of dissolving polylactide acid in dimethylformamide which is a good solvent for polylactide acid but also has a high solubility in water. Upon contact with water, the homogeneous solution of polylactide acid in dimethylformamide promotes the precipitation of polylactide acid around the thyme oil core. The quality of encapsulated oil has shown that apolar compounds of thyme oil are preferentially encapsulated over polar compounds. The overall encapsulation efficiency of thyme oil was found to be 30.5%.

Araújo, Jr. et al. (2013) applied an extrusion-spheronization method to improve the photostability of 4-nerolidylcatechol from *Pothomorphe umbellata* (L.) Miq. synonym *Piper umbellatum* L., Piperaceae. This species has been extensively used in Brazilian folk medicine and is well known for its strong antioxidant properties. However, its main active constituent, 4-nerolidylcatechol, is sensitive to ultraviolet and visible light, which can limit the use of the intermediate and final herbal preparations of this species. Using coated multiparticulate solid dosage forms of *P. umbellata*, the authors obtained a formulation with the purpose of increasing the stability of 4-nerolydilcatechol. *Piper umbellatum* extract was used as a wetting liquid for the preparation of pellets via extrusion-spheronization. The pellets were coated in a fluidized bed with three different polymers; hydroxypropyl methylcellulose, polyvynilpirrolidone K-30 and polyvinyl alcohol-polyethylene glycol graft-copolymer. The photoprotection was higher in pellets coated with polyvynilpirrolidone K30 and polyvinyl alcohol-polyethylene glycol graft-copolymer. The polyvinyl alcohol-polyethylene glycol graft-copolymer-coated pellets with a weight increase of 6.0 and 9.0%, respectively, resulted in a final concentration of 4-nerolydilcatechol that was approximately five times higher than the uncoated pellets or liquid extracts, suggesting the potential of this formulation as a multiparticulate solid dosage form for *Piper umbellatum* extracts.

Cecropia glaziovii Snethl., Urticaceae, is a vegetal species that is commonly used by the South American population as an antihyperlipidaemic agent. Several pharmacological studies have recently reported the potential of *C. glaziovii* as a hypotensive, anti-asthmatic and anxiolytic agent. Although dry herbal extracts are more stable in comparison to the aqueous form, they have a complex composition and contain a number of active components and hydrophilic ingredients with a tendency to absorb moisture, leading to microbial growth and hydrolysis. The highly hydrophilic characteristics of these compounds may facilitate the production of highly concentrated herbal extracts in aqueous and hydroalcoholic solutions. However, this characteristic also presents a challenge due to the significant effects on physicochemical and technological properties as well as on the biopharmaceutical properties of the extracts. The strict requirements regarding the quality, safety and effectiveness of phytopharmaceutical products represent an enormous challenge in the search for products with a high level of uniformity, reproducibility and stability. The incorporation of dry extracts into multiparticulate dosage forms, such as pellets produced using extrusion-spheronization technology, is a suitable alternative to overcome the lack of technological options for dry extracts because they are associated with low flowability and high hygroscopicity. The authors also optimized dry extract of *C. glaziovii* production into pellets to decrease the sorption of moisture and increase the stability, safety and percentage of the extract in the final product. Pellets containing approximately 50.0% dry extract *C. glaziovii* were considered to be the most technologically viable, offering a significant improvement in flowability and compressibility properties, and a decrease in moisture compared with the dry extract of *C. glaziovii*. This formulation could be used as an intermediate dosage form to fill capsules or be compressed into tablets. In conclusion, pellets containing a high dose of *C. glaviovii* extract were successfully prepared and achieved degrees of quality, physical stability and feasibility compatible with the desirable characteristics of a phytopharmaceutical product (Beringhs et al. 2013).

A modified starch was evaluated by Dukic´ et al. (2007) as an alternative excipient to microcrystalline cellulose for pellets prepared via extrusion-spheronization using theophylline as a model drug. They concluded that the modified starch in that study had good potential as the main excipient in formulations intended for extrusion-spheronization based on the high process yield, good pellet sphericity and immediate release properties of the pellets. This modified starch could represent an alternative to microcrystalline cellulose in the processing of formulations using this specific method.

Deshmukh and Amin (2013) developed pellets that melt in the mouth (registered under the trademark of *Meltlets*), containing a 40% herbal extract of soy isoflavones that provide antioxidant activity in menopausal women. The process of extrusion-spheronization was optimized in terms of the extruder speed, extruder screen size, spheronization speed, and time. The process parameters were optimized to afford a melt-in-the-mouth dosage form that dissolves within 60 seconds without the aid of water. The antioxidant activity of the *Meltlets*® was comparable to the original extract, indicating that the antioxidant activity of the extract was not altered after being subjected to harsh processing conditions and that the utilized excipients did not contribute to the antioxidant activity of the formulation.

In industry, researchers encounter the universal problem of incorporating medicinal herbal extracts into safe, efficacious, consistent and stable oral dosage forms (for example, tablets, granules, capsules and pellets). The highly hygroscopic nature of the herbal materials may be attributed to their hydrophilic ingredients. The sorption of moisture can exert significant effects on physicochemical and technological properties as well as on the biopharmaceutical parameters of the extracts. Chen et al. (2010) investigated the moisture-proofing effect and its underlying mechanism in herbal extracts using extrusion-spheronization together with hot-melt coating. They concluded that this method is valuable for producing moisture-proof pellets. They also concluded that this technique is solvent-free and can save time, protect the environment and reduce production costs.

Various synthetic and herbal drugs are incorporated into liposomes to improve efficacy. These are used as a good delivery vehicle for plant extracts. Various extracts can be incorporated into liposomes, including turmeric, carrot extract, papaya extract, *Aloe vera*, and green tea extract (Singh et al. 2014). These extracts are incorporated to increase the rate of permeation into the skin and to decrease the adverse side effects of the extract (Patel et al. 2014).

The formulation of liposomes containing green tea and *Gaultheria procumbens*, which has antibacterial properties, decreased the minimum inhibitory concentration against *Micrococcus luteus*, which causes acne. The dose was decreased to avoid overdosing and consequent intoxication (Singh and Sankar 2012).

The bioavailability of lipophilic drugs such as curcumin is very poor when they are administered orally or applied topically. This poor absorption through the intestine and skin necessitates high doses to reach therapeutic levels, which is sometimes inconvenient. When a gel containing curcumin liposomes was applied to the skin of mice, there was significant protection of the connective tissue, collagen bundles and elastin fibres at the level of the deep dermis. In addition, skin moisture was restored (Gupta and Dixit 2011).

When these formulations are applied to the skin, the liposomes are deposited on the skin surface and begin to merge with the cell membranes. These liposomal preparations reduce skin roughness due to interactions with corneocytes and to the presence of intercellular lipids that result in skin softening and smoothening (Singh et al. 2014).

An equivalent problem occurs with silymarin, a hepatoprotective agent that is obtained from *Silybum marianum* (L.) Gaertn.; it is poorly absorbed (20.0–50.0%) from the gastrointestinal tract. Kumar et al. (2014) showed that silymarin in a liposomal carrier system provided better *in vitro* and *in vivo* hepatoprotection.

The size of a liposome is approximately 200 nm, which allows efficient uptake into the intestine, thus bypassing metabolism in the liver, and prolonged contact with the intestinal wall due to the adhesive properties of liposomes (Takahashi et al. 2009).

Liposomes have been used as an effective delivery system to the brain because the particles entrap compounds and prevent rapid elimination or degradation. They also promote penetration through the blood-brain barrier and dispersion throughout the brain tissue (Tong-Un et al. 2010).

Tong-Un et al. (2010) confirmed the efficacy of quercetin liposomes as anxiolytic and antidepressive agents via nasal administration. The observed effects might be associated with the ability of liposomes to traverse the blood-brain barrier.

The encapsulation of *Myrtus communis* L. extract in liposomes has a more intense antioxidant effect (25.0% higher) than the same extract in its pure form. The methanolic extract of *Myrtus communis* L. has also been studied for its antimicrobial activity before and after encapsulation in liposomes. The extract appeared to be active against most of the studied microorganisms. After its encapsulation, enhanced antimicrobial activity was detected (Gortzi et al. 2008).

Thymol is a volatile phenolic compound that is extracted from thyme for use as an antimicrobial agent, and it shows a high level of activity against *Salmonella typhimurium in vitro*; however, it has poor water solubility and palatability due to its unpleasant taste and smell (Nieddu et al. 2014). Nieddu et al. (2014) produced an inclusion complex with thymol that increased the dissolution rate of thymol, which dissolves slowly in the gastrointestinal tract. The observed effect was due to the presence of β-cyclodextrin as a solubilizer of substances that are poorly soluble in water.

Hsu et al. (2013) studied rhubarb, a traditional Chinese medicinal plant with hydrophobic compounds, which is widely used as a laxative or as a liver cleanser for the treatment of indigestion. The inclusion complex demonstrated a greater solubility than the rhubarb extract, which indicated that the insoluble components of the rhubarb extract had been included with the 2-hydroxypropyl-β-cyclodextrin, thus improving its aqueous solubility.

Teixeira et al. (2013) evaluated the antioxidant and antimicrobial activity of black pepper (*Piper nigrum* L.) oleoresin encapsulated in hydroxypropyl-β-cyclodextrin. The encapsulated black pepper had significantly increased antioxidant activity; this complex improved the water solubility of the black pepper and increased its contact with hydrophilic 2,2-diphenyl-1-picrylhydrazyl radicals, thus elevating the antioxidant activity. The complex also enhanced the antimicrobial activity efficacy against *Salmonella typhimurium* at lower concentrations than the free black pepper oleoresin.

Colon-specific drug delivery based on natural polysaccharides has highly been acclaimed in recent years. A colon-specific drug delivery system should prevent drug release in the stomach as well as in the small intestine. Based on this approach, Pachuau and Mazumder (2013) reviewed the use of different natural polysaccharides as colonic drug delivery systems due to their safety and availability in a variety of structures.

All of the different technologies presented in this chapter have demonstrated effectiveness in reaching one or more targets for the preservation of sensory characteristics and prevention of sticking (for example, during the spray drying process). The administration of medication via microparticulate systems is advantageous because microparticles can be ingested or injected. In addition, they can be tailored to achieve desired release profiles, and when they are applied via site-specific delivery, they can even provide, in some cases, organ-targeted release. To date, a series of herbal ingredients such as rutin, camptothecin, essential oils, seed oils, tetrandrine, quercetin and herbal extracts and their derivatives such as dried extracts, fluid extracts, and tinctures, has been developed into microparticles. In addition, there have been numerous recent reports investigating immune and magnetic microspheres. Immune microspheres possess immune competence as a result of the coating or adsorption of antibody and antigen on polymer microspheres (Saraf 2010).

Conclusion

Herbal formulations in the world market have experienced a large growth in recent years. Extensive research is currently investigating novel pharmacotechnical products using herbal ingredients; however, this research remains in the exploratory stage. Many problems associated with research and development stages must be resolved. For example, new wall (coat/shell) materials should be evaluated to discover more suitable and cheaper polymers that can reduce toxicity, enhance bioavailability and improve the overall quality of the herbal ingredients, which have huge therapeutic potential. Many herbal extracts and phytocompounds, despite having excellent bioactivity *in vitro*, are inactive *in vivo* because of their poor absorption. When administered via novel pharmacotechnical systems, several phytocompounds show much better absorption, enabling their enhanced the bioavailability. Hence there is a great potential in the development of novel pharmacotechnical systems for the herbal ingredients.

In this chapter, we have ventured to provide the reader with a short overview of some techniques that are employed in the development of proper manufacturing processes for the generation of a desirable herbal product with high quality standards and presentation. Currently, this promising market and business niche has an increased product value, mostly due to the quality and functionality of the products. Key trends predicted for this type of approach continue to grow. The methods described in this chapter were selected mostly based on the simplicity, cost and target of the technique and do not represent a general opinion.

References

Abascal, K., Ganora, L. and Yarnell, E. 2005. The effect of freeze-drying and its implications for botanical medicine: a review. Phytother. Res. 19: 655–660.

Abdul-Fattah, A.M. and Truong, V.L. 2010. Drying process methods for biopharmaceutical products: an overview. pp. 705–738. *In*: F. Jameel and S. Hershenson (eds.). Formulation and Process Development Strategies for Manufacturing Biopharmaceuticals. John Wiley & Sons, Inc., Hoboken, New Jersey.

Adams, G. 2007. The principles of freeze-drying. pp. 15–38. *In*: J.G. Day and G.R.C. Stacey (eds.). Cryopreservation and Freeze-Drying Protocols. 2nd edition. Humana Press, New York.

Akhtar, N. 2014. Vesicles: a recently developed novel carrier for enhanced topical drug delivery. Curr. Drug Del. 1: 87–97.

Allen, T.M. and Cullis, P.R. 2004. Drug delivery systems: entering the mainstream. Science 303(5665): 1818–1822.

Ameri, M. and Maa, Y.F. 2006. Spray drying of biopharmaceuticals: stability and process considerations. Dry. Technol. 24: 763–768.

Araújo, Jr., C.A., Costa, F.S.O., Taveira, S.T., Marreto, R.N., Valadares, M.C. and Lima, E.M. 2013. Preparation of pellets containing *Pothomorphe umbellata* extracts by extrusion-spheronization: improvement of 4-nerolidylcatechol photostability. Braz. J. Pharmacog. 23(1): 169–174.

Araújo, R.R., Teixeira, C.C.C. and Freitas, L.A.P. 2010. The Preparation of Ternary Solid Dispersions of an Herbal Drug via Spray Drying of Liquid Feed. Dry. Technol. 28: 412–421.

Araya-Farias, M. and Ratti, C. 2008. Dehydration of foods: general concepts. pp. 01–36. *In*: C. Ratti (ed.). Advances in Food Dehydration. CRC Press, New York.

Azevedo, J.C.S., Fujita, A., Oliveira, E.L., Genovese, M.I. and Correia, R.T.P. 2014. Dried camu-camu (*Myrciaria dubia* H.B.K. McVaugh) industrial residue: a bioactive-rich Amazonian powder with functional attributes. Food Res. Int. 62: 934–940.

Beringhs, A.O., Souza, F.M., Campos, A.M., Ferraz, H.G. and Sonaglio, D. 2013. Technological development of *Cecropia glaziovii* extract pellets by extrusion-spheronization. Braz. J. Pharmacog. 23(1): 160–168.

Bonifácio, B.V., Silva, P.B., Ramos, M.A.S., Negri, K.M.S., Bauab, T.M. and Chorilli, M. 2013. Nanotechnology-based drug delivery systems and herbal medicines: a review. Int. J. Nanomed. 9(1): 1–15.

Bruttini, R. and Liapis, A.I. 2006. Freeze drying. pp. 257–271. *In*: A.S. Mujumdar (ed.). Handbook of Industrial Drying. 3rd Edition. CRC Press, New York.

Caliskan, G. and Dirim, N. 2013. The effects of the different drying conditions and the amounts of maltodextrin addition during spray drying of sumac extract. Food Bioprod. Process. 91(4): 539–548.

Chen, Q., Bi, J., Zhou, Y., Liu, X., Wu, X. and Chen, R. 2014. Multi-objective Optimization of Spray Drying of Jujube (*Zizyphus jujuba* Miller) Powder Using Response Surface Methodology. Food Bioprocess Technol. 7: 1807–1818.

Chen, X., Deng, Y., Xue, Y. and Liang, J. 2012. Screening of bioactive compounds in Radix Salviae Miltiorrhizae with liposomes and cell membranes using HPLC. J. Pharmaceut. Biomed. 70: 194–201.

Cragg, G.M. and Newman, D.J. 2013. Natural products: a continuing source of novel drug leads. Biochim. Biophys. Acta 1830: 3670–3695.

Deshmukh, K. and Amin, P. 2013. Meltlets® of soy isoflavones: process optimization and the effect of extrusion spheronization process parameters on antioxidant activity. Indian J. Pharm. Sci. 75: 450–456.

Dewettinck, K. and Huyghebaert, A. 1998. Top-spray fluidized bed coating: effect of process variables on coating efficiency. Lebensm. Wiss. Technol. 31: 568–575.

Dewettinck, K. and Huyghebaert, A. 1999. Fluidized bed coating in food technology. Trends Food Sci. Tech. 10: 163–168.

Dima, C., Cotârlet, M., Alexe, P. and Dima, S. 2014. Microencapsulation of essential oil of pimento (*Pimenta dioica* (L.) Merr.) by chitosan/k-carrageenan complex coacervation method. Innov. Food Sci. Emerg. 22: 203–211.

Dukic´, A., Mens, R., Adriaensens, P., Foreman, P., Gelan, J., Remon, J.P. and Vervaet, C. 2007. Development of starch-based pellets via extrusion/spheronisation. Eur. J. Pharm. Biopharm. 66: 83–94.

Dukic´-Ott, A., Thommes, M., Remon, J.P., Kleinebudde, P. and Vervaet, C. 2009. Production of pellets via extrusion–spheronisation without the incorporation of microcrystalline cellulose: a critical review. Eur. J. Pharm. Biopharm. 71: 38–46.

Elnaggar, W.S.R., El-Refaie, W.M., El-Massik, M.A. and Abdallah, O.Y. 2014. Lecithin-based nanostructured gels for skin delivery: an update on state of art and recent applications. J. Control. Release 180: 10–24.

Erkoboni, D.F. 2003. *Extrusion/Spheronization*. pp. 277–322. *In*: I. Ghebre-Sellassie and C. Martin (eds.). Pharmaceutical Extrusion Technology. Marcel Dekker Inc., New York.

Ezhilarasi, P.N., Indrani, D., Jenaa, B.S. and Anandharamakrishnan, C. 2014. Microencapsulation of Garcinia fruit extract by spray drying and its effect on bread quality. J. Sci. Food Agr. 94: 1116–1123.

Filková, I., Huang, L.X. and Mujumdar, A.S. 2006. Industrial spray drying systems. pp. 215–257. *In*: Arun S. Mujumdar (ed.). Handbook of Industrial Drying. CRC Press, New York.

Fissore, D. and Velardi, S. 2012. Freeze drying: basic concepts and general calculation procedures. pp. 47–72. *In*: R.H. Mascheroni (ed.). Operations in Food Refrigeration. CRC Press, New York.

Gortzi, O., Lalas, S., Chinou, I. and Tsaknis, J. 2008. Reevaluation of bioactivity and antioxidant activity of *Myrtus communis* extract before and after encapsulation in liposomes. Eur. Food Res. Technol. 226: 583–590.

Gupta, N.K. and Dixit, V.K. 2011. Development and evaluation of vesicular system for curcumin delivery. Arch. Dermatol. Res. 303: 89–101.

Holm, Y. and Hiltunen, R. 2002. Plant-derived drugs and extracts. pp. 23–44. *In*: K.M. Oksman-Caldentey and W.H. Barz (eds.). Plant Biotechnology and Transgenic Plants. Marcel Dekker, New York.

Hsu, C.M., Yu, S.C., Tsai, F.J. and Tsai, Y. 2013. Enhancement of rhubarb extract solubility and bioactivity by 2-hydroxypropyl—cyclodextrin. Carbohyd. Polym. 98: 1422–1429.

Kasper, J.C. and Friess, W. 2011. The freezing step in lyophilization: physico-chemical fundamentals, freezing methods and consequences on process performance and quality attributes of biopharmaceuticals. Eur. J. Pharm. Biopharm. 78: 248–263.

Kerdudo, A., Dingas, A., Fernandez, X. and Faure, C. 2014. Encapsulation of rutin and naringenin in multilamellar vesicles for optimum antioxidant activity. Food Chem. 159: 12–19.

Krishnan, S., Bhosale, R. and Singhal, R.S. 2005. Microencapsulation of cardamom oleoresin: Evaluation of blends of gum arabic, maltodextrin and a modified starch as wall materials. Carbohyd. Polym. 61: 95–102.

Kudra, T. and Mujumdar, A.S. 2009a. Spray-Freeze-Drying. pp. 337–341. *In*: Kudra, T. and Mujumdar, A.S. (eds.). Advanced Drying Technologies. CRC Press, New York.

Kudra, T. and Mujumdar, A.S. 2009b. Innovation and trends in drying technologies. pp. 19–26. *In*: Kudra, T. and Mujumdar, A.S. (eds.). Advanced Drying Technologies. CRC Press, New York.

Kumar, N., Rai, A., Reddy, N.D., Raj, P.V., Jain, P., Deshpande, P., Mathew, G., Kutty, N.G., Udupa, N. and Rao, C.M. 2014. Silymarin liposomes improves oral bioavailability of silybin besides targeting hepatocytes and immune cells. Pharmacol. Rep. 116: 1–11.

Lin, Y.C., Kuo, J.Y., Hsu, C.C., Tsai, W.C., Li, W.C., Yu, M.C. and Wen, H.W. 2013. Optimizing manufacture of liposomal berberine with evaluation of its antihepatoma effects in a murine xenograft model. Int. J. Pharm. 441: 381–388.

Loksuwan, J. 2007. Characteristics of microencapsulated b-carotene formed by spray drying with modified tapioca starch, native tapioca starch and maltodextrin. Food Hydrocolloid. 21: 928–935.

Maa, Y.F. and Costantino, H.R. 2004. Spray freeze drying of biopharmaceuticals: applications and stability consideration. pp. 519–561. *In*: H.R. Costantino and M. Pikal (eds.). Lyophilization of Biopharmaceuticals. American Association of Pharmaceutical Scientists Press, Arlington, VA.

Marcolino, V.A., Zanin, G.M., Durrant, L.R., Benassi, M.T. and Matioli, G. 2011. Interaction of Curcumin and Bixin with β-Cyclodextrin: Complexation Methods, Stability, and Applications in Food. J. Agr. Food Chem. 59: 3348–3357.

Marques, H.M.C. 2010. A review on cyclodextrin encapsulation of essential oils and volatiles. Flavour Frag. J. 25: 313–326.

Martí, M., Maza, A., Parra, J.L. and Coderch, L. 2014. Role of liposomes in textile dyeing lipid. pp. 401–412. *In*: G.P.N. Kučerka, M.P. Nieh and J. Katsaras (eds.). Liposomes, Bilayers and Model Membranes. CRC Press, New York.

Martina, K. and Cravotto, G. 2012. Cyclodextrins. pp. 593–606. *In*: L.M.L. Nollet and F. Toldrá (eds.). Handbook of Analysis of Active Compounds in Functional Foods. CRC Press, New York.

Martins, I.M., Rodrigues, S.N., Barreiro, F. and Rodrigues, A.E. 2008. Microencapsulation of thyme oil by coacervation. J. Microencapsul. 26(8): 667–675.

Materska, M. 2014. Bioactive phenolics of fresh and freeze-dried sweet and semi-spicy pepper fruits (*Capsicum annuum* L.). J. Funct. Foods 7: 269–277.

McChesney, J. 1999. Commercialization of plant-derived natural products as pharmaceuticals: a view from the trenches. pp. 258–269. *In*: S.J. Cutler and H.G. Cutler (eds.). Biologically Active Natural Products Pharmaceuticals. CRC Press, New York.

Mukherji, G. and Wilson, C.G. 2002a. Enteric coating for colonic delivery. pp. 223–232. *In*: M.J. Rathbone, J. Hadgraft and M.S. Roberts (eds.). Modified-Release Drug Delivery Technology. CRC Press, New York.

Mukherji, G. and Wilson, C.G. 2002b. Biopolymers and colonic delivery. pp. 233–242. *In*: M.J. Rathbone, J. Hadgraft and M.S. Roberts (eds.). Modified-Release Drug Delivery Technology. CRC Press, New York.

Munin, A. and Edwards-Lévy, F. 2011. Encapsulation of natural polyphenolic compounds; a review. Pharmaceutics 3: 793–829.

Mura, P. 2014. Analytical techniques for characterization of cyclodextrin complexes in aqueous solution: a review. J. Pharm. Biomed. Anal. 101: 238–50.

Nail, S.L. 2005. Physical and chemical stability considerations in the development and stress testing of freeze-dried pharmaceuticals. pp. 261–291. *In*: S.W. Baertschi (ed.). Pharmaceutical Stress Testing Predicting Drug Degradation. CRC Press, New York.

Nieddu, M., Rassu, G., Boatto, G., Bosi, P., Trevisi, P., Giunchedi, P., Carta, A. and Gavinia, E. 2014. Improvement of thymol properties by complexation with cyclodextrins: *In vitro* and *in vivo* studies. Carbohydr Polym. 102: 393–399.

Nii, T. and Ishii, F. 2005. Encapsulation efficiency of water-soluble and insoluble drugs in liposomes prepared by the microencapsulation vesicle method. Int. J. Pharm. 298: 198–205.

Nori, M.P., Favaro-Trindade, C.S., Alencar, S.M., Thomazini, M., Balieiro, J.C.C. and Castillo, C.J.C. 2011. Microencapsulation of propolis extract by complex coacervation. Lebensm. Wiss. Technol. 44: 429–435.

Oliveira, W.P., Souza, C.R.F., Kurozawa, L.E. and Park, K.J. 2010. Spray drying of food and herbal products. pp. 113–156. *In*: M.W. Woo, A.S. Mujumdar and W.R.W. Daud (eds.). Spray Drying Technology. V. 1. Wai's Research Group, Singapore.

Pachuau, L. and Mazumder, B. 2013. Colonic drug delivery systems based on natural polysaccharides and their evaluation. Mini Rev. Med. Chem. 13: 1982–1991.

Pang, S.F., Yusoff, M.M. and Gimbun, J. 2014. Assessment of phenolic compounds stability and retention during spray drying of *Orthosiphon stamineus* extracts. Food Hydrocolloid. 37: 159–165.

Park, S.N., Lee, H.J. and Gu, H.A. 2014. Enhanced skin delivery and characterization of rutin-loaded ethosomes. The Korean Korean J. Chem. Eng. 31(3): 485–489.

Patel, N.A., Patel, N.J. and Patel, R.P. 2009. Formulation and Evaluation of Curcumin Gel for Topical Application. Pharm. Dev. Technol. 14: 80–89.

Pinho, E., Grootveld, M., Soares, G. and Henriques, M. 2014. Cyclodextrins as encapsulation agents for plant bioactive compounds. Carbohydr. Polym. 101: 121–135.

Piroyan, A., Koshkaryev, A., Riehle, R.D. and Torchilin, V.P. 2014. Polymer-modified liposomes. pp. 341–360. *In*: G.P.N. Kučerka, M.P. Nieh and J. Katsaras (eds.). Liposomes, Bilayers and Model Membranes. CRC Press, New York.

Rashidinejad, A., Birch, E.J., Waterhouse, D.S. and Everett, D.W. 2014. Delivery of green tea catechin and epigallocatechin gallate in liposomes incorporated into low-fat hard cheese. Food Chem. 156: 176–183.

Rodrigues, R.A.F., Rodrigues, M.V.N., Oliveira, T.I.V., Bueno, C.Z., Souza, I.M.O., Sartoratto, A. and Foglio, M.A. 2011. Docosahexaenoic acid ethyl esther (DHAEE) microcapsule production by spray-drying: optimization by experimental design. Ciênc. Tecnol. Aliment. 31(3): 589–596.

Rowe, R.C., Sheskey, P.J. and Owen, S.C. 2006. Handbook of Pharmaceutical Excipients. 5th edition. Pharmaceutical Press, London.

Sadikoglu, H. 2010. Spray freeze drying. pp. 157–182. *In*: M.W. Woo, A.S. Mujumdar and W.R.W. Daud (eds.). Spray Drying Technology. Wai's Research Group, Singapore.

Sahoo, N., Manchikanti, P. and Dey, S. 2010. Herbal drugs: Standards and regulation. Fitoterapia 81: 462–471.

Sanguansri, L. and Augustin, M.A. 2006. Microencapsulation and delivery of omega-3 fatty acids. pp. 297–327. *In*: John Shi (ed.). Functional Food Ingredients and Nutraceuticals Processing Technologies. CRC, New York.

Saraf, A. 2010. Applications of novel drug delivery system for herbal formulations. Fitoterapia 81: 680–689.

Şerbetçi, T., Ozsoy, N., Demirci, B., Can, A., Kültür, S., Başer, K.H.C. 2012. Chemical composition of the essential oil and antioxidant activity of methanolic extracts from fruits and flowers of *Hypericum lydium* Boiss. Ind. Crop. Prod. 36: 599–606.

Shao, P., Zhang, J., Fang, Z. and Sun, P. 2014. Complexing of chlorogenic acid with β-cyclodextrins: Inclusion effects, antioxidative properties and potential application in grape juice. Food Hydrocolloid. 41: 132–139.

Shu, B., Yu, W., Zhao, Y. and Liu, X. 2006. Study on microencapsulation of lycopene by spray-drying. J. Food Eng. 76: 664–669.

Singh, A., Vengurlekar, P. and Rathod, S. 2014. Design, development and characterization of liposomal neem gel. International Journal of Pharma Sciences and Research 5(4): 140–148.

Singh, R. and Sankar, C. 2012. Formulation of herbal liposomes containing green tea and *Gaultheria procumbens* for anti-acne activity. IJAPR 10: 1211–1216.

Siriamornpun, S., Kaisoon, O. and Meeso, N. 2012. Changes in colour, antioxidant activities and carotenoids (lycopene, b-carotene, lutein) of marigold flower (*Tagetes erecta* L.) resulting from different drying processes. J. Funct. Foods 4: 757–766.

Souza, T.P., Gómez-Amoza, J.L., Pacheco, R.M. and Petrovick, P.R. 2009. Development of granules from *Phyllanthus niruri* spray-dried extract. Braz. J. Pharm. Sci. 45(4): 669–675.

Stoner, G.D., Wang, L.S., Zikri, N., Chen, T., Hecht, S.S., Huang, C., Sardo, C. and Lechner, J.F. 2007. Cancer prevention with freeze-dried berries and berrys components. Sem. Canc. Biol. 17: 403–410.

Takahashi, M., Uechi, S., Takara, K., Asikin, Y. and Wada, K. 2009. Evaluation of an oral carrier system in rats: bioavailability and antioxidant properties of liposome-encapsulated curcumin. J. Agr. Food Chem. 57: 9141–9146.

Teixeira, B.N., Ozdemir, N., Hill, L.E. and Gomes, C.L. 2013. Synthesis and characterization of nano-encapsulated black pepper oleoresin using hydroxypropyl beta-cyclodextrin for antioxidant and antimicrobial applications. J. Food Sci. 12: 1913–1920.

Teunou, E. and Poncelet, D. 2002. Batch and continuous fluid bed coating—review and state of the art. J. Food Eng. 53: 325–340.

Teunou, E. and Poncelet, D. 2005. Fluid-bed coating. pp. 197–212. *In*: C. Onwulata (ed.). Encapsulated and Powdered Foods. CRC Press, New York.

Tong-Un, T., Wannanon, P., Wattanathorn, J. and Phachonpai, W. 2010. Quercetin Liposomes via Nasal Administration Reduce Anxiety and Depression-Like Behaviors and Enhance Cognitive Performances in Rats. Am. J. Pharmacol. Toxicol. (5), 2: 80–88.

Trivedi, N.R., Rajan, M.G., Johnson, J.R. and Shukla, A.J. 2007. Pharmaceutical Approaches to Preparing Pelletized Dosage Forms Using the Extrusion-Spheronization Process. Crit. Rev. Ther. Drug. Carrier Syst. 24(1): 1–40.

Umesalma, S., Nagendraprabhu, P. and Sudhandiran, G. 2014. Antiproliferative and apoptotic-inducing potential of ellagic acid against 1,2-dimethyl hydrazine-induced colon tumorigenesis in Wistar rats. Mol. Cel. Biochem. 388: 157–172.

Vervaet, C., Baert, L. and Remon, J.P. 1995. Extrusion-spheronisation A literature review. Int. J. Pharm. 116: 131–146.

Villacrez, J.L., Carriazo, J.G. and Osorio, C. 2014. Microencapsulation of Andes Berry (*Rubus glaucus* Benth.) Aqueous Extract by Spray Drying. Food Bioprocess Technol. 7: 1445–1456.

Wang, Y., Xie, Y., Xu, D., Lin, X., Feng, Y. and Hong, Y. 2014. Hydroxypropyl Methylcellulose Reduces Particle Adhesion and Improves Recovery of Herbal Extracts During Spray Drying of Chinese Herbal Medicines. Dry. Technol. 32: 557–566.

Welti-Chanes, J., Bermudez, D., Valdez-Fragoso, A., Mujica-Paz, H. and Alzamora, S.M. 2005. Principles and applications of freeze-concentration and freeze-drying. pp. 106/2–106/8. *In*: Y.H. Hui (ed.). Handbook of Food Science, Technology, and Engineering. CRC Press, New York.

Werner, S.R.L., Jones, J.R., Paterson, A.H.J., Archer, R.H. and Pearce, D.L. 2007. Air-suspension particle coating in the food industry: Part I—state of the art. Powder Technol. 171: 25–33.

Williams, C. 2013. *Asteraceae* daisies of the apothecary. pp. 47–86. *In*: Williams, C. (ed.). Medicinal Plants in Australia: An Antipodean Apothecary. Rosenberg Publishing, Australia.

Wilson, C.G. 2002. Colonic drug delivery. pp. 217–222. *In*: M.J. Rathbone, J. Hadgraft and M.S. Roberts (eds.). Modified-Release Drug Delivery Technology. CRC Press, New York.

Wu, K.G., Chai, X.H. and Chen, Y. 2005. Microencapsulation of fish oil by simple coacervation of hydroxypropyl methylcellulose. Chinese J. Chem. 23: 1569–1572.

Xiao, J.X., Yu, H.Y. and Yang, J. 2011. Microencapsulation of sweet orange oil by complex coacervation with soybean protein isolate/gum Arabic. Food Chem. 125: 1267–1272.

Yaseen, A.D.B. and Mo'ala, A. 2014. Spectral, thermal, and molecular modeling studies on the encapsulation of selected sulfonamide drugs in β-cyclodextrin nano-cavity. Spectrochim. Acta Mol. Biomol. Spectrosc. 131C: 424–431.

Cultivation and Utilization of Medicinal Plants for Development of Products

Pedro Melillo de Magalhães

IIIIIIIIIIIIIIIIIIIIIIIIIIIIIIIIIIIIIIIIIIIIIIIIIIIIIIIIIIIIIIIIIIIIIIIIIIIIIIIIIIIIIIIIIIIIIIIIIIIIIIIIIIIIIIIIIIIIIIIIIIIIIIIIIIIIIIIIIIIIIIIIIIIIIIIIIIIIIIIIIIIIIIIIIII

Introduction

Medical prescriptions based on many cultures of medicinal plants, used since ancient times, are once again arousing the interest of people, and some have recently been put on the public health medicine list in Brazil as a result of scientific evidence of their therapeutic properties.

This same scenario can be seen throughout the world and also in the pharmaceutical industry, which is dedicating itself to the development of safe, effective medicines based on medicinal plants, with a view of increasing the demand for such products. This interest in medicinal plants is due, in part, to advances in scientific research which have revealed their therapeutic properties, toxicity and composition with greater precision, and, in part, due to the valorization of natural products in the context of overall health. For both industry and public health, agricultural production activities and agronomical research into medicinal plants are of fundamental importance for the supply and quality of the products, principally when the species are used *in natura* or with limited processing, as in the case of infusions, capsules and extracts.

The use of medicinal plants and their preparations are coming back into prominence, not as a retrocession to primitive practices, but as 'new generation' medicines, taking advantage of a high level of technology in all its aspects, even when relatively simple, mainly with respect to the raw material, so that it can be obtained in a regular, standardized form. In the case of medicinal plants, agronomical research is also monitored by chemical and biological evaluations. In fact, once the active complex related to the desired biological activity is known, the agro-technological development and species improvement are directed at specific targets, allowing for the selection of superior, standardized genotypes. Thus the process of developing the cultivation of medicinal plants includes two main lines: (a) the selection of genotypes that unite the characteristics of interest within the existing genetic variability in nature, promoting improvement of the species with respect to the active principles and in certain agronomical characteristics, and (b) the development of cultivation technology to bring the species from its wild state into the condition of a cultivated plant (domestication), and/or its acclimatization to a determined region. These two lines must be carried out in parallel, constructing protocols capable of being followed by agriculturalists, who ultimately will be co-responsible for the quality of the phytomedicines and should receive information concerning the techniques of cultivation, harvesting and post-harvest processing, apart from having access to genotypes superior with respect to the characteristics of interest.

The selection and cultivation processes for six medicinal species are dealt with in this chapter, chosen for representing highly successful cases and being at different technological levels. The 'history' of the research into these species serves as a model for many other plants which, despite being of interest from the point of view of their medicinal properties, are still in the wild state, making their commercial production on a large scale difficult, and consequently impeding the amplifying of their therapeutic benefits.

This case study deals with the following species:

- *Artemisia annua* L.
- *Pfaffia glomerata* (Spreng.) Pedersen
- *Mikania laevigata* Sch. Bip. ex Baker
- *Phyllanthus amarus* Schumach
- *Cordia verbenacea* DC.
- *Baccharis dracunculifolia* DC.

Artemisia annua

Description

Artemisia annua L. (Asteraceae) is an annual, erect herbaceous species with a height varying from 0.80 to 1.70 m. The inflorescence contains eight to 10 feminine flowers and 13 to 23 hermaphrodite flowers and the achene is light brown and striped and about 1 mm in length. *A. annua* stores essential oil with a balsamic aroma (camphor) in its trichomes. The number of chromosomes is 2n = 18.

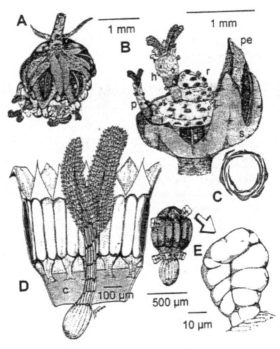

Figure 12.1. Floral morphology of *A. annua*: (A) Capitulum. (B) Expanded capitulum showing the calyx with its sepals (*s*) and petals (*pe*), receptacle (*r*), marginal pistil (*p*) and internal hermaphrodite flowers (*h*). Numerous trichomes can be found in the receptacle and flowers of the capitulum. (C) Cross sectional cut of the involucre. (D) Details of the hermaphrodite flowers. (E) Trichome (Ferreira 1994).

Ecology

Habitat: *A. annua* is a native of Asia, China and Vietnam, occurring at a latitude of 40°

Altitude: 700 to 1500 meters

Climate: Temperate, humid

Soil: Not too demanding with respect to type of soil, but responds favorably to correction of the pH to about 7.0

Properties and applications

Infusions made with *A. annua* leaves were cited for the treatment of fevers in general in China more than 2000 years ago. The Chinese researcher Dr. To Youyou isolated the principal active compound with anti-malarial activity, the sesquiterpene lactone artemisinin, in the 1970s. This molecule, and other derivatives obtained later by semi-synthesis such as sodium artesunate, artemeter and arteeter, became the most efficient drugs for making the therapy for cerebral malaria, destroying *Plasmodium falciparum*, including the strains resistant to the anti-malarial drugs previously used. With an aim to long-lasting efficiency so as to avoid the selection of an artemisinin resistant parasite, the strategy used was to administer artemisinin or its derivatives together with other anti-malarial drugs with distinct action modes which remain in the blood stream, such as amodiaquine, lumefantrine and mefloquine.

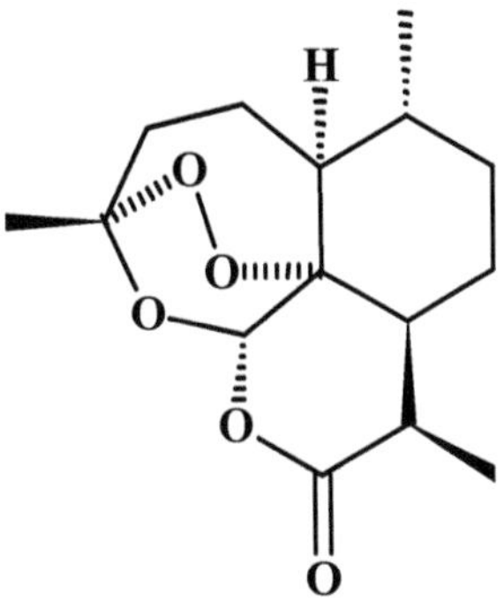

Figure 12.2. Artemisinin molecule.

In addition to the recent use of drugs obtained from artemisinin, traditional tea made from *A. annua* leaves is also being considered as therapeutic for malaria, since the leaves used come from an improved strain of the plant, expressively richer in artemisinin as well as in other compounds synthesized by the plant, according to the promising therapeutic activity presented by Mueller et al. 2000. This form of its use is being studied as a means of obtaining an efficient and safe anti-malarial effect in locations where the disease occurs, which are normally far from the cities and difficult to access, making the treatment with more elaborated drugs not always feasible. The studies aim to arrive at a simple, although judicious form of treatment with respect to the dosage and efficacy. It is interesting to note that the dose of tea that promotes the same therapeutic effect as artemisinin when administered in the pure form, only contains 1/3 of the amount of artemisinin used, admitting the occurrence of synergism between the artemisinin and other plant components. However, in both treatments, tea or pure artemisinin, the efficiency is unsatisfactory, a later parasitological or clinical failure occurring around day 21 to day 28 after the start of the therapy. Other pharmacological and agronomic uses have also been discovered, such as an anti-ulcer activity (Dias et al. 2000) and an inhibitory activity against the germination of invading species (Delabays et al. 1998). The essential oil also presents anti-microbial activity (Foglio et al. 2000) and is used in phytosanitary products.

State of the culture

The genetic improvement program carried out by MEDIPLANT (Switzerland) and also adopted by CPQBA-UNICAMP managed to explore the natural variability of the species and arrive at hybrid strains and artemisinin rich (1%) populations, as well as the standardization of biomass development and late flowering (Magalhães 1996, Delabays 1997, Magalhães et al. 1997). The main genotypes selected and developed for the region of Campinas-SP-Brazil were the hybrids CPQBA 5 x 2/39 and CPQBA 3M x POP. Apart from the large scale cultivation being carried out at CPQBA in Campinas-SP-Brazil, there is no other cultivation of the species in Brazil except for some experimental areas for research into the physiology of the plant or acclimatization of the species to other regions where malaria is endemic. The main difficulty for cultivation is early flowering, which occurs because of short days with less than 12 hours of daylight. At altitudes above 1000 m *A. annua* rarely flowers, but this condition does not coincide with the areas where malaria is endemic. Viability of the culture away from the 22° latitude requires a new selection of genotypes, obtaining, in the natural variability of the species, individuals that do not flower in a determined region.

Figure 12.3. Cultivation of *Artemisia annua* at CPQBA-UNICAMP (the present author in front of the cultivation of the hybrid CPQBA 5 x 2/39).

Multiplication

A. annua propagates itself by seed dispersal, and, in fact, is considered to be a weed species due to the large amount of seeds produced and the ease of dissemination. However, the species also accepts vegetative propagation by micro-cuttings or by meristem culture, which allows for fixation of the genetic characteristics of interest and accelerates multiplication of the parents for the crossing and production of hybrid seeds. The seeds are harvested at the end of the plant cycle, when the plant is in senescence. The inflorescences are collected at this point and pressed against sieves of various sized mesh to obtain the seed-rich material. The definitive separation from the impurities is done in a delicate manner by consecutive fanning operations. Seeding is carried out superficially in small tubes containing vegetable substrate (earth, matured manure, worm humus) and placed under 50% shade netting. After 15 days or when the plants have grown to a height of 5 cm, they are transplanted, leaving just one plant in each tube. The seedlings are ready to be transferred to the field when they are approximately 10 cm in height, which should coincide with 60 days after seeding. Vegetative propagation by micro-cuttings is carried out using terminal segments of matrix plants, taking 2 cm sections of the stalk with their young leaves. In this case the environmental humidity must be saturated until the cuttings have produced functional roots. The tissue culture micro-propagation method is for the *in vitro* maintenance of the genotypes selected as the parents to obtain the hybrids, as described by Nopper 1993.

Spacing and seeding time

The recommended spacing for the hybrid CPQBA-1 is of 1 meter between lines and 0.60 m between plants, and the ideal seeding time in the Campinas-SP region is between September and October. The harvest is defined by the phonological state of the pre-florescence, coinciding with the months of February and March.

Fertilization

The plant *A. annua* responds well to nitrogenated manuring based on 90 Kg of N/ha. In addition, liming to correct acid soils promotes a significant increase in biomass, without altering the artemisinin content.

Cultivation care

This species has a fasciculated surface radicular system which demands a constant water supply principally during the first month in the field. Normally two hoeing sessions are sufficient to maintain the culture clean up to the end, taking special care to control climbing weeds such as morning glory and mucuna.

Insects and disease

Although by many insects and bees are attracted to the culture, no case of critical prejudice has been observed in the region.

Harvest

The point of harvest to obtain artemisinin is when the leaves start to differentiate for the formation of inflorescences. The leaves appear to be smaller and more spaced out on the stem, giving the appearance of having a smaller number of leaves. Although mechanical cereal harvesters adapted to harvest *A. annua* do exist (Laughlin 2002), manual harvesting using pruning shears is preferred for smaller areas or a mower attached to a tractor. In both cases the stalks are cut at the point where the leaves start, usually about 50 cm from the ground. The thicker stalks at the base, with few green leaves, serve as a place to carry out the pre-drying of the harvested material, and after this operation, are broken up with a hoe and incorporated into the soil with a plough or leveler. Harvesting should be carried out on a clear, hot day, favoring the subsequent operations of pre-drying and separation of the leaves and stems.

However, to obtain the essential oil, the harvest should be carried out in different physiological states, depending on the compound of interest in the oil (Table 12.1). Nevertheless the greatest yield in oil is obtained in the full flowering or florescence phase.

Drying

Drying starts in the field. The cut branches bearing green leaves are placed in a horizontal position on the plant from which they were cut. After two days of drying they are transferred to tracks or left between the lines, where they are spread over screens and are exposed to sunlight. At night they roll up into a large 'roly-poly' and are covered with a plastic tarpaulin. The next day they are unrolled and exposed to the

Figure 12.4. Harvesting of *Artemisia annua* at CPQBA-UNICAMP using pruning shears. The remains of the stalks that sustained the material harvested can be seen at the front, and the plants still to be harvested, with a height of approximately 2.0 m, can be seen behind them.

Table 12.1. Yield and composition of the essential oil from *Artemisia annua* in three phonological stages of the culture: (1) Vegetative in the eminence of florescence; (2) Initial florescence; (3) In flower and with seeds, close to senescence (Magalhães et al. 2001).

$T_r/'$	Compounds	(1)	(2)	(3)
7.04	α-pineno	1.89	1.84	2.38
7.55	Canfeno	2.86	2.67	4.24
8.37	Sabineno	4.14	4.58	3.97
9.00	ß-mirceno	7.89	12.41	4.09
9.93	α-terpineno	0.41	0.39	0.53
10.37	1-metil-4-(1-metileil) benzeno	1.23	-	-
10.73	1,8 cineol	17.06	21.88	28.76
11.64	δ-terpineno	1.03	1.56	0.73
12.03	Terpin-4-ol	0.66	0.57	0.98
12.88	α-terpinoleno	0.11	0.10	0.16
15.87	Cânfora	28.44	14.89	30.87
17.66	α-terpineol	1.46	1.31	1.15
19.97	Carvona	0.10	0.12	0.13
25.12	2-metoxi-4-(2-propenil) fenol	0.24	0.16	-
28.08	*Trans*-cariofileno	3.28	2.97	1.68
29.53	ß-farneseno	3.18	2.92	2.40
30.84	ß-germacreno	5.85	5.30	0.34
	Yield E.O. (%) ⟹	**0.40**	**0.30**	**0.21**

sunlight and at night rolled up again, and so on. This procedure of exposing the tissues to sunlight while they still have water activity can increase the artemisinin content by up to 15% (Simonnet et al. 2001). These authors suggested the occurrence of a photo-oxidative reaction of the precursors of artemisinin, principally of artemisininic acid, resulting in an increase in the content of the active principle. If the leaves are still moist after two weeks, complementary drying in a dryer at 40°C to constant weight is used. Once the material has been well dried the procedure of separating the leaves from the finer stems starts, first manually by lightly beating the material, and then using sieves, also separating other fractions, such as the larger petioles.

Yields

The mean yields of these hybrids, developed and cultivated at CPQBA-UNICAMP with a spacing of 1.0 x 0.6 m were:

> Dry leaves: 1.8 to 2.2 ton/ha
> Artemisinin content of the dry leaves: 0.9 to 1.10%
> Mean quantity of artemisinin/ha: 20 Kg
> Finer stems (dry): 2 ton/ha
> Thick dry stems (in the field): 13 ton/ha

Pfaffia glomerata

Description

The Pfaffia, *Pfaffia glomerata* (Spreng.) Pederson is a perennial shrub, also known by the popular name of Brazilian ginseng. The genus *Pfaffia* Mart. belongs to the family *Amarantaceae*, and there are 33 species distributed throughout Central and South America. There are 21 species in Brazil, found in the forest and field forms (Siqueira 1988). *Pfaffia glabrata* Mart. was established as a species type by Martius in 1826, the name of the species being in honor of Cristian H. Pfaff (1774–1852), a professor of medicine in Germany. In 1972, Smith and Downs presented the following description for *P. glomerata*: perennial herb up to 2 m in height, with erect slender stalks. The leaves are lanceolate in shape, 5–12 cm in length and 1~2 cm in width.

The flowers are monoecious and arranged on a cob. *P. glomerata* (Spreng.) Pedersen and also *P. iresinoides* (H.B.K.) Spreng., belong to the section *Sertunera* (Mart.), both being of sub-shrub or shrub size and showing very similar morphological characteristics, leading to doubts as to whether they should be classified as distinct species. However the first is widely distributed occupying various climatic and soil conditions. On the other hand, *P. iresinoides* has a much more limited distribution than *P. glomerata*, only occurring in the Brazilian Amazon, in the North of Brazil, and in adjacent countries.

Ecology

Habitat: *P. glomerata* occurs in different Brazilian states, being rarer in the southern states (Santa Catarina and Rio Grande do Sul). *Pfaffia* is commonly found on riverbanks and at the edge of wasteland where it can get a lot of light.

Altitude: up to 800–1000 meters.

Climate: Tropical or sub-tropical and moist. It cannot tolerate low temperatures.

Soil: Occurs in light soils although the yield is greater in clayey or more fertile soils.

Figure 12.5. *P. glomerata* cultivated at CPQBA-UNICAMP.

Properties and applications

The interest shown in *P. glomerata* is due to the popular use of its roots, used as a tonic, stimulant and aphrodisiac, also known as 'Brazilian ginseng'. In fact the composition of *P. glomerata* roots, according to a study by Shiobara et al. 1993, showed the presence of hormones such as β-ecdison, rubrosterone, oleanólic acid and b-glucopyranosyl oleanolate with characteristics of the adaptogenic effect (The adaptogenic effect is the pharmacological effect of promoting an increase in immunity). The authors also isolated two new compounds: glomeric acid (a triterpenic acid) and pfameric acid (a nortriterpenic acid).

Nishimoto et al. 1987, 1988 and Shiobara et al. 1992, analyzed the composition of the roots of *P. iresinoides*, a species close to *P. glomerata*, isolating large amounts of ecdysterone together with polipodine and pterosterone. The authors also isolated new steroidal glycosides as well as a new yellow pigment called iresinoside.

The β-ecdysone content can be determined by HPLC using the following methodology developed at CPQBA:

Take 1 gram of dry-ground sample and extract with 130 mL of methanol in a soxhlet apparatus for four hours. Concentrate the extract to 10 mL, add 50 mL of H_2O and centrifuge at 3000 rpm for 10 minutes. Filter the supernatant through a 0.45 μm Millipore filter, since some solids remain in the supernatant. Prepare a standard curve using 40, 120 and 200 ppm of the standard (β-ecdysone) dissolved in methanol. The content is determined by HPLC with a Waters 510 pump, Michrosorb RP 18 column (4.6 mm x 250 mm, 10 μm), 20 μl loop and UV detector set at 245 nm. The methanolic extraction is selective for β-ecdysone by polarity, and it is detected in the UV range at 245 nm.

Figure 12.6. The structure of β-ecdysone, present in *P. glomerata* roots.

State of the culture

Although there is an increasing national and worldwide demand for *P. glomerata* roots, mainly in Japan, where the species has been widely studied from the phytochemical and pharmacological points of view, the entire production comes from Brazil, as a function of the ecological requirements of the plant in terms of the climate of the inter-tropical regions.

However, until very recently, no techniques were available for the cultivation of the species, and the only option for the supply of the raw material was the extractivism process, which became more intense after studies for the scientific validation of its therapeutic properties (Alcântara 1994, Galvão et al. 1996, Dias et al. 1996).

Approximately 30 tons of *Pfaffia* sp. roots are exported to Japan every month coming from the River Paraná basin–PR–Brazil and the municipality of Mogi das Cruzes–SP–Brazil, where the extractivist suppliers receive about US$ 2.00 per kilo of fresh *P. glomerata* roots. Such severe uncontrolled extractivism, which also occurs in other regions, puts the variability of the populations and the continuity of the supply of the raw material at risk, as well as representing an aggression to the environment.

Thus its cultivation has become very urgent, and could indeed be an interesting alternative agricultural activity. Studies carried out by CPQBA, located in Paulínia-SP-Brazil, 22°48'Lat. South, 47°07'Long., have defined the yields and some basic parameters for the cultivation of this species.

Multiplication

P. glomerata produces fertile seeds with a germinating power between 50 and 77% (Magalhães et al. 1994, Ribeiro[a] and Pereira 1994). The species also accepts vegetative propagation using cuttings and tissue culture (Alves et al. 1996). Since the roots are the final product, propagation by seeds has the advantage of promoting deeper, thicker roots (pivoting), as compared with the roots of plants propagated using cuttings, where the radicular system is fasciculate. The commercial acceptance of thicker roots is higher. On the other hand, when dealing with a species in the domestication phase, propagation by seeds promotes populations with a certain genetic variability. Calago (pers. comm.) showed that both complete and polygamous flowers occurred on the same *P. glomerata* plant, making it difficult to determine the rate of crossing of the species. Nevertheless, relative uniformity is possible if one collects seeds from a single plant or from similar plants (half-brother populations). Montanari, Jr. (pers. comm.) has been studying the genetic variability of *P. glomerata*, selecting genotypes as a function of their biomass and β-ecdysone yields, apart from the 'aerial part: roots' ratio and the correlation between the active principle content and the root color.

Figure 12.7. *P. glomerata* inflorescences.

Spacing and planting time

According to Ribeiro[b] and Pereira 1994, Magalhães and Pereira 1996, when planted in clayey soils, *P. glomerata* should be grown on top of furrows in order to make the harvesting of the roots easier, which would thus be located near the soil surface. The furrows can be made using large plows such as those used in sugarcane plantations, which, on making the grooves, leave a 'furrow' between one groove and another (Magalhães 1997). The spacing promoting the best productivity was that of 0.5 m between plants on the furrow, and 1.5 m between the furrows (Montanari, Jr. et al. 1997). The best time for transplanting is at the end of winter or beginning of spring, such that harvesting of the roots, carried out after one, two or three years, coinciding with the end of the winter, when the reserves synthesized by the plants have been transferred to the roots.

Fertilization

Although no recommendations for the fertilization of *Pfaffia* were found in the literature, trials carried out in different soils, both sandy and clayey, showed that the productivity was considerably greater in clayey soils, which are normally more fertile. However, these were only preliminary studies, since until the population is genetically standardized, investigations on the effect of the environment will be masked by genetic variability.

Cultivation care

The culture *Pfaffia* requires no special treatment during development, just the control of weeds and irrigation as required. In some cases it may be necessary to protect the plants from being bent over by the wind, by planting hedges around the plantation.

Pests and diseases

P. glomerata is susceptible to rust and to nematodes. Important studies were developed by Mattos 1993, Araújo 1994, who, while exploring the variability, found genotypes resistant to both.

While no resistant varieties are available on a commercial scale, the use of nematode-repelling plants, such as: *Sesame* (*Pedaliaceae*) and *Tagetes* (*Asteraceae*), are recommended.

Harvest

Pfaffia roots can be harvested at the end of winter as from one year after planting. The operation is made easier by using a winged sub-soiler or plow, which passes at the base of the furrow and pulls out the roots. The previous cutting and removal of the aerial part of the plants is recommended to facilitate the operation (In the case of a selection program in a seed propagated population, the aerial part could be used to vegetative propagation of the plants of greater interest. Thus care must be taken to catalog the aerial part with its respective roots).

The roots from plants cultivated in clayey soil require cleaning, which can be carried out using pressurized water.

Drying

After allowing the washing water to drain off, the roots should be cut into approximately 1 cm^3 pieces or 2–3 mm slices before feeding into the dryer. The drying temperature should be between 40 and 50°C and the material must be well distributed inside the dryer to avoid localized humidity which could favor the development of fungi.

Yields

The trial carried out with the species *P. glomerata* in Paulínia-SP-Brazil showed rapid growth when compared with other trials carried out in the lowlands of the State of Mato Grosso-Brazil, where the species occurs naturally. It showed satisfactory production of about 1.9 tons dry roots/ha at 12 months of age, 3.2 tons/ha at 24 months of age and 4.1 tons/ha at 36 months of age. These productivities were obtained considering a mean spacing between plants of 0.75 m on the furrow. The β-ecdysone content did not suffer significant variations as a function of the age of the plant, varying from 0.67 to 0.71% (Montanari, Jr., pers. comm.).

It was concluded that an annual harvest resulted in the highest yields, since the sum of three yields equal to that of the first year was higher than that obtained at 36 months of age. On the other hand, to make such a comparative analysis, one should also consider the cost of planting annually.

The great variability observed in the population with respect to the morphological parameters and yields (biomass and β-ecdysone content) can be explained due to the fact that the plants were obtained from seeds and from wild matrixes.

Mikania laevigata

Description

Mikania laevigata Schultz Bip., ex Baker belongs to the family *Asteraceae* and is popularly known as Guaco in Brazil. It is a climbing species with a perennial cycle and is native to South America. In Brazil, other *Mikania* species are also known as Guaco, principally *M. glomerata*. Moraes 1997 presented the following description of *M. laevigata* Sch. Bip. ex Baker: liana, striped hairless branches, petioles measuring 1.5–4.2 cm, lanceolate, strictly ovate leaves, sometimes slightly lobate with an obtuse base, sharp tip and hairless on both faces. The panicle inflorescences are foliaceous and bracteal and the achene from 3 to 3.3 mm. It flowers from August to October and the flowers are visited by *Apis mellifera* and various Lepdoptera species.

M. laevigata is similar to *M. glomerata* Spreng., but is distinct from the latter because its leaves are non-lobate to slightly lobate whereas those of the latter are distinctly lobate. The leaves of *M. laevigata* are also thicker, shinier and more aromatic (similar to that of vanilla) than those of *M. glomerata*. Nevertheless for many years *M. laevigata* was treated as *M. glomerata*, indicating that the literature concerning these species should be considered with care. In fact the species considered in the Brazilian Pharmacopeia as *M. glomerata* is really *M. laevigata*.

Oliveira et al. 1992a described the histological differences between the various species of *Mikania*. From the chemistry-taxonomical point of view, the genus *Mikania* can be divided into species that produce sesquiterpenic lactones and those that produce diterpenes (Núñez and Roque 1997).

Figure 12.8. *M. laevigata* vine at the start of florescence cultivated in direct sunlight at CPQBA.

Ecology

Habitat: *M. laevigata* is native to the South American continent growing on the edge of wild coastline vegetation (under partially shady conditions), a common condition on the slopes of coastal mountain ridges. It can be found from the state of São Paulo to the state of Rio Grande do Sul.

Altitude: 0 to 800 meters.

Climate: Subtropical, hot and humid. It can be cultivated in direct sunlight or in partially shady regions, avoiding cold climates.

Soil: Undemanding with respect to the type of soil, but it adapts better to well-drained clayey-sandy or clayey soils, with an elevated amount of organic matter.

Properties and applications

M. hirsutissima and *M. glomerata*, also known as Guaco, can be found in the Brazilian Pharmacopeia. The syrup, made with the leaves and twigs of the plant, is popularly indicated in cases of asthma, coughs and as an anti-ophidian (Ruppelt et al. 1991). Pharmacological studies have confirmed the anti-inflammatory properties of crude Guaco extracts (Oliveira et al. 1996, Soares de Moura et al. 1996). Criddle et al. 1996 and Carvalho et al. 1996 found evidence that fractions of the hydro-alcoholic extract of Guaco showed potent inhibitory action against smooth respiratory muscle, in contrast to the slight dilatory activity against smooth vascular muscles. Oliveira et al. 1992b developed a chromatographic method for the fluid extract of *M. glomerata*, characterizing more than 18 substances, including: coumarin, cinnamoyl grandifloric acid and kaurenoic acid. Santos 1992, while studying the composition of *M. glomerata*, isolated and characterized coumarin and estigmast-22-en-3-ol. The chromatographic fractionation of the hexane extract allowed for the isolation of coumarin, kaurenoic acid, isobutyryl kaurenoic acid and lupeol (Santos et al. 1996a). The concentrations of coumarin were much higher than those of the other constituents, and the antimicrobial activity of coumarin is well known (Santos et al. 1996b). With respect to the volatile oil composition of *M. laevigata*, Suyenaga et al. 1996 found 0.65% total oils, the following sesquiterpenes standing out: β-caryophyllene (20.9%), germacrene (29.8%) and bicyclogermacrene (13.4%).

According to Lopes et al. 1997, coumarin is the only substance present in the hydro-alcoholic tincture of *Mikania glomerata* that presents significant relaxing activity on the smooth respiratory musculature. In a report by the Brazilian Medications Center—CEME, of the Ministry of Health, it is stated that Guaco syrup is completely innocuous and safe, and shows broncho-dilatory action and an anti-coughing effect.

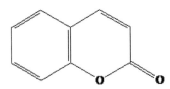

Figure 12.9. The structure of coumarin.

Evidence for other therapeutic activities has been found in phytochemical and pharmacological studies carried out on *M. laevigata* at CPQBA-UNICAMP (Rehder 2002). Amongst these activities, the following were found: anti-cancer effect (*in vitro*), anti-gastric ulcer effect, and a caries and bacterial plaque preventative effect for the teeth.

State of the culture

Although Guaco has long been used in Brazil, large scale cultivation and the use of selected genotypes are still rare. Considering coumarin as the main bioactive compound, advances were made on selection, and some genotypes were characterized for their biomass production and coumarin content at CPQBA-UNICAMP. The fact that *M. laevigata* has a perennial cycle and accepts vegetative propagation (cuttings) made it easier to obtain homogeneous populations by cloning. No papers were found in the literature concerning improvement aimed at obtaining seeds, probably due to the fact that vegetative propagation is so easy and promotes good rooting of the seedlings. Analytical methods for the dosage of coumarin and kaurenoic acid have been described by Vilegas et al. 1997a, 1997b.

Figure 12.10. Detail of the insertion of *M. laevigata* leaves.

Multiplication

M. laevigata is easily multiplied using cuttings from healthy, relatively thick segments from the plant matrix, chosen as a function of their good agronomical characteristics and elevated coumarin content. The segments used were from 15 to 20 cm long, cut on the slant with a very sharp knife. The cuttings should only have one pair of leaves, and these should be cut transversely, leaving just 50% of the foliar area. The cuttings can be treated with fungicide when they are stuck into the substrate in little tubes. The ideal time to take cuttings of Guaco is at the beginning of spring (Magalhães 1997). Dechamps et al. 1997 showed that the treatment of *M. glomerata* cuttings by immersing them in water for an hour, significantly increased the percentage of rooting. Pereira et al. 1997 arrived at the conclusion that the formation of *M. glomerata* seedlings using cuttings was twice as quick as the formation of seedlings by the *in vitro* micropropagation technique, probably due to problems with contamination by endophytic bacteria and fungi in the latter case. Scheffer et al. 1997 studied the influence of temperature and storage on the germination of *M. glomerata* seeds, and concluded that after three months of storage the seeds stored in the cold chamber had lost 50% of their germination power, whereas those stored under room conditions lost 87%. They also found that of the three temperatures tested in the germination tests, 20°C, 25°C and 20–30°C, the latter inhibited germination.

Spacing and time of planting

A 45 to 60 day period should be considered for the formation of seedlings in the greenhouse when multiplication is by cuttings, and in Brazil, transplantation to the field should occur at the beginning of November, when the environmental conditions in the Southeast of Brazil are hot and humid. Since the species is a climber, the cultivation of Guaco should be supported by some kind of trellis. Wooden or concrete stakes can be used at a distance of 3 m from each other, with three wires stretched from one to another at heights of 50, 100 and 150 cm from the ground. At the time of transplanting, bamboo stakes or transversal wires can be used to guide the seedlings in the direction of the three wires. The recommended spacing is of 1.0 m in the line and 2.5 m between lines, resulting in a plant density of 4,000 plants per hectare. The spacing between lines can be altered in function of the tread of the available machines. The alignment of the stakes should be in a north-south direction so as to promote greater incidence of the sun on the plants.

Fertilization

Correa Júnior et al. 1991 recommended the following fertilization: on planting apply 5 kg stable manure or organic compound or 3 kg poultry manure + 1.5 kg calcareous rocks +300 to 500 g natural phosphate per seed hole. Repeat the organic fertilization every year.

Cultivation care

Cultivating *M. laevigata* in direct sunlight increases the production of coumarin (Rehder et al. 1998). On the other hand, direct exposition of the plants to the sun induced intense fluorescing with a great expenditure of energy and a decline in plant vigor, which results in a progressive decrease in productivity in terms of twigs and leaves every harvest. To avoid this process, pruning of twigs presenting gaps at the start of florescence, is recommended.

Pests and diseases

The culture of *M. laevigata* at CPQBA-UNICAMP showed symptoms of a severe attack of the roots and neck by pests, causing the death of the smitten plants. The borers and larvae of *Migdorus* were found to be responsible for the damage. Wilting of the roots caused by bacteria was also observed, resulting in the complete drying out of the sick plants. Some plants presented the symptom of wilted leaves, caused by mites and not virus.

   All these cases can be controlled by spraying the vine with natural insecticides, and observing the holding time before harvesting the leaves. Culture rotation is recommended in the case of severe infestation by soil pests.

Harvest

The Guaco is harvested manually using pruning shears. It can be cut twice a year in mid spring and at the start of autumn, removing the leaves and twigs together. The pruning method is that of only cutting the secondary twigs, those that come from the main branches, which should remain intact on the sustaining wires. Although no differences have been found between samples harvested in the morning and those harvested in the afternoon with respect to the active principle contents (Rehder et al. 1998), harvesting in the morning is recommended so that the material can be processed fresh.

Drying

M. laevigata loses a large amount of coumarin (50%) during the drying process, even when low temperatures are used (40°C). Nevertheless, when it is impossible to process the fresh material, it should be dried quickly in dryers with air circulation at 40°C. The drying of leaves and twigs together makes the operation more difficult, since the leaves dry well before the twigs.

Yields

The *M. laevigata* cultivated in direct sunlight at CPQBA-UNICAMP with a spacing of 2 x 2.5 meters, provided 1.95 tons of dry leaves and twigs/ha (Figueira et al. 1991). The coumarin content of the *M. laevigata* clone was 1.13% when cultivated in direct sunlight and 0.68% under partially shady conditions. Also, the greatest coumarin content was obtained in the months with longer days, culminating in the summer solstice (longest day) on December 22 in the Southern Hemisphere. In fact, coumarin and the flavonoids are incited by ultra-violet-UV radiation, characterizing an important ecological function of protection against these rays. It is said that in this case the ultra-violet rays illicit coumarin, and, from the practical point of view, harvesting of the leaves should be done in the months with longer days (summer) in order to obtain the best coumarin yields, and, consequently the greatest therapeutic activity.

Phyllanthus amarus

Description

In Brazil, the common name of many species of *Phyllanthus* is 'Quebra-Pedra' or Stonebreaker. The species *P. niruri* L. (*Euphorbiaceae*) is considered to be a ruderal plant, and is more common in the southeastern states of Brazil, found in moist, shady places, whereas *P. amarus* Schumach is found in the North and Northeast of Brazil. Correa, Jr. et al. 1991 presents the following botanical description: herbaceous plant with a height of 10–50 cm, perennial or semi-perennial (three–four years), with rounded twigs, small flowers (3-4 mm in diameter), monoecious and with seeds measuring from 0.8 to 1.0 in diameter.

Figure 12.11. (a) Details of the whitish leaves of *P. niruri* seedlings; (b) the seeds (1 mm).

Ecology

Habitat: This plant occurs from the north of Mexico to Argentina. It is very common in moist, shady soils.

Altitude: from 0 to 800 m.

Climate: It is a plant that is acclimatized for the whole of Brazil, mainly in the tropical and equatorial regions. It shows some restrictions for growth in altitudes above 1000 m or in cold climates.

Soil: It is undemanding with respect to soil type, preferring clayey soils with elevated amounts of organic matter.

Properties and applications

As its name suggests, the popular indication of Stonebreaker is related to problems with kidney stones. The use of tea made with the entire plant (some popular indications even include the roots) has been recommended to favor the expulsion of kidney or bile stones (calculus), and as an antispasmodic agent. Studies supported by the Medication Center (CEME) in Brazil have reported that Stonebreaker has no acute toxic effect, shows an uricosuric effect and increases glomerular filtration (GFR), suggesting its potential use not only for its lithic effect and/or prevention of the formation of kidney stones, but also its possible use in hyperuricemic patients and those with kidney failure. Other research has demonstrated various medicinal properties for certain Stonebreaker species. Kumar et al. 1989 showed a sugar reducing effect for the alcoholic extract of *P. niruri* leaves administered orally at 250 mg/Kg in rabbits. In clinical trials, Srividya et al. 1995 showed that a preparation made from the whole *P. amarus* (cited as synonymous with *P. niruri*) plant and administered orally to diabetic patients for 10 days, had a diuretic, hypotensive and sugar reducing effect. Zhou-Shiwen et al. 1997 elucidated the hepatic-protective activity of *P. urinaria*, while Prakash et al. 1995 compared this effect in *P. urinaria, P. niruri,* and *P. simplex.* Sane et al. 1995 showed that rat liver cells damaged by the action of CCl_4, were regenerated by treatment with *P. amarus* and principally with *P. debilis*, when administered orally in the form of ground leaves at 0.66 g/kg. Species of the genus *Phyllanthus* have also been investigated for their antiviral action. Wang et al. 1995 found evidence of the efficacy of *P. urinaria* and *P. amarus* in a clinical trial with 123 patients with chronic hepatitis B. In another clinical trial Jayaram et al. 1997 demonstrated that hepatitis B patients treated with *P. amarus* showed a greater percent recovery (86.9%) than those treated with the conventional drug (50%) and the response was also quicker. Since the hepatitis B virus is similar to the AIDS virus, *P. amarus* extracts have also been tested for the control of AIDS, and showed promising results. Ogata et al. 1992 showed that and aqueous extract of *P. amarus* inhibited the transcriptase of type 1 HIV, and identified repandusinic acid as the active principle.

Using a methanolic extract of *P. niruri*, Qian-Cutrone et al. 1996 isolated another inhibitor of HIV called niruside at the genus expression level (HIV REV/RRE). The phytochemical study of *P. niruri* isolated various compounds, such as: gallic acid, epicatechin, gallocatechin, epigallocatechin, epicatechin 3-O-gallate and epigallocatechin 3-O gallate (Ishimaru et al. 1992). However, it was the lignans phyllanthin, hypophyllanthin and nirtetralin, isolated from a methanolic extract of the aerial part (fresh) of *P. niruri* that presented inhibitory action against the endothelin receptor, an important arterial vasoconstrictor (Hussain et al. 1995). The lignans phyllanthin and hypophyllanthin were tested in cancer cells, but did not present any significant cytotoxic effect on any tumor cell line (Srinivasulu 1992).

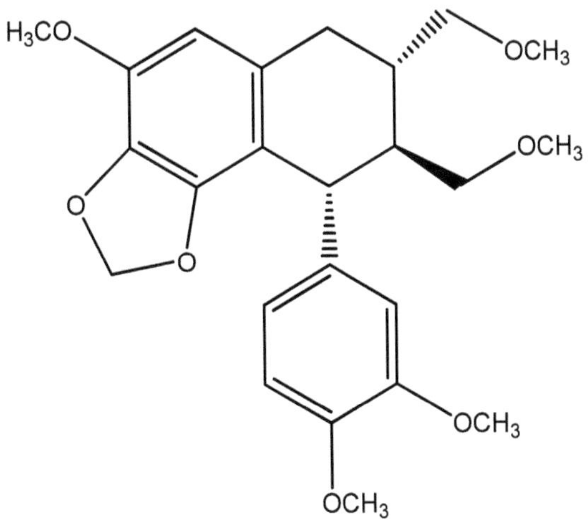

Figure 12.12. The structure of hypophyllanthin, the most active of the *Phyllanthus lignans* as an endothelin receptor inhibitor.

In Brazil, various validation studies concerning the pharmacological activities of *P. niruri* and *P. amarus* have demonstrated antispasmodic (Calixto et al. 1998, Rae et al. 1987, Dias et al. 1995) and analgesic actions (Santos et al. 1993a, 1993b, 1994a, 1994b, 1995, Cechinel filho et al. 1996).

State of the culture

P. amarus, like other Stonebreakers, is still a wild species presenting genetic variability amongst populations from regions with distinct environmental conditions, although not presenting significant variability or segregation amongst the same population. These accesses, when cultivated under the same environmental conditions maintain their characteristics, giving evidence of their distinct genetic bases. So far there are no improved strains or varieties. In trials carried out at CPQBA-UNICAMP, 14 genotypes of *P. amarus* were evaluated showing great differences between them with respect to germination, resprouting, productivity and hypophyllanthin contents.

Multiplication

The most indicated form for propagating Stonebreaker is by seeds, although the species can be multiplied vegetatively using cuttings, treating them with 100 ppm of Indole Butyric Acid (IBA) (Oliveira et al. 1996). In the case of propagation by seeds, the fruits are harvested when ripe. In practice this is an empirical way, observing the fruit and seeds with respect to size and coloration. Since ripening is irregular, one can use a comparative criterion concerning the parameters, taking the larger fruits and/or the seeds with a more intense color (brown). Brown seeds show a significantly higher percent germination than yellow seeds (Oliveira et al. 1996).

The fruits are left in the sun for a few hours with some kind of covering (netting, for example) to avoid loss of the seeds, since on losing moisture they tend to spring out of the fruits. The fruits are exposed to the sun for a time sufficient for the expulsion of most of the seeds.

Studies carried out by Figueira et al. 1998 with six *Phyllanthus* species (*P. niruri, P. tenellus, P. amarus, P. urinaria, P. carolinensis* and *Phyllanthus* sp. 'Pantanal') showed that the *Phyllanthus* seeds remained dormant for the first four months after harvesting, but the dormancy of recently harvested seeds can be broken by treatment with 2000 ppm gibberellic acid (GA_3) for six hours.

With the exception of *P. niruri*, the other species cited in this study reacted as positive photoblastic types, demanding light for germination. Unander et al. 1995 also cited the dormancy of recently harvested seeds and their light-dependent behavior for germination. They concluded that Stonebreaker should be seeded very near the surface with just a very fine covering of soil, to allow for the passage of light. The seedlings should be formed in a shady greenhouse using an organic substrate composed of equal parts of soil, cattle manure and worm humus. Seedling formation takes from 45 to 60 days.

Spacing and time of planting

The best time to transplant Stonebreaker in Brazil is from October to January, coinciding with the hot, rainy season. Spacing can be highly dense in the lines, from 15 to 20 cm, and 50 cm between lines. Thus with this spacing the number of seedlings will be 133 thousand/ha. To provide this number of seedlings in the greenhouse, one should plant an additional 20% to allow for eventual germination failures or even to permit a selection of the best seedlings.

In the field, formation of the furrows can be mechanized with the help of a tracer adapted to penetrate up to 20 cm in depth.

Fertilization

Correa Júnior et al. 1991 recommend organic composting with 5 kg/m^2 stable manure or 3 kg/m^2 chicken manure. Fertilization should be carried out after every cutting.

Cultivation care

Stonebreaker demands frequent irrigation to maintain the soil humidity in the range from 60 to 80%. Cultivation can be carried out in direct sunlight, although it vegetates better in shady locations. With respect to the temperature, it prefers higher temperatures, typical of the Brazilian summer. During the first months after transplantation to the field hoeing must be carried out periodically between the lines to maintain them free of weeds. After the first cutting the need for hoeing decreases, since the Stonebreaker sprouts and spreads out, occupying the spaces between the lines itself.

Pests and diseases

Stonebreaker is very susceptible to attack by cutting ants, which can destroy the cultivated area in a few hours. Thus ant control should be intense especially during the first days after transplantation to the field. This can be done using a commercial ant killer or, in a natural way, by planting lines of sesame (*Sesamum indicum*) around the cultivated area as a protective barrier.

   The ants will prefer to cut the sesame leaves, which will intoxicate the fungus that serves as food for the ants in the anthill. Another fatal pest for the culture is the occurrence of the military caterpillar (larvae of the geometric moth), which attacks the neck of the plant, leading to its death. The caterpillars prefer *P. carolinensis.*

Harvest

Stonebreaker is cut manually with the aid of pruning shears, cutting it from 5 to 10 cm from the soil, or using a back-carried hoe with a three-pointed star shaped disk which rotates at approximately 2,000 rpm. The yield from the latter equipment is 1 ha/day. The harvested material should be placed on a cloth or fine netting to avoid the loss of leaves and fruits. The first cut should be made three months after transplanting, and subsequent cuts should be done as a function of resprouting, up to four cuts per year.

Figure 12.13. Harvesting of *P. amarus* at CPQBA-UNICAMP.

Drying

Most of the pharmacological studies were carried out with the fresh, recently harvested plant.

However, the moisture content must be reduced by drying in order to store the drug. The drying temperature should not go above 40°C in order to minimize losses of the active principle during drying, and drying should be quick, that is, passing a large volume of unsaturated air over the material in a well distributed way. The Stonebreaker leaves and fruits should be distributed in the dryer, placed on trays lined with fine mesh netting to avoid physical losses. The material is ready for storage when the moisture content reaches 10–12% (dry weight basis).

Yields

The yield of the aerial part (leaves, stems and fruits of *P. niruri*) with 12% moisture content is estimated as 1.5 ton/ha per cut. The yield normally increases after the first cut due to sprouting of the plant.

It must be remembered that the populations under study are genotypes in a selection process and hence these yields should improve. For the same reason there is no definitive characterization of the active constituent contents.

Cordia verbenacea

Description

C. verbenacea DC., Boraginaceae, synonym for *Cordia curassavica* auctt. Bras. Ex Fresen, is an aromatic shrub reaching 2.5 to 3 meters in height and 2 m in breadth (projection of the crown).

The base of the trunk is rigid and thick (5 to 8 cm in diameter), whilst the numerous upper branches are fine and tender, of a dusky brown color with whitish tips (lenticels). The leaves are found predominantly on the terminal parts of the branches, the interior of the plant consisting mainly of leafless branches. The leaves are lance-shaped or oblong-lanceolate, measuring from 5 to 14 cm in length and from 1.5 to 4 cm in width, sessile or sub-sessile with a pointed apex and attenuated base, dented edges, the upper face being dark green and rough (wrinkled) whereas the back of the leaves is lighter and covered with hairs. A mixture of yellow and green leaves is common, this phenomenon being a physiological characteristic of the species. The leaves have a typical aroma which is similar to seasoning (such as Knorr/Maggi broth). The flowers are hermaphrodite and the inflorescences are arranged along 2 to 8 cm long ears. The calyx is from 2.5 to 3.5 mm in length and the corolla from 5 to 7 mm in length.

C. verbenacea leaves can be confounded with *C. corymbosa* leaves, but the latter are wider and do not have the characteristic aroma of *C. verbenacea*. As its name suggests, the inflorescences of *C. corymbosa* are in the form of a corymb (the flowers come from different heights, but reach the same level at the top), and are therefore different from those of *C. verbenacea* which has inflorescences on an ear.

Ecology

Habitat: *C. verbenacea* vegetates preferentially in sandy soil such as sandbanks, dunes and beaches. In Brazil it flowers during the hottest eight months of the year (September to April). Flowering and fructification are irregular and the flowers attract African bees, hymenopteras, flies and butterflies. The small fruits are surrounded by a red aril, are edible and highly sought out by various species of birds, which, on defecating, promote seed dispersion. The incidence of cutting ants is common and coleopterous larvae of the family *Carabidae*, which feed on the leaf blade, causing serious damage to the plant (Montanari 2000).

Altitude: from 0 to 800 m.

Climate: It accepts a wide range of climatic conditions so long as the temperature is not too low. Ideally it prefers hot regions with a mean temperature around 25°C.

Soil: Although basically undemanding with respect to the type of soil, it responds favorably to well-drained, fertile soils.

Strategy for launching the anti-inflammatory product

The essential oil stored in the leaves of *C. verbenacea* (popular name: 'Erva Baleeira' = 'Whaler herb') show anti-inflammatory action, a fact well known by coastal populations, who use it for the treatment of contusions and muscular pains. Pharmacological research validated the anti-inflammatory property of this essential oil, identifying an alpha-humulene-rich fraction as being the most efficient.

In addition to its use in humans, the Whaler herb is also used in veterinary treatments, specifically for the treatment of mastitis in cattle.

Based on scientific data and a study of the market for anti-inflammatory agents, the laboratory Aché contracted the State University of Campinas-UNICAMP to carry out the large-scale sustainable production of the raw material, with a view of launching an anti-inflammatory medication based on Whaler herb essential oil onto the market. The amount of essential oil required to launch the product into the market was estimated at 100 Kg per year. In 2001 they increased the area planted to 12 ha, and installed steam-stripping essential oil extractors with a capacity to obtain 1 Kg Whaler herb oil/day. The product was only put into the market in mid-2003 when sufficient raw material was available for the predicted demand. However in the first year of commercialization the sales surpassed the estimates by a considerable margin, and the product became the topical anti-inflammatory medication most sold in Brazil. Due to the rapid growth in commercialization, other production strategies were organized, including the supply of the raw material by family farmers and producers with a higher technological level. In parallel with the production areas, where the cultivation technology and processing was being constantly improved, UNICAMP also installed a genetic selection program to obtain, as from the natural variability, plants producing more essential oil which containing more alpha humulene.

State of the culture

The main domestication studies have still not resulted in cultivars that could be registered in the Brazilian Ministry of Agriculture, showing the cultures still with a high degree of genetic variability. CPQBA-UNICAMP is the institute that mostly supplies propagation material due to advances in the selection of more productive genotypes. The direct harvest in locations where the species occurs, mainly at the coast, results in heterogeneous populations due to the high degree of crossing of the species.

Multiplication

Sexual propagation by seeds is completely feasible but the species also accepts vegetative propagation by cuttings or by meristem culture. In the case of propagation by seeds, these should be harvested in the red fruit state, since soon after this, generalized oviposition of hymenoptera occurs, whose larvae destroy the embryo. Seeding can be carried out in small tubes after removal of the aril, which prevents germination. According to Montanari 2000, germination occurs 20 to 50 days after seeding. Since this is a species at the start of domestication, there are still no cultivars available, and even the cultivation technology is subject to constant adjustments.

Cultivations carried out at CPQBA-UNICAMP indicated that the species requires fertile, slightly acid soil and good drainage Montanari 2000.

Seedling formation can be carried out as from seeds in greenhouses with shade netting (50%).

Although the species accepts vegetative propagation from branch cuttings, the formation of seedlings from seeds is of greater interest due to the large amount of seeds produced and also to promote genetic variability, which is important in this phase of species domestication. The seeds present germination power of approximately 70%, reaching this value between 15 and 20 days after seeding, and they take about two months to be ready for transplantation to the cultivation areas. Seeding should preferably take place in the hot months, from October to January, to allow for transference to the field while still in the rainy, hot season.

The Whaler herb seeds reach maturity in a highly irregular way. The ears, which contain the reproductive part of the plant, present flowers and fruits in different states at the same time. In fact this is a typical process in wild plants, as is their florescence, without a defined time and occurring throughout the year. However, due to greater growth of the plant in summer, greater seed formation was also found in summer. The seeds should be harvested when the aril is red (bright, brilliant), since in this stage the seeds show the best germination conditions and are usually healthy. In later phases the fruits are normally destroyed by larvae which develop inside the seeds and damage the embryo. It should be mentioned that the red seeds on the ears (the best for harvest) are also a target for birds. A regular, daily harvest is necessary when a large amount of seeds is required, considering the competition with the birds.

Figure 12.14. Floral ear of *C. verbenacea* showing mature fruit (red).

Spacing and time of planting

Seeding can be done in any season of the year in Brazil, especially if the cultivation is irrigated.

However if there is no irrigation, seeding should ideally be in mid-September, such that transference to the field after about 60 days coincides with the rains. The spacing recommended in the line is 0.5 m between plants, and 1.8 m between the furrows. The spacing between the furrows may vary according to the gage of the tractor and of the other implements that will be used.

Plant cycle (Phenology)

C. verbenacea is a perennial species with the following growing habits: in the first developmental stage the plant presents a single principle stalk bearing alternate leaves and a foliar apex responsible for vegetative growth. As the foliar apex differentiates itself into florescence (ear I or phase I), three to four leaf bearing branches appear from the region just below the florescence, also showing their respective foliar apexes. These branches develop in a vegetative way whilst ear I continues its formation until it produces fruits and attains senescence. As from the point where ear I dries out, each of the branches below develops more, and at the same time differentiates itself, and its apex forms fruit-producing ears. These inflorescences are called ears II or phase II. Ears II will also continue their formation and attain senescence, promoting the development of other three-four branches which, for their part, will terminate in ears III, and so on, conferring a highly ramified architecture to the plant. This behavior is important for spacing and harvesting. The growth of *C. verbenacea* is quick, and in six to eight months plants cultivated in direct sunlight can be found in phase III or even more advanced.

Phase I

Phase II

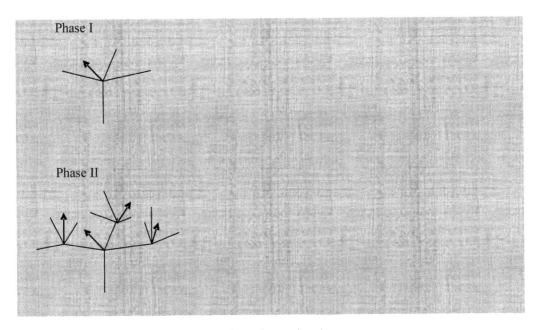

Figure 12.15. Characteristics of the development of *C. verbenacea* branches.

Fertilization

N-P-K, 10-10-10 fertilization is recommended after each harvest at 50 g/plant.

Cultivation care

Two weeding sessions are sufficient to keep the spaces between lines free of invaders, one about 30 days after transference of the seedlings to the field, and the other before the end of cultivation.

Special care must be taken concerning the removal of aromatic invaders present in the cultivation line. Invading plants containing essential oil will contaminate the principle product on being harvested and processed together in the distiller. The plantation of wild peanut (*Arachis pintoi*) is recommended between lines.

Pests and diseases

The recently-transferred seedlings must be checked for attack by ants. Sesame, when planted every 10 lines of *C. verbenacea*, is able to control the anthills.

Whaler herb attracts a great variety of insects, and its leaves are even used to capture insects in traps. At UNICAMP, while cultivating on a small scale, up to 200–500 m², the insects caused no damage. However, when larger areas were installed, 8–12 ha, an extremely aggressive pest appeared on the Whaler herb leaves, identified as the larva of *Chrysomelideae*, sub-family *Cassidinae*, a close relative of the glow-worm. Oviposition occurred on the leaves and the larvae hatched and fed voraciously on the leaves. Cultivated areas of 2 ha were completely devastated in about 10 days at CPQBA-UNICAMP. Many plants failed to recover, and, in fact, the literature reports that this pest is used to eliminate the species in Malaysia, where *C. verbenacea* is an invader of the coconut cultivation. Fortunately these larvae are easy to control using organic insecticides such as Tracer® (Spinosad® in USA, from DowAgroscience). This insecticide is manufactured from a toxin produced by a bacterium and is safe for the environment. As soon as the pest is detected, applications of Tracer® must be made immediately using a dose of 75 mL/ha in a solution of 400 L/ha, using sprays with a conical nozzle. The application must be repeated every seven days, and three applications are normally sufficient. Another option that also controls these larvae and is safe for the environment is a product based on neonicotinoids, trade name Capypso®. This must be used with a dosage of 100 mL/ha in a solution of 400 L/ha. After four years of cultivation at UNICAMP, the natural enemy of these larvae was found—a sucking insect similar to one popularly known as 'Stinking Maria' (Maria Fedida). This insect is highly aggressive and sucks up the larvae.

Figure 12.16. Young *C. verbenacea* plant attacked by *Chrysomelideae* larvae.

Harvest

The *C. verbenacea* leaves which, in this case, represent the vegetable drug since they contain the essential oil, occur predominantly at the periphery of the branches. Thus it is recommended not to allow the plant to grow too much, and it appears that the best harvesting point is when most of the plants are in phase II of development. At CPQBA, once the culture is established (after one year in the field), three harvests are made per year: one in March, another in June and the last one in November. However these dates will probably be adjusted as a function of the times of the year and the states of development.

Studies evaluating the production of leaves and essential oil by the Whaler herb, carried out monthly at CPQBA for a whole year, indicated that the greatest yields coincided with the months from January to May, as shown in Frame 1, but the best time of the day in terms of yield is still unknown. Depending on the scale, the Whaler herb can be harvested using pruning shears, pneumatic shears, reapers or 'tarup' type harvesters. Pruning shears are recommended for small areas, making the cuts at the periphery of the plant in order to harvest predominant leaves, inflorescences and fine twigs. The fine twigs have an important role of providing a uniformly impermeable mass for the steam inside the distiller. After making a peripheral harvest, a second cut is recommended to lower the height of the plants, such that the recovery of the foliage occurs at a height adequate for the next cut. When a reaper is used, the cut will already be lower, resulting in a material including thicker branches. In this case the second cut can be made using pruning shears just to prevent the 'skirt' from getting too wide. If a harvester is used, a second cut is unnecessary since the ideal height for cutting can be regulated with a view to foliage recovery in accordance with the plant architecture.

When dealing with raw material for the extraction of essential oil, the material harvested must proceed rapidly to processing in the distiller. It is known that fresh leaves give higher oil yields since the loss of volatile fractions is reduced, although in the present case, the compound of greatest interest, alpha humulene, showed no significant loss, even when the material was dried at 40°C. When using steam stripping distillation, the alpha humulene is extracted in the second half hour of distillation. In order to maintain the integrity of all the constituents of the oil, and not just of the marker, cutting at times when immediate processing will not be possible, should be avoided.

Figure 12.17. *C. verbenacea* plantation at harvesting point (CPQBA-UNICAMP).

Distillation of the fresh leaves is the ideal procedure to extract the oil from the Whaler herb. The material harvested is placed in distillation vats which, in the case of CPQBA, consist of two 1500 liter vats with a single condenser and separator flask. The entire equipment was constructed in stainless steel and the steam is produced by a gas boiler with a capacity for 200 Kg steam per hour. The installation also has a rolling bridge with a vessel and dynamometer to control the yields and facilitate loading and unloading of the mass. The vats are filled and the mass compacted by workers stamping on top of it, before closing with lids that contain a water seal. The steam passes through the mass with an upward flow at 50 L steam/m^3/hour and distillation is completed in 90 minutes. The oil is then removed from the separating flask and kept aside in a separating funnel before filtering through a carbon column. The filtered oil is stored in a fridge in amber glass jars with screw tops containing Teflon seals.

Figure 12.18. Steam stripping distillation vats for the processing of *C. verbenacea* leaves.

Expected yields

The Whaler herb yield is composed of the following parameters: number of plants per area; leaf biomass per plant; the number of cuts per year; the essential oil content of the leaves; and the alpha humulene content of the oil. The ideal parameter which translates into the yield of interest is the weight of essential oil per area and per year. Table 12.2 shows the variations in biomass and essential oil content of the Whaler herb evaluated, month by month, during one year.

Table 12.2. Evaluation of the essential oil yields and compositions of fresh *C. verbenacea* leaves.

Months Parameters	July 2001	Aug	Sept	Oct	Nov	Jan/2	Mar	Apr	May	Jun	July 2002
Leaves*Kg/pl.	0,30	0,33	0,36	0,64	0,44	1,55	1,24	0,98	0,67	0,74	0,30
Oil % p/p	0,10	0,11	0,14	0,16	0,16	0,11	0,16	0,19	0,19	0,17	0,18
Oil(g)/plant	0,30	0,36	0,5	1,03	0,71	1,71	1,99	1,86	1,27	1,25	0,54
1.8-cineol	3,34	5,15	4,3	4,39	4,20	3,50	4,31	4,28	5,48	7,02	5,37
Trans-caryophyllene	19,9	26,9	18,9	22,6	17,5	19,0	23,3	21,1	10,3	8,73	13,4
a-humulene	3,88	3,06	2,69	3,97	4,01	4,74	4,80	3,73	1,39	1,04	1,54

The amount of oil (g)/plant varied from 0.30 to 1.55, correlating positively with the rainiest and hottest months of the year. The markers 1,8 cineole (from 3.34 to 5.15%), trans-caryophyllene (17.51 to 26.9%) and a-humulene (from 2.69 to 4.8%) did not present these variations in a way correlated with the times of year. The variation in the a-humulene contents correlated negatively with the variation observed in the 1,8 cineole contents.

These yields indicate the ideal times of the year for cutting. However, in an intensive harvesting system, when three to four cuts are carried out per year in the same area, it is not always possible to carry out the cuts at the ideal time. In order to determine the cutting point, it is more important to consider the recovery of the plant in terms of biomass, than to go after the best oil yield. The flow-sheet for the processing of Whaler herb at CPQBA-UNICAMP is shown below.

The smaller scale maintains proportionality with this example at the level of mass processed and steam used.

Descriptive flow-sheet of the production of *Cordia verbenacea* essential oil

As from one hectare of Whaler herb plantation under the conditions defined below,* the area had the potential to produce 7.5 tons of fresh biomass** at each cutting (three cuts per year) and was processed in steam stripping extractors in 400 Kg batches. In order to process the amount of biomass referring to the three cuttings from one hectare, estimated as 22 tons of fresh material, the extractor system received 3.5 tons of steam plus 123.2 m^3 of water for cooling the condenser, and generated 26.4 Kg of essential oil. The extractor system also generated about 22 tons of moist biomass exhausted of its oil (vegetable cake), which returned to the plantations as organic manure; 123.2 m^3 of water from the condenser and 3.5 m^3 of hydrolyte.

An average yield in essential oil of 833.3 Kg fresh biomass: 1 Kg essential oil was considered.

---

* a 'formed plantation' is considered to be an area which has already undergone formation pruning and where the architecture of the plants is defined—this normally occurs after one year of cultivation. The cultivation conditions considered in this flow-sheet were: spacing of 3.20 m between lines and 0.5 m between plants, organic handling area with pH and nutrient correction of the soil, periodic irrigation and mechanical control of invaders.

**fresh biomass manually harvested is constituted of the aerial part, that is, leaves, inflorescences and branches of up to 1.5 cm in diameter. These fractions alter as a function of the stage of development and treatment of the plant, a large amount of leaves being of greater interest, since the essential oil is in the leaves.

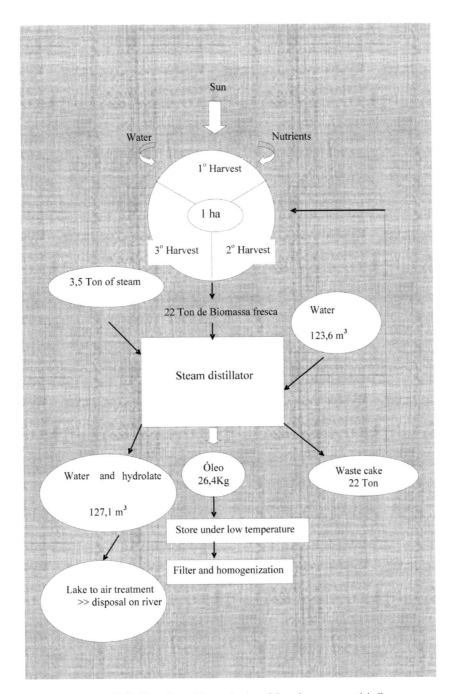

Figure 12.19. Flow-sheet of the production of *C. verbenacea* essential oil.

It can be seen that the most productive harvest was done in March, and that with the three cuttings per year system; fertilizer replacement is required after each harvest.

The yields varied from 1.77 to 5.08 Kg of oil/ha between the areas cultivated, demonstrating the potential to increase the mean yield (3.29 Kg/ha) by replacing the nutrients.

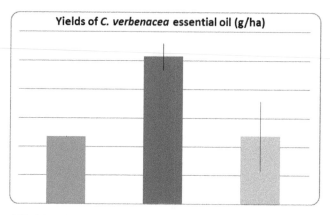

Figure 12.20. Average yield of *C. verbenacea* essential oil in commercial areas at CPQBA-UNICAMP harvested in a three cuttings/year system for two years (average of nine areas).

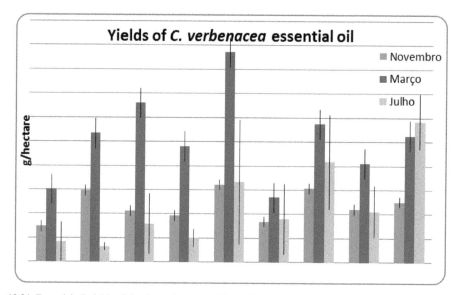

Figure 12.21. Essential oil yields of *Cordia verbenacea* cultivated in nine commercial areas at CPQBA-UNICAMP during the years of 2007 and 2008 (means of two cuttings).

Storage

If the final product is the dried Whaler herb leaves, they should be thoroughly dried and stored in double Kraft paper bags with a polypropylene lining. Storage should be at mild temperatures with a dehumidifier. If the final product is the essential oil, this should be stored in amber glass recipients with screw tops and Teflon seals. The flasks should be stored in the fridge at temperatures between 5 and 10°C.

Commercialization

For commercialization of the dried leaves, whenever possible one should negotiate with the potential buyer before obtaining the product. When this is not possible it is of interest to work with a scheme of samples that truly represents the available batch. One must take care with the packaging of the samples to be sent to the companies interested in the product, such that the color (green) can be seen, the aroma appreciated and the general aspect of the product be clearly visualized (purity, leaf integrity, absence of insects, etc.).

The label should include the scientific name, the part of the plant (dried leaves) and the date obtained. Do not include the therapeutic use on the label, even if the product is destined for drug manufacture. The same instructions apply to commercialization of the oil, that is, negotiate beforehand with potential buyers or send samples in order to determine their interest. The estimated price for dried Whaler herb leaves is between three and five US$/kg, and for the essential oil approximately US$ 1,500.00/kg. As a quality standard, the oil should have a minimum of 2% w/w of alpha humulene. To obtain 1 Kg of oil one must process approximately 800 Kg of fresh Whaler herb biomass (considering an oil content of 0.12% w/w).

Table 12.3. Cost estimate to obtain 1 L of *C. verbenacea* oil at CPQBA.

Cost of Items	$ unit	Total in Reais $
Harvest and process post-harvest on fresh leaves (700 Kg, without stems) 4 people + taxes, 5 days > 4.(50,00 + 40,00).5	2,57/Kg	1.800,00
Facilities rent/5 days	60,00/day	300,00
Steam generator (1500 Kg of steam/36 hours = 41,5 Kg/hour)		500,00
Management (transport, steam operation)	150,00/day	750,00
Taxes (20%)		670,00
Total in Reais $		**3.980,00**

Baccharis dracunculifolia

Description

Baccharis dracunculifolia DC. (Asteraceae) is a perennial shrub species which is aromatic and very common in open fields and pastures. It is a dioecious species, the flowers of the female plants liberating their achenes when mature. On the other hand, the flowers of the male plants do not open completely. The primordial foliage is densely covered by glandular trichomes and tectors but these are rare in the adult leaves. This explains why bees prefer to visit the new shoots to collect the resin for the formation of green propolis. The adult leaf is amphi-stomatal and the mesophyll is composed of palisade parenchyma and small amounts of spongy parenchyma and secretory canals associated with the phloem.

The occurrence of 'ambrosia galls' is common, which are induced by dipteras (*Cecidomyiidae*) and have no nutritive tissue, since the inductor larvae feed on fungal hyphae.

The *B. dracunculifolia* ambrosia galls are constituted of a single larval chamber containing one inductor. The hyphae are confined to the larval chamber in the *B. dracunculifolia* galls and the palisade parenchyma cells are elongated. The chlorenchyma cells close to the larval chamber are slightly elongated. The pericyclic fibers of the vascular system lose their secondary walls.

Figure 12.22. Apical branch of *B. dracunculifolia* being visited by *Apis mellifera*.

Ecology: The species presents the characteristics of an invasive colonizing plant and is indicated for the recovery of degraded areas.

Habitat: Open dry fields

Altitude: from 0 to 1200 m

Climate: Temperate

Soil: Predominantly in sandy soil, but can be found in a wide range of soil conditions

Properties and applications

The epidermis of *B. dracunculifolia* leaves is the main constituent of the green propolis produced by *Apis mellifera*. Thus the properties and applications of the plant are confounded with those of the propolis, whose composition depends on the time of year, genotype and collection location.

Phytochemical analyses of propolis have shown a high proportion of artepillin C and other cinnamic acid derivatives. The *B. dracunculifolia* DC essential oil is considered to be an exotic essence, and is exported as a raw material for the manufacture of perfumes, the sesquiterpenes E-nerolidol and spathulenol being the major components (Queiroga 2014). E-nerolidol is a sesquiterpene present in the essential oil of many plants and has been approved by the Food & Drug Administration (FDA, USA) for use as a flavoring agent in foods (Arruda et al. 2005). According to Wattenberg 1991 (*apud* Arruda et al. 2005), this compound also shows antineoplastic properties. Studies with the compound have shown an inhibitory effect on the growth of *Plasmodium falciparum*, the causal agent of malaria (Macedo et al. 2002), and on that of *Leishmania amazonensis*, causal agent of American tegumentary leishmaniasis (Arruda et al. 2005). The compound spathulenol shows important biological activity with antibacterial properties and moderate cytotoxic activity (Limberger et al. 2004). Another component of the *B. dracunculifolia* essential oil, although not a major one, is the sesquiterpene caryophyllene oxide, which the literature reports as having anti-carcinogenic activity (Zheng et al. 1992).

The essential oil produced by *B. dracunculifolia* inhibited the growth of the following cariogenic bacterial strains: *Streptococcus mutans* (ATCC 2575); *S. sobrinus* (ATCC 27607); *S. sanguis* (ATCC 10557) and *Lactobacillus casei* (ATCC 4646) (Ferronato et al. 2007).

The aqueous extracts of *B. dracunculifolia* showed evidence of allelopathic potential, reducing the germination of seeds and the growth of the aerial part and root system of the following species: *Brassica campestris*, *B. oleracea* cv. Capitata, *Citrullus lanatus*, *Eruca sativa*, *Lactuca sativa* cv. Branca Boston, *L. sativa* cv. Grand Rapids, *L. sativa* cv. Simpson, *Lycopersicum esculentum*, *Raphanus sativus* and *Zea mays* L. (Gusman et al. 2008).

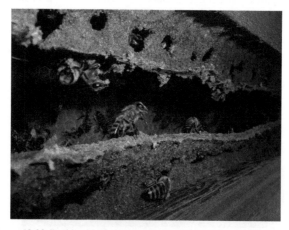

Figure 12.23. Beehive highlighting a thread of *B. dracunculifolia* propolis.

State of the culture

Unicamp, together with the Brazilian organization Sebrae, are developing the improvement of the species in order to attract the interests of green propolis producers in the Brazilian state of Minas Gerais, selecting individuals visited by the bees and showing high artepillin C contents.

However, with the exception of this program, the species is only found in the wild state throughout the whole of the large distribution area, presenting important genetic variability. Other characteristics considered in the Unicamp improvement program are the responses to pruning and the yields in biomass and essential oil.

Multiplication

The achenes germinate in the light between 15°C and 20°C, and should be seeded as soon as harvested since they lose their germinating power to a drastic degree during storage.

Spacing and time of planting

The spacing with a view to annual pruning is of 1 x 1 m, and in non-irrigated plantations the time to transplant should coincide with the rainy season, October-November.

Figure 12.24. Cultivated area of *B. dracunculifolia* at CPQBA-UNICAMP.

Fertilization

The species is relatively undemanding with respect to soil fertility, manuring and liming being carried out as a function of an analysis of the soil and the intensity of extraction of the nutrients by the cultures.

Cultivation care

During cultivation manual hoeing (weeding) is recommended as from the transplant of the seedlings to the field up to finalizing of the culture. The techniques for harvesting the leaves and annual pruning are still in the experimental phase.

Pests and diseases

No disease or pest was observed in *B. dracunculifolia*, but 'ambrosia galls' occur, with no prejudice to plant development. The hypothesis that the galls influence an increase in artepillin C content is under study.

Harvest

Harvesting of the aerial parts, leaves and branches, is the result of pruning, the first pruning being done when the plant reaches approximately 1.5 m in height. Pruning can be done by manual cutting using shears, or mechanically using hedge trimmers.

Drying

Wherever possible, distillation of the leaves to obtain the essential oil should be done with fresh material, but if the material is stored for subsequent distillation, drying should be done in dryers with air circulation at temperatures below 50°C.

Yields

The yield in dry leaves from an irrigated plantation, using a half-brother population with one year of development, is between 1500 and 2000 Kg/ha and from 15 to 20 Kg of essential oil.

Conclusions

All these six species results in an agricultural opportunity since the commercial demand for them are increasing. The main points to finally consider are:

Artemisia annua new varieties with very high content of artemisinin and flavonoids have now brought a new version for traditional use, since the tea made with the selected plant produce an important effect for the treatment of malaria. *Pfaffia glomerata*, also called Brazilian Ginseng due to its similar effect as an stimulant has a significant space in the market. The genetic selection increased the content of β-ecdisone in the roots promoting an interesting material to growers. On *Mikania laevigata* researches, the focus were done to its content of coumarin which is linking with some therapeutic properties of the plant to respiratory diseases. Coumarin was monitoring to define the better season and region to cultivate. Also high temperatures were avoid during the drying process. In *Phyllanthus niruri* a great genetic variation for lignin content was found, which is associated with hepatitis B treatment. The agro-technology advances showed that the species accept even four cuts by cycle. While cultivating Whaler herb (*Cordia verbenacea*) it was technically and economically feasible to produce the amount of raw material destined for the production of the anti-inflammatory medication launched into the market. Naturally the installation of a new culture as from a wild population demands continuous domestication procedures in order to carry out its cultivation on a large scale. For anti-inflammatory properties the yield of essential oil and its content of α-humulene was the goal in terms of plant breeding. The interest in *Baccharis dracunculifolia* has grown due to its ability to promote and enhance green propolis production, to obtain Nerolidol to treat malaria and also to its use in aromas and cosmetics. The plant breeding program is helped by bees who visit the better genotypes that contain higher content of artepellin C, nerolidol, and flavonoids.

References

Abreu Mattos, J.K. 1993. Biologia da Ferrugem (*Uromycesplatensis* SPEG.) da *P. glomerata* PEDERSEN. Dissertação de Mestrado. Universidade de Brasília.

Alcântara, M.F.A. de. 1994. Atividade antimicrobiana de *Pfaffia glomerata* (Spreng.) Pedersen. XIII Simpósio de Plantas Medicinais do Brasil. Fortaleza-CE. p-072.

Alves, M.N., Magalhães, P.M. and Ricoy, M. 1996. Tissue culture of *Pfaffia* sp. Kuntze (*Amarantaceae*) aiming vegetative micropropagation and the induction of friable callus for future cell suspension culture. XIV Simpósio de Plantas Medicinais do Brasil. Florianópolis-SC. A-025.

Araujo, W., Mattos, J. and Souza, R. 1994. Fontes de resistência a Meloidogynejavanica entre procedências de *Pfaffia glomerata*. Fitopatologia Brasileira, 19(nsupl).

Arruda, D.C., D'Alexandri, A. and Uliana, S. 2005. Antileishmanial activity of the terpene nerolidol. Antimicrob. Agents and Chemother. 49: 1.679–1.687.

Calixto, J.B., Santos, A.R.C., Cechinel-Filho, V.C. and Yunes, R.A. 1998. A review of the plants of the genus *Phyllanthus*: their chemistry, pharmacology and therapeutic potential. Medicinal Research Reviews 18(4): 225–258.

Carvalho, L.C.R.M., Soares de Moura, R., Lopes, C.S., Criddle, D.N., Pinto, A.C., Jansen, J.M., Alves Pereira, S. and Freitas, M.C. 1996. Efeito do Guaco (*Mikania glomerata*) na musculatura lisa respiratória. XIV Simpósio de Plantas Medicinais do Brasil. F-126.

Cechinel Filho, V., Santos, A.R.S., Campos, R.O.P., Miguel, O.G., Yunes, R.A., Ferrari, F., Messana, J. and Calixto, J.B. 1996. Chemical and pharmacological studies of *Phyllanthus caroliniensis* in mice. Journal of Pharmacy and Pharmacology, EstadosUnidos 48: 1231–1236.

Correa Júnior, C., Scheffer, M. and Ming, L.C. 1991. Cultivo de plantas medicinais, condimentares e aromáticas. EMATER-Paraná, 162 p.

Criddle, D.N., Soares de Moura, R., Lopes, C.S. and Carvalho, L.C.R.M. 1996. Efeito do Guaco (*Mikania glomerata*) no leito vascular mesentérico (LVM) do rato. XIV Simpósio de Plantas Medicinais do Brasil. F-109.

Dechamps, C., Boeing, C., Scheffer, M. and DoniFilho, L. 1997. Influencia del tiempo de inmersión en agua y de la aplicación de fitorreguladores en la propagación vegetativa de guaco (*Mikania glomerata* Spreng.). WOCMAP II. Mendoza-Argentina. P-083.

Delabays, N. 1997. Biologie de la reproductionchez L′*Artemisia annua* L. et genetique de la production en Artemisinine: Contribuition à la domestication et à l′amelorationgénétique de l′especie. These de doctorat. Faculte des Sciences de L′Universite de Lausanne, Suisse, 169 p.

Delabays, N., Ançay, A. and Mermillod, G. 1998. Recherche d′espèces végétales à propriétés allélopathiques. RevuesuisseVitic. Arboric. Hortic. 30(6): 383–387.

Dias, M.A., Cechinel Filho, V., Yunes, R.A. and Calixto, J.B. 1995. Análise do mecanismo envolvido na resposta contrátil para o extrato hidroalcoólicodo *Phyllanthus urinaria* na veia porta isolada de rato. *In*: Anais da X ReuniãoAnual da Federação de Sociedadesde Biologia Experimental, 23 a 26Agosto, Serra Negra, SP, pp. 260.

Dias, P.C., Foglio, M.A., Rehder, V.L.G. and Carvalho, J.E. 2000. "Atividade antiulcerogenica da dihidro-epideoxiartenuína b isolada d e *Artemisia annua* L.", XVI Latin American Congress of Pharmacology—XXXII Brazilian Congress of Pharmacology and Experimental Therapeutics—II Ibero American Congress of Pharmacology—VII Interamerican Congress of Clinical Pharmacology and Therapeutics-Águas de Lindoia, SP, 13–17 de setembro de 2000.

Dias, R.F., Espínola, E.B., Galvão, S.M.P., Marques, L.C., Mattei, R. and Carlini, E.A. 1996. Avaliação farmacológica de plantas medicinais brasileiras com possível efeito adaptógeno—II. Determinação da atividade motora e da esquiva passiva. XIV Simpósio de Plantas Medicinais do Brasil. Florianópolis-SC. F-212.

Ferreira, J.F. da S. 1994. Production and detection of artemisinin in *Artemisia annua* L. Thesis of Doctor of Philosophy. Purdue University, 125 p.

Ferronatto, R., Marchesan, E.D., Pezenti, E., Bednarski, F. and Onofre, S.B. 2007. Atividade antimicrobiana de óleos essenciais produzidos por *Baccharis dracunculifolia* D.C. e *Baccharis uncinella* D.C. (Asteraceae). Revista Brasileirade Farmacognosia, João Pessoa 17: 224–230.

Figueira, G.M., Montanari Junior, I., Magalhães, P.M., Pereira, B. and Archangelo Junior, U. 1991. Técnicas de cultivo de guaco (*Mikania glomerata*, Spreng.). XXXI Congresso Brasileiro de Olericultura. Belo Horizonte. Hort. bras. 9(1): 37.

Figueira, G.M., Pereira, B. and Magalhães, P.M. 1998. Estudos sobre a germinação de Quebra-Pedra. XV Simpósio de Plantas medicinais do Brasil, 02.025.

Foglio, M.A., Duarte, M.C.T., Belarmelina, C. and Silva, E.F. 2000 "Antimicrobial activity of sesquiterpenes isolated from *Artemisia annua* L.". 22º IUPAC International Symposiumon The Chemistry of Natural Products—Universidade Federal de São Carlos, SP.

Galvão, S.M.P., Dias, R.F., Espínola, E.B., Marques, L.C., Mattei, R. and Carlini, E.A. 1996. Avaliação farmacológica de plantas medicinais brasileiras com possível efeito adaptógeno—I. Estudos preliminares. XIV Simpósio de Plantas Medicinais do Brasil. Florianópolis-SC. F-209.

Gusman, G.S., Bittencourt, A.H.C. and Vestena, S. 2008. Alelopatia de *Baccharis dracunculifolia* DC. sobre a germinação e desenvolvimento de espécies cultivadas. Acta Sci. Biol. Sci. Maringá 30(2): 119–125.

Hussain, R., Dickey, J.K., Rosser, M.P., Matson, J.A., Kozlowski, M.R., Brittain, R.J., Webb, M.L., Rose, P.M. and Fernandes, P. 1995. A novel class of non-peptidicendothelin antagonists isolated from medicinal herb *Phyllanthus niruri*. J. Nat. Prod. 58(10): 1515–1520.

Ishimaru, K., Yoshimatsu, K., Yamakawa, T., Kamada, H. and Shimomura, K. 1992. Phenolic constituents in tissue cultures of *Phyllanthus niruri*. Phytochemistry 31(6): 2015–2018.

Jayaram, S., Thyagarajan, S.P., Sumati, S., Manjula, S., Malathi, S. and Madangopal, N. 1997. Efficacy of *Phyllanthus amarus* treatment in acute viral hepatitis A, B and non A non B: an open clinical trial. Indian J. Virol. 13(1): 59–64.

Kumar, N.G., Nair, A.M.C., Raghunandanan, V.R. and Rajagopalan, M.K. 1989. Hypoglycemic effect of *Phyllanthus niruri* leaves in rabbits. Kerala J. Vet. Sci. 20(1): 77–80.

Laughlin, J.C., Heazlewood, G.N. and Beattie, B.M. 2002. Cultivation of *Artemisia annua* L. pp. 159–193. *In*: C.W. Wright (ed.). Artemisia—Medicinal & Aromatic Plants-Industrial Profiles Series. University Brandford, UK, Taylor & Francis, London.

Limberger, R.P., Sobral, M. and Henriques, A. 2004. Óleos voláteis de espécies de *Myrcia* nativas do Rio Grande do Sul. Química Nova, São Paulo 27(6): 916–919.

Lopes, C.S., Carvalho, L.C.R.M., Pinto, A.C., Jansen, J.M. and Soares de Moura, R. 1997. Efecto de *Mikania glomerata* (Guaco) enla musculatura lisa respiratória. WOCMAP II. Mendoza-Argentina. P-325.

Macedo, C.S., Uhrig, M.L., Kimura, E.A. and Katzin, A.M. 2002. Characterization of the isoprenoid chain of coenzyme Q in *Plasmodium falciparum*. FEMS Microbiol. Lett. 207: 13–20.

Magalhães, P.M. 1993. A experimentação agrícola com plantas medicinais e aromáticas. Atualidades Científicas-CPQBA03, pp. 31–56.

Magalhães, P.M. 1997. O Caminho Medicinal das Plantas: aspectos sobre o cultivo. Ed. RZM. Campinas, SP, 120 p.

Magalhães, P.M., Figueira, G.M., Pereira, B. and Rodrigues, J.A. 1994. Propagação de algumas espécies do ginseng do Brasil. XIII. Simpósio de Plantas Medicinais do Brasil. Fortaleza, CE. P. 110.

Magalhães, P.M. de. 1996. Seleção, Melhoramento e Nutrição da *Artemisiaannua* L., para cultivo em região intertropical. Tese de doutorado. Instituto de Biologia, UNICAMP/Mediplant (Suíça), 117 p.

Magalhães, P.M. de and Pereira, B. 1996. Observações do desenvolvimento de *Pfaffia paniculata* em três composições de solo. XIV Simpósio de Plantas Medicinais do Brasil. Florianópolis-SC. A-12.

Magalhães, P.M. de, Delabays, N. and Sartoratto, A. 1997. New hybrid lines of antimalarial species *Artemisia annua* L. guarantee its growth in Brazil. Ciência e Cultura Journal of the Brazilian Association for the Advancement of Science 49(5/6): 413–415.

Magalhães, P.M. de, Pereira, B. and Sartoratto, A. 2001. Rendimentos da espécie antimalárica *Artemisia annua* L. X Congreso Ítalo-Latino americano de Etnomedicina. Islã Margaita—Venezuela, pp. 137–138.

Martius, C.F.P. von. 1826. Nova Generaet Species Plantarum. Monachii, Caroli Wolf. 2: 21–46.

Montanari, Jr. 1997. Influences of plantation density and cultivation cycle in root productivity and tenors of β-ecdisone in *Pfaffia glomerata* (Spreng.) Pedersen. Wocmap II. Mendoza.

Montanari Jr., Ilio. 2000. Cultivo comercial da erva-baleeira. Agroecologia Hoje 1(3): 14–15.

Moraes, M.D. de. 1997. A família *Asteraceae* na planície litorânea de Picimguaba Município de Ubatuba-São Paulo. Tese de Mestrado. UNICAMP-IB.

Mueller, M.S., Karhagomba, I.B., Hirt, H.M. and Wemalor, E. 2000. The potential of *Artemisia annua* L. as a locally produced remedy for malaria in the tropics: agricultural, chemical and clinical aspects. J. Etnopharmacol. 73: 487–493.

Nishimoto, N., Shiobara, Y., Fujino, M., Inoue, S.S., Takemoto, T., De Oliveira, F. and Matsuura, H. 1987. Ecdysteroids from *Pfaffia iresinoides* and reassignment of some ¹³CNMR chemical shifts. Phytochemistry 26(9): 2505–2507.

Nishimoto, N., Shiobara, Y., Inoue, S.-S., Fujino, M., Takemoto, T., Yeoh, C.L., Oliveira, F., Akisue, G., Akisue, M.K. and Hashimoto, G. 1988. Three ecdysteroid glycosides from *Pfaffia iresinoides*. Phytochemistry 27(6): 1665–1668.

Nopper, M.A. and Magalhães, P.M. 1993. Micropropagação de *Artemisia annua* visando obter clones de plantas com florescimento tardio. Revista Brasileira de Fisiologia Vegetal 5: 108.

Núñez, C.V. and Roque, N.F. 1997. Terpenos de *Mikania* sp. (Asteraceae). WOCMAP II. Mendoza-Argentina. P-164.

Ogata, T. et al. 1992. HIV-1 reverse transcriptase inhibitor from *Phyllanthusniruri*. AIDS Res. Hum. Retroviruses 8(11): 1937–1944.

Oliveira, D.M. 1996. Efeito dos extratos brutos de *Cephaelis ipecacuanha* e de *Mikania glomerata* na pleurisia induzida pelo Zimosan em ratos. XIV Simpósio de Plantas Medicinais do Brasil. F-005.

Oliveira, E. and Rand, A.M. 1996. Polimorfismo em sementes de algumas espécies de *Phyllanthus* (Euforbiaceae) e enraizamento de *P. niruri*. XIV Simpósio de Plantas Medicinais do Brasil. p. 48.

Oliveira, F., Saito, M.L. and Garcia, L.O. 1992a. Anatomia foliar das espécies brasileiras de *Mikania willdenow* seção *Globosae* Robinson—visão farmacognóstica. XII Simpósio de Plantas Medicinais do Brasil. P-179.

Oliveira, F., Saito, M.L. and Garcia, L.O. 1992b. Caracterização cromatográfica do extrato fluido de *Mikania glomerata* Sprengel. XII Simpósio de Plantas Medicinais do Brasil. P-096.

Pereira, A.M.S., Câmara, A.M.S. and França, S.C. 1997. Vegetative propagation in *Mikania glomerata*: Micropropagation and Cutting. WOCMAP II. Mendoza—Argentina. P-226.

Prakash, A., Satyan, K.S., Wahi, S.P. and Singh, R.P. 1995. Comparative hepatoprotective activity of three *Phyllanthus* species, *P. urinaria, P. niruri* and *P. simplex*, on carbon tetrachloride induced liver injury in rat. Phytother. Res. 9(8): 594–596.

Qian-Cutrone, J., Huang, S., Trimble, J., Li, H., Lin, P.F., Alam, M., Klohr, S.E. and Kadow, K.F. 1996. Niruside, a new HIV REV/RRE binding inhibitor from *Phyllanthus niruri*.J. Nat. Prod. 59: 196–199.

Queiroga, C.L., Cavalcante, M.Q., Ferraz, P.C., Coser, R.N., Sartoratto, A. and Magalhães, P.M. 2014. High-speed countercurrent chromatography as a tool to isolate nerolidol from the volatile oil. J. Essent. Oil Res. 1: 1–4.

Rae, G. et al. 1987. Avaliação farmacológica pré-clínica de extratos de *Phyllanthus niruri* em alguns modelos experimentais *in vivo*. SBPC, 17-G.1.7.

Rehder, V.L.G. 2002. Amplo espectro: O Guaco, planta nativa da mata atlântica, tem mais propriedades terapêuticas do que se supunha. Pesquisa FAPESP. Abril, pp. 43–45.

Rehder, V.L.G., Sartoratto, A., Magalhães, P.M., Figueira, G.M., Montanari Junior, I. and Lourenço, C. 1998. Variação fenológica do teor de cumarina em *Mikania laevigata* Schultz Bip., ex Baker. III Workshop de Plantas Medicinais de Botucatu, p. 26.

Ribeiro[a], P.G.F. and Pereira, E.F. 1994. Estudo da germinação de sementes de *Pfaffiaglomerata*. XIII Simpósio de Plantas Medicinais do Brasil, p. 207.

Ribeiro[b], P.G.F. and Pereira, E.F. 1994. Influência do método de propagação e tipos de solo na produção de raízes de Pfaffia (*Pfaffia glomerata*). XIII Simpósio de Plantas Medicinais do Brasil, p. 206.

Ruppelt, B.M., Pereira, E.F., Gonçalves, L.C. and Pereira, N.A. 1991. Pharmacological screening of plants recommended by folk medicine as anti-snake venon - I. Analgesic and anti-inflammatory activities. Mem. Inst. Oswaldo Cruz. 86(Suppl. II): 203–205.

Sane, R.T., Kuber, V.V., Chalissery, M.S. and Menon, S. 1995. Hepatoprotection by *Phyllanthus amarus* and *Phyllanthus debilis* in CCl$_4$-induced liver dysfunction. Curr. Sci. 68(12): 1243–1246.

Santos, A.R.S., Niero, R., Cechinel Filho, V., Yunes, R.A., Moneche, F.D., Calixto, J.B. and Santos, A.R.S. 1993a. Ações antinociceptivas de extratos de espécies de *Phyllanthus*. FESBE, p. 187.

Santos, A.R.S., Niero, R., Cechinel Filho, V., Yunes, R.A., Moneche, F.D. and Calixto, J.B. 1993b. Atividade analgésica de frações e compostos de *Phyllanthus corcovadensis*. FESBE, p. 176.

Santos, A.R.S., Cechinel Filho, V., Yunes, R.A. and Calixto, J.B. 1994a. Análise do mecanismo de ação analgésica de espécies de *Phyllanthus*. FESBE, p. 197.

Santos, A.R., Valdir filho, C., Niero, R., Viana, A.M., Moreno, F.N., Campos, M.M. and Calixto, J.B. 1994b. Analgesic effects of callus culture extracts from selected species of *Phyllanthus* in mice. J. Pharm. Pharmacol. 6(9): 755–759.

Santos, A.R., Valdir Filho, C., Yunes, R.A. and Calixto, J.B. 1995. Analysis of the mechanisms underlying the antinociceptive effect of extracts of plants from genus *Phyllanthus*. Gen. Pharmac. 26(7): 1499–1506.

Santos, T.C. dos. 1992. *Mikania glomerata*, Srengel—Estudo químico de seus componentes. XII Simpósio de Plantas Medicinais do Brasil. P-149.

Santos, T.C., Cabral, L.M. and Tomassini, T.C.B. 1996a. Contribuição para o estudo fitoquímico de *Mikania glomerata* Sprengel. XIV Simpósio de Plantas Medicinais do Brasil. Q-034.

Santos, T.C., Tomassini, T.C.B., Sanchez, E. and Cabral, L.M. 1996b. Estudo da atividade antimicrobiana da *Mikaniaglomerata* Sprengel. XIV Simpósio de Plantas Medicinais do Brasil. F-173.

Scheffer, M.C., Rodrigues, C., Bello, M. and Doni Filho, L. 1997. Germinación y almacenamiento de semillas de guaco (*Mikania glomerata* Spreng.—*Asteraceae*). WOCMAP II. Mendoza-Argentina. P-084.

Shiobara, Y., Inoue, S.-S., Nishiguchi, Y., Takemoto, T., Nishimoto, N., Oliveira, F. and Akisue, G. 1992. Iresinoide, a yellow pigment from *Pfaffiairesinoides*. Phytochemistry 31(3): 953–956.

Shiobara, Y., Inoue, S.-S., Kato, K., Nishiguchi, Y., Nishimoto, N., de Oliveira, F. and Hashimoto, G. 1993. Pfaffane-type nortriterpenoids from *P. pulverulenta*. Phytochemistry 33(4): 897–899.

Simonnet, X., Gaudin, M., Hausammann, H. and Vergeres, Ch. 2001. Le fanage au champ d'*Artemisia annua* L.: élever la teneur en artémisinine et abaisser les coûts de production. Revue Suisse Vitic. Arboric. Hortic. 33(5): 263–266.

Siqueira, J.C. 1988. Considerações taxonômicas sobre as espécies brasileiras do gênero *Pfaffia*. Acta Biológica Leopoldensia 10(2): 269–278.

Smith, L.B. and Downs, R.J. 1972. Amaranthaceas de Santa Catarina. Flora Ilustrada Catarinense. Itajaí, SC., HBR, pp. 35–50.

Soares de Moura, R., Carvalho, L.C.R.M., Lopes, C.S. and Vieira de Souza, M.A. 1996. Efeitos do Guaco (*Mikaniag lomerata*) sobre o edema podal em camundongos induzido pelo veneno da *Bothrops jararaca*. XIV Simpósio de Plantas Medicinais do Brasil. F-025.

Srinivasulu, C. 1992. Phyllanthin and hypophyllanthin lack anticancer activity. Indian J. Pharm. Sci. 54(6): 253–254.

Srividya, N.A. and Periwal, S. 1995. Diuretic, hypotensive and hypoglycaemic effect of *Phyllanthus amarus*. Indian journal of experimental biology 33: 11, 861–864.

Suyenaga, E.S., Limberger, R., Menut, C., Chaves, C.G. and Henriques, A.T. 1996. Composição química de óleos essenciais de três espécies de *Mikania* de ocorrência no sul do Brasil. XIV Simpósio de Plantas Medicinais do Brasil. Q-010.

Unander, D.W., Bryan, H.H., Lance, C.J. and Mcmillan, R.T. 1995. Factors affecting germination and stand establishment of *Phyllanthus amarus* (Euphorbiaceae). Econ. Bot. 49(1): 49–55.

Vilegas, J.H., de Marchi, E. and Lanças, F.M. 1997a. Determination of coumarin and kaurenoic acid in *Mikania glomerata* (guaco) leaves by capillary gas chromatography. Phytochem. Anal. 8(2): 74–77.

Vilegas, J.H.Y., De Marchi, E. and Lancas, F.M. 1998b. Extraction of low-polarity compounds (with emphasis on coumarin and kaurenoic acid) from *Mikania glomerata* ('guaco') leaves. Phytochem. Anal. 8(5): 266–270.

Wang, M., Cheng, H., Meng, L., Zhao, G. and Mai, K. 1995. Herbs of the genus *Phyllanthus* in the treatment of chronic hepatitis B: Observations with three preparations from different geographic sites. J. Lab. Clin. Med. 126(4): 350–352.

Zheng, G.Q., Kenney, P.M. and Lam, L.K.T. 1992. Sesquiterpenes from clove (*Eugeniacaryophyllata*) as potential anticarcinogenic agents. J. Nat. Prod. 55: 999–1003.

Zhou-Shiwen, Xu-Chuanfu, Zhou-Ning, Huang-Yongping, Huang-Linqing, Chen-Xiaohong, Hu-Youmei and Liao-Yaqin. 1997. Mechanism of protective action of *Phyllanthus urinaria* L. against injuries of liver cells. ZhongguoZhongyaoZazhi 22(2): 109–111, 129.

Essential Oils and their Products as Antimicrobial Agents: Progress and Prospects

Janne Rojas* and Alexis Buitrago

Introduction

Plants produce a great variety of secondary metabolites that are used for protection against predators and to attract the attention of pollinators. Among these secondary metabolites, essential oils constitute a significant group; are defined as complex mixtures of terpenes, terpenoids, aromatic and aliphatic compounds biosynthesized by plants and all are characterized by possessing low molecular weight (Guenther 1950).

The word *essential oil* was defined by Paracelsus von Hohenheim, for the first time, in the 16th century, referring to it as *Quinta essentia* (Pichersky et al. 2006). However, the Ebers papyrus is the first document that describes the use of essential oils for therapeutic reasons.

Essential oils are a mixture of volatile constituents produced by the secondary metabolism of aromatic and other variety of plants (Bassolé et al. 2012). Generally, these compounds have a lower density than that of water (Bakkali et al. 2008) and although only represent a small fraction of plant's composition; they confer the characteristics by which aromatic plants are used in food, cosmetics and pharmaceutical industries (Pourmortazavi et al. 2007).

Nowadays, there are approximately 17,500 aromatic plant species and about 3,000 essential oils are known out of which 300 are commercially important for pharmaceuticals, cosmetics and perfume industries (Bakkali et al. 2008) apart from pesticides (Bajpai et al. 2011, Chang and Cheng 2002). The ability of plants to accumulate essential oils is quite high in both Gymnosperms and Angiosperms, although the most commercially important essential oil plant sources are related to Angiosperms (Anwar et al. 2009a, Hussain et al. 2008, Celiktas et al. 2007). However, a limited number of families are considered large producers of essential oils. These include: Myrtaceae, Lauraceae, Rutaceae, Lamiaceae, Asteraceae, Apiaceae, Cupressaceae, Poaceae, Zingiberaceae and Piperaceae (Tripathi et al. 2009, Hussain et al. 2008, Cava et al. 2007, Sood et al. 2006). Currently, essential oils are still widely used in traditional medicine for their multiple activities, including antimicrobial effects. Thus, a number of investigations are continuously carried out to learn more about the chemistry and biological potential of these types of components.

Organic Biomolecular Research Group, Research Institute, Faculty of Pharmacy and Bioanalysis. University of Los Andes. Mérida 5101, Venezuela.
* Corresponding author: janner@ula.ve

The present chapter aims to summarize the most important aspects of essential oils, their potential as antimicrobials and to describe some of the market products commercialized and used as traditional medicines.

Role of essential oils in nature

Many years of investigation have revealed that essential oils play an important role in plants. Attraction of different insects in order to promote the dispersion of pollen and seeds, allopathic communication between plants (Bakkali et al. 2008), feeding deterrent and repelling effects against herbivores, are among the mechanisms plants use for survival. Thus, the detection of some of biological properties needed for the survival of plants has also been the base for searching similar properties for the combat of several microorganisms responsible for some infectious diseases in humans and animals (Bajpai et al. 2011).

On the other hand, essential oils may be synthesized by all plant organs (flowers, buds, seeds, leaves, twigs, barks, herbs, woods, fruits and roots) and are mainly stored in secretory cells, cavities, canals, epidermic cells or glandular trichomes (Miguel 2010, Bakkali et al. 2008, Burt 2007, Burt 2004). Moreover, volatile components of essential oils can be classified into four main groups: terpenes, benzene derivatives, hydrocarbons and other miscellaneous compounds (Hussain et al. 2008, Cava et al. 2007, Sood et al. 2006).

Biosynthesis of chemical compounds present in essential oils

Different biosynthetic pathways are used by plants to synthesize terpenes and phenylpropane type of components which are the main constituents of essential oils (Bassolé et al. 2012). The biosynthesis begins with the formation of a molecule of isopentenyl diphosphate. Two molecules of isopentenyl diphosphate form a prenyldiphosphate unit, precursor of various classes of terpenes. Modification of allyliprenyl diphosphate by terpene specific synthetases forms the terpene skeleton and finally, secondary enzymatic modification of the skeleton (redox reaction) attributes the functional properties of these secondary metabolites (Fig. 13.1) (Bajpai et al. 2011, Bakkali et al. 2008, Dewick 2002a, Dewick 2002b, Marcano and Hasegawa 2002).

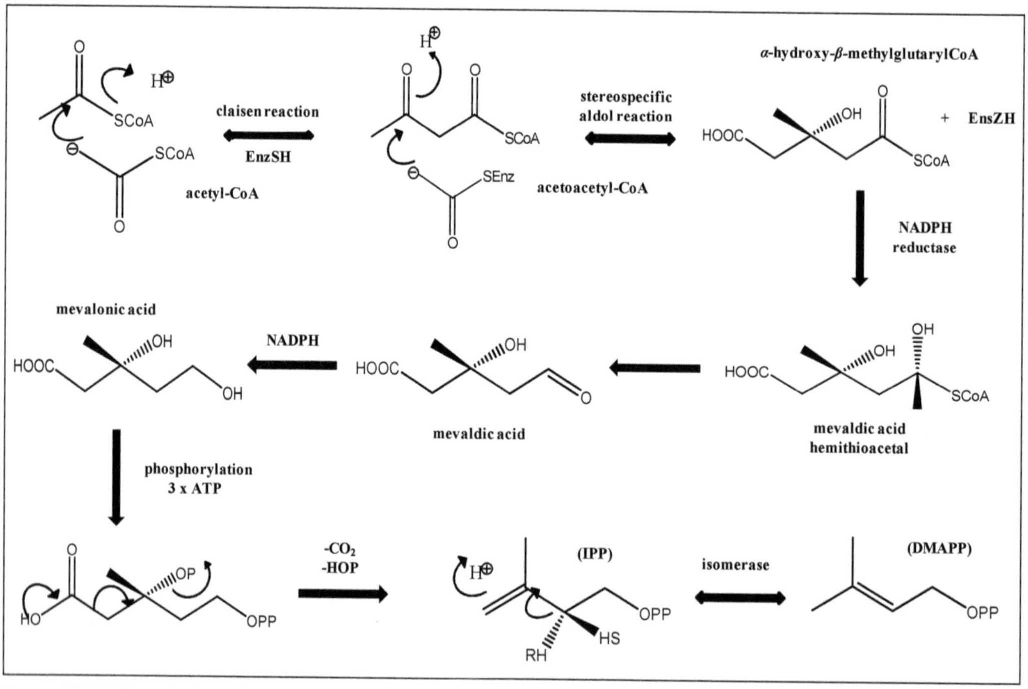

Figure 13.1. Biosynthetic pathway of isopentenyl diphosphate (**IPP**) and dimethylallyl diphosphate (**DMAPP**) (Dewick 2002a).

Terpenes, present in essential oils, are made from combinations of isoprene units and are classified according to the number of these units in monoterpenes (C10, two isoprene units), sesquiterpenes (C15, three isoprene units) and diterpenes (C20, four isoprene units), although hemiterpenes (C5, one isoprene unit), sesterpenes (C25, five isoprene units), triterpenes (C30, six isoprene units) and tetraterpenes (C40, eight isoprene units) are also part of terpene type components even though they are not commonly isolated as part of essential oils (Bajpai et al. 2011, Tripathi et al. 2009). A terpene containing oxygen is called a terpenoid (Fig. 13.2) (Bakkali et al. 2008).

Monoterpenes are the most representative molecules constituting 90% of the essential oils. These allow a variety of structures and may consist of several functions (Tripathi et al. 2009, Bakkali et al. 2008). In case of sesquiterpenes, the extension of the chain increases the number of cyclizations, which also allows a great variety of structures (Bajpai et al. 2011, Adams 2007).

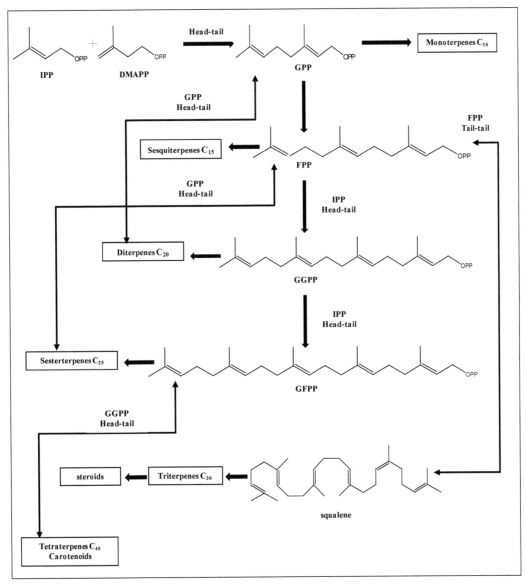

Figure 13.2. Biosynthesis of terpenes (Marcano and Hasegawa 2002, Dewick 2002a). **IPP:** isopentenyl diphosphate, **DMAPP:** dimethylallyl diphosphate, **GPP:** geranyl diphosphate, **FPP:** farnesyl diphosphate, **GGPP:** geranyl geranyl diphosphate, **GFPP:** geranyl farnesyl diphosphate.

On the other hand, the aromatic compounds derived from phenylpropane, occur less frequently than terpenes. Although biosynthetic pathways concerning terpenes and phenylpropanic derivatives generally are separated in plants. These may coexist in some, with one major pathway taking over. Aromatic compounds including aldehyde, alcohol, phenols, methoxy derivatives and ethylene dioxy compounds are derived by the shikimate pathway that provides an alternative route to these types of components (Bajpai et al. 2011, Bakkali et al. 2008). A central intermediate in the pathway is shikimic acid from which a number of derivatives are formed (Fig. 13.3) (Dewick 2002a).

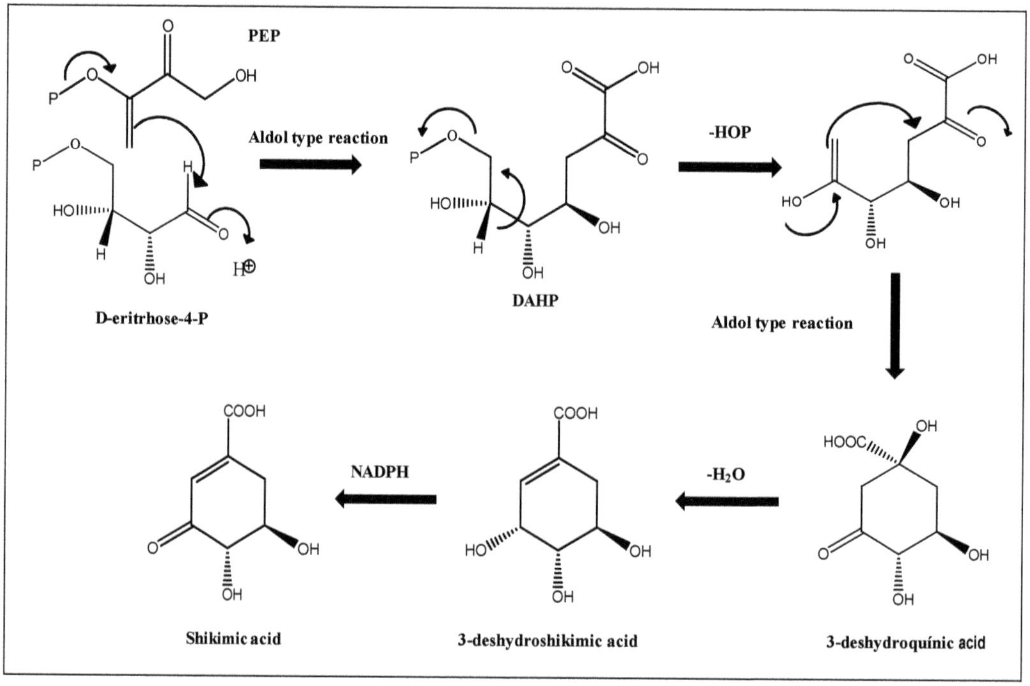

Figure 13.3. Biosynthesis of Shikimic acid, precursor of some aromatic compounds present in essential oils (Dewick 2002a). **PEP:** phosphoenolpyruvate, **DAHP:** 3-deoxy-D-arabino-heptulosonic acid 7-phosphate, **NADPH:** nicotinamide adenine dinucleotide phosphate (reduced).

In addition, several investigations have been conducted to obtain more detailed information about the biosynthetic pathways of terpenoids and phenylpropanoids, in order to reveal the enzymes and enzyme mechanisms involved in the process and to learn more about genes encoding for these enzymes (Dewick 2002a, Dewick 2002b). The interest in this topic is based on the idea that genetic engineering of metabolic pathways might be promising for improving the production of essential oils. On this matter a review article was published showing several results of different authors concerning the production of volatile metabolites by transgenic microorganisms and genetically engineered plants (Gounaris 2010). For that purpose, bacteria, yeasts and plants have been genetically altered either for the production of terpenoids or shikimic acid-derived volatiles. Although, some authors have concluded that this type of approach could be used successfully to generate noticeable levels of a particular group of terpenoids, engineering of some classes of this group of compounds is quite difficult since owning the pool of terpenoid precursors may not be sufficient for the production of substantial quantities of the desired compound (Asaph et al. 2006).

Factors affecting quality and quantity of essential oils

There are numerous factors that determine the composition and yield of essential oils (Terblanche 2000). These variables may include seasonal and maturity variation, geographical origin, genetic variation, growth stages, part of plant utilized, extraction methods and post-harvest drying and storage (Anwar et al. 2009a, Tripathi et al. 2009, Hussain et al. 2008, Terblanche 2000).

Seasonal and maturity variations

Essential oil yields may vary considerably depending on the time of the year where the plant is collected and is also influenced by the environmental conditions at the harvesting time. Also the place where the plant is growing (sun or shade) affects the composition and yield of the oil (Juliani et al. 2002). There are many reports in literature regarding the variation in the chemical profile of essential oils from a variety of plants collected during different seasons (Hussain et al. 2008, Celiktas et al. 2007, Van Vuuren et al. 2007, Angioni et al. 2006, McGimpsey et al. 1994).

A good example of this variation was presented by (McGimpsey et al. 1994). They reported seasonal variation in the essential oil composition and yield of *Thymus vulgaris.* In this investigation a higher yield was observed in spring; regarding the composition, thymol, observed as the major compound, was found to vary from 31.5 to 52.4% (Atti-Santos et al. 2004). With respect to the harvesting time, in the same study essential oils were richer in oxygenated compounds in the spring, followed by summer, autumn and winter (Atti-Santos et al. 2004). The maximum yields and phenol content of the oil were exhibited from crop harvested after flowering in December (McGimpsey et al. 1994).

On the other hand, many reports are available in the literature regarding the variation in the yield and quality of essential oils with respect to maturity stages (Anwar et al. 2009a, Yildirim et al. 2004). A study of *Tagetes minuta*, before flowering, reported higher oil yields compared to the same plant collected at full flowering stages (Skoula and Harborne 2002, Chalchat et al. 1995). *Origanum vulgare* ssp. *hirtum* is known to produce less essential oil during the cool and wet vegetative period, and more oil during the warmer and drier flowering stage. After flowering, the oil yield decreases, as the leaves age and become drier. In addition, thymol and/or carvacrol contents decrease during autumn. Mohammadreza et al. 2008 and Skoula and Harborne 2002; reported the variation in the quality and quantity of the essential oil obtained from *Artemisia annua* at different developmental growth stages including pre-flowering, flowering and post-flowering.

Lippia origanoides, collected in two different seasons (dry and rainy) from same location in Mérida, Venezuela showed several differences in the composition of the essential oil. The major compound identified as thymol was observed in higher concentrations in the rainy season (61.9%) compared to (44.7%) in the oil collected in dry season. The concentration of carvacrol also varied considerably (7.9%, rainy season and 16.8%, dry season). On the other hand, linalool was only observed in the sample collected in the rainy season while carvacryl acetate and caryophyllene oxide were only present in the oil obtained in the dry season (Rojas et al. 2006).

Geographical conditions

Variation in the yield and chemical composition of the essential oil with respect to geographical regions has also been reported (Celiktas et al. 2007, Van Vuuren et al. 2007). Viljoen et al. 2006, described variations in the yield and chemical profile of essential oils from *Mentha longifolia* (L.) and *Tagetes minuta* populations, collected from different geographical locations. Such differences could be linked to the varied soil textures and possible adaptation response of different populations, resulting in different chemical products being formed, without morphological differences being observed in the plants (Hussain et al. 2008).

Altitude seems to be another important environmental factor influencing the essential oil content and chemical composition (Vokou et al. 1993). A good example is a study on the composition of the essential oil from leaves of *Vismia baccifera* var. *dealbata* collected from two different locations, Chiguará (1250 m) and La Hechicera (1800 m), Mérida-Venezuela. Several differences were observed: germacrene-D (15.8%), α-cadinol (14.5%), epi-α-cadinol (11.9%), β-caryophyllene (10.1%) and δ-cadinene (7.5%) were the major constituents in the Chiguará sample while β-caryophyllene (45.7%), valencene (12.3%), β-elemene (10.7%), α-humulene (8.9%) and germacrene-D (6.3%) were observed as major components in the La Hechicera sample, components such as viridiflorene (4.5%), germacrene-A (2.5%), germacrene-B

(2.4%), α-muurolol (2.1%), khusimonene (2.1%), γ-muurolene (2.1%) and γ-cadinene (2.0%) were present in the Chiguará sample but absent in the La Hechicera oil, whereas, nerolidol (1.2%) and selinelol (1.2%) were observed only in the La Hechicera sample (Buitrago et al. 2009).

Genetic variation

Genotype is usually defined as the genetic make-up of an organism, as characterized by its physical appearance or phenotype, while chemotype is defined as a group of organisms that produce the same chemical profile for a particular class of secondary metabolites. Genetic variations have been observed in several studies carried out with oils produced from specimens collected from the same population and location, thus demonstrating the presence of different chemotypes within the same species (Contreras et al. 2014, Ahmad et al. 2006, Wink 2003). An investigation carried out by (Galambosi and Peura 1996) also showed significant differences between the oil composition and yields of wild and cultivated caraway types in a population grown under the same conditions.

Other factors affecting yield and composition of essential oil

There are a number of factors that might affect the yield and composition of essential oils. Several investigations have revealed that the part of the plant used (Santos-Gomes et al. 2001); post-harvest drying (Skoula and Harborne 2002); length of exposure to sunlight (Clark et al. 1979), availability of water, height above sea level (Galambosi and Peura 1996), plant density (Graven et al. 1990), time of sowing (Galambosi and Peura 1996) and the presence of fungal diseases and insects (Margina and Zheljazkov 1994), have a strong influence in the production of essential oil in the plant.

In addition, the oil composition and yield may also change as a result of the harvesting methods used (Bonnardeaux 1992), the isolation techniques employed (Moates and Reynolds 1991), the moisture content of the plants at the time of harvest (Burbott and Loomis 1957) and the prevailing steam distillation conditions.

There are different opinions regarding post-harvest drying of material, using methods that include exposure to natural air in the shade, sun-drying, as well as drying by blowing warm air over the material before extracting the essential oil. Some people believe this practice improves oil yield and accelerates distillation, by improving heat transfer and providing increased loading capacity, due to loss of plant moisture. Furthermore, the reduction of microbial growth and the inhibition of some biochemical reactions in dried material is also consider as an advantage of this technique (Hernandez-Arteseros et al. 2003). However, some amount of the oil may be lost during such post-harvest treatment due to volatilization and mechanical damage to oil glands (Skoula and Harborne 2002).

In some plants, such as *Eucalyptus* species and tea tree (*Melaleuca alternifolia*), post-harvest drying, with leaves kept intact on the branches, resulted in an increased oil yield. This has been shown to be a result of continued post-harvest metabolic activity within the plant, perhaps during which oil is moved from the stem to the leaves (Whish and Williams 1996). For the production of peppermint oil, *Mentha piperita*, leaves are commonly cut and allowed to partially dry (35% moisture) in the field (Mactavish and Harris 2002), however, this practice is known to affect the chemical composition of the oil since chemical transformation in its monoterpenoids occur when leaves are allowed to dry (Asekun et al. 2007). Another report found no significant differences in the oil yield of fresh bay leaves (*Laurus nobilis*) compared to that of leaves dried at 40, 50 and 60°C, as well as in sunny and shady conditions, however, the oil composition of the dried material differed significantly from that of the fresh material (Demir et al. 2004).

The composition of essential oils from different parts of the same plant can also vary widely. For example, oil obtained from the seeds of coriander (*Coriandrum sativum* L.) has a quite different composition to the oil of cilantro, which is obtained from the immature leaves of the same plant (Delaquis et al. 2002). Regarding, antibacterial activity, some authors believe that essential oils produced from herbs harvested during or immediately after flowering, might possess the strongest antimicrobial activity (Marino et al. 1999).

Extraction methods commonly used to isolate essential oils

Composition of oil varies to a large extent depending on the isolation method used. The chemical profile of the essential oil products differs not only in the number of molecules but also in the stereochemical types of molecules extracted (Tripathi et al. 2009). Steam distillation and hydrodistillation are the procedures most frequently used to isolate essential oils by Clevenger-type apparatus from the botanical material. Although, enfleurage, solvent extraction, supercritical fluid extraction and microwave assisted extraction are also methods widely utilized for the extraction of essential oils (Tripathi et al. 2009, Abad et al. 2007, Masango 2004, Augusto et al. 2003).

Steam distillation

This method enables a mixture of compounds to be distilled at a temperature substantially below boiling point of individual constituents, since in the presence of steam; these substances are volatilized at a temperature lower than 100°C, at atmospheric pressure. The mixture of hot vapors will pass through a cooling system that allows condensing to form a liquid in which the oil and water comprise two distinct layers. Knowing that most essential oils are lighter than water, these are placed on the top layer and therefore will be easy to separate (Donelian et al. 2009).

The disadvantage of this process is that the hydrolyzable compounds such as esters, as well as thermally labile components, may be decomposed during the distillation process (Houghton and Raman 1998). In addition, partial loss of more polar constituents of the oil, due to their affinity for water, may also occur (Masango 2004). Although steam distillation is very popular for the isolation of essential oils on commercial scale and 93% of the oils are produced by this process, it is not a preferred method in research laboratories (Masango 2004). Most studies which focus on the essential oil of herbs, use the hydrodistillation method in a Clevenger-type apparatus (Rojas et al. 2011a, Buitrago et al. 2011, Rojas et al. 2011b, Rojas et al. 2010, Hussain et al. 2008, Sokovic and Van Griensven 2006).

Hydrodistillation

Hydrodistillation is quite similar to steam distillation but in the hydrodistillation procedure, the material is immersed in water, which is heated to boiling point using an external heat source. Care must be taken to ensure efficient condensation of steam, thereby preventing the loss of the more volatile oil components. The recovery of the oil is similar to the steam distillation where the vapors are allowed to condense and the oil is then separated from the aqueous phase (Houghton and Raman 1998).

Some authors affirm that hydrodistillation produces a finer, more complete product than steam distillation (Ackerman 2001), however, it has been suggested that steam distillation is more efficient in removing oil from plant material (Charles and Simon 1990). The authors have stated that steam distillation gave higher yields of oil, although the same compounds were obtained (Khanavi et al. 2004). Nevertheless, hydrodistillation has been accepted as a simpler and more rapid method for oil isolation and better yields has been reported by using this method (Sefidkon et al. 2007, Ackerman 2001).

The duration of this process is also an important parameter that might have some influence on the yield and composition of essential oils (Masango 2004). Some authors believe that long distillation cycles may influence the composition of the oil obtained. Oxygenated compounds are released earlier from intact plant material than non-oxygenated compounds (Baker et al. 2000). This might be due to the fact that boiling water (steam) penetrates the oil glands dissolving part of the oil present in it. Once released from the glands, the oil components are immediately vaporized. Since polar oxygenated compounds are more water-soluble than non-oxygenated compounds they diffuse faster and are distilled first (Baker et al. 2000).

On the other hand, sesquiterpenes, probably because of their large molecular size, tend to distill later than oxygenated monoterpenes. It is important to consider that sabinene, *cis*-sabinene hydrate and *trans*-sabinene hydrate are known to undergo thermal transformation to terpinen-4-ol, α-terpinene, γ-terpinene and terpinolene, during hydro and steam distillation (Baker et al. 2000).

Enfleurage

For this method, highly purified and odorless vegetable or animal fat is placed on top of the plant material that is usually composed of petals and flowers freshly picked just before the extraction goes. Glass plates in a frame are used to place the whole material and then pressure is applied. Hereafter, essential oil is extracted by using solvents such as petroleum ether, ethanol or hexane. Finally, solvents are distilled leaving a waxy aromatic mixture which is used for aromatherapy (Baker et al. 2000).

Simultaneous distillation-extraction

For this technique an amount of sample is placed along with distilled water in a container; a suitable volume of an extracting solvent denser than water (dichloromethane, chloroform) is placed in another separated container, both devices are heated and vapors from water and solvent pass through the material carrying the essential oil present in this, then vapors are allowed to condense over the surface of a cold tube. Both, water and solvent are collected after their condensation and return to the corresponding flasks, allowing continuous reflux. A modified apparatus allows the use of solvents lighter than water, such as pentane or ethyl acetate, as extractors (Augusto et al. 2003).

Supercritical fluid extraction

This method is based on the fact that, near the critical point of the solvent, its properties change rapidly with only slight variations of pressure. Thus, supercritical fluids can be used to extract analytes from samples. Carbon dioxide (CO_2) is the reagent widely used as the supercritical solvent, since it diffuses very easily, has a low viscosity, low surface tension and extracts analytes faster and more environmentally friendly than organic solvents. Besides, CO_2 becomes hypercritical at 33 degrees Celsius; in this state it is neither gas nor liquid, but has qualities of both, and is an excellent solvent to use in the extraction of essential oils. Thus, is inert and therefore does not chemically interact with the essence that is being extracted. To remove the solvent is easy, one just needs to remove the pressure under which it is kept. The disadvantage of this method is the special equipment required because this process has to take place in a closed chamber for the hypercritical pressure required (200 atmospheres for CO_2); this is 200 times the pressure of normal atmosphere (Augusto et al. 2003).

On the other hand, several authors have compared the composition of essential oils obtained by hydro/steam distillation and the product obtained by super critical fluid extraction. They found that hydro/steam-distilled oil contained higher percentages of terpene hydrocarbons. In contrast, the super critical extracted oil contained a higher percentage of oxygenate compounds (Donelian et al. 2009).

Microwave assisted extraction

In case of microwave assisted extraction, heating occurs in a targeted and selective manner with practically no heat being lost to the environment as the heating occurs in a closed system. This unique heating mechanism can significantly reduce the extraction time (usually less than 30 minutes) as compared to Soxhlet (Huie 2002). The principle of heating using a microwave is based on its direct impact with polar materials/solvents and is governed by two phenomena: ionic conduction and dipole rotation, which in most cases occurs simultaneously (Letellier et al. 1999). Ionic conduction refers to the electrophoretic migration of ions under the influence of the changing electric field. The resistance offered by the solution to the migration of ions generates friction, which eventually heats up the solution (Mandal et al. 2007).

Plant cells contain minute microscopic traces of moisture that serve as the target for microwave heating. The moisture, heated up inside the plant cell due to the microwave effect, evaporates and generates pressure on the cell wall (Wang and Weller 2006). This pressure pushes the cell wall from inside, stretching and ultimately rupturing it, which facilitates leaching out of the active constituents. This phenomenon can even be more intensified if the plant matrix is impregnated with solvents with higher heating efficiency under a microwave (Wang and Weller 2006).

In terms of time, the microwave assisted extraction exhibit a considerable reduction of time (15 minutes) and solvent consumption (10 mL) comparing to hydrodistillation/steam distillation techniques that require an extraction time of 270 minutes for heating 6 kg of water and 500 g of plant material (Lucchesi et al. 2004). Although the supercritical fluid extraction method only requires about 30 minutes for heating the plant and evaporation of the *in situ* water and essential oil, still requires double the time than microwave extraction, so this is consider an advantage of this method (Talebi et al. 2004).

Reduction in extraction time has been reported by several authors using the microwave assisted extraction compared to conventional techniques (Zhou and Liu 2006, Zhang et al. 2005). However, several differences have been noticed in the composition of essential oils extracted with hydrodistillation and microwave extraction (Silva et al. 2004).

Characterization of essential oils

Capillary Gas Chromatography (GC) with Flame Ionization Detection (FID) is the method of choice for quantitative determinations. However, many researchers use Mass Spectrometers (MS), coupled with GC, to determine the identities of components (Rojas et al. 2013a, Rojas et al. 2013b, Buitrago et al. 2011, Rojas et al. 2011a, Rojas et al. 2011b, Rojas et al. 2010, Anwar et al. 2009b, Hussain et al. 2008, Cavaleiro et al. 2004). Alternatively, Kovats indices, determined by co-injection of the oil with a homologous series of *n*-alkanes, are also widely used to identify compounds, where authentic standards are not available (Burt 2007, Abad et al. 2007, Juliani et al. 2002).

The capillary columns frequently used are: HP-5ms, DB-5 (cross-linked 5% diphenyl/95% dimethyl siloxane) or DB-1, also known as SE-30 (polydimethyl siloxane). These non-polar stationary phases are often complimented by the use of a more polar stationary phase, such as polyethylene glycol (Cavaleiro et al. 2004).

The complexity of essential oils makes the quantification of their components difficult. There are at least four widely used approaches: relative percentage abundance, internal standard normalized percentage abundance, 'absolute' or true quantification of one or more components using internal and/or external standards, and quantification by a validated method. These methods have been explained in the literature (Tadeg et al. 2005).

Importance of essential oils for the treatment of several human diseases

Essential oils are widely used to prevent and treat a wide range of human diseases, including the prevention and treatment of human cancer such as glioma, colon cancer, gastric cancer, human liver tumor, pulmonary tumors, breast cancer and leukemia (Lin et al. 2012, De Angelis 2001). Furthermore, cardiovascular diseases, atherosclerosis and thrombosis, as well as antibacterial, antifungal, antiviral, antioxidants and antidiabetic properties have also been attributed to essential oils (Vizcaya et al. 2014, Araujo et al. 2013, Morales et al. 2013, Torres et al. 2013, Edris 2007).

Regarding anticancer activity, a number of investigations have been conducted to evaluate this property. Nutmeg (*Myristica fragrans*) has shown potent hepatoprotective activity that might be correlated to myristicin, the major constituent (Morita et al. 2003). Citral, a monoterpene present in many essential oils, has been investigated as a cancer chemopreventive agent towards inflammation-related carcinogenesis such as skin cancer (Henderson et al. 1998) and colon cancer (Edris 2007, Mulder et al. 1995).

The presence of phenolic and monoterpene compounds might be related to the hepatoprotective activity in essential oils, for instance, thymoquinone present in black cumin (*Nigella sativa*) (Mansour et al. 2001); *d*-limonene in orange (*Citrus sinencis*) essential oil (Bodake et al. 2002); *d*-limonene and β-myrcene in sweet fennel (*Foeniculum vulgare*) (Ozbek et al. 2003). In addition, *d*-limonene has shown chemopreventive efficacy in preclinical hepatocellular carcinoma models (Parija and Das 2003) and antiangiogenic and proapoptotic effects on human gastric cancer implanted in nude mice, thus inhibiting tumor growth and metastasis (Guang et al. 2004). Garlic essential oil is a rich source of volatile organosulfur components also recognized as a group of potential cancer chemopreventive agents (Edris 2007, Chen et al. 2004).

The sesquiterpene α–bisabolol, present in Chamomile (*Matricaria chamomilla*) essential oil, could be considered as a possible apoptosis inducer in malignant glioma cells (Cavalieri et al. 2004). Other investigations have revealed that elemene, which is found in many essential oils, could possibly be effective in the treatment of glioma (Tan et al. 2000). Eucalyptol has also shown induction of apoptosis on HL-60 cells indicating that may be used for the treatment of human leukemia (Moteki et al. 2002). Other investigations *in vitro* have revealed that lemon balm (*Melissa officinalis* L.) essential oil is effective against a series of human cancer cell lines (A549, MCF-7, Caco-2, HL-60, K562) (Edris 2007, De Sousa et al. 2004).

Another important and interesting activity observed in essential oils is, antioxidant, since these are natural sources of phenolic components. Oxidative stress is a phenomena produced when an imbalance between free radical production on human body and their removal by the antioxidant system is produced (Abdollahi et al. 2004). It is well known that free radicals and other reactive oxygen species cause oxidation of proteins, amino acids, unsaturated lipids and DNA, producing molecular alterations related to aging, atherosclerosis, cancer (Gardner 1997), Alzheimer's disease (Butterfield and Lauderback 2002), Parkinson's disease, diabetes and asthma (Zarkovic 2003).

The essential oils of *Thymus serpyllus* and *Thymus spathulifolius* have shown free radical scavenging activity close to that of butylated hydroxytoluene (BHT); this activity was attributed to the high content of thymol and carvacrol (Tepe et al. 2005, Sokmen et al. 2004). *Origanum vulgare* essential oil also showed antioxidant activity compared to that of α-tocopherol and BHT, but less effective than ascorbic acid (Kulisic et al. 2004). The activity was also attributed to the high content of thymol and carvacrol (Botsoglou et al. 2004). Other essential oils, such as, *Salvia cryptantha*, *Salvia multicaulis* (Tepe et al. 2004), *Achillea millefolium* subsp. *Millefolium* (Candan et al. 2003), *Curcuma zedoaria* (Mau et al. 2003), *Mentha aquatic*, *Mentha longifolia* and *Mentha piperita* have also been proven to reduce DPPH radicals. In case of *M. aquatic*, 1,8-cineole was the major component present on the oil, while for *M. longifolia* and *M. piperita*, menthone and isomenthone were present in higher concentrations (Mimica-Dukic et al. 2003). *Melaleuca alternifolia* oil has been suggested as a natural antioxidant alternative for BHT (Kim et al. 2004); the antioxidant activity is mainly attributed to the α-terpinene, γ-terpinene and α-terpinolene content. It is clear that essential oils may be considered as potential natural antioxidants and could perhaps be formulated as a part of daily supplements or additives to prevent oxidative stress that may well contribute with several degenerative diseases (Edris 2007).

Atherosclerosis ailment, is a process in which deposits of plaque accumulates in the artery's inner layer. This plaque is composed by low density lipoproteins (LDLs) under cholesterol form and can eventually reduce blood flow causing serious health problems (Barter 2005). However, atherosclerosis can be inhibited by preventing the oxidation of LDLs using a daily intake of antioxidants. Essential oils and their aroma volatile constituents, especially terpinolene, eugenol and γ-terpinene (Takahashi et al. 2003) have shown antioxidant activity, thus, can effectively inhibit the oxidation of both the lipid and the protein part of LDL (Naderi et al. 2004). Additionally, thrombosis and several cardio-circulatory disorders may also be prevented by the use of natural products. The essential oil of lavender (*Lavandula hybrida*) composed by about 36% of linalyl acetate, showed a broad spectrum antiplatelet effect and was able to inhibit platelet aggregation with no prohemorrhagic properties (Ballabenia et al. 2004). Onion (*Allium cepa*) is also well known for promoting cardiovascular health and many people use this vegetable to avoid atherosclerosis and thrombotic diseases (Kendler 1987). Allicin and ajoene are organosulfur components isolated from garlic essential oil; both have shown potent inhibition of platelet aggregation (Calvey et al. 1994). However, ajoene, a metabolite may be formed from allicin during steam or water distillation of the essential oil, proved to be even more effective as antithrombotic than allicin itself (Edris 2007, Ibrel et al. 1990).

On the other hand, essential oils are also considered as potential alternatives to synthetic antiviral drugs since they have demonstrated virucidal properties, with the advantage of low toxicity compared to synthetic antiviral drugs (Baqui et al. 2001, Primo et al. 2001). *Herpes simplex* virus (type I, II) causes some of the most common viral infections in humans; incorporation of *Artemisia arborescens* essential oil in multilamellar liposomes significantly improved its activity against intracellular *Herpes simplex* virus type-1 (HSV-1) (Sinico et al. 2005). *Melissa officinalis* essential oil can inhibit the replication of HSV-2, due to the presence of citral and citronellal (Allahverdiyev et al. 2004). *Cymbopogon citratus* essential oil also possesses a potent anti-HSV-1 activity even at a concentration of 0.1% (Minami et al. 2003). *Mentha piperita* essential oil revealed high levels of virucidal activity against HSV-1, HSV-2 (Schuhmacher et al. 2003). The essential oil of *Lippia junelliana* and *Lippia turbinate* showed a potent inhibition against Junin virus (Garcia et al. 2003), while *Santolina insularis* showed an antiviral activity against HSV-1 and HSV-2 *in vitro* (De Logu et al. 2000).

Essential oils as natural antimicrobial agents

In vitro investigations have demonstrated that essential oils may act as antibacterial agents against a wide spectrum of pathogenic bacterial strains including *Listeria monocytogenes* (Gaysinsky et al. 2005), *Listeria innocua, Salmonella typhimurium, Salmonella enteritidis, Salmonella choleraesuis, Escherichia coli, Shigella dysenteria* (Santoyo et al. 2006, Penalver et al. 2005), *Bacillus cereus* (Ultee et al. 2002), *Staphylococcus aureus* and *Salmonella typhimurium* (Cai et al. 2012, Schmidt et al. 2005, Jirovetz et al. 2005); among others. This activity has been attributed to phenolic components, such as carvacrol, thymol, eugenol present in many essential oils (Edris 2007).

Essential oils antibacterial effect has also been evaluated against multidrug-resistant pathogens such as *Pseudomonas aeruginosa* and *Escherichia coli*. Among the oils tested, oregano (*Origanum vulgare*), proved to be highly effective (Bozin et al. 2006), *Ocimum gratissimum* can also inhibit multidrug-resistant strains of *Shigellae* (Iwalokun et al. 2003), peppermint (*Mentha piperita*) and spearmint (*Mentha spicata*) essential oils showed strong activity as well against *Staphylococcus aureus* (Imai et al. 2001) and *Achillea clavennae* oil demonstrated strong activity against *Klebsiella pneumoniae, Streptococcus pneumoniae, Haemophilus influenzae* and *Pseudomonas aeruginosa* (Edris 2007, Skocibusic et al. 2004).

Regarding traditional uses, tea tree (*Melaleuca alternifolia*) oil, has been used as a skin-wash formula, primarily to eliminate bacteria from hands, since it does not lead to dermatological problems, nor affect the original protective bacterial flora of the skin (Carson and Riley 1995). In order to prove this property an evaluation was carried out on these products containing tea tree oil as well as on pure tea tree oil against *Staphylococcus aureus, Acinetobacter baumannii, Escherichia coli* and *Pseudomonas aeruginosa* (Messager et al. 2005a, 2005b) showing antibacterial activity depending on the concentration used (Carson et al. 2006). On the other hand, a cream formulation containing 10% of tea tree oil has been used in veterinary therapy for the treatment of canine localized acute and chronic dermatitis (Reichling et al. 2004). Some investigations related to this essential oil have concluded that the antimicrobial activity might be attributed to the major component, terpinen-4-ol, present in high concentrations (Edris 2007, Dryden et al. 2004, May et al. 2000).

In addition, oral applications such as mouth washes containing essential oils are also considered to prevent bacterial aggregation and avoid plaque formation (Seymour 2003). A controlled clinical survey demonstrated that a mouth rinse containing essential oils showed a comparable antiplaque and antigingivitis activity to that containing the synthetic antibacterial agent, chlorhexidine (Charles et al. 2004). *Croton cajucara* essential oil was proved to be toxic for some pathogenic bacteria associated with oral cavity disease (Alviano et al. 2005) thus, it could be incorporated into rinses or mouth washes for general improvement of oral health and to control oral malodor (Edris 2007, Yengopal 2004).

It is well known that the use of plants for treating diseases is a very ancient practice and is based on popular observations, they are frequently prescribed, even if their chemical constituents are not always completely acknowledged. On the other hand, infectious diseases represent an important cause of morbidity and mortality among the general population, particularly in developing countries. Therefore, pharmaceutical companies have been motivated to develop new antimicrobial drugs in recent years, especially due to the constant emergence of microorganisms resistant to conventional antimicrobials (Abad et al. 2007).

As a result of the increasing resistance to antibacterial and antifungal compounds and the reduced number of available drugs, the number of investigations aiming for therapeutic alternatives among aromatic plants and their essential oils have been conducted (Akin et al. 2012, Mahmoodi et al. 2012, Polatoglu et al. 2012, Vasudeva and Sharma 2012, Zellagui et al. 2012, Teimouri 2012, Kateryna and Mahendra 2012,

Adeleye et al. 2011, Nuzu et al. 2011, Abad et al. 2007, Mahmoud and Croteau 2002). Several researches have revealed that the composition, structure and functional groups of essential oils play an important role in determining their antimicrobial activity (Gudrun and Buchbauer 2012, Celiktas et al. 2007). Those compounds containing phenolic groups are usually the most effective (Burt 2007, Omidbeygi et al. 2007, Dorman and Deans 2000). However, since essential oils are complex mixtures of numerous molecules, their biological effects might be the result of a synergism of all molecules or the reflection of only those of the main molecules present at the highest levels according to gas chromatographic analysis (Holley and Patel 2005). Some authors consider that it is also possible that the activity of the main components be modulated by the minor molecules (Tajkarimi et al. 2010, Burt 2007, Ipek et al. 2005, Hoet et al. 2006, Santana-Rios et al. 2001).

Moreover, it has been stated that several components of essential oils are important elements in defining the fragrance, density, texture, color, cell penetration (Marino et al. 2001), lipophilic or hydrophilic attraction, fixation on cell walls and membranes, and cellular distribution. This last characteristic is very important because the distribution of the oil in the cell, determines the different types of radical reactions produced. In that sense, for biological purposes, researchers believe that it is more informative to study the entire oil rather than some of its components (Bakkali et al. 2008).

In vitro methods for antimicrobial assessment of essential oil

A number of methods used for evaluation of antimicrobial activity of essential oils have been reported in literature (Celiktas et al. 2007, Cal 2006, Bozin et al. 2006). Different analyses like disk diffusion assay, well diffusion assay and microdilution assay are often used for determining this activity on essential oils (Da Silveira et al. 2012, Savita and Goval 2012, Yong-Wei et al. 2012, Bachir-Raho and Benali 2012, Schelz et al. 2012, Selim 2011, Nuran et al. 2011, Kelen and Tepe 2008, Bakkali et al. 2008, Burt 2004, Kalemba and Kunicka 2003).

Nevertheless, testing and evaluating the antimicrobial activity of essential oils is difficult because of their volatility, water insolubility and complexity. Essential oils are of hydrophobic nature and high viscosity; these properties may reduce the dilution capability or cause unequal distribution of the oil through the medium used (Giweli et al. 2012). Thus, researchers have adapted different experimental protocols to better represent the applications in their particular field (Burt 2007). On the other hand, since the results of a test can be affected by different factors such as the method used to extract the essential oils from plant material, the volume of inoculum, growth phase, culture medium used, pH of the media and incubation time and temperature (Rios et al. 1998), comparison of published data results are very complicated (Burt 2004, Kalemba and Kunicka 2003, Friedman et al. 2002). However, many investigations are carried out every day in the constant search for antimicrobial properties in essential oils and the following methods are the most frequently used to develop these studies:

Disk and well diffusion assay (agar diffusion method)

This is the most common technique for antimicrobial activity assessment and is recognized as precise and reliable, even though it produces semi-quantitative results, and according to some authors, only qualitative and not always repeatable (Kalodera et al. 1997). The procedure of this method is as follow: Petri dishes of 5–12 cm diameter (usually 9 cm) are filled with 10–20 mL of agar broth and inoculated with microorganisms. Then, the essential oils are incorporated in two possible ways; on a paper disk (Elansary et al. 2012, Jian-Yu et al. 2012, Kačániová et al. 2012, Serban et al. 2011, Mickiené et al. 2011, Rondón et al. 2008, Rojas et al. 2008, Velasco et al. 2007) or into a well prepared agar medium (Kahriman et al. 2012, Lara-Júnior et al. 2012, Benites et al. 2011, Mihailovi et al. 2011, Marković et al. 2011, Ascensão et al. 1998, Lis-Balchin et al. 1998). However, the most important parameters for every variation of this method are the diameter of Whatman's paper disk or the well, the amount of essential oil, as well as the sort of dispersing solvent (Pauli and Schilcher 2010, Rondón et al. 2005, Friedman et al. 2002).

To get the most reliable results, solutions of different concentrations of essential oils are used (Buitrago et al. 2015, Rondón et al. 2006a). After incorporation of the oil, plates are stored at room temperature for

30 minutes to allow all the oil components to diffuse in the agar broth, and then are incubated at 37°C for 24 hours (Buitrago et al. 2015, Rojas et al. 2007, Rondón et al. 2006b). The effectiveness of essential oil is demonstrated by the size of the zone of microorganism growth inhibition around the disk or well, and it is usually expressed as the diameter of this zone (mm or cm), including the disk or well diameter (Sharopova et al. 2012, Zomorodian et al. 2012, Gholamreza et al. 2012, Mary Helen et al. 2012, Mothana 2012, Mickiené et al. 2011, Djenane et al. 2011, Burt 2007, Kalemba and Kunicka 2003, Griffin et al. 2000).

Although, the agar diffusion method is considered inappropriate for essential oils, as their volatile components may possibly evaporate with the dispersing solvent during the incubation time and the components might not diffuse well in the agar broth (Soković et al. 2010), still it is the most common technique for the evaluation of antibacterial and antifungal activity of essential oils because it is easy to perform and requires only small amounts of testing material (Haghighat et al. 2012, Mahboubi et al. 2012, Nascimento et al. 2011, Ortega-Nieblas et al. 2011, Mahboubi et al. 2011, Oladosu et al. 2011). Some authors consider that this method may be recommended as a pre-screening technique for testing a large amount of samples. Once selected the most active samples, more sophisticated methods may be carefully chosen for further analysis (Burt 2007, Kalemba and Kunicka 2003).

The dilution method

This method is used for both, bacteria and fungi (Gao et al. 1999). Agar broth cultures are carried out in Petri dishes or tubes, while liquid broth cultures (used mostly for fungi) are cultivated in conical flasks filled with 100 mL medium (used for molds) (Hili et al. 1997) or test tubes with 2.5–5 mL medium (bacteria and molds); a modification using microtitre plates is also possible (Deans et al. 1994, Rios et al. 1988). For the liquid broth in conical flasks, the inhibitory growth index is computed (% changes in the mold's biomass compared to the control culture) (Lambert et al. 2001). The inhibitory effect of essential oil in the test tubes cultures and microtitre plates is measured turbidimetrically (Kim et al. 1998, Deans et al. 1995), by optical density or with the count plate method (Pintore et al. 2002, Deans et al. 1994). The activity of essential oils against molds can also be investigated by checking sporulation inhibition (Ultee et al. 2000) and toxin productivity (Paster et al. 1995). This method has been claimed by some authors as very accurate compared to the agar diffusion method, nevertheless, it is not often used due to higher costs and laborious handling (Basilico and Basilico 1999).

In both, agar diffusion and dilution methods, the most important parameter to measure the antomicrobial performance of essential oils is the Minimal Inhibitory Concentration (MIC), that is defined as the lowest essential oil concentration in the broth resulting in the lack of visible microorganism growth changes (Cal 2006). This MIC factor gives information about the accurate, precise and reproducible results of an experiment. Some researchers also use the Minimum Bactericidal Concentration (MBC) parameter that refers to the lowest concentration at which no growth is observed after subculturing into fresh broth; or the Bacteriostatic Concentration (BC) that indicates the lowest concentration at which bacteria fail to grow in broth, but are cultured when broth is plated onto agar (Burt 2007).

A different microdilution method for determining the MIC of essential oils based on the use of resazurin as a visual indicator has also been proposed. Resazurin, an oxidation-reduction indicator, is used for the assessment of cell growth. It is a violet/blue non-fluorescent and non-toxic dye that turns purple/pink and fluorescent when reduced to resorufin by oxidoreductases enzyme within viable cells. Resorufin can further reduce to hydroresorufin, a non-fluorescent and uncolored compound (Varela et al. 2008).

Important aspects to consider in the evaluation of antimicrobial activity of essential oils

Microorganisms used in antimicrobial assays are usually derived from internationally recognized pure culture collections, but they can also be isolates taken from clinical patients (Rojas et al. 2008, Sarker et al. 2007) or some other different natural environments (Khallouki et al. 2000). Before testing, these microorganisms have to reach an appropriate growth phase and a specified number of cells have to be used for the test. It is very important that all parameters which may affect the experimental results have to be strictly defined (Kalemba and Kunicka 2003).

Experiments are always repeated at least two or three times and negative (solvents) as well as positive controls (standard antibiotics) are also usually verified to evaluate the susceptibility of tested strains (Rojas et al. 2008, Rondón et al. 2006a, Rondón et al. 2005) and possibly to compare the essential oil efficiency with the positive controls (Rojas et al. 2011a, Rojas et al. 2011b, Rondón et al. 2008, Kalodera et al. 1997). However, it is worth noting that it is very difficult to have an appropriate comparison of the antimicrobial activity of essential oils and antibiotics just by using these standard controls (Kalemba and Kunicka 2003).

A main concern in all experiments is the solvent used to dissolve the essential oils. Several substances have been used for this purpose: menthanol/ethanol (Marino et al. 2001, Packiyasothy and Kyle 2002, Nielsen and Rios 2000), dimethyl sulfoxide (Rojas et al. 2011a, Rojas et al. 2011b, Cárdenas et al. 2012, Baldovino et al. 2009), n-hexane (Senatore et al. 2000), Tween-20/Tween-80 (Wilkinson et al. 2003, Bassole et al. 2003) and agar (Burt et al. 2003, Gill et al. 2002). However, a number of researchers found it unnecessary to use an additive (Cimanga et al. 2002, Mejlholm et al. 2002, Lambert et al. 2001, Elgayyar et al. 2001).

Other methods used to assess antimicrobial activity of essential oils

Vapor phase test (Pauli and Schilcher 2010), air washer coupled with air sampler (Sato et al. 2006) and direct bioautography (Choma and Grzelak 2011) are among other methods used by researchers to evaluate the antimicrobial activity of essential oils. In the vapor phase test a seeded agar plate is placed upside down onto a basin which comprises a certain amount of volatile oil. An inhibition zone observed in the agar plate is considered as antimicrobial activity (Pauli and Schilcher 2010). The air washer coupled with an air sampler method, is used to determine the antimicrobial activity of essential oils against air-borne microbes. An air washer machine is filled with diluted essential oil which is vaporized into it. By the air sampler, air-borne microorganisms are fixed on agar strips.

After incubating these strips the number of microbes in the air can be counted. The comparison between the amount of bacteria before and after the application of essential oil will be the measure of antimicrobial activity (Sato et al. 2006). For the direct bioautography method a developed thin-layer chromatography plate which contains the test substances is immersed into a microbially contaminated liquid. After direct incubation of the plates the zones of inhibition are exposed with tetrazolium salts. These molecules are transformed by the microbial enzyme dehydrogenase into colored products (purple formazan). As a consequence, colorless inhibition zones are visible in regions without microbial growth since there is no dehydrogenase activity (Choma and Grzelak 2011).

Mechanism of action of essential oils in relation to antibacterial activity

Previous investigations have indicated that essential oils possess a higher antibacterial effect against Gram-positive than Gram-negative bacteria (Bigos et al. 2012, Lang and Buchbauer 2012, Harpaz et al. 2003, Deans et al. 1995). It has also been observed that *Pseudomonas*, in particular *P. aeruginosa*, appears to be least sensitive to the action of essential oils (Senatore et al. 2000, Deans et al. 1995). The negative response of the Gram-negative bacteria might be explained since the outer cell membrane surrounding the cell wall restricts the diffusion of hydrophobic compounds through its lipopolysaccharide covering (Saei-Dehkordi et al. 2010, Harpaz et al. 2003).

On the other hand, a number of possible mechanisms have been used to explain the mechanism of action of essential oils against microorganisms and mainly against Gram-positive bacteria. Degradation of the cell wall (Helander et al. 1998), damage to cytoplasmic membrane (Ultee et al. 2002), damage to membrane proteins (Ultee et al. 1999), leakage of cell contents (Celikel et al. 2008, Lambert et al. 2001) and coagulation of cytoplasm (Gustafson et al. 1998) are some of these (Burt 2007, Sandri et al. 2007).

Recent interest for the use of essential oils as pesticides

The excessive use of pesticides has caused several environmental problems and has been a matter of concern for both scientists and the public in recent years. Thus, to reduce negative impacts to human health and the environment, there has recently been, a great interest for green chemistry, since natural products are an excellent alternative as green pesticides because they are eco-friendly, economic and biodegradable (Manindra et al. 2011).

On the other hand, essential oils, usually, show a broad spectrum of activity against pest insects and plant pathogenic fungi ranging from insecticidal, antifeedant, repellent, oviposition deterrent, growth regulatory and antivector activities. In addition, resistance might develop more slowly to essential oil-based pesticides due to the complex mixture of constituents present in these natural products and have proved to be well received by consumers for use against home and garden pests, thus, could also be effective in farms particularly for organic food production. These features indicate that pesticides based on plant essential oils could be used in a variety of ways to control a large number of pests (Manindra et al. 2011).

Some examples of essential oils used as green pesticides are: lemon grass (*Cymbopogon flexuosus*), eulcalyptus (*Eulcalyptus globules*), rosemary (*Rosmarinus officinalis*), vetiver (*Vetiveria zizanioides*), clove (*Eugenia caryophyllus*) and thyme (*Thymus vulgaris*); these have been found effective as animal repellents, antifeedants, insecticides, miticides and antimicrobials (Zaridah 2003). Furthermore, peppermint (*Mentha piperita*) is well known to repel ants, flies, lice and moths, while spearmint (*Mentha spicata*) and basil (*Ocimum basilicum*) are also effective in warding off flies. Similarly, *Artemesia vulgaris*, *Melaleuca leucodendron*, *Pelargonium roseum*, *Lavandula angustifolia*, *Mentha piperita* and *Juniperus virginiana* are also effective against various insects and fungal pathogens (Kordali 2005, Manindra et al. 2011).

Domestically, people use a combination of citronella (*Citronella* sp.), lemon (*Citrus limon*), rose (*Rosa damascena*), lavender (*Lavandula angustifolia*) and basil (*Ocimun basilicum*) essential oils to protect homes against indoor insect pests. Nepetalactone, for example, is an active constituent present in catnip (*Nepeta cataria*) and has been proved to repel mosquitoes 10 times more than DEET and is also effective against *Aedes aegypti*, a vector for yellow fever virus (Manindra et al. 2011). Despite all these qualities only few pest control products based on plant essential oils have appeared in the market. This may be a consequence of several factors, regulatory barriers to commercialization and the fact that essential oils require greater application rates and may necessitate frequent reapplication when used out-of-doors, among others. Additionally, availability of sufficient quantities of plant material, standardization and refinement of pesticide products and protection of technology is also required for these natural pesticides. Another issue to be considered is that the chemical profile of plant species may vary naturally depending on geographic, genetic, climatic, annual or seasonal factors, thus, manufacturers must work on these elements to ensure the consistency of their products (Manindra et al. 2011, Isman and Machial 2006).

Nonetheless, some U.S. companies have introduced pesticides, based on essential oils, in recent years. EcoSMART Technologies has introduced insecticides containing eugenol and 2-phenethyl propionate to control crawling and flying insects, under the brand name EcoPCO®. The advantage about EcoSMART products is that active ingredients are exempt from Environmental Protection Agency registration and are approved as direct food additives or classifies as GRAS (generally recognized as safe) by the Food and Drug Administration. EcoTrol®, insecticide-miticide containing rosemary oil as the active ingredient, has recently been introduced for use on horticultural crops. Another product based on rosemary oil is Sporan®, used as a fungicide, while Matran®, a formulation of clove oil based mainly in eugenol, is used for weed control. In terms of technology, green pesticides prepared as oil-in-water micro emulsions is being recently developed to replace the traditional emulsifiable concentrates, in order to reduce the use of organic solvents and increase the penetration properties of the droplets. The advantages of using these micro emulsions are the efficacy improvement and the lower dosage required (Manindra et al. 2011).

Essential oils global market

Essential oils have long been used in flavoring, pharmaceutical and fragrances industries for manufacturing products such as beverages, detergents, perfumes, medicines among others (Department of Agriculture,

Forestry and Fisheries 2012, Juliani et al. 2004, Williams 1997). Over the last 50 years, the demand for essential oil products has gradually increased, perhaps due to the public consciousness for health plus the fact that natural products are assumed to have a superior quality, thus, there has been a progressive urge for incorporating natural materials into a wide range of foods, medicines and lifestyle products such as fragrances, culinary herbs, aromatic teas, among others (Department of Agriculture, Forestry and Fisheries 2012, Juliani et al. 2004, Simon 1990). It has been estimated that around 50% of the world production of essential oils is used in flavors (predominantly beverages), 5 to 10% in phytotherapy and aromatherapy, 20 to 25% in fragrances and 20 to 25% is used by chemical industries for the isolation of aroma compounds (Department Trade and Industry Republic of South Africa 2004).

Besides, the global market of pharmaceutical products represents 33% of world-wide chemical compounds production (Newman et al. 2003), now a days the most important pharmaceutical companies (Seidl 2002) are investing in research to find novel compounds in order to synthesize new lead medicines (Silva et al. 2008). Concerning essential oils, the major world market is the United States, followed by Japan and Europe. However, production is mostly concentrated in Europe (Department of Agriculture, Forestry and Fisheries 2012); among crops the most predominant are: citrus such as orange, lemon, grapefruit (produced by US, Brazil, Mexico); mint oils such as peppermint (US), spearmint (US) and cornmint (China, India, South America); and lemon fragrance oils such as citronella, lemongrass and listsea cubeba (China, India, South America) (Department Trade and Industry Republic of South Africa 2004, Lawrence 1993).

However, the most used essential oils are: citrus, mints and condimentous (sesame, clove, ginger, pepper, garlic), citrous essential oils (orange, lemon, grapefruit and bergamot), mints essential oils (peppermint and spearmint), essential oils extracted from pine and eucalyptus and floral essential oils (rose and rosemary) while menthol and limonene are among isolated compounds from essential oils with more commercial uses in dental and medicinal preparations containing organic active ingredients (Silva et al. 2008). In relation to therapeutic uses, dermatological, antibiotic, analgesic, anti-inflammatory, neurological, gastrointestinal, heart diseases and anticancer are the most predominant areas (Silva et al. 2008).

Conclusions

A wide range of investigations have been achieved concerning essential oils and a variety of biological properties such as, antibacterial, antifungal, antiviral, insecticidal and antioxidant have also been revealed. Despite this, these potential, essential oils have been primarily used in food preservation, aromatherapy and fragrance industries. However, the increasing tolerance of several microorganisms against commonly used antibiotic drugs have encouraged researchers to find alternative ways for the treatment of such infections, and aromatic plants, particularly essential oils, seems promising.

A significant problem that researchers have to deal in their investigations is the discordance between data obtained from same species studied. To understand this variability, a number of factors have to be considered: climatic, seasonal and geographic conditions, harvest period and distillation technique are among these. Besides, the plant maturity at the time of oil production and the existence of chemotypic differences can also drastically affect the oil composition. Fortunately, capillary-GC experiments can provide the exact composition of essential oils. Nevertheless, it is important to consider that essential oils are composed by a heterogeneous mixture of substances, thus, biological activities are possibly due to the synergism of these components, although antagonism might also occur.

Knowledge of methods for testing essential oils is therefore necessary to discover the spectrum of action of these natural products, their modes of action and their therapeutic applications. It is also important to emphasize that over the last 50 years, the demand for essential oil products has gradually increased, perhaps due to the public consciousness for health plus the fact that natural products are assumed to have superior quality, thus, there has been a progressive urge for incorporating natural materials into a wide range of foods, medicines and lifestyle products such as fragrances, culinary herbs, aromatic teas, among others. Pharmaceutical companies have also been motivated to develop new antimicrobial drugs in recent years, especially due to the constant emergence of microorganisms resistant to conventional antimicrobials, therefore, they are sponsoring research to find novel compounds in order to synthesize new lead medicines.

References

Abad, M.J., Ansuategui, M. and Bermejo, P. 2007. Active antifungal substances from natural sources. ARKIVOC 7: 116–145.

Abdollahi, M., Ranjbar, A., Shadnia, S., Nikfar, S. and Rezaie, A. 2004. Pesticides and oxidative stress: a review. Med. Sci. Monit. 10: 141–147.

Ackerman, D. 2001. In the begning was smell. p. 27. *In*: Suzanne Catty (ed.). Hydrosols: The Next Aromatherapy. Inner Traditions, Rochester, Vermont.

Adams, R. 2007. Identification of Essential Oil Components by Gas Chromatography/Mass Spectrometry. 4th Ed. Allured Publishing Corporation, Carol Stream IL, USA, p. 804.

Adeleye, I.A., Omadime, M.E. and Daniels, F.V. 2011. Antimicrobial activity of essential oil and extracts of *Gongronema latifolium* decne on bacterial isolates from blood stream of HIV infected patients. J. Pharmacol. Toxicol. 6: 312–320.

Ahmad, M.M., Salim-ur-Rehman, Iqbal, Z., Anjum, F.M. and Sultan, J.I. 2006. Genetic variability to essential oil composition in four citrus fruit species. Pak. J. Bot. 38(2): 319–324.

Akın, M., Saraçoğlu, H.T., Demirci, B., Başer, K.H.C. and Küçüködük, M. 2012. Chemical composition and antibacterial activity of essential oils from different parts of *Bupleurum rotundifolium* L. Rec. Nat. Prod. 6(3): 316–320.

Allahverdiyev, A., Duran, N., Ozguven, M. and Koltas, S. 2004. Antiviral activity of the volatile oils of *Melissa officinalis* L. against Herpes simplex virus type-2. Phytomedicine 11: 657–661.

Alviano, W.S., Mendonça-Filho, R.R., Alviano, D.S., Bizzo, H.R., Souto-Padrón, T., Rodrigues, M.L., Bolognese, A.M., Alviano, C.S. and Souza, M.M. 2005. Antimicrobial activity of *Croton cajucara* Benth linalool-rich essential oil on artificial biofilms and planktonic microorganisms. Oral Microbiol. Immunol. 20: 101–105.

Angioni, A., Barra, A., Coroneo, V., Dessi, S. and Cabras, P. 2006. Chemical composition, seasonal variability, and antifungal activity of *Lavandula stoechas* L. ssp. stoechas essential oils from stem/leaves and flowers. J. Agr. Food Chem. 54: 4364–4370.

Anwar, F., Hussain, A.I., Sherazi, S.T.H. and Bhanger, M.I. 2009a. Changes in composition and antioxidant and antimicrobial activities of essential oil of fennel (*Foeniculum vulgare* Mill.) fruit at different stages of maturity. J. Herbs Spices Med. Plants 15: 1–16.

Anwar, M.A., Hussain, A.I. and Shahid, M. 2009b. Antioxidant and antimicrobial activities of essential oil and extracts of fennel (*Foeniculum vulgare* Mill.) seeds from Pakistan. Flavour Frag. J. 24: 170–176.

Araujo, L., Moujir, L.M., Rojas, J., Carmona, J. and Rondón, M. 2013. Chemical composition and biological activity of *Conyza bonariensis* (L.) Cronquist essential oil collected in Mérida-Venezuela. Nat. Prod. Commun. 8(8): 1175–1178.

Asaph, A., Jongsma, M.A., Kim, T.Y., Ri, M.B., Giri, A.P., Verstappen, W.A., Schwab, W. and Bouwmeester, H.J. 2006. Metabolic engineering of terpenoid biosynthesis in plants. Phytochem. Rev. 5: 49–58.

Ascensão, L., Figueiredo, C.A., Barosso, J.G., Pedro, L.G., Schripsema, J., Deans, S.G. and Scheffer, J.J.C. 1998. *Plectranthus madagascariensis*: morphology of the glandular trichomes, essential oil composition, and its biological activity. Int. J. Plant Sci. 159: 31–38.

Asekun, O.T., Grierson, D.S. and Afolayan, A.J. 2007. Effects of drying methods on the quality and quantity of the essential oil of *Mentha longifolia* L. sunsp. Capensis, Food Chem. 101: 995–998.

Atti-Santos, A.C., Pansera, M.R., Paroul, N., Atti-Serafini, L. and Moyna, P. 2004. Seasonal variation of essential oil yield and composition of *Thymus vulgaris* L. (Lamiaceae) from South Brazil. J. Essent. Oil Res. 16: 294–295.

Augusto, F., Lopes, A.L. and Zini, C.A. 2003. Sampling and sample preparation for analysis of aromas and fragrances. Trac-Trend Anal. Chem. 22: 160–169.

Bachir-Raho, G. and Benali, M. 2012. Antibacterial activity of the essential oils from the leaves of *Eucalyptus globulus* against *Escherichia coli* and *Staphylococcus aureus*. Asian Pac. J. Trop. Biomed. 2: 739–742.

Bajpai, V.K., Kang, S., Xu, H., Lee, S.G., Baek, K.H. and Kang, S.C. 2011. Potential role of essential oils on controlling plant pathogenic bacteria *Xanthomonas* species: a review. Plant. Pathol. J. 27(3): 207–224.

Baker, G.R., Lowe, R.F. and Southwell, I.A. 2000. Comparison of oil recovered from tea tree by ethanol extraction and steam distillation. J. Agr. Food Chem. 48: 4041–4043.

Bakkali, F., Averbeck, S., Averbeck, D. and Idaomar, M. 2008. Biological effects of essential oils. A review. Food Chem. Toxicol. 46: 446–475.

Baldovino, S., Rojas, J., Rojas, L.B., Lucena, M., Buitrago, A. and Morales, A. 2009. Chemical composition and antibacterial activity of the essential oil of *Monticalia andicola* (Asteraceae) collected in Venezuela. Nat. Prod. Commun. 4(11): 1601–1604.

Ballabenia, V., Tognolini, M., Chiavarini, M., Impicciatore, Bruni, R., Bianchi, A. and Barocelli, E. 2004. Novel antiplatelet and antithrombotic activities of essential oil from Lavandula hybrida Reverchon 'grosso'. Phytomedicine 11: 596–601.

Baqui, A., Kelley, J., Jabra-Rizk, M., Depaola, L., Falkler, W. and Meiller, T. 2001. *In vitro* effects of oral antiseptics human immunodeficiency virus-1 and herpes simplex virus type 1. J. Clin. Periodontol. 28: 610–616.

Barter, P. 2005. The inflammation: Lipoprotein cycle. Atherosclerosis Suppl. 6(2): 15–20.

Basilico, M.Z. and Basilico, J.C. 1999. Inhibitory effects of some spice essential oils on *Aspergillus ochraceus* NRRL 3174 growth and ochratoxin a production. Lett. Appl. Microbiol. 29: 238–241.

Bassolé, I.H.N. and Juliani, H.R. 2012. Essential oils in combination and their antimicrobial properties. Molecules 17(4): 3989–4006.

Bassole, I.H.N., Ouattara, A.S., Nebie, R., Ouattara, C.A.T., Kabore, Z.I. and Traore, S.A. 2003. Chemical composition and antibacterial activities of the essential oils of *Lippia chevalieri* and *Lippia multiflora* from Burkina Faso. Phytochem. 62: 209–212.

Benites, J., Bravo, F., Rojas, M., Fuentes, R., Moiteiro, C. and Venâncio, F. 2011. Composition and antimicrobial screening of the essential oil from the leaves and stems of *Senecio atacamensis* Phil. from Chile. J. Chil. Chem. Soc. 56: 712–714.

Bigos, M., Wasiela, M., Kalemba, D. and Sienkiewicz, M. 2012. Antimicrobial activity of geranium oil against clinical strains of *Staphylococcus aureus*. Molecules 17: 10276–10291.

Bodake, H., Panicker, K., Kailaje, V. and Rao, V. 2002. Chemopreventive effect of orange oil on the development of hepatic preneoplastic lesions induced by N-nitrosodiethylamine in rats: an ultrastructural study. Indian J. Exp. Biol. 40: 245–251.

Bonnardeaux, J. 1992. The effect of different harvesting methods on the yield and quality of basil oil in the Ord River irrigation area. J. Essent. Oil Res. 4: 65–69.

Botsoglou, N., Florou-Paneri, P., Christaki, E., Giannenas, I. and Spais, A. 2004. Performance of rabbits and oxidative stability of muscle tissues as affected by dietary supplementation with oregano essential oil. Arch. Anim. Nutr. 58: 209–218.

Bozin, B., Mimica-Dukic, N., Simin, N. and Anackov, G. 2006. Characterization of the volatile composition of essential oils of some *Lamiaceae* spices and the antimicrobial and antioxidant activities of the entire oils. J. Agric. Food. 54: 1822–1828.

Buitrago, A., Rojas, L.B., Rojas, J., Buitrago, D., Usubillaga, A. and Morales, A. 2009. Comparative study of the chemical composition of the essential oil from fresh leaves of *Vismia baccifera* var. dealbata (Guttiferae) collected in two different locations in Mérida-Venezuela. J. Essent. Oil Bear. Pl. 12(6): 651–655.

Buitrago, A., Rojas, J., Rojas, L., Morales, A., Aparicio, R. and Rodríguez, L. 2011. Composition of the essential oil of leaves and roots of *Allium shoenosprasum*, BLACPMA 10(3): 218–221.

Buitrago, A., Rojas, J., Rojas, L., Velasco, J., Morales, A., Peñaloza, Y. and Díaz, C. 2015. Essential oil composition and antimicrobial activity of *Vismia macrophylla* leaves and fruits collected in Táchira-Venezuela. Nat. Prod. Commun. 10(2): 375–377.

Burbott, A.J. and Loomis, W.D. 1957. Effects of light and temperature on the monoterpenes of peppermint. J. Plant Physiol. 42: 20–28.

Burt, S. 2004. Essential oils: their antimicrobial properties and potential application in foods—a review. Int. J. Food Microbiol. 94: 223–253.

Burt, S.A. 2007. Antibacterial activity of essential oils: potential applications in food. Ph.D. Thesis, Institute for Risk Assessment Sciences, Division of Veterinary Public Health Utrecht University, Netherlands.

Burt, S.A. and Reinders, R.D. 2003. Antibacterial activity of selected plant essential oils against *Escherichia coli* O157:H7. Lett. Appl. Microbiol. 36(3): 162–167.

Butterfield, D. and Lauderback, C. 2002. Lipid peroxidation and protein oxidation in Alzheimer's disease brain: potential causes and consequences involving amyloid *beta* peptideassociated free radical oxidative stress. Free Radic. Biol. Med. 32: 1050–1060.

Cai, Y., Hu, X., Huang, M., Sun, F., Yang, B., He, J., Wang, X., Xia, P. and Chen, J. 2012. Characterization of the antibacterial activity and the chemical components of the volatile oil of the leaves of *Rubus parvifolius* L. Molecules 17: 7758–7768.

Cal, K. 2006. Skin penetration of terpenes from essential oils and topical vehicles. Planta Med. 72: 311–316.

Calvey, E., Roach, J. and Block, E. 1994. Supercritical fluid chromatography of garlic (*Allium sativum*) extracts with mass spectrometric identification of allicin. J. Chromatogr. Sci. 32: 93–96.

Candan, F., Unlu, M., Tepe, B., Daferera, D., Polissiou, M., Sökmen, A. and Akpulat, H.A. 2003. Antioxidant and antimicrobial activity of the essential oil and methanol extracts of *Achillea millefolium* subsp. *millefolium Afan*. (Asteraceae). J. Ethnopharmacol. 87: 215–220.

Cárdenas, J., Rojas, J., Rojas-Fermin, L., Lucena, M. and Buitrago, A. 2012. Essential oil composition and antibacterial activity of *Monticalia greenmaniana* (Asteraceae). Nat. Prod. Commun. 7(2): 243–244.

Carson, C. and Riley, T. 1995. Toxicity of the essential oil of *Melaleuca alternifolia* or tea tree oil. J. Toxicol. Clin. Toxicol. 33: 193–195.

Carson, F., Hammer, A. and Riley, V. 2006. *Melaleuca alternifolia* (Tea Tree) oil: a review of antimicrobial and other medicinal properties. Clin. Microbiol. Rev. 19: 50–62.

Cava, R., Nowak, E., Taboada, A. and Marin-Iniesta, F. 2007. Antimicrobial activity of clove and cinnamon essential oils against *Listeria monocytogenes* in pasteurized milk. J. Food Protect 70(12): 2757–2763.

Cavaleiro, C., Salgueiro, L.R., Miguel, M.G. and Proenca da Cunha, A. 2004. Analysis by gas chromatography-mass spectrometry of the volatile components of *Teucrium lusitanicum* and *Teucrium algarbiensis*. J. Chromatogr. 1033: 187–190.

Cavalieri, E.S., Mariotto, Fabrizi, C., de Prati, A.C., Gottardo, R., Leone, S., Berra, L.V., Lauro, G.M., Ciampas, A.R. and Suzuki, H. 2004. α-Bisabolol, a nontoxic natural compound, strongly induces apoptosis in glioma cells. Biochem. Biophys. Res. Commun. 315: 589–594.

Celikel, N. and Kavas, G. 2008. Antimicrobial properties of some essential oils against some pathogenic microorganisms. Czech J. Food Sci. 26: 174–181.

Celiktas, O.Y.E., Kocabas, E.H., Bedir, E., Sukan, F.V., Ozek, T. and Baser, K.H.C. 2007. Antimicrobial activities of methanol extracts and essential oils of *Rosmarinus officinalis*, depending on location and seasonal variations. Food Chem. 100: 553–559.

Chalchat, J.C., Garry, R.P. and Muhayimana, A. 1995. Essential oil of *Tagetes minuta* from Rwanda and France: chemical composition according to harvesting location, growth stage and part of plant extracted. J. Essent. Oil Res. 7: 375–386.

Chang, S.T. and Cheng, S.S. 2002. Antitermitic activity of leaf essential oil and components from *Cinnamomum osmophleum*. J. Agric. Food Chem. 50: 1389–1392.

Charles, C., Mostler, K., Bartels, L. and Mankodi, S. 2004. Comparative antiplaque and antigingivitis effectiveness of a chlorhexidine and an essential oil mouthrinse: 6-month clinical trial. J. Clin. Periodontol. 31: 878–884.

Charles, D.J. and Simon, J.E. 1990. Comparison of extraction methods for the rapid determination of essential oil content and composition. J. Am. Soc. Hortic. Sci. 115: 458–462.

Chen, C., Pung, D., Leong, V., Hebbar, V., Shen, G., Nair, S., Li, W. and Kong, A.N. 2004. Induction of detoxifying enzymes by garlic organosulfur compounds through transcription factor NRF2: effect of chemical structure and stress signals. Free Radic. Biol. Med. 37: 1578–1590.

Chen, H., Yang, Y., Xue, J., Wei, J., Zhang, Z. and Chen, H. 2011. Comparison of compositions and antimicrobial activities of essential oils from chemically stimulated agarwood, wild agarwood and healthy *Aquilaria sinensis* (Lour.) gilg trees. Molecules 16: 4884–4896.

Choma, I.M. and Grzelak, E.M. 2011. Bioautography detection in thin-layer chromatography. J. Chromatogr. A. 1218(19): 2684–2691.

Cimanga, K., Kambu, K., Tona, L., Apers, S., De Bruyne, T., Hermans, N., Totte, J., Pieters, L. and Vlietinck, A.J. 2002. Correlation between chemical composition and antibacterial activity of essential oils of some aromatic medicinal plants growing in the Democratic Republic of Congo. J. Ethnopharmacol. 79: 213–220.

Clark, R.J. and Menary, R.C. 1979. The importance of harvest date and plant density on the yield and quality of *Tasmanian peppermint* oil. J. Am. Soc. Hortic. Sci. 104: 702–706.

Contreras, B., Rojas, J., Celis, M., Rojas, L., Méndez, L. and Landrum, L. 2014. Componentes volátiles de las hojas de *Pimenta racemosa* var. *racemosa* (Mill.) J.W. Moore (Myrtaceae) de Táchira–Venezuela. BLACPMA 13(3): 305–310.

Da Silveira, S.M., Júnior, A.C., Scheuermann, G.N., Secchi, F.L. and Werneck, C.R. 2012. Chemical composition and antimicrobial activity of essential oils from selected herbs cultivated in the South of Brazil against food spoilage and foodborne pathogens. Ciênc Rural. 42: 1300–1306.

De Angelis, L. 2001. Brain tumors. N. Engl. J. Med. 344: 114–123.

De Logu, A., Loy, G., Pellerano, M., Bonsignore, L. and Schivo, M. 2000. Inactivation of HSV-1 and HSV-2 and prevention of cellto-cell virus spread by *Santolina insularis* essential oil. Antiviral Res. 48: 177–185.

De Sousa, A., Alviano, D., Blank, A., Alves, P., Alviano, C. and Gattass, C. 2004. *Melissa officinalis* L. essential oil: antitumoral and antioxidant activities. J. Pharm. Pharmacol. 56: 677–681.

Deans, S.G., Kennedy, A.I., Gundidza, M.G., Mavi, S., Waterman, P.G. and Gray, A.I. 1994. Antimicrobial activities of the volatile oil of *heteromorpha trifoliata* (wendl.) eckl. & zeyh. (apiaceae). Flavour Fragr. J. 9(5): 245–248.

Deans, S.G., Noble, R.C., Hiltunen, R., Wuryani, W. and Penzes, L.G. 1995. Antimicrobial and antioxidant properties of *Syzygium aromaticum*. Flavour Fragr. J. 10(5): 323–328.

Delaquis, P.J., Stanich, K., Girard, B. and Mazza, G. 2002. Antimicrobial activity of individual and mixed fractions of dill, cilantro, coriander and eucalyptus essential oils. Int. J. Food Microbiol. 74: 101–109.

Demir, V., Gunhan, T., Yagcioglu, A.K. and Degirmencioglu, A. 2004. Mathematical modeling and the determination of some quality parameters of air-dried bay leaves. Biosyst. Eng. 88: 325–335.

Department of Agriculture, Forestry and Fisheries. 2012. A guide to essential oil crops [on line article]. Revised on January 27th 2014, source: www.daff.gov.za/docs/.../guideessoilcrops.pdf.

Department Trade and Industry Republic of South Africa. 2004. Study into the establishment of an aroma and fragrance fine chemicals value chain in South Africa (Tender number T79/07/03. [on line article]. Revised on January 15th 2014, source: http://www.thedti.gov.za/industrial_development/docs/fridge/Aroma_Part1.pdf.

Dewick, P. 2002a. Medicinal Natural Products. John Wiley & Sons, Nottingham, UK.

Dewick, P.M. 2002b. The biosynthesis of C_5-C_{25} terpenoid compounds. Nat. Prod. Rep. 19: 181–222.

Djenane, D., Yangüela, J., Montañés, L., Djerbal, M. and Roncalés, P. 2011. Antimicrobial activity of *Pistacia lentiscus* and *Satureja montana* essential oils against *Listeria monocytogenes* CECT 935 using laboratory media: Efficacy and synergistic potential in minced bee. Food Control 22(7): 1046–1053.

Donelian, A., Carlson, L.H.C., Lopes, T.J. and Machado, R.A.F. 2009. Comparison of extraction of patchouli (*Pogostemon cablin*) essential oil with supercritical CO_2 and by steam distillation. J. Supercrit. Fluids 48: 15–20.

Dorman, H.J. and Deans, S.G. 2000. Antimicrobial agents from plants: antibacterial activity of plant volatile oils. J. Appl. Microbiol. 88: 308–316.

Dryden, M., Dailly, S. and Crouch, M. 2004. A randomized, controlled trial of tea tree topical preparations versus a standard topical regimen for the clearance of MRSA colonization. J. Hosp. Infect. 58: 86–87.

Edris, A.E. 2007. Pharmaceutical and therapeutic potentials of essential oils and their individual volatile constituents: A Review. Phytother. Res. 21(4): 308–323.

Elansary, H.O., Salem, M.Z.M., Ashmawy, N.A. and Yacout, M.M. 2012. Chemical composition, antibacterial and antioxidant activities of leaves essential oils from *Syzygium cumini* L., *Cupressus sempervirens* L. and *Lantana camara* L. from Egypt. J. Agri. Sci. 4(10): 144–152.

Elgayyar, M., Draughon, F.A., Golden, D.A. and Mount, J.R. 2001. Antimicrobial activity of essential oils from plants against selected pathogenic and saprophytic microorganisms. J. Food Protect. 64(7): 1019–1024.

Friedman, M., Henika, P.R. and Mandrell, R.E. 2002. Bactericidal activities of plant essential oils and some of their isolated constituents against *Campylobacter jejuni, Escherichia coli, Listeria monocytogenes,* and *Salmonella enteric.* J. Food Protect. 65(10): 1545–1560.

Galambosi, B. and Peura, P. 1996. Agrobotanical features and oil content of wild and cultivated forms of caraway. J. Essent. Oil Res. 8: 389–397.

Gao, Y., Van Belkum, M.J. and Stiles, M.E. 1999. The outer membrane of gram negative bacteria inhibits antibacterial activity of brochocin-C. Appl. Environ. Microbiol. 65: 4329–4333.

Garcia, C., Talarico, L., Almeida, N., Colombres, S., Duschatzky, C. and Damonte, B. 2003. Virucidal activity of essential oils from aromatic plants of San Luis, Argentina. Phytother. Res. 17: 1073–1075.

Gardner, P. 1997. Superoxide-driven aconitase FE-S center cycling. Biosci. Rep. 17: 33–42.

Gaysinsky, S., Davidson, P., Bruce, D. and Weiss, J. 2005. Growth inhibition of *Escherichia coli* O157:H7 and *Listeria monocytogenes* by carvacrol and eugenol encapsulated in surfactant micelles. J. Food Prot. 68: 2559–2566.

Gholamreza, A., Jalali, M. and Sadoughi, E. 2012. Antimicrobial activity and chemical composition of essential oil from the seeds of *Artemisia aucheri* Boiss. J. Nat. Pharm. Prod. 6: 11–15.

Gill, A.O., Delaquis, P., Russo, P. and Holley, R.A. 2002. Evaluation of anti-listerial action of cilantro oil on vacuum packed ham. Int. J. Food. Microbiol. 73: 83–92.

Giweli, A., Džamić, A.M., Soković, M., Ristić, M.S. and Marin, P.D. 2012. Antimicrobial and antioxidant activities of essential oils of *Satureja thymbra* growing wild in Libya. Molecules 17: 4836–4850.

Gounaris, Y. 2010. Biotechnology for the production of essential oils, flavours and volatile isolates. A review. Flavour Fragr. J. 25(5): 367–386.

Graven, E.H., Webber, L., Venter, M. and Gardner, J.B. 1990. The development of *Artemisia afra* (Jacq.) as a new essential oil crop. J. Essent. Oil Res. 2: 215–220.

Griffin, S.G., Markham, J.L. and Leach, D.N. 2000. An agar dilution method for the determination of the minimum inhibitory concentration of essential oils. J. Essent. Oil Res. 12: 249–255.

Guang, L., Li-Bin, Z., Bing-An, F., Ming-Yang, Q., Li-Hua, Y. and Ji-Hong, X. 2004. Inhibition of growth and metastasis of human gastric cancer implanted in nude mice by d-limonene. World J. Gastroenterol. 10: 2140–2144.

Gudrun, L. and Buchbauer, G. 2012. A review on recent research results (2008–2010) on essential oils as antimicrobials and antifungals. Flavour Fragr. J. 27: 13–32.

Guenther, E. 1950. The Essential Oils. Van Nostrand Co., Inc., New York, USA.

Gustafson, J.E., Liew, Y.C., Chew, S., Markham, J.L., Bell, H.C., Wyllie, S.G. and Warmington, J.R. 1998. Effects of tea tree oil on *Escherichia coli*. Lett. Appl. Microbiol. 26: 194–198.

Haghighat, M.H., Alizadeh, A. and Nouroznejadfard, M.J. 2012. Essential oil composition and antimicrobial activity in Iranian *Salvia mirzayanii* Rech. & Esfand. Adv. Environ. Biol. 6: 1985–1989.

Harpaz, S., Glatman, L., Drabkin, V. and Gelman, A. 2003. Effects of herbal essential oils used to extend the shelf-life of freshwater reared Asian sea bass (Lates calcarifer). J. Food Protect 66: 410–417.

Helander, I.M., Alakomi, H.L., Latva-Kala, K., Mattila-Sandholm, T., Pol, I., Smid, E.J., Gorris, L.G.M. and Von Wright, A. 1998. Characterization of the action of selected essential oil components on *Gram-negative* bacteria. J. Agr. Food Chem. 46: 3590–3595.

Henderson, C., Smith, A., Ure, J., Brown, K., Bacon, E. and Wolf, C. 1998. Increased skin tumorigenesis in mice lacking π-class glutathione S-transferase. Proc. Natl. Acad. Sci. USA 95: 5275–5280.

Hernandez-Arteseros, J.A., Vila, R., Cruz, S.M., Caceres, A. and Canigueral, S. 2003. Composition of the essential oil of *Lippia chiapasensis*. *In*: Proceedings of the 5th European Colloquium on Ethnopharmacology, Valencia, España.

Hili, P., Evans, C.S. and Veness, R.G. 1997. Antimicrobial action of essential oils: the effect of dimethylsulphoxide on the activity of *cinnamon* oil. Lett. Appl. Microbiol. 24(4): 269–275.

Hoet, S., Stévigny, C., Hérent, M.F. and Quetin-Leclercq, J. 2006. Anti-trypanosomal compounds from the leaf essential oil of *Strychnos spinosa*. Planta Med. 72: 480–482.

Holley, R.A. and Patel, D. 2005. Improvement in shelf-life and safety of perishable food by plant essential oils and smoke antimicrobials. Food Microbiol. 22: 273–292.38: 319–324.

Houghton, P.J. and Raman, A. 1998. Laboratory Handbook for the Fractionation of Natural Extracts, 1st Ed. Chapman and Hall, London, UK.

Huie, C.W. 2002. A review of modern sample-preparation techniques for the extraction and analysis of medicinal plants. Anal. Bioanal. Chem. 373: 23–30.

Hussain, A.I., Anwar, F., Sherazi, S.T.H. and Przybylski, R. 2008. Chemical composition antioxidant and antimicrobial activities of basil (*Ocimum basilicum*) essential oils depends on seasonal variations. Food Chem. 108: 986–995.

Iberl, B., Winkler, G. and Knobloch, K. 1990. Products of allicin transformation: Ajoene and dithiins. Characterization and their determination by HPLC. Planta Med. 56: 202–211.

Imai, H., Osawa, K., Uasuda, H., Hamashima, H., Arai, T. and Sasatsu, M. 2001. Inhibition by the essential oils of peppermint and spearmint of the growth of pathogenic bacteria. Microbios. 106(S1): 31–39.

Ipek, E., Zeytinoglu, H., Okay, S., Tuylu, B.A., Kurkcuoglu, M. and Husnu-Can Baser, K. 2005. Genotoxicity and antigenotoxicity of *Origanum* oil and *carvacrol* evaluated by Ames *Salmonella*/microsomal test. Food Chem. 93: 551–556.

Isman, M.B. and Machial, C.M. 2006. Pesticides based on plant essential oils: from traditional practice to commercialization. pp. 29–44. *In*: M. Rai and M.C. Carpinella (eds.). Elsevier, New York, USA.

Iwalokun, B., Gbenle, G., Adewole, T., Smith, S., Akinsinde, K. and Omonigbehin, E. 2003. Effects of *Ocimum gratissimum* L. essential oil at subinhibitory concentrations on virulent and multidrug-resistant *Shigella* strains from Lagos, Nigeria. APMIS 111: 477–482.

Jian-Yu, S., Zhu, L. and Ying-Juan, T. 2012. Chemical composition and antimicrobial activities of essential oil of *Matricaria songarica*. Int. J. Agric. Biol. 14: 107–110.

Jirovetz, L., Buchbauer, G. and Denkova, Z. 2005. Antimicrobial testing and gaschromatographic analysis of pure oxygenated monoterpenes 1,8-cineol, alpha-terpineol, terpene-4-ol and camphor as well as target compounds in essential oils of pine (*Pinus pinaster*) rosemary (*Rosmarinus officinalis*) and tea tree (*Melaleuca alternifolia*). Sci. Pharm. 73: 27–39.

Juliani, H.R., Karoch, A.R., Juliani, H.R., Trippi, V.S. and Zygadlo, J.A. 2002. Intraspecific variation in the leaf oils of *Lippia junelliana* (mold.) tronc. Ecology 30: 163–170.

Juliani, H.R., Koroch, A., Simon, J.E., Biurrun, F.N., Castellano, V. and Zygadlo, J.A. 2004. Essential oils from argentinean aromatic plants. Acta Horticulturae 624: 491–498.

Kačániová, M., Vukovič, N., Hleba, L., Bobková, A., Pavelková, A., Rovná, K. and Arpášová, H. 2012. Antimicrobial and antiradicals activity of *Origanum vulgare* L. and *Thymus vulgaris* essential oils. J. Microb. Biotech. Food Sci. 2: 263–271.

Kahriman, N., Yaylı, B., Yücel, M., Alpay Karaoglu, S. and Yaylı, N. 2012. Chemical Constituents and antimicrobial activity of the essential oil from *Vicida dianorum* extracted by hydro and microwave distillations. Rec. Nat. Prod. 6(1): 49–56.

Kalemba, D. and Kunicka, A. 2003. Antibacterial and antifungal properties of essential oils. Curr. Med. Chem. 10(17): 813–829.

Kalodera, Z., Pepeljnak, S., Blazevic, N. and Petrak, T. 1997. Chemical composition and antimicrobial activity of *Tanacetum parthenium* essential oil. Pharmazie. 52: 885–886.

Kateryna, K. and Mahendra, R. 2012. Antibacterial activity of *Thymus vulgaris* essential oil alone and in combination with other essential oils. Bioscience 4(2): 50–56.

Kelen, M. and Tepe, B. 2008. Chemical composition, antioxidant and antimicrobial properties of the essentials oils of three *Salvia* species from Turkish flora. Bioresource Technol. 99: 4096–4104.

Kendler, B. 1987. Garlic (*Allium sativum*) and onion (*Allium cepa*): a review of their relationship to cardiovascular disease. Prevent. Med. 16: 670–685.

Khallouki, F., Hmamouchi, M., Younos, C., Soulimani, R., Bessiere, J.M. and Essassi, E.M. 2000. Antibacterial and molluscicidal activities of the essential oil of *Chrysanthemum viscidehirtum*. Fitoterapia. 71(5): 544–546.

Khanavi, M., Hadjiakhoondi, A., Amin, G., Amanzadeh, Y., Rustaiyan, A. and Shafiee, A. 2004. Comparison of the volatile composition of *Stachys persica* Gmel. and *Stachys byzantina* C. Koch. Oils obtained by hydrodistillation and steam distillation. Z. Naturforsch. 59(7/8): 463–467.

Kim, H., Moon, K.H., Ryu, S.Y., Moon, D.C. and Lee, C.K. 1998. Screening and isolation of antibiotic resistance inhibitors from herb materials IV-resistance inhibitors from *Anetheum graveolens* and *Acorus gramineus*. Arch. Pharm. Res. 21(6): 734–737.

Kim, H., Chen, F., Wu, C., Wang, X., Chung, H. and Jin, Z. 2004. Evaluation of antioxidant activity of Australian tea tree (*Melaleuca alternifolia*) oil and its components. J. Agric. Food. Chem. 52: 2849–2854.

Kordali, S., Cakir, A., Mavi, A., Kilic, H. and Yildirim, A. 2005. Screening of chemical composition and antifungal and antioxidant activities of the essential oils from three turkish Artemisia species. J. Agric. Food Chem. 53: 1408–1416.

Kulisic, T., Radonic, A., Katalinic, V. and Milos, M. 2004. Analytical, nutritional and clinical methods use of different methods for testing antioxidative activity of oregano essential oil. Food Chem. 85: 633–640.

Lambert, R.J.W., Skandamis, P.N., Coote, P.J. and Nychas, G.J.E. 2001. A study of the minimum inhibitory concentration and mode of action of oregano essential oil, thymol and carvacrol. J. Appl. Microbiol. 91(3): 453–462.

Lang, G. and Buchbauer, G. 2012. A review on recent research results (2008–2010) on essential oils as antimicrobials and antifungals. A review. Flavour Fragr. J. 27(1): 13–39.

Lara-Júnior, C.R., Lopes de Oliveira, G., Ferreira Mota, B.C., Gonçalves Fernandes, M.F., Figueiredo, L.S., Martins, E.R., Moreira, D.L. and Coelho Kaplan, M.A. 2012. Antimicrobial activity of essential oil of *Piper aduncum* L. (Piperaceae). J. Med. Plants Res. 6: 3800–3805.

Lawrence, B.M. 1993. A planning scheme to evaluate new aromatic plants for the flavour and fragrance industries. pp. 620–627. *In*: J. Janick and J.E. Simon (eds.). New Crops. John Wiley & Sons, New York, USA.

Letellier, M., Budzinski, H., Charrier, L., Capes, S. and Dorthe, A.M. 1999. Optimization by factorial design of focused microwave assisted extraction of polycyclic aromatic hydrocarbons from marine sediment. J. Anal. Chem. 364: 228–237.

Lin, J., Dou, J., Xu, J. and Aisa, H. 2012. Chemical composition, antimicrobial and antitumor activities of the essential oils and crude extracts of *Euphorbia macrorrhiza*. Molecules 17(5): 5030–5039.

Lis-Balchin, M., Deans, S.G. and Eaglesham, E. 1998. Relationship between bioactivity and chemical composition of commercial essential oils. Flavour Fragr. J. 13: 98–104.

Lucchesi, M., Chemat, E.F. and Smalja, J. 2004. Solvent-free microwave extraction of essential oil from aromatic herbs: comparison with conventional hydro-distillation. J. Chromatogr. A. 1043: 323–327.

Mactavish, H. and Harris, D. 2002. An economic study of essential oil production in the UK. ADAS R & D contract report (M137/62) to the Government-Industry Forum on Non Food Uses of Crops.

Mahboubi, M. and Kazempour, N. 2011. Chemical composition and antimicrobial activity of *Satureja hortensis* and *Trachyspermum copticum* essential oil. IJM 3(4): 194–200.

Mahboubi, M., Bokaee, S., Dehdashti, H. and Feizabadi, M.M.I. 2012. Antimicrobial activity of *Mentha piperitae, Zhumeria majdae, Ziziphora tenuior* oils on ESBLs producing isolates of *Klebsiella pneumonia*. Biharean Biol. 6(1): 5–9.

Mahmoodi, A., Roomiani, L., Soltani, M., Basti, A.A., Kamali, A. and Taheri, S. 2012. Chemical composition and antibacterial activity of essential oils and extracts from *Rosmarinus officinalis, Zataria multiflora, Anethum graveolens* and *Eucalyptus globules*. Global Veterinaria. 9(1): 73–79.

Mahmoud, S.S. and Croteau, R.B. 2002. Strategies for transgenic manipulation of monoterpene biosynthesis in plants. Trends Plant Sci. 7(8): 366–373.

Mandal, V., Mohan, Y. and Hemalatha, S. 2007. Microwave assisted extraction. An innovative and promising extraction tool for medicinal plant research. Phcog. Rev. 1: 7–18.

Manindra, M., S., Zafar, H., Harish-Chandra, A. and Vijay-Kant, P. 2011. Essential oils as green pesticides: for sustainable agricultura. Res. J. Pharm. Biol. Chem. 2(4): 100–106.

Mansour, M., Ginawi, O., El-Hadiyah, T., El-Khatib, A., Al-Shabanah, O. and Al-Sawaf, H. 2001. Effects of volatile oil constituents of *Nigella sativa* on carbon tetrachloride-induced hepatotoxicity in mice: evidence for antioxidant effects of thymoquinone. Res. Commun. Mol. Pathol. Pharmacol. 110: 239–251.

Marcano, D. and Hasegawa, M. 2002. Fitoquímica Orgánica, Consejo de Desarrollo Científico y Humanístico de la Universidad Central de Venezuela, Caracas, Venezuela.

Margina, A. and Zheljazkov, V. 1994. Control of mint rust (*Puccinia menthae* Pers.) on mint with fungicides and their effect on essential oil content. J. Essent. Oil Res. 6: 607–615.

Marino, M., Bersani, C. and Comi, G. 1999. Antimicrobial activity of the essential oils of *Thymus vulgaris* L. measured using a bioimpedometric method. J. Food Protect. 62(9): 1017–1023.

Marino, M., Bersani, C. and Comi, G. 2001. Impedance measurements to study the antimicrobial activity of essential oils from Lamiacea and Compositae. Int. J. Food Microbiol. 67: 187–195.

Marković, T., Chatzopoulou, P., Šiljegović, J., Nikolić, M., Glamočlija, J., Ćirić, A. and Soković, M. 2011. Chemical analysis and antimicrobial activities of the essential oils of *Satureja thymbra* L. and *Thymbra spicata* L. and their main components. Arch. Biol. Sci. 63: 457–464.

Mary Helen, P.A., Susheela Gomathy, K., Jayasree, S., Nizzy, A.M., Rajagopal, B. and Jeeva, S. 2012. Phytochemical characterization and antimicrobial activity of *Curcuma xanthorrhiza* Roxb. Asian Pac. J. Trop. Biomed. 2: S637–S640.

Masango, P. 2004. Cleaner production of essential oil by steam distillation. J. Clean Prod. 13: 833–839.

Mau, J., Laib, E., Wang, N., Chen, C., Chang, C. and Chyau, C. 2003. Composition and antioxidant activity of the essential oil from *Curcuma cedoaria*. Food Chem. 82: 583–591.

May, J., Chan, C., King, A., Williams, L. and French, G. 2000. Time-kill studies of tea tree oils on clinical isolates. J. Antimicrob. Chemother. 45: 639–643.

McGimpsey, J.A., Douglas, M.H., Van Klink, J.L., Beauregard, D.A. and Perry, N.B. 1994. Flavour Fragr. J. 9347–352.

Mejlholm, O. and Dalgaard, P. 2002. Antimicrobial effect of essential oils on the seafood spoilage micro-organism *Photobacterium phosphoreum* in liquid media and fish products. Lett. Appl. Microbiol. 34: 27–31.

Messager, S., Hammer, K., Carson, C. and Riley, T. 2005a. Assessment of the antibacterial activity of tea tree oil using the European EN 1276 and EN 12054 standard suspension tests. J. Hosp. Infect. 59: 113–125.

Messager, S., Hammer, K., Carson, C. and Riley, T. 2005b. Effectiveness of hand-cleansing formulations containing tea tree oil assessed *ex vivo* on human skin and *in vivo* with volunteers using European standard EN 1499. J. Hosp. Infect. 59: 220–228.

Mickienė, R., Ragažinskienė, O. and Bakutis, B. 2011. Antimicrobial activity of *Mentha arvensis* L. and *Zingiber officinale* R. essential oils. Biologija. 57(2): 92–97.

Miguel, M.G. 2010. Antioxidant and anti-Inflammatory activities of essential oils: a short review. Molecules 15(12): 9252–9287.

Mihailovi, V., Vukovi, N., Niciforovi, N., Soluji, S., Mladenovi, M., Maškovi, P. and Stankovi, M.S. 2011. Studies on the antimicrobial activity and chemical composition of the essential oils and alcoholic extracts of *Gentiana asclepiadea* L. J. Med. Plants Res. 5: 1164–1174.

Mimica-Dukic, N., Bozin, B., Sokovic, M., Mihajlovic, B. and Matavulj, M. 2003. Antimicrobial and antioxidant activities of three Mentha species essential oils. Plant Med. 69: 413–419.

Minami, M., Kita, M., Nakaya, T., Yamamoto, T., Kuriyama, H. and Imanishi, J. 2003. The inhibitory effect of essential oils on herpes simplex virus type-1 replication *in vitro*. Microbiol. Immunol. 47: 681–684.

Moates, G.K. and Reynolds, J. 1991. Comparison of rose extracts produced by different extraction techniques. J. Essent. Oil Res. 3: 289–294.

Mohammadreza, V.R. 2008. Variation in the essential oil composition of *Artemisia annua* L. of different growth stages cultivated in Iran. African J. Plant Sci. 2: 16–18.

Morales, A., Rojas, J., Moujir, L.M., Araujo, L. and Rondón, M. 2013. Chemical composition, antimicrobial and cytotoxic activities of *Piper hispidum* Sw. essential oil collected in Venezuela. J. App. Pharm. Sci. 3(6): 16–20.

Morita, T., Jinno, K., Kawagishi, H., Arimoto, Y., Suganuma, H., Inakuma, T. and Sugiyama, K. 2003. Hepatoprotective effect of myristicin from nutmeg (*Myristica fragrans*) on lipopolysaccharide/d-galactosamine-induced liver injury. J. Agric. Food. Chem. 51: 1560–1565.

Moteki, H., Hibasami, H., Yamada, Y., Katsuzaki, H., Imai, K. and Komiya, T. 2002. Specific induction of apoptosis by 1,8-cineole in two human leukemia cell lines, but not a in human stomach cancer cell line. Oncol. Rep. 9: 757–760.

Mothana, R.A. 2012. Chemical composition, antimicrobial and antioxidant activities of the essential oil of *Nepeta deflersiana* Growing in Yemen. Rec. Nat. Prod. 6(2): 189–193.

Mulder, T.P., Verspaget, H.W., Sier, C.F., Roelofs, H.M., Ganesh, S., Griffioen, G. and Peters, W.H. 1995. Glutathione Stransferase-pi in colorectal tumors is predictive for overall. Cancer Res. 55: 2696–2702.

Naderi, G., Asgary, S., Ani, M., Sarraf-Zadegan, N. and Safari, M. 2004. Effect of some volatile oils on the affinity of intact and oxidized low-density lipoproteins for adrenal cell Surface receptors. Mol. Cell. Biochem. 267: 59–66.

Nascimento, J.C., Barbosa, L.C.A., Paula, V.L.F., David, J.M., Fontana, R., Silva, L.A.M. and França, R.S. 2011. Chemical composition and antimicrobial activity of essential oils of *Ocimum canum* Sims. and *Ocimum selloi* Benth. An. Acad. Bras. Cienc. 83(3): 787–799.

Newman, D.J., Cragg, G.M. and Snader, K.M. 2003. Natural products as sources of new drugs over the period 1981–2002. J. Natl. Prod. 66(7): 1022–1037.

Nielsen, P.V. and Rios, R. 2000. Inhibition of fungal growth on bread by volatile components from spices and herbs, and their possible application in active packaging, with specia emphasis on mustard essential oil. Int. J. Food Microbiol. 60: 219–229.

Nuran, K., Tosun, G., Terzioglu, S., Karaoglu, S.A. and Nurettin, Y. 2011. Chemical composition and antimicrobial activity of the essential oils from the flower, leaf, and stem of *Senecio pandurifolius*. Rec. Nat. Prod. 5: 82–91.

Nuzu, W.M., Salleh, H.W., Ahmad, F., Yen, K.H. and Sirat, H.M. 2011. Chemical compositions, antioxidant and antimicrobial activities of essential oils of *Piper caninum* Blume. Int. J. Mol. Sci. 12(11): 7720–7731.

Oladosu, I.A., Usman, L.A., Olawore, N.O. and Atata, R.F. 2011. Antibacterial Activity of rhizome essential oils of two types of *Cyprerus articulatus* growing in Nigeria. Adv. Biol. Resear. 5: 179–183.

Omidbeygi, M., Barzegar, M., Hamidi, Z. and Naghdibadi, H. 2007. Antifungal activity of *thyme, summer savory* and clove essential oils against *Aspergillus flavus* in liquid medium and tomato paste. Food Control. 18: 1518–1523.

Ortega-Nieblas, M.M., Robles-Burgueño, M.R., Acedo-Félix, E., González-León, A., Morales-Trejo, A. and Vázquez-Moreno, L. 2011. Chemical composition and antimicrobial activity of oregano (*Lippia palmeri* S. WATS) essential oil. Rev. Fitotec. Mex. 34: 11–17.

Ozbek, H., Uğraş, S., Dülger, H., Bayram, I., Tuncer, I., Oztürk, G. and Oztürk, A. 2003. Hepatoprotective effect of *Foeniculum vulgare* essential oil. Fitoterapia. 74: 317–319.

Packiyasothy, E.V. and Kyle, S. 2002. Antimicrobial properties of some herb essential oils. Food Australia 54: 384–387.

Parija, T. and Das, B. 2003. Involvement of YY1 and its correlation with c-myc in NDEA induced hepatocarcinogenesis, its prevention by d-limonene. Mol. Biol. Rep. 30: 41–46.

Paster, N., Menasherov, M., Ravid, U. and Juven, B.J. 1995. Antifungal activity of *oregano* and *thyme* essential oils applied as fumigant against fungi attacking stored grain. J. Food Prot. 58: 81–85.

Pauli, A. and Schilcher, H. 2010. *In vitro* antimicrobial activities of essential oils monographed in the European Pharmacopoeia, 6th Ed. pp. 353–547. *In*: K.H.C. Baser, G. Buchbauer (eds.). Handbook of Essential Oils: Science, Technology, and Applications. Taylor & Francis, Boca Raton, USA.

Penalver, P., Huerta, B., Borge, C., Astorga, R., Romero, R. and Perea, A. 2005. Antimicrobial activity of five essential oils against origin strains of the Enterobacteriaceae family. APMIS 113: 1–6.

Pichersky, E.J., Noel, P. and Dudareva, N. 2006. Biosynthesis of plant volatiles: nature's diversity and ingenuity. Science 311: 808–811.

Pintore, G., Usai, M., Bradesi, P., Juliano, C., Boatto, G., Tomi, F., Chessa, M., Cerri, R. and Casanova, J. 2002. Chemical composition and antimicrobial activity of *Rosmarinus officinalis* L. oils from Sardinia and Corsica. Flavour Fragr. J. 17: 15–19.

Polatoglu, K., Demirci, F., Demirci, B., Gören, N. and Can Baser, K.H. 2012. Essential oil composition and antimicrobial activities of *Tanacetum chiliophyllum* (Fisch. & Mey.) Schultz Bip. var. *monocephalum* Grierson from Turkey. Rec. Nat. Prod. 6(2): 184–188.

Pourmortazavi, S.M. and Hajimirsadeghi, S.S. 2007. Supercritical fluid extraction in plant essential and volatile oil analysis. J. Chromatogr. A. 1163: 2–24.

Primo, V., Rovera, M., Zanon, S., Oliva, M., Demo, M., Daghero, J. and Sabini, L. 2001. Determination of the antibacterial and antiviral activity of the essential oil from *Minthostachys verticillata* (Griseb.) Epling. Rev. Argent. Microbiol. 33: 113–117.

Reichling, J., Fitzi, J., Hellmann, K., Wegener, T., Bucher, S. and Saller, R. 2004. Topical tea tree oil effective in canine localised pruritic dermatitis–a multi-centre randomized double-blind controlled clinical trial in the veterinary practice. Dtsch. Tierarztl. Wochenschr. 111: 408–414.

Rios, J.L., Recio, M.C. and Villar, A. 1988. Screening methods for natural products with antimicrobial activity: A review of the literature. J. Ethnopharmacol. 23: 127–149.

Rojas, J., Morales, A., Pasquale, S., Márquez, A., Rondón, M., Imré, M. and Veres, K. 2006. Comparative study of the chemical composition of the essential oil of *Lippia oreganoides* L. collected in two different seasons. Nat. Prod. Commun. 1(3): 205–207.

Rojas, J., Velasco, J., Rojas, L.B., Díaz, T., Carmona, J. and Morales, A. 2007. Chemical composition and antibacterial activity of the essential oil of *Baccharis latifolia* Pers. and *B. prunifolia* H. B. & K. (Asteraceae). Nat. Prod. Commun. 2(12): 1245–1248.

Rojas, J., Velasco, J., Morales, A., Rojas, L., Díaz, T.M. Rondón and Carmona, J. 2008. Chemical composition and antibacterial activity of the essential oil of *Baccharis trinervis* (Lam.) Pers. (Asteraceae) collected in Venezuela. Nat. Prod. Commun. 3(3): 369–372.

Rojas, J., Buitrago, A., Rojas, L.B., Morales, A. and Baldovino, S. 2010. Chemical composition of the essential oil of leaves and roots of *Ottoa oenanthoides* (Apiaceae) from Mérida, Venezuela. Nat. Prod. Commun. 5(7): 1115–1116.

Rojas, J., Mender, T., Rojas, L., Guillen, E., Buitrago, A., Lucena, M. and Cárdenas, N. 2011. Estudio comparativo de la composición química y actividad antibacteriana del aceite esencial de *Ruta graveolens* L. recolectada en los estados Miranda y Mérida. Avances en Química. 6(3): 89–93.

Rojas, J., Buitrago, A., Rojas, L., Morales, A., Lucena, M. and Baldovino, S. 2011a. Essential oil composition and antibacterial activity of *Vismia baccifera* fruits collected from Mérida, Venezuela. Nat. Prod. Commun. 6(5): 699–700.

Rojas, J., Buitrago, A., Rojas, L.B. and Morales, A. 2011b. Essential oil composition of *Vismia macrophylla* Leaves (Guttiferae). Nat. Prod. Commun. 6(1): 85–86.

Rojas, J., Buitrago, A., Rojas, L. and Morales, A. 2013a. Chemical composition of *Hypericum laricifolium* Juss essential oil collected from Mérida-Venezuela. JMAPS 2(5): 1–3.

Rojas, J., Buitrago, A., Rojas, L., Cárdenas, J. and Carmona, J. 2013b. Chemical composition of the essential oil of *Croton huberi* Steyerm. (Euphorbiaceae) collected from Táchira-Venezuela. JEOBP 16(3): 646–650.

Rondón, M., Velasco, J., Morales, A., Rojas, J., Carmona, J., Gualtieri, M. and Hernández, V. 2005. Composition of the essential oil and antibacterial activity of *Salvia leucantha* cav. cultivated in Venezuela Andes. Rev. Latinoamer. Quím. 33(2): 55–59.

Rondón, M., Velazco, J., Hernández, J., Pecheneda, M., Morales, A., Rojas, J., Carmona, J. and Díaz, T. 2006a. Chemical composition and antibacterial activity of the essential oil of *Tagetes patula* (Asteraceae) collected in the Venezuela Andes. Rev. Latinoamer. Quím. 34(1): 32–36.

Rondón, M., Araque, M., Morales, A., Gualtieri, M., Rojas, J., Veres, K. and Máthé, I. 2006b. Chemical composition and antibacterial activity of the essential oil of *Lasiocephalus longipenicillatus* (*Senecio longipenicillatus*). Nat. Prod. Commun. 1(2): 113–115.

Rondón, M.E., Delgado, J., Velasco, J., Rojas, J., Rojas, L.B., Morales, A. and Carmona, J. 2008. Chemical composition and antibacterial activity of the essential oil of *Porophyllum ruderale* (Jacq.) Cass. Collected from the Venezuela Andes. Revista de la Facultad de Ciencia. 16: 5–9.

Saei-Dehkordi, S.S., Tajik, H., Moradi, M. and Khalighi-Sigaroodi, F. 2010. Chemical composition of essential oils in *Zataria multiflora* Boiss. from different parts of Iran and their radical scavenging and antimicrobial activity. Food Chem. Toxicol. 48: 1562–1567.

Sandri, I.G., Zacaria, J., Fracaro, F., Delamare, A.P.L. and Echeverrigaray, S. 2007. Antimicrobial activity of the essential oils of Brazilian species of the genus *Cunila* against foodborne pathogens and spoiling bacteria. Food Chem. 103: 823–828.

Santana-Rios, G., Orner, G.A., Amantana, A., Provost, C., Wu, S.Y. and Dashwood, R.H. 2001. Potent antimutagenic activity of white tea in comparison with green tea in the *Salmonella* assay. Mutat. Res. 495: 61–74.

Santos-Gomes, P.C. and Fernandes-Ferreira, M. 2001. Organ and season dependent variation in the essential oil composition of *Salvia officinalis* L. cultivated at two different sites. J. Agric. Food Chem. 49: 2908–16.

Santoyo, S., Cavero, S., Jaime, L., Ibanez, E., Senorans, J. and Reglero, G. 2006. Supercritical carbon dioxide extraction of compounds with antimicrobial activity from *Origanum vulgare* L.: determination of optimal extraction parameters. J. Food Prot. 69: 369–375.

Sarker, S.D., Nahar, L. and Kumarasamy, Y. 2007. Microtitre plate-based antibacterial assay incorporating resazurin as an indicator of cell growth, and its application in the *in vitro* antibacterial screening of phytochemicals. Methods 42(4): 321–324.

Sato, K., Krist, S. and Buchbauer, G. 2006. Antimicrobial effect of transcinnamaldehyde, (-)-perillaldehyde, (-)-citronellal, citral, eugenol, and carvacrol on airborne microbes using an air-washer. Biol. Pharm. Bull. 29: 2292–2294.

Savita, A. and Goyal, S. 2012. Comparative analysis of antimicrobial activity of essential oil of *Ocimum kilimandscharium*. Asian J. Pharm. Clin. Res. 5: 53–55.

Schelz, Z., Hohmann, J. and Molnar, J. 2012. Recent advances in research of antimicrobial effects of essential oils and plant derived compounds on bacteria. Ethnomedicine 1: 179–201.

Schmidt, E., Jirovetz, L., Buchbauer, G., Denkova, Z., Stoyanova, A., Murgov, I. and Geissler, M. 2005. Antimicrobial testing and gas chromatographic analysis of aroma chemicals. JEOBP 8: 99–106.

Schuhmacher, A., Reichling, J. and Schnitzler, P. 2003. Virucidal effect of peppermint oil on the enveloped viruses herpes simplex virus type 1 and type 2 *in vitro*. Phytomedicine 10: 504–510.

Sefidkon, F., Abbasi, K., Jamzad, Z. and Ahmad, S. 2007. The effect of distillation methods and stage of plant growth on the essential oil content and composition of *Setureja rechingeri* Jamzad. Food Chem. 100: 1054–1058.

Seidl, P.R. 2002. Pharmaceuticals from natural products: current trends. An. Acad. Bras. Cienc. 74(1): 145–150.

Selim, S.A. 2011. Composición química, actividad antioxidante y antimicrobiana del aceite esencial y de extractos de metanol de limoncillo egipcio (*Cymbopogon proximus* Stapf.-S.A. Selim). Grasas y Aceites. 62(1): 55–61.

Senatore, F., Napolitano, F. and Ozcan, M. 2000. Composition and antibacterial activity of the essential oil from *Crithmum maritimum* L. (Apiaceae) growing wild in Turkey. Flavour Fragr. J. 15: 186–189.

Şerban, E.S., Ionescu, M., Matinca, D., Maier, C.S. and Bojiţă, M.T. 2011. Screening of the antibacterial and antifungal activity of eight volatile essential oils. FARMACIA 59(3): 440–446.

Seymour, R. 2003. Additional properties and uses of essential oils. J. Clin. Periodontol. 30(S5): 19–21.

Sharopova, F.S., Kukanieva, M.A., Thompson, R.M., Satyalb, P. and Setzer, W.N. 2012. Composition and antimicrobial activity of the essential oil of *Hyssopus seravschanicus* growing wild in Tajikistan. Der. Pharma. Chemica. 4: 961–966.

Silva, M.G.V., Matos, F.J.A., Lopes, P.R.O., Silva, F.O. and Holanda, M.T. 2004. Composition of essential oils from three *Ocimum* species obtained by steam and microwave distillation supercritical CO_2 extraction. ARKIVOC 6: 66–71.

Silva-Santos, A., Antunes, A.M.S., Bizzo, H.R., D'Avila, L.A., Souza-Santos, L.C. and Souza, R.C. 2008. Analysis of uses of essential oils and terpenics/terpenoids compounds by pharmaceutical industry through USPTO granted patents. Rev. Bras. Pl. Med. 10(1): 8–15.

Simon, J.E. 1990. Essential oils and culinary herbs. pp. 472–483. *In*: J. Janick and J.E. Simon (eds.). Advances in New Crops. Timber Press, Portland, USA.

Sinico, C., De Logu, A., Lai, F., Valenti, D., Manconi, M., Loy, G., Bonsignore, L. and Fadda, A.M. 2005. Liposomal incorporation of *Artemisia arborescens* L. essential oil and *in vitro* antiviral activity. Eur. J. Pharm. Biopharm. 59: 161–168.

Skocibusic, M., Bezic, N., Dunkic, V. and Radonic, A. 2004. Antibacterial activity of *Achillea clavennae* essential oil against respiratory tract pathogens. Fitoterapia. 75: 733–736.

Skoula, M. and Harborne, J.B. 2002. The taxonomy and chemistry of *Origanum*. *In*: S.E. Kintzios (ed.). *Oregano*: The Genera *Origanum* and *Lippia*. Taylor & Francis, London, UK.

Sokmen, A., Gulluce, M., Akpulat, A., Dafererad, D., Bektas, T., Moschos, P., Münevver, S. and Fikrettin, S. 2004. The in vitro antimicrobial and antioxidant activities of the essential oils and methanol extracts of endemic *Thymus spathulifolius*. Food Control. 15: 627–634.

Sokovic, M. and Van Griensven, L.J.L.D. 2006. Antimicrobial activity of essential oils and their components against the three major pathogens of the cultivated button mushroom *Agaricus bisporus*. Eur. J. Plant Pathol. 116: 211–224.

Soković, M., Glamočlija, J., Marin, D.P., Brkić, D. and Van Griensven, J.L.D.L. 2010. Antibacterial effects of the essential oils of commonly consumed medicinal herbs using an *in vitro* model. Molecules 15: 7532–7546.

Sood, S., Vyas, D. and Nagar, P.K. 2006. Physiological and biochemical studies during flower development in two rose species. Sci. Hortic. 108: 390–396.

Tadeg, H., Mohammed, E., Asres, K. and Gebre, T. 2005. Antimicrobial activities of some selected traditional Ethiopian medicinal plants used in the treatment of skin disorders. J. Ethnopharmacol. 100: 168–175.

Tajkarimi, M.M., Ibrahim, S.A. and Cliver, D.O. 2010. Antimicrobial herb and spice compounds in food. Food Control 21(9): 1199–1218.

Takahashi, Y., Inaba, N., Kuwahara, S. and Kuki, W. 2003. Antioxidative effect of Citrus essential oil components on human lowdensity lipoprotein *in vitro*. Biosci. Biotechnol. Biochem. 67: 195–197.

Talebi, M., Ghassempour, A., Talebpour, Z., Rassouli, A. and Dolatyari, L. 2004. Optimization of the extraction of paclitaxel from *Taxus baccata* L. by the use of microwave energy. J. Sep. Sci. 27: 1130–36.

Tan, P., Zhong, W. and Cai, W. 2000. Clinical study on treatment of 40 cases of malignant brain tumor by elemene emulsion injection. Chin. J. Integ. Trad. Western Med. 20: 645–648.

Teimouri, M. 2012. Antimicrobial activity and essential oil composition of *Thymus daenensis* Celak from Iran. J. Med. Plants Res. 6(4): 631–635.

Tepe, B., Donmez, E., Unlub, M., Candanc, F., Dafererad, D., Vardar-Unlub, G., Polissioud, M. and Sokmena, A. 2004. Antimicrobial and antioxidative activities of the essential oils and metanol extracts of *Salvia cryptantha* (Montbret et Aucher ex Benth.) and *Salvia multicaulis* (Vahl). Food Chem. 84: 519–525.

Tepe, B., Sokmen, M., Akpulat, A., Daferera, D., Polissiou, M. and Sokmen, A. 2005. Antioxidative activity of the essential oils of *Thymussipyleus* subsp. *sipyleus* var. *Sipyleus* and *Thymus sipyleus* subsp. *sipyleus* var. *Rosulans*. J. Food. Eng. 66: 447–454.

Terblanche, F.C. 2000. Ph.D. Thesis, The characterization, utilization and manufacture of products recovered from *Lippia scaberrima* Sond, Pretoria, University, South Africa.

Torres, L., Rojas, J., Morales, A., Rojas, L., Lucena, M. and Buitrago, A. 2013. Chemical composition and evaluation of antibacterial activity of essential oils of *Ageratina jahnii* and *Ageratina pichinchensis* collected in Mérida-Venezuela. BLACPMA 12(1): 92–98.

Tripathi, A.K., Upadhyay, S., Bhuiyan, M. and Bhattacharya, P.R. 2009. A review on prospects of essential oils as biopesticides in insect pest management. J. Pharmacognosy Phyther. 1: 52–63.

Ultee, A., Kets, E.P.W. and Smid, E.J. 1999. Mechanisms of action of carvacrol on the food-borne pathogen *Bacillus cereus*. Appl. Microbiol. Applied 65: 4606–4610.

Ultee, A., Bennik, M. and Moezelaar, R. 2002. The phenolic hydroxyl group of carvacrol is essential for action against the foodborne pathogen *Bacillus cereus*. Appl. Environ. Microbiol. 68: 1561–1568.

Ultee, A., Kets, E.P.W., Alberda, M., Hoekstra, F.A. and Smid, E.J. 2000. Adaptation of the food-borne pathogen *Bacillus cereus* to carvacrol. Arch. Microbiol. 174: 233–238.

Van Vuuren, S.F., Viljoen, A.M., Ozek, T., Demirici, B. and Baser, K.H.C. 2007. Seasonal and geographical variation of *Heteropyxis natalensis* essential oil and the effect thereof on the antimicrobial activity. S. Afr. J. Bot. 73: 441–448.

Varela, N.P., Friendship, R., Dewey, C. and Valdivieso, A. 2008. Comparison of agar dilution and E-test for antimicrobial susceptibility testing of *Campylobacter coil* isolates recovered from 80 Ontario swine farms. Can. J. Vet. Res. 72(2): 168–174.

Vasudeva, N. and Sharma, T. 2012. Chemical composition and antimicrobial activity of essential oil of *Citrus limettioides* Tanaka. J. Pharm. Technol. Drug Res. 1: 1–7.

Velasco, J., Rojas, J., Salazar, P., Rodríguez, M., Díaz, T., Morales, A. and Rondón, M. 2007. Antibacterial activity of the essential oil of *Lippia oreganoides* against multiresistant bacterial strains of nosocomial origin. Nat. Prod. Commun. 2(1): 85–88.

Viljoen, A.M., Petkar, S., Van-Vuuren, S.F., Figueiredo, A.C., Pedro, L.G. and Barroso, J.G. 2006. Chemo-geographical variation in essential oil composition and the antimicrobial properties of "Wild Mint" *Mentha longifolia* subsp. Polyadena (Lamiaceae) in Southern Africa. J. Essent. Oil Res. 18: 60–65.

Vizcaya, M., Pérez, C., Rojas, J., Rojas, L., Plaza, C., , A. and Perez, P. 2014. Composición química y evaluación de la actividad antifúngica del aceite esencial de corteza de *Vismia baccifera* var. *dealbata*. RSVM 34: 86–90.

Vokou, D., Kokkini, S. and Bessiere, J.M. 1993. Geographic variation of Greek oregano (*Origanum vulgare* ssp. *hirtum*) essential oils. Biochem. Syst. Ecol. 21: 287–295.

Wang, L. and Weller, C.L. 2006. Recent advances in extraction of nutraceuticals from plants. Trends Food Sci. Technol. 17: 300–312.

Whish, J.P.M. and Williams, R.R. 1996. Effects of post-harvest drying on the yield of tea tree oil (*Melaleuca alternifolia*). J. Essent. Oil Res. 8: 47–51.

Wilkinson, J.M, Hipwell, M., Ryan, T. and Cavanagh, H.M.A. 2003. Bioactivity of *Backhousia citriodora*: antibacterial and antifungal activity. J. Agric. Food Chem. 51: 76–81.

Williams, D.G. 1997. The Chemistry of Essential Oils. Micelle Press, New York, USA.

Wink, M. 2003. Evolution of secondary metabolites from an ecological and molecular phylogenetic perspective. Phytochem. 64: 3–19.

Yengopal, V. 2004. Preventative dentistry: essential oils and oral malodour. SADJ 59: 204–206.

Yildirim, A., Cakir, A., Mavi, A., Yalcin, M., Fauler, G. and Taskesenligil, Y. 2004. The variation of antioxidant activities and chemical composition of essential oils of *Teucrium orientale* L. var. *orientale* during harvesting stages. Flavour Fragr. J. 19: 367–372.

Yong-Wei, W., Wei-Cai, Z., Pei-Yu, X., Ya-Jia, L., Rui-Xue, Z., Kai, Z., Yi-Na, H. and Hong, G. 2012. Chemical composition and antimicrobial activity of the essential oil of Kumquat (*Fortunella crassifolia* Swingle) Peel. Int. J. Mol. Sci. 13: 3382–3393.

Zaridah, M.Z., Nor-Azah, M.A., Abu-Said, A. and Mohd-Faridz, Z. 2003. Larvicidal properties of citronella (*Cymbopogon nardus*) essential oils from two different localities. Trop. Biomd. 20(2): 169–174.

Zarkovic, N. 2003. 4-Hydroxynonenal as a bioactive marker of pathophysiological processes. Mol. Aspects. Med. 24: 281–291.

Zellagui, A., Gherraf, N., Ladjel, S. and Hameurlaine, S. 2012. Chemical composition and antibacterial activity of the essential oils of *Ferula vesceritensis* Leaves, endemic in Algeria. Org. Medicinal Chem. Lett. 2: 2–4.

Zhang, F., Chen, B., Xiao, S. and Yao, S. 2005. Optimization and comparison of different extraction techniques for sanguinarine and chelerythrine in fruits of *Macleaya cordata* (Willd) R. Br. Sep., Purif. Technol. 42: 283–290.

Zhou, H. and Liu, C. 2006. Microwave assisted extraction of solanesol from tobacco leaves. J. Chromatogr. A. 1129: 135–39.

Zomorodian, K., Saharkhiz, M.J., Shariati, S., Pakshir, K., Rahimi, M.J. and Khashei, R. 2012. Chemical composition and antimicrobial activities of essential oils from *Nepeta cataria* L. against common causes of food-borne infections. ISRN Pharm. 1: 1–6.

Some Latin American Plants Promising for the Cosmetic, Perfume and Flavor Industries

Amner Muñoz-Acevedo,[1,*] Erika A. Torres,[1] Ricardo G. Gutiérrez,[1] Sandra B. Cotes,[1] Martha Cervantes-Díaz[2] and Geovanna Tafurt-García[3]

Introduction

Natural resources have been studied, used, and exploited by mankind as food sources, for health applications, cosmetics, and others. It is estimated that a quarter of pharmaceutical products are derived from plants and in developing countries, which are rich in biodiversity but not technological leaders, a high percentage of the population depends on traditional botanical medicine for their primary care (Moran et al. 2001). This chemical, biological, and cultural richness provides opportunities: (i) knowledge can be used and be allowed to endure over time; (ii) technological processes can be developed to obtain and secure raw materials in sufficient quantities and in sustainable ways; and (iii) the environment and the communities that live in the surroundings can be protected and conserved (Cotes et al. 2012).

Latin America and the Caribbean (LAC) is the region with the greatest biological/genetic diversity of the planet with six of the so-called megadiverse countries. The region accounts for nearly half of the tropical forest of the world and 32% of the global biodiversity in vascular plants (Arroyo et al. 2010). Latin American ecosystems are valuable sources of new useful models for foods, active ingredients for pharmaceutical products, and potential industrial applications (chemical substances) (UNEP 2010). This biodiversity represents an abundant resource for the LAC region and its countries, which are willing to exploit this potential resource to promote growth, social and economic equities, protected by government policies and regulatory frameworks for defense and sustainable uses (Jaramillo-Castro 2012).

A reliable indicator of these trends has been the increase of commercialization of products derived from biodiversity: transactions in the order of US$750 billions per year are invested in medicinal plants, natural ingredients, rubber, latex, resins and natural dyes; many of these raw materials are employed in the development of cosmetic, food and cosmeceutical industries (Gómez and Mejía 2010).

[1] Chemistry and Biology Research Group – Universidad del Norte, Barranquilla, Colombia.
[2] Environmental Research Group for Sustainable Development – Universidad Santo Tomás, Bucaramanga, Colombia.
[3] Orinoquía's Science Research Group – Universidad Nacional de Colombia, Sede Orinoquía, Arauca, Colombia.
* Corresponding author: amnerm@uninorte.edu.co

This chapter discusses the potential of a group of Latin American plants and some of its applications in the cosmetic, fragrance and flavor industries and is organized in six parts: (1) text mining of scientific publications on Latin American plants, its application areas and the countries investigating them; (2) botanical description of the species; (3) chemical and biological aspects; (4) exploitation and sustainability of natural resources; (5) technological development of patents; and (6) formulations and applications.

Text mining

Text mining was conducted to identify the state of the scientific and inventive activities about promising Latin American species and their applications in cosmetic, flavor and perfume areas. The criterion taken into account was the number of scientific papers indexed in the *Scopus* database (*Elsevier*, BV 2014). The records obtained were processed using the VantagePoint software (*Search Technology,* http://www. thevantagepoint.com/), which is specialized for Text mining[1] and can identify the dynamics, trends, progress of the publications in time, principal authors, countries, and institutions with interests in this subject and the most important application areas. For this search, the name of each plant family was used as a 'keyword' and the consultation period was set between the years 2000–2014.

The scientific trend in publications dealing with botanical families of promising Latin American species is presented in Fig. 14.1. From the structured search equation, the families with the highest number of records were Asteraceae, Fabaceae, Lamiaceae and Euphorbiaceae with 300614, 11947, 5185 and 3985 records, respectively. It is very important to emphasize that some botanical families of Latin American plants had a low number of publications according to the search criteria used. For example, only eight records were recovered regarding the Quillajaceae family.

The degree of applications of medicinal plants or medicinal and aromatic plants was determined by the equation: (*TITLE-ABS-KEY ('medicinal plant*' OR 'medicinal and aromatic plant*') AND TITLE-ABS-KEY (application*ORuse*ORproduct*)) AND DOCTYPE(ar) AND PUBYEAR>1999.* 24532 documents were found.

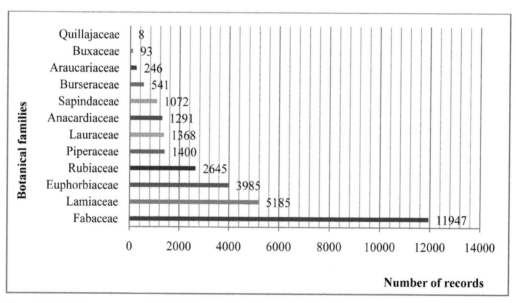

Figure 14.1. Distribution of the number of publications related to the botanical families of promising Latin American species. Source: Information calculated in accordance with Scopus database (Elsevier, B.V. 2014) and processed with *VantagePoint* (Version 8.0, *Search Technology*).

---

[1] Although the data mining could be used, the processing unit itself is not data but 'text' could taken as 'data'. In text mining, the 'text' term is treated as values or data allowing their statistical or scientometric analysis.

Also, the specific applications of the natural products from promising plants were determined by the number of publications based on the search equation: *(TITLE-ABS-KEY ('medicinal plant*' OR 'medicinal and aromatic plant*' OR 'essential oil*' OR 'natural extract*') AND TITLE-ABS-KEY (cosmetic*ORperfume*ORflavor*)) AND DOCTYPE(ar) AND PUBYEAR>1999.* 1829 articles were collected.

The number of scientific publications per year was also monitored as shown in Fig. 14.2; it can be seen that there is a growing interest on the applications of these plants in the cosmetic, perfume and flavor sectors. The year 2013 had the highest number of publications on the subject.

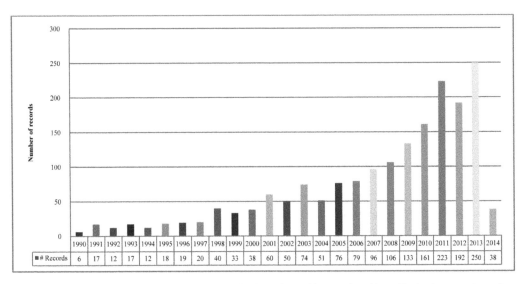

	1990	1991	1992	1993	1994	1995	1996	1997	1998	1999	2000	2001	2002	2003	2004	2005	2006	2007	2008	2009	2010	2011	2012	2013	2014
■# Records	6	17	12	17	12	18	19	20	40	33	38	60	50	74	51	76	79	96	106	133	161	223	192	250	38

Figure 14.2. Dynamics of the number of scientific publications of promising species with application in cosmetic, perfume and flavor industries.
Source: Information calculated in accordance with Scopus database (Elsevier, B.V. 2014) and processed with *VantagePoint* (Version 8.0, *Search Technology*).

Medicinal plants have been used for a long time because of their therapeutical and preservative properties, and as flavoring agents for foods. Some of these properties are partially attributed to Essential Oils (EO). In the 16th century, the 'essential' term was coined by *Paracelsus von Hohenheim*, who appointed the effective component of a drug as '*Quinta essentia*'. Since the 20th century, EO have been preferentially used as ingredients in perfumes, cosmetics and as flavoring agents in foods (Edris 2007).

The number of publications obtained from using the scientific names of promising plants as keywords is presented in Fig. 14.3; the most studied plants have been *Araucaria angustifolia* (327), *Bixa orellana* (302), *Lippia alba* (198) and *Quillaja saponaria* (181); in contrast, the least studied were *Croton malambo* (5) and *Spilanthes oleracea* (2).

The distribution of plant-related scientific publications by countries (worldwide) was also considered. Brazil appeared first with 969 articles, followed by the U.S.A. with 159 articles and India with 86 articles. Within Latin American countries, Brazil was the country with the highest number of publications related to promising plants (838), followed by México (73), Argentina (59), Colombia (58), and Chile (33) (see Fig. 14.4); species that topped the list were *A. angustifolia*, *L. alba*, and *B. orellana* with 262, 158 and 150 records on papers respectively. Most of these papers were produced by Brazil, which is the country with the highest plant diversity in the world.

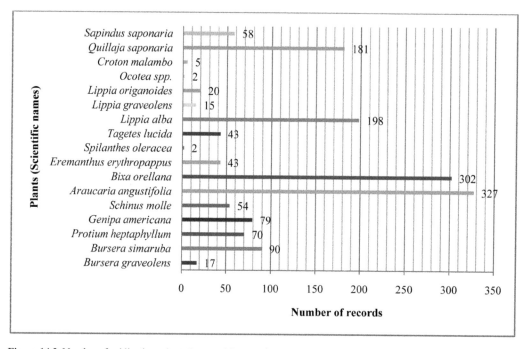

Figure 14.3. Number of publications about the promising species.
Source: Information calculated in accordance with Scopus database (Elsevier, B.V. 2014) and processed with *VantagePoint* (Version 8.0, *Search Technology*).

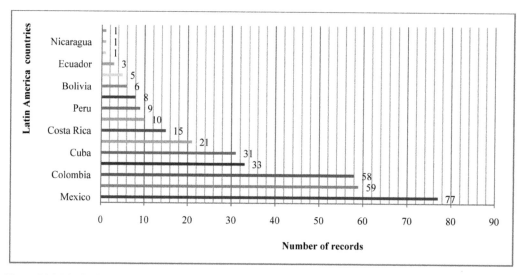

Figure 14.4. Distribution of publications about the plants in Latin American countries.
Source: Information calculated in accordance with Scopus database (Elsevier, B.V. 2014) and processed with *VantagePoint* (Version 8.0, *Search Technology*).

Botanical description of promising species

Plants described below were selected based on review of scientific publications, patents and existing ethnobotanical information. Most of them belong to different families of angiosperms (flowering plants) such as Anacardiaceae, Asteraceae, Bixaceae, Burseraceae, Euphorbiaceae, Lamiaceae, Lauraceae,

Quillajaceae, Rubiaceae and Sapindaceae; and only one is a gymnosperm (Araucariaceae). All these plants are distributed mainly in the dry, wet and gallery forests of Latin America.

An overview on the families formerly mentioned is presented in Table 14.1 and this includes promising plants (Fig. 14.5). Below, botanical aspects of each species are briefly described.

A.
Source:
http://dendrome.ucdavis.edu/treegenes/species/oracjp
g/araucaria_angustifolia_cone.jpg

B.
Source: Taken by S.A. Meyer

C.
Source: Taken by Denise Sasaki

D.
Source: Taken by P.H. Nobre

E.
Source: Taken by María Ignez Calhau

F.
Source: Taken by W. Ariza

G.
Source: Taken by INBio

Figure 14.5. Pictures of some promising Latin American plants.
(A) *Araucaria angustifolia*; (B) *Bursera graveolens*; (C) *Protium heptaphyllum*; (D) *Eremanthus erythropappus*; (E) *Lippia origanoides*; (F) *Ocotea caparrapi*; (G) *Genipa americana*.

Table 14.1. General information of the families related to promising Latin America species—origin, genera/species, metabolites and uses.

Family	Origin	Genera-Species number	Type vegetation	Metabolites	Uses or application	Promising species	Reference
Anacardiaceae	Tropic: South America, Africa	Gen: 80 Spp.: 600–800	Trees/shrubs evergreen/deciduous	Volatile oils, fixed oils; resins	Traditional medicine, food	*Schinus molle* L.	(Cabrera 2005, Medina-Lemos and Fonseca 2009, Muñoz-Garmendia and navarro 2011, Pell et al. 2011)
Araucariaceae	Austral hemisphere (Australia, Oceania), South America	Gen: 4 Spp.: 32	Monoecious or dioecious evergreen trees	Resins	Timber, chemical industry	*Araucaria angustifolia*	(Page 1990, Salazar et al. 2000a, López-González 2007)
Asteraceae	Eurasia, Australasia, South America	Gen: 1600 Spp.: 23000	Herbs/shrubs, annual, biennial, perennial	Volatile oils, resins and latex	Traditional medicine, food	*Eremanthus erythropappus, Spilanthes oleracea, Tagetes lucida*	(Cronquist 1980, Anderberg et al. 2007, Funk et al. 2009)
Bixaceae	Tropic, West Indian Islands	Gen: 4 Spp.: 23	Trees/shrubs-herbs	Volatile oils, yellow/red Savia, dyes	Ethnomedicine, magic-religious	*Bixa orellana*	(Lozada-Pérez, 2003, Poppendieck 2003, Cabrera 2005)
Burseraceae	Tropic: Asia, Africa, America	Gen: 19 (8 in America, 6 endemics) Spp.: 700	Trees/deciduous shrubs	Volatile oils and resins	Traditional medicine, magic-religious, timber	*Bursera graveolens, B. simaruba, Protium heptaphyllum*	(Cuatrecasas 1957, Harrar and Harrar 1962, Weeks et al. 2005, Daly et al. 2011)
Euphorbiaceae	Tropic-Subtropic	Gen: 299 Spp.: 8000	Trees/shrubs-herbs, monoecious or dioecious, annual or perennial	Volatile oils, fixed oils resins, latex	Traditional medicine, food, timber	*Croton malambo*	(Clarke and Lee 1987, Hickey and King 1997, Webster 2014)
Lamiaceae	Cosmopolita	Gen: 236 Spp.: 7173	Trees, shrubs/subshrubs annual, biennial and perennial	Volatile oils	Traditional medicine, food	*Lippia alba, L. graveolens, L. origanoides*	(Hickey and King 1998, Harley et al. 2004)

Lauraceae	Tropic, subtropic	Gen: 50 Spp.: 2500–3500	Trees, shrubs some monoecious	Volatile oils	Timber traditional medicine, food	*Ocotea caparrapi, O. quixos*	(Hickey and King 1998, Rohwer 2011)
Quillajaceae	South America	Gen: 1 Spp.: 2	Trees, shrubs	Saponins	Timber, cleaning	*Quillaja saponaria*	(Kubitzki 2007)
Rubiaceae	Tropic, cosmopolita	Gen: 600 Spp.: 13000	Trees, shrubs, hermaphrodite herbs, dioecious	Volatile oils, natural dyes	Food, traditional medicine, magic-religious	*Genipa americana*	(Hickey and King 1998)
Sapindaceae	Tropic	Gen: 140 Spp.: 1750	Trees, shrubs deciduous vines, dioecious, hermaphrodite	Latex, resins, saponins	Food, timber, cleaning	*Sapindus saponaria*	(Richardson 1995, Acevedo-Rodriguez et al. 2011)

***Araucaria angustifolia* (Bertol.) Kuntze.**—synonyms: *A. brasiliensis* London; *A. brasiliana* Richard; *A. dioica* (Vell.) Stellfeld; *A. elegans* Carriere; *A. ridolfiana* Pi. Savi. Tree (10–35 m) with stratified crown, horizontal foliage clustered at the terminal; straight bole free of branches in almost all its extension (Salazar et al. 2000a). Lanceolate simple leaves and pseudofruits grouped in cone-shaped. Species native to South America distributed in Brazil, Chile and Argentina (Quattrocchini 2012). The plant has a resin that is used in traditional medicine and in the chemical industry. Common names: Cori, Curi, Pinheiro Macaco, Pinheiro branco, Candelabra tree, Missionary Pino (Brazil), Kuntze, kuri´y (Paraguay), Pino Paraná (Argentina) (Salazar et al. 2000b, Quattrocchini 2012, Grandtner and Chevrette 2013).

***Schinus molle* L.**—synonyms: *Sch. angustifolius* Sessé & Moc.; *Sch. areira* L.; *Sch. bituminosus* Salisb; *Sch. occidentalis* Sessé & Moc. Tree (3–16 m) branched and bushy, dark brown color bark, with imparipinnate leaves, lanceolate leaflets and terminal inflorescence; its fruits are reddish globose drupes, fragrant like pepper (Correa and Bernal 1989, Gupta 1995). Native to Perú, distributed from México to Chile (Peter 2012). The plant is characterized by a high content of volatile oils, along with resins and tannins; It is considered a promising tree for 'Países del Convenio Andrés Bello' (Correa and Bernal 1989). Common names: Tree of Life, Molle, Mulli (Bolivia), Aroeira, cures all Balm, Pirul, Aguaribai, Pimenteira, California pepper, Terebinto (Colombia, Brasil, México), Muelle (Chile), Lentisco, Peruvian pepper, Uchu, Turbinto (Perú) (García-Barriga 1992, Gupta 1995, Pérez-Arbeláez 1996, Seidemann 2005, Duke et al. 2009, Bernal et al. 2011, Peter 2012).

***Bixa orellana* Linneo**—synonyms: *B. odorata* Ruiz & Pavon ex G Don, *Orellana americana* Kuntze, *B. acuminata* Poir., *B. orleana* Noronha, *B. tinctoria* Salisb (Gupta 1995, Duke et al. 2009). Perennial shrub (2–10 m) with oval leaves and flowers in paniculate-thyrsi, ovoid fruits with orange-red seed (Quattrocchinni 2012, Grandtner and Chevrette 2013). Native to Central America and distributed from México to Brazil. The seeds are used as dyes and for their digestive properties in traditional medicine. This species is considered promising in 'Países del Convenio Andrés Bello' (Bernal and Correa 1989, García-Barriga 1992). Common names: Achiote, Achote, Uruco, Açafrão, Unoto, Bixa (Bernal and Correa 1989, Gupta 1995, Pérez-Arbeláez 1996, Seidemann 2005, Bernal et al. 2011, Quattrocchini 2012, Grandtner and Chevrette 2013).

***Bursera graveolens* (Kunth) Triana & Planch**—synonyms: *B. tatamaco* Triana & Planch, *B. pilosa* (Engl.) L. Riley, *Amyris graveolens* Spreng, *Terebinthus graveolens* (Kunth) Rose, *Elaphrium graveolens* Kunth. Tree (3–15 m) with branch and bark with reddish-brown color, alternate leaves, unisexual flowers and subglobose fruits. Distributed from México to Perú (García-Barriga 1992, McMullen 1999). The plant has a fragrant resin with a pleasant smell and is considered a tree promising in 'Países del Convenio Andrés Bello' (Correa and Bernal 1990, Gupta 1995). Common names: Caraña, Bija, Palo Santo or Tatamaco (Colombia), Caragana (Bolivia), Caraño (Nicaragua), Copalillo (Honduras), Sassafras (Cuba), Crispin and Huacor (Perú) (García-Barriga 1992, Duke et al. 2009, Bernal et al. 2011, Grandtner and Chevrette 2013).

***Bursera simaruba* (L.) Sarg.**—synonyms: *B. arboreas* (Rose) L. Riley, *B. pilosa* (Engl.) L. Riley, *Pistacia simaruba* L., *Terebinthus simaruba* (L.) W. Wight ex Rose, *Elaphrium simaruba* (L.) Rose (García-Barriga 1992, Gupta 1995). Tree (5–20 m) with a bright reddish-brown bark, shedding into thin strips; fragrant, deciduous and oblong/elliptical leaves (Correa and Bernal 1990, Gupta 1995, Pérez-Arbeláez 1996). This plant is known as Almácigo, in Central and South America. Common names: Indio desnudo (Venezuela), Aceitero (Cuba), Gommier (Antilles), Gumbo-limbo (Belize), Chacajiota (México), Chicchica (Guatemala), Resbala-mono, Caratero, Guácimo (Duke et al. 2009, Grandtner and Chevrette 2013).

***Protium heptaphyllum* (Aubl.)**—synonyms: *Icica heptaphyllum* Aublet, *Pr. aromaticum* Engl., *Pr. insigne* (Triana & Planchon) Engler., *Pr. microphyllum* H.B.K., *Pr. tacamahaca* M. Tree (4–10 m) with pubescent branches of ferruginous color, alternate and imparipinnate leaves, with flowers, and reddish drupes as fruits. Its bark is fragrant and exudes a transparent semi-viscous resin. It is widely distributed in Latin America from Colombia to northern Argentina and Paraguay. Common names: Anime, Breu, Breu branco, Árvoredo-incense (Brazil), Currucay (Perú), Incense, Ysy (Paraguay) (Gupta 1995, Pérez-Arbeláez 1996, Duke et al. 2009, Bernal et al. 2011, Grandtner and Chevrette 2013).

***Eremanthus erythropappus* (DC.) MacLeisch**—synonyms: *Albertinia candolleana* Garner, *Al. claussenni* Sch.Bip. ex Baker, *Vanillosmopsis candolleana* (Gardner) Sch. Bip., *Vernonia glomerata* Sch.Bip. Tree (6–10 m) deciduous with white flowers (Fern 2012). It is distributed in South America, mainly in Brazil. Its bark contains an essential oil of importance for the cosmetic industry. Common names: Candeia-do-serra and Candle tree (Brazil) (Grandtner and Chevrette 2013).

***Spilanthes oleracea* (L.)**—synonyms: *Acmella oleracea* (L.) R.K.; *Sp. fusca* Lam.; *Bidens fervida* Lam.; *Cotula pyrethraria* L.; *Pyrethrum spilanthus* Medik.; *Sp. radicans* Schrad. (Duke et al. 2009, Quattrocchini 2012). Annual herb, succulent, semi-erect (50 cm), with ovate leaves and yellow flowers. Distributed mainly in the tropic particularly Perú-Brazil (Grubben and Denton 2004). The leaves and flowers have a spicy flavor that gives its medicinal and flavoring properties (Vega 2001). Common names: Toothache plant, Pará cress, Pimenteira do Pará, Jambú (Brazil), Chisacá, Yerba del espanto, Botoncillo (Perú), Cabrito (Cuba) (Grubben and Denton 2004, Seidemann 2005, Duke et al. 2009).

***Tagetes lucida* Cav.**—synonyms: *T. lucida* (Sweet) Voss, *T. lucida* f. *lucida*, *T. lucida* subsp. *schiedeana* (Less) Neher. Perennial grass, with yellow flowers and opposite and lanceolate leaves, which have oil glands (with essential oil) that when crushed release an aniseed odor (Raghavan 2007, Perdomo-Roldán and Mondragón-Pichardo 2009). Native to Central América (Gupta 1995, Perdomo-Roldán and Mondragón-Pichardo 2009) and widely distributed through Mesoamerica and South America (Cicció 2005, Bernal et al. 2011). Common names: Pericón, Yerba anís, San Juan Herb, Periquillo (México), Tarragón, Santa María herb, Winter tarragon (Colombia) (Duke et al. 2009, Bernal et al. 2011).

***Croton malambo* H. Karst.**—synonym: *Oxydectes malambo* (H. Karts.) Kuntze (Quattrocchini 2012; Web 1). Tree (3–8 m) with alternate leaves and oily glands in all its parts, fragrant yellowish bark (García-Barriga 1992). The plant is distributed in Colombia and Venezuela and its leaves and bark are used in traditional medicine. Common names: Malambo, palo Matías, palomitas, carnapire and torco (García-Barriga 1992, Pérez-Arbeláez 1996, Pollak-Eltz 2001, Bernal et al. 2011, Quattrocchini 2012).

***Lippia alba* (Mill.) N.E.Br. ex Britton & Wills.**—synonyms: *L. geniculate* H.B.K., *Lantana alba* Mill., *Verbena odorata* (Pers.) Steud, *Zaponia odorata* Pers., *Lippea asperifolia* A. Rich. (Seidemann 2005, Duke et al. 2009, Quattrocchini 2012). Perennial shrub (1–2 m) with opposite and acuminate leaves, fragrant with a lime or mint-like odor; axillary solitary flowers and small fruit inside the chalice. Plant native to America and distributed from Central America to South America. Common names: Pronto alivio, American mint, American salvia, Verbena anise, Quita-dolor, Pennyroyal, Juanislama, Hierbaluisa, Orozul, Sonora, Salsa branca, Erva cidreira, Salsa Limao (Colombia, Perú, Cuba, Brazil, Uruguay) (Gupta 1995, DeBaggio and Tucker 2009, Dellacassa 2010, Bernal et al. 2011, Quattrocchini 2012).

***Lippia graveolens* Kunth.**—synonyms: *L. berlandieri* Schauer, *Lantana origanoides* Martens & Galeotti, *Goniostachyum graveolens* (H.B.K.) Small, *L. graveolens* H.B.K. f. *macrophylla* Moldenke, *L. amentacea* H.B.K. (Seidemann 2005, Sánchez et al. 2007, DeBaggio and Tucker 2009). Perennial fragrant shrub (1–3 m), hairy branches, decussate-elliptic leaves and white flowers (Gupta 1995, Willmann et al. 2000). Distribution: southern North America and Central America (Willmann et al. 2000, DeBaggio and Tucker 2009). Plant used in cooking and traditional medicine (Dellacassa 2010). Common names: Epazote, Orégano, Mexican oregano, Oreganón verde, Orégano de la tierra, Orégano de Monte, Salvia (Bernal et al. 2011, Duke et al. 2009).

***Lippia origanoides* H.B.K.**—synonym: *L. berteroi* Spreng, *L. microphyla* Benth, *L. schomburgkiana* Shauer (Web 1). Shrub (1–2 m) with ovate leaves and glandular trichomes containing essential oils; axillary inflorescences with small white flowers sweetly fragrant. Distribution: Colombia, Venezuela and Brazil (Staples and Kristiansen 1999). Species used in cooking and traditional medicine. Common names:

Orégano, Orégano cimarrón, Orégano de monte, Orégano ancho, Orégano de burro, Culantro cimarrón (García-Barriga 1992, Staples and Kristiansen 1999, Bernal et al. 2011).

***Ocotea caparrapi* (Nates) Dugand**—synonyms: *Nectandra caparrapi* Sandino Groot ex Nates, *N. oleifera* Posada Arango ex Nates, *O. oleifera* Posada Arango, *Oreodaphne okifera* Posada Arango (Bernal and Correa 1994, Pérez-Arbeláez 1996). Tree (15–25 m) perennial, lateral branches with tapered foliage and thick trunk with medicinal resin; abundant, alternate and lanceolate leaves; terminal cluster inflorescences consisting of fragrant small flowers; fruit type ovoid berry (Gupta 1995, Bernal and Correa 1994). This species is characterized by an exuding resin (oil) of the trunk and is endemic to Colombia. Common names: Caparrapí, Palo de Caparrapí, Caparrapí oil, Palo de Aceite, Aceituno, Aguarrás, Arenillo, Canelo real, Laurel canelo (Bernal and Correa 1994, Gupta 1995, Pérez-Arbeláez 1996, Naranjo and Mideros 1997, Bernal et al. 2011).

***Ocotea quixos* (Lam.) Kostern.**—synonyms: *Laurus quixos* Lam., *Licaria quixos* (Lam.) Kostern., *Borbonia peruviana* Juss. ex Steud., *N. cinnamomoides* (Kunth) Nees (Seidemann 2005; Web 1). Perennial, branched and leafy tree (2–5 m) with elliptic leaf-blades and cuneate base; greenish-white flowers and oval fruit with a large seed (Gupta 1995, Rios et al. 2007). Distribution: South America, particularly Ecuador, Colombia and Brazil (Seidemann 2005). It is characterized by its cinnamon-like fragrance. The plant is used for timber, as a condiment and in traditional medicine. Common names: Canelón, Quixos cinnamon, Ispingu, Quichua, Ishpingu, Ispungu, American cinnamon, Ishipingo, Canelo de los Andaquíes (Ecuador, Colombia) (León 1968, Gupta 1995, Seidemann 2005, Cárdenas and Salinas 2006, Rios et al. 2007, Noriega and Dacarro 2008, Ríos et al. 2008).

***Quillaja saponaria* Molina**—synonyms: *Q. molinae* DC., *Q. poeppigii* Walp, *Q. saponaria* Poir, *Q. smegmadermos* D.C., *Smegmadermos emarginatus* Ruíz et Pav. (Gupta 1995, Duke et al. 2009). Monoecious tree (5–10 m) with overhanging branches and greenish-white flowers (Quattrocchini 2012). Endemic to Chile and Perú (Gupta 1995, Kubitzki 2007). Common names: Quillay, Kallay, Soapbark tree, Madera Panamá, Jabón de palo (Chile, Perú, Panamá) (Duke et al. 2009, Quattrocchini 2012).

***Genipa americana* L.**—synonyms: *G. barbata* Presl., *G. caruto* Kunth, *G. excelsa*, *G. onlongifolia*, *G. pubescens* DC., *G. cymosa* Spruce (Gupta 1995, Duke et al. 2009, Quattrocchini 2012, Grandtner 2005). Dioecious tree (10–25 m) with smooth bark, thick foliage, opposite and elliptic leaves, large yellow flowers and fruit type ovoid berry (Gupta 1995, Salazar et al. 2000b). Distribution: southern México, the Caribbean and northern South America. A dye is isolated from the unripe fruit that is used in foods, for tattoos and to stain objects (Gupta 1995, Pérez-Arbeláez 1996, Salazar et al. 2000b). Common names: Carcarutoto (Venezuela), Uito, Huito, Jagua, Ink tree, Guate, Kipará (Colombia), Bicito, Totumillo (Bolivia), Genipapa, Nandipá, Genipapo (Brazil), Maluco (México), White guayatil, Crayo (Guatemala), Irayol, Canuto (El Salvador) (Gupta 1995, Pérez-Arbeláez 1996, Braham et al. 2000, Salazar et al. 2000b, Grandtner 2005, Bernal et al. 2011, Quattrocchini 2012).

***Sapindus saponaria* L.**—synonyms: *S. inaequalis* DC., *Cupaina saponarioides* SW., *S. abruptus* Loureiro, *S. detergens* Roxb., *S. indica* Poir., *S. stenopterus* DC. (García-Barriga 1992, Quattrocchini 2012)]. Deciduous tree (6–18 m) with pinnate alternate and elliptic leaves, small flowers and globose fruit. Distribution: from southern North America to South America (Grandtner and Chevrette 2013). The fruits are traditionally used as soap and the bark and roots, in traditional medicine; the seeds contain edible oil (García-Barriga 1992, Pérez-Arbeláez 1996). Common names: Jaboncillo, Angelina, Chumbimbo, Michú, Chambimbe, Pepo, Majagua (Colombia), Sululu (Bolivia), Choloque (Perú), Para-para (Venezuela), Jurupe (Panama), Sulluco (Ecuador), Saboeiro (Brazil) (García-Barriga 1992, Pérez-Arbeláez 1996, Moreno-Amado 2003, Duke et al. 2009, Bernal et al. 2011).

Biological and chemical aspects

Other important trends were found from the analysis of text mining when focusing on the chemical composition and biological activities of the promising species. Figure 14.6 shows the distribution on the number of publications related to the promising plants and the biological activities evaluated.

It was found that the most important properties for the application in cosmetic, perfume and flavor industries were antimicrobial, antioxidant and cytotoxicity. The cytotoxicity (54 records), antioxidant (51 records), antimicrobial/antifungal (48 records) and anti-inflammatory (44 records) were those biological activities most evaluated on species. Species with the highest number of records on bioactivity assessments were *L. alba* (66), *B. orellana* (55) and *Q. saponaria* (32). The species that had a lesser number of records on biological evaluation were *C. malambo*, *L. origanoides* and *L. graveolens*, with the equal number of records (5).

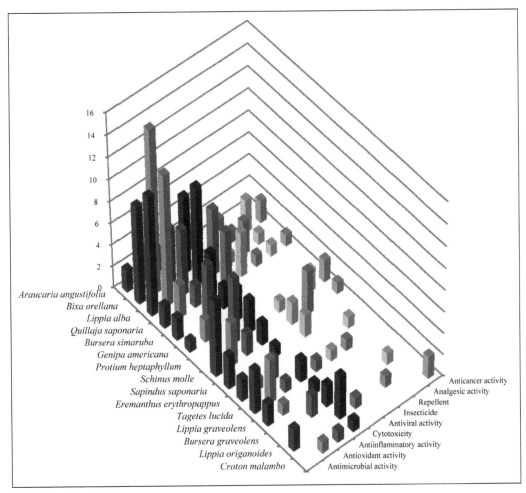

Figure 14.6. Distribution of the number of publications related to promising Latin American species and evaluation of some biological activity.
Source: Information calculated in accordance with Scopus database (Elsevier, B.V. 2014) and processed with *VantagePoint* (Version 8.0, *Search Technology*).

Moreover, based on the search criterion used in the text mining, the chemical compositions of the essential oils and oleoresins were related to the promising species. The compounds found in the EO and oleoresins of these plants were mostly carvone, citral and caryophyllene, each one with 39 records, followed by myrcene (38), linalool (37), germacrene D (26), sabinene (23), and *p*-cymene (22).

Finally, in accordance with the scientometric analysis and in compliance with the selected records by the consulted database and looking for the correlation between biological activity and chemical components, it was found that the greatest number of records was for the antimicrobial activity related to caryophyllene (8), followed by cytotoxicity associated to citral (4), and antioxidant related to β-carotene and thymol (2).

The chemical and biological aspects of each species were investigated from the existing ethnobotanical information. Leaves, bark and resin of **Schinus molle** prepared in different forms of beverages (juice and infusion) and applied topically have been used in ethnomedicine to treat opthalmia, rheumatism, and diarrhea (Hernández et al. 2003, Goldstein and Coleman 2004, Tene et al. 2007). In addition, they were useful to treat amenorrhea, bronchitis, gingivitis, tuberculosis, ulcers, urethritis, urogenital diseases and intestinal parasites. Furthermore, other properties were attributed: astringent, balsamic, diuretic, expectorant, stomachic and tonic (Johnson 1999, Cerón 2006, Orwa et al. 2009, Dellavalle et al. 2011, Huamantupa et al. 2011). The tree produces a fragrant resin and the fruits are used to make beverages; due to the taste of their seeds, they are used to adulterate the pepper (Kramer 1957). The fruits contain EO with a smell similar to a mixture of fennel-pepper. Chemical composition, biological activities, parts studied and place of origin (particularly Latin America) for the EO and extracts of Aroeria are showed in Table 14.2.

The components with greater presence in the EO of the different parts of the plant were α-phellandrene, β-phellandrene, myrcene, and limonene. Further, the bioactivities with the best results were antioxidant (β-carotene/linoleic acid—57 and 19% inhibition of oxidation of the leaves and fruit, compared with 1% ascorbic acid), antimicrobial (*Staphylococcus epidermidis*—MIC 63 µg/mL and 125 µg/mL, of the leaves and fruits), and toxicity (lower in mice) (Martins et al. 2014). Other promising results were the anticancer (Díaz et al. 2008, Bendaoud et al. 2010, Simionatto et al. 2011), repellency and insecticidal activities (Abdel-Sattar et al. 2010, Deveci et al. 2010) that would permit other applications in the cosmeceutical and agrochemical areas. Irritability, phototoxicity, sensitization and acute toxicity data, reported for the EO of *Sch. molle* fruits (Opdyke 1979), have allowed their use in the flavor and fragrance industries. Scheme 1 shows the structures of the most abundant components found in EO from *Sch. molle* according to Table 14.2.

Information about traditional medicine related to **A. angustifolia** was based on the use of the tinctures, brew and syrup obtained from the nodes, leaves, bark and resin, administered topically or orally for the treatment of rheumatism, kidney disease, sexually transmitted diseases, varicose veins, muscle injuries, respiratory infections, fatigue, anemia, dry skin, herpes zoster, and wounds (Freitas et al. 2009).

The secondary metabolites isolated and chemically characterized from the wood and resin have been type lignans, e.g., (+)-pinoresinol, (+)-lariciresinol, (−)-secoisolariciresinol and derivatives; norlignans, e.g., criptoresinol, 2,3-bis-(p-hydroxyphenyl)-2-cyclopentene-1-one and 4-4'-dihydroxychalcone (Fonseca et al. 1979, Ohashi et al. 1992). The study of the chemical composition of the calluses obtained by *in-vitro* cultures of this species allowed isolating the octadecyl p-coumarate and octadecyl ferulate isomers, three biflavonoids type amentoflavone, trans-communic acid, vanillin, p-hydroxybenzaldehyde, coniferaldehyde, cabreuvine and irisolidone (Fonseca et al. 2000). Other biflavonoid compounds type amentoflavone were isolated from the leaves of the species in adult stage (Yamaguchi et al. 2005) (Scheme 2).

The scientific validation of the traditional uses of *A. angustifolia* has proven the antiviral (herpes simplex type 1—selectivity index value of 38–58, attributed to proanthocyanidins from leaves), antioxidant (Fenton reaction, lipid peroxidation, Folin-Ciocalteu reagent, DPPH· and enzymatic methods—biflavonoids type amentoflavone, polyphenols and flavonoids are responsible for the activity), antigenotoxicity (against MRC-5 human lung fibroblast cell), antidepressant and anti-inflammatory (lecithin responsible for activities) activities and their capacity as photoprotective agent against UV radiation (amentoflavones responsible for photoprotection) (Yamaguchi et al. 2005, Mota et al. 2006, Freitas et al. 2009, Vasconcelos et al. 2009, Yamaguchi et al. 2009, Koehnlein et al. 2012, Shahzad Aslam et al. 2013, Souza et al. 2014).

In general, the leaves of *B. orellana* have been used in botanical medicine to treat skin problems and liver diseases; they have also been prepared as a poultice to heal or in decoction for throat and stomach pains or as a diuretic and mild laxative; in infusion, these have been used in the treatment of jaundice,

Table 14.2. Chemical compositions and bioactivities of EO and extracts of *Sch. Molle.*

Components (Relative amount, %)	Part used	Yield EO-Extract, %	Origin	Bioactivity	Reference
Myrcene (20%) α-phellandrene (17%)	Fruits	2.8	U.S.A.	----	(Bernhard et al. 1983)
β-Pinene (31%) α-pinene (23%)	Leaves	1.2	Costa Rica	Cytotoxicity (leukemic cell line K-562–IC$_{50}$: 79 µg/mL, 48 hours) antioxidant (DPPH–IC$_{50}$: 36 µg/mL)	(Díaz et al. 2008)
epi-α-Cadinol (27%) caryophyllene oxide (10%)	Leaves	----	Brazil	Cytotoxicity (human tumor cell line: K-562–IC$_{50}$: 0.9 µg/mL; NCI-ADR/RES–IC$_{50}$: 3.4 µg/mL) antimicrobial antioxidant allelopathic	(Simionatto et al. 2011)
Limonene (41%) β-caryophyllene (16%) bicyclogermacrene (12%)	Mix leaves twig	---- (SFE extract)		----	(Barroso et al. 2011)
Bicyclogermacrene (29%)	Leaves	----	Uruguay		(Rossini et al. 1996)
Myrcene (40%) p-cymene (20%)	Fruits	----	México	Antibacterial (*St. pneumoniae*, MIC 125 mg/mL)	(Pérez-López et al. 2011)
Sabinene (45%) α-pinene (12%)	Mix fruits, leaves steam	----	Argentina	Antioxidant (DPPH–EC$_{50}$: 29.1 ± 0.4 g mtra/g DPPH)	(Guala et al. 2009)
epi-Bicyclosesquiphellandrene (19%), β-pinene (15%) α-pinene (12%)	Mix leaves branch	----		----	(Chamorro et al. 2012)
α-Phellandrene (26%) β-phellandrene (12%)	----	----	Algeria	Antibacterial	(Belhamel et al. 2009)
α-Phellandrene (21%) β-phellandrene (11%)	Aereal parts	2.0	Ethiopia	Antibacterial	(Ljalem and Unnithan 2014)
Myrcene + α-phellandrene (10–29%), limonene + β-phellandrene (16–24%) β-caryophyllene (11%)	Leaves	0.7–2.3	India	Fungitoxicity	(Dikshit et al. 1986)
α-Phellandrene (55%) β-phellandrene (15%) limonene (14%)	Fruits	----	Italy		(Maffei and Chialva 1990)
α-Phellandrene (30%) elemol (13%)	Leaves				
α-Phellandrene (26%) β-phellandrene + limonene (21%), elemol (11%)	----	---- (SFE extract)			(Marongiu et al. 2004)

Table 14.2. contd....

Table 14.2. contd.

Components (Relative amount, %)	Part used	Yield EO-Extract, %	Origin	Bioactivity	Reference
α-Phellandrene (26%) limonene (12%) myrcene (11%) β-phellandrene (10%)	Leaves	1.1	Portugal	Antimicrobial (Gram+, Gram−, fungi) antioxidant (DPPH, linoleic acid/β-carotene) toxicity (*Artemia salina*, in mice)	(Martins et al. 2014)
Myrcene (51%) limonene (14%) α-phellandrene (14%) β-phellandrene (11%)	Fruits	0.9			
p-Cymene (33%) β-pinene (19%)	Fruits	4.3	Saudi Arabia	Insecticidal repellent	(Abdel-Sattar et al. 2010)
p-Cymene (69%)	Leaves	2.1			
α-Phellandrene (47%) β-phellandrene (28%)	Leaves	1.1	Tunisia	---	(Ennigrou et al. 2011)
α-Phellandrene (36%) β-phellandrene (29%) β-pinene (16%)	Fruits	0.8		Antimicrobial (Gram+, Gram−, fungi)	(Hayouni et al. 2008)
α-Phellandrene (46%) β-phellandrene (21%)	Fruits	2.7		Anticancer (MCF-7) antioxidant (DPPH, ABTS$^+$)	(Bendaoud et al. 2010)
α-Phellandrene (35–40%) limonene + β-phellandrene (22–37%) myrcene (8–25%)	Fruits	1.2–5.2		Antimicrobial Antioxidant	(Hosni et al. 2011)
α-Phellandrene (46%) β-phellandrene (14%) limonene (13%)	Leaves	1.7	Turkey	---	(Baser et al. 1997)
α-Phellandrene (22–38%) β-phellandrene (10–12%) limonene (10–12%)	Fruits	3.1–4.3			
δ-Cadinene (11%) α-cadinol (11%)	Leaves fruits	3.0–3.8		Antimicrobial Repellent	(Deveci et al. 2010)
Germacrene D (21%) β-caryophyllene (14%)	Leaves fruits	0.5–2.7 (hexane extract)			

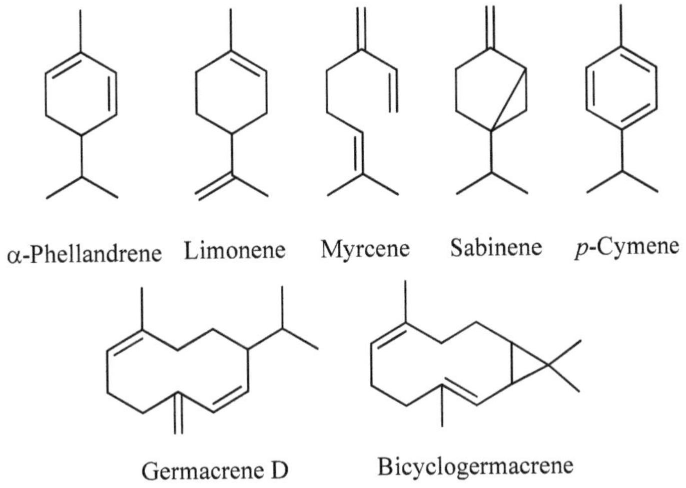

α-Phellandrene   Limonene   Myrcene   Sabinene   *p*-Cymene

Germacrene D   Bicyclogermacrene

Scheme 1. Structures of some components present in the EO from *Sch. molle* L.

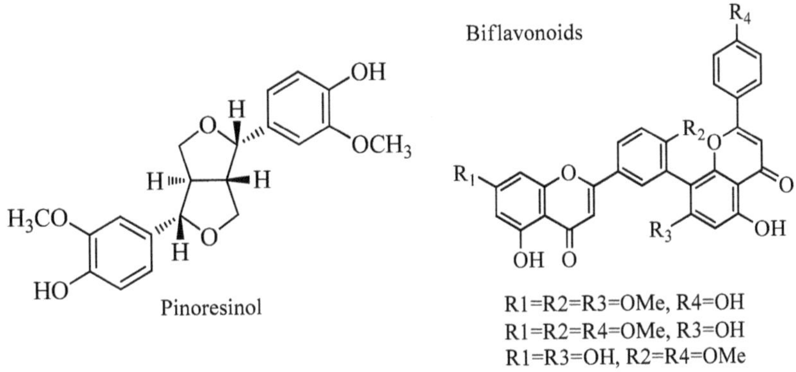

Biflavonoids

Pinoresinol

R1=R2=R3=OMe, R4=OH
R1=R2=R4=OMe, R3=OH
R1=R3=OH, R2=R4=OMe

Scheme 2. Structures of pinoresinol and biflavonoids from *A. angustifolia*.

dysentery and diarrhea (Radhika et al. 2010, Shilpi et al. 2006). The seeds are used as astringent or antipyretic and their dye as a repellent (Bernal and Correa 1989, Johnson 1999, Shilpi et al. 2006). The chemistry of non-polar extracts of the leaves (petroleum ether, chloroform and ethyl acetate) is related to compounds type steroid, terpenoids, phenols and tannins. Polar extracts (methanol and water) are rich in flavonoids, terpenoids, phenols, tannins, alkaloids, glycosides, carbohydrates and proteins. The yields are between 1–23% (Radhika et al. 2010). Resinous coat of the seed is a source of some terpenoids (*trans*-geranylgeraniol, farnesylacetone, etc.) and of apocarotenoids (annatto colorings) (Jondiko and Pattenden 1989, Mercadante et al. 1996). *cis*-Bixin (main compound) represents 80% (Scheme 3); *trans*-bixin, *cis*-norbixin, *trans*-norbixin, resins, orellin (yellow dye), volatile compounds and fatty acids complete the total 100% (Mercadante et al. 1997, Mercadante et al. 1999, Rodrígues et al. 2014).

It has been proved that ethanol extract of Annatto has protective-preventive effects dose-dependent on Albino rats (50 mg/kg body weight—best dose) related to the antimalarial activity and monitored by TBARS, GSH and CAT (Conrad et al. 2013). Besides, polar extracts (methanol/ethanol) of the leaves and seeds have been evaluated as antimicrobial (Gram+, Gram– and fungi); leaves gave the best results (Fleischer et al. 2003, Shilpi et al. 2006). Other bioactivities evaluated of the leaf extracts were antioxidant (DPPH—IC_{50}: 22 µg/mL), anti-diarrheal (induced by castor oil), anticonvulsive (induced by strychnine) and gastrointestinal motility (Shilpi et al. 2006). Extracts of different polarities of the seeds showed high

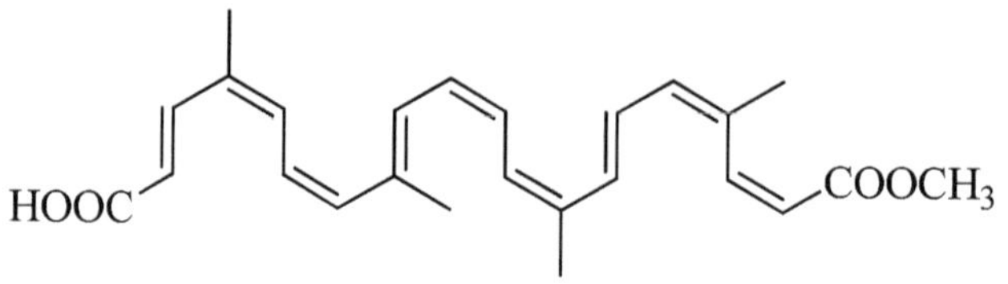

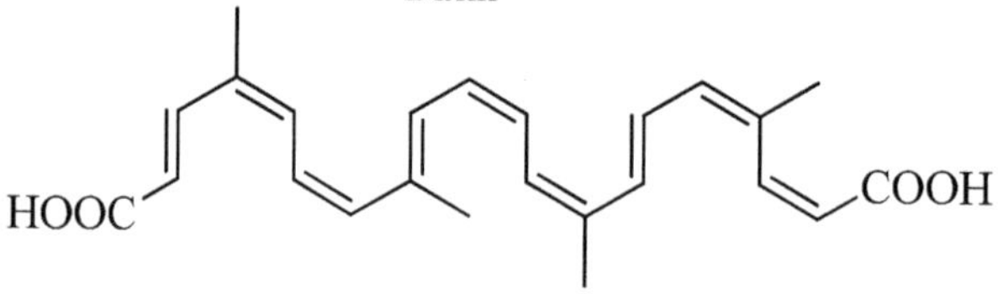

Bixin

Norbixin

Scheme 3. Apocarotenoids found in *B. orellana.*

antioxidant capacity in different model systems (Kiokias and Gordon 2003, Campos Chisté et al. 2011). The most current applications of these apocarotenoids are by the food industry and there is a great potential for the textile and tannery, cosmetic and pharmaceutical industries (Selvi et al. 2013).

Different parts of the species *B. graveolens, B. simaruba,* and *P. heptaphyllum,* e.g., leaves, stems, bark and resin, are used in herbal medicine in some Latin American countries such as Peru, Costa Rica, Nicaragua, Guatemala, Cuba, Colobia, Brazil, Bolivia, Mexico and Paraguay. These plants are prepared as infusions, decoctions, smokes, baths, poultices and compresses, due to some therapeutic effects as healing, antipyretic, anti-inflammatory, antitumor, antidiarrheal, antiseptic, astringent, ananlgesic, tonic, diaphoretic, expectorant insecticide, mosquito-repellent and to treat skin problems (wounds, rashes), urinary tract infections, respiratory diseases, blood purification and anemia, toothache and abdominal pain, caries, colics, cold (cold and cough), influenza, dermatitis, rheumatism, and swelling. Also, these plants are used as fragrances. In the same way, the resins of these plants have been used for healing, as an analgesic or a narcotic, during religious rituals (Soukup 1970, Correa and Bernal 1990, Álvarez 1991, García-barriga 1992, Gupta 1995, Sosa et al. 2002, Sánchez et al. 2006, Acero 2007, Tene et al. 2007, Rüdiger et al. 2007, Alonso-Castro et al. 2011, Bernal et al. 2011). Table 14.3 presents the chemical compositions and the biological activities associated to the EO and extracts of the studied parts from *Bursera* spp. and *Protium* sp.

Components with greater presence in the EO and extracts of the different parts of *Bursera* spp. and *Protium* sp. were α-pinene, β-pinene, terpinolene, *p*-cymene, myrcene, limonene, α-terpinene, 1,8-cineole, mintlactone, *iso*-mintlactone, germacrene D, viridiflorol, guaiol, α-amyrin, β-amyrin, burseranin, lupeol, and brein (Scheme 4). Also, the most important bioactivities found were cytotoxicity, antimicrobial, anti-inflammatory, antitumor, antioxidant, gastroprotective and repellent.

Bark/stems extracts from *B. graveolens* showed bioactivities as antimicrobial (against Gram+ bacteria, 50–250 mg/mL), anti-inflammatory (% inhibition: 50–72% (Robles et al. 2005) and % inflammation: 14% (Manzano et al. 2009)) and cytotoxicity (HT1080, ED_{50}: 60 mg/L (Nakanishi et al. 2005)); antiproliferative properties on breast cancer cells (MCF-7, IC_{50}: 49 mg/L) for the EO were reported, antiparasitic (*L. amazonensis,* IC_{50}: 37 mg/L), cytotoxicity (BALB, IC_{50}: 104 mg/L) (Monzote et al. 2012), and antibacterial activities (Gram+ bacteria, MIC: 1–7 mg/mL; MBC: 2–9 mg/mL) (Luján-Hidalgo et al. 2012). The EO from *B. simaruba* showed anticancer activity in the cell lines A-549 (GI_{50}: 42 mg/L) and DLD-1 (GI_{50}: 48 mg/L) (Sylvestre et al. 2007), and antibacterial (6 strains Gram+ and Gram–) (Junor et al. 2007). Finally, the EO form *P. heptaphyllum* showed cytotoxicity (neoplastic cell, % inhibition: 59–67%) (Siani et al. 2011), repellency (mites, 2–10 µl EO/L air) (Pontes et al. 2007), antioxidant, anti-inflammatory (Siani et al. 1999, Bandeira et al. 2006) and antinociceptive activities (Rao et al. 2007) and gastroprotective effect (Araujo et al. 2011). The extracts also showed antimicrobial activity (Gram+, Gram– and fungi, MIC: 62–500 mg/L) (Violante et al. 2012), anti-inflammatory and gastoprotector effects (lesion area, 2–9 mm², dosage 200–400 mg/kg) (Oliveira et al. 2004).

Table 14.3. Chemical compositions and biological activities of EO and extracts from *Bursera* spp. and *Protium* sp.

Components (Relative amount, %)	Part used	Yield EO-Extract, %	Origin	Bioactivity	Reference
***Bursera graveolens* (Kunth) Triana & Planch**					
Germacrene D (21%) β-caryophyllene (18%)	Leaves	SDE (NR)	Colombia	----	(Muñoz-Acevedo et al. 2013)
Mintlactone (43–45%) iso-mintlactone (6–7%) 3-hydroxymintlactone (6–16%)	Branches and bark				
Limonene (23%) mintlactone (16%) mintlactone derivative (15%) Pulegone (12%)	Resin				
Limonene (48%) myrcene (20%) caryophyllene oxide (14%)	Leaves	0.1		----	(Leyva et al. 2007)
Limonene (42%) menthofuran (15%)	Stems	0.1			
Tetracyclic terpenes β-elemonic acid α-elemolic acid	Bark	0.7 (different extracts)		Antimicrobial anti-inflammatory	(Robles et al. 2005)
Limonene (26%)	----	----	Cuba	Antitumor (MCF-7) anti-leshmaniasis cytotoxicity	(Monzote et al. 2012)
Limonene (31%) β-ocimene (21%) β-elemene (11%)	Leaves	1.1		----	(Carmona et al. 2007)
Limonene (59%) α-terpineol (11%)	Stems	3.7	Ecuador		(Young et al. 2007)
Viridiflorol (71%)	Branches	0.05		Anti-inflammatory	(Manzano et al. 2009)
Limonene (43%) β-ocimene (17%) β-elemene (12%)	Leaves	0.1	México	Antimicrobial (Gram+)	(Luján-Hidalgo et al. 2012)
Burseranin picropolygamain, lupeol epi-lupeol	Stems	4.8 (methanol extract)		Cytotoxicity	(Nakanishi et al. 2005)

Table 14.3. contd....

Table 14.3. contd.

Components (Relative amount, %)	Part used	Yield EO-Extract, %	Origin	Bioactivity	Reference
3-Hydroxymintlactone 2,3-didehydromintlactone Mintlactone, *iso*-mintlactone 10-*epi*-γ-eudesmol, juneol *iso*-dihydroagarofuran	Branches	32 (diethyl ether extract)	Perú	---	(Yukawa and Iwabuchi 2003, Yukawa et al. 2004a, Yukawa et al. 2004b)
***Bursera simaruba* (L.) Sarg.**					
α-Terpinene (26%) δ-terpinene (20%) α-pinene (18%) *p*-cymene (16%)	Fruits	2.0	Costa Rica	---	(Rosales-Ovares and Ciccó-Alberti 2002)
Limonene (47%) β-caryophyllene (15%) α-humulene (13%)	Leaves	---	Guadeloupe	Anticancer (A-549; DLD-1)	(Sylvestre et al. 2007)
α-Amyrin β-amyrin, lupeol epi-lupeol, epiglutinol picropolygamain	Resin	(CHCl$_3$ extract)	México	---	(Peraza-Sánchez et al.1995)
Yatein hinokinin burschernin	Bark	(ether, CHCl$_3$, methanol)	Belize	---	(Maldini et al. 2009)
α-Pinene (10%) *trans*-cadina-1(6),4-diene (10%) β-caryophyllene (9%)	Leaves	0.07	Jamaica	Antibacterial	(Junor et al. 2007, Junor et al. 2008)
α-Pinene (32%) β-pinene (14%)	Bark	0.03			
α-Pinene (28%) β-pinene (24%) terpinen-4-ol (13%)	Fruits	0.3			
Neophytadiene Phytosterols, α-amyrin methyl-β-peltatin A methyl-β-peltatin B	Leaves	5.2 (hexane extract) 3.9 (methanol extract)	Venezuela	Anti-inflammatory	(Noguera et al. 2004, Carretero et al. 2008)

Protium heptaphyllum (Aubl.)

Composition	Part	Yield	Country	Activity	Reference
Terpinolene (42%) p-cymen-8-ol (14%) limonene (12%) p-cymene (5–40%) tetradecane (13%) dihydro-4-carene (12%)	Resin	8.6–11.3	Brazil	---	(Marques et al. 2010)
Terpinolene (22%) Dillapiole (16%) p-cymene (11%) p-cymen-8-ol (11%)	Resin	10		Anti-inflammatory antinociceptive toxicity hemolytic cell proliferation	(Siani et al. 1999)
Myrcene (35%) α-pinene (27%) sabinene (11%) terpinolene (28%) p-cymene (16%)	Resin	---		Cytotoxicity	(Siani et al. 2011)
Myrcene (19%) β-caryophyllene (18%)	Leaves	0.3		Antimicrobial antioxidant	(Bandeira et al. 2001, Bandeira et al. 2006)
α-Pinene (71%)	Fruits	3.0			
Terpinolene (28%) α-phellandrene (17%) limonene (17%) α-pinene (10%)	Resin	11.0			
1,8-Cineole (59%) α-terpinene (14%) α-phellandrene (10%)	Resin	---		Antinociceptive	(Rao et al. 2007)
9-*epi*-Caryophyllene (21%) 14-hydroxi-9-epi-caryophyllene (17%) *trans*-isolongifolanone (11%)	Leaves	0.7		Toxicity and repellent on mite	(Pontes et al. 2007)
α-Terpinene (48%)	Fruits	1.3			
β-Elemene (22%) terpinolene (15%) β-caryophyllene (11%)	Leaves	0.2		---	(Zoghbi et al. 1995)
Terpinolene (40%)	Stems	0.06			

Table 14.3. contd....

Table 14.3. contd.

Components (Relative amount, %)	Part used	Yield EO-Extract, %	Origin	Bioactivity	Reference
p-Menth-3-ene-1,2,8-triol α-amyrin; β-amyrin; brein quercetin; catechin scopoletin	Leaves Fruits Resin	---- 1.5–3.0 11.0		Antiacetyl-cholinesterase	(Bandeira et al. 2002)
β-Caryophyllene (32%) germacrene B (17%) β-ocimene (16%) Ledene (15%)	Leaves	0.02		---	(Citó et al. 2006)
Limonene (93%)	Fruits	0.5			
----	Bark	---- (ethanol, ethyl acetate, CH$_2$Cl$_2$, water)		Antimicrobial	(Violante et al. 2012)
α-Pinene (40%) p-mentha-1,4(8)-diene (12%) α-phellandrene (10%)	Leaves stems	----		Gastroprotective effect antioxidant, toxicity	(Araujo et al. 2011)
Lupeol, phytosterols	Leaves	10 (ethanol extract)		---	(Almeida et al. 2002)
Coumarins phytosterols	Stems bark	12 (ethanol extract)			
α-Amyrin β-amyrin malinadiol brein	Resin	0.1		Gastroprotective effect anti-inflammatory toxicity	(Susunaga et al. 2001, Oliveira et al. 2004)
Germacrene D (14%) germacrene B (13%) bicyclogermacrene (12%)	Flowers	SDE (NR)	Colombia	---	(Tafurt-García and Muñoz-Acevedo 2012)
Guaiol (14%) 1,10-di-epi-cubenol (8%) guaiol (7%)	Leaves				
Germacrene D (28%) germacrene B (13%) bicyclogermacrene (12%)	Bark				
p-Cymene (30%) α-pinene (22%) limonene (14%)	Resin				

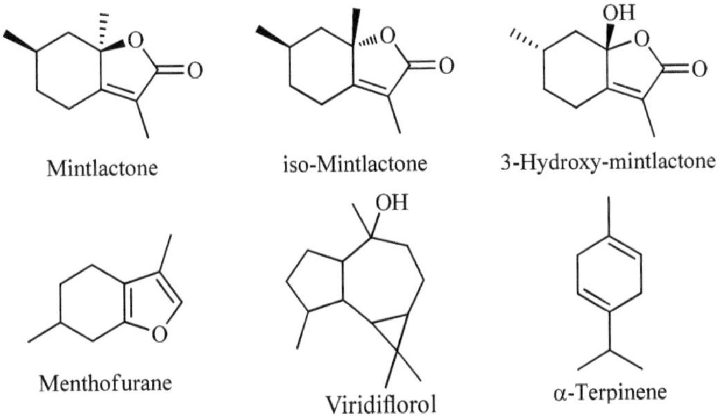

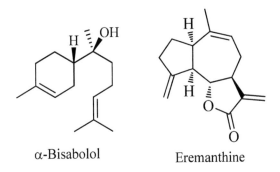

Mintlactone · iso-Mintlactone · 3-Hydroxy-mintlactone

Menthofurane · Viridiflorol · α-Terpinene

Scheme 4. Terpenoids found in *Bursera* spp. and *Protium* sp.

In Brazilian traditional medicine, ***Eremanthus erythropappus*** has been utilized as antimicrobial and antiphlogistic agents (Silvério et al. 2013). Validation of botanical medicine allowed the characterization of the EO and extracts and the evaluation of the anti-inflammatory, antinociceptive and antimicrobial properties (Silvério et al. 2008, Sousa et al. 2008, Soares and Fabri 2011). Thus, the chemical compositions of the EO of the different parts (flowers, leaves and branches) were represented by α-bisabolol (93%—branches, Scheme 5), β-caryophyllene (18–22%), germacrene D (12–17%), α-copaene (9–10%), α-muurolol (10%), and β-pinene (9%).

α-Bisabolol · Eremanthine

Scheme 5. Structures of α-bisabolol and eremanthine from *E. erythropappus*.

The yields of EO were 0.2–0.4%, and antimicrobial (Gram+, Gram– and fungi, MIC: 10–500 mg/L) and antioxidant (DPPH·, IC_{50}: 39–102 mg/L; Fe^{+3} reducing power, IC_{50}: 110–170 mg/L) activities were promising (Silvério et al. 2013). In the hexane, dichloromethane, and ethanol extracts from branches were isolated eremanthine (sesquiterpene-lactone) and other α-methylene-γ-lactones that had activity against *Schistosoma mansoni* (50–75 mg/L) (Alves 2011, Almeida et al. 2012); other ethanol extracts also showed anti-inflammatory, antinociceptive and anti-ulcer activities (Silvério et al. 2008); and the leaf extracts, of different chemical nature, were evaluated using the DPPH· (IC_{50}: 29–34 mg/L—antioxidant) and *Artemia salina* (LC_{50}: 64 mg/L—toxicity) assays (Soares and Fabri 2011).

Species belonging to the genus *Spilanthes* are commonly known as 'plants for toothache' referring to the analgesic action of the N-alkylamide constituents present in their parts (Paulraj et al. 2013). ***Spilanthes oleracea*** (Jambú) is used in cooking for typical preparations of the Amazon region of Brazil (Lovera 2005) or as a flavoring agent for salads, soups and meat (Grubben and Denton 2004). Traditionally, the decoction of flowers or leaves are used to treat stomachache and toothache, gums and throat infections, for rheumatism and dysentery, blood parasites (malaria); as anesthetic, stimulant, diuretic, antiseptic, bacteriostatic, anti-inflammatory (Hind and Biggs 2003, Grubben and Denton 2004, Quattrocchini 2012,

Dubey et al. 2013, Srinath and Laksmi 2014). The EO of flowers and leaves of Jambu were chemically characterized and presented as main compounds to β-caryophyllene (11–30%), (E)-2-hexenol (26%), limonene (24%), thymol (18%), Z-β-ocimene (14%), γ-cadinene (13%), 2-tridecanone (13%), germacrene D (11%), hexanol (11%), and myrcene (10%). The yields EO were 0.04–0.4% (Lemos et al. 1991, Baruah and Leclercq 1993, Jirovetz et al. 2005).

The yields of hexane, SFE-CO_2 and ethanol extracts obtained from flowers, leaves and stems were 1–39% and contained spilanthol in about 2–65% of the total, which is the N-alkyl amide representative of this genus. These extracts have shown antinociceptive (10–100 mg/kg reduced inflammatory phase), antioxidant (β-carotene/linolenic acid –16–72% inhibition; DPPH·—38–96% inhibition, dose 333 mg/L), anti-inflammatory (high lipoxygenase inhibition), diuretic, larvicidal (*A. aegyptii*—LD_{100}: 12 mg/L), antifeedant (*H. zea*—66% inh., dose 250 mg/L), toxicity (*A. salina*—LC_{50}: 10–16 mg/L), cytotoxicity (neoplastic cells HEp-2—IC_{50}: 513 mg/L), insecticidal (*Tuta absoluta* (Meyrick)—LD_{50}: 0.13 mg/g), antimalarial (*Plasmodium falciparum*—IC_{50}: 6–41 mg/L), and antimicrobial activities (Gram+, Gram–, MIC: 64–256 mg/L) (Ramsewak et al. 1999, Ratnasooriya et al. 2004, Prachayasittikul et al. 2009, Spelman et al. 2011, Dias et al. 2012, Moreno et al. 2012, Abeysiri et al. 2013, Nomura et al. 2013, Pacheco Soares et al. 2014).

In Mesoamerica, the leaves and flowers of ***Tagetes lucida*** were used during religious ceremonies as smokes or beverage mixed with other plants, for their apparent hallucinogenic properties. Also, it has been useful as a disinfectant and as a spice in cooking. The flowers of *T. lucida* were applied to treat diarrhea and dysentery. Nowadays, the principal application given to their leaves and flowers is as a seasoning because of its aniseed flavor and as a tea to treat colic, gastrointestinal pain, headache, colds, and fever (Olivas Sánchez 1999, Osuna Torres et al. 2005, Pennacchio et al. 2010, Lim 2014). *T. lucida* has some therapeutic effects as anesthetic, fumigant, repellent, anti-inflammatory, diuretic, laxative and stimulant (Etkin 1986, Osuna Torres et al. 2005, Cerón 2006, Duke et al. 2009). The chemical compositions of the EO of leaves and flowers of tarragon were characterized by containing estragole (12–97%) as the main component; nonetheless, also have been reported anethole (24–74%), methyleugenol (17–80%) and nerolidol (40%). Yields of EO were 0.07–0.9% (Bicchi et al. 1997, Marotti et al. 2004, Cicció 2005, Muñoz-Acevedo et al. 2007, Can Başer and Buchbauer 2010). In addition, coumarins and flavonoids were identified in the polar extracts (Abdala 1999, Xu et al. 2012). The EO and ethanol, ether and aqueous extracts showed antifungal (*C. albicans*), antibacterial (Gram–, Gram+), spasmolytic, antioxidant (DPPH·, TBARS, $ABTS^+$, lipidic peroxidation), antiplasmodial (*P. berghei*), repellent (*Tribolium castaneum*—5 µL/g; *Sitophilus zeamais* —48–79% dose, 0.06–0.5 µL/cm^2), antidepressant (using forced swimming test) and nematocidal activities (Vasudevan et al. 1997, Céspedes et al. 2006, Guadarrama-Cruz et al. 2008, Muñoz-Acevedo et al. 2009, Nerio et al. 2009, Hooks et al. 2010, Regalado et al. 2011, Olivero-Verbel et al. 2013).

The leaves and bark of ***Croton malambo*** have been traditionally prepared as poultices, infusions or tinctures to treat colic, colitis, diarrhea, rheumatism (García-Barriga 1992) and the therapeutic effects showed were sedative, analgesic, anti-inflammatory and antinociceptive (Suárez et al. 2003, Quattrocchini 2012). The main constituent identified in the EO of the bark and leaves from Malambo has been methyleugenol (64–94%) (Suárez et al. 2005, Suárez et al. 2008, Jaramillo et al. 2010). The yields of these EO were 0.6–1.2% for species from Venezuela and Colombia. Additionally, the EO and extracts of *C. malambo* showed bioactivities as antioxidant ($ABTS^+$, TAA—2.3 mmol/kg EO), antibacterial (Gram+, Gram–), toxicity (*Artemia salina*, LC_{50}: 120 mg/L) and cytotoxicity (cell lines MCF-7, PC-3 and LoVo; moderate activity on MCF-7) (Jaramillo et al. 2007, Jaramillo et al. 2010, Suárez et al. 2008).

Scheme 6 shows the structures of spilanthol, estragol, anethole and methyleugenol, principal components from *Sp. oleracea*, *T. lucida* and *C. malambo*.

Lippia species (***L. alba***, ***L. graveolens*** and ***L. origanoides***) have been used in ethnomedicine for the treatment of different disorders: respiratory (e.g., cold, flu, cough, bronchitis); gastrointestinal (indigestion, abdominal pain, diarrhea, dysentery); skin (burns, wounds, ulcers); and in cooking, as a condiment. Therapeutic effects attributed to these species were antispasmodic, carminative, emmenagogue, analgesic, anti-inflammatory, antipyretic, stimulating appetite, antidiabetes, antiseptic and antiparasitic (Pascual et al. 2001a). The chemical compositions and biological activities associated to the EO of the studied parts from *Lippia* spp. are presented in Table 14.4.

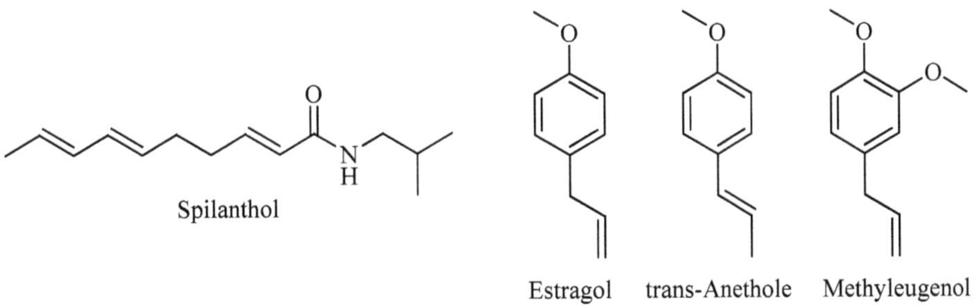

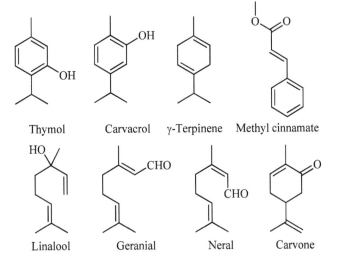

Spilanthol

Estragol   trans-Anethole   Methyleugenol

Scheme 6. Structures of spilanthol, estragol, anethole and methyleugenol, principal components from *Sp. oleracea*, *T. lucida* and *C. malambo*.

Linalool (68–79%), myrcene (26%), and neral (10%) were the principal constituents found in the EO of leaves and flowers from *Lippia alba* of Indian origin. The yields of these essential oils were 0.7–1.2% (Mishra et al. 2010, Singh et al. 1999).

The chemical compositions of the EO of leaves and flowers from *Lippia* species were represented by linalool (55–92%), carvacrol (8–71%), thymol (7–64%), carvone (17–60%), nerolidol (1–47%), geranial (10–41%), *p*-cymene (10–26%), a-pinene (5–19%), limonene (10–44%), γ-terpinene (8–17%), 1,8-cineole (2–35%), neral (10–30%), myrcene (4–26%), β-caryophyllene (2–16%), methyl cinnamate (8–27%), and piperitone (31%), as main components (Scheme 7). Yields of EO were 0.2–4.6%. Compounds type iridoids (3 iridoids) and flavonoids (3–23 flavonoids) were identified in the polar (methanol, ethanol, water) and no-polar (SFE-CO_2) extracts (Rastrelli et al. 1998, Barbosa et al. 2006, Lin et al. 2007, Stashenko et al. 2013).

The EO and ethanol, ether, and aqueous extracts showed some biological activities: antimicrobial (fungi, yeasts and Gram+, inhibition halo diameter—20–40 mm; MIC < 2 mg/mL), antispasmolytic (contractile concentration-response curves of acetylcholine and calcium on intestine portion—7–37 mg EO/mL), antioxidant (ABTS+; linoleic acid oxidation; DPPH·; β-carotene/linoleic acid), insecticidal (*Myzus persicae* Sulzer, 0.1–0.5% evaluated concentration), antiprotozoal (*Leishmania chagasi* and *Trypanozoma cruzi*, IC_{50}: 4–12 µg/mL), antigenotoxicity (DNA protective effect against bleomycin-induced genotoxicity), antimigraine (50% pain reduction), repellent (*Tribolium castaneum*—24–48 hours, 19–93% inh., dose 0.00002–0.2 µL/cm²), cytotoxicity (MTT—Selective index > 7 (7–1241)), analgesic, antinociceptive (tail immersion method, 80% response at 60 minutes), anti-inflammatory (carrageenan induced paw edema

Thymol   Carvacrol   γ-Terpinene   Methyl cinnamate

Linalool   Geranial   Neral   Carvone

Scheme 7. Some structures of the main components from *Lippia* spp.

Table 14.4. Chemical compositions and biological activities of EO from *Lippia* spp.

Components (Relative amount, %)	Part used	Yield EO-Extract, %	Origin	Bioactivity	Reference
Lippia origanoides					
Carvacrol (39%), thymol (18%), p-cymene (10%)	Leaves	1.0	Brazil	Antimicrobial	(Oliveira et al. 2007)
Carvacrol (33–43%), p-cymene (12–16%), γ-terpinene (8–10%)	Leaves	4.6			(Dos Santos et al. 2004)
E-Nerolidol (1–47%), E-methyl cinnamate (8–27%), p-cymene (1–26%), 1,8-cineole (2–25%), carvacrol (1–24%), α-pinene (5–19%), E-caryophyllene (2–16%)	Leaves	2.1–4.6		---	(Ribeiro et al. 2014)
Carvacrol (42%), p-cymene (18%), γ-terpinene (17%)	Leaves	---		Antioxidant (DPPH, β-carotene/linoleic acid), insecticidal	(Teixeira et al. 2014)
Limonene (12%), α-phellandrene (11%), p-cymene (10%)	Leaves	0.4–0.5	Colombia	---	(Stashenko et al. 2010)
Thymol (60%), Thymol (34%), carvacrol (26%)	Leaves	1.4–2.2		Antigenotoxicity	(Vicuña et al. 2010)
Carvacrol (44–52%), thymol (7–14%), p-cymene (9–10%), γ-terpinene (1–10%), p-Cymene (11–16%), 1,8-cineole (7–11%)				Antioxidant (ABTS$^+$) antiprotozoal antitubercular repellent	(Stashenko et al. 2008, Bueno-Sánchez et al. 2009, Escobar et al. 2010, Caballero-Gallardo et al. 2012)
Thymol (64%)	Leaves	2.5	Venezuela	Antimicrobial (fungi, yeast, Gram+, Gram−; MIC: <2 mg/mL; MFC: 0.5–1 mg/mL)	(Chataing et al. 2012)
Thymol (48–62%), carvacrol (8–17%)		---		---	(Rojas et al. 2006)

Lippia graveolens					
Thymol (32%)	Leaves	0.3	Guatemala	---	(Senatore and Rigano 2001)
Carvacrol (35–71%)	Leaves	---	El Salvador	---	(Vernin et al. 2001)
Thymol (30%)	Leaves	1.8	México	---	(Uribe-Hernández et al. 1992)
Carvacrol (62%)	Leaves	--		---	(Calvo-Irabien et al. 2009)
Lippia alba					
Linalool (64%)	Leaves	1.2	Argentina	Antispasmodic	(Blanco et al. 2013)
Geranial (20%), carvone (17%) neral (12%)	Leaves	0.7			
Citral (65%) geranial (37%), neral (28%)	Leaves	1.0	Brazil	Antifungal	(Fun and Svendsen 1990)
Linalool (77–92%)	Leaves	--		---	(Frighetto et al. 1998, Forero-Peñuela et al. 2013)
Geranial (36–41%) neral (27–30%)	Leaves	1.1–1.6		---	(Matos et al. 1996)
Carvone (42–55%)	Leaves				
1,8-Cineole (35%) limonene (18%)	Leaves	--		---	(Zoghbi et al. 1998)
Limonene (32%) carvone (32%), myrcene (11%)					
Germacrene D (25%) geranial (22%), neral (14%) β-caryophyllene (10%)					
Carvone (25–60%) limonene (10–18%) myrcene (6–12%) germacreno D (8–10%)	Leaves	0.8–2.2		---	(Teles et al. 2012)
Citral (55%), myrcene (10%)	Leaves	1.1–1.6		Behavioral effect	(Vale et al. 1999)
Citral (63%), limonene (23%)					
Carvone (55%) limonene (12%)					

Table 14.4. contd....

Table 14.4. contd.

Components (Relative amount, %)	Part used	Yield EO-Extract, %	Origin	Bioactivity	Reference
Geranial (31%), neral (30%), β-elemene (16)	Leaves	0.5		Antibacterial	(Veras et al. 2011)
Geranial (13–21%), neral (9–16%), myrcene (4–15%), nerol (4–10%)	Leaves	0.2–0.3		Antimicrobial	(Oliveira et al. 2006)
Geranial (25%), carvenone (21%)	Leaves	1.6		Antimigraine	(Conde et al. 2011)
Carvone (25–50%), limonene (22–36%)	Leaves	0.4–3.0	Colombia	Antifungal	(Mesa-Arango et al. 2010, Durán et al. 2007)
Geranial (10–30%), neral (10–24%)	Leaves				
Carvone (40–57%), limonene (24–32%)	Leaves	0.7		Antioxidant (linoleic acid oxidation)	(Stashenko et al. 2004)
Carvone (29%), β-guaiene (12%)	Leaves/ stems	2.1	Cuba	---	(Pino et al. 1996)
Carvone (40%), β-guaiene (10%)	Leaves	2.2		Antibacterial (Gram+, Gram–)	(Pino-Alea et al. 1996)
Myrcenone-Z-ocimenone 1,8-cineol, neral and geranial	Leaves	---	Guatemala	---	(Fischer et al. 2004)
Limonene (44%), piperitone (31%)	Leaves	0.2		---	(Senatore and Rigano et al. 2001)
Linalool (55%)	Leaves	---	Uruguay	---	(Lorenzo et al. 2001)
Camphor 1,8-cineole, β-cubebene	Leaves	---		---	(Dellacassa et al. 1990)

method, 20% response at 240 minutes), antiviral, antiulcerogenic (rat gastric mucosa) and anesthetic (*Rhamdia quelen*, 100–500 mg/L induced anethesia) (Fun and Svendsen 1990, Pino-Alea et al. 1996, Vale et al. 1999, Pascual et al. 2001b, Stashenko et al. 2004, Oliveira et al. 2006, Durán et al. 2007, Oliveira et al. 2007, Hennebelle et al. 2008, Stashenko et al. 2008, Da Cunha et al. 2010, Escobar et al. 2010, Mesa-Arango et al. 2010, Vicuña et al. 2010, Conde et al. 2011, Veras et al. 2011, Caballero-Gallardo et al. 2012, Chataing et al. 2012, Haldar et al. 2012, Blanco et al. 2013, Teixeira et al. 2014, Zapata et al. 2014).

The leaves and bark from **Ocotea caparrapi** have been used to treat insect bites, sexually transmitted diseases, ulcers and other epithelial lesions, and as mosquito repellents. Also, they have showed therapeutic effects as antirheumatic, antivenom and antitumor (García-Barriga 1992, Pérez-Arbeláez 1996, Bernal et al. 2011). In the case of *O. quixos*, the main traditional application has been as spices replacing the Ceylon cinnamon. In botanical medicine, this species was useful to treat stomach pain and toothache, spasms, arthritis, colds and hydropsy (Gupta 1995, Cerón 2006, Cárdenas and Salinas 2007). Table 14.5 shows the chemical compositions and biological activities of the EO from *Ocotea* spp.

The oil from *O. caparrapi* was characterized by having a high percentage of nerolidol (Scheme 8) and only one publication related to biological activity (cytotoxicity) of *O. caparrapi* was found in the scientific literature reviewed (Palomino et al. 1996).

The EO of the calices of *O. quixos* showed some biological activities as antiplatelet (inhibited collagen-induced platelet aggregation in human and rodent plasma), antithrombotic (prevented acute thrombosis induced on mice, dose 30–100 mg/kg.day), antioxidant (DPPH, 52% quenching), antimicrobial (*S. aureus* subsp. *aureus*, MIC: 0.1 mg/mL; *E. foecalis*, MIC: 0.2 mg/mL; *P. aeruginosa*, MIC: 0.05 mg/mL; *E. coli*,

Table 14.5. Chemical compositions and bioactivities of EO from *Ocotea* spp.

Components (Relative amount, %)	Part used	Yield EO-Extract, %	Origin	Bioactivity	Reference
Ocotea caparrapi					
Nerolidol (92%)	Oil	----	Colombia	----	(Borges del Castillo et al. 1966)
Caparrapi oxide (43%)	Oil	----		----	(Brooks and Campbell 1969, Ohloff 1978)
Caparratriene	Oil	----	----	Cytotoxicity (CEM leukemic cell—IC$_{50}$: 3.0 x 10^{-6} M)	(Palomino et al. 1996)
Ocotea quixos					
β-Caryophyllene (15%) cinnamyl acetate (11%) sabinene (8%)	Leaves	0.1 (1.6 mL/kg)	Ecuador	----	(Sacchetti et al. 2006)
trans-Cinnamaldehyde (28%) methyl cinnamate (22%)	Calices	1.9		Antiplatelet antithrombotic antioxidant (DPPH) antimicrobial (Gram+,Gram–) antiinflammatory	(Bruni et al. 2004, Ballabeni et al. 2007, Ballabeni et al. 2010)

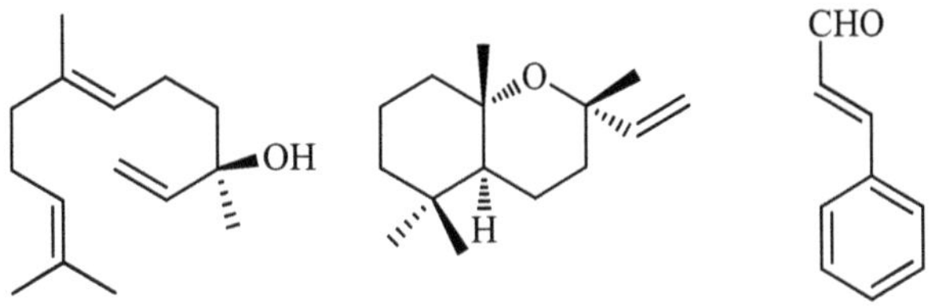

Nerolidol   Caparrapi oxide   trans-Cinnamaldehyde

Scheme 8. Principal constituents of the oils from *Ocotea* spp.

MIC: 0.1 mg/mL; *Saccharomyces cerevisiae*, MIC: 0.02 mg/mL; *C. albicans*, MIC: 0.02 mg/mL) and anti-inflammatory (carrageenan induced rat paw edema method, 30–100 mg/kg showed effect) (Bruni et al. 2004, Ballabeni et al. 2007, Ballabeni et al. 2010).

The leaves, bark and root of ***Quillaja saponaria*** were used in herbal medicine as an expectorant, diuretic, emulsifying agent or detergent, to treat bronchitis, and hairlessness; moreover, they have shown anti-inflammatory, sedative, analgesic, antiviral, antinociceptive effects. The bark extract is an excellent skin stimulating agent and has been used as a hair tonic. However, there is a potential risk if this extract is swallowed due to the toxicity associated to saponins (*ca.* 9% of content), attempting to dissolve the blood corpuscles (Gupta 1995, Johnson 1999, Montenegro et al. 2001, Gonsalves 2010, Quattrocchini 2012).

Approximately, 60 saponins (glycosides of high molecular weight) have been recognized and partially identified in the bark extract of Quillay (Van Setten et al. 1995, Nord and Kenne 1999, Guo et al. 2000, Nord and Kenne 2000, Nyberg et al. 2000, Thalhamer and Himmelsbach 2014). The base structure of these saponins is the molecule of quillaic acid (triterpenoid or steroid aglycone) bound to different substituted oligosaccharides (e.g., pentose, rhamnose) at positions C-3 and C-28, or a dimeric acyl group (C_9) linked to one or more terminal monosaccharides (Scheme 9) (Nyberg et al. 2000, Yang et al. 2013).

Saponins isolated or present in the extracts of *Q. saponaria* have shown larvicidal (100% larval mortality in *A. aegypti* and *Culex pipiens*, 800–1000 mg/L, one–five days), immunomodulatory (antibodies type IgG), antiviral (Herpes viruses, HIV, reovirus and Rhesus rotavirus (RRV)—concentration < 0.1 mg/L obstruct to VIH, 15 µg extract reduced 79–11% diarrhea induced by RRV), antifungal (*Botrytis cinerea*), antinociceptive (showed activity in murine thermal models in a dose-dependent manner), cytotoxicity (100 mg/L, not cytotoxic), toxicity (dose 500 µg—80% mortality), anti-inflammatory activities and capability to improve the conditions of greasy/oily skin; also, they function as an adjuvant in the formulation of vaccines (Wu et al. 1992, Soltysik et al. 1995, Barr and Mitchell 1996, Marciani et al. 2001, Pelah et al. 2002, Kirk et al. 2004, Roner et al. 2007, Arrau et al. 2011, Rodríguez-Díaz et al. 2011, Tam and Roner 2011, Rigano et al. 2012, Zuñiga et al. 2012).

Traditionally, the flowers, leaves, fruits, bark and resin of ***Genipa americana*** have been used for medicinal purposes for treating bronchitis, inflammation, diarrhea, ulcers, anemia, high blood pressure, intestinal worms, and colic. In addition, insecticide, fungicide, repellency, antimicrobial, diuretic, vermifuge, laxative, purgative, astringent and tonic properties were attributed to this species. During ceremonial rites, the indigenous communities use the dye of *G. americana* for facial and body painting (Núñez Meléndez 1982, Duke and Vásquez-Martínez 1994, Santos-Granero and Barclay 1998, Johnson 1999, UNCTAD 2005, Chizmar et al. 2009, Quattrocchini 2012). The compound responsible for the coloring function (blue-black) is genipin (compound type iridoid). The fruit also contains two antibiotic compounds (cyclopentoid monoterpenes—genipic and genipinic acids) (Tallent 1964, Ueda et al. 1991, Wrigley et al. 2000, Ono et al. 2005, UNCTAD 2005).

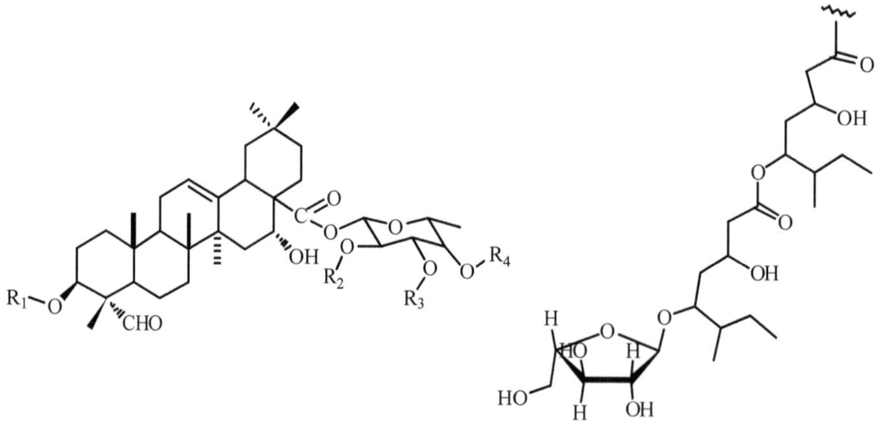

Scheme 9. Base structure of the saponins isolated from *Q. saponaria*.

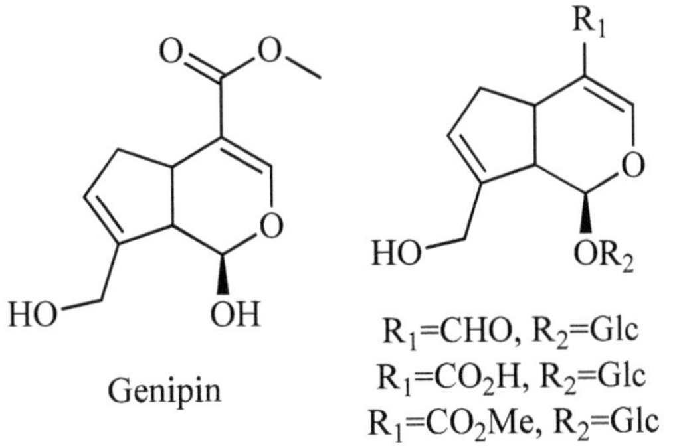

Scheme 10. Structure of genipin and their derivatives glycosides from *G. americana*.

The pericarp of the fruit of **Sapindus saponaria** has been traditionally used for washing because of its detergent properties, and as a hair tonic. However, in natural medicine it has been used along with leaves, flowers and seeds as antidotal, antipyretic, antirheumatic, astringent, expectorant, soporific, and to treat conjunctivitis and inflammation; additionally, the insecticidal, piscicide, molluscicide (snail; LC_{90}: 90 mg/L, 24 hours exposure), espermicidal (50% immobilization, dose 0.5% p/p), acaricide, antiprotozoal, antiulcer (significant reduction of lesion index on rats), antimicrobial (dental pathogens Gram+, the inhibitory activity was dose dependent), antifungal (*Candida* spp.—MIC: 70–500 mg/L on *C. parapsilosis*), antiviral (VIH using the syncytia formation assay; methanol 80% extract was the most active agent with Therapeutic Index (TI) value higher than 11), repellent, spasmolytic, cytotoxic activities were determined (Albiero et al. 2002, Abdel-Gawad et al. 2004, Kashanipour and McGee 2004, Guirado 2005, Álvarez-Gómez et al. 2007, Tsuzuki et al. 2007, Pelegrini et al. 2008, Gonsalves 2010, Quattrocchini 2012, Thota et al. 2012, Damke et al. 2013, Iannacone et al. 2013, Rashed et al. 2013a,b). Sapogenins responsible for the detergent action are hederagenin derivatives (triterpenoid mono- and di-acetylated) (Núñez Meléndez 1990, Díaz González 2010). Their seeds are a potential source of fatty acids C_{16}–C_{22}, with a yield of 43% (Lovato et al. 2014).

The study of the volatile fractions of the fruits of Huito allowed to determine that octanoic acid (34%), hexanoic acid (18%), 2-methylbutyric acid (9%), hexadecanoic acid, octadecanoic acid, tetradecanoic acid, linalool, limonene and esters of low boiling point were the components responsible for aroma (Borges and Rezende 2000, Pino et al. 2005, Pinto et al. 2006). The ethanol extracts of fruits, leaves and flowers, rich in iridoid glycosides, tannins and flavonoids have shown antitumor, antimitotic (*Allium* test), cytotoxic (cell line NCTC-929; sheep corneal epithelial cells, 98–100% viability), antioxidant (ABTS$^{+\cdot}$—TEAC: 0.4–1.1; DPPH$\cdot$—RSA%: 10–15; FRAP—TEAC: 0.7–1.0; CUPRAC—TEAC: 0.8–1.3; Total phenolic content –62–188 mg AGE/g extract; lipid peroxidation), antiacetylcholinesterase (TLC bioassay, 0.8 cm—AChE inhibition zone similar to the positive control), antihelmintic activities (LC_{90}, 80–29 mg/mL) (Ueda et al. 1991, Omena et al. 2012, Nogueira et al. 2014).

Exploitation and sustainable aspects

In most cases, the medicinal and aromatic plants of interest are cultivated or collected in the wild to be used as renewable sources of EO, colorants, dyes, biocides and active principles, with potential utility to a commercial level and in developing products (cosmetics, perfumes, paints, foods and drugs) (Lubbe and Verpoorte 2011). However, when collecting the plant material is carried out without considering the established legislation by the appropriate control agencies for the management of biodiversity (regional/national/international levels), the population of exploited plants could be threatened (i.e., become vulnerable, endangered or even critically endangered) or in the worst cases be driven to extinction (Cárdenas and Salinas 2006). Because of this, and given the need to meet commercial or industrial demands, generated from natural products, initiatives have been devised for the sustainable management of natural resources. One of these initiatives is biotrade, an economic activity around crops with a direct impact on the local communities, which has led to the improvement of their quality of life. However, for the success of large-scale cultivation, an investment policy is necessary for the production of high quality material for an international market (Lubbe and Verpoorte 2011).

Several sustainable biotrade programs have been developed in Latin America as an initiative of UNCTAD (United Nations Conference on Trade and Development) along with the support of public and state institutions. For example, Colombia initiated the Sustainable Development Program in 1998 (Ministry of Environment and the Alexander von Humboldt Institute). In 2001, Ecuador launched its program sponsored by the Ministry of the Environment. In 2003, Perú started with the National BioTrade Promotion Programme (NBPP) under the direction of the Ministry of Trade and Tourism. In the same year, Bolivia began its program (Friends of Nature Foundation) (Jaramillo-Castro 2012). Among the successful cases of biotrade in Latin America the Jambi Kiwa (Ecuador), Natura (Brazil) and Nativa (Colombia) projects stand out.

Table 14.6 presents, *grosso modo*, the aspects related to the categories established for the management-exploitation of biodiversity (extinct, endangered or at risk) along with initiatives established for sustainable management of the promising species according to the Red List of the IUCN (International Union for Conservation of Nature).

Schinus molle is cultivated for restoration of gardens and riverbanks (Perú) (Whaley et al. 2010). Also, in Bolivia the plant is categorized as an agroforestry promising species due to its socio-cultural and economic value (Brandt et al. 2012).

Although **Bixa orellana** is a species from America, some African countries have joined efforts to reduce deforestation and ensure the sustainability of this species, collecting the ethnobotanical and biological information, and the methods used to obtain dyes (MacFoy 2004). Additionally, this species has been useful for the sustainable management by incorporating nutrients to soil (Schroth et al. 2001, Elias et al. 2002, Dinkelmeyer et al. 2003). Regarding the sustainability of **Genipa americana**, in Costa Rica studies were conducted to understand their capacity to accumulate biomass in crops mixed with other species (Stanley and Montagnini 1999). In Colombia, *G. americana* and *B. orellana* crops are being developed for obtaining sustainable black, blue and red natural dyes (Web 2), in a joint effort between private (Ecoflora) and environmental institutions, that involve Afro-communities in the Colombian Pacific region. These communities learn the steps and practices of collection, distribution, production capacity, and post-harvest handling (Assunção and Kutsch 2006).

Table 14.6. Analysis of the species of interest according to the Red List of the IUCN.

Species	Red List IUCN	
	YES (category)	NO
Schinus molle	----	X
Bixa orellana	----	X
Genipa americana	----	X
Bursera graveolens	----	X
Bursera simaruba	----	X
Protium heptaphyllum	----	X
Araucaria angustifolia	----	X
Quillaja saponaria	----	X
Sapindus saponaria	----	X
Ocotea spp.	----	X
Lippia spp.	----	X
Eremanthus erythropappus	----	X

Lippia alba is prioritized by Nativa Organization of Colombia for exploitation through a sustainable biotrade (Web 3). On the other hand, *Ocotea quixos* is classified as a threatened species in Colombia because of the exhaustive timber exploitation by the population located in the Amazonian piedmont of the country's southwest.

The program of sustainable biotrade of Ecuador chose *Bursera graveolens* as the priority species. This program has allowed them to analyze the sustainability of the resource (vulnerability, exploited part of the plant, reproductive ecology, population) and its exploitation (harvesting rate), the 'good' exploitation practices (setting up of nurseries, manual for collectors and gatherers, traceability records) and the implementation of corporate environmental policies (Web 4). In the case of *Bursera simaruba*, Nicaragua has used this species in studies of sustainable management of soil in riverbanks, using the bio-engineering techniques (Petrone and Preti 2008).

The overexploitation of *Quillaja saponaria* to obtain bark has caused a decrease in this resource along with considerable ecological damages in Chile. The Chilean National Forestry Corporation (CONAF) has a plan for forest management, in which every five to 10 years, a maximum of 35% of basal area of the tree can be cut; but despite these controls the population of old trees has decreased and prices have increased dramatically (San Martin and Briones 1999, San Martin 2000).

Lippia graveolens an endemic species in México, has been cultivated and marketed as a source of economic income for rural families for a long period of time. Thus, strategies have been established with producers, traders and brokers, to design an alternative management to ensure the sustainability of the activity (Granados-Sánchez et al. 2013). Unfortunately, the species in this country is categorized as threatened due to anthropic activities, scarcity and nonexisting management policies (Arellanes et al. 2013).

Araucaria angustifolia is the most important timber species of Brazil and produces 3400 ton per year of fruit and seeds for human consumption. The Brasilian federal government is promoting some initiatives for the protection of this species (Thomas 2013). However, conservation efforts seem to be efficient against the rapid loss of their diversity, despite the unsustainable exploitation (Stefenon et al. 2009). Also, a reforestation strategy is being implemented and improvements in soil properties and the capacity of absorption of nutrients (Ilany et al. 2010, Fagotti et al. 2012).

Due to the importance of α-bisabolol found in the crude extract or EO from *Eremanthus erythropappus* some conservation plans have been developed (Pérez et al. 2004). Wood is mainly bought by industries that extract the EO and then it is sold to the cosmetic industries (Osuna Torres et al. 2005). Monocrop conditions or intercropped with maize, have also been studied, taking into account the economic feasibility of the demand in a sustainable way to reduce the impact on native fragments of the species (Silva et al. 2012).

Technological development patents related to Latin American plants

It is considered that more than 75000 plants are used in herbal medicine and only 1% has scientific studies that support their possible use for commercial purposes. The worldwide market for herbal medicinal products is estimated at *ca.* US$60 billion and the cost to develop a product or drug, is on average *ca.* US $ 80 million and in some cases, can reach several billion dollars (Kartal 2007). When the benefits and the possible commercial exploitation of some of the plants are scientifically verified, they are included as ingredients in formulations that end in technological development, these are protected through patents. For example, pharmaceutical patents are granted if they satisfy any of the following requirements: (1) discovery of new chemical substances, (2) know-how in development and manufacturing of product, and (3) patents for commercialization.

The European Patent Office, in Article 53 (b), does not grant any patent protection for plant varieties or biological essential processes for the production of plants (Kartal 2007). The protection of inventive activity in natural products can occur in the following cases: (1) patents for new chemical compounds (isolation, biological activity and synthesis), (2) patents for known compounds (structural modification to enhance or potentialize some properties), and (3) patents for traditional herbal medicines or herbal medicinal products (formulations and processes).

The IPC (International Patent Classification) is the code (a combination of letters and numbers) assigned to a patent and provides a short description of the area of application in which the invention is registered (Oldham 2006); its main purpose is to facilitate the identification of patents. The IPC classification joins together to the patents in eight sections, of which the section A (human necessities) along with its sub-sections, classes, subclasses, groups and subgroups, are related to applications in cosmetics, cosmeceuticals, perfumes and flavors. According to the eighth edition of the IPC (IPC 8[2]), the distribution of the epigraphs that characterize to the patents related to skin care would be:

SECTION A	Human necessities
A61	Medical or veterinary science; hygiene
A61K	Preparations for medical, dental, or toilet purposes
A61K8/00	Cosmetics or similar toilet preparations
A61K9/00	Medicinal preparations characterised by special physical form

In order to establish the inventive activity in natural products (EO and/or extracts) obtained from promising plants and their application as ingredients for cosmetic, perfume and flavor (fragrances and flavorings) products, a search equation was defined using A61K epigraph, the scientific name of each species and the period 1970–2014. Of the 18 promising species only 13 plants reported some inventive activity measured by the number of requested patents.

In accordance with the search equation established, the distribution of the number/family of patents for each plant is shown in Fig. 14.7. The species *Quillaja saponaria* had the highest number of patents request (293) distributed in 95 families of patents, followed by *Genipa americana* with 36 patents (17 families), *Sapindus saponaria* with 15 patents (nine families), and *Bixa orellana* with 14 patents (11 families).

Examples of some patents for application related to promising plants are contained in Table 14.7, which includes the code, title, IPC, assignee and publication date of the patent.

Based on the analysis of patents by VantagePoint software, *Quillaja saponaria* was the plant with the highest number of patent applications, of which 33 patent families were found in the WIPO (World

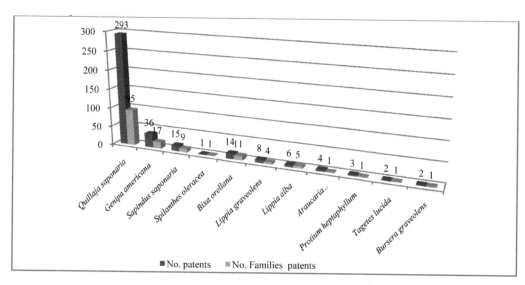

Figure 14.7. Distribution of the number of patents of promising species with application in cosmetic, perfume and flavor industries.

Source: Information calculated in accordance with *Thomson Innovation* patents database (*Thomson Reuters*) and processed with *VantagePoint* (Version 8.0, *Search Technology*). Data provided by Professor Fernando Palop, Triz XXI—Spain.

---

[2] http://www.wipo.int/classifications/ipc/ipc8/?lang=en.

Table 14.7. Patent related to Latin America promising plants for applications in the cosmetic, fragrance and flavor industries.

Plant	Patent code	Title of patent	IPC	Assignee	Date of publication
Araucaria angustifolia	FR 2885049 A1	Topical cosmetic and/or dermatological composition, useful to protect the skin and to fight against the climatic and environmental effect, comprises a seed extract of Araucaria	A61K36/00, A61K8/9, A61P17/00, A61Q19/00	Laboratoire Nuxe Société anonyme	03-nov-06
Bixa orellana	JP2005082561	Cosmetic	A61K7/00, A61K7/42	Broad KK	31-mar-05
	ES2192997A1	New cosmetic formulation with extract of natural origin (plant) obtained of the annatto (*Bixa orellana* L.) For topical application as a sunscreen	A61K-0007/42	Valefarma, S.L.	16-oct-03
Genipa americana	JP2000319120A	Cosmetic composition including vegetable extract having moisture retaining property	A61K07/50, A61K07/48, A61K07/16, A61K07/08, A61K07/075, A61K07/00	Ichimaru Pharcos	21-nov-00
	WO 2010105320 A1	A process for obtaining insoluble substances from genipap-extract precipitates, substances from genipap-extract precipitates and their uses	A23L1/27, A61K36/74, A61K8/97	Natura Cosméticos S.A.	23-sep-10
Protium heptaphyllum	WO2009082797A1	Active formulations based on essential oil of plants of the genus *Protium, Guatteria, Cyperus* and the mixture thereof	A61Q90/00, A61K8/99	Inst Nac De Pesquisa Da Amazon, 8 más	09-jul-09
Lippia graveolens	WO2001015680A1	Composition for treatment of infections of humans and animals	A61P31/00, B01F15/06, A61K31/05, B01F7/22, A61K36/75, A61K36/53, A61K36/23, A61K36/85, A61K36/534	Beek Global Ninkov L L C Van	08-mar-01
Tagetes lucida	WO 2007100567 A2	Method and composition for improved meat quality	A61K-0036/00	Greger Douglas L.	07-sep-07
Spilanthes oleracea	WO 2010010394 A2	Local pharmaceutical compositions	A61K36/28	Neurosolutions Ltd.	28-ene-10
	WO 2013185060 A2	Oral stimulatory product	A61K36/81	Foundation Brands, Llc	12-dic-13

Table 14.7. contd....

Table 14.7. contd.

Plant	Patent code	Title of patent	IPC	Assignee	Date of publication
Lippia alba	JP2001322941A	Skin care preparation	A61K-0008/00, A61K-0008/96, A61K-0036/18, A61K-0036/48, A61P-0017/16, A61Q-0001/00, A61Q-0019/00	Kose Corp	15-may-00
	JP11050050A	Antioxidant	A23L-0003/3472, A61K-0008/00, A61K-0008/06, A61K-0008/97, A61K-0036/18	Lion Corp	06-ago-97
Quillaja saponaria	US 5443829 A	Modified saponins isolated from *Quillaja saponaria*	A61K47/28, C07J53/00, A61K31/70, A61K38/28, A61K9/00, A61K39/39	Cambridge Biotech Corporation	22-ago-95
	US 2010021583 A1	Method for producing extract of *Quillaja saponaria* Molina saponins and use to stabilize beer foam	C12H1/15	Universidad Mayor, Chile	28-ene-10
Sapindus saponaria	FR2456515A1	Composition cosmetique capillaire notamment pour le lavage et/ou le demelage des cheveux, a base d'un extrait de plantes contenant des saponosides	A61K-0008/00, A61K-0008/34, A61K-0008/63, A61K-0008/81, A61K-0008/88, A61K-0008/97	L'Oreal	12-dic-80
	EP 2568813 A1	Method for extraction of material from a sapindacea family fruit	A01P7/00, A01P3/00, A01N65/00, A01N65/08	Restrepo Jaime Toro, 3 más	20-mar-13

Intellectual Property Organization—short name: WO), 17 in the United States Patent Office and 14 in the Japan Patent Office. Most of the Assignees (patent owners) were pharmaceutical companies like Pharmexa A/S (Denmark), Aquila Biopharmaceuticals Inc. (U.S.A.), Glaxosmithkline Biologicals SA (Belgium).

For *Genipa americana*, six patent families were registered in the Japan Patent Office, five in the WIPO, and three in the United States Patent Office. For this plant, the most important assignees were cosmetic companies such as L'Oreal (France), Ichimaru Pharcos Co. Ltd. (Japan) and Cavezza Alexandre (France).

Many of these patents are used for formulations and applications of cosmetic products based on Latin American promising plants.

Formulations and applications

The most important application of Latin American plants is its use as an ingredient in cosmetic products, fragrances for perfumes, aromas and flavors. In particular cases, it is also added in pharmaceutical products.

An interesting case is *Schinus molle* or 'pink pepper' (Lim 2012) which has been incorporated as an ingredient of a homeopathic drug as **Poconeol No. 26**® (*Pierre Fabre Médicaments*) and it is prescribed for the treatment of atherosclerosis in adults (Web 5, Web 6). The fruit, the 'pink berry' (Web 7) is marketed directly as a spice and their EO is an active ingredient in perfumes like '**28 La pausa**®'—Chanel, Pleasures—Estée Lauder and Arsène Lupin—Guerlain (Web 8–10); exfoliating soaps and body creams as **Bodhi & Birch Ylang-Ylang Incensa Sensual Bath & Shower Therapy**®, **Pink YUM Exfoliating Solid Lotion Bar**, **Body Cream with YUZU Essential (M)**, by Sophia's Choice, Pink YUM and Botanicus, respectively (Web 11–13); and, balms for revitalizing the skin as **Revitality Balm**® by Delarom Paris (Web 14). Most of the ingredients (13 constituents) of this latter product are EO of different plants. The European Commission Health and Consumers (ECHC) considered that the EO and extract of the fruits of *S. molle* (CAS. 68917-52-2/94334-31-3; EC.-/305-104-2) can be applied as masking, perfuming, humectant, and skin protecting and conditioning (Web 15–18).

According to the ECHC, the extract and powder of the seeds from *A. angusifolia* (CAS. 8015-68-7) can be useful as an emollient, skin conditioning, abrasive and humectant (Web 19–21); for this reason, Nuxe company (Natural Origin Cosmetics—Paris) uses it for the manufacture of some products **Nuxuriance**®: Crème Nuxuriance® Jour, Skin Deep® Nuxe, Crènmme Nuxuriance® Nuit, Emulsion Nuxuriance® Jour, and so on (Web 22–25).

Some body creams (e.g., **Shea Butter & Baobab**® **Antioxidant Cream**) manufactured by Shea Radiance company (Columbia), TanOrganic Original Try Me® ultra-moisturizing sunscreen and AgeLoss® First Day cream contain extracts from *B. orellana* as natural coloring ingredient (Web 26–28). The ECHC declares that oil, seed and leaf extracts and pulp of the fruits from *B. orellana* (CAS. 89957-43-7; EC. 289-561-2) fulfill a function as an emollient, humectant, skin protecting, antioxidant, masking and skin conditioning (Web 29–33) and can be used in the cosmetic industry. There is a no-cosmetic product (Bixa®, dietary supplement) where the main ingredient is the dried extract (1.3 g) of seeds from *B. orellana* (Web 34).

The EO from *B. graveolens* are found as ingredients in some products for hair care, e.g., **Rahua**® **shampoo**, **Rahua**® **conditioner**, **Rahua**® **finishing treatment**, and **Rahua**® **hair wax**, manufactured and marketed by Amazon Beauty (New York) (Web 35–38). The ECHC reports that the EO obtained from fruits and wood of *B. graveolens* can be used as antiperspirant, flavoring, masking, and perfuming (Web 39; Web 40). Despite having no commercial cosmetic product related to *B. simaruba*, the ECHC proclaims that the bark extract of *B. simaruba* has applications as an antioxidant (Web 41).

The resin of *P. heptaphyllum* is the most used part in the cosmetic industry and products have been developed such as massage oils for babies (**Centifolia Bio**—France); products for hair care (**Shampooing Detox**®, **Shampooing ultra brillance** and **masque punch**—Hip—Paris) and skin (**Crème hydratante Protectrice**, **Emulsion hydratante Equilibre**—Experelle—France) (Web 42–47). In accordance with the ECHC, the resin of *P. heptaphyllum* (CAS. 92704-59-1; EC. 296-493-7) could be used as binding and masking (Web 48).

The EO of branches and leaves from *E. erythopappus* blended with other EO are marketed under the name of **MelaCare-Oil (ECO)** by SOCRI s.r.l. (Italy) as raw material and skin conditioning for the cosmetic industry (Web 49; Web 50). ThaiDham Allianze Co. Ltd. (Bangkok) uses the EO of *E. erythopappus* as an active ingredient for the manufacture of skin lightening cream as **Anti-Melasma Cream** (Web 51).

The ECHC states that the extract of the aerial parts of *Sp. oleracea* can be used as skin protectant (Web 52). This is why some commercially available products (**A. Vogel**), with cosmetic/food purposes (mouthwash/supplement), use extracts (tinctures) of the leaves from *Sp. oleracea* to treat problems of the skin or nails caused by fungi (Web 53). Etat Pure and Jean D'Arcel Cosmetique companies offer **Actif Pur A04** (**Acmella**) and the serum '**Advanced Face Lift**', respectively, which are products containing extracts from flowers and stems of *Sp. oleracea* acting as a muscle relaxant to treat expression lines (Web 54; Web 55).

Essential oils from leaves and stems of *L. alba* (CAS. 8024-12-2) could be applied as an antioxidant, masking and perfuming, according to ECHC (Web 56). The EO of *L. alba* is used in the manufacture of cosmetic products for men face care (**Thrive**, face wash, shave oil and face balm—San Francisco), shampoos and soaps (**Herbia, Belverde**—Brazil) (Web 57–61).

L. graveolens is an ingredient in the formulation of a cosmetic product type shampoo (**Neem Aloe Body Shampoo**—Neem Tree Farms—Florida) (Web 62).

The essential oil obtained from leaves, flowers and stems of *L. origanoides* can be used in the cosmetic industry as antimicrobial agent, antioxidant, flavoring, masking, and preservative (Web 63). However, there is not a cosmetic product based on the EO in the market but it is sold as raw material by the Colombian company **Neyber**® **S.A.S.** for the purpose of exploiting their antimicrobial, antioxidant and flavoring properties (Web 64).

The EO of *O. quixos* is used in making of cosmetic products for skin care of face (**A.R.T.**® **Beauty Masque** and **A.R.T.**® **Creme Masque**—Young Living Essential Oil, Australia) (Web 65; Web 66), toothpastes (**Thieves**® **AromaBright Toothpaste**—Young Living Essential Oil, USA) (Web 67) and external use products that reduce the stress and nervous tension (**Roll-On-Stress Away**—Young Living Essential Oil, Australia) (Web 68). Likewise, the leaves of *O. quixos* are used for fabrication of beverages (**Slique tea**—Young Living Essential Oil, USA) (Web 69).

The extracts of the wood, bark and roots obtained from *Q. saponaria* (CAS. 68990-67-0; EC. 273-620-4) function as skin conditioning, masking, antidandruff, cleansing, emulsifying, foaming, moisturizing, and surfactant (Web 70–72). GreenShield Organic® (USA) manufactures many of its products, using to the bark of *Q. saponaria* as raw material: '**Laundry detergent—free & clear**', and '**Pet stain & odor remover**' (Web 73–75). The wood/bark extracts are used in the manufacture of dental care products (**Thieves**® **Fresh Essence Plus Mouthwash**—Young Living Essential Oil, USA) and shampoos (Web 76; Web 77).

The juice obtained of the fruit from *G. americana* is used for masking and perfuming in the cosmetics industry, according to the ECHC (Web 78). The fruit extract of *G. americana* is marketed as a cosmetic dye and used for the manufacture of exfoliating creams and hair dyes as **ExfoliKate**® **Acne Clearing Treatment Cream** and **Surya Henna** (USA) (Web 79–81).

There are patents relating to the active compounds (saponins) of *S. saponaria*; however, nowadays trademarks associated or existing with the manufacturing or marketing of products containing the plant do not exist.

List of functions proposed by the European Commission Health and Consumers (Web 82):

Abrasive—Removes materials from various body surfaces or aids mechanical tooth cleaning or improves gloss.

Antidandruff—Helps control dandruff.

Antiperspirant—Reduces perspiration.

Binding—Provides cohesion in cosmetics.

Cleansing—Helps to keep the body surface clean.

Emollient—Softens and smooths the skin.

Emulsifying—Promotes the formation of intimate mixtures of non-miscible liquids by altering the interfacial tension.

Film forming—Produces, upon application, a continuous film on skin, hair or nails.

Flavoring—Gives flavor to the cosmetic product.

Foaming—Traps numerous small bubbles of air or other gas within a small volume of liquid by modifying the surface tension of the liquid.

Hair conditioning—Leaves the hair easy to comb, supple, soft and shiny and/or imparts volume, lightness, gloss, etc.

Humectant—Holds and retains moisture.

Masking—Reduces or inhibits the basic odor or taste of the product.

Moisturizing—Increases the water content of the skin and helps keep it soft and smooth.

Perfuming—Used for perfume and aromatic raw materials (Section II).

Skin conditioning—Maintains the skin in good condition.

Skin protecting—Helps to avoid harmful effects to the skin from external factors.

Surfactant—Lowers the surface tension of cosmetics as well as aids the even distribution of the product when used.

Conclusion

Some plants that have been discussed in this chapter are being incorporated as active ingredients in the way of extracts, essential oils or pure substances for cosmetic, perfumes and flavor formulations (protected by patents/technological development by well known multinational companies or individuals) and these commercial products have been largely accepted in the global market. However, other plants that have not been yet commercially exploited (e.g., *Lippia origanoides, Croton malambo, Ocotea caparrapi, Bursera simaruba, Sapindus saponaria*) have a high potential due to the chemical compositions and their demonstrated biological activities. Moreover, companies and countries are aware of the need for a controlled exploitation of renewable vegetable resources through sustainable management strategies regardless of the commercial purposes for these Latin American plants. If the policies of sustainable usage are applied correctly, it will contribute to the development of the Latin American countries that are directly involved.

Acknowledgements

The authors would like to thank: the Universidad del Norte, for its financial support through Strategic Area 'Biodiversidad, Servicios Ecosistémicos y Bienestar Humano'; Colciencias for their support through Program 'Jóvenes Investigadores e Innovadores 2013'; Bibliometrics Unit of the Universidad de Santo Tomás de Aquino, Bucaramanga. In addition, a special thanks goes to Professor Fernando Palop, Universidad Politécnica de Valencia (Spain) and finally, to the Instituto de Estudios de la Orinoquia de la Universidad Nacional de Colombia.

References

Abdala, L.R. 1999. Flavonoids of the aerial parts from *Tagetes lucida* (Asteraceae). Biochem. Syst. Ecol. 27: 753–754.

Abdel-Gawad, M.M., El-Sayed, M.M., El-Nahas, H.A. and Abdel-Hameed, E.S. 2004. Laboratory evaluation of the molluscicidal, miracidicidal and cercaricidal properties of two egyptian plants. Bull. Pharm. Sci. 27: 331–339.

Abdel-Sattar, E., Zaitoun, A.A., Farag, M.A., El Gayed, S.H. and Harraz, F.M.H. 2010. Chemical composition, insecticidal and insect repellent activity of *Schinus molle* L. leaf and fruit essential oils against *Trogoderma granarium* and *Tribolium castaneum*. Nat. Prod. Res. 24: 226–235.

Abeysiri, G.R.P.I., Dharmadasa, R.M., Abeysinghe, D.C. and Samarasinghe, K. 2013. Screening of phytochemical, physico-chemical and bioactivity of different parts of *Acmella oleraceae* Murr. (Asteraceae), a natural remedy for toothache. Ind. Crops. Prod. 50: 852–856.

Acero, L.E. 2007. Plantas útiles de la cuenca del Orinoco. Segunda edición. Corporinoquia. Ecopetrol, Bogotá.

Acevedo-Rodríguez, P., Van Welzen, P.C., Adema, F. and Van Der Ham, R.W.J.M. 2011. Sapindaceae. pp. 357–407. *In*: K. Kubitzki (ed.). Flowering Plants. Eudicots: Sapindales, Cucurbitales, Myrtaceae. Series title—The Families and Genera of Vascular Plants. Vol. X. Springer-Verlag, Berlin.

Albiero, A.L.M., Sertié, J.A.A. and Bacchi, E.M. 2002. Antiulcer activity of *Sapindus saponaria* L. in the rat. J. Ethnopharmacol. 82: 41–44.

Almeida, E.X., Conserva, L.M. and Lemos, R.P.L. 2002. Coumarins, coumarinolignoids and terpenes from *Protium heptaphyllum*. Biochem. Syst. Ecol. 30: 685–687.

Almeida, L.M.S., Farani, P.G.S., Tosta, L.A., Silvério, M.S., Sousa, O.V., Mattos, A.C.A., Coelho, P.M.Z., Vasconcelos, E.G. and Faria-Pinto, P. 2012. *In vitro* evaluation of the schistosomicidal potential of *Eremanthus erythropappus* (DC) McLeisch (Asteraceae) extracts. Nat. Prod. Res. 26: 2137–2143.

Alonso-Castro, A.J., Villarreal, M.L., Salazar-Olivo, L.A., Gómez-Sánchez, M., Domínguez, F. and García-Carranca, A. 2011. Mexican medicinal plants used for cancer treatment: pharmacological, phytochemical and ethnobotanical studies. J. Ethnopharmacol. 133: 945–972.

Álvarez, O.L. 1991. Recuperación de la medicina indígena. Memorias del I Encuentro de Médicos Tradicionales. Santiago—Putumayo. Cartilla de Educación Popular No. 8. Servicio Colombiano de Comunicación, Bogotá.

Álvarez-Gómez, A.M., Cardona-Maya, W.D., Castro-Álvarez, J.F., Jiménez, S. and Cadavid, A. 2007. Nuevas opciones en anticoncepción: posible uso espermicida de plantas colombianas. Actas Urol. Esp. 31: 372–381.

Alves, J.C.F. 2011. A review on the chemistry of Eremanthine: a sesquiterpene lactone with relevant biological activity. Org. Chem. Internat. Article ID 170196.

Anderberg, A.A., Baldwin, B.G., Bayer, R.G., Breitwieser, J., Jeffrey, C., Dillon, M.O., Eldenäs, P., Funk, V., Garcia-Jacas, N., Hind, D.J.N., Karis, P.O., Lack, H.W., Nesom, G., Nordenstam, B., Oberprieler, C., Panero, J.L., Puttock, C., Robinson, H., Stuessy, T.F., Susanna, A., Urtubey, E., Vogt, R., Ward, J. and Watson, L.E. 2007. Compositae. pp. 61–588. *In*: K. Kubitzki, J.W. Kadereit and C. Jeffrey (eds.). Flowering Plants. Eudicots: Asterales. Series title—The Families and Genera of Vascular Plants. Vol. VIII. Springer-Verlag, Berlin.

Araujo, D.A.O.V., Takayama, C., de-Faria, F.M., Socca, E.A.R., Dunder, R.J., Manzo, L.P., Luiz-Ferreira, A. and Souza-Brito, A.R.M. 2011. Gastroprotective effects of essential oil from *Protium heptaphyllum* on experimental gastric ulcer models in rats. Brazil J. Pharmacogn. 21: 721–729.

Arellanes, Y., Casas, A., Arellanes, A., Vega, E., Blancas, J., Vallejo, M., Torres I., Rangel-Landa, S., Moreno, A.I., Solis, L. and Pérez-Negrón, E. 2013. Influence of traditional markets on plant management in the Tehuacan Valley. J. Ethnobiol. Ethnomed. 9: 38.

Arrau, S., Delporte, C., Cartagena, C., Rodríguez-Díaz, M., González, P., Silva, X., Casseles, B.K. and Miranda, H.F. 2011. Antinociceptive activity of *Quillaja saponaria* Mol. saponin extract, quillaic acid and derivatives in mice. J. Ethnopharmacol. 133: 164–167.

Arroyo, M.K.T., Dirzo, R., Castillas, J.C., Cejas, F. and Joly, C.A. 2010. Biodiversity in Latin America and the Caribbean: an assessment of knowledge, research scope and priority areas. Volume 1. Science for a better life: developing regional scientific programs in priority areas for Latin America and the Caribbean. ICSU-LAC/CONACYT. Rio de Janeiro and Mexico City.

Assunção, L. and Kutsch, R. UNCTAD. 2006. Primeras experiencias en el apoyo a cadenas de valor de productos de Biocomercio. Documento informativo con motivo de la Octava Conferencia de las Partes del Convenio de Diversidad Biológica, Iniciativa Biotrade, Programa de Facilitación del Biocomercio, Ginebra, 31 pp.

Ballabeni, V., Tognolini, M., Bertoni, S., Bruni, R., Guerrini, A., Moreno Ruega, G. and Barocelli, E. 2007. Antiplatelet and antithrombotic activities of essential oil from wild *Ocotea quixos* (Lam.) Kosterm. (Lauraceae) calices from Amazonian Ecuador. Pharmacol. Res. 55: 23–30.

Ballabeni, V., Tognolini, M., Giorgio, C., Bertoni, S., Bruni, R. and Barocelli, E. 2010. *Ocotea quixos* Lam. essential oil: *in vitro* and *in vivo* investigation on its anti-inflammatory properties. Fitoterapia. 81: 289–295.

Bandeira, P.N., Machado, M.I.L., Cavalcanti, F.S. and Lemos, T.L.G. 2001. Essential oil composition of leaves, fruits and resin of *Protium heptaphyllum* (Aubl.) March. J. Essent. Oil Res. 13: 33–34.

Bandeira, P.N., Pessoa, O.D.L., Trevisan, M.T.S. and Lemos, T.L.G. 2002. Metabólitos secundários de *Protium heptaphyllum* March. Quim. Nova. 25: 1078–1080.

Bandeira, P.N., Fonseca, A.M., Costa, S.M.O., Lins, M.U.D.S., Pessoa, O.D.L., Monte, F.J.Q., Nogueira, N.A.P. and Lemos, T.L.G. 2006. Antimicrobial and antioxidant activities of the essential oil of resin of *Protium heptaphyllum* (Aubl.) March. Nat. Prod. Comm. 1: 117–120.

Barbosa, F.G., Lima, M.A.S., Braz-Filho, R. and Silveira, E.R. 2006. Iridoid and phenylethanoid glycosides from *Lippia alba*. Biochem. Syst. Ecol. 34: 819–821.

Barr, I.G. and Mitchell, G.F. 1996. ISCOM (immunostimulating complexes): the first decade. Immunol. Cell Biol. 74: 8–25.

Barroso, M.S.T., Villanueva, G., Lucas, A.M., Pérez, G.P., Vargas, R.M.F., Brun, G.W. and Cassel, E. 2011. Supercritical fluid extraction of volatile and non-volatile compounds from *Schinus molle* L. Braz. J. Chem. Eng. 28: 305–312.

Baruah, R.N. and Leclercq, P.A. 1993. Characterization of the essential oil from flower heads of *Spilanthes acmella*. J. Essent. Oil Res. 5: 693–695.

Baser, K.H.C., Kürkçüoglu, M., Demirçakmak, B., Uülker, N. and Beis, S.H. 1997. Composition of the essential oil of *Schinus molle* L. grown in Turkey. J. Essent. Oil Res. 9: 693–696.

Belhamel, K., Abderrahim, A. and Ludwig, R. 2009. Chemical composition and antibacterial activity of the essential oil of *Schinus molle* L. growth in Algeria. Acta Hort. (ISHS) 826: 201–204. Available from: http://www.actahort.org/books/826/826_27.htm (March, 2014).

Bendaoud, H., Romdhane, M., Souchard, J.P., Cazaux, S. and Bouajila, J. 2010. Chemical composition and anticancer and antioxidant activities of *Schinus molle* L. and *Schinus terebinthifolius* Raddi berries essential oils. J. Food Sci. 75: C466–C472.

Bernal, H.Y. and Correa, J.M. 1989. *Bixa orellana* L. pp. 260–289. *In*: Especies vegetales promisorias de los países del Convenio Andrés Bello. Tomo II. Programa de Recursos Vegetales del Convenio Andrés Bello. Editora Secretaria Ejecutiva del Convenio Andrés Bello, Bogotá.

Bernal, H.Y. and Correa, J.M. 1994. *Ocotea caparrapi* (N.) Dug. pp. 227–261. *In*: Especies Vegetales Promisorias de los países del Convenio Andrés Bello. Tomo X. Programa de Recursos Vegetales del Convenio Andrés Bello. Editora Secretaria Ejecutiva del Convenio Andrés Bello, Bogotá.

Bernal, H.Y., García-Martínez, H. and Quevedo-Sánchez, G.F. 2011. Pautas para el conocimiento, conservación y uso sostenible de las plantas medicinales nativas en Colombia: Estrategia nacional para la conservación de plantas. Instituto de Investigación de Recursos Biológicos Alexander von Humboldt, Bogotá.

Bernhard, R.A., Shibamoto, T., Yamaguchi, K. and White, E. 1983. The volatile constituents of *Schinus molle* L. J. Agric. Food Chem. 31: 463–466.

Bicchi, C., Fresia, M., Rubiolo, P., Monti, D., Franz and Goehler, I. 1997. Constituents of *Tagetes lucida* Cav. ssp. *lucida* essential oil. Flavour Frag. J. 12: 47–52.

Blanco, M.A., Colareda, G.A., van Baren, C., Bandoni, A.L., Ringuelet, J. and Consolini, A.E. 2013. Antispasmodic effects and composition of the essential oils from two South American chemotypes of *Lippia alba*. J. Ethnopharmacol. 149: 803–809.

Borges, E.S. and Rezende, C.M. 2000. Main aroma constituents of genipap (*Genipa americana* L.) and bacuri (*Platonia insignis* M.). J. Essent. Oil Res. 12: 71–74.

Borges del Castillo, J., Brooks, C.J.W. and Campbell, M.M. 1966. Caparrapidiol and Caparrapitriol: Two new acyclic sesquiterpene alcohols. Tetrahedron Lett. 31: 3731–3736.

Braham, W.C., Pinzón, C.A., Pachón, M.E., Rojas, L.F. and Aragón, J.C. 2000. Estudio de las especies promisorias productoras de colorantes en el trapecio amazónico. Universidad Distrital Francisco José de Caldas, Bogotá.

Brandt, R., Zimmermann, H., Hensen, I., Mariscal, J.C. and Rist, S. 2012. Agroforestry species of the Bolivian Andes: an integrated assessment of ecological, economic and socio-cultural plant values. Agroforest. Syst. 86: 1–16.

Brooks, C.J.W. and Campbell, M.M. 1969. Caparrapi oxide, a sesquiterpenoid from caparrapi oil. Phytochemistry 8: 215–218.

Bruni, R., Medici, A., Andreotti, E., Fantin, C., Muzzoli, M., Dehesa, M., Romagnoli, C. and Sacchetti, G. 2004. Chemical composition and biological activities of Ishpingo essential oil, a traditional Ecuadorian spice from *Ocotea quixos* (Lam.) Kosterm. (Lauraceae) flower calices. Food Chem. 85: 415–421.

Bueno-Sánchez, J.G., Martínez-Morales, J.R., Stashenko, E.E. and Ribón, W. 2009. Anti-tubercular activity of eleven aromatic and medicinal plants occurring in Colombia. Biomedica. 29: 51–60.

Caballero-Gallardo, K., Olivero-Verbel, J. and Stashenko, E.E. 2012. Repellency and toxicity of essential oils from *Cymbopogon martinii*, *Cymbopogon flexuosus* and *Lippia origanoides* cultivated in Colombia against *Tribolium castaneum*. J. Stored Prod. Res. 50: 62–65.

Cabrera, I. 2005. Las plantas y sus usos en las Islas de Providencia y Santa Catalina. Universidad del Valle Programa Editorial, Cali.

Calvo-Irabien, L.M., Yam-Puc, J.A., Dzib, G., Escalante-Erosa, F. and Peña-Rodriguez, L.M. 2009. Effect of postharvest drying on the composition of mexican oregano (*Lippia graveolens*) essential oil. J. Herb. Spic. Med. Plants. 15: 281–287.

Campos Chisté, R., Zerlotti Mercadante, A., Gomes, A., Fernandes, E., da Costa Lima, J.L.F. and Bragagnolo, N. 2011. *In vitro* scavenging capacity of annatto seed extracts against reactive oxygen and nitrogen species. Food Chem. 127: 419–426.

Can Başer, K.H. and Buchbauer, G. 2010. Handbook of Essential Oils: Science, Technology, and Applications. CRC Press and Taylor & Francis Group, Boca Raton.

Cárdenas, D. and Salinas, N.R. 2006. Libro rojo de plantas de Colombia. Especies maderables amenazadas: I Parte. Instituto Amazónico de Investigaciones Científicas (SINCHI), Ministerio de Ambiente, Vivienda y Desarrollo Territorial, Bogotá, pp. 35–135.

Cárdenas, D. and Salinas, N.R. 2007. Libro rojo de plantas de Colombia. Volumen 4. Especies maderables amenazadas: I Parte. Serie libros rojos de especies amenazadas de Colombia. Instituto Amazónico de Investigaciones Científicas SINCHI, Ministerio de Ambiente, Vivienda y Desarrollo Territorial. Bogotá, pp. 115–117.

Carmona, R., Quijano-Celís, C.E. and Pino, J.A. 2007. Leaf oil composition of *Bursera graveolens* (Kunth) Triana et Planch. J. Essent. Oil Res. 21: 387–389.

Carretero, M.E., López-Pérez, J.L., Abad, M.J., Bermejo, P., Tillet, S., Israel, A. and Noguera-P, B. 2008. Preliminary study of the anti-inflammatory activity of hexane extract and fractions from *Bursera simaruba* (Linneo) Sarg. (Burseraceae) leaves. J. Ethnopharmacol. 116: 11–15.

Cerón, C.E. 2006. Plantas medicinales de los Andes ecuatorianos. pp. 285–293. *In*: M. Moraes, B. Øllgaard, L.P. Kvist, F. Borchsenius and H. Balslev (eds.). Botánica económica de los Andes Centrales. Universidad Mayor de San Andrés, La Paz.

Céspedes, C.L., Avila, J.G., Martínez, A., Serrato, B., Calderón-Mugica, J.C. and Salgado-Garciglia, R. 2006. Antifungal and antibacterial activities of Mexican tarragon (*Tagetes lucida*). J. Agric. Food Chem. 54: 3521–3527.

Chamorro, E.R., Zambón, S.N., Morales, W.G., Sequeira, A.F. and Velasco, G.A. 2012. Study of the chemical composition of essential oils by gas chromatography. pp. 307–324. *In*: B. Salih (ed.). Gas Chromatography in Plant Science, Wine Technology, Toxicology and some Specific Applications. InTech.

Chataing, B., Rojas, L., Usubillaga, A. and Mora, D. 2012. Chemical composition and bioactivity on bacteria and fungi of the essential oil from *Lippia origanoides* H.B.K. J. Essent. Oil Bear. Pl. 15: 454–460.

Chizmar, C., Lu, A. and Correa, M. 2009. Plantas de uso floclórico y tradicional de Panamá. Instituto Nacional de Biodiversidad—INBio, Santo Domingo de Heredia, pp. 103–105.

Cicció, J.F. 2005. Source of almost pure methyl chavicol: volatile oil from the aerial parts of *Tagetes lucida* (Asteraceae) cultuvated in Costa Rica. Rev. Biol. Trop. 52: 853–857.

Citó, A.M.G.L., Costa, F.B., Lopes, J.A.D., Oliveira, V.M.M. and Chaves, M.H. 2006. Identificação de constituintes voláteis de frutos e folhas de *Protium heptaphyllum* Aubl (March). Rev. Bras. Pl. Med. 8: 4–7.

Clarke, I. and Lee, H. 1987. Name that Flower: The Identification of Flowering Plants. Melbourne University Press, Melbourne.

Conde, R., Corrêa, V.S.C., Carmona, F., Contini, S.H.T. and Pereira, A.M.S. 2011. Chemical composition and therapeutic effects of *Lippia alba* (Mill.) N. E. Brown leaves hydro-alcoholic extract in patients with migraine. Phytomedicine 18: 1197–1201.

Conrad, O.A., Dike, I.P. and Agbara, U. 2013. *In vivo* antioxidant assessment of two antimalarial plants—*Allamamda cathartica* and *Bixa orellana*. Asian Pac. J. Trop. Biomed. 23: 388–394.

Correa, J.M. and Bernal, H.Y. 1989. *Schinus molle* L. pp. 169–183. *In*: Especies Vegetales promisorias de los países del Convenio Andrés Bello. Tomo I. Programa de Recursos Vegetales del Convenio Andrés Bello. Editora Secretaria Ejecutiva del Convenio Andrés Bello, Bogotá.

Correa, J.M. and Bernal, H.Y. 1990. *Bursera graveolens* (H.B.K.) Triana & Planchon. *Bursera simaruba* (L.) Sargent. *Protium heptaphyllum* M.B.K. pp. 57–78. *In*: Especies Vegetales Promisorias de los países del Convenio Andrés Bello. Tomo III. Programa de Recursos Vegetales del Convenio Andrés Bello. Editora Secretaria Ejecutiva del Convenio Andrés Bello, Bogotá.

Cotes, A.M., Barrero, L.S., Rodríguez, F., Zuluaga, M.V. and Arévalo, H. 2012. Bioprospección para el desarrollo del sector agropecuario de Colombia. Corpoica, Bogotá.

Cronquist, A. 1980. Vascular Flora of the Southeastern United States. Volume I: Asteraceae. The University of North Carolina Press, Raleigh.

Cuatrecasas, J. 1957. Prima flora colombiana: 1. Burseracea. Webbia. 12: 375–441.

Da Cunha, M.A., de Barros, F.M.C., Garcia, L.O., Veeck, A.P.L., Heinzmann, B.M., Loro, V.L., Emanuelli, T. and Baldisserotto, B. 2010. Essential oil of *Lippia alba*: a new anesthetic for silver catfish, *Rhamdia quelen*. Aquaculture 306: 403–406.

Daly, D.C., Harley, M.M., Martínez-Habibe, M.C. and Weeks, A. 2011. Burseraceae. pp. 76–104. *In*: K. Kubitzki (ed.). Flowering Plants. Eudicots: Sapindales, Cucurbitales, Myrtaceae. Series title—The Families and Genera of Vascular Plants. Vol. X. Springer-Verlag, Berlin.

Damke, E., Tsuzuki, J.K., Chassot, F., Cortez, D.A.G., Ferreira, I.C.P., Mesquita, C.S.S., da Silva, V.R.S., Svidzinski, T.I.E. and Consolaro, M.R.L. 2013. Spermicidal and anti-*Trichomonas vaginalis* activity of Brazilian *Sapindus saponaria*. BMC Complem. Altern. Med. 13: 196.

DeBaggio, T. and Tucker, A.O. 2009. The Encyclopedia of Herbs: A Comprehensive Reference to Herbs of Flavor and Fragrance. Timber Press, Portland.

Dellacassa, E. 2010. Normalización de productos naturales obtenidos de especies de la flora aromática latinoamericana. Proyecto CYTED IV.20. Editora Universitaria da PUCRS, Porto Alegre.

Dellacassa, E., Soler, E., Menéndez, P. and Moyna, P. 1990. Essential oils from *Lippia alba* (Mill.) N.E. Brown and *Aloysia chamaedrifolia* Cham. (verbenaceae) from Uruguay. Flavour Frag. J. 5: 107–108.

Dellavalle, P.D., Cabrera, A., Alem, D., Larrañaga, P., Ferreira, F. and Dalla Rizza, M. 2011. Antifungal activity of medicinal plant extracts against phytopathogenic fungus *Alternaria* spp. Chil. J. Agr. Res. 71: 231–239.

Deveci, O., Sukan, A., Tuzun, N. and Hames Kocabas, E.E. 2010. Chemical composition, repellent and antimicrobial activity of *Schinus molle* L. J. Med. Plants Res. 4: 2211–2216.

Dias, A.M.A., Santos, P., Seabra, I.J., Júnior, R.N.C., Braga, M.E.M. and de Sousa, H.C. 2012. Spilanthol from *Spilanthes acmella* flowers, leaves and stems obtained by selective supercritical carbon dioxide extraction. J. Supercrit. Fluid. 61: 62–70.

Díaz, C., Quesada, S., Brenes, O., Aguilar, G. and Cicció, J.F. 2008. Chemical composition of *Schinus molle* essential oil and its cytotoxic activity on tumor cell lines. Nat. Prod. Res. 22: 1521–1534.

Díaz González, G.J. 2010. Plantas tóxicas de importancia en salud y producción animal en Colombia. Editorial Universidad Nacional de Colombia, Bogotá, pp. 188–189.

Dikshit, A., Naqvi, A.A. and Husain, A. 1986. *Schinus molle*: a new source of natural fungitoxicant. Appl. Environ. Microbiol. 51: 1085–1088.

Dinkelmeyer, H., Lehmann, J., Renck, A., Trujillo, L., da Silva, J.P., Gebauer, G. and Kaiser, K.. 2003. Nitrogen uptake from N-15-enriched fertilizer by four tree crops in an Amazonian agroforest. Agroforest. Syst. 57: 213–224.

Dos Santos, F.J.B., Lopes, J.A.D., Cito, A.M.G.L., de Oliveira, E.H., de Lima, S.G. and Reis, F.A.M. 2004. Composition and biological activity of essential oils from *Lippia origanoides* H.B.K. J. Essent. Oil Res. 16: 504–506.

Dubey, S., Maity, S., Singh, M., Saraf, S.A. and Saha, S. 2013. Phytochemistry, pharmacology and toxicology of *Spilanthes acmella*: a review. Adv. Pharmacol. Sci. Article ID 423750.

Duke, J.A. and Vásquez-Martínez, R. 1994. Amazonian Ethnobotanical Dictionary. CRC Press LLC, Boca Raton.

Duke, J.A., Bogenschutz-Godwin, M.J. and Ottesen, A.R. 2009. Duke's Handbook of Medicinal Plants of Latin America. CRC Press, Boca Raton.

Durán, D.C., Monsalve, L.A., Martínez, J.R. and Stashenko, E.E. 2007. Estudio comparativo de la composición química de aceites esenciales de *Lippia alba* provenientes de diferentes regiones de Colombia, y efecto del tiempo de destilación sobre la composición del aceite. Sci. Techn. 33: 435–438.

Edris, A.E. 2007. Pharmaceutical and therapeutic potentials of essential oils and their individual volatile constituents: A review. Phytother. Res. 21: 308–323.

Elias, M.E.A., Schroth, G., Macedo, J.L.V., Mota, M.S.S. and D'Angelo, S.A. 2002. Mineral nutrition, growth and yields of annatto trees (*Bixa orellana*) in agroforestry on an Amazonian ferralsol. Exp. Agr. 38: 277–289.

Ennigrou, A., Hosni, K., Casabianca, H., Vulliet, E. and Smiti, S. 2011. Leaf volatile oil constituants of *Schinus terebinthifolius* and *Schinus molle* from Tunisia. FOODBALT-Conference Proceeding. pp. 90–92. Available from: http://llufb.llu.lv/conference/foodbalt/2011/FOODBALT-Proceedings-2011-90-92.pdf (January, 2014).

Escobar, P., Leal, S.M., Herrera, L.V., Martinez, J.R. and Stashenko, E. 2010. Chemical composition and antiprotozoal activities of Colombian *Lippia* spp. essential oils and their major components. Mem. Inst. Oswaldo Cruz. 105: 184–190.

Etkin, N.L. 1986. Plants in Indigenous Medicine & Diet: Biobehavioral Approaches. Routledge—Taylor & Francis Group, Abingdon.

Fagotti, D.S.L., Miyauchi, M.Y.H., Oliveira, A.G., Santinoni, I.A., Eberhardt, D.N., Nimtz, A., Ribeiro, R.A., Paula, A.M., Queiroz, C.A.S., Andrade, G., Zangaro, W. and Nogueira, M.A. 2012. Gradients in N-cycling attributes along forestry and agricultural land-use systems are indicative of soil capacity for N supply. Soil Use Manage. 28: 292–298.

Fern, K. 2012. *Eremanthus erythropappus*. Tropical species database. Available from: http://theferns.info/tropical/viewtropical.php?id=Eremanthus+erythropappus (January, 2014).

Fischer, U., Lopez, R., Pöll, E., Vetter, S., Novak, J. and Franz, C.M. 2004. Two chemotypes within *Lippia alba* populations in Guatemala. Flavour Frag. J. 19: 333–335.

Fleischer, T.C., Ameade, E.P.K., Mensah, M.L.K. and Sawer, I.K. 2003. Antimicrobial activity of the leaves and seeds of *Bixa orellana*. Fitoterapia. 74: 136–138.

Fonseca, F.N., Ferreira, A.J.S., Sartorelli, P., Lopes, N.P., Floh, E.I.S., Handro, W. and Kato, M.J. 2000. Phenylpropanoid derivatives and biflavones at different stages of differentiation and development of *Araucaria angustifolia*. Phytochemistry 55: 575–580.

Fonseca, S.F., Nielsen, L.T. and Rúveda, E.A. 1979. Lignans of *Araucaria angustifolia* and ¹³C NMR analysis of some phenyltetralin lignans. Phytochemistry 18: 1703–1708.

Forero-Peñuela, L.Y., Biasi, L.A., Bizzo, H.R., de Souza, M.S. and Deschamps, C. 2013. Potential of *Lippia alba* (Mill.) N.E. Br. ex Britt. & P. Wilson, as available source of linalool in southern Brazil. J. Essent. Oil Res. 25: 464–467.

Freitas, A.M., Almeida, M.T.R., Andrighetti-Fröhner, C.R., Cardozo, F.T.G.S., Barardi, C.R.M., Farias, M.R. and Simones, C.M.O. 2009. Antiviral activity-guided fractionation from *Araucaria angustifolia* leaves extract. J. Ethnopharmacol. 126: 512–517.

Frighetto, N., de Oliveira, J.G., Siani, A.C. and das Chagas, K.C. 1998. *Lippia alba* Mill. N.E. Br. (Verbenaceae) as a source of linalool. J. Essent. Oil Res. 10: 578–570.

Fun, C.E. and Svendsen, A.B. 1990. The essential oil of *Lippia alba* (Mill.) N.E.Br. J. Essent. Oil Res. 2: 265–267.

Funk, V.A., Susanna, A., Stuessy, T.F. and Bayer, R.J. 2009. Systematics, Evolution, and Biogeography of Compositae. International Association for Plant Taxonomy. Smithsonian Institute, Washington.

García-Barriga, H. 1992. Flora medicinal de Colombia. Botánica médica. Tomo II. Tercer Mundo Editores, Bogotá.

Goldstein, D.J. and Coleman, R.C. 2004. *Schinus molle* L. (Anacardiaceae) Chicha production in the Central Andes. Econ. Bot. 58: 523–529.

Gómez, J.A. and Mejía, D.G. 2010. Biodiversidad y desarrollo: una oportunidad para el sector cosmético natural en Colombia. Rev. Cosmet. Versión 97(1): 7.

Gonsalves, J. 2010. Economic Botany and Ethnobotany. International Scientific Publishing Academy, New Delhi.

Granados-Sánchez, D., Martínez-Salvador, M., López-Ríos, G.F., Borja-De la Rosa, A. and Rodríguez-Yam, G.A. 2013. Ecology, harvesting and marketing of oregano (*Lippia graveolens* H.B.K.) in Mapimi, Durango. Rev. Chapingo Serie Cie. Forest. Amb. 19: 305–321.

Grandtner, M.M. 2005. Elsevier's Dictionary of Trees. Volume I. Elsevier B.V., Amsterdam.

Grandtner, M.M. and Chevrette, J. 2013. Dictionary of Trees. South America. Nomenclature, Taxonomy and Ecology. Volume 2. Academic Press (Elsevier), USA.

Grubben, G.J.H. and Denton, O.A. 2004. Plant Resource of Tropical Africa 2. Vegetables. PROTA Foundation, Wageningen.

Guadarrama-Cruz, G., Alarcon-Aguilar, F.J., Lezama-Velasco, R., Vazquez-Palacios, G. and Bonilla-Jaime, H. 2008. Antidepressant-like effects of *Tagetes lucida* Cav. in the forced swimming test. J. Ethnopharmacol. 120: 277–281.

Guala, M.S., Elder, H.V., Pérez, G. and Chiesa, A. 2009. Evaluación del poder antioxidante de fracciones de aceite esencial crudo de *Schinus molle* L. obtenidas por destilación al vacío. Inf. Tecnol. 20: 83–88.

Guirado, O.A.A. 2005. Potencial medicinal del género *Sapindus* L. (Sapindaceae) y de la especie *Sapindus saponaria* L. Rev. Cubana Plant. Med. 10: 1–10.

Guo, S., Falk, E., Kenne, L., Ronnberg, B. and Sundquist, B.G. 2000. Triterpenoid saponins containing an acetylated branched D-fucosyl residue from *Quillaja saponaria* Molina. Phytochemistry 53: 861–868.

Gupta, M.P. 1995. 270 plantas medicinales iberoamericanas. CYTED-SECAB. Editora Presencia Ltda, Bogotá, pp. 21–566.

Haldar, S., Kar, B., Dolai, N., Kumar, R.B.S., Behera, B. and Haldar, P.K. 2012. *In vivo* anti-nociceptive and anti-inflammatory activities of *Lippia alba*. Asian Pac. J. Trop. Dis. S667–S670.

Harley, R.M., Atkins, S., Budantsev, A.L., Cantino, P.D., Conn, B.J., Grayer, R., Harley, M.M., De Kok, R., Krestovskaja, T., Morales, R., Paton, A.J., Ryding, O. and Upson, T. 2004. Labiatae. pp. 167–228. *In*: K. Kubitzki and J.W. Kadereit (eds.). Flowering Plants dicotyledons. Lamiales (except Acanthaceae including Avicenniaceae). Series title—The Families and Genera of Vascular Plants. Vol. 7. Springer-Verlag, Berlin.

Harrar, E.S. and Harrar, J.G. 1962. Guide to Southern Trees. McGraw-Hill, New York.

Hayouni, A.E., Chraief, I., Abedrabba, M., Bouixe, M., Leveaue, J.Y., Mohammed, H. and Hamdi, M. 2008. Tunisian *Salvia officinalis* L. and *Schinus molle* L. essential oils: their chemical compositions and their preservative effects against Salmonella inoculated in minced beef meat. Int. J. Food Microbiol. 125: 242–251.

Hennebelle, T., Sahpaz, S., Joseph, H. and Bailleul, F. 2008. Ethnopharmacology of *Lippia alba*. J. Ethnopharmacol. 116: 211–222.

Hernández, T., Canales, M., Avila, J.G., Duran, A., Caballero, J., Romo de Vivar, A. and Lira, R. 2003. Ethnobotany and antibacterial activity of some plants used in traditional medicine of Zapotitlán de las Salinas, Puebla (México). J. Ethnopharmacol. 88: 181–188.

Hickey, M. and King, C. 1997. Common Families of Flowering Plants. Cambridge University Press, Cambridge.

Hickey, M. and King, C. 1998. 100 Families of Flowering Plants. Cambridge University Press, Cambridge.

Hind, N. and Biggs, N. 2003. Plate 460. *Acmella oleracea*. Compositae. Curtis's BotMag. 20: 31–39.

Hooks, C.R.R., Wang, K.H., Ploeg, A. and McSorley, R. 2010. Using marigold (*Tagetes* spp.) as a cover crop to protect crops from plant-parasitic nematodes. Appl. Soil Ecol. 46: 307–320.

Hosni, K., Jemli, M., Dziri, S., M'rabet, Y., Ennigrou, A., Sghaier, A., Casabianca, H., Vulliet, E., Ben Brahim, N. and Sebei, H. 2011. Changes in phytochemical, antimicrobial and free radical scavenging activities of the Peruvian pepper tree (*Schinus molle* L.) as influenced by fruit maturation. Ind. Crop. Prod. 34: 1622–1628.

Huamantupa, I., Cuba, M., Urrunaga, R., Paz, E., Ananya, N., Callalli, M., Pallqui, N. and Coasaca, H. 2011. Riqueza, uso y origen de plantas medicinales expendidas en los mercados de la ciudad del Cusco. Rev. Peru. Biol. 18: 283–291.

Iannacone, J., La Torre, M.I., Alvariño, L., Cepeda, C., Ayala, H. and Argota, G. 2013. Toxicidad de los bioplaguicidas *Agave americana*, *Furcraea andina* (Asparagaceae) y *Sapindus saponaria* (Sapindaceae) sobre el caracol invasor *Melanoides tuberculata* (Thiaridae). Neotrop. Helminthol. 7: 231–241.

Ilany, T., Ashton, M.S., Montagnini, F. and Martinez, C.U. 2010. Using agroforestry to improve soil fertility: effects of intercropping on *Ilex paraguariensis* (yerba mate) plantations with *Araucaria angustifolia*. Agroforest. Syst. 80: 399–409.

Jaramillo, B.E., Olivero, J.T. and Muñoz, K. 2007. Composición química volátil y toxicidad aguda (CL$_{50}$) frente a *Artemia salina* del aceite esencial del *Croton malambo* colectado en la costa norte Colombiana. Sci. Techn. 33: 299–302.

Jaramillo, B.E., Duarte, E., Muñoz, K. and Stashenko, E. 2010. Composición química volátil del aceite esencial de *Croton malambo* H. Karst. colombiano y determinación de su actividad antioxidante. Rev. Cubana Plant. Med. 15: 133–142.

Jaramillo-Castro, L. 2012. Trade and biodiversity: the BioTrade experiences in Latin America. United Nations Conference on Trade and Development. UNCTAD. United Nations Publications, New York and Geneva.

Jirovetz, L., Buchbauer, G., Wobus, A., Shafi, M.P. and Abraham, G.T. 2005. Essential oil analysis of *Spilanthes acmella* Murr. Fresh plants from southern India. J. Essent. Oil Res. 17: 429–431.

Johnson, T. 1999. CRC Ethnobotany Desk Reference. CRC Press LLC, Boca Raton.

Jondiko, I.J.O. and Pattenden, G. 1989. Terpenoids and an apocarotenoid from seed of *Bixa orellana*. Phytochemistry 28: 3159–3162.

Junor, G.A.O., Porter, R.B.R., Facey, P.C. and Yee, T.H. 2007. Investigation of essential oil extracts from four native jamaican species of *Bursera* for antibacterial activity. West Indian Med. J. 56: 22–25.

Junor, G.A.O., Porter, R.B.R. and Yee, T.H. 2008. The chemical composition of the essential oils from the leaves, bark and fruits of *Bursera simaruba* (L.) Sarg. from Jamaica. J. Essent. Oil Res. 20: 426–429.

Kartal, M. 2007. Intellectual property protection in the natural product drug discovery, traditional herbal medicine and herbal medicinal products. Phytother. Res. 21: 113–119.

Kashanipour, R.A. and McGee, R.J. 2004. Northern Lacandon Maya medicinal plant use in the communities of Lacanja Chan Sayab and Naha', Chiapas, Mexico. J. Ecol. Anthropol. 8: 47–66.

Kiokias, S. and Gordon, M.H. 2003. Antioxidant properties of annatto carotenoids. Food Chem. 83: 523–529.

Kirk, D.D., Rempel, R., Pinkhasov, J. and Walmsley, A.M. 2004. Application of *Quillaja saponaria* extracts as oral adjuvants for plant-made vaccines. Expert Opin. Biol. Ther. 4: 947–958.

Koehnlein, E.A., Santos Carvajal, A.E., Koehnlein, E.M., da Silva Coelho-Moreira, J., Inácio, F.D., Castoldi, R., Bracht, A. and Peralta, R.M. 2012. Antioxidant activities and phenolic compounds of raw and cooked Brazilian pinhão (*Araucaria angustifolia*) seeds. Afr. J. Food Sci. 6: 512–518.

Kramer, F.L. 1957. The pepper tree, *Schinus molle* L. Econ. Bot. 11: 322–326.

Kubitzki, K. 2007. Quillajaceae. pp. 407–408. *In*: K. Kubitzki (ed.). Flowering Plants. Eudicots. Berberidopsidales, Buxales, Crossosomatales, Fabales p.p., Geraniales, Gunnerales, Myrtales p.p., Proteales, Saxifragales, Vitales, Zygophyllales, Clusiaceae Alliance, Passifloraceae Alliance, Dilleniaceae, Huaceae, Picramniaceae, Sabiaceae. Series title—The Families and Genera of Vascular Plants. Vol. IX. Springer-Verlag, Berlin.

Lemos, T.L.G., Pessoa, O.L.D., Matos, F.J.A., Alencar, J.W. and Craveiro, A.A. 1991. The essential oil of *Spilanthes acmella* Murr. J. Essent. Oil Res. 3: 369–370.

León, J. 1968. Fundamentos botánicos de los cultivos tropicales. Serie: Textos y Materiales de Enseñanza No. 18. Editorial Instituto Interamericano de Ciencias Agrícola de la OEA, Lima.

Leyva, M.A., Martínez, J.R. and Stashenko, E.E. 2007. Composición química del aceite esencial de hojas y tallos de *Bursera graveolens* (Burseraceae) de Colombia. Sci. et Techn. 33: 201–202.

Lim, T.K. 2012. *Schinus molle*. pp. 153–159. *In*: Edible Medicinal and Non-Medicinal Plants. Vol. 1. Fruit. Springer-Verlag, Heidelberg.

Lim, T.K. 2014. Edible medicinal and non-medicinal plants, Volume 7—Flowers. Springer Science + Business Media, Berlin.

Lin, L.Z., Mukhopadhyay, S., Robbins, R.J. and Harnly, J.M. 2007. Identification and quantification of flavonoids of Mexican oregano (*Lippia graveolens*) by LC-DAD-ESI/MS analysis. J. Food Compos. Anal. 20: 361–369.

Ljalem, H.A. and Unnithan, C.R. 2014. Chemical composition and antibacterial activity of essential oil of *Schinus molle*. Unique J. Pharm. Biol. Sci. 2: 9–12.

López-González, G.A. 2007. Guía de los árboles y arbustos de la Península Ibérica y Baleares (especies silvestres y las cultivadas más comunes). 3a edición. Ediciones Mundi-Prensa, Madrid.

Lorenzo, D., Paz, D., Davies, P., Vila, R., Cañigueral, S. and Dellacassa, E. 2001. Composition of a new essential oil type of *Lippia alba* (Mill.) N.E. Brown from Uruguay. Flavour Frag. J. 16: 356–359.

Lovato, L., Pelegrini, B.L., Rodrigues, J., de Oliveira, A.J.B. and Ferreira, I.C.P. 2014. Seed oil of *Sapindus saponaria* L. (Sapindaceae) as potential C_{16} to C_{22} fatty acids resource. Biomass Bioenerg. 60: 247–251.

Lovera, J.R. 2005. Food Culture in South America. Greenwood Press, Westport.

Lozada-Pérez, L. 2003. Bixaceae. pp. 3–5. *In*: N. Diego-Pérez and R.M. Fonseca (eds.). Flora de Guerrero. Prensas de Ciencias, México.

Lubbe, A. and Verpoorte, R. 2011. Cultivation of medicinal and aromatic plants for specialty industrial materials. Ind. Crops. Prod. 34: 785–801.

Luján-Hidalgo, M.C., Gutiérrez-Miceli, F.A., Ventura-Canseco, L.M.C., Dendooven, L., Mendoza-López, M.R., Cruz-Sánchez, S., García-Barradas, O. and Abud-Archila, M. 2012. Composición química y actividad antimicrobiana de los aceites esenciales de hojas de *Bursera graveolens* y *Taxodium mucronatum* de Chiapas, México. Gayana Bot. 69: 7–14.

MacFoy, C. 2004. Ethnobotany and sustainable utilization of natural dye plants in Sierra Leone. Econ. Bot. 58: S66–S76.

Maffei, M. and Chialva, F. 1990. Essential oils from *Schinus molle* L. berries and leaves. Flavour Frag. J. 5: 49–52.

Maldini, M., Montoro, P., Piacente, S. and Pizza, C. 2009. Phenolic compounds from *Bursera simaruba* Sarg. bark: phytochemical investigation and quantitative analysis by tandem mass spectrometry. Phytochemistry 70: 641–649.

Manzano, P., Miranda, M., Gutiérrez, Y., García, G., Orellana, T. and Orellana, A. 2009. Efecto antiinflamatorio y composición química del aceite de ramas de *Bursera graveolens* Triana & Planch (palo santo) de Ecuador. Rev. Cubana Plant. Med. 14: 45–53.

Marciani, D.J., Pathak, A.K., Reynolds, R.C., Seitz, L. and May, R.D. 2001. Altered immunomodulating and toxicological properties of degraded *Quillaja saponaria* Molina saponins. Int. Immunopharmacol. 1: 813–818.

Marongiu, B., Porcedda, A.P.S., Casu, R. and Pierucci, P. 2004. Chemical composition of the oil and supercritical CO_2 extract of *Schinus molle* L. Flavour Frag. J. 19: 554–558.

Marotti, M., Piccaglia, R., Biavati, B. and Marotti, I. 2004. Characterization and yield evaluation of essential oils from different *Tagetes* species. J. Essent. Oil Res. 16: 440–444.

Marques, D.D., Sartori, R.A., Lemos, T.L.G., Machado, L.L., de Souza, J.S.N. and Monte, F.J.Q. 2010. Chemical composition of the essential oils from two subspecies of *Protium heptaphyllum*. Acta Amaz. 40: 227–230.

Martins, M.R., Arantes, S., Candeias, F., Tinoco, M.T. and Cruz-Morais, J. 2014. Antioxidant, antimicrobial and toxicological properties of *Schinus molle* L. essential oils. J. Ethnopharmacol. 151: 485–492.

Matos, F.J.A., Machado, M.I.L., Craveiro, A.A. and Alencar, J.W. 1996. Essential oil composition of two chemotypes of *Lippia alba* grown in northeast Brazil. J. Essent. Oil Res. 8: 695–698.

McMullen, C.K. 1999. Flowering Plants of the Galápagos. Cornell University Press, New York.

Medina-Lemos, R. and R.M. Fonseca. 2009. Anacardiaceae. Flora del Valle de Tehuacan-Cuicatlan. 1a. Instituto de Biología, Universidad Nacional Autónoma de México. D. F, México. 71: 54 pp.

Mercadante, A.Z., Steck, A., Rodriguez-Amaya, D., Pfander, H. and Britton, G. 1996. Isolation of methyl 9'Z-apo-6'-lycopenoate from *Bixa orellana*. Phytochemistry 41: 1201–1203.

Mercadante, A.Z., Steck, A. and Pfander, H. 1997. Isolation and structure elucidation of minor carotenoids from annatto (*Bixa orellana* L.). Phytochemistry 46: 1379–1383.

Mercadante, A.Z., Steck, A. and Pfander, H. 1999. Three minor carotenoids from annatto (*Bixa orellana* L.) seeds. Phytochemistry 52: 135–139.

Mesa-Arango, A.C., Betancur-Galvis, L., Montiel, J., Bueno, J.G., Baena, A., Durán, D.C., Martínez, J.R. and Stashenko, E.E. 2010. Antifungal activity and chemical composition of the essential oils of *Lippia alba* (Miller) N.E. Brown grown in different regions of Colombia. J. Essent. Oil Res. 22: 568–574.

Mishra, R.K., Chaudhary, S., Pandey, R., Gupta, S., Mallavarapu, G.R. and Kumar, S. 2010. Analysis of linalool content in the inflorescence (flower) essential oil and leaf oil of *Lippia alba* cultivar 'Kavach'. J. Essent. Oil Res. 22: 3–7.

Montenegro, G., Peña, R.C. and Timmermann, B.N. 2001. La corteza de quillay (*Quillaja saponaria* Mol.), un recurso de la farmacopea internacional. Rev. Acad. Colomb. Cienc. Exact. Fis. Nat. 25: 420–427.

Monzote, L., Hill, G.M., Cuellar, A., Scull, R. and Setzer, W.N. 2012. Chemical composition and anti-proliferative properties of *Bursera graveolens* essential oil. Nat. Prod. Commun. 7: 1531–1534.

Moran, K., King, S.R. and Carlson, T.J. 2001. Biodiversity prospecting lessons and prospects. Annu. Rev. Anthropol. 30: 505–526.

Moreno, S.C., Carvalho, G.A., Picanço, M.C., Morais, E.G. and Pereira, R.M. 2012. Bioactivity of compounds from *Acmella oleracea* against *Tuta absoluta* (Meyrick) (Lepidoptera: Gelechiidae) and selectivity to two non-target species. Pest Manag. Sci. 68: 386–393.

Moreno-Amado, M. 2003. Guía para procesos de cereria, jaboneria y cremas. Serie Ciencia y Tecnología No. 117. Convenio Andrés Bello, Bogotá.

Mota, M.R.L., Criddle, D.N., Alencar, N.M.N., Gomes, R.C., Meireles, A.V.P., Santi-Gadelha, T., Gadelha, C.A.A., Oliveira, C.C., Benevides, R.G., Cavada, B.S. and Assreuy, A.M.S. 2006. Modulation of acute inflammation by a chitin binding lectin from *Araucaria angustifolia* seeds via mast cells. Naunyn-Schmiedeberg's Arch. Pharmacol. 374: 1–10.

Muñoz-Acevedo, A., Bottia, E.J., Cárdenas, C.Y., Patiño, J.G., Díaz, O.L., Martínez, J.R., Kouznetsov, V.V. and Stashenko, E.E. 2007. Estudio comparativo sobre la capacidad de atrapamiento del catión radical ABTS$^+$ por los aceites esenciales de especies aromáticas con alto contenido de trans-anetol y estragol. Sci. Techn. 33: 117–120.

Muñoz-Acevedo, A., Kouznetsov, V.V. and Stashenko, E.E. 2009. Composición y capacidad antioxidante *in vitro* de aceites esenciales ricos en Timol, Carvacrol, trans-Anetol o Estragol. Salud UIS 41: 287–294.

Muñoz-Acevedo, A., Serrano-Uribe, A., Parra-Navas, X.J., Olivares-Escobar, L.A. and Niño-Porras, M.E. 2013. Análisis multivariable y variabilidad química de los metabolitos volátiles presentes en las partes aéreas y la resina de *Bursera graveolens* (Kunth) Triana & Planch. de Soledad (Atlántico, Colombia). Bol. Latinoam. Caribe. 12: 322–337.

Muñoz-Garmendia, F. and Navarro, C. 2011. CXIV. Anacardiaceae. *In*: S. Castroviejo, C. Aedo, M. Laínz, F. Muñoz Garmendia, G. Nieto Feliner, J. Paivaand and C. Benedí (eds.). Flora ibérica. Real Jardín Botánico. CSIC. Madrid. 9: 1–16.

Nakanishi, T., Inatomi, Y., Murata, H., Shigeta, K., Iida, N., Inada, A., Murata, J., Farrera, M.A., Linuma, M., Tanaka, T., Tajima, S. and Oku, N. 2005. A new and known cytotoxic aryltetralin-type lignans from stems of *Bursera graveolens*. Chem. Pharm. Bull. 53: 229–231.

Naranjo, P. and Mideros, R. 1997. Etnomedicina: progresos italo-latinoamericanos. Volumen II. Editorial Abya-Yala, Quito.

Nerio, L.S., Olivero-Verbel, J. and Stashenko, E.E. 2009. Repellent activity of essential oils from seven aromatic plants grown in Colombia against *Sitophilus zeamais* Motschulsky (Coleoptera). J. Stored Prod. Res. 45: 212–214.

Nogueira, F.A., Nery, P.S., Morais-Costa, F., Oliveira, N.J.F., Martins, E.R. and Duarte, E.R. 2014. Efficacy of aqueous extracts of *Genipa americana* L. (Rubiaceae) in inhibiting larval development and eclosion of gastrointestinal nematodes of sheep. J. Appl. Anim. Res. 42: 356–360.

Noguera, B., Díaz, E., García, M.V., San Feliciano, A., López-Perez, J.L. and Israel, A. 2004. Anti-inflammatory activity of leaf extract and fractions of *Bursera simaruba* (L.) Sarg (Burseraceae). J. Ethnopharmacol. 92: 129–133.

Nomura, E.C.O., Rodrigues, M.R.A., da Silva, C.F., Hamm, L.A., Nascimento, A.M., de Souza, L.M., Cipriani, T.R., Baggio, C.H. and Werne, M.F.P. 2013. Antinociceptive effects of ethanolic extract from the flowers of *Acmella oleracea* (L.) R.K. Jansen in mice. J. Ethnopharmacol. 150: 583–589.

Nord, L.I. and Kenne, L. 1999. Separation and structural analysis of saponins in a bark extract from *Quillaja saponaria* Molina. Carbohyd. Res. 320: 70–81.

Nord, L.I. and Kenne, L. 2000. Novel acetylated triterpenoid saponins in a chromatographic fraction from *Quillaja saponaria* Molina. Carbohyd. Res. 329: 817–829.

Nyberg, N.T., Kenne, L., Ronnberg, B. and Sundquist, B.G. 2000. Separation and structural analysis of some saponins from *Quillaja saponaria* Molina. Carbohyd. Res. 323: 87–97.

Ohashi, H., Kawai, S., Sakurai, Y. and Yasue, M. 1992. Norlignan from the knot resin of *Araucaria angustifolia*. Phytochemistry 31: 1371–1373.

Ohloff, G. 1978. Recent development in the field of naturally-occurring aroma components. pp. 431–527. *In*: W. Gerz, H. Grisebach and G.W. Kirby (eds.). Fortschritte der Chemie Organischer Naturstoffe/Progress in the Chemistry of Organic Natural Products, Volume 35. Springer-Verlag, Vienna.

Oldham, P. 2006. Biodiversity and the patent system: an introduction to research methods. Lancaster-Cardiff University, CESAGen. Centre for Economic and Social Aspects of Genomics, Lancaster.

Olivas Sánchez, M.P. 1999. Plantas medicinales del Estado de Chihuahua. Volumen 1. Universidad Autónoma de Ciudad Juárez, Ciudad Juárez, pp. 97–98.

Oliveira, D.R., Leitão, G.G., Santos, S.S., Bizzo, H.R., Lopes, D., Alviano, C.S., Alviano, D.S. and Leitão, S.G. 2006. Ethnopharmacological study of two *Lippia* species from Oriximina, Brazil. J. Ethnopharmacol. 108: 103–108.

Oliveira, D.R., Leitão, G.G., Bizzo, H.R., Lopes, D., Alviano, D.S., Alviano, C.S. and Leitão, S.G. 2007. Chemical and antimicrobial analyses of essential oil of *Lippia origanoides* H.B.K. Food Chem. 101: 236–240.

Oliveira, F.A., Vieira-Júnior, G.M., Chaves, M.H., Almeida, F.R.C., Florêncio, M.G., Lima Jr., R.C.P., Silva, R.M., Santos, F.A. and Rao, V.S.N. 2004. Gastroprotective and anti-inflammatory effects of resin from *Protium heptaphyllum* in mice and rats. Pharmacol. Res. 49: 105–111.

Olivero-Verbel, J., Tirado-Ballestas, I., Caballero-Gallardo, K. and Stashenko, E.E. 2013. Essential oils applied to the food act as repellents toward *Tribolium castaneum*. J. Stored Prod. Res. 55: 145–147.

Omena, C.M.B., Valentim, I.B., Guedes, G.S., Rabelo, L.A., Mano, C.M., Bechara, E.J.H., Sawaya, A.C.H.F., Trevisan, M.T.S., da Costa, J.G., Ferreira, R.C.S., Sant'Ana, A.E.G. and Goulart, M.O.F. 2012. Antioxidant, anti-acetylcholinesterase and cytotoxic activities of ethanol extracts of peel, pulp and seeds of exotic Brazilian fruits. Antioxidant, anti-acetylcholinesterase and cytotoxic activities in fruits. Food Res. Int. 49: 334–344.

Ono, M., Ueno, M., Masuoka, C., Ikeda, T. and Nohara, T. 2005. Iridoid glucosides from the fruit of *Genipa americana*. Chem. Pharm. Bull. 10: 1342–1344.

Opdyke, D.L.J. 1979. Monographs on Fragrance Raw Materials. Pergamon Press Ltd., Oxford. 732p.

Orwa, C., Mutua, A., Kindt, R., Jamnadass, R. and Anthony, S. 2009. *Schinus molle* L. *In*: Agroforestree Database: a tree reference and selection guide version 4.0. World Agroforestry Centre, Kenya. 1-5 p. Available from: http://www.worldagroforestry.org/treedb/AFTPDFS/Schinus_molle.pdf (January, 2014).

Osuna Torres, L., Tapia Pérez, M.E. and Aguilar Contreras, A. 2005. Plantas medicinales de la medicina tradicional mexicana para tratar afecciones gastrointestinales: estudio etnobotánico, fitoquímico y farmacológico. Edicions Universitat Barcelona, Barcelona.

Pacheco Soares, C., Lemos, V.R., da Silva, A.G., Campoy, R.M., da Silva, C.A.P., Menegon, R.F., Rojahn, I. and Joaquim, W.M. 2014. Effect of *Spilanthes acmella* hydroethanolic extract activity on tumour cell actin cytoskeleton. Cell Biol. Int. 38: 131–135.

Page, C.N. 1990. Araucariaceae. pp. 294–299. *In*: K.U. Kramer and P.S. Green (eds.). Pteridophytes and Gymnosperms. Series title—The Families and Genera of Vascular Plants. Vol. 1. Springer-Verlag, Berlin.

Palomino, E., Maldonado, C., Kempff, M.B. and Ksebati, M.B. 1996. Caparratriene, an active sesquiterpene hydrocarbon from *Ocotea caparrapi*. J. Nat. Prod. 59: 77–79.

Pascual, M.E., Slowing, K., Carretero, E., Sanchez, D. and Villar, A. 2001a. *Lippia*: traditional uses, chemistry and pharmacology: a review. J. Ethnopharmacol. 76: 201–214.

Pascual, M.E., Slowing, K., Carretero, M.E. and Villar, A. 2001b. Antiulcerogenic activity of *Lippia alba* (Mill.) N. E. Brown (Verbenaceae). Farmaco. 56: 501–504.

Paulraj, J., Govindarajan, R. and Palpu, P. 2013. The genus *Spilanthes* ethnopharmacology, phytochemistry, and pharmacological properties: a review. Adv. Pharmacol. Sci. Article ID 510298.

Pelah, D., Abramovich, Z., Markus, A. and Wiesman, Z. 2002. The use of comercial saponin from *Quillaja saponaria* bark as a natural larvicidal agent against *Aedes aegypti* and *Culex pipiens*. J. Ethnopharmacol. 81: 407–409.

Pelegrini, D.D., Tsuzuki, J.K., Amado, C.A.B., Cortez, D.A.G. and Ferreira, I.C.P. 2008. Biological activity and isolated compounds in *Sapindus saponaria* L. and other plants of the genus *Sapindus*. Lat. Am. J. Pharm. 27: 922–927.

Pell, S.K., Mitchell, J.D., Millerand, A.J. and Lobova, T.A. 2011. Anacardiaceae. pp. 7–50. *In*: K. Kubitzki (ed.). Flowering Plants. Eudicots: Sapindales, Cucurbitales, Myrtaceae. Series title—The Families and Genera of Vascular Plants. Vol. X. Springer-Verlag, Berlin.

Pennacchio, M., Jefferson, L. and Havens, K. 2010. Uses and abuses of plant-derived smoke: its ethnobotany as hallucinogen, perfume, incense, and medicine. Oxford University Press, Oxford.

Peraza-Sánchez, S.R., Salazar-Aguilar, N.E. and Peña-Rodríguez, L.M. 1995. A new triterpene from the resin of *Bursera simaruba*. J. Nat. Prod. 58: 271–274.

Perdomo-Roldán, F. and Mondragón-Pichardo, J. 2009. Malezas de México. *Tagetes lucida* Cav. Available from: http://www.conabio.gob.mx/malezasdemexico/asteraceae/tagetes-lucida/fichas/ficha.htm (January, 2014).

Pérez, J.F.M., Scolforo, J.R.S., de Oliveira, A.D., de Mello, J.M., Borges, L.F.R. and Camolesi, J.F. 2004. Management system for native candeia forest (*Eremanthus erythropappus* (DC.) MacLeish)—the option for selective cutting, Cerne. 10: 257–273.

Pérez-Arbeláez, E. 1996. Plantas útiles de Colombia. Quinta edición. Fondo FEN Colombia, Bogotá.

Pérez-López, A., Torres Cirio, A., Rivas-Galindo, V.M., Salazar Aranda, R. and Waksman de Torres, N. 2011. Activity against *Streptococcus pneumoniae* of the essential oil and δ-cadinene isolated from *Schinus molle* Fruit. J. Essent. Oil Res. 23: 25–28.

Peter, K.V. 2012. Handbook of herbs and spices. Volume 2. Woodhead Publishing Limited, Cambridge, pp. 561–562.

Petrone, A. and Preti, F. 2008. Suitability of soil bioengineering techniques in Central America: a case study in Nicaragua. Hydrol. Earth Syst. Sci. 12: 1241–1248.

Pino, J., Marbot, R. and Vazquez, C. 2005. Volatile constituents of genipap (*Genipa americana* L.) fruit from Cuba. Flavour Fragr. J. 20: 583–586.

Pino, J.A., Ortega, A. and Rosado, A. 1996. Chemical composition of the essential oil of *Lippia alba* (Mill.) N.E. Brown from Cuba. J. Essent. Oil Res. 8: 445–446.

Pino Alea, Jorge A et al. Composición y propiedades antibacterianas del aceite esencial de Lippia alba (Mill.) n. e. Brown. Rev Cubana Farm [online]. 1996, vol. 30, n.1 [citado 2015-07-02], pp. 0-0 . Disponible en: <http://scielo.sld.cu.

Pinto, A.B., Guedes, C.M., Moreira, R.F.A. and Maria, C.A.B.D. 2006. Volatile constituents from headspace and aqueous solution of genipap (*Genipa americana*) fruit isolated by the solid-phase extraction method. Flavour Frag. J. 21: 488–491.

Pollak-Eltz, A. 2001. La medicina tradicional venezolana. Primera edición. Universidad Católica Andrés Bello, Caracas.

Pontes, W.J.T., de Oliveira, J.C.G., da Camara, C.A.G., Lopes, A.C.H.R., Gondim, M.G.C., Jr., de Oliveira, J.V., Barros, R. and Schwartz, M.O.E. 2007. Chemical composition and acaricidal activity of the leaf and fruit essential oils of *Protium heptaphyllum* (Aubl.) Marchand (Burseraceae). Acta Amaz. 37: 103–110.

Poppendieck, H.H. 2003. Bixaceae. pp. 33–35. *In*: K. Kubitzki and C. Bayer (eds.). Flowering Plants. Dicotyledons. Malvales, Capparales and Non-betalain Caryophyllales. Series title—The Families and Genera of Vascular Plants. Vol. V. Springer-Verlag, Berlin.

Prachayasittikul, S., Suphapong, S., Worachartcheewan, A., Lawung, R., Ruchirawat, S. and Prachayasittikul, V. 2009. Bioactive metabolites from *Spilanthes acmella* Murr. Molecules 14: 850–867.

Quattrocchini, U. 2012. CRC World Dictionary of medicinal and poisonous plants: common names, scientific names, eponyms, synonyms, and etymology. CRC Press-Taylor and Francis Group, Boca Raton, pp. 56–3319.

Radhika, B., Begum, N. and Srisailam, K. 2010. Pharmacognostic and preliminary phytochemical evaluation of the leaves of *Bixa orellana*. Pharmacogn. J. 2: 132–136.

Raghavan, S. 2007. Handbook of Spices, Seasoning, and Flavorings. CRC Press LLC and Taylor & Francis Group, Boca Raton.

Ramsewak, R.S., Erickson, A.J. and Nair, M.G. 1999. Bioactive N-isobutylamides from the flower buds of *Spilanthes acmella*. Phytochemistry 51: 729–732.

Rao, V.S., Maia, J.L., Oliveira, F.A., Lemos, T.L.G., Chaves, M.H. and Santos, F.A. 2007. Composition and antinociceptive activity of the essential oil from *Protium heptaphyllum* resin. Nat. Prod. Commun. 2: 1199–1202.

Rashed, K., Luo, M.T., Zhang, L.T. and Zheng, Y.T. 2013a. Phytochemical content and anti-HIV-1 activity of *Sapindus saponaria* L. J. Forest. Prod. Ind. 2: 22–26.

Rashed, K.N., Ćirić, A., Glamočlija, J., Calhelha, R.C., Ferreira, I.C.F.R. and Soković, M. 2013b. Antimicrobial activity, growth inhibition of human tumour cell lines, and phytochemical characterization of the hydromethanolic extract obtained from *Sapindus saponaria* L. aerial parts. BioMed. Res. Internat. Article ID 659183.

Rastrelli, L., Caceres, A., Morales, C., De Simone, F. and Aquino, R. 1998. Iridoids from *Lippia graveolens*. Phytochemistry 49: 1829–1832.

Ratnasooriya, W.D., Pieris, K.P.P., Samaratunga, U. and Jayakody, J.R.A.C. 2004. Diuretic activity of *Spilanthes acmella* flowers in rats. J. Ethnopharmacol. 91: 317–320.

Regalado, E.L., Fernández, M.D., Pino, J.A., Mendiola, J. and Echemendia, O.A. 2011. Chemical composition and biological properties of the leaf essential oil of *Tagetes lucida* Cav. from Cuba. J. Essent. Oil Res. 23: 63–67.

Ribeiro, A.F., Andrade, E.H.A., Salimena, F.R.G. and Maia, J.G.S. 2014. Circadian and seasonal study of the cinnamate chemotype from *Lippia origanoides* Kunth. Biochem. Syst. Ecol. 55: 249–259.

Richardson, A. 1995. Plants of the Rio Grande Delta. University of Texas Press, Austin.

Rigano, L., Bonfigli, A. and Walther, R. 2012. Bioactivity evaluations of *Quillaja saponaria* (soap bark tree) saponins in skin and scalp sebaceous imbalances. SOFW-Journal. 138: 13–21.

Rios, M., Koziol, M.J., Borgtoft Pedersen, H. and Granda, G. 2007. Plantas útiles del Ecuador: aplicaciones, retos y perspectivas. Ediciones Abya-Yala, Quito.

Ríos, M., De la Cruz, R. and Mora, A. 2008. Conocimiento tradicional y plantas útiles del Ecuador: saberes y prácticas. Ediciones Abya-Yala, Quito.

Robles, J., Torrenegra, R., Gray, A.I., Piñeros, C., Ortíz, L. and Sierra, M. 2005. Triterpenos aislados de corteza de *Bursera graveolens* (Burseraceae) y su actividad biológica. Rev. Bras. Farmacog. 15: 283–286.

Rodrígues, L.M., Alcázar-Alay, S.C., Petenate, A.J. and Meireles, M.A.A. 2014. Bixin extraction from defatted annatto seeds. C. R. Chimie. 17: 268–283.

Rodríguez-Díaz, M., Delporte, C., Cartagena, C., Cassels, B.K., González, P., Silva, X., León, F. and Wessjohann, L.A. 2011. Topical anti-inflammatory activity of quillaic acid from *Quillaja saponaria* Mol. and some derivatives. J. Pharm. Pharmacol. 63: 718–724.

Rohwer, J.G. 2011. Lauraceae. pp. 366–391. *In*: K. Kubitzki, J.G. Rohwer and V. Bittrich (eds.). Flowering Plants. Dycotiledons. Magnoliid, Hamamelid and Caryophyllid Families. Series title—The Families and Genera of Vascular Plants. Vol. 2. Springer-Verlag, Berlin.

Rojas, J., Morales, A., Pasquale, S., Márquez, A., Rondón, M., Veres, K. and Máthé, I. 2006. Comparative study of the chemical composition on the essential oil of *Lippia oreganoide* L. collected in two different seasons. Nat. Prod. Commun. 1: 205–207.

Roner, M.R., Sprayberry, J., Spinks, M. and Dhanji, S. 2007. Antiviral activity obtained from aqueous extracts of the Chilean soapbark tree (*Quillaja saponaria* Molina). J. Gen. Virol. 88: 275–285.

Rosales-Ovares, K.M. and Ciccio-Alberti, J.F. 2002. The volatile oil of the fruits of *Bursera simaruba* (L.) Sarg. (Burseraceae) from Costa Rica. Ing. Cienc. Quim. 20: 60–61.

Rossini, C., Menéndez, P., Dellacassa, E. and Moyna, P. 1996. Essential oils from leaves of *Schinus molle* and *S. lentiscifolius* of uruguayan origin. J. Essent. Oil Res. 8: 71–73.

Rüdiger, A.L., Siani, A.C. and Veiga Junior, V.F.. 2007. The chemistry and pharmacology of the South America genus *Protium* Burm. f. (Burseraceae). Phcog. Rev. 1: 93–104.

Sacchetti, G., Guerrini, A., Noriega, P., Bianchi, A. and Bruni, R. 2006. Essential oil of wild *Ocotea quixos* (Lam.) Kosterm. (Lauraceae) leaves from Amazonian Ecuador. Flavour Frag. J. 21: 674–676.

Salazar, R., Méndez, J.M. and Soihet, C. 2000a. *Araucaria angustifolia* (Bertol.) Kuntze. Nota Técnica No. 76. *In*: Manejo de semillas de 100 especies forestales de América Latina. Volumen 1. Serie técnica. Manual técnico CATIE No. 41. Centro Agronómico Tropical de Investigación y Enseñanza. Proyecto de Semillas Forestales. CATIE, Turrialba, pp. 151–152.

Salazar, R., Méndez, J.M. and Soihet, C. 2000b. *Genipa americana* L. Nota Técnica No. 72. In: Manejo de semillas de 100 especies forestales de América Latina. Volumen 1. Serie técnica. Manual técnico CATIE No. 41. Centro Agronómico Tropical de Investigación y Enseñanza. Proyecto de Semillas Forestales. CATIE, Turrialba, pp. 143–144.

San Martin, R. 2000. Sustainable production of *Quillaja saponaria* Mol. Saponins: saponins in food, feedstuffs and medicinal plants. Proc. Phythochem. Soc. Europe. 45: 271–279.

San Martin, R. and Briones, R. 1999. Industrial uses and sustainable supply of *Quillaja saponaria* (Rosaceae) saponins. Econ. Bot. 53: 302–311.

Sánchez, O., Kvist, L.P. and Aguirre, Z. 2006. Bosques secos en Ecuador y sus plantas útiles. pp. 188–204. *In*: M. Morales, B. Øllgaard, L.P. Kvist, F. Borchsenius and H. Balslev (eds.). Botánica Económica de los Andes Centrales. Universidad Mayor de San Andrés, La Paz.

Sánchez, O., Medellín, R., Aldama, A., Goettsch, B., Soberón-Mainero, J. and Tambutti, M. 2007. Método de evaluación del riesgo de extinción de especies silvestres en México (MER). INE-SEMARNAT; CONABIO, México.

Santos-Granero, F. and Barclay, F. 1998. Guía etnográfica de la alta amazonía. Volumen 3. Smithsonian Tropical Research Institute/Editorial Abya Yala, Quito.

Schroth, G., Elias, M.E., Uguen, K., Seixas, R. and Zech, W. 2001. Nutrient fluxes in rainfall, throughfall and stemflow in tree-based land use systems and spontaneous tree vegetation of central Amazonia. Agric. Ecosyst. Environ. 87: 37–49.

Seidemann, J. 2005. World Spice Plants. Economic Usage, Botany, Taxonomy. Springer-Verlag, Heidelberg.

Selvi, A.T., Aravindhan, R., Madhan, B. and Rao, J.R. 2013. Studies on the application of natural dye extract from *Bixa orellana* seeds for dyeing and finishing of leather. Ind. Crop. Prod. 43: 84–86.

Senatore, F. and Rigano, D. 2001. Essential oil of two *Lippia* spp. (Verbenaceae) growing wild in Guatemala. Flavour Frag. J. 16: 169–171.

Shahzad Aslam, M., Choudhary, B.A., Uzair, M. and Subhan Ijaz, A. 2013. Phytochemical and ethno-pharmacological review of the genus *Araucaria*—review. Trop. J. Pharm. Res. 12: 651–659.

Shilpi, J.A., Taufiq-Ur-Rahman, Md., Uddin, S.J., Alam, Md.S., Sadhu, S.K. and Seidel, V. 2006. Preliminary pharmacological screening of *Bixa orellana* L. leaves. J. Ethnopharmacol. 108: 264–271.

Siani, A.C., Ramos, M.F.S., Menezes-de-Lima, O., Jr., Ribeiro-dos-Santos, R., Fernandez-Ferreira, E., Soares, R.O.A., Rosas, E.C., Susunaga, G.S., Guimarães, A.C., Zoghbi, M.G.B. and Henriques, M.G.M.O. 1999. Evaluation of anti-inflammatory-related activity of essential oils from the leaves and resin of species of *Protium*. J. Ethnopharmacol. 66: 57–69.

Siani, A.C., Ramos, M.F.S., Monteiro, S.S., Ribeiro-dos-Santos, R. and Soa, R.O.A. 2011. Essential oils of the oleoresins from *Protium heptaphyllum* growing in the brazilian southeastern and their cytotoxicity to neoplasic cells lines. J. Essent. Oil Bear. Pl. 14: 373–378.

Silva, C.P.D., Coelho, L.M., de Oliveira, A.D., Scolforo, J.R.S., de Rezende, J.L.P. and Lima, I.C.G. 2012. Economic analysis of agroforestry systems with candeia. Cerne. 18: 585–594.

Silvério, M.S., Sousa, O.V., Del-Vechio-Vieira, G., Miranda, M.A., Matheus, F.C. and Kaplan, M.A.C. 2008. Propriedades farmacológicas do extrato etanólico de *Eremanthus erythropappus* (DC.) McLeisch (Asteraceae). Rev. Bras. Farmacogn. 18: 430–435.

Silvério, M.S., Del-Vechio-Vieira, G., Pinto, M.A.O., Alves, M.S. and Sousa, O.V. 2013. Chemical composition and biological activities of essential oils of *Eremanthus erythropappus* (DC) McLeisch (Asteraceae). Molecules 18: 9785–9796.

Simionatto, E., Chagas, M.O., Peres, M.T.L.P., Hess, S.C., da Silva, C.B., Ré-Poppi, N., Gebara, S.S., Corsino, J., Morel, A.F., Stukerc, C.Z., Matos, M.F.C. and de Carvalho, J.E. 2011. Chemical composition and biological activities of leaves essential oil from *Schinus molle* (Anacardiaceae). J. Essent. Oil Bear. Pl. 14: 590–599.

Singh, G., Pandey, S.K., Leclercq, P.A. and Sperkova, J. 1999. Studies on essential oils. Part 15. GC/MS analysis of chemical constituents of leaf oil of *Lippia alba* (Mill.) from North India. J. Essent. Oil Res. 11: 206–208.

Soares, T.V. and Fabri, R.L. 2011. Composição química e avaliação do potencial antioxidante e citotóxico das folhas de *Eremanthus erythropappus* (DC) Mcleish (Candeia). Rev. Eletrôn. Farm. 8: 41–52.

Soltysik, S., Wu, J.Y., Recchia, J., Wheeler, D.A., Newman, M.J., Coughlin, R.T. and Kensil, C.R. 1995. Structure/function studies of QS-21 adjuvant: assessment of triterpene aldehyde and glucuronic acid roles in adjuvant function. Vaccine. 13: 1403–1410.

Sosa, S., Balick, M.J., Arvigo, R., Esposito, R.G., Pizza, C., Altinier, G. and Tubaro, A. 2002. Screening of the topical anti-inflammatory activity of some Central America plants. J. Ethnopharmacol. 81: 211–215.

Soukup, J. 1970. Vocabulario de los nombres vulgares de la flora peruana. Imprenta Salesiana, Lima, Perú.

Sousa, O.V., Silverio, M.S., Del-Vechio-Vieira, G., Matheus, F.C., Yamamoto, C.H. and Alves, M.S. 2008. Antinociceptive and anti-inflammatory effects of the essential oil from *Eremanthus erythropappus* leaves. J. Pharm. Pharmacol. 60: 771–777.

Souza, M.O., Branco, C.S., Sene, J., DallAgnol, R., Agostini, F., Moura, S. and Salvador, M. 2014. Antioxidant and antigenotoxic activities of the brazilian pine *Araucaria angustifolia* (Bert.) O. Kuntze. Antioxidants 3: 24–37.

Spelman, K., Depoix, D., McCray, M., Mouray, E. and Grellier, P. 2011. The traditional medicine *Spilanthes acmella*, and the alkylamides spilanthol and undeca-2*E*-ene-8,10-diynoic acid isobutylamide, demonstrate *in vitro* and *in vivo* antimalarial activity. Phytother. Res. 25: 1098–1101.

Srinath, J. and Laksmi, T. 2014. Therapeutic potential of *Spilanthes acmella*—a dental note. Int. J. Pharm. Sci. Rev. Res. 25: 151–153.

Stanley, W.G. and Montagnini, F. 1999. Biomass and nutrient accumulation in pure and mixed plantations of indigenous tree species grown on poor soils in the humid tropics of Costa Rica. Forest. Ecol. Manag. 113: 91–103.

Staples, G. and Kristiansen, M.S. 1999. Ethnic Culinary Herbs: A Guide to Identification and Cultivation in Hawaii. University of Hawaii Press, Hawaii.

Stashenko, E.E., Jaramillo, B.E. and Martínez, J.R. 2004. Comparison of different extraction methods for the analysis of volatile secondary metabolites of *Lippia alba* (Mill.) N.E. Brown, grown in Colombia, and evaluation of its *in vitro* antioxidant activity. J. Chromatogr. 1025: 93–103.

Stashenko, E.E., Ruiz, C., Muñoz, A., Castañeda, M. and Martínez, J. 2008. Composition and antioxidant activity of essential oils of *Lippia origanoides* H.B.K. grown in Colombia. Nat. Prod. Commun. 3: 563–566.

Stashenko, E.E., Martínez, J.R., Ruíz, C.A., Arias, G., Durán, C., Salgar, W. and Cala, M. 2010. *Lippia origanoides* chemotype differentiation based on essential oil GC-MS and principal component analysis. J. Sep. Sci. 33: 93–103.

Stashenko, E.E., Martínez, J.R., Cala, M.P., Durán, D.C. and Caballero, D. 2013. Chromatographic and mass spectrometric characterization of essential oils and extracts from *Lippia* (Verbenaceae) aromatic plants. J. Sep. Sci. 36: 192–202.

Stefenon, V.M., Steiner, N., Guerra, M.P. and Nodari, R.O. 2009. Integrating approaches towards the conservation of forest genetic resources: a case study of *Araucaria angustifolia*. Biodivers. Conserv. 18: 2433–2448.

Suárez, A.I., Compagnone, R.S., Salazar-Bookaman, M.M., Tillett, A., Delle Monache, F., Di Giulio, C. and Bruges, G. 2003. Antinociceptive and anti-inflammatory effects of *Croton malambo* bark aqueous extract. J. Ethnopharmacol. 88: 11–14.

Suárez, A.I., Vásquez, L.J., Manzano, M.A. and Compagnone, R.S. 2005. Essential oil composition of *Croton cuneatus* and *Croton malambo* growing in Venezuela. Flavour Frag. J. 20: 611–614.

Suárez, A.I., Vásquez, L.J., Taddei, A., Arvelo, F. and Compagnone, R.S. 2008. Antibacterial and cytotoxic activity of leaf essential oil of *Croton malambo*. J. Essent. Oil Bear. Pl. 11: 208–213.

Susunaga, G.S., Siani, A.C., Pizzolatti, M.G., Yunes, R.A. and Delle Monache, F. 2001. Triterpenes from the resin of *Protium heptaphyllum*. Fitoterapia 72: 709–711.

Sylvestre, M., Longtin, A.P.A. and Legault, J. 2007. Volatile leaf constituents and anticancer activity of *Bursera simaruba* (L.) Sarg. essential oil. Nat. Prod. Commun. 12: 1273–1276.

Tafurt-García, G. and Muñoz-Acevedo, A. 2012. Metabolitos volátiles presentes en *Protium heptaphyllum* (Aubl.) March. colectado en Tame (Arauca–Colombia). Bol. Latinoam. Caribe. 11: 223–232.

Tallent, W.H. 1964. Two new antibiotic cyclopentanoid monoterpenes of plant origin. Tetrahedron. 20: 1781–1787.

Tam, K.I. and Roner, M.R. 2011. Characterization of *in vivo* anti-rotavirus activities of saponin extracts from *Quillaja saponaria* Molina. Antivir. Res. 90: 231–241.

Teixeira, M.L., Cardoso, M.G., Figueiredo, A.C.S., Moraes, J.C., Assis, F.A., de Andrade, J., Nelson, D.L., Gomes, M.S., de Souza, J.A. and de Albuquerque, L.R.M. 2014. Essential oils from *Lippia origanoides* Kunth. and *Mentha spicata* L.: chemical composition, insecticidal and antioxidant activities. American J. Plant Sci. 5: 1181–1190.

Teles, S., Pereira, J.A., Santos, C.H.B., Menezes, R.V., Malheiro, R., Lucchese, A.M. and Silva, F. 2012. Geographical origin and drying methodology may affect the essential oil of *Lippia alba* (Mill) N.E. Brown. Ind. Crop. Prod. 37: 247–252.

Tene, V., Malagón, O., Vita Finzi, P., Vidari, G., Armijos, C. and Zaragoza, T. 2007. An ethnobotanical survey of medicinal plants used in Loja and Zamora-Chinchipe, Ecuador. J. Ethnopharmacol. 111: 63–81.

Thalhamer, B. and Himmelsbach, M. 2014. Characterization of quillaja bark extracts and evaluation of their purity using liquid chromatography-high resolution mass spectrometry. Phytochem. Lett. 8: 97–100.

Thomas, P. 2013. *Araucaria angustifolia*. *In*: IUCN Red List of Threatened Species. Version 2013.2. Available: www. iucnredlist.org (April 2014).

Thota, P., Praveen Kumar, B., Narasimha Rao, V.B., Shirish Reddy, G. and Sushma Kumari, C. 2012. *In vitro* studies on antimicrobial screening of leaf extracts of *Sapindus saponaria* against common dental pathogens. Plant Sci. Feed. 2: 15–18.

Tsuzuki, J.K., Svidzinski, T.I.E., Shinobu, C.S., Silva, L.F.A., Rodrigues-Filho, E., Cortez, D.A.G. and Ferreira, I.C.P. 2007. Antifungal activity of the extracts and saponins from *Sapindus saponaria* L. An. Acad. Bras. Cienc. 79: 577–583.

Ueda, S., Iwahashi, Y. and Tokuda, H. 1991. Production of anti-tumor-promoting iridoid glucosides in *Genipa americana* and its cell cultures. J. Nat. Prod. 54: 1677–1680.

United Nations Conference on Trade and Development (UNCTAD). 2005. Market Brief in the European Union for selected natural ingredients derived from native species. *Genipa americana*, Jagua—Huito. UNCTAD/BioTrade Facilitation Programme, Netherlands and Switzerland.

United Nations Environment Programme (UNEP). 2010. State of biodiversity in Latin America and the Caribbean. Regional Office for Latin America and the Caribbean—UNEP, Panamá. Available from: http://www.unep.org/delc/Portals/119/ LatinAmerica_StateofBiodiv.pdf (March, 2014)

Uribe-Hernández, C.J., Hurtado-Ramos, J.B., Olmedo-Arcega, E.R. and Martinez-Sosa, M.A. 1992. The essential oil of *Lippia graveolens* H.B.K. from Jalisco, Mexico. J. Essent. Oil Res. 4: 647–649.

Vale, T.G., Matos, F.J.A., de Lima, T.C.M. and Viana, G.S.B. 1999. Behavioral effects of essential oils from *Lippia alba* (Mill.) N.E. Brown chemotypes. J. Ethnopharmacol. 167: 127–133.

Van Setten, D.C., van de Werken, G., Zomer, G. and Kersten, G.F.A. 1995. Glycosyl compositions and structural characteristics of the potential immuno-adjuvant active saponins in the *Quillaja saponaria* Molina extract Quil A. Rapid Commun. Mass Sp. 9: 660–666.

Vasconcelos, S.M.M., Lima, S.R., Soares, P.M., Assreuy, A.M.S., de Sousa, F.C.F., Lobato, R.F.G., Vasconcelos, G.S., Santi-Gadelha, T., Bezerra, E.H.S., Cavada, B.S. and Patrocínio, M.C.A. 2009. Central action of *Araucaria angustifolia* seed lectin in mice. Epilepsy Behav. 15: 291–293.

Vasudevan, P., Kashyap, S. and Sharma, S. 1997. *Tagetes*: a multipurpose plant. Bioresour. Technol. 62: 29–35.

Vega, M. 2001. Etnobotánica de la Amazonía peruana. Ediciones Abya-Yala, Quito.

Veras, H.N.H., Campos, A.R., Rodrigues, F.F.G., Botelho, M.A., Coutinho, H.D.M. and da Costa, J.G.M. 2011. *Lippia alba* (mill.) N.E. essential oil interfere with aminoglycosides effect against *Staphylococcus aureus*. J. Essent. Oil Bear. Pl. 14: 574–581.

Vernin, G., Lageot, C., Gaydou, E.M. and Parkanyi, C. 2001. Analysis of the essential oil of *Lippia graveolens* HBK from El Salvador. Flavour Frag. J. 16: 219–226.

Vicuña, G.C., Stashenko, E.E. and Fuentes, J.L. 2010. Chemical composition of the *Lippia origanoides* essential oils and their antigenotoxicity against bleomycin-induced DNA damage. Fitoterapia. 81: 343–349.

Violante, I.M.P., Hamerski, L., Garcez, W.S., Batista, A.L., Chang, M.R., Pott, V.J. and Garcez, F.R. 2012. Antimicrobial activity of some medicinal plants from the Cerrado of the Central-Western region of Brazil. Braz. J. Microbiol. 43: 1302–1308.

Web 1. The plant list. 2010. Version 1. Published on the Internet; http://www.theplantlist.org/ (February 2014).

Web 2. Q & B Colorantes Naturales Ltda. Online: http://www.qbcolornatural.com/responsabilidad.php (February, 2014).

Web 3. Gobernación del Valle. Online: http://sisav.valledelcauca.gov.co/CADENAS_PDF/AROMATICAS/c08.pdf (February, 2014).

Web 4. Nativa Ecuador. Online: http://www.nativaecuador.org (February, 2014).

Web 5. La Sante. Médicament. Forme et vitalité. Compléments alimentaires. Poconéol No. 26—Artériosclérose—Solution buvable 15 mL. Online: http://lasante.net/nos-medicaments/forme-et-vitalite/medicaments-complements-alimentaires/poconeol-n26-arteriosclerose.html (January, 2014).

Web 6. Pierre Fabre. Reference product name/A.P.I. wording. Poconéol—*Schinus molle*. Online: http://www.pierre-fabre.com/sites/default/files/liste_dci_pierre_fabre_nov13.pdf (January, 2014).

Web 7. Osmoz—Share your fragrances. Baie rose (*Schinus molle*). Online: http://www.osmoz.fr/encyclopedie/matieres-premieres/epice/124/baie-rose-schinus-molle (January, 2014).

Web 8. Chanel. Fragance. Les exclusifs de Chanel. 28 La pausa. Description. Online: http://www.chanel.com/en_GB/fragrance-beauty/Fragrance-Les-Exclusifs-de-CHANEL-28-LA-PAUSA-EAU-DE-TOILETTE-SPRAY-95428 (January, 2014).

Web 9. Estée Lauder. Fragance. Estée Lauder pleasures—Eau de Parfum Spray. Product Details. Online: http://www.esteelauder.com/product/571/2094/Product-Catalog/Fragrance/Este-Lauder-Pleasures/Eau-de-Parfum-Spray (January, 2014).

Web 10. Guerlain. Arsène Lupin—Eau de parfum. Fragrance. Online: http://www.guerlain.com/int/en-int/exclusive-collections/les-parisiens/arsene-lupin-eau-de-parfum-bottle (January, 2014).

Web 11. Sophia's Choise. Men. Body Care. Bodhi & Birch Ylang-Ylang Incensa Sensual Bath & Shower Therapy. Online: http://www.sophiaschoice.co.uk/bodhi-birch-ylang-ylang-incensa-bath-shower-therapy-200ml.html (January, 2014).

Web 12. Bodylish. Our Ingredients. Pink Peppercorn Essential Oil. Online: http://www.bodylish.com/our-ingredients/ (January, 2014).

Web 13. Botanicus USA. Products. Body Care. Body Cream with YUZU Essential. Online: http://www.botanicususa.com/searchquick-submit.sc?keywords=schinus+molle (January, 2014).

Web 14. Urban Retreat – The ultimate in beauty and lifestyle. Delarom Paris. Revitality Balm. Online: http://www.urbanretreat.co.uk/beautique/product/delarom/revitality_balm.aspx (January, 2014).

Web 15. European Commission Health and Consumers. Cosmetics—CosIng. Ingredient: *Schinus molle* oil. Online: http://ec.europa.eu/consumers/cosmetics/cosing/index.cfm?fuseaction=search.details&id=58294 (January, 2014).

Web 16. European Commission Health and Consumers. Cosmetics—CosIng. Ingredient: *Schinus molle* fruit extract. Online: http://ec.europa.eu/consumers/cosmetics/cosing/index.cfm?fuseaction=search.details_v2&id=41101 (January, 2014).

Web 17. European Commission Health and Consumers. Cosmetics—CosIng. Ingredient: *Schinus molle* extract. Online: http://ec.europa.eu/consumers/cosmetics/cosing/index.cfm?fuseaction=search.details_v2&id=82584 (January, 2014).

Web 18. European Commission Health and Consumers. Cosmetics—CosIng. Ingredient: *Schinus molle* seeds extract. Online: http://ec.europa.eu/consumers/cosmetics/cosing/index.cfm?fuseaction=search.details_v2&id=85556 (January, 2014).

Web 19. European Commission Health and Consumers. Cosmetics—CosIng. Ingredient: *Araucaria angustifolia* seed extract. Online: http://ec.europa.eu/consumers/cosmetics/cosing/index.cfm?fuseaction=search.details&id=82192 (January, 2014).

Web 20. European Commission Health and Consumers. Cosmetics—CosIng. Ingredient: *Araucaria angustifolia* seed powder. Online: http://ec.europa.eu/consumers/cosmetics/cosing/index.cfm?fuseaction=search.details_v2&id=88005 (January, 2014).

Web 21. European Commission Health and Consumers. Cosmetics—CosIng. Ingredient: *Araucaria angustifolia* seedcoat extract. Online: http://ec.europa.eu/consumers/cosmetics/cosing/index.cfm?fuseaction=search.details_v2&id=90168 (January, 2014).

Web 22. Nuxe Paris. Crème Nuxuriance® Jour—REPLUMPS, RESTORES RADIANCE. Online: http://americas.nuxe.com/en/creme-nuxuriance-jour-40/nuxuriance#composition (January, 2014).

Web 23. Environmental Working Group (EWG) is Skin Deep®. NUXE Nuxuriance Anti-Aging Re-Densifying Care, Normal to Dry Skin—Day, Age 55+. http://www.ewg.org/skindeep/product/518180/NUXE_Nuxuriance_Anti-Aging_Re-Densifying_Care%2C_Normal_to_Dry_Skin_-_Day%2C_Age_55%2B/(January, 2014).

Web 24. Nuxe Paris. Crème Nuxuriance® Nuit—REPLUMPS, SMOOTHES. Online: http://americas.nuxe.com/en/creme-nuxuriance-nuit-120/nuxuriance#composition (January, 2014).

Web 25. Nuxe Paris. Emulsion Nuxuriance® Jour—REPLUMPS, TIGHTENS PORES. Online: http://americas.nuxe.com/en/emulsion-nuxuriance-jour-41/nuxuriance (January, 2014).

Web 26. Shea Radiance®—Women's Gold. Online: http://www.shearadiance.com/shea-butter-baobab-oil-body-cream-pure/ (January, 2014).

Web 27. TanOrganic. Product. TanOrganic original Trye Me. Online: http://www.tanorganic.com/product/tanorganic-original-try-me-30ml/(January, 2014).

Web 28. Nature's Product®. The Energy Supplements®. AgeLoss® First day cream. Online: http://www.naturesplus.com/products/productdetail.php?productNumber=80038&criteria=keywordSearchResults&category= (January, 2014).

Web 29. European Commission Health and Consumers. Cosmetics—CosIng. Ingredient: *Bixa orellana* seed oil. Online: http://ec.europa.eu/consumers/cosmetics/cosing/index.cfm?fuseaction=search.details_v2&id=54725 (January, 2014).

Web 30. European Commission Health and Consumers. Cosmetics—CosIng. Ingredient: *Bixa orellana* leaf extract. Online: http://ec.europa.eu/consumers/cosmetics/cosing/index.cfm?fuseaction=search.details_v2&id=90857 (January, 2014).

Web 31. European Commission Health and Consumers. Cosmetics—CosIng. Ingredient: *Bixa orellana* seed extract. Online: http://ec.europa.eu/consumers/cosmetics/cosing/index.cfm?fuseaction=search.details_v2&id=74509 (January, 2014).

Web 32. European Commission Health and Consumers. Cosmetics—CosIng. Ingredient: *Bixa orellana* pulp extract. Online: http://ec.europa.eu/consumers/cosmetics/cosing/index.cfm?fuseaction=search.details_v2&id=54724 (January, 2014).

Web 33. European Commission Health and Consumers. Cosmetics—CosIng. Ingredient: *Bixa orellana* seed. Online: http://ec.europa.eu/consumers/cosmetics/cosing/index.cfm?fuseaction=search.details_v2&id=86341 (January, 2014).

Web 34. Experelle, Vinicosm-Ethique®. Les ingrédients. Naturels et bio. Bixa. http://www.dplantes.com/les-ingredients/les-ingredients-naturels/bixa-annatto.html (January, 2014).

Web 35. Rahua. Hair. Shampoo. Online: http://www.rahua.com/us/hair/shampoo.html (January, 2014).

Web 36. Rahua. Hair. Rahua Conditioner. Online: http://www.rahua.com/us/hair/rahua-conditioner.html (January, 2014).

Web 37. Rahua. Hair. Rahua Finishing Treatment. Online: http://www.rahua.com/us/hair/rahua-finishing-treatment.html (January, 2014).

Web 38. Rahua. Hair. Rahua Hair Wax. Online: http://www.rahua.com/us/hair/rahua-hair-wax.html (January, 2014).

Web 39. European Commission Health and Consumers. Cosmetics—CosIng. Ingredient: *Bursera graveolens* fruit oil. Online: http://ec.europa.eu/consumers/cosmetics/cosing/index.cfm?fuseaction=search.details_v2&id=83625 (January, 2014).

Web 40. European Commission Health and Consumers. Cosmetics—CosIng. Ingredient: *Bursera graveolens* wood oil. Online: http://ec.europa.eu/consumers/cosmetics/cosing/index.cfm?fuseaction=search.details_v2&id=83624 (January, 2014).

Web 41. European Commission Health and Consumers. Cosmetics—CosIng. Ingredient: *Bursera simaruba* bark extract. Online: http://ec.europa.eu/consumers/cosmetics/cosing/index.cfm?fuseaction=search.details_v2&id=88209 (January, 2014).

Web 42. Centifoia Bio Nature. Baby. Organic massaje oil. Massaje oil. Online: http://www.centifoliabio.fr/en/organic-massage-oil/6-massage-oil-for-baby.html (January, 2014).

Web 43. HIP. Detox shampoo. Online: http://www.hip.fr/shampooing-detox.html (January, 2014).

Web 44. HIP. Ultra shine shampoo. Online: http://www.hip.fr/shampooing-ultra-brillance.html (January, 2014).

Web 45. HIP. Punch hair mask. Online: http://www.hip.fr/masque-punch-1.html (January, 2014).

Web 46. Experelle, Vinicosm-Ethique®. Cosmétique naturalle. Crème Hydratante Protectrice. Online: http://www.experelle.fr/cosmetique-naturelle/15-creme-hydratante-protectrice-3760227600090.html (January, 2014).

Web 47. Experelle, Vinicosm-Ethique®. Cosmétique naturalle. Emulsion Hydratante Equilibre. Online: http://www.experelle.fr/cosmetique-naturelle/8-emulsion-hydratante-equilibre-3760227600083.html (January, 2014).

Web 48. European Commission Health and Consumers. Cosmetics—CosIng. Ingredient: *Protium heptaphyllum* resin. Online: http://ec.europa.eu/consumers/cosmetics/cosing/index.cfm?fuseaction=search.details_v2&id=83225 (January, 2014).

Web 49. SOCRI. MelaCar-Oil (ECO). Online: http://www.socri.it/Component.aspx?ComponentID=77 (January, 2014).

Web 50. Australian Certified Organic. Cosmos raw materials data base. MelaCare-Oil. Online: http://aco.net.au/index.php/product-search/natural-cosmetics-ingredients/2013-06-17-01-40-37 (January, 2014).

Web 51. Thaidham Allianze Ltd.—The innovation of health & beauty. Anti-Melasma Cream. Online: http://www.thaidham.com/th_new/index.php?page=shop.product_details&product_id=184&flypage=flypage.tpl&pop=0&option=com_virtuemart&Itemid=80 (January, 2014).

Web 52. European Commission Health and Consumers. Cosmetics—CosIng. Ingredient: *Acmella oleracea* extract. Online: http://ec.europa.eu/consumers/cosmetics/cosing/index.cfm?fuseaction=search.details_v2&id=85212 (January, 2014).

Web 53. A. Vogel. *Spilanthes oleracea*—Tincture of freshly harvested *Spilanthes* leaves. Online: http://www.avogel.ie/herbal-remedies/spilanthes/ (January, 2014).

Web 54. ETAT PUR—Pure and active solutions for every skin. *Acmella*. Online: http://www.etatpur.co.uk/index.php/acmella-sheet (January, 2014).

Web 55. Beauty Pure—German luxury cosmetics. Jean D'Arcel Cosmetique. Miratense®. Advanced face—La Cure Delux. Online: http://www.beautypurecosmetics.com/BEAUTYPURE/Jean__D_Arcel_Cosmetique/Premium_Skincare_line/Jean_D_Arcel_Miratense/ADVANCED_FACE_LIFT_/advanced_face_lift_.html (January, 2014).

Web 56. European Commission Health and Consumers. Cosmetics—CosIng. Ingredient: *Lippia alba* leaf/stem oil. Online: http://ec.europa.eu/consumers/cosmetics/cosing/index.cfm?fuseaction=search.details_v2&id=86877 (January, 2014).

Web 57. Thrive care. Face wash ingredients. Online: http://www.thrivecare.co/ingredients-face-wash (January, 2014).

Web 58. Thrive care. Shave oil ingredients. Online: http://www.thrivecare.co/ingredients-shave-oil (January, 2014).

Web 59. Thrive care. Face balm ingredients. Online: http://www.thrivecare.co/ingredients-face-balm (January, 2014).

Web 60. Belverde biocosméticos, produtos naturais, orgânicos e sustentáveis. Shampoo Orgânico *Lippia alba*. Online: http://www.lojabelverde.com.br/#!product/prd1/2633320631/shampoo-org%C3%A2nico-lippia-alba (January, 2014).

Web 61. Belverde biocosméticos, produtos naturais, orgânicos e sustentáveis. Sabonete Líquido Orgânico *Lippia alba*. Online: http://www.lojabelverde.com.br/#!product/prd1/2633320591/sabonete-l%C3%ADquido-org%C3%A2nico-lippia-alba (January, 2014).

Web 62. Neem Tree Farms©. Neem Aloe Body Shampoo. Online: http://neemtreefarms.com/neem-aloe-body-shampoo-p-140.html (January, 2014).

Web 63. European Commission Health and Consumers. Cosmetics—CosIng. Ingredient: *Lippia origanoides* leaf/flower/stem oil. Online: http://ec.europa.eu/consumers/cosmetics/cosing/index.cfm?fuseaction=search.details_v2&id=89648 (January, 2014).

Web 64. Neyber® S.A.S.. Aceites esenciales. *Lippia origanoides* Neyber ™. Online: http://www.neyber.co/index.php/es/aceites-esenciales (January, 2014).

Web 65. Young Living Essential Oil©. A.R.T.® Beauty Masque. Online: http://www.youngliving.com/en_AU/products/beauty/facial-care/art-beauty-masque (January, 2014).

Web 66. Young Living Essential Oil©. A.R.T.® Cream Masque. Online: http://www.youngliving.com/en_AU/products/beauty/facial-care/art-creme-masque (January, 2014).

Web 67. Young Living Essential Oil©. Thieves® AromaBirght Tootpaste. Online: http://www.youngliving.com/en_US/products/beauty/oral-care/thieves-aromabright-toothpaste (January, 2014).

Web 68. Young Living Essential Oil©. Roll-On-Stress Away. Online: http://www.youngliving.com/en_AU/products/essential-oils/roll-ons/stress-away-roll-on (January, 2014).

Web 69. Young Living Essential Oil©. Slique tea—Ocotea Oolong Cacao. Online: http://www.youngliving.com/en_US/products/wellness/weight-management/slique-tea-ocotea-oolong-cacao (January, 2014).

Web 70. European Commission Health and Consumers. Cosmetics—CosIng. Ingredient: *Quillaja saponaria* wood extract. Online: http://ec.europa.eu/consumers/cosmetics/cosing/index.cfm?fuseaction=search.details_v2&id=82372 (January, 2014).

Web 71. European Commission Health and Consumers. Cosmetics—CosIng. Ingredient: *Quillaja saponaria* root extract. Online: http://ec.europa.eu/consumers/cosmetics/cosing/index.cfm?fuseaction=search.details_v2&id=59043 (January, 2014).

Web 72. European Commission Health and Consumers. Cosmetics—CosIng. Ingredient: *Quillaja saponaria* bark extract. Online: http://ec.europa.eu/consumers/cosmetics/cosing/index.cfm?fuseaction=search.details_v2&id=59042 (January, 2014).

Web 73. GreenShield Organic®. Search: *Quillaja saponaria*. http://www.greenshieldorganic.com/search?q=quillaja+saponaria (January, 2014).

Web 74. GreenShield Organic®. 100 oz. Laundry Detergent—Free & Clear. http://www.greenshieldorganic.com/products/100-oz-laundry-detergent-free-clear#.U1QstVV5OSp (January, 2014).

Web 75. GreenShield Organic®. 32 oz. Pet Stain & Odor Remover. http://www.greenshieldorganic.com/products/32-oz-pet-stain-odor-remover#.U1Qsw1V5OSp (January, 2014).

Web 76. Young Living Essential Oil©. Thieves® Fresh Essence Mouthwash. Online: http://www.youngliving.com/en_US/products/beauty/oral-care/thieves-fresh-essence-mouthwash (January, 2014).

Web 77. Soignée Inc©. Botanical Shampoo. Online: http://www.soignee.com/botanical-shampoo.html (January, 2014).

Web 78. European Commission Health and Consumers. Cosmetics—CosIng. Ingredient: *Genipa americana* fruit juice. Online: http://ec.europa.eu/consumers/cosmetics/cosing/index.cfm?fuseaction=search.details&id=82899 (January, 2014).

Web 79. Premier Specialities, Inc©. Products. Cosmetic. Cosmeceuticals with attributes. Premier *Genipa* Blue Extract. Online: http://www.premierfragrances.com/products/cosmetic/cosmeceuticals.html (January, 2014).

Web 80. Nordstrom, Inc©. Women. Beauty & Fragance. Skincare. All Skincare. ExfoliKate® Acne Clearing Treatment. Online: http://shop.nordstrom.com/s/kate-somerville-exfolikate-acne-clearing-exfoliating-acne-treatment/3221630 (January, 2014).

Web 81. SURYA Brasil. Surya Henna Cream Ash Blonde. Online: http://shopsuryabrasil.com/suryahennacreamashblonde231floz70ml.aspx (January, 2014).

Web 82. European Commission Health and Consumers. Cosmetics—CosIng. List of functions. Online: http://ec.europa.eu/consumers/cosmetics/cosing/index.cfm?fuseaction=ref_data.functions (January, 2014).

Webster, G.L. 2014. Euphorbiaceae. pp. 51–216. *In*: K. Kubitzki (ed.). Flowering Plants. Eudicots. Malpighiales. Series title—The Families and Genera of Vascular Plants. Vol. 11. Springer-Verlag, Berlin.

Weeks, A., Daly, D.C. and Simpson, B.B. 2005. The phylogenetic history and biogeography of the frankincense and myrrh family (Burseraceae) based on nuclear and chloroplast sequence data. Mol. Phylogen. Evol. 35: 85–101.

Whaley, O.Q., Beresford-Jones, D.G., Milliken, W., Orellana, A., Smyk, A. and Leguía, J. 2010. An ecosystem approach to restoration and sustainable management of dry forest in southern Peru. Kew Bull. 65: 613–641.

Willmann, D., Schmidt, E.M., Heinrich, M. and Rimpler, H. 2000. Flora del valle de Tehuacán-Cuicatlán: Fascículo 27. Verbenaceae J.St.-Hil. Universidad Nacional Autónoma de México, México.

Wrigley, S.K., Hayes, M.A., Thomas, R., Chrystal, E.J.T. and Nicholson, N. 2000. Biodiversity: new Leads for the pharmaceutical and agrochemical industries. Royal Society of Chemistry, Cambridge.

Wu, J.Y., Gardner, B.H., Murphy, C.I., Seals, J.R., Kensil, C.R., Recchia, J., Beltz, G.A., Newman, G.W. and Newman, M.J. 1992. Saponin adjuvant enhancement of antigen-specific immune responses to an experimental HIV-1 vaccine. J. Immunol. 148: 1519–1525.

Xu, L.W., Chen, J., Qi, H.Y. and Shi, Y.P. 2012. Phytochemicals and their biological activities of plants in *Tagetes* L. Chinese Herb. Med. 4: 103–117.

Yamaguchi, L.F., Vassão, D.G., Kato, M.J. and Di Mascio, P. 2005. Biflavonoids from brazilian pine *Araucaria angustifolia* as potentials protective agents against DNA damage and lipoperoxidation. Phytochemistry 66: 2238–2247.

Yamaguchi, L.F., Kato, M.J. and Di Mascio, P. 2009. Biflavonoids from *Araucaria angustifolia* protect against DNA UV-induced damage. Phytochemistry 70: 615–620.

Yang, Y., Leser, M.E., Sher, A. and McClements, D.J. 2013. Formation and stability of emulsions using a natural small molecule surfactant: quillaja saponin (Q-Naturale®). Food Hydrocolloid. 30: 589–596.

Young, D.G., Chao, S., Casabianca, H., Bertrand, M.C. and Minga, D. 2007. Essential oil of *Bursera graveolens* (Kunth) Triana et Planch from Ecuador. J. Essent. Oil Res. 19: 525–526.

Yukawa, C. and Iwabuchi, H. 2003. Terpenoids of the volatile oil of *Bursera graveolens*. J. Oleo Sci. 52: 483–489.

Yukawa, C., Iwabuchi, H., Kamikawa, T., Komemushi, S. and Sawabe, A. 2004a. Terpenoids of the volatile oil of *Bursera graveolens*. Flavour Frag. J. 19: 565–570.

Yukawa, C., Iwabuchi, H., Komemushi, S. and Sawabe, A. 2004b. Eudesmane-type sesquiterpenoids in the volatile oil from *Bursera graveolens*. J. Oleo Sci. 53: 343–348.

Zapata, B., Betancur-Galvis, L., Duran, C. and Stashenko, E. 2014. Cytotoxic activity of Asteraceae and Verbenaceae family essential oils. J. Essent. Oil Res. 26: 50–57.

Zoghbi, M.G.B., Maia, J.G.S. and Luz, A.I.R. 1995. Volatile constituents from leaves and stems of *Protium heptaphyllum* (Aubl.) March. J. Essent. Oil Res. 7: 541–543.

Zoghbi, M.G.B., Andrade, E.H.A., Santos, A.S., Silva, M.H.L. and Maia, J.G.S. 1998. Essential oils of *Lippia alba* (Mill.) N.E. Br. growing wild in the Brazilian Amazon. Flavour Frag. J. 13: 47–48.

Zúñiga, G.E., Junqueira-Gonçalves, M., Pizarro, M., Contreras, R., Tapia, T. and Silva, S. 2012. Effect of ionizing energy on extracts of *Quillaja saponaria* to be used as an antimicrobial agent on irradiated edible coating for fresh strawberries. Radiat. Phys. Chem. 81: 64–69.

The Therapeutic Potential of Products based on Polyphenols from Wine Grapes in Cardiovascular Diseases

Raúl Vinet,[1,2,*] José Luis Martínez[3] and Leda Guzmán[4]

Introduction

Cardiovascular diseases (CVD) have become a major worldwide cause of morbidity and mortality (Teo et al. 2013). CVD are a result of genetic and environmental factors such as smoking, obesity, lack of exercise, overweight, high blood pressure, elevated blood cholesterol levels, Insulin Resistance (IR) and diabetes (Potvin et al. 2000, Lusis et al. 2008). The physiopathological mechanism implicates an inflammatory process that might contribute to the endothelial dysfunction (Zhang 2008). Tumor Necrosis Factor-α (TNF-α), IL-6 the pro-coagulant Plasminogen Activator Inhibitor-1 (PAI) and macrophages contribute to CVD (Matsuzawa 2005). Oxidative stress has been proposed to be a potential pathogenic mechanism linking obesity and IR with the development of the CVD where an oxidative stress occurs in the vessel wall, resulting in a pathogenic feature of vascular atherosclerosis (Van Gaal et al. 2006). Increased oxidative stress is a primary cause of endothelial dysfunction by reducing Nitric Oxide (NO.) production, promoting inflammation, and through its involvement in the activation of intracellular signal transduction pathways influencing ion channel activation, intercellular communication, and gene expression to, eventually, produce the cessation of cell division and premature senescence (Elnakish et al. 2013, Victor 2013).

Oxidative stress describes a physiological states characterized by elevated Reactive Oxygen Species (ROS) levels. ROS are reactive chemical units involving two main categories: free radicals such as superoxide ($O_2^{-\cdot}$), hydroxyl (OH$\cdot$) and NO$\cdot$, and non-radical derivatives of O_2 such as hydrogen peroxide (H_2O_2) and peroxynitrite (ONOO$^-$). ROS control numerous physiological processes such as host defense,

[1] Faculty of Pharmacy, Universidad de Valparaíso, Valparaíso 2360102, Chile.
[2] Regional Center for Studies in Healthy Foods (CREAS), Valparaíso 2362696, Chile.
[3] Vicerrectoría de Investigación, Desarrollo e Innovación, Universidad de Santiago de Chile, Santiago 9160000, Chile.
[4] Faculty of Sciences, Pontificia Universidad Católica de Valparaíso, Valparaíso 2373223, Chile.
* Corresponding author: raul.vinet@uv.cl

biosynthesis of hormones, fertilization, and cellular signaling. Normally, oxidative stress results in damage to proteins, lipids and DNA, causing cellular dysfunction (Elnakish et al. 2013, Zamora and Villamena 2013). In particular, an overload of $O_2^{-\cdot}$ may decrease NO· availability, resulting in endothelial dysfunction and a decrease in endothelium-dependent vasodilation; furthermore, $O_2^{-\cdot}$ is implicated in the generation of oxidized LDLs (OxLDLs), a key initiator of atherosclerosis (Peluso et al. 2012).

Evidence-based strategies focusing on increasing fitness, weight control, improving food quality intake - including nutraceuticals, vitamins, antioxidants and minerals - should be considered as a key element in cardiovascular disease prevention (Folta and Nelson 2010, Houston 2010).

This chapter describes the cardiovascular effects of stilbene phytoalexins with focus on *trans*-resveratrol (*t*-RV) and its derivatives. The reason that these secondary metabolites have drawn the attention of many researchers is due to their ability to modulate various biological systems, including the cardiovascular system.

Resveratrol, an important stilbene phytoalexin

A MEDLINE search using the MeSH terms 'resveratrol' in the title field yielded 3,899 articles (1988 to June, 2014) with a clear exponential growth (Fig. 15.1). Resveratrol, is one of the most important stilbenes synthesized in grapevines, and specifically *t*-RV, has been the matter of intensive research over the past two decades due to their health promoting properties, as evidenced by *in vitro* and *in vivo* bioassays (Pervaiz 2003, Opie and Lecour 2007, Bishayee 2009, Subramanian et al. 2010, Rayalam et al. 2011, Fernandez-Mar et al. 2012, Hector et al. 2012).

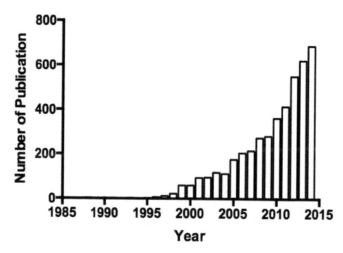

Figure 15.1. MEDLINE search terms 'resveratrol'. Search was carried out using the MeSH terms 'resveratrol' in the title field from 1988 to December 2014.

Resveratrol belongs to a family of compounds termed stilbene and more specifically hydroxystilbenes (Fig. 15.2) which are secondary metabolites found in plants. Those derivatives are also part of a broad class of plant metabolites, commonly known as phytoalexins (Shen 2000). Phytoalexins are produced by plants, in response to biotic and abiotic stress protecting them from the attack of fungi and insects (Ahuja et al. 2012).

The hydroxystilbenes are important components of human and animal diet, having a broad spectrum of beneficial effects, such as: (i) to protect the skin against UV radiation (Weisshaar and Jenkins 1998, Chen et al. 2006); (ii) to protect plants against fungal and bacterial attack (Tropf et al. 1995, Weisshaar and Jenkins 1998). Resistance of *Vitis* species to fungal infection is usually correlated with their capability

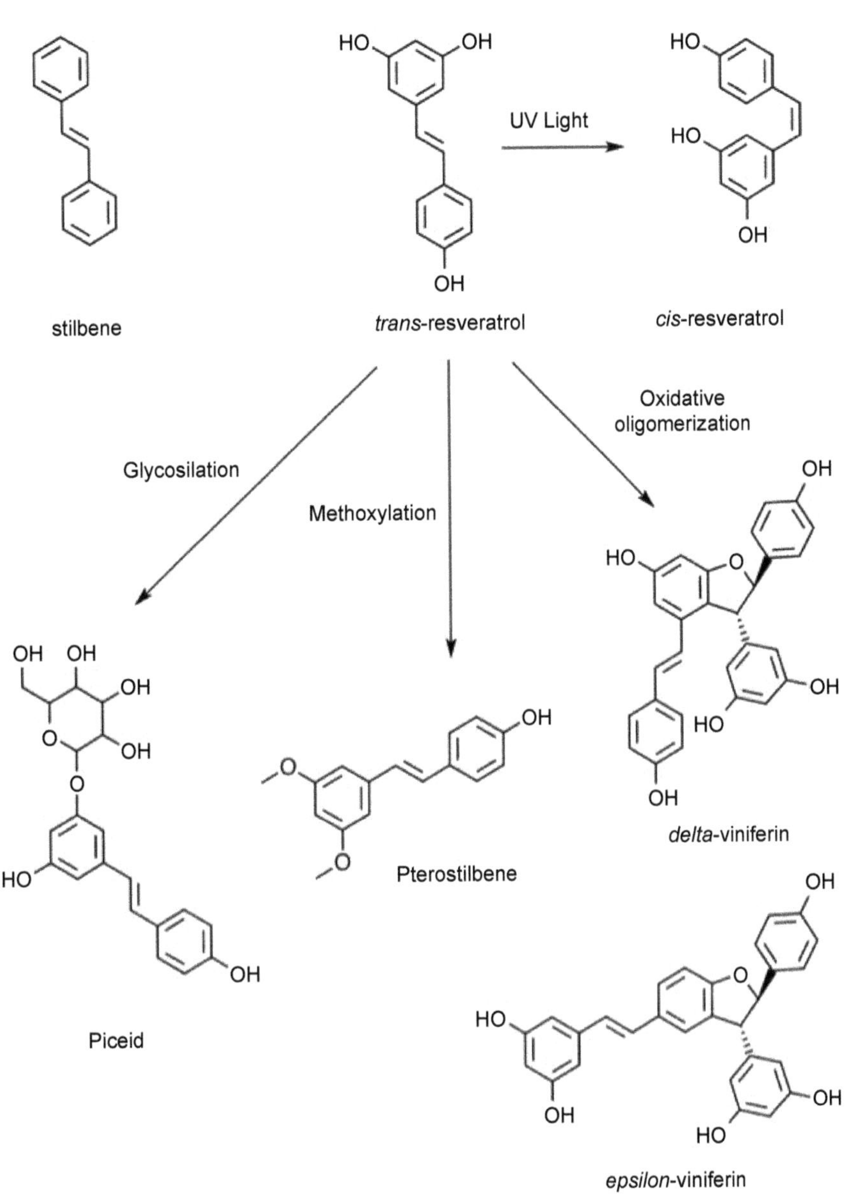

Figure 15.2. Resveratrol and related compound belonging to a family termed hydroxystilbenes. Trans-resveratrol is a precursor molecule of piceid, pterostilbene and viniferins by glycosylation, methoxylation, and oxidative oligomerization, respectively.

to produce stilbene phytoalexins (Douillet-Breuil et al. 1999, Malacarne et al. 2011). As indicated above, phytoalexins and related compounds have been studied for their beneficial effects on human diseases. Among these compounds are biologically active molecules with the potential to be used as ingredients in nutraceuticals and functional foods (Boue et al. 2009). As pointed out in Fig. 15.1 resveratrol has been one of the most studied stilbene phytoalexins. Resveratrol exists as two isomers, *cis*-resveratrol (*cis*-3,5,4'-trihydroxystilbene) and *trans*-resveratrol (*t*-RV, *trans*-3,5,4'-trihydroxystilbene), where the latter can suffer isomerization to the *cis*- form when exposed to ultraviolet irradiation (Lamuelaraventos et al. 1995).

While red wines are the best-known dietary sources of *t*-RV and its derivatives, these compounds are produced by a multiplicity of phylogenetically diverse plant species, including Japanese knotweed (*Fallopia japonica*), peanut (*Arachishypogaea*), sorghum (*Sorghum bicolor*), and *Picea* and *Pinus* species (Parage et al. 2012, Dai et al. 2013). Figure 15.2 also shows *t*-RV acting as a precursor molecule that is converted into a wide diversity of compounds including piceid, pterostilbene, and viniferins. Phytoalexins are produced in grapevines constitutively at low levels and are present in root, stem, leaf, and grape tissues (Wang et al. 2010). Nevertheless, production and accumulation of phytoalexins within leaf and grape tissues is intensely induced by stress stimuli as UV light, ozone, heavy metals, and fungal infection (Jeandet et al. 2002).

Table 15.1. Phenolic compounds in different parts of the grapevine and its products.

Resource	Phenolic compounds
Seed	Gallic acid, (+)-catechin, epicatechin, dimeric procyanidin, proanthocyanidins
Skin	Proanthocyanidins, ellagic acid, myricetin, quercetin, kaempferol, *trans*-resveratrol
Leaf	Myricetin, ellagic acid, kaempferol, quercetin, gallic acid
Stem	Rutin, quercetin 3-O-glucuronide, *trans*-resveratrol, astilbin
Raisin	Hydroxycinnamic acid, hydroxymethylfurfural
Red wine	Malvidin-3-glucoside, peonidin-3-glucoside, cyanidin-3-glucoside, petunidin-3-glucoside, catechin, quercetin, resveratrol, hydroxycinnamic acid

Taken from Xia et al. (2010).

The higher levels of *t*-RV in red wines are explained by the long fermentation process that involves contact with the grape skins. This permits highly hydrophobic stilbene compounds to be extracted from grape skin into the forming ethanol. Although red wines are a comparatively rich dietary source of *t*-RV and its derivatives, the absolute concentrations of these compounds are nevertheless low, ranging from untraceable to 14.3 mg/L (62.7 μM) and fluctuating with wine variety and location (Naugler et al. 2007, Stervbo et al. 2007).

As shown diagrammatically in Fig. 15.3, *t*-RV is the initial stilbene product of *p*-coumaroyl-CoA and three molecules of malonyl-CoA in a reaction catalyzed by stilbene synthase. While most plants are able to generate malonyl-CoA and CoA esters of cinnamic acid, the capability to synthesize stilbenes is restricted to only a few plant species. In grapevine, stilbene synthase is encoded by between 20 to 40 stilbene synthase genes, showing that the genetic diversity of this plant is considerably higher than found in other plant species (Chong et al. 2009, Parage et al. 2012, Dai et al. 2013).

The biosynthesis of resveratrol and flavonoids shares the following enzymes of the phenylpropanoid pathway: phenylalanine-ammonia lyase (PAL) (EC 4.3.1.5), cinnamate 4-hydroxylase (C4H) (EC 1.14.13.11) and 4-coumarate:coenzymeA ligase (4CL) (EC 6.2.1.12). PAL catalyzes the starting reaction; the phenylalanine is deaminated to cinnamic acid. Then, the cinnamic acid is hydroxylated at 4th position by C4H. Previous to the final stages of resveratrol biosynthesis, the coummaric acid must be activated by esterification to coenzyme A, catalyzed by 4CL. The coummaroyl-CoA produced could be used in resveratrol synthesis. In some plants PAL, i.e., *Zea mays*, also has tyrosine-ammonia lyase activity (TAL), rendering coummaric acid directly, bypassing C4H (Schoppner and Kindl 1984, Rosler et al. 1997). Finally, coummaroyl-CoA and malonyl-CoA converge in the last step of resveratrol biosynthesis, reaction catalyzed by the stilbene synthase (EC 2.3.1.95), a member of the broad class of polyketide synthases. The stilbene synthases (STS) are found in a broad range of plant species, such as peanut, grape and pine trees (Schoppner and Kindl 1984, Schanz et al. 1992, Sparvoli et al. 1994, Raiber et al. 1995). All the STS catalyze the synthesis of resveratrol from three units of malonyl-CoA and one molecule of coummaroyl-CoA. The reaction starts (hypothetically) by binding of coummaroyl residue to the enzyme, followed by three cycles of decarboxylating condensation with malonyl-CoA, as is indicated in Fig. 15.3 (Parage et al. 2012, Dai et al. 2013).

Biological activity of resveratrol on the cardiovascular system

The beneficial effects of *t*-RV were initially associated with the bioactive component of red wines responsible for the 'French Paradox'. The French Paradox is the observation of low Coronary Heart Disease (CHD)

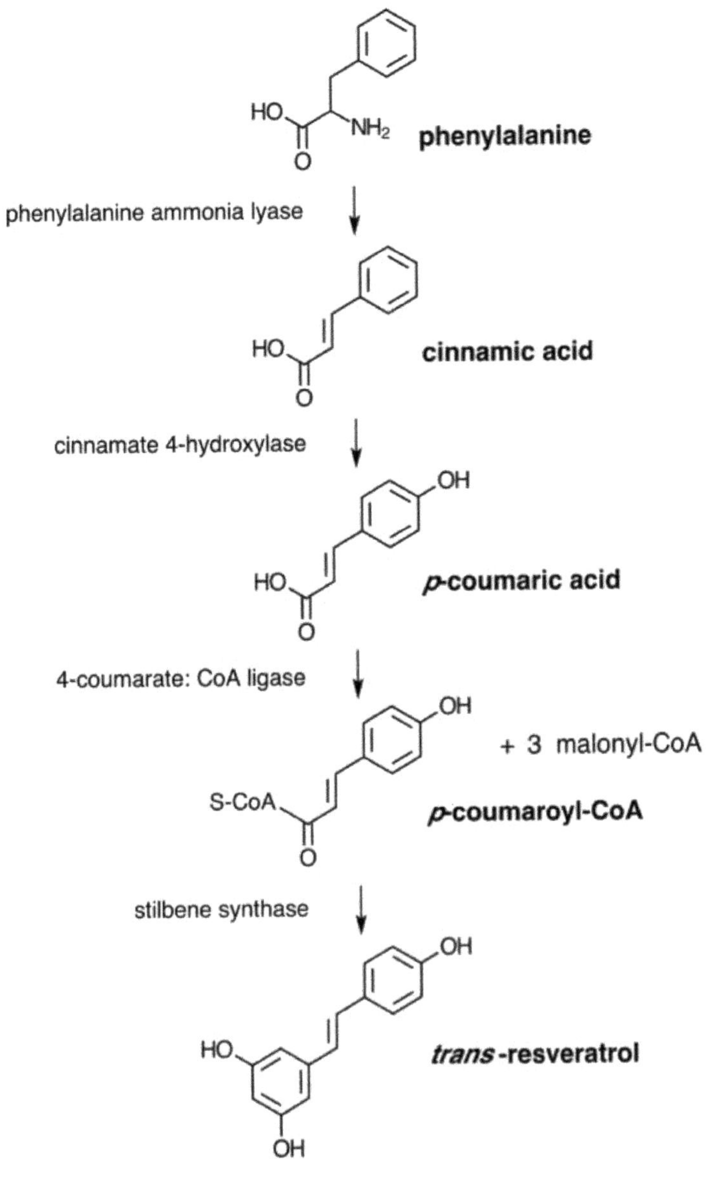

Figure 15.3. Biosynthetic pathway of *trans*-resveratrol.

death rates despite high intake of dietary cholesterol and saturated fat (Renaud and de Lorgeril 1992). French epidemiologists formulated the French Paradox concept in the 1980s (Richard et al. 1981).

The first assays with *t*-RV in order to evaluate its bioactivity potential were focused on its effect upon the contractibility of isolated aorta (Chen and Pace-Asciak 1996) and its interactions with vascular endothelium and platelets aggregation (Bertelli 1998, Ferrero et al. 1998). The cardioprotective effects attributed to *t*-RV include its antiatherogenic effects, improvement in lipid profile and reduction of damage following a myocardial infarction (Petrovski et al. 2011, Wang et al. 2012).

While the observed cardiovascular effects were originally attributed to the chemical antioxidant capacity of *t*-RV, more recent research has focused on its ability to upregulate the expression of endogenous antioxidant enzymes, inhibit the inflammatory activity of cyclo-oxygenases, and promote NO. signaling and vasodilation by activating Nitric Oxide Synthase (eNOS) (Cianciulli et al. 2012, Mohar and Malik 2012, Qureshi et al. 2012).

Lipid oxidation in LDL has been suggested as a potential mechanism in the pathogenesis of atherosclerotic processes. Even, OxLDLs promote the transformation of macrophages into foam cells (Henriksen et al. 1981). The possibility of inhibiting LDL oxidation has been studied by using many antioxidants including vitamin E (Williams et al. 1992), α-tocopherol (Reaven et al. 1993), vitamin C, and β-carotene (Diaz et al. 1997).

Phenolic substances contained in red wine have been found to inhibit oxidation of human LDL, especially *t*-RV (Frankel et al. 1993). Evidence showed that red wine consumption reduces the susceptibility of human plasma and LDL to lipid peroxidation (Fuhrman et al. 1995). However, it is hard to attribute a decrease of atherosclerotic processes and consequently, protection from coronary artery disease to only the inhibition of LDL oxidation since several vascular effects of antioxidants are not related to the resistance of LDL to oxidation (Diaz et al. 1997).

Since, after oral administration of *t*-RV, low plasma concentrations are reached, it was not possible to explain an antioxidant protection of LDL. On this basis Bertelli (1998) and (Ferrero et al. 1998) hypothesized that *t*-RV should act for other pathways. They showed that *t*-RV, at concentrations as low as 1 μM and 100 nM, significantly inhibited ICAM-1 and VCAM-1 expression by TNF-α-stimulated Human Umbilical Vein Endothelial Cells (HUVEC) and lipopolysaccharide-stimulated Human Saphenous Vein Endothelial Cells (HSVEC), respectively. ICAM-1 and VCAM-1 are two adhesion molecules usually expressed in atherogenesis and metastatic and inflammatory processes.

Using the productive isolated aorta model (Vinet et al. 2012), both resveratrol and quercetin, in a dose-dependent manner, inhibited the contractile response to noradrenaline (NA), effect that was blocked by pretreatment with N^G-Nitro-L-arginine (L-NNA), a eNOS inhibitor (Chen and Pace-Asciak 1996). The same work showed that *t*-RV at $> 3 \times 10^{-5}$ M, produced relaxation of the phenylephrine pre-contracted endothelium-intact aorta, effect that was also reversed by L-NNA; at higher concentrations ($> 6 \times 10^{-5}$ M), *t*-RV and quercetin also relaxed the endothelium-denuded aortic rings. The same effect was observed with angiotensin II, where *t*-RV also inhibited the contraction induced by this peptide hormone in rat aorta (Isbir-Soylemez 2006).

A research that evaluated the effect of *t*-RV on isolated epicardial porcine coronary arteries strips, which have been described to be pathologically prone for vasospastic contractions, displays relaxant effect of *t*-RV (Jager and Nguyen-Duong 1999). The results show that the tonic component of the biphasic contractions induced by histamine, together with the contractions induced by F$^-$ ions (10 mM), which activate G proteins downstream of the receptors, could dose-dependently be inhibited by *t*-RV (0.1–100 μM) (Jager and Nguyen-Duong 1999).

Investigating the mode of action of *t*-RV-induced vasorelaxation in aortic rings with denuded endothelium, K$^+$ channels were evaluated (Novakovic et al. 2006). Since 4-aminopiridine, a non-selective blocker of voltage-gated K$^+$ (Kv) channels and margatoxin that inhibits Kv1 channels abolished the relaxation induced by *t*-RV, the authors suggested that 4-aminopyrine and margatoxin-sensitive K$^+$ channels located in the smooth muscle of rat aorta mediated this relaxation. Studies carried out in endothelium-dependent relaxation induced by *t*-RV similarly proposed that smooth muscle Kv channels are included in this relaxation (Gojkovic-Bukarica et al. 2007). In conclusion, *t*-RV produces endothelium-dependent and -independent relaxation of vascular smooth muscle, and in both cases K$^+$ channels were involved.

A study indicated that *t*-RV affected vascular smooth muscle and endothelium in different ways. While in vascular smooth muscle *t*-RV decreased the Ca^{2+} sensitivity but did not affect the KCl-stimulated intracellular Ca^{2+} concentration ([Ca^{2+}]$_i$), in the endothelial cells *t*-RV augmented the agonist-stimulated [Ca^{2+}]$_i$ that might trigger NO synthesis from these cells (Buluc and Demirel-Yilmaz 2006). In addition, *t*-RV has recently been shown to decrease both, extracellular calcium influx and intracellular calcium release, which results in vasorelaxation (Shen et al. 2013). Together, these studies clearly provide evidence that *t*-RV affects ionic channels present in endothelial cells. The effect of *t*-RV on Ca^{2+}-activated K$^+$ currents in an endothelial cell line derived from HUVEC was also studied with the patch-clamp technique (Li et al.

2000). This work showed that *t*-RV directly stimulates K_{Ca} channels, which may underlie the mechanism through which *t*-RV produces endothelium-dependent vasorelaxation.

An important effect of *t*-RV is related to its action on Silencing information regulator (SirT1), a NAD-dependent histone deacetylase, and an essential mediator of longevity in normal cells by caloric restriction. SirT1 has many biological functions, including transcription regulation, cell differentiation inhibition, cell cycle regulation, and anti-apoptosis (Stunkel and Campbell 2011). Evidence shows that *t*-RV induced SirT1 activation, improving endothelial dysfunction and suppressing vascular inflammation (Hu et al. 2011).

It has been suggested that SirT1 could be a key factor in the endothelial cells of atherosclerotic coronary artery bypass graft surgery patients and aging rats. The study shows that *t*-RV is a potential candidate for preventing oxidative stress-induced aging in endothelial cells. The same work shows that *t*-RV may also prevent ROS-induced damage via increased endothelial SirT1 expression in HUVEC (Kao et al. 2010).

Chronic effect of *t*-RV, evaluated in Spontaneously Hypertensive Rats (SHRs), showed that *t*-RV administrated in drinking water, during 28 days at a dosage that mimic moderated red wine consumption, enhanced endothelium-dependent relaxation but did not alter endothelial eNOS levels in aorta (Rush et al. 2007). A similar study evaluated whether the relaxation of rat aorta to estrogen was differently modified in male and female rats after long-term *t*-RV treatment (Soylemez et al. 2008). The results showed that *t*-RV treatment increased both, endothelium-dependent and -independent relaxations to estrogen, especially in aorta from males, an effect that was associated to an elevated NO and/or decreased superoxide production, possibly mediated by classical estrogen receptors (Soylemez et al. 2008).

Furthermore, a chronic study with *t*-RV was conducted in pigs fed an atherogenic diet (AD) (Azorin-Ortuno et al. 2012). Pigs fed the AD for four months showed early atherosclerotic lesions in the thoracic aorta such as degeneration and fragmentation of elastic fibers, increase of intima thickness, subendothelial fibrosis, and accumulation of fatty cells and anion superoxide radicals. *t*-RV-rich grape extract (GE-*t*-RV) was an effective treatment and prevented the early atherosclerotic lesions in aortic tissue, suggesting that the consumption of this GE-*t*-RV nutraceutical, in a dietary prevention context, could prevent early atherosclerotic events (Azorin-Ortuno et al. 2012).

The cardioprotective effects associated to *t*-RV include the ability to reduce the severity of damage that follow a myocardial infarction, antiatherogenic effects and positive effects on blood lipid profiles (Das and Maulik 2006, Penumathsa and Maulik 2009).

Using two experimental models Bradamante et al. (2003) proved that *t*-RV reduced cardiac ischemic-reperfusion injury by means of a NO- and adenosine-dependent mechanism; first, in acute *ex vivo* rat hearts, *t*-RV (10 µM, 10 minutes) significantly increased adenosine release and coronary flow compared with baseline; second, in chronic *in vivo,* where rats received 25 mg/L *t*-RV for 15 days, hearts showed better functional recovery at reperfusion and significant vasodilation suggesting that long-term moderate *t*-RV consumption could play an important role in late cardioprotective effects (Bradamante et al. 2003).

Another research, that also evaluated the potential role of *t*-RV in protecting the isolated rat myocardium, from the deleterious effects of Ischemia-Reperfusion (I/R) injury, resulted in similar conclusions (Dernek et al. 2004). The results from this study indicated that *t*-RV has potent cardioprotective properties against I/R injury in rat hearts, and also its administration, as pretreatment, amplified the beneficial effects over the standard treatment (Dernek et al. 2004).

It is noteworthy that *t*-RV in combination with statin induces cardioprotection against myocardial infarction in hypercholesterolemic rats (Penumathsa et al. 2007). Results showed that acute as well as chronic protection allowed by combination treatment with statin and *t*-RV may be due to pro-angiogenic, antihyperlipidemic and anti-apoptotic effects and long-term effects may be caused by increased neo-vascularization of the myocardial infarction zone leading to less ventricular remodeling (Penumathsa et al. 2007).

To this point, studies show that *t*-RV pretreatment can protect the heart by inducing pharmacological preconditioning. Xi et al. (2009) attempted to show whether *t*-RV protects the heart when applied at reperfusion. They observed that *t*-RV reduced infarct size and prevented cardiac mitochondrial swelling by a mechanism that involves glycogen synthase kinase 3 beta and mitochondrial permeability transition pore (Xi et al. 2009).

Whereas, autophagy is shown to protect the myocardium and cardiac cells against IR injury (Hamacher-Brady et al. 2006), and considering that *t*-RV induces preconditioning-like effects, the possibility that *t*-RV could induce autophagy was evaluated. Effectively, *t*-RV at lower doses (0.1 and 1 µM in cardiac myoblast cells and 2.5 mg/kg/day in rats) induced cardiac autophagy and then cell survival (Gurusamy et al. 2010).

On the other hand, *t*-RV and melatonin—also found in red wine—protect the heart in an experimental model of mouse myocardial infarction, via the Survivor Activating Factor Enhancement (SAFE) prosurvival signaling pathway, that involves the activation of TNF-α and the signal transducer and activator of transcription 3 (STAT3) (Lamont et al. 2011).

Rapid cardioprotection induced by *t*-RV has been associated to Na^+-H^+ exchanger (Thuc et al. 2012). In this work infarct size and functional recovery in rat isolated perfused hearts followed by reperfusion were measured. *t*-RV elicited cardioprotection by reducing ROS and preserving mitochondrial function; the authors proposed that PKC-α-dependent inhibition of Na^+-H^+ exchanger and subsequent attenuation of $[Ca^{2+}]_i$ overload may be a cardioprotective mechanism.

Other stilbenes

There are only few data describing the cardiovascular effects of the other most abundant stilbenes present in red wines. Park et al. (2010) investigated the effects of pterostilbene on Platelet-Derived Growth Factor (PDGF)-BB-induced vascular smooth muscle cells proliferation and the molecular mechanisms. The results showed that the inhibition of pterostilbene to the cell proliferation and DNA synthesis of PDGF-BB-stimulated vascular smooth muscle cells seem to be mediated by the suppression of Akt kinase. Furthermore, they proposed that pterostilbene could be a potential anti-proliferative agent for the treatment of atherosclerosis and angioplasty restenosis. Since piceid showed pterostilbene like cardiovascular effects, it has been suggested that it may also be cardioprotective (Park et al. 2010).

In addition, pterostilbene inhibited OxLDLs-induced apoptosis and effectively induced autophagosome formation in HUVECs (Zhang et al. 2013). Autophagy occurred via a rapid elevation in $[Ca^{2+}]_i$ and subsequent AMP-activated protein kinase-α1-subunit (AMPKα1), which in turn inhibited mammalian target of rapamycin, a potent inhibitor of autophagy. Pterostilbene also stimulated cytoprotective autophagy helping in the removal of OxLDLs and inhibition of apoptosis in HUVECs. The study indicates that pterostilbene could serve as a potential lead compound for developing a class of autophagy regulator as autophagy-related diseases therapy (Zhang et al. 2013).

Recently, Zghonda et al. (2012) compared the effects of *t*-RV and its dimer ε-viniferin on endothelial cells functions, and on blood pressure and cardiac mass of SHRs. Treatment of endothelial cells with these compounds enhanced cell proliferation via NO production and protected them from oxidative stress by suppressing increases in intracellular ROS. ε-Viniferin was more potent than *t*-RV in most of these effects. ε-Viniferin, but not *t*-RV inhibited angiotensin-converting enzyme activity *in vitro*. Three weeks of ε-viniferin treatment reduced the systolic blood pressure and improved the whole cardiac mass and left ventricle mass indexes in SHRs, suggesting that both, ε-viniferin and *t*-RV, may be involved in protecting cardiovascular functions (Zghonda et al. 2012).

In our laboratory we were interested in evaluating the biological activity of *p*-coumaric acid (*p*-CA), an intermediary metabolite in the synthesis of *t*-RV. Experiments carried out recently show that *p*-CA induces endothelium-dependent relaxation of rat aortic rings (Vinet et al. 2014). In addition, *p*-CA protected endothelial-dependent relaxation significantly affected in vessels pre-incubated with high glucose (25 mM).

Conclusions

Research carried out during the last two decades has generated strong evidence regarding the potential beneficial effect of stilbene phytoalexins on health. *Trans*-resveratrol and its derivatives have played a key role in understanding the mechanisms of action associated with its protective effect on the cardiovascular system. It is expected that future research and development, can be transferred to the market as new nutraceuticals- and functional foods-containing stilbene phytoalexins derivatives, capable of decreasing the risk of cardiovascular disease.

Acknowledgements

This work was supported by Grants DI PIA 037.430/12 and 037.301/2013 from the DIPUCV, P. Universidad Católica de Valparaíso, Chile, Grants DIPUV-56/2011 and DIPUV-57/2011 from the DIUV, Universidad de Valparaíso, Chile, and Grant CREAS R12C1001 from the CONICYT-REGIONAL, GORE Región de Valparaíso, CHILE.

References

Ahuja, I., Kissen, R. and Bones, A.M. 2012. Phytoalexins in defense against pathogens. Trends Plant Sci. 17: 73–90.

Azorin-Ortuno, M., Yanez-Gascon, M.J., Pallares, F.J., Rivera, J., Gonzalez-Sarrias, A., Larrosa, M., Vallejo, F., Garcia-Conesa, M.T., Tomas-Barberan, F. and Espin, J.C. 2012. A dietary resveratrol-rich grape extract prevents the developing of atherosclerotic lesions in the aorta of pigs fed an atherogenic diet. J. Agr. Food Chem. 60: 5609–5620.

Bertelli, A.A.E. 1998. Modulatory effect of resveratrol, a natural phytoalexin, on endothelial adhesion molecules and intracellular signal transduction. Pharm. Biol. 36: 44–52.

Bishayee, A. 2009. Cancer prevention and treatment with resveratrol: from rodent studies to clinical trials. Cancer Prev. Res. 2: 409–418.

Boue, S.M., Cleveland, T.E., Carter-Wientjes, C., Shih, B.Y., Bhatnagar, D., McLachlan, J.M. and Burow, M.E. 2009. Phytoalexin-enriched functional foods. J. Agric. Food Chem. 57: 2614–2622.

Bradamante, S., Barenghi, L., Piccinini, F., Bertelli, A.A.E., De Jonge, R., Beemster, P. and De Jong, J.W. 2003. Resveratrol provides late-phase cardioprotection by means of a nitric oxide- and adenosine-mediated mechanism. Eur. J. Pharmacol. 465: 115–123.

Buluc, M. and Demirel-Yilmaz, E. 2006. Resveratrol decreases calcium sensitivity of vascular smooth muscle and enhances cytosolic calcium increase in endothelium. Vascul. Pharmacol. 44: 231–237.

Chen, C.K. and Pace-Asciak, C.R. 1996. Vasorelaxing activity of resveratrol and quercetin in isolated rat aorta. Gen. Pharmacol. 27: 363–366.

Chen, M.L., Li, J., Xiao, W.R., Sun, L., Tang, H., Wang, L., Wu, L.Y., Chen, X. and Xie, H.F. 2006. Protective effect of resveratrol against oxidative damage of UVA irradiated HaCaT cells. J. Cent. South Univ. (Med. Sci.) 31: 635–639.

Chong, J.L., Poutaraud, A. and Hugueney, P. 2009. Metabolism and roles of stilbenes in plants. Plant Sci. 177: 143–155.

Cianciulli, A., Calvello, R., Cavallo, P., Dragone, T., Carofiglio, V. and Panaro, M.A. 2012. Modulation of NF-kappa B activation by resveratrol in LPS treated human intestinal cells results in downregulation of PGE2 production and COX-2 expression. Toxicol. *In Vitro* 26: 1122–1128.

Dai, R., Gao, F., Gassmann, W. and Qiu, W. 2013. Functional analysis of grapevine stilbene synthase genes. Phytopathology 103: 32–32.

Das, D.K. and Maulik, N. 2006. Resveratrol in cardioprotection: a therapeutic promise of alternative medicine. Mol. Interv. 6: 36–47.

Dernek, S., Ikizler, M., Erkasap, N., Ergun, B., Koken, T., Yilmaz, K., Sevin, B., Kaygisiz, Z. and Kural, T. 2004. Cardioprotection with resveratrol pretreatment: improved beneficial effects over standard treatment in rat hearts after global ischemia. Scand Cardiovasc. J. 38: 245–254.

Diaz, M.N., Frei, B., Vita, J.A. and Keaney, J.F., Jr. 1997. Antioxidants and atherosclerotic heart disease. N. Engl. J. Med. 337: 408–416.

Douillet-Breuil, A.C., Jeandet, P., Adrian, M. and Bessis, R. 1999. Changes in the phytoalexin content of various *Vitis* spp. in response to ultraviolet C elicitation. J. Agric. Food Chem. 47: 4456–4461.

Elnakish, M.T., Hassanain, H.H., Janssen, P.M., Angelos, M.G. and Khan, M. 2013. Emerging role of oxidative stress in metabolic syndrome and cardiovascular diseases: important role of Rac/NADPH oxidase. J. Pathol. 231: 290–300.

Fernandez-Mar, M.I., Mateos, R., Garcia-Parrilla, M.C., Puertas, B. and Cantos-Villar, E. 2012. Bioactive compounds in wine: resveratrol, hydroxytyrosol and melatonin: a review. Food Chem. 130: 797–813.

Ferrero, M.E., Bertelli, A.A.E., Fulgenzi, A., Pellegatta, F., Corsi, M.M., Bonfrate, M., Ferrara, F., De Caterina, R., Giovannini, L. and Bertelli, A. 1998. Activity *in vitro* of resveratrol on granulocyte and monocyte adhesion to endothelium. Am. J. Clin. Nutr. 68: 1208–1214.

Folta, S.C. and Nelson, M.E. 2010. Reducing cardiovascular disease risk in sedentary, overweight women: strategies for the cardiovascular specialist. Curr. Opin. Cardiol. 25: 497–501.

Frankel, E.N., Kanner, J., German, J.B., Parks, E. and Kinsella, J.E. 1993. Inhibition of oxidation of human low-density lipoprotein by phenolic substances in red wine. Lancet 341: 454–457.

Fuhrman, B., Lavy, A. and Aviram, M. 1995. Consumption of red wine with meals reduces the susceptibility of human plasma and low-density lipoprotein to lipid peroxidation. Am. J. Clin. Nutr. 61: 549–554.

Gojkovic-Bukarica, L., Novakovic, A., Lesic, A. and Bumbasirevic, M. 2007. Endothelium-dependent relaxation of rat aorta induced by resveratrol. Acta Vet-Beograd 57: 123–132.

Gurusamy, N., Lekli, I., Mukherjee, S., Ray, D., Ahsan, M.K., Gherghiceanu, M., Popescu, L.M. and Das, D.K. 2010. Cardioprotection by resveratrol: a novel mechanism via autophagy involving the mTORC2 pathway. Cardiovasc. Res. 86: 103–112.

Hamacher-Brady, A., Brady, N.R. and Gottlieb, R.A. 2006. Enhancing macroautophagy protects against ischemia/reperfusion injury in cardiac myocytes. J. Biol. Chem. 281: 29776–29787.

Hector, K.L., Lagisz, M. and Nakagawa, S. 2012. The effect of resveratrol on longevity across species: a meta-analysis. Biol. Letters 8: 790–793.

Henriksen, T., Mahoney, E.M. and Steinberg, D. 1981. Enhanced macrophage degradation of low density lipoprotein previously incubated with cultured endothelial cells: recognition by receptors for acetylated low density lipoproteins. Proc. Natl. Acad. Sci. USA 78: 6499–6503.

Houston, M.C. 2010. The role of cellular micronutrient analysis, nutraceuticals, vitamins, antioxidants and minerals in the prevention and treatment of hypertension and cardiovascular disease. Ther. Adv. Cardiovasc. Dis. 4: 165–183.

Hu, Y., Liu, J., Wang, J.F. and Liu, Q.S. 2011. The controversial links among calorie restriction, SIRT1, and resveratrol. Free Radical Bio. Med. 51: 250–256.

Isbir-Soylemez, S. 2006. Resveratrol inhibits contractions to angiotensin II in rat aorta. Acta Pharmacol. Sin 27: 347–347.

Jager, U. and Nguyen-Duong. 1999. Relaxant effect of trans-resveratrol on isolated porcine coronary arteries. Arzneimittel-Forsch 49: 207–211.

Jeandet, P., Douillt-Breuil, A.C., Bessis, R., Debord, S., Sbaghi, M. and Adrian, M. 2002. Phytoalexins from the vitaceae: biosynthesis, phytoalexin gene expression in transgenic plants, antifungal activity, and metabolism. J. Agric. Food Chem. 50: 2731–2741.

Kao, C.L., Chen, L.K., Chang, Y.L., Yung, M.C., Hsu, C.C., Chen, Y.C., Lo, W.L., Chen, S.J., Ku, H.H. and Hwang, S.J. 2010. Resveratrol protects human endothelium from H_2O_2-Induced oxidative stress and senescence via SirT1 Activation. J. Atheroscler. Thromb. 17: 970–979.

Lamont, K.T., Somers, S., Lacerda, L., Opie, L.H. and Lecour, S. 2011. Is red wine a SAFE sip away from cardioprotection? Mechanisms involved in resveratrol- and melatonin-induced cardioprotection. J. Pineal Res. 50: 374–380.

Lamuelaraventos, R.M., Romeroperez, A.I., Waterhouse, A.L. and Delatorreboronat, M.C. 1995. Direct HPLC analysis of cis-resveratrol and trans-resveratrol and piceid isomers in Spanish red Vitis-Vinifera Wines. J. Agric. Food Chem. 43: 281–283.

Li, H.F., Chen, S.A. and Wu, S.N. 2000. Evidence for the stimulatory effect of resveratrol on Ca2+-activated K+ current in vascular endothelial cells. Cardiovasc. Res. 45: 1035–1045.

Lusis, A.J., Attie, A.D. and Reue, K. 2008. Metabolic syndrome: from epidemiology to systems biology. Nat. Rev. Genet. 9: 819–830.

Malacarne, G., Vrhovsek, U., Zulini, L., Cestaro, A., Stefanini, M., Mattivi, F., Delledonne, M., Velasco, R. and Moser, C. 2011. Resistance to plasmoparaviticola in a grapevine segregating population is associated with stilbenoid accumulation and with specific host transcriptional responses. BMC Plant Biol. 11: 114.

Matsuzawa, Y. 2005. White adipose tissue and cardiovascular disease. Best Pract. Res. Clin. Endocrinol. Metab. 19: 637–647.

Mohar, D.S. and Malik, S. 2012. The sirtuin system: the holy grail of resveratrol? J. Clin. Exp. Cardiolog. 3:216. doi:10.4172/2155-9880.1000216.

Naugler, C., McCallum, J.L., Klassen, G. and Strommer, J. 2007. Concentrations of trans-resveratrol and related stilbenes in Nova Scotia wines. Am. J. Enol. Viticult. 58: 117–119.

Novakovic, A., Bukarica, L.G., Kanjuh, V. and Heinle, H. 2006. Potassium channels-mediated vasorelaxation of rat aorta induced by resveratrol. Basic Clin. Pharmacol. 99: 360–364.

Opie, L.H. and Lecour, S. 2007. The red wine hypothesis: from concepts to protective signalling molecules. Eur. Heart J. 28: 1683–1693.

Parage, C., Tavares, R., Rety, S., Baltenweck-Guyot, R., Poutaraud, A., Renault, L., Heintz, D., Lugan, R., Marais, G.A., Aubourg, S. and Hugueney, P. 2012. Structural, functional, and evolutionary analysis of the unusually large stilbene synthase gene family in grapevine. Plant Physiol. 160: 1407–1419.

Park, E.S., Lim, Y., Hong, J.T., Yoo, H.S., Lee, C.K., Pyo, M.Y. and Yun, Y.P. 2010. Pterostilbene, a natural dimethylated analog of resveratrol, inhibits rat aortic vascular smooth muscle cell proliferation by blocking Akt-dependent pathway. Vasc. Pharmacol. 53: 61–67.

Peluso, I., Morabito, G., Urban, L., Ioannone, F. and Serafini, M. 2012. Oxidative stress in atherosclerosis development: the central role of LDL and oxidative burst. Endocr. Metab. Immune Disord. Drug Targets 12: 351–360.

Penumathsa, S.V. and Maulik, N. 2009. Resveratrol: a promising agent in promoting cardioprotection against coronary heart disease. Can. J. Physiol. Pharm. 87: 275–286.

Penumathsa, S.V., Thirunavukkarasu, M., Koneru, S., Juhasz, B., Zhan, L.J., Pant, R., Menon, V.P., Otani, H. and Maulik, N. 2007. Statin and resveratrol in combination induces cardioprotection against myocardial infarction in hypercholesterolemic rat. J. Mol. Cell Cardiol. 42: 508–516.

Pervaiz, S. 2003. Resveratrol: from grapevines to mammalian biology. Faseb J. 17: 1975–1985.

Petrovski, G., Gurusamy, N. and Das, D.K. 2011. Resveratrol in cardiovascular health and disease. Ann. N Y Acad. Sci. 1215: 22–33.

Potvin, L., Richard, L. and Edwards, A.C. 2000. Knowledge of cardiovascular disease risk factors among the Canadian population: relationships with indicators of socioeconomic status. CMAJ 162: S5–11.

Qureshi, A.A., Guan, X.Q., Reis, J.C., Papasian, C.J., Jabre, S., Morrison, D.C. and Qureshi, N. 2012. Inhibition of nitric oxide and inflammatory cytokines in LPS-stimulated murine macrophages by resveratrol, a potent proteasome inhibitor. Lipids Health Dis. 11: 76.

Raiber, S., Schroder, G. and Schroder, J. 1995. Molecular and enzymatic characterization of two stilbene synthases from Eastern white pine (Pinusstrobus). A single Arg/His difference determines the activity and the pH dependence of the enzymes. Febs Lett. 361: 299–302.

Rayalam, S., Della-Fera, M.A. and Baile, C.A. 2011. Synergism between resveratrol and other phytochemicals: Implications for obesity and osteoporosis. Mol. Nutr. Food Res. 55: 1177–1185.

Reaven, P.D., Khouw, A., Beltz, W.F., Parthasarathy, S. and Witztum, J.L. 1993. Effect of dietary antioxidant combinations in humans. Protection of LDL by vitamin E but not by beta-carotene. Arterioscler. Thromb. 13: 590–600.

Renaud, S. and de Lorgeril, M. 1992. Wine, alcohol, platelets, and the French paradox for coronary heart disease. Lancet 339: 1523–1526.

Richard, J.L., Cambien, F. and Ducimetiere, P. 1981. Epidemiologic characteristics of coronary disease in France. Nouv Presse Med. 10: 1111–1114.

Rosler, J., Krekel, F., Amrhein, N. and Schmid, J. 1997. Maize phenylalanine ammonia-lyase has tyrosine ammonia-lyase activity. Plant Physiol. 113: 175–179.

Rush, J.W.E., Quadrilatero, J., Levy, A.S. and Ford, R.J. 2007. Chronic resveratrol enhances endothelium-dependent relaxation but does not alter eNOS levels in aorta of spontaneously hypertensive rats. Exp. Biol. Med. 232: 814–822.

Schanz, S., Schroder, G. and Schroder, J. 1992. Stilbene synthase from Scots pine (*Pinussylvestris*). Febs Lett. 313: 71–74.

Schoppner, A. and Kindl, H. 1984. Purification and properties of a stilbene synthase from induced cell suspension cultures of peanut. J. Biol. Chem. 259: 6806–6811.

Shen, B. 2000. Biosynthesis of aromatic polyketides. pp. 1–51. *In*: F.J. Leeper and J.C. Vederas (eds.). Topics in Current Chemistry. Springer, Berlin.

Shen, M., Zhao, L., Wu, R.X., Yue, S.Q. and Pei, J.M. 2013. The vasorelaxing effect of resveratrol on abdominal aorta from rats and its underlying mechanisms. Vasc. Pharmacol. 58: 64–70.

Soylemez, S., Gurdal, H., Sepici, A. and Akar, F. 2008. The effect of long-term resveratrol treatment on relaxation to estrogen in aortae from male and female rats: Role of nitric oxide and superoxide. Vasc. Pharmacol. 49: 97–105.

Sparvoli, F., Martin, C., Scienza, A., Gavazzi, G. and Tonelli, C. 1994. Cloning and molecular analysis of structural genes involved in flavonoid and stilbene biosynthesis in grape (*Vitis vinifera* L.). Plant Mol. Biol. 24: 743–755.

Stervbo, U., Vang, O. and Bonnesen, C. 2007. A review of the content of the putative chemopreventive phytoalexin resveratrol in red wine. Food Chem. 101: 449–457.

Stunkel, W. and Campbell, R.M. 2011. Sirtuin 1 (SIRT1): the misunderstood HDAC. J. Biomol. Screen 16: 1153–1169.

Subramanian, L., Youssef, S., Bhattacharya, S., Kenealey, J., Polans, A.S. and van Ginkel, P.R. 2010. Resveratrol: challenges in translation to the clinic—a critical discussion. Clin. Cancer Res. 16: 5942–5948.

Teo, K., Lear, S., Islam, S., Mony, P., Dehghan, M., Li, W., Rosengren, A., Lopez-Jaramillo, P., Diaz, R., Oliveira, G., Miskan, M., Rangarajan, S., Iqbal, R., Ilow, R., Puone, T., Bahonar, A., Gulec, S., Darwish, E.A., Lanas, F., Vijaykumar, K., Rahman, O., Chifamba, J., Hou, Y., Li, N., Yusuf, S. and Investigators, P. 2013. Prevalence of a healthy lifestyle among individuals with cardiovascular disease in high-, middle- and low-income countries: The Prospective Urban Rural Epidemiology (PURE) study. JAMA 309: 1613–1621.

Thuc, L.C., Teshima, Y., Takahashi, N., Nishio, S., Fukui, A., Kume, O., Saito, S., Nakagawa, M. and Saikawa, T. 2012. Inhibition of Na+-H+ exchange as a mechanism of rapid cardioprotection by resveratrol. Br. J. Pharmacol. 166: 1745–1755.

Tropf, S., Karcher, B., Schroder, G. and Schroder, J. 1995. Reaction mechanisms of homodimeric plant polyketide synthase (stilbenes and chalcone synthase). A single active site for the condensing reaction is sufficient for synthesis of stilbenes, chalcones, and 6'-deoxychalcones. J. Biol. Chem. 270: 7922–7928.

Van Gaal, L.F., Mertens, I.L. and De Block, C.E. 2006. Mechanisms linking obesity with cardiovascular disease. Nature 444: 875–880.

Victor, V.M. 2013. Editorial (hot topic: new insights into insulin resistance, cardiovascular diseases and oxidative stress: pathophysiological and clinical consequences). Curr. Pharm. Des. 19: 5661–5662.

Vinet, R., Knox, M., Mascher, D., Paredes-Carbajal, C. and Martínez, J.L. 2012. Isolated aorta model and its contribution to phytopharmacology. Bol. Latinoam Caribe Plant Med. Aromat. 11: 35–45.

Vinet, R., Araos, P., Gentina, J.C., Knox, M. and Guzmán, L. 2014. p-Coumaric acid reduces high glucose-mediated impairment of endothelium-dependent relaxation in rat aorta. Bol. Latinoam Caribe Plant Med. Aromat. 13: 232–237.

Wang, H., Yang, Y.J., Qian, H.Y., Zhang, Q., Xu, H. and Li, J.J. 2012. Resveratrol in cardiovascular disease: what is known from current research? Heart Fail Rev. 17: 437–448.

Wang, W., Tang, K., Yang, H.R., Wen, P.F., Zhang, P., Wang, H.L. and Huang, W.D. 2010. Distribution of resveratrol and stilbene synthase in young grape plants (*Vitis vinifera* L. cv. Cabernet Sauvignon) and the effect of UV-C on its accumulation. Plant Physiol. Bioch. 48: 142–152.

Weisshaar, B. and Jenkins, G.I. 1998. Phenylpropanoid biosynthesis and its regulation. Curr. Opin. Plant Biol. 1: 251–257.

Williams, R.J., Motteram, J.M., Sharp, C.H. and Gallagher, P.J. 1992. Dietary vitamin E and the attenuation of early lesion development in modified Watanabe rabbits. Atherosclerosis 94: 153–159.

Xi, J.K., Wang, H.H., Mueller, R.A., Norfleet, E.A. and Xu, Z.L. 2009. Mechanism for resveratrol-induced cardioprotection against reperfusion injury involves glycogen synthase kinase 3 beta and mitochondrial permeability transition pore. Eur. J. Pharmacol. 604: 111–116.

Xia, E.Q., Deng, G.F., Guo, Y.J. and Li, H.B. 2010. Biological activities of polyphenols from grapes. Int. J. Mol. Sci. 11: 622–646.

Zamora, P.L. and Villamena, F.A. 2013. Pharmacological approaches to the treatment of oxidative stress-induced cardiovascular dysfunctions. Future Med. Chem. 5: 465–478.

Zghonda, N., Yoshida, S., Ezaki, S., Otake, Y., Murakami, C., Mliki, A., Ghorbel, A. and Miyazaki, H. 2012. epsilon-Viniferin is more effective than its monomer resveratrol in improving the functions of vascular endothelial cells and the heart. Biosci. Biotechnol. Biochem. 76: 954–960.

Zhang, C. 2008. The role of inflammatory cytokines in endothelial dysfunction. Basic Res. Cardiol. 103: 398–406.

Zhang, L., Cui, L., Zhou, G., Jing, H., Guo, Y. and Sun, W. 2013. Pterostilbene, a natural small-molecular compound, promotes cytoprotective macroautophagy in vascular endothelial cells. J. Nutr. Biochem. 24: 903–911.

16

From Plant Collection to Lab and Market—Traditional Medicine in Northern Peru

Rainer W. Bussmann* and Douglas Sharon

Introduction

Traditional medicine is used globally and is of rapidly growing economic importance. In developing countries, traditional medicine is often the only accessible and affordable treatment available. In Uganda, for instance, the ratio of traditional practitioners to the population is between 1:200 and 1:400, while the availability of western doctors is typically 1:20,000 or less. Moreover, doctors are mostly located in cities and other urban areas, and are therefore inaccessible to rural populations. In Africa, up to 80% of the population uses traditional medicine as the primary healthcare system. In Latin America, the WHO Regional Office for the Americas (AMRO/PAHO) reports that 71% of the population in Chile and 40% of the population in Colombia use traditional medicine. In many Asian countries traditional medicine is widely used, even though western medicine is often readily available. In Japan, 60–70% of allopathic doctors prescribe traditional medicines for their patients. In China, traditional medicine accounts for about 40% of all healthcare, and is used to treat roughly 200 million patients annually. The number of visits to providers of Complementary-Alternative Medicine (CAM) now exceeds by far the number of visits to all primary care physicians in the US (WHO 1999a,b, 2002).

Complementary-Alternative Medicine is becoming more and more popular in many developed countries. Forty-eight percent of the population in Australia, 70% in Canada, 42% in the US, 38% in Belgium and 75% in France, have used Complementary-Alternative Medicine at least once (WHO 1998, Fisher and Ward 1971, Health Canada 2001). A survey of 610 Swiss doctors showed that 46% had used some form of CAM, mainly homeopathy and acupuncture (Domenighetti et al. 2000). In the United Kingdom, almost 40% of all general allopathic practitioners offer some form of CAM referral or access (Zollman and Vickers 2000). In the US, a national survey reported the use of at least one of 16 alternative therapies increased from 34 in 1990 to 42% in 1997 (Eisenberg et al. 1998, UNCTD 2000).

The expenses for the use of traditional and complementary-alternative medicine are exponentially growing in many parts of the world. In Malaysia, an estimated US$500 million is spent annually on traditional medicine, compared to about US$300 million on allopathic medicine. The 1997 out-of-pocket

---

William L. Brown Center, Missouri Botanical Garden, P.O. Box 299, St. Louis, MO 63166-0299, USA.
* Corresponding author: rainer.bussmann@mobot.org

Complementary-Alternative Medicine expenditure was estimated at US$2,700 million in the USA. In Australia, Canada, and the United Kingdom, annual Complementary-Alternative Medicine expenditure is estimated at US$80 million, US$2,400 million and US$2300 million, respectively. The world market for herbal medicines based on traditional knowledge was estimated at US$60,000 million in the late 1990s (Breevort 1998). A decade later it was around US$60 billion (Tilbert and Kaptchuk 2008) with estimates for 2015 at US$90 billion (GIA 2012). The sales of herbs and herbal nutritional supplements in the US increased 101% between May 1996 and May 1998. The most popular herbal products included ginseng (*Ginkgo biloba* L.), garlic (*Allium sativum* L.), *Echinacea* spp. and St. John's Wort (*Hypericum perforatum* L.) (Breevort 1998).

Traditional and Complementary-Alternative Medicine are gaining more and more respect by national governments and health providers. Peru's National Program in Complementary Medicine and the Pan American Health Organization recently compared complementary medicine to allopathic medicine in clinics and hospitals operating within the Peruvian Social Security System. A total of 339 patients—170 being treated with Complementary-Alternative Medicine and 169 with allopathic medicine—were followed for one year. Treatments for osteoarthritis; back pain; neurosis; asthma; peptic acid disease; tension and migraine headache; and obesity were analyzed. The results, with 95% significance, showed that the cost of using Complementary-Alternative Medicine was less than the cost of western therapy. In addition, for each of the criteria evaluated—clinical efficacy, user satisfaction, and future risk reduction—Complementary-Alternative Medicine's efficacy was higher than that of conventional treatments, including fewer side effects, higher perception of efficacy by both the patients and the clinics, and a 53–63% higher cost efficiency of Complementary-Alternative Medicine over that of conventional treatments for the selected conditions (EsSalud 2000).

According to WHO (1998), the most important challenges for traditional medicine/complementary-alternative medicine for the next years are:

- Research into safe and effective traditional medicine and complementary alternative medicine treatments for diseases that represent the greatest burden, particularly among poorer populations.
- Recognition of the role of traditional medicine practitioners in providing healthcare in developing countries.
- Optimized and upgraded skills of traditional medicine practitioners in developing countries.
- Protection and preservation of the knowledge of indigenous traditional medicine.
- Sustainable cultivation of medicinal plants.
- Reliable information for consumers on the proper use of traditional medicine and complementary-alternative medicine therapies and products.

A policy report, 'Biodiversity, Traditional Knowledge and Community Health: Strengthening Linkages', published by the United Nations University—Institute of Advanced Studies in Yokohama, Japan addresses many of the issues discussed above (Unnikrishnan and Suneetha 2012). Building on the WHO Alma Ata Declaration of 1978, relating to traditional medicine and primary health care, the UN Convention on Biological Diversity of 1992, and the UN's Middle Development Goals (MDGs) of 2011.

Antecedents—Medicinal plant research and traditional medicine in Peru

In all Peruvian ethnic groups, plant knowledge is invaluable because it reinforces national identity and values, which are being lost in the complementary processes of modernization and globalization. In the current situation the emerging recognition and incipient application of these resources and associated knowledge emphatically underscores the critical need for ethnobotanical research in light of the following facts:

- Absorption and devaluing of native culture due to modernization and globalization.
- At the same time, recuperation/revalorization of traditional knowledge of Peruvian flora.
- Emerging 'first world' awareness of the therapeutic potential of medicinal plants.
- Recent ethnobotanical research by a growing number of Peruvian scholars (de Zandra 1988).

In 'Sinopsis histórica de la Etnobotánica en el Perú', La Torre and Alban (2006) outline the history of formal floristic studies in Peru starting in 1778 with the work of Hipólito Ruiz, José Pavón and Joseph Dombey followed by Alexander von Humbolt and Aime Bonpland.

Considerable progress has been made in the overall taxonomic treatment of the flora of Peru over the last few decades (Valdizan and Maldonado 1922). However, while the Amazon rainforests have received a great deal of scientific attention, the mountain forests and remote highland areas are still relatively unexplored. Until the late 1990s little work had been done on vegetation structure, ecology, and ethnobotany in the mountain forests and coastal areas of the north.

Northern Peru represents the 'health axis' of the old Central Andean cultural area stretching from Ecuador to Bolivia (Camino 1992/1999). The traditional use of medicinal plants in this region, which encompasses in particular the departments of Piura, Lambayeque, La Libertad, Cajamarca, Amazonas, and San Martin possibly dates as far back as the first millennium B.C. (north coastal Cupisnique Culture) or at least to the Moche period (A.D. 100–800), with healing scenes and healers frequently depicted in ceramics.

Precedents for our research have been established by early colonial period chroniclers (Monardes 1574, Acosta 1590, Cobo 1653/1956); the plant collections that Bishop Baltasar Jaime Martínez Compañón sent to the Palacio Real de Madrid along with cultural materials in 1789 under the title 'Trujillo del Perú' in nine illustrated volumes (Martínez Compañon 1789, Schjellerup 2009, Sharon and Bussmann 2006); the travel journals of H. Ruiz from 1777–88; the work of the Italian naturalist Antonio Raimondi (1857); ethnoarchaeological analysis of the psychedelic San Pedro cactus (Sharon 2000); *curandera* depictions in Moche ceramics (Glass-Coffin et al. 2000), and research on the medicinal plants of southern Ecuador (Béjar et al. 1997, 2000, Bussmann 2006, Bussmann and Sharon 2007a).

Two decades of ethnobotanical research in southern Ecuador and northern Peru

Work up to 2012—besides developing a database of 510 medicinal plants (Bussmann and Sharon 2006a,b, 2007a,b) and 974 remedies of mixtures (Bussmann et al. 2010a) has demonstrated that herbal commerce in Peru is a major economic resource (Bussmann et al. 2007a), which, although used alongside modern pharmaceutical products, is showing signs of diminished popular knowledge of applications (Bussmann et al. 2007a,b, 2009a). Laboratory research on most of the database has ranged from minimum inhibition concentrations (Bussmann et al. 2010b) to toxicity screening (Bussmann et al. 2011) as well as bioassays to determine antibacterial activity (Bussmann et al. 2009b, Bussmann et al. 2010a) and phytochemical analysis (Bussmann et al. 2009c). An ethnography of peasant herbalists which documented aspects of the market supply chain showed that suppliers are not adequately remunerated and revealed threats posed by lack of conservation measures and overharvesting (Revene et al. 2008, Carrillo 2012) criticized the scientific reductionism of laboratory research in attempting to appropriately verify traditional remedies. Anthropological studies of traditional *curanderos* and their curing altars (*mesas*) include articles by Sharon (2009) and Glass-Coffin et al. (2000).

Ethnobotanical data were collected from plant sellers while purchasing plant materials in local markets; by accompanying local healers (*curanderos*) to the markets when they purchased plants for curing sessions, and into the field when they were harvesting. In addition, plants were collected by the project members in the field, and—together with the material purchased in the markets—taken to the homes of *curanderos* to discuss the plants' healing properties, applications, harvesting methodology, and origins. At the *curanderos*' homes the authors also observed the preparation of remedies and participated in healing rituals. Plant uses were discussed in detail with informants, after seeking prior informed consent from each respondent. Following a semi-structured interview technique, respondents were asked to provide detailed information about the vernacular plant name in Spanish or Quechua; plant properties (hot/cold); harvesting region; ailments for which a plant was used; best harvesting time and season; plant parts used as well as mode of preparation and application; and specific instructions for the preparation of remedies, including the addition of other plant species. Data on plant species, families, vernacular names, plant parts used, traditional uses, and modalities of use were recorded.

Many of the species reported from northern Peru are widely known by *curanderos* and herb vendors as well as the general population of the region, and are used for a large number of medical conditions. One hundred fifty to two hundred plant species, including most of the introductions, are commonly sold in the local markets (Bussmann et al. 2007b). Rare indigenous species were either collected by the healers themselves, or were ordered from special collectors or herb vendors. The same plants were frequently used by a variety of healers for the same purposes, with only slight variations in recipes. However, different healers might give preference to different species for the treatment of the same medical condition. All species found were well known to the healers and herb vendors involved in the study, even if they themselves did not use or sell the species in question. Many species were often easily recognized by their vernacular names by other members of the population. This indicates that these remedies have been in use for a long time by many people. The use of some species, most prominently San Pedro (*Echinopsis pachanoi* Britton & Rose) Friedrich & G.D. Rowley, Maichil (*Thevetia peruviana* K. Schum.) and Ishpingo (various species of *Nectandra*), can be traced back to the Moche culture (AD 100–800). Representations of these plants are frequently found on Moche ceramics, and the remains of some were found in a variety of burials of high-ranking individuals of the Moche elite, e.g., the tomb of the Lord of Sipán (Bussmann and Sharon 2009).

Changing markets

Exotics played an important role amongst all plants sold in northern Peruvian markets. Fifty-nine species (15%) found in all markets were exotics. However, amongst the species most commonly encountered in the inventories, 40–50% were exotics. *Matricaria recutita* L. (chamomile) was found in the inventory of approximately 70% of vendors. The next most popular species sold in these markets included *Equisetum giganteum* L., *Phyllanthus urinaria* L., *Phyllanthus stipulatus* (Raf.) G.L. Webster, *Phyllanthus niruri* L. (chanca piedra—stone breaker), *Eucalyptus globulus* Labill. (eucalyptus), *Piper aduncum* L., *Uncaria tomentosa* (Willd.) DC. (cat's claw), *Rosmarinus officinalis* L. (rosemary), *Peumus boldus* Molina, *Bixa orellana* L. (achiote) and *Buddleja utilis* Kraenzl. However, when taking sales volume into account, *Croton lechleri* Müll.-Arg. (dragon's blood), *Uncaria tomentosa* (Willd.) DC., and *Eucalyptus globulus* Labill. were clearly the most important species (Bussmann and Sharon 2009b).

While it was very easy for all the vendors to name their most important and frequently sold species, it proved impossible to get detailed information about species that vendors observed as 'rare' or 'disappearing'. In most cases, vendors mentioned species as rare because they themselves did not sell them, in many cases these plants were very common outside the market, or because demand was so low, that it would not have made sense to carry them in their inventories. Very small vendors had inventories that represented the most common medicinal plants available, and excluded most species in the large 'witchcraft' segment of the pharmacopoeia. On the other hand, well-established large stands specialized in supplies for healers (including 'magical' plants).

All researched markets had inventories containing more than 50% of all listed plant species, but lacked many of the 'generalist' plants sold by other vendors. The portfolio of these stands focused almost entirely on 'magical' species that are needed to cure illnesses like '*susto*' (fright), '*mal aire*' (evil wind), '*daño*' (damage), '*envidia*' (envy) and other 'magical' or psychosomatic ailments. At the same time, all four vendors catered also to the esoteric tourism crowd that tends to frequent the large markets, and carried a variety of plants that were not used by *curanderos*, but instead were sold to meet tourist demand.

A look on sustainability—how much plant and for which price?

More than two thirds of all species sold in northern Peruvian were claimed to originate from the highlands (*sierra*), above the timberline, which represents areas often heavily used for agriculture and livestock grazing. The overall value of medicinal plants in these markets reaches a staggering US$1.2 million a year. This figure only represents the share of market vendors, and does not include the amount local healers charge for their cure. Thus, medicinal plants contribute significantly to the local economy. Such an immense market raises questions of the sustainability of this trade, especially because the market analysis does not take into account any informal sales.

Most striking was the fact that seven indigenous and three exotic species, i.e., 2.5% of all species traded, accounted for more than 40% of the total sales volume (with 30 and 12% respectively). Moreover, 31 native species accounted for 50% of all sales, while only 16 introduced plants contributed to more than a quarter of all material sold. This means that little over 11% of all plants in the market accounted for about three fourths of all sales. About one third of this sale volume includes all exotic species traded. None of these are rare or endangered. However, the rising market demand might lead to increased production of these exotics, which in turn could have negative effects on the local flora (Bussmann et al. 2010a).

A look at the indigenous species traded highlights important conservation threats. *Croton lechleri* Müll.-Arg. (dragon's blood), and *Uncaria tomentosa* (Willd.) DC. (cat's claw) are immensely popular at a local level and each contributes to about 7% to the overall market value. Both species are also widely traded internationally. The latex of *Croton* is harvested by cutting or debarking the whole tree. *Uncaria* is mostly traded as a bark, and again the whole plant is normally debarked. *Croton* is a pioneer species, and apart from *C. lechleri* a few other species of the genus have found their way in the market. Sustainable production of this genus seems possible, but the process has to be closely monitored, and the current practice does not appear sustainable because most *Croton* is harvested wild. The cat's claw trade is so immense, that in fact years ago collectors of this primary forest liana started complaining about a lack of resources (Cabieses Molina 2000) and during the years of this study other *Uncaria* species, or even *Acacia* species have appeared in the market as 'cat's claw' (pers. observation). As such, the *Uncaria* trade is clearly not sustainable.

Some of the other 'most important' species are either common weeds (e.g., *Desmodium molliculum* (Kunth) DC), or have large populations (e.g., *Equisetum giganteum* L.). However, a number of species are very vulnerable. *Tillandsia cacticola* L.B. Sm. for example grows in small areas of the coast as epiphyte. The habitat, coastal dry forest and shrub, is heavily impacted by urbanization and mechanized agriculture, the impact of the latter worsened by the current bio-fuel boom.

Gentianella alborosea (Gilg.) Fabris, *Gentianella bicolor* (Wedd.) Fabris ex J.S. Pringle, *Gentianella graminea* (Kunth) Fabris, *Geranium ayavacense* Willd. ex Kunth and *Laccopetalum giganteum* (Wedd.) Ulbr. are all high altitude species with very limited distribution. Their large-scale collection is clearly unsustainable, and in case of *Laccopetalum* collectors indicate that supply is harder and harder to find. The fate of a number of species with similar habitat requirements raises comparable concern. The only species under cultivation at this point are exotics, and a few common indigenous species.

Does traditional medicine work? A look at antibacterials used in northern Peru

Plants with potential medicinal activity have recently come to the attention of western scientists, and studies have reported that some are bioactive (Perumal Samy and Ignacimuthu 2000). Potentially active compounds have been isolated from a few of the plants tested (D'Agostini et al. 1995a,b, Okuyama et al. 1994, Rodriguez et al. 1994, Umana and Castro 1990).

In order to evaluate the antibacterial activity of species used in TM in northern Peru, 525 plant samples of at least 405 species were tested in simple agar-bioassays for antibacterial activity against *Staphylococcus aureus, Escherichia coli, Salmonella enterica typhi* and *Pseudomonas aeruginosa* (Bussmann et al. 2010b). A much larger number of ethanolic plant extracts showed antibacterial activity compared to water extracts for all antibacterial activity. One-hundred-ninety-three ethanolic extracts and 31 water extracts were active against *S. aureus*. In 21 cases only the water extract showed activity (for all bacterial species) compared to ethanol only. None of the aqueous extracts were active against the other three bacteria, with the activity of the ethanolic extracts also greatly reduced, as only 36 showed any activity against *E. coli*, and three each against *S. enterica typhi* and *P. aeruginosa*. Eighteen ethanol extracts were effective against both *E. coli* and *S. aureus*, while in two cases the ethanol extract showed activity against *E. coli* and the water extract against *S. aureus*. The ethanol extract of *Dioscorea trifida* L.f. was effective against *E. coli, S. aureus* and *P. aeruginosa*. *Caesalpinia spinosa* (Feuilleé ex Molina) Kuntze was the only species that showed high activity against all bacteria, including *Salmonella enterica typhi* and *Pseudomonas aeruginosa*, when extracted in ethanol.

Two hundred twenty-five extracts came from plant species that are traditionally used against bacterial infections. One hundred sixty-six (73.8%) of these were active against at least one bacterium. Of the 300 extracts from plants without traditional antibacterial use, only 96 (32%) showed any activity. This shows clearly that plants traditionally used as antibacterial had a much higher likelihood to be antibacterially active than plants without traditional antibacterial use. However, the efficacy of plants used traditionally for antibacterial related applications did vary, which underlines the need for studies aiming to clearly understand traditional disease concepts. Plants used for respiratory disorders, inflammation/infection, wounds, diarrhea, and to prevent post-partum infections were efficacious in 70–88% of the tests. Plants used for 'kidney inflammation' had a much lower efficacy against bacteria, and fell within the range of species that are traditionally used to treat other bodily disorders. Only species used for spiritual/ritual treatments scored worse. Of these only 22% showed some antibacterial; activity. However, amongst the 'spiritual' plants 38% of the species used for cleansing baths did in fact show activity, while only 15% of the plants often used in protective amulets (mostly species with the families of Lycopodiaceae and Valerianaceae) showed limited antibacterial activity.

A variety of species showed higher efficacy than the control antibiotics employed. However, extracts were often highly inconsistent in their efficacy.

Extracts of the same species traditionally used to treat infections often produced vastly diverging results when collected from different localities.

Almost all remedies are traditionally prepared as water extracts, although ethanol (in the form of sugarcane spirit) is readily available. This might at a first glance seem astonishing, given the low efficacy of water extraction found in this study. However, initial results from Brine-Shrimp toxicity assays indicate that the ethanolic extracts are by far more toxic than water extracts of many species, and thus ethanolic extraction might in many cases not be suitable for application in patients. This again indicates the considerable sophistication and care with which traditional healers in northern Peru chose their remedies for a specific purpose.

If the botanical documentation of Peruvian medicinal plants has been neglected, investigations of the phytochemical composition of useful plants is lagging even further behind. Most studies on the phytochemistry of Peruvian plants concentrate on a few 'fashionable' species that have been marketed heavily on a global scale, especially Maca (*Lepidium meyenii* Walp.), Sangre del Drago or del Grado (*Croton lechleri* Müll.-Arg.), and Uña de Gato (*Uncaria tomentosa* (Willd.) DC. and *Uncaria guianensis* (Aubl.) J.F. Gmel.). The number of other Peruvian plants for which at least some phytochemical studies exist is still miniscule, and most efforts are fuelled by the fads and fashions of the international herbal supplement market. Studies involving multiple species were initiated as late as the 1990s (Oblitas 1992).

Minimum inhibitory concentrations found for Peruvian plant extracts ranged from 0.008 to 256 mg/ml. The ethanolic extracts exhibited stronger activity and a much broader spectrum of action than the water extracts. Most MIC values reported in this work were largely higher than those obtained for South American species (Bastos et al. 2009, Jiménez et al. 2001, Meléndez et al. 2006, Zampini et al. 2009) and African studies (Kirira et al. 2006). However, they were in range or lower than concentrations reported by Kloucek et al. (2007) and Nascimento et al. (2000). Most species effective against *S. aureus* are traditionally used to treat wound infection, throat infections, serious inflammations, or are post partum infections. Interestingly many species used in cleansing baths also showed high activity against this bacterium. Many of these species are either employed topically, or in synergistic mixtures, so that possible toxicity seems not to be an issue. The species effective against *E. coli* were mostly used in indications that traditional healers identified as 'inflammation'.

Most of the plants used by the healers have antibacterial activity, but only eight of the 141 plants (5.6%) examined in this study show any MIC values of 200 or less mg/ml of extract. Of these eight plants five are used to treat diseases believed to be in bacterial origin by TM, one is a disease not believed to be caused by bacteria and one is used for undefined treatment purposes.

Nine out of 141 plants (6.3%) tested that were not used for diseases believed to be bacterial in origin by TM, five showed high antibacterial activity with MIC values below 16 mg/ml. Four of these were among the most potent plants tested with MIC values of 2 or less mg/ml including the hallucinogen and extracts used to treat diabetes and epilepsy. Diseases such as diabetes often compromise the health of the individual and antibacterial treatments can be warranted for secondary complications of the disease. In addition, TM does determine sometimes that diseases not originally believed to be bacterial in origin, such as ulcers, are actually caused by bacteria. Currently TM is seriously looking at the role of inflammation (which can certainly be bacterial in origin) in heart disease.

Toxicity in traditional medicine

Crude medicinal activities have been investigated for a wide variety of plants (e.g., Hammond et al. 1998). But while toxicity assays are available for many other regions of the world, no data exists on the potential toxicity of Peruvian medicinal species (Bussmann et al. 2011).

Brine shrimp (*Artemia*) are frequently used as agents in laboratory assays to determine toxicity values by estimating LC_{50} values (median lethal concentration). The Brine shrimp lethality activity of 501 aqueous and ethanolic extracts of 341 plant species belonging to 218 genera of 91 families used in Peruvian traditional medicine was tested (Bussmann et al. 2011). The aqueous extracts of 55 species showed high toxicity values (LC_{50} below 249 µg/ml), 18 species showed median toxicity (LC_{50} 250–499 µg/ml) and 18 low toxicity (LC_{50} 500–1000 µg/ml). The alcoholic extracts proved to be much more toxic: 220 species showed high toxicity values (LC_{50} below 249 µg/ml, with 37 species having toxicity levels of >1 µg/ml), 43 species showed median toxicity (LC_{50} 250–499 µg/ml) and 23 species low toxicity (LC_{50} 500–1000 µg/ml). Over 24% of the aqueous extracts and 76% of the alcoholic extracts showed elevated toxicity levels to brine-shrimp. Traditional preparation methods are taking this into account—most remedies are multiple extracts from different collections of the same species showed in most cases very similar toxicity values. However, in some cases the toxicity of extracts from different collections of the same species varied from non-toxic to highly toxic.

Conclusions

Current research indicates that the composition of the local pharmacopoeia in northern Peru and southern Ecuador has changed since colonial times (Martinez Compañon 1789, Sharon and Bussmann 2006, Bussmann and Sharon 2009a). However, in northern Peru the overall number of medicinal plants employed seems to have remained at a comparable level, while plant use in southern Ecuador has decreased. This indicates that the northern Peruvian health tradition is still going strong, and that the healers and public are constantly experimenting with new remedies. One example of this is the sudden appearance of Noni (*Morinda citrifolia* L.) fruits and products in large quantities in plant pharmacies and markets in the region since 2005. This plant was not available before, but is now largely marketed worldwide. Peruvian sellers are clearly reacting to a global market trend and are trying to introduce this new species to their customers. This indicates that local herbalists and herb merchants are carefully watching international health trends to include promising species in their own repertoire. In southern Ecuador, healers were not able to experiment with new remedies due to persecution and legal restrictions. As a result, the pharmacopoeia in this region remained on an early colonial level, with loss of significant knowledge.

The knowledge of medicinal plants is still taught by word of mouth, with no written record. Illustrated identification guides for the medicinal plants of northern Peru and southern Ecuador and their uses (Bussmann and Sharon 2007b) will hopefully help to keep the extensive traditional knowledge of this area alive. However, traditional medicine is experiencing increasing demand, especially from a Peruvian perspective, as indicated by the fact that the number of herb vendors, in particular in the markets of Trujillo, has increased in recent years. Also, a wide variety of medicinal plants from northern Peru can be found in the global market. While this trend might help to maintain traditional practices and to give traditional knowledge the respect it deserves, it poses a serious threat, as signs of over-harvesting of important species are becoming increasingly common.

It is apparent that the respondents used medicinal herbs more often than pharmaceutical medicines, but only to a small degree. Bussmann et al. (2007b, 2009a) showed in their studies that patients both at western and herbalist clinics often had a preference for pharmaceutical medicines only to a small degree. People generally assumed that plants are healthier and better to use because they are natural and are thought to not have any side effects. It is difficult to determine if the knowledge of the use of medicinal plants is growing or decreasing, but the indications are that the last generation knows more than the present. However, most of the present generation does teach their children about the use of medicinal plants. This study also showed what medicinal plants the respondents used for which purposes. It would be interesting to evaluate the properties of the species used in bioassays. Similarly, the plant knowledge of patients at both facilities was largely identical, with an essentially overlapping selection of common, mostly introduced, species, and basically the same number of medicinal plants mentioned overall. This indicates that traditional medicinal knowledge is a major part of a people's culture that is being maintained while patients are also embracing the benefits of western medicine.

This attitude does however lead to profound challenges when it comes to the safety of the plants employed, in particular for applications that require long-term use. Bussmann et al. (2013) found that various species were often sold under the same common names. Some of the different fresh species were readily identifiable botanically, but neither the collectors nor the vendors made a direct distinction between species. Often material is sold in finely powdered form, which makes the morphological identification of the species in the market impossible, and greatly increases the risk for the buyer. The best way to ensure correct identification would be DNA bar-coding. The necessary technical infrastructure is however not available locally. The use of DNA bar-coding as a quality control tool to verify species composition of samples on a large scale would require to carefully sample every batch of plant material sold in the market. The volatility of the markets makes this is an impossible logistical task. Often the same or closely related species mentioned in literature sell under wide variety of common names. Worse, one species might be sold, e.g., as 'Hercampuri' in one location or market stand, while selling under a different name at a neighboring stand. As expected there is no consistency in the dosage of plants used, nor do vendors agree on possible side effects.

Studies indicate that the plant use in northern Peru, although footing on a millennial tradition, has changed considerably even during the last decades. Even in case of plant species used for very clearly circumscribed applications, patients run a considerable risk when purchasing plants for their remedies in the local markets, and the possible side effects can be serious. Much more control, and a much more stringent identification of the material sold in public markets, and entering the global supply chain via Internet sales, would be needed.

References

Bastos, A., Lima, M.L.F., Conserva, M.R.M., Andrade, L.S., Rocha, V.M. and Lemos, E.M. 2009. Studies in the antimicrobial activity and brine shrimp toxicity of *Zeyheria tuberculosa* (Vell.) Bur. (Bignoniaceae) extracts and their main constituents. Ann. Clin. Micr. Antimicr. 8(16).

Béjar, E., Bussmann, R.W., Roa, C. and Sharon, D. 1997. Pharmacological search for active ingredients in medicinal plants of Latin America. pp. 63–81. *In*: T. Shuman, M. Garrett and L. Wozniak (eds.). International Symposium on Herbal Medicine, A Holistic Approach. SDSU International Institute for Human Resources Development, San Diego.

Béjar, E., Bussmann, R.W., Roa, C. and Sharon, D. 2001. Herbs of Southern Ecuador—Hierbas del Sur Ecuatoriano. Latin Herbal Press, San Diego.

Breevort, P. 1998. The Booming U.S. Botanical Market, A New Overview. HerbalGram, vol. 44.

Bussmann, R.W. 2006. Manteniendo el balance de naturaleza y hombre, La diversidad.florística andina y su importancia por la diversidad cultural—ejemplos del Norte de.Perú y Sur de Ecuador. Arnaldoa 13(2): 382–397.

Bussmann, R.W. and Sharon, D. 2006a. Traditional plant use in Southern Ecuador. J. Ethnobiol. Ethnobiomed. 2(44).

Bussmann, R.W. and Sharon, D. 2006b. Traditional plant use in Northern Peru, Tracking two thousand years of health culture. J. Ethnobiol. Ethnobiomed. 2(47).

Bussmann, R.W. and Sharon, D. 2007a. Plants of longevity—The medicinal flora of Vilcabamba. Plantas de la longevidad—La flora medicinal de Vilcabamba. Graficart, Trujillo.

Bussmann, R.W. and Sharon, D. 2007b. Plants of the four winds—The magic and medicinal flora of Peru. Plantas de los cuatro vientos—La flora mágica y medicinal del Perú. Graficart, Trujillo.

Bussmann, R.W., Sharon, D., Vandebroek, I., Jones, A.A. and Revene, Z. 2007a. Health for sale, the medicinal plant markets in Trujillo and Chiclayo, Northern Peru. J. Ethnobiol. Ethnobiomed. 3(37).

Bussmann, R.W., Sharon, D. and Lopez, A. 2007b. Blending traditional and Western medicine, medicinal plant use among patients at Clinic Anticona in El Porvenir, Peru. Ethnobot. Res. Appl. 5: 185–199.

Bussmann, R.W. and Sharon, D. 2009a. Shadows of the colonial past-diverging plant use in Northern Peru and Southern Ecuador. J. Ethnobiol. Ethnobiomed. 5(4).

Bussmann, R.W. and Sharon, D. 2009b. From collection to market and cure-Traditional medicinal use in Northern Peru. pp. 184–207. *In*: U. Albuquerque (ed.). Recent Development and Case Studies in Ethnobotany. Nupea, Recife.

Bussmann, R.W., Sharon, D. and Garcia, M. 2009a. From Chamomile to Aspirin? Medicinal Plant use among clients at Laboratorios Beal in Trujillo, Peru. Ethnobot. Res. Appl. 7: 399–407.

Bussmann, R.W., Glenn, A., Meyer, K., Rothrock, A., Townesmith, A., Sharon, D., Castro, M., Cardenas, R., Regalado, S., Toro, R., Chait, G., Malca, G. and Perez, F. 2009b. Antibacterial activity of medicinal plants of Northern Peru-Part II. Arnaldoa 16(1): 93–103.

Bussmann, R.W., Glenn, A., Meyer, K., Rothrock, A., Townesmith, A., Sharon, D., Castro, M., Cardenas, R., Regalado, S., Toro, R., Chait, G., Malca, G. and Perez, F. 2009c. Phyto-chemical analysis of Peruvian medicinal plants. Arnaldoa 16(1): 105–110.

Bussmann, R.W., Glenn, A., Meyer, K., Kuhlman, A. and Townesmith, A. 2010a. Herbal mixtures in traditional medicine in Northern Peru. J. Ethnobiol. Ethnobiomed. 6(10).

Bussmann, R.W., Malca, G., Glenn, A., Sharon, D., Chait, G., Diaz, D., Pourmand, K., Jonat, B., Somogny, S., Guardado, G., Aguirre, C., Chan, R., Meyer, K., Kuhlman, A., Townesmith, A., Effio, J., Frias, F. and Benito, M. 2010b. Minimum inhibitory concentrations of medicinal plants used in Northern Peru as antibacterial remedies. J. Ethnopharmacol. 132: 101–108.

Bussmann, R.W., Malca, G., Glenn, A., Sharon, D., Nilsen, B., Parris, B., Dubose, D., Ruiz, D., Saleda, J., Martinez, M., Carillo, L., Kuhlman, A. and Townesmith, A. 2011. Toxicity of medicinal plants used in traditional medicine in Northern Peru. J. Ethnopharmacol. 137: 121–140.

Bussmann, R.W., Paniagua Zambrana, N., Rivas Chamorro, M., Molina Moreira, N., Cuadros Negri, M.L. and Olivera, J. 2013. Peril in the market—classification and dosage of species used as anti-diabetics in Lima, Peru. Journal of Ethnobiology and Ethnomedicine 9(37).

Cabieses Molina, F. 2000. *La Uña de Gato u su entorno. De la Selva a la farmacia.* Universidad de San Martin De Porres, Lima.

Camino, L. 1992/1999. *Cerros, plantas y lagunas ponderosas—la medicina al norte del Perú.* Lluvia Editores, Lima.

Carrillo, L. 2012. Scientific validation? How bioprospecting laboratory practices contribute to the devaluation of traditional medicinal knowledge. Berkl. Mc. Nair. Res J., vol. 19.

Cobo, B. 1653. Historia del Nuevo Mundo, 2 tomos. Sevilla.

Cobo, B. 1956. Historia del Nuevo Mundo. F. Mateos (ed.). Ediciones Atlas, Madrid.

D'Agostino, M., De Simone, F., Tommasi, N. and Pizza, C. 1995a. Constituents of *Culcitium canescens.* Fitoterapia 66: 550–551.

D'Agostino, M., Pizza, C. and De Simone, F. 1995b. Flavone and flavonol glycosides from *Desmodium mollicum.* Fitoterapia 66: 384–385.

De Acosta, J. 1590. *Historia natural y moral de las Indias.* Sevilla.

De Zandra, A. Alarco. 1988. *Perú, el libro de las plantas mágicas.* Concytec, Lima.

Domenighetti, G., Grilli, R., Gutzwiller, F. and Quaglia, J. 2000. Usage personnel de pratiques relevant des médecines douce sou alternatives parmi les médecins suisses. Med. Hyg. 58: 22–91.

Eisenberg, D.M., Davis, R.B., Ettner, S.L., Appel, S., Wilkey, S., van Rompay, M. and Kessler, R.C. 1998. Trends in alternative medicine use in the United States, 1990–1997, results of a follow-up national survey. J. Am. Med. Ass. 280(18): 1569–1575.

EsSalud/Organización Panamericana de Salud. 2000. Estudio Costo-Efectividad, Programa Nacional de Medicina Complementaria. Seguro Social de EsSalud (Study of Cost Effectiveness, National Program in Complementary Medicine. Social Security of EsSalud). Lima, EsSalud/Organización Panamericana de Salud.

Fisher, P. and Ward, A. 1971. Medicine in Europe, complementary medicine in Europe. Brit. Med. J. 309: 107–111.

Glass-Coffin, B., Sharon, D. and Uceda, S. 2004. Curanderos a la sombra de la Huaca de la luna. Bull. Inst. fr. Et. And. 33(1): 81–95.

Global Industry Analysts Inc. 2012. Herbal Supplements and Remedies, A Global Strategic Business Report. Global Industry Analysts, San Jose.

Hammond, G.B., Fernández, I.D., Villegas, L. and Vaisberg, A.J. 1998. A survey of traditional medicinal plants from the Callejón de Huaylas, Department of Ancash, Perú. J. Ethnopharm. 61: 17–30.

Health Canada. 2001. Perspectives on Complementary and Alternative Health Care. A Collection of Papers Prepared for Health Canada. Ottawa, Health Canada.

Jiménez, G., Hasegawa, M., Rodríguez, M., Estrada, O., Méndez, J., Castillo, A., Gonzalez-Mujica, F., Motta, N., Vázquez, J. and Romero-Vecchione, E. 2001. Biological screening of plants from the Venezuelan Amazon. J. Ethnopharm. 77: 77–83.

Kirira, P.G., Rukunga, G.M., Wanyonyi, A.W., Gathirwa, J.W., Mathaura, C.N., Omar, S.A., Tolo, F., Mungai, G.M. and Ndiege, I.O. 2006. Anti-plasmodial activity and toxicity of extracts of plants used in traditional malaria therapy in Meru and Kilifi Districts in Kenya. J. Ethnopharm. 106: 403–407.

Kloucek, P., Svoboda, B., Polesny, Z., Langrova, I., Smrcek, S. and Kokoska, L. 2007. Antimicrobial activity of some medicinal barks used in Peruvian Amazon. J. Ethnopharm. 111: 427–429.

La Torre, M.M. and Alban, J. 2006. Etnobotánica en los Andes del Perú. pp. 239–245. *In*: M. Morales, L. Ollgaard, L. Kvist, F. Borchsenius and H. Balslev (eds.). Botánica Económica de los Andes Centrales.

Martínez Compañon, D.B. 1789. *Razón de las especies de la naturaleza y del arte del obispado de Trujillo del Perú*. Tomos III-V, Sevilla.

Meléndez, P.A. and Capriles, V.A. 2006. Antibacterial properties of tropical plants from Puerto Rico. Phytomed. 13: 272–276.

Monardes, N. 1574. Primera y segunda y tercera partes de la história medicinal de las cosasque se traen de nuestras Indias Occidentales, que sirven en medicina; Tratado de la piedra bezaar, y de la yerva escuerçonera; Diálogo de las grandezas del hierro, y de sus virtudes medicinales; Tratado de la nieve, y del beuer frio. Alonso Escrivano, Seville.

Nascimento, G.F., Locatelli, J., Freitas, P.C. and Silva, G.L. 2000. Antibacterial activity of plant extracts and phytochemicals on antibiotic resistan bacteria. Braz. J. Microbiol. 31: 247–256.

Oblitas, E. 1992. Plantas medicinales de Bolivia. Editorial Los Amigos del Libro, La Paz.

Okuyama, E., Umeyama, K., Ohmori, S., Yamazaki, M. and Satake, M. 1994. Pharmacologically active components from a Peruvian medicinal plant, Huira-Huira (*Culcitium canescens* H. & B.). Chem. Pharm. Bull. 42: 2183–2186.

Perumal Samy, R. and Ignacimuthu, S. 2000. Antibacterial activity of some medicinal plants used by tribals in Western Ghats, India. J. Ethnopharm. 69: 63–71.

Raimondi, A. 1857. Elementos de Botánica aplicada a la medicina y la industria. Lima.

Revene, Z., Bussmann, R.W. and Sharon, D. 2008. From Sierra to Coast: tracing the supply of medicinal plants in northern Peru—A plant collector´s tale. Ethnobot. Res. Appl. 6: 15–22.

Rodriguez, J., Pacheco, P., Razmilic, I., Loyola, J.I., Schmeda-Hirschmann, G. and Theoduloz, C. 1994. Hypotensive and diuretic effect of Equisetum bogotense and Fuchsia magellanica and micropropagation of *E. bogotense*. Phytother. Res. 8: 157–160.

Ruiz, H. 1998. Relación del viaje hecho a los reynos del Perú y Chile. 1777–1788. Translated by R.E. Schultes and M.J. Nemry von Thenen de Jaramillo-Arango as "The Journals of Hipólito Ruiz". Timber Press, Portland.

Schjellerup, I. 2009. Razon de las Especies de la Naturaleza y del Arte del Obispado de Trujillo del Peru del Obispo D. Balazar Martinez Compagñon. *In*: E. Vergara and R. Vásquez (eds.). Medicina Tradicional, Conocimento Milenario. Ser. Anthropol. 1: 128–152.

Sharon, D. 2000. Shamanismo y el Cacto Sagrado—Shamanism and the Sacred Cactus. San Diego Mus. Pap., vol. 37.

Sharon, D. 2009. Tuno y sus colegas: Notas comparativas. pp. 255–267. *In*: C. Galvez (ed.). Medicina Tradicional Conocimiento Milenario. *Ser. Antropol.*, no 1. Museo de Arquelogía, Antropología e Historia, Facultad de Ciencias Sociales, Universidad Nacional de Trujillo.

Sharon, D. and Bussmann, R.W. 2006. *Plantas Medicinales en la Obra del Obispo Don Baltasar Jaime Martínez Compagñon (Siglo XVIII)*. pp. 147–165. *In*: L. Millones and T. Kato (eds.). *Desde el exterior, El Perú y sus estudios*. Tercer Congreso Internacional de Peruanistas, Nagoya, 2005, UNMSM, Lima.

Tilbert, J.C. and Kaptchuk, T.J. 2008. Herbal medicine research and global health, an ethical analysis. Bull. WHO 86: 594–599.

Umana, E. and Castro, O. 1990. Chemical constituents of *Verbena littoralis*. Int. J. Crude Drug Res. 28: 175–177.

United Nations Conference on Trade and Development. 2000. Systems and National Experiences for Protecting Traditional Knowledge, Innovations and Practices. Background Note by the UNCTAD Secretariat Geneva, United Nations Conference on Trade and Development, (document reference TD/B/COM.1/EM.13/2).

Unnikrishnan, P.M. and Suneetha, M.S. 2012. Biodiversity, Traditional Knowledge and Community Health, Strengthening Linkages. Yokohama, United Nations University-Institute of Advanced Studies.

Valdizan, H. and Maldonado, Y.A. 1922. *La medicina popular peruana*. Torres Aguirre, Lima.

Villegas, L.F., Fernandez, I.D., Maldonado, H., Torres, R., Zavaleta, A., Vaisberg, A.J. and Hammond, G.B. 1997. Evaluation of the wound-healing activity of selected traditional medicinal plants from Peru. J. Ethnopharm. 55: 193–200.

World Health Organization. 1998. Technical Briefing on Traditional Medicine. Forty-ninth Regional Committee Meeting, Manila, Philippines, 18 September 1998, Manila, WHO Regional Office for the Western Pacific.

World Health Organization. 1999a. Consultation Meeting on TM and Modern Medicine, Harmonizing the Two Approaches. Geneva, World Health Organization (document reference. (WP)TM/ICP/TM/001/RB/98–RS/99/GE/32(CHN)), World Health Organization, Geneva.

World Health Organization. 1999b. Traditional, Complementary and Alternative Medicines and Therapies. Washington DC, WHO Regional Office for the Americas/Pan American Health Organization (Working group OPS/OMS).

World Health Organization. 2002. WHO Traditional Medicine Strategy 2002–2005. World Health Organization, Geneva.

Zampini, I.C., Cuello, S., Alberto, M.R., Ordoñez, R.M., Alameida, R.D., Solorzano, E. and Isla, M.I. 2009. Antimicrobial activity of selected plant species from the "Argentine Puna" against sensitive and multi-resistant bacteria. J. Ethnopharm. 124: 499–505.

Zollman, C. and Vickers, A.J. 2000. ABC of Complementary Medicine. BMJ Books, London.

An Update on Plant Originated Remedies for the Medication of Hyperlipidemia

Mert Ilhan, Ipek Süntar and Esra Küpeli Akkol*

Introduction

Lipids are a heterogeneous group of substances including fats, mono/di/tri-glycerides, phospholipids, waxes, fat-soluble vitamins and sterols which are soluble in organic solvents, but, insoluble in water. There are various functions of lipids, for instance, they serve as structural components of membranes and provide protection as an outer coating of the organism. They are storage and transport forms of metabolic fuel. The lipoprotein has hydrophobic structures inside, such as cholesterol esters and triacylglycerol (TAG), which has a surface coat of polar lipids including unesterified cholesterol and phospholipids as well as apolipoproteins (Trinick and Duly 2005).

Glycerol molecules esterified with three fatty acid molecules are called TAG, which are the storage form of fatty acids. They constitute the main energy storage form in mammals. Cholesterol is found in the membranes as free cholesterol, and in the plasma mainly as esterified. The excess cholesterol is eliminated through bile as cholesterol or after conversion to bile acids some of which are absorbed back into the enterohepatic circulation to be used again (Trinick and Duly 2005).

The major function of the lipoproteins is to transport lipids from one organ to the other (Reckless and Lawrence 2003). In lipoprotein metabolism, a series of receptors, transporters, and enzymes take place. Several diseases in particular, coronary heart disease, stroke, and peripheral vascular diseases are caused due to the circulating high lipid concentrations (Trinick and Duly 2005).

Lipoproteins are classified according to their density as chylomicrons, Very Low-Density Lipoprotein (VLDL), Immediate-Density Lipoprotein (IDL), Low-Density Lipoprotein (LDL), High-Density Lipoprotein (HDL) (Trinick and Duly 2005).

VLDL are large in size, triglyceride-rich lipoproteins. Their size depends on their triglyceride (TG) content. They function in the transportation of endogenous TGs, which are derived from dietary carbohydrate or from plasma. These large VLDL may be a poorer substrate for the usual path of metabolism and thus stay longer in the blood. By hydrolyzation of the TG, with the loss of surface components, VLDL particles

Gazi University, Faculty of Pharmacy, Department of Pharmacognosy, Etiler 06330, Ankara, Turkey.
* Corresponding author: esrak@gazi.edu.tr

shrink to HDL and catabolism to LDL through IDL. Therefore, IDL occur as an intermediate step in the metabolism of VLDL. The level of IDL in the plasma is usually low due to rapid catabolization (Trinick and Duly 2005).

LDL particles can be classified into LDL1, LDL2, and LDL3 according to their size and density. A high amount of LDL3 is associated with a high risk of coronary heart disease (Trinick and Duly 2005). Indeed, LDL is the main cholesterol carrier towards tissues having atherogenic potential (Liu et al. 2012). On the other hand, HDL, which are small and high dense lipoproteins, are involved in the transportation of the cholesterol from the peripheral tissues to the liver (Reckless and Lawrence 2003). Therefore, HDL provides protection against many cardiac problems and obesity (Liu et al. 2012).

Hypercholesterolemia has become the primary problem for human health (Sharma et al. 2008, Gao et al. 2013). A cholesterol rich diet is the main contributor to an imbalanced lipoprotein metabolism (Luo et al. 2008) resulting in the increases of serum TG, TC (total cholesterol), LDL and decrease of HDL levels (Kamesh and Sumathi 2012). Since high concentrations of LDL accumulate in the extracellular subendothelial space of arteries and are highly atherogenic and toxic to vascular cells, the increase of serum total cholesterol and LDL concentrations can be considered as the major risk factors in the initiation and progression of atherosclerotic impasse and cardiovascular diseases including atherosclerosis and hypertension (Ghule et al. 2009, Vijayaraj et al. 2013). Insulin resistance is also associated with a high risk of cardiovascular diseases and is often accompanied by low levels of HDL (Gliozzi et al. 2013).

The insulin deficiency in diabetes mellitus also causes lipid breakdown in adipose tissues and enhances the level of free fatty acids (Sridevi et al. 2011). Enhancement in the mobilization of fatty acids from adipose tissue and elevation of free fatty acid concentration in serum result in a disorder in lipid metabolism (Ravi et al. 2005). Therefore, diabetes is associated with alterations in carbohydrate, fat and protein metabolisms which lead to several secondary complications and an increased risk of hyperlipidemia, coronary artery disease, renal failure, stroke, neuropathy, retinopathy and blindness (Kasetti et al. 2010, Latha and Daisy 2011). Being more prone to hypercholesterolemia and hypertriglyceridemia (Ravi et al. 2005) diabetic patients also display abnormal antioxidant status (Sedaghat et al. 2011). Enzymatic and non-enzymatic antioxidative defence systems are altered leading to Reactive Oxygen Species (ROS) mediated damage (Vijayaraj et al. 2013) such as lipid peroxidation, stellate cell activation, steatosis and precirrhotic steatohepatitis in the liver (Matsuzawa et al. 2007, Küpeli Akkol et al. 2009). Due to the oxidative stress, LDL turns into oxidized LDL, a reactive form (Chang et al. 2010). Oxidized LDL is considered highly atherogenic. Because it increases lipid accumulation in macrophages (Sparow et al. 1989), stimulates chemotaxis of monocytes (Berliner et al. 1990), and modulates expression of various cytokines (Nomura et al. 2001). On the other hand, HDL can be considered as the anti-atherogenic lipoproteins and plays an important role in protecting LDL against oxidation (Berrougui et al. 2006).

The occurrence and progression of atherosclerotic lesions can be effectively prevented by providing a reduction in the elevated concentrations of LDL cholesterol in the plasma (Kumar et al. 2012b). To decrease the blood cholesterol TAG levels, drug therapy is highly beneficial (Miettinen et al. 1995, Marinangeli et al. 2006). Especially the use of LDL receptor and proprotein convertase subtilisin/kexin type 9 (PCSK9) and 3-hydroxy-3-methylglutaryl coenzyme A (HMG CoA) reductase inhibitors, also known as statins, ezetimibe, bile acid sequestrants, nicotinic acid and fibric acids play important roles in regulating cholesterol homeostasis (Marinangeli et al. 2006, Chong et al. 2011). However, safety concerns regarding the long-term use of these agents have recently occurred (de Denus et al. 2004, Marinangeli et al. 2006), since statins sometimes cause severe side effects (Magalhaes 2005) and a group of the population responds poorly to statin drugs (Jia et al. 2008). To overcome these problems screening the novel compounds is important. Plant-based pharmaceuticals are obviously active and produce minimal side effects in clinical practice when compared to oral synthetic agents (Sedaghat et al. 2011). In addition, an increasing number of patients with hyperlipidemia have been searching for natural products to regulate their serum lipid concentrations as an alternative to cholesterol-lowering drugs (Jia et al. 2008). Combination of plant sterols and endurance training is recommended for patients with hyperlipidemia (Marinangeli et al. 2006) Although varieties of synthetic drugs are used in the treatment, natural products that have the capacity to regulate serum lipid concentrations as an alternative to cholesterol-lowering drugs are still being searched (Kumar et al. 2012b).

There are a lot of well-documented studies regarding the protective effect of natural products in lowering plasma cholesterol concentrations. In the light of this information, the objective of this study is to present preclinical research on plant based remedies against hyperlipidemia.

Experimental researches on medicinal plants against hyperlipidemia

In vitro and *in vivo* experimental studies on medicinal plants traditionally used for the treatment of hyperlipidemia are presented according to the family names of the plants in alphabetic order.

Amaranthaceae

A previous study was carried out by Rajesh et al. for the evaluation of the effect of methanol (MEAL) and aqueous extracts (AEAL) of the aerial parts of *Aerva lanata* Linn. Juss in streptozotocin-induced diabetic rats. The streptozotocin induced diabetic rats were orally treated with vehicle (Normal saline), reference drug (glibenclamide), MEAL and AEAL. MEAL and AEAL significantly decreased the blood glucose level, lipid profile, increased body weight and reduced Serum Glutamate-Oxaloacetate Transaminase (SGOT), Serum Glutamate-Pyruvate Transaminase (SGPT), creatinine, alkaline phosphatase (ALP), Blood Urea Nitrogen (BUN) and total bilirubin levels. The findings suggested that MEAL and AEAL possess antihyperlipidemic effect in addition to the antidiabetic activity (Rajesh et al. 2012).

The objective of the previous study by Kumar et al. was to search the potential antidiabetic, antihyperlipidemic and antioxidant effects of the methanolic extract of whole plant of *Amaranthus viridis* Linn. in alloxan (ALX)-induced diabetic albino Wistar rats. *A. viridis* and glibenclamide (as the reference drug) orally administered daily for 15 days. At the end of the experiment, rats were sacrificed and blood was collected for determining HDL, LDL, VLDL, TC, TG and Total Protein (TP). For the evaluation of *in vivo* antioxidant activity of *A. viridis*, blood malondialdehyde (MDA), glutathione (GSH), catalase activity (CAT) and total thiols were investigated in the liver tissues. When compared to the control group *A. viridis* exhibited notable lowering effect of blood glucose, lipid levels as well as remarkable improvement in MDA, GSH, CAT and total thiols, thus concluding that *A. viridis* displayed antidiabetic, antihyperlipidemic and antioxidant activities (Kumar et al. 2012a).

Apocynaceae

Alstonia scholaris Linn. R. Br., native of India, possesses several pharmacological activities such as astringent, thermogenic, laxative, antipyretic, anthelmintic, galactogoguic and cardiotonic (Nadkarni 1976, Kirtikar and Basu 2002). It has been used for the treatment of diabetes in traditional medicine. Arulmozhi et al. evaluated the activity of ethanol extract of the leaves of *A. scholaris* in streptozotocin-induced diabetic rats. The blood glucose level, body weight, glycosylated haemoglobin, muscle and liver glycogen, lipid profile, lipid peroxidation, antioxidant status were investigated. Ethanol extract of the leaves of *A. scholaris* was reported to reduce the blood glucose level, glycosylated haemoglobin and lipid peroxidation, whereas they increased body weight, liver and muscle glycogen and antioxidant status. According to the results of the study, it was suggested that *A. scholaris* exerted antidiabetic, antihyperlipidemic and antioxidant activities in the streptozotocin-induced diabetic model (Arulmozhi et al. 2010).

Aquifoliaceae

Mate tea is prepared as an infusion made from the leaves of the tree *Ilex paraguariensis* A. St. Hil. It is widely consumed as a beverage in South America and contains different bioactive components including polyphenols, alkaloids and triterpenoid saponins (Heck and de Mejia 2007, Bracesco et al. 2011). A previous research was designed to assess the activity of the aqueous extract of Mate on body weight, serum lipids, antioxidant enzyme activity, lipoprotein metabolism enzyme activity and gene expression involved in lipid metabolism in hyperlipidemia induced hamsters by a high-fat diet. Administration of Mate enhanced antioxidant enzyme activity, improved lipoprotein lipase (LPL) and hepatic lipase (HL) effects in serum and liver, upregulated mRNA expression of peroxisome proliferator-activated receptor α and LDL receptor, and downregulated mRNA expression of sterol regulatory element-binding protein 1c and acetyl CoA carboxylase in the liver, which indicated that Mate tea ameliorates hyperlipidemia (Gao et al. 2013).

Asteraceae

The flower of *Chrysanthemum morifolium* Ramat. (CM) is widely used in China for its antioxidant, cardiovascular protective and anti-inflammatory functions. Lii et al. explored the activities of HCM (a hot water extract of CM), ECM (an ethanol extract of CM), and the flavonoids apigenin and luteolin in CM on the oxidized LDL (oxLDL)-induced expression of ICAM-1 and E-selectin in human umbilical vein endothelial cells. HCM, ECM, apigenin, and luteolin were found to have the ability to inhibit ICAM-1 and E-selectin expression and adhesion of HL-60 by oxLDL. Moreover, HCM, ECM, apigenin and luteolin reversed the inhibition of phosphorylation of Akt and CREB by oxLDL. The ROS scavenging capacity of HCM, ECM, apigenin, and luteolin was observed in a dose-dependent manner in the presence of oxLDL. The results of the present research indicated that cardiovascular-protective potential of *C. morifolium* could be attributed to its antioxidant activity and modulation of the PI3K/Akt signalling pathway (Lii et al. 2010).

Eclipta prostrata Linn. is used as a tonic and deobstruent and to treat hepatic disorders, spleen enlargement, skin diseases, obesity and hypercholesterolemia in Indian traditional medicine (Anonymous 1952). The total alcoholic extract of the plant was tested for antihyperlipidemic potential. The results showed that the extract of *E. prostrata* displayed a dose-dependent effect in albino rats when compared to reference drugs, which supported the traditional utilization of the plant in the treatment of hyperlipidemia (Santhosh Kumari et al. 2006).

Vernonia anthelmintica (L.) A Willd. was reported to be used for antidiabetic activity in folk medicine (Nagaraju and Rao 1990). Ethanol extract prepared from the seeds of *V. anthelmintica* was investigated for its antihyperglycemic activity in streptozotocin (STZ)-induced diabetic rats according to biological activity-guided fractionation assay. Administration of ethanol extract provided reduction in the blood glucose concentration in diabetic rats. Ethanol extract was further fractionationated using silica gel column chromatography. Among the five fractions, polyphenolic fraction displayed the maximum antihyperglycemic effect. Administration of the active fraction for 45 days resulted in notable reduction in plasma glucose (PG), HbA1C, cholesterol, TG, LDL, VLDL, free fatty acids, phospholipids and HMG-CoA reductase. Moreover, remarkable decrease in plasma insulin, protein, HDL and hepatic glycogen were normalized with the treatment of the active fraction. Therefore, it was suggested that the seeds of *V. anthelmintica* could be used as a beneficial agent for the management of diabetes and hyperlipidemia without evident toxic effects (Fatima et al. 2010).

Biognoniaceae

Hypolipidemic and antihyperlipidemic activities of the water extract of *Pachyptera hymenaea* (DC.) A. H. Gentry. were investigated. Serum lipid profiles were estimated after the administration of the extract to normal and diet-induced hypercholesterolemic rats for 28 days and 200 and 400 mg/kg doses exhibited a notable reduction in plasma LDL-cholesterol, TG and TC levels compared to normal rats. Diallyldisulphide and diallyltrisulphide were determined as the two main organosulphur compounds of the volatile oil by Gas Chromatography-Mass Spectroscopy analysis. The potential antihyperlipidemic effect of the extract was attributed to the presence of organosulphur compounds, flavonoids and polyphenols (Verma et al. 2012b).

Caesalpiniaceae

The extracts of *Caesalpinia bonducella* (Linn.) Flem. were shown to have antihyperlipidemic actions in diabetes-induced hyperlipidemia. *C. bonducella* demonstrated a significant reduction in the total cholesterol and the triglyceride concentrations as well as a remarkable increase in the HDL by suppressing the LDL level when compared to the standard drug (Kannur 2011).

Cassia auriculata L., a common plant in Asia, has been reported to have antioxidant potential (Kumaran and Karunakaran 2007). *C. auriculata* also reported to have the ability to control blood glucose (Surana et al. 2008), dyslipidemia, cardiovascular risk (Javekar and Halade 2006), rheumatism, and conjunctivitis (Joshi 2000). The phytochemical data on *C. auriculata* revealed the presence of flavonoids, β-sitosterol-β-D-glucoside, polysaccharides, anthracene, dimeric procyanidins and myristyl alcohol. In a previous study by Vijayaraj et al. the possible antihyperlipidemic and antioxidative effects of *C. auriculata* flower (CAF) were examined in Triton WR 1339 induced-hyperlipidemic rats. Ethanol extract prepared from CAF (Et-CAF) was administered to normal and hyperlipidemic rats. Serum and liver tissue were analyzed for lipid profile, lipid peroxidation products and antioxidants enzymes and were compared to the reference drug lovastatin. After the treatment with Et-CAF, parameters which changed during hyperlipidemia reverted back to near normal values. In Et-CAF treated rats, lipid peroxidation reduced while the activities of superoxide dismutase (SOD), GSH peroxidase and CAT increased. The results suggested that Et-CAF could be used for its beneficial effects in treating hyperlipidemia and ROS (Vijayaraj et al. 2013).

Combretaceae

In order to assess the antidiabetic, antihyperlipidemic and antioxidant potential of *Anogeissus latifolia* (Roxb. ex DC.) Wall. ex. Guill. & Perr. in STZ-nicotinamide (STZ-NIN)-induced type 2 diabetic rats, aqueous extract prepared from the barks of the plant was orally administered for four weeks. With the treatment *A. latifolia*, increase in body weight, haemoglobin and decreased blood glucose, glycosylated haemoglobin was reported. Altered lipid profiles and antioxidant levels were reversed to near normal by *A. latifolia* administration at 200 mg/kg dose. Aqueous extract of the barks of *A. latifolia* exhibited significant antidiabetic, antihyperlipidemic and antioxidant activity in type 2 diabetic rats (Ramachandran et al. 2012).

In Indian folk medicine *Terminalia bellerica* (Gaertn.) Roxb. is utilized to treat various kinds of diseases including diabetes. A study was conducted to determine the antidiabetic effect of the fruit rind of *T. bellerica* and identify its active compounds in STZ-induced diabetic male wistar rats. According to bioassay guided fractionation assay gallic acid was isolated as the active compound. Gallic acid was administered at different doses to the rats. PG, TC, TG, LDL, urea, uric acid, creatinine levels notably reduced and at the same time plasma insulin, C-peptide and glucose tolerance significantly increased when compared to the control. The results demonstrated that gallic acid was the active constituent of fruit rind of *T. bellerica* responsible for normalizing the biochemical parameters related to the patho-biochemistry of diabetes mellitus (Latha and Daisy 2011).

Convolvulaceae

In some countries, leaves of *Ipomoea batatas* L. (sweet potato) are consumed as a fresh vegetable (Villareal et al. 1982, Nwinyi 1992). Sweet potato leaves were found to be rich in polyphenols according to recent research (Islam et al. 2002a, 2002b, Islam et al. 2003, Taira et al. 2007). The phenolic compounds were identified as mainly caffeoylquinic acid (CQA) derivatives (Cliffold et al. 2003, Ishiguro et al. 2007) having DPPH-radical scavenging, antimutagen and antitumour activities (Oki et al. 2002, Yoshimoto et al. 2002). The anti-oxidative activity of the *I. batatas* leaves was investigated in a LDL oxidation induction system. Indeed, the oxidative modification of the LDL was reported to play an important role in the pathogenesis of atherosclerosis (Fogelman et al. 1980, Rosenfeld et al. 1991). Therefore, protection against LDL-oxidation could be a target study for preventing atherosclerosis. According to the results, all the caffeoylquinic acid (CQA) derivatives from the leaves of *I. batatas* demonstrated anti-LDL oxidation activity. The antioxidant activity of sweet potato leaves was correlated with the amounts of CQA derivatives. It was concluded that sweet potato leaves might prevent developing atherosclerosis caused by the oxidation of LDL (Taira et al. 2013).

Cucurbitaceae

A study was undertaken to explore the activities of methanol extract of *Lagenaria siceraria* (Molina) Standl. in experimentally induced hyperlipidemia in rats. In order to evaluate its antihyperlipidemic effect methanol extract of *L. siceraria* fruits (LSFE) at doses of 100, 200 and 300 mg/kg was orally administered to the high fat-diet-induced hyperlipidemic rats for 30 days. Atorvastatin (10 mg/kg; p.o.) was applied as a reference drug. On the day 30, a significant reduction in total cholesterol, LDL, VLDL, TG concentrations were observed in the LSFE treated rats as compared to the rats fed with high-fat diet. Furthermore, LSFE also displayed notable enhancement in excretion of bile acids, which demonstrated that the LSFE had a definite antihyperlipidemic activity potential (Ghule et al. 2009).

Antihyperlipidemic activity of crude ethanolic extract of *Melothria maderaspatana* (L.) M. Roem. leaves was investigated in deoxycorticosterone acetate (DOCA)-salt hypertensive rats. The plant was shown to have antihypertensive effect at the dose of 200 mg/kg in hypertensive rats (Veeramani et al. 2010). The same researchers also investigated the activity of *M. maderaspatana* on the lipid metabolism on DOCA-salt induced hypertensive rats. The plasma and tissue concentration of TC, TG, free fatty acid, phospholipids (PL), LDL and VLDL notably increased in DOCA-salt hypertensive rats. On the other hand, administration of *M. maderaspatana* brought these parameters to normality, which proved its antihyperlipidemic activity. Histopathology of liver, kidney and heart showed reduced damages and revealed that *M. maderaspatana* protected the liver, kidney and heart due to its antihyperlipidemic activity against DOCA-salt administration (Veeramani et al. 2012).

Dioscoreaceae

Saponins from the rhizome of *Dioscorea nipponica* Makino, a traditional Chinese herb, possess various pharmacological activities (Liu et al. 2004) and is prescribed for improving blood circulation. In a study by Wang et al. anti-hyperlipidemic effect of trillin, a steroidal saponin from *D. nipponica* was studied in hyperlipidemia-induced rats by using different biochemical assays. The blood levels of TC, TG, LDL and HDL were increased in rats fed with high-fat diet. The intra-peritoneal administration of trillin at 0.5 mg/kg dose significantly restored the levels of TC, TG, LDL and HDL to a normal condition. Moreover, the administration of trillin in rats exhibited beneficial effects in improving the levels of lipid peroxidation and superoxide dismutase activity. The results suggested that trillin could be used as a therapeutic drug against hyperlipidemia and cardiovascular diseases in future (Wang et al. 2012).

Euphorbiaceae

Phyllanthus buxifolius (Blume) Müll. Arg. is a medicinal plant that has been utilized for the treatment of various types of diseases especially for its antilipidemic and anticholesterolemic effects by Indonesian people. The aim of the study by Wardah et al. was to investigate the activities of *P. buxifolius* leaf powder supplementation in broiler feed on intracellular lipid accumulation, serum leptin and meat cholesterol levels. Group I was fed commercial feed and Group II was fed commercial feed with 5% powdered leaves of *P. buxifolius*. The results showed that the accumulation of intracellular lipids, serum leptin levels, fat and cholesterol of meat and abdominal fat weight of chickens significantly decreased with *P. buxifolius* leaf powder supplementation. The leaves of *P. buxifolius* were reported to have various secondary metabolites such as polyphenols, alkaloids, quinones, steroids and triterpenoids (Wardah et al. 2007). Among these compounds flavonoids are known to be good antioxidants (Gonzalez-Paramas et al. 2004) and suppress the synthesis of fatty acids (Rodrigues et al. 2005) and adipogenesis (Kuppusamy and Das 1994). A previous research revealed that the presence of polyphenols as well as flavonoids in chicken diet significantly reduced hyperlipidemia (Xia et al. 2010). Moreover, saponins are known to inhibit the absorption of fat by the intestine (Dong et al. 2007) and tannins inhibit the digestion and absorption of protein (Matsui et al. 2006). The potential effect of *P. buxifolius* could be attributed to the presence of these compounds (Wardah et al. 2012).

Fabaceae

Antihyperlipidemic activity of methanol extract of *Bauhinia variegata* (Linn.) leaves was evaluated in Triton WR-1339 induced hyperlipidemic rats. According to the biological activity fractionation assay, butanol fraction of methanol extract was further fractionated with column chromatography. By comparison with standard drug fenofibrate, subfraction D showed significant decrease in serum cholesterol, TG, LDL and VLDL levels as well as increase in HDL concentration at the dose of 65 mg/kg (Kumar et al. 2012b).

Luo et al. investigated the potential effects of the stilbenes containing extract-fraction obtained from *Cajanus cajan* L. (sECC) in diet-induced hypercholesterolemic Kunming mice. Simvastatin was used as a reference drug. The activities of sECC were evaluated by monitoring serum and liver lipid profile and serum SOD activity. For the further exploration of the mechanism of action, hepatic HMG-CoA reductase, cholesterol 7α-hydroxylase (CYP7A1), and LDL receptor expressions in cholesterol homeostasis were analyzed by reverse transcription PCR. After the treatment procedure, serum and hepatic TC, TG levels of serum and liver were significantly decreased by sECC. Serum LDL cholesterol decreased and the activities of serum superoxide dismutase increased. Atherogenic index and body weight were also significantly reduced. The mRNA expressions of HMG-CoA reductase, CYP7A1, and LDL-receptor were significantly enhanced with sECC treatment, while those expressions were suppressed by the hypercholesterolemic diet. The results indicated that sECC at 200 mg/kg/day reduced the atherogenic properties of dietary cholesterol in mice (Luo et al. 2008).

According to ethnobotanical data, *Mucuna pruriens* L. (Velvet bean), one of the tropical legumes, was empirically used to improve the cholesterol profile. After previous findings regarding the antioxidant activity of *M. pruriens*, its potential anticholesterolemic effect was investigated by Ratnawati and Widowati. For the determination of the anticholesterol activity, crude extract and ethyl acetate fraction was prepared. Simvastatin and vitamin E were used as reference drugs. After the administration of the test samples, the plasma levels of TC, LDL, TG and HDL were measured in hypercholesterolemic rats induced by high-fat diet. The treatment groups were administered by crude extract and ethyl acetate fraction, simvastatin and vitamin E for 10 days. Ethyl acetate fraction was found to decrease total cholesterol at dose of 15 mg/kg, while LDL-cholesterol reduced at 60 mg/kg dose. Two hundred mg/kg of crude extract increased HDL-cholesterol and decreased the TG levels (Ratnawati and Widowati 2011).

Flacourtiaceae

Casearia sylvestris S.W. was reported to be used in folk medicine as an antiseptic, cicatrizant in skin diseases and as a topical anesthetic (Hoehne 1939). The methanolic extract was investigated for its antihyperlipidemic activity at doses of 125–500 mg/kg in olive oil-loaded mice. Acute treatment provided inhibition in the TG and serum lipase (Schoenfelder et al. 2008).

Lamiaceae

Dracocephalum kotschyi Boiss. is a herbaceous plant native to Iran (Rechinger 1982). The ethnobotanical data revealed that hydroalcoholic extract and polyphenolic fraction of *D. kotschyi* have been used to decrease the plasma lipid levels. Administration of the hydroalcoholic extract (120 mg/kg) and polyphenolic fraction (40 mg/kg) of the leaves of *D. kotschyi* to the rats provided a notable decrease in blood triglyceride, total cholesterol and LDL-cholesterol concentrations and increase in HDL-cholesterol level (Sajjadi et al. 1998).

The aim of another study by Berrougui et al. was to show the beneficial effects of the water extracts of *Marrubium vulgare* L. for the treatment of cardiovascular diseases by preventing the lipid peroxidation of human-LDL and promoting HDL-mediated cholesterol efflux by using *in vitro* techniques. The outcome of the present research provided that *M. vulgare* could be a good source of natural antioxidants, which inhibit LDL oxidation and improve reverse cholesterol transport. Therefore, it could be used to prevent the development of cardiovascular diseases (Berrougui et al. 2006).

The chemical composition and hyperlipidemic and antioxidant activities of total flavonoids of the leaves of *Perilla frutescens* (L.) Britton (TFP) were investigated in the hyperlipidemia rats induced by a high-fat diet. According to HPLC analysis, TFP was found to be rich in apigenin with a smaller amount of luteolin. Oral administration of TFP (50–200 mg/kg) was highly effective in decreasing the serum TC, TG, LDL levels, and the lipid accumulation of the adipose tissue, increasing the serum HDL levels and antioxidant enzyme effect. The results suggested that TFP could be used as a potential food additive for the prevention of hyperlipidemic diseases (Feng et al. 2011).

Based on its traditional utilization, antidiabetic, antihyperlipidemic and antioxidant effects of the methanol extract of *Tectona grandis* Linn. f. flowers were investigated in STZ-induced diabetic rats. Administration of *T. grandis* markedly decreased blood glucose concentrations and increased body weight, serum insulin, haemoglobin and total protein levels in STZ-induced diabetic rats as well as reversed lipid profiles and antioxidants levels to near normal (Ramachandran et al. 2011).

The leaves of *Thymbra spicata* var. *spicata* L. are very popular as a remedy to combat hypercholesterolaemia (Baser et al. 1986). In order to evaluate the antihypercholesterolaemic, antioxidant and antisteatohepatitic activities of the plant, diethyl ether, ethyl acetate and aqueous extracts were prepared. The effects on the plasma TC, HDL, LDL, TG and glucose; MDA and reduced GSH; erythrocyte SOD and CAT activity of the plant extracts were evaluated using *in vivo* models in mice fed with high-fat diet which induces increase in plasma TC, TG, LDL, MDA levels. Administration of diethyl ether extract with a high-fat diet decreased the levels of TC, LDL, TG and MDA, while increased HDL levels along with GSH, SOD and CAT activities compared to the high-fat diet group. The activity of the other plant extract, remaining an aqueous extract, was similar to that of the diethyl ether extract. According to the results, it is suggested that the diethyl ether extract and partially remaining aqueous extract of *T. spicata* var. *spicata* exhibit significant antihypercholesterolaemic, antioxidant and antisteatohepatitic activities. The activity potential of the extract was attributed to the presence of carvacrol according to the HPLC analysis (Küpeli Akkol et al. 2009).

Malvaceae

The flower extract of *Hibiscus sabdariffa* L. was reported to reduce blood pressure in both man and rats (Haji Faraji and Haji Tarkhani 1999, Onyenekwe et al. 1999). Chang et al. assessed the antioxidant effect of *Hibiscus* anthocyanins by evaluating their activities on LDL oxidation and anti-apoptotic abilities *in vitro*. The results obtained from the 3-(4,5-dimethylthiazol-2-yl)-2,5-diphenyltetrazolium bromide (MTT) assay, Leukostate staining analysis and Western blotting demonstrated that *Hibiscus* anthocyanins could inhibit oxLDL-induced apoptosis and might be used as a chemopreventive agent (Chang et al. 2006).

Previous research demonstrated that *H. sabdariffa* leaves possess hypoglycemic (Sachdewa et al. 2001), antioxidant (Ochani and D'Mello 2009), and hypolipidemic effects (Ochani and D'Mello 2009, Gosain et al. 2010). Chen et al. explored the anti-atherosclerotic effect of a flavonoid enriched extract of *H. sabdariffa* leaves (HLP). The inhibitory effect of HLP on oxidation and lipid peroxidation of LDL was investigated by using *in vitro* methods. HLP showed potential in reducing intracellular lipid accumulation in oxidized LDL-induced macrophage J774A.1 cells. These activities of HLP were reported to be probably mediated via liver-X receptor a/ATP-binding cassette transporter A1 pathway. The results indicated that HLP could be used for the development of an anti-atherosclerotic agent (Chen et al. 2013).

Moraceae

In Chinese medicine mulberry (*Morus alba* L.) leaves, bark and branches have been utilized for the treatment of fever, protecting the liver, improving eyesight, strengthening joints, facilitating discharge of urine and reducing blood pressure (Zhishen et al. 1999). In a previous study, Doi et al. (2000) reported that 1-butanol extract of mulberry leaves had antioxidant activity. Katsube et al. investigated the LDL antioxidant effect of the extracted compounds of mulberry leaves. Quercetin-3-(6''-malonyl)-glucoside, rutin and isoquercitrin were identified as the major LDL antioxidant compounds by LC-MS and NMR. The amounts of these compounds were evaluated by HPLC. The results demonstrated that quercetin-3-(6"-malonyl)-glucoside and rutin were the predominant flavonol glycosides in the mulberry leaves (Katsube et al. 2006).

Another study was carried out by Ma et al. for the investigation of the effect and possible mechanism of flavonoids extracted from *Morus indica* L. (FMI) on blood lipids and glucose in high-fat diet (HFD) and STZ-induced hyperlipidemia-diabetic rats. FMI treatment reduced TC, TG, and LDL, increased HDL and significantly decreased the atherosclerosis index. FMI also downregulated the elevation of blood glucose induced by STZ. Moreover, it increased hepatic SOD activity and reduced hepatic MDA content. The expression of hepatic CYP2E1 was significantly reduced while the expression of GLUT-4 in skeletal muscles was increased by the administration of FMI. The results suggested that FMI might be used for the prevention of hyperlipidemia and hyperglycemia (Ma et al. 2012).

Moringaceae

Leaves of *Moringa oleifera* Lam. were examined for its hypolipidemic, antioxidant, anticoagulant, platelet anti-aggregatory and anti-inflammatory activities in hyperlipidemic male Wistar rats. The hydroalcoholic extract of *M. oleifera* was applied orally to the rats for a period of 28 days. The results of the present study demonstrated that *M. oleifera* showed remarkable reduction in increased levels of body weight, TC, TG, LDL, VLDL and notable increase in HDL at 100 and 200 mg/kg/b.wt doses. Therefore, it was concluded that *M. oleifera* could be prescribed for the treatment of coronary artery diseases (Rajanandh et al. 2012).

Myrtaceae

The seeds of *Eugenia jambolana* Lam. were reported to have hypoglycemic, anti-inflammatory, neuropsychopharmacological, antibacterial, anti-HIV and antidiarrhoeal effects (Bhatia and Bajaj 1975, De Lima et al. 1998) and contain various active components including flavonoids, gallic acid, ellagic acid, glycosides, triterpenoids and saponins. Anti-hyperlipidemic efficacy of *E. jambolana* seed kernel (EJs-kernel) was investigated in STZ-induced diabetic rats by Ravi et al. In EJs-kernel treated animals restored elevated levels of TC, PL, TG and free fatty acid to near normal. The plasma lipoproteins (HDL, LDL, VLDL-cholesterol) and fatty acid composition were also reverted back to near normal by EJs-kernel administration. Consequently, it was concluded that, EJs-kernel provided hypolipidemic effect, which could be attributed to the presence of flavonoids, saponins, glycosides and triterpenoids in the extract (Ravi et al. 2005).

The study by Kasetti et al. was undertaken to determine potent antihyperglycemic and antihyperlipidemic effects of the aqueous extract of the seeds of *Syzygium alternifolium* (Wight) Walp. by using bioassay guided fractionation assay. Administration of the phenolic fraction at a dose of 50 mg/kg/day for 30 days to STZ diabetic rats resulted in a significant decrease in blood glucose, glycosylated haemoglobin with a remarkable rise in plasma insulin concentration. Fraction C also demonstrated antihyperlipidemic activity by decreasing the serum TC, TG, LDL, VLDL levels along with the enhancement of HDL level in diabetic rats. A notable reduction in the effects of SGOT, SGPT, creatinine, ALP and low levels of serum urea and creatinine in diabetic rats revealed the protective role of the fraction C against liver and kidney damage (Kasetti et al. 2010).

Pinaceae

A Mongolian medicinal plant, *Abies sibirica* Ladeb, was shown to have an inhibitory effect on lipase activity and oxidation of LDL, which are preventative factors for arteriosclerosis. The methanol extract was subjected to silica gel column chromatography for fractionation and its active constituents were screened. Twenty terpenoids were isolated from the lipid soluble fractions and triterpenes were demonstrated to have moderate lipase inhibitory and LDL anti-oxidative activities (Handa et al. 2013).

Plantaginaceae

Kamesh and Sumathi, attempted to assess the activity of the alcoholic extract of *Bacopa monniera* (L.) Pennell on high cholesterol diet-induced (HCD) rats by the evaluation of lipid and lipoprotein status, antioxidant status, cardiac marker enzyme and histological changes of aorta. *B. monniera* treatment (40 mg/kg) significantly reduced the levels of TC, TG, PL, LDL, VLDL, atherogenic index, LDL/HDL ratio, and TC/HDL ratio and notably enhanced the concentration of HDL as compared with HCD induced rats. Effects on the liver antioxidant status were significantly increased with reduction in the level of lipoproteins in *B. monniera* treated group. Administration of *B. monniera* notably reduced the effect of serum cardiac marker enzymes SGOT, LDH and creatine phosphokinase (CPK). The results of the study indicated that *B. monniera* may be useful for the treatment of hypercholesterolemia (Kamesh and Sumathi 2012).

Poaceae

Lemongrass oil, obtained from the leaves of *Cymbopogon citratus* (DC.) Stapf, was reported to reduce cholesterol in hypercholesterolemic subject (Jones et al. 2007). For the assessment of the antihyperlipidemic activity of lemongrass oil, dexamethasone induced hyperlipidemic rats were used. Lemongrass oil (at doses of 100 and 200 mg/kg) was administered to the rats and demonstrated remarkable inhibition in the serum levels of cholesterol and triglycerides. Atherogenic index was found to be near to normal levels and the antihyperlipidemic effect of the lemongrass oil was comparable with the reference drug atorvastatin. The possible mechanism was reported to be associated with the decrease in lecithin cholesterol acetyl transferase effect (Kumar et al. 2011).

Ramjiganesh et al. analyzed the hypocholesterolemic mechanisms of corn (*Zea mays* L.) husk oil (CoHO) in male Hartley guinea pigs fed diets containing increasing doses of CoHO and cholesterol. Plasma LDL cholesterol concentrations were found to be decreased with increasing doses of CoHO. Moreover, intake of CoHO resulted in lower hepatic total and esterified cholesterol and TAG concentrations compared with the control group. Hepatic β-hydroxy-β-methylglutaryl-coenzyme A reductase activity was not modified by CoHO intake while cholesterol 7α-hydroxylase was upregulated. On the other hand, CoHO intake caused hepatic acyl coenzyme A cholesterol acyltransferase activity to be downregulated in a dose-dependent manner. CoHO intake resulted in increased faecal cholesterol. The data suggested that CoHO exerted its hypocholesterolemic activity by decreasing cholesterol absorption and increasing bile acid output (Ramjiganesh et al. 2000).

Polygonaceae

A study was conducted to assess the hypolipidemic activities of *Fagopyrum esculentum* Moench (Buckwheat) leaf and flower mixture (BLF) in male Sprague Dawley rats fed a high-fat diet. The plasma TC and TG levels were notably lower in the BLF group than in the other groups. Hepatic cholesterol and TG concentrations of the BLF group were found to be similar to those of the control group. BLF enhanced the faecal TG and acidic sterol concentrations in the BLF group. The results suggested that the beneficial effect of this buckwheat on plasma and hepatic lipid profiles in high-fat fed rats is partly mediated by higher excretion of faecal lipids. This was probably due to the synergistic effect of phenolic compounds and fibre present in the BLF (Lee et al. 2010).

The activity on serum glucose and lipid profile of *Rumex patientia* L. seeds was investigated in STZ-diabetic Wistar rats. Serum glucose notably decreased in *R. patientia* seed fed diabetic rats when compared to untreated diabetic animals. Although serum HDL-cholesterol markedly increased and LDL-cholesterol significantly reduced, serum TC and TG levels did not exhibit remarkable reductions in diabetic rats treated with *R. patientia*. *R. patientia* also attenuated the elevated MDA content and reduced activity of SOD in hepatic tissue (Sedaghat et al. 2011).

Rosaceae

The antihyperlipidemic and antioxidant activities of *Crataegus pinnatifida* Bunge (Chinese hawthorn) pectic oligosaccharide (HPOS) were investigated using a model of hyperlipidemia induced by a high-fat diet in mice by Li et al. HPOS remarkably reduced the serum levels of TC and TG and inhibited the accumulation of body fat. It notably increased the effect of superoxide dismutase to suppress oxidative reactions and inhibited the synthesis and accumulation of MDA in serum. It was concluded that HPOS could be a valuable agent in the development of nutritional and drug therapies to combat cardiovascular diseases (Li et al. 2010).

Pyrus biossieriana Buhse grows wild in northern Iran and Turkmenistan. The leaves are used for treatment of inflammation of the bladder, bacteriuria, high blood pressure and urinary stones and as a diuretic (Zargari 1996). The leaves were reported to contain high amounts of arbutin, a naturally occurring derivative of hydroquinone (Azadbakht et al. 2004). The antihyperglycemic, antihyperlipidemic and antioxidant effects of *P. biossieriana* leaf extract were evaluated by using alloxan-induced rat model of hyperglycaemia. Administration of the the extract markedly reduced serum glucose, TG and cholesterol levels and increased serum insulin levels. Serum antioxidant levels were significantly higher in rats treated with *P. biossieriana* extracts at doses of 500 and 1000 mg/kg/day (Shahaboddin et al. 2011).

Rutaceae

Recently, the effects of *Citrus bergamia* Risso (bergamot)-derived polyphenolic fraction (BPF) was investigated by using animal models of diet-induced hyperlipidemia as well as in patients suffering from metabolic syndrome. BPF was found to reduce serum cholesterol, TG and glycaemia (Mollace et al. 2011). Besides these results, Gliozzi et al. designed a prospective, open-label, parallel group, placebo-controlled study to explore the mechanism of action of bergamot-derived extract in comparison with statins in 77 patients with metabolic syndrome. Patients were randomly divided to a control group (n = 15), two reference groups receiving orally administered rosuvastatin, a group receiving BPF and a group receiving BPF plus rosuvastatin. Both Rosuvastatin and BPF decreased TC, LDL, the LDL/HDL ratio and urinary mevalonate in hyperlipidemic patients when compared to the control group. The cholesterol lowering effect was accompanied by reductions of MDA, oxyLDL receptor LOX-1 and phosphoPKB. Combination of BPF and rosuvastatin markedly increased rosuvastatin-induced effect on serum lipemic profile compared to rosuvastatin alone (Gliozzi et al. 2013).

In vitro antioxidant and antihyperlipidemic activities of *Toddalia asiatica* (L.) Lam. leaves were investigated in Triton WR-1339 and high fat diet-induced hyperlipidemic rats. Ethyl acetate extract showed significant scavenging activity on DPPH, hydroxyl and nitric oxide radicals, as well as high reducing power. This extract at the doses of 200 and 400 mg/kg was also shown to decrease the levels of serum TC, TG, LDL and significantly increase HDL in comparison with hexane and methanol extracts (Irudayaraj et al. 2013).

Solanaceae

Kou et al. (2009) analyzed the total phenol contents of *Cyphomandra betacea* Sendt., a subtropical plant, and examined the antioxidant activity of the ethanol extract of the fruit and its ethyl acetate, butanol and water fractions on 2,2-diphenyl-1-picrylhydrazyl (DPPH) and 2,2'-azino-bis(3-ethylbenzthiazoline-6-sulphonic acid) (ABTS) cation radical scavenging activity as well as the protective effect against LDL oxidation. The ethyl acetate fraction displayed the highest antioxidant activity and total phenol content. The phenolic compounds in ethyl acetate fraction inhibited copper-induced LDL oxidation (Kou et al. 2009).

Another study by Sridevi et al. was conducted to evaluate the antihyperlipidemic activity of alcoholic leaf extract of *Solanum surattense* Burm. f. in STZ-induced diabetic male albino Wistar rats. Administration of *S. surattense* to diabetic rats for 45 days significantly reversed the increase of the levels of TC, TG, PL and free fatty acids, LDL and VLDL in the plasma, liver and kidney and reduction of HDL. The antihyperlipidemic effect was attributed to the presence of alkaloids, flavonoids, tannins, triterpenoids and sterols in the extract (Sridevi et al. 2011).

Saponins from *Solanum anguivi* Lam. fruits were screened for the hypoglycemic, antiperoxidative and antihyperlipidemic activities in alloxan-induced diabetic rats. The outcome of the research demonstrated that administration of saponins remarkably decreased the elevated levels of glucose, TC, TG, LDL and increased HDL in the serum. Furthermore, saponins displayed strong inhibition of lipid peroxidation and increased the levels of antioxidant enzymes in the serum, liver and pancreas. The results suggested that saponins of *S. anguivi* fruit could improve the hypoglycemic, hypolipidemic and antioxidant conditions in alloxan-induced diabetic rats (Elekofehinti et al. 2013).

Theaceae

Preclinical research in experimental animals also demonstrated potential hypocholesterolemic activity of *Camellia sinensis* (L.) Kuntze. (green tea) or green tea extracts enriched in catechins. The mechanism of cholesterol lowering activity of green tea was proposed as the inhibition of cholesterol absorption. As faecal excretion of total lipids and cholesterol were determined to be high in the animals consuming green tea extracts (Muramatsu et al. 1986, Matsuda et al. 1986, Yang and Koo 1997, Chan et al. 1999, Yang and Koo 2000). In another study, the EGCG was shown to inhibit the uptake of cholesterol from the intestine (Chisaka et al. 1988). However, little is known about the activities of green tea on cholesterol biosynthesis. To determine the activity mechanism of green tea on plasma cholesterol concentration, a study was conducted by Bursill et al. in the cholesterol-fed rabbit. Administration of the green tea catechin extract notably lowered the cholesterol concentrations in plasma, liver and aorta compared to the control group. Cholesterol synthesis index decreased remarkably while hepatic LDL receptor activity increased. On the other hand, intrinsic capacity of cholesterol absorbtion from the intestines did not change. Consequently it was concluded that catechins of green tea decreased plasma, liver and aortic cholesterol by reducing cholesterol synthesis and upregulating the hepatic LDL receptor in the cholesterol-fed rabbit (Bursill et al. 2007). Epigallocatechin gallate is one of the important polyphenols of green tea, *Camellia sinensis* (L.) O. Kuntze. (Theaceae). Epidemiological studies have shown that drinking green tea is associated with lower levels of plasma TG (Kono et al. 1992, Stensvold et al. 1992, Imai et al. 1995).

Thymelaeaceae

According to the *in vitro* and *in vivo* experiments *Phaleria macrocarpa* (Scheff.) Boerl was shown to reduce cholesterol level. The study by Chong et al. investigated the anti-hypercholesterolemic effect of *P. macrocarpa* fruit. In the *in vivo* study, *P. macrocarpa* extract was administered to the cholesterol enriched-diet-induced hypercholesterolemic male Sprague Dawley rats. Simvastatin was used as a reference drug. In the *in vitro* study, in the presence of *P. macrocarpa* extract or simvastatin liver hepatocellular cells (HepG2) cells were cultured in serum-free RPMI supplemented with 0.2% BSA with or without LDL. *P. macrocarpa* extract (20 mg/kg) markedly decreased the body weight gain, total cholesterol, TG, HDL and LDL levels and upregulated hepatic LDL receptor along with PCSK9 proteins of hypercholesterolemic rats. The study provided evidence to the potential use of *P. macrocarpa* fruit in controlling the body weight of obese people and treating hypercholesterolemia (Chong et al. 2011).

Ulvaceae

In some countries, *Ulva pertusa* Kjellman, green alga, is used as an important food source because of its components including vitamins, trace elements and dietary fibres (Lahaye and Jegon 1993). Furthermore, it has been used in traditional Chinese medicine for the treatment of hyperlipidemia, sunstroke and urinary diseases. Polysaccharides of *U. pertusa* displayed antihyperlipidemic actions when it was degraded into low molecular weights fractions (Yu et al. 2003). In another study by Qi et al. antihyperlipidemic activity of ulvan and high sulphate content ulvan (HU) was investigated in mice. By the administration of both ulvan and HU, TG and LDL levels significantly reduced. However, obvious differences in antihyperlipidemic effects between natural ulvan and HU were detected. The antihyperlipidemic activity of HU was higher, when compared to natural ulvan (Qi et al. 2012).

Recently, polysaccharide type compounds were reported to have antihyperlipidemic activity including the polysaccharides of *U. pertusa* and *U. lactuca* L. (Sathivel et al. 2008, Qi et al. 2012), chitosan and chitosan derivatives (Muzzarelli and Muzzarelli 2006). In light of this knowledge, *in vitro* antioxidant and *in vivo* antihyperlipidemic effects of polysaccharides of Sea cucumber, *Apostichopus japonicus* Selenka, prepared by protease hydrolysis method, were evaluated by Liu et al. The antioxidant activitiy of *A. japonicus* was evaluated by using the assays of scavenging DPPH, hydroxyl and superoxide radicals, and reducing power. The results indicated that *A. japonicus* had potent antioxidant activity. By administration of *A. japonicus* serum TC, TG and LDL significantly decreased and HDL notably increased in hyperlipidemic Wistar rats. Phytochemical analysis revealed that *A. japonicus* was mainly composed of glucosamine, galactosamine, glucuronic acid, mannose, glucose, galactose and fucose. These results suggested that *A. japonicus* may be used as a potential candidate for a therapeutic agent in the treatment of hyperlipidemia (Liu et al. 2012).

Urticaceae

A study was conducted to test the antidiabetic effect of the flavonoid rich fraction of *Pilea microphylla* (L.) Liebm. (PM1). *P. microphylla* inhibited dipeptidyl peptidase IV (DPP-IV) *in vitro* also provided dose-dependent reductions in glucose excursion, body weight, PG, TG and TC content in high-fat STZ-induced diabetic mice. HPLC analysis of *P. microphylla* revealed the presence of chlorogenic acid, rutin, luteolin-7-*O*-glucoside, isorhoifolin, apigenin-7-*O*-glucoside, and quercetin. Consequently, *P. microphylla* exhibited an antidiabetic effect and improved antioxidant levels probably due to the presence of its phenolic ingredient (Bansal et al. 2012).

Zingiberaceae

According to previous reports, Zingiberaceae plants, as well as their isolated constituents have been demonstrated to possess hypolipidemic activities. For instance, *Alpinia zerumbet* (Pers.), *Alpinia officinarum* Hance and curcumin from *Curcuma longa* L. are considered potential hypolipidemic agents (Ramirez-Tortosa et al. 1999, Shin et al. 2004, Olszanecki et al. 2005, Lin et al. 2008). The effect of *Alpinia pricei* Hayata on lipid profiles of hamsters fed a chow-based hypercholesterolemic diet (HCD) was investigated. Seventy per cent ethanol extract of *A. pricei* was prepared and 250 and 500 mg/kg doses were administered to hamsters fed HCD. The results showed that *A. pricei* exhibited suppressive and preventive potencies against hypercholesterolemia by lowering serum TC and LDL concentrations, reducing thiobarbituric acid reactive substances (TBARS) and alanine aminotransferase (ALT) activities of hamsters fed a HCD diet (Chang et al. 2010).

Nadeem et al. (2012) studied the activity of ethanol extract of *A. calcarata* Roscoe rhizomes on the serum lipid and leptin concentrations of hyperlipidemia induced-male wistar rats by high-fat diet. One hundred, 200 and 300 mg/kg doses of ethanol extract of *A. calcarata* were administered to the high fat-diet-induced hyperlipidemic rats for 30 days. Atorvastatin (10 mg/kg; p.o.) was used as a reference drug. The results revealed that *A. calcarata* decreased rat weight gain, TG, TC, LDL, VLDL, TP and leptin concentrations. The results suggested that *A. calcarata* had good pharmacological potential for the prevention of hyperlipidemia (Nadeem et al. 2012).

Zygophyllaceae

Ghoul et al. (2012) investigated the antihyperglycemic, antioxidant and antihyperlipidemic effects of the aqueous extract of *Zygophyllum album* Linn. in STZ-induced diabetic mice. Oral administration of the aqueous extract of *Z. album* at the doses of 100 and 300 mg/kg reduced the blood glucose, TC, TG, LDL and VLDL concentrations. According to the findings obtained in this study, aqueous extract of *Z. album* displayed the antidiabetic and antihypercholesterolemic activities through its antioxidant properties (Ghoul et al. 2012).

Plant constituents having antihyperlipidemic activity

Flavonoids are the compounds which exhibit a wide range of biological activities (Bruneton 1995). The anti-atherosclerosis effects of flavonoids have been confirmed (Fuhrman et al. 1997, Vaya et al. 1997, Renaud and de Lorgeril 1992, Ruiz-Larrea et al. 1997). Moreover, their inhibitory actions on LDL oxidation *in vitro* and ability to decrease *ex vivo* LDL susceptibility to oxidation were reported (Vaya et al. 2003). For the evaluation of the relationship of the flavonoids' structure to the inhibition of LDL oxidation was assessed on 20 flavonoid compounds by using *in vitro* methods including copper ion, 2,2'-azobis (2-amidino-propane) dihydrochloride induction. The most effective inhibitors were found to be flavonols and/or flavonoids with two adjacent hydroxyl groups at ring B. Isoflavonoids were more active inhibitors than the other flavonoid subclasses with a similar structure. A remarkable inhibitory effect was achieved by the substitution of ring B with hydroxyl group(s) at 2' position (Vaya et al. 2003).

A study by Srinivasan and Pari (2013), was conducted to assess the antihyperlipidemic effect of diosmin (DS), a flavone, in STZ-induced diabetic male albino Wistar rats. DS was applied intragastrically at 100 mg/kg dose for 45 days. The concentration of plasma and tissue lipids reduced, while HDL cholesterol increased. The altered effects of lipid metabolic enzymes were restored to near normal. It was suggested that DS could ameliorate lipid abnormalities in experimental diabetes (Srinivasan and Pari 2013).

The effects of cinnamaldehyde, an active and major compound in cinnamon, was investigated on glucose metabolism and insulin resistance in C57BLKS/J db/db mice. By the application of cinnamaldehyde reduction in the serum levels of fasting blood glucose and insulin with elevated concentration of serum HDL-C level was observed. It was concluded that cinnamaldehyde has antihyperglycemic and antihyperlipidemic actions in db/db mice and could be used as a beneficial agent for the treatment of type-2 diabetes (Li et al. 2012).

Berberine (BBR), an alkaloid, has been used for years as a folk medicine to treat many illnesses including bacterial infection and gastro-intestinal disorders (Zeng et al. 2003, Jagetia et al. 2004, Kong et al. 2004, Cicero et al. 2007). Previous studies showed the significant cholesterol-lowering efficacy of BBR compared to statin drugs by upregulating LDL receptor mediated cholesterol clearance. Moreover, BBR has also been reported to decrease plasma TAG concentrations in humans and animals (Kong et al. 2004, Abidi et al. 2005 and 2006, Tang et al. 2006). Since BBR and phytosterols reduce cholesterol through different mechanisms, it was hypothesized that the combination of BBR and Plant Stanols (PS) would synergistically lower the serum cholesterol and TAG levels. In order to determine the beneficial effects of the combination of BBR and PS on plasma lipid profiles an *in vivo* study was conducted in male Sprague Dawley rats. Animals were divided into four groups and fed a cornstarch–casein–sucrose-based high-cholesterol and high-fat diet. Three treatment groups were administered either BBR, PS, or the combination of both (BBRPS). At the end of six weeks, the animals were sacrificed and blood and organ samples were collected. Lipid analysis demonstrated that PS and combination treatment of BBRPS lowered TC, non-HDL-cholesterol and liver cholesterol when compared to the control group, while BBR had no effect on both TC and non-HDL-cholesterol. BBR decreased the plasma TG levels, while PS had no activity. On plasma TAG, BBRPS displayed an additive activity of BBR and PS. BBR and PS, either alone or in combination, were devoid of any toxic effect as evaluated by plasma levels of hepatic biochemical parameters. The outcome of the present research revealed that combination of BBR and PS synergistically reduced plasma cholesterol levels and remarkably decreased liver cholesterol, without any toxic effect (Jia et al. 2008).

The study by Wang et al. was designed to evaluate the activity and mechanism of BBR, PS and their combination on plasma lipids in Male Golden Syrian hamsters fed a cornstarch–casein–sucrose-based diet containing cholesterol and fat. Three treatment groups were supplemented with BBR, PS, or a combination of both (BBRPS). At the end of four weeks, plasma lipids, cholesterol absorption and synthesis, and gene and protein expressions in the liver and small intestine were analyzed. According to the results, BBR and PS markedly reduced plasma TC and non HDL-cholesterol levels, and BBRPS notably improved cholesterol-lowering activity compared to BBR or PS alone. Further examinations indicated that both BBR and PS inhibited the absorption of cholesterol but increased its synthesis when combined synergistic action was observed. According to PCR analysis, BBR upregulated sterol 27-hydroxlase gene expression and BBRPS enhanced both cholesterol-7α-hydroxylase and sterol 27-hydroxlase gene expressions. BBR and PS also synergistically reduced plasma TAGs. The outcome of the study suggested that the cholesterol-lowering action of BBR might involve a combination of inhibition of cholesterol absorption and stimulation of bile acid synthesis (Wang et al. 2010).

According to previous findings, foods added with PS/sterol esters decreased LDL-cholesterol level (Katan et al. 2003, Berger et al. 2004, Demonty et al. 2009). Indeed, when added to fat based food, phytosterol esters reduces LDL-cholesterol (Miettinen et al. 1995, Amundsen et al. 2002, Katan et al. 2003). In a previous research, anti-hypercholesterolemic effect of softgel capsules containing phytosterols (2 g) was investigated in a double-blinded, randomized, placebo-controlled crossover study. The subjects were randomized into two four-weeks intervention periods. At the end of this period, no significant difference in total- or LDL-cholesterol between the phytosterol and the placebo period were observed. Daily intake of softgel capsules did not decrease total- or LDL-cholesterol significantly which may emphasize the importance of choosing a suitable dosage-delivery system to achieve optimal cholesterol lowering activity (Ottestad et al. 2013).

A randomized, placebo-controlled, crossover trial was conducted for the determination of the lipid-altering activity of a softgel capsule dietary supplement, providing esterified plant sterols/stanols 1.8 g/d, in 28 participants with primary hypercholesterolemia. After a five week period of National Cholesterol Education Program (NCEP) Therapeutic Lifestyle Changes (TLC) diet, subjects received double-blinded placebo or sterol/stanol softgel capsules for six weeks and then crossed over to the opposite product for six weeks. The incorporation of softgel capsules into the NCEP TLC diet produced favourable changes in atherogenic lipoprotein cholesterol levels in these subjects with hypercholesterolemia (Maki et al. 2013).

Herbal formulations having antihyperlipidemic activity

The antihyperlipidemic activity potential of Ogi-Keishi-Gomotsu-To-Ka-Kojin (OKGK), a Kampo medicine, containing Astragali radix, Cinnamomi cortex, Paeoniae radix, Zingiberis rhizoma, Zizyphi fructus, Ginseng radix rubra, was investigated in experimentally-induced hyperlipidemic rats by a cholesterol and fat-enriched diet. Oral administration of OKGK at the doses of 0.69 or 1.38 g/kg/day remarkably decreased the high levels of serum TG and phospholipids (PL). OKGK was also given as a nutritional supplement (1.25%) in the diet. It was found to suppress the increase of serum TG and PL in rats with hypercholesterolemia. OKGK decreased acetic acid and oleic acid incorporation into TG and PL, which suggests OKGK suppresses TG and PL syntheses in the liver. Moreover, OKGK enhanced the activities of LPL and hepatic TG lipase (HTGL). The results suggested the significant activity of OKGK in the treatment of hypertriglyceridemia by inhibiting triglyceride synthesis in the liver and stimulating the hydrolysis of TG in lipoprotein (Wu et al. 1997).

In a previous study, Nomura et al. assessed the activities of Saiko-ka-ryukotsu-borei-to (CHAIHU-JIA-LONGGU-MOLI-TANG), a Japanese herbal medicine, containing Bupleuri radix, Pinelliae tuber, Hoelen, Cinnamomi cortex, Scutellariae radix, Zizyphi fructus, Ginseng radix, Fossilia ossis mastodi, Zingiberis rhizoma, Ostreae testa, in patients with elevated triglyceride levels (Mizoguchi et al. 2002). After the administration of Saiko-ka-ryukotsu-borei-to, levels of CD62P, PMPs, sE-selectin, and anti-oxidized LDL antibody significantly decreased. Concentrations of TG, TC and MMPs also reduced. These results indicated that Saiko-ka-ryukotsu-borei-to could be used to prevent the development of vascular complications in hyperlipidemia patients (Nomura et al. 2001).

In a recent study the aqueous extracts of a traditional prescription, Dae-Shiho-Tang, containing *Paeonia lactiflora* Pall (Paeoniaceae), exhibited a lowering activity on the total cholesterol level in experimentally-induced hyperlipidemic rats. Moreover, methanol extract obtained from *P. lactiflora* displayed the lowering effect of TC, TG and LDL levels. For the identification of the active principle responsible from the activity, methanol extract of *P. lactiflora* was fractionated according to biological activity-guided fractionation assay. The component isolated from the active fraction was elucidated as paeoniflorin, a monoterpene glycoside. Paeoniflorin also showed a lowering effect on total cholesterol, LDL and triglyceride levels in experimentally-induced hyperlipidemic rats at the dose of 200 and 400 mg/kg (Yang et al. 2004).

An ayurvedic herbal formulation called Arborium Plus containing *Hippophae rhamnoides* L. (Elaeagnaceae) and *Rhododendron arboreum* Smith (Ericaceae) was reported for its hepatoprotective (Verma et al. 2011) and hypolipidemic activity in experimentally induced hypercholesterolemic rabbits (Murty et al. 2010). However, no study was reported on the antihyperlipidemic activity of *R. arboreum*. Therefore, a study was carried out by Verma et al. on the the ethanol extract of *R. arboreum* flowers. Normal and STZ-induced diabetic rats were treated with the fractions of ethanol extract. In a short term study, fraction 3 provided a remarkable fall in blood glucose level. Oral application of fraction 3 once daily for 30 days resulted in a significant reduction in blood glucose level, haemoglobin A1C, serum urea and creatinine accompanied with a notable increase in insulin concentration. Moreover, fraction 3 exhibited antihyperlipidemic activity by lowering the serum TC, TG, LDL cholesterol and VLDL cholesterol levels with elevation of HDL cholesterol in the diabetic rats (Verma et al. 2012a).

Conclusion

Hypercholesterolemia is a common health problem, that generally results in atherosclerosis and coronary heart disease. Diabetes is also associated with alterations in the plasma lipid profile and with an increased risk of heart diseases (Srinivasan et al. 2013). The present chapter reveals that treatment with herbal drugs has an important role in lipid metabolism. Previous studies have shown that plants provide a significant decrease in the concentrations of TC, LDL-cholesterol, TBARS, ALT and many other parameters associated with hyperlipidemia. The beneficial effect of plants is sometimes mediated by their antioxidant, antidyslipidemic and hypoglycemic activities (Chang et al. 2010). Indeed, some plants reduce lipid peroxidation, improve endothelial function, enhance lipolysis and contribute a remarkable decrease in serum lipid levels. Statins, fibrates, niacin, bile acid sequestrants (resins), Zetia, Orlistat are some allopathic hypolipidemic drugs, however many side effects have been reported including hyperuricemia, diarrhoea, nausea, myopathy, gastric irritation, flushing, dry skin and abnormal liver function (Irudayaraj et al. 2013). In recent years there has been a strong desire for natural hypolipidemic substances derived from medicinal plants. A number of plants have been utilized traditionally to control hyperlipidemia and thereby for the treatment of various cardiovascular diseases. Furthermore, many studies demonstrated that constituents isolated from plants such as polyphenols, alkaloids, stilbenes and plant stanols possess antihyperlipidemic activity potential (Chen et al. 2013). Consequently, the use of herbal medicines and their active components has been considered as a novel strategy for the management of metabolic diseases and maintenance of a healthy life.

References

Abidi, P., Zhou, Y., Jiang, J.D. and Liu, J. 2005. Extracellular signal-regulated kinasedependent stabilization of hepatic low-density lipoprotein receptor mRNA by herbal medicine berberine. Arterioscler. Thromb. Vasc. Biol. 25: 2170–2176.
Abidi, P., Chen, W., Kraemer, F.B., Li, H. and Liu, J. 2006. The medicinal plant goldenseal is a natural LDL-lowering agent with multiple bioactive components and new action mechanisms. J. Lipid Res. 47: 2134–2147.
Amundsen, A.L., Ose, L., Nenseter, M.S. and Ntanios, F.Y. 2002. Plant sterol ester-enriched spread lowers plasma total and LDL cholesterol in children with familial hypercholesterolemia. Am. J. Clin. Nutr. 76(2): 338–344.
Anonymous. 1952. The Wealth of India. Raw Materials, Vol. III. Council of Scientific and Industrial Research, New Delhi, p. 127.
Arulmozhi, S., Mazumder, P.M., Lohidasan, S. and Thakurdesai, P. 2010. Antidiabetic and antihyperlipidemic activity of leaves of *Alstonia scholaris* Linn. R.Br. Eur. J. Integr. Med. 2: 23–32.
Azadbakht, M., Marston, A., Hostettmann, K., Ramezani, M. and Jahromi, M. 2004. Biological activity of leaf extract and phenolglycoside arbutin of *Pyrus boissieriana* Buhse. J. Med. Plants 3: 9–14.
Bansal, P., Paul, P., Mudgal, J., Nayak, P.G., Pannakal, S.T., Priyadarsini, K.I. and Unnikrishnan, M.K. 2012. Antidiabetic, antihyperlipidemic and antioxidant effects of the flavonoid rich fraction of *Pilea microphylla* (L.) in high fat diet/streptozotocin-induced diabetes in mice. Exp. Toxicol. Pathol. 64: 651–658.
Baser, K.H.C., Honda, G. and Miki, W. 1986. Herb Drugs and Herbalists in Turkey, Tokyo Publishing & Printing Co., Ltd., Tokyo.
Berger, A., Jones, P.J. and AbuMweis, S.S. 2004. Plant sterols: factors affecting their efficacy and safety as functional food ingredients. Lipids Health Dis. 3: 5.
Berliner, J.A., Territo, M.C., Sevanian, A., Ramin, S., Kim, J.A., Bamshad, B., Esterson, M. and Fogelman, A.M. 1990. Minimally modified low density lipoproteins stimulates monocyte endothelial interactions. J. Clin. Invest. 85: 1260–1266.

Berrougui, H., Isabelle, M., Cherki, M. and Khalil, A. 2006. *Marrubium vulgare* extract inhibits human-LDL oxidation and enhances HDL-mediated cholesterol efflux in THP-1 macrophage. Life Sci. 80: 105–112.

Bhatia, I.S. and Bajaj, K.L. 1975. Chemical constituents of the seeds and bark of *Syzygium cumini*. Planta Med. 28: 346–352.

Bracesco, N., Sanchez, A.G., Contreras, V., Menini, T. and Gugliucci, A. 2011. Recent advances on *Ilex paraguariensis* research: minireview. J. Ethnopharmacol. 136: 378–384.

Bruneton, J. 1995. Pharmacognosy, Phytochemistry, Medicinal Plants. Lavoisier Publishing, Paris, France, pp. 275–294.

Bursill, C.A., Abbey, M. and Roach, P.D.A. 2007. Green tea extract lowers plasma cholesterol by inhibiting cholesterol synthesis and upregulating the LDL receptor in the cholesterol-fed rabbit. Atherosclerosis 193: 86–93.

Chan, P.T., Fong, W.P., Cheung, Y.L., Huang, Y., Ho, W.K. and Chen, Z.Y. 1999. Jasmine green tea epicatechins are hypolipidemic in hamsters (*Mesocricetus auratus*) fed a high fat diet. J. Nutr. 129: 1094–1101.

Chang, N.W., Wu, C.T., Wang, S.Y., Pei, R.J. and Lin, C.F. 2010. *Alpinia pricei* Hayata rhizome extracts have suppressive and preventive potencies against hypercholesterolemia. Food Chem. Toxicol. 48: 2350–2356.

Chang, Y.C., Huang, K.X., Huang, A.C., Ho, Y.C. and Wang, C.J. 2006. *Hibiscus* anthocyanins-rich extract inhibited LDL oxidation and oxLDL-mediated macrophages apoptosis. Food Chem. Toxicol. 44: 1015–1023.

Chen, J.H., Wang, C.J., Wang, C.P., Sheu, C.Y., Lin, C.L. and Lin, H.H. 2013. *Hibiscus sabdariffa* leaf polyphenolic extract inhibits LDL oxidation and foam cell formation involving up-regulation of LXRα/ABCA1 pathway. Food Chem. 141: 397–406.

Chisaka, T., Matsuda, H., Kubomura, Y., Mochizuki, M., Yamahara, J. and Fujimura, H. 1988. The effect of crude drugs on experimental hypercholesteremia: mode of action of (−)-epigallocatechin gallate in tea leaves. Chem. Pharm. Bull. 36: 227–233.

Chong, S.C., Dollah, M.A., Chong, P.P. and Maha, A. 2011. *Phaleria macrocarpa* (Scheff.) Boerl fruit aqueous extract enhances LDL receptor and PCSK9 expression *in vivo* and *in vitro*. J. Ethnopharmacol. 137: 817–827.

Cicero, A.F., Rovati, L.C. and Setnikar, I. 2007. Eulipidemic effects of berberine administered alone or in combination with other natural cholesterol-lowering agents. A single-blind clinical investigation. Arznei-Forschung 57: 26–30.

Cliffold, M.N., Johnson, K.L., Knight, S. and Kuhnert, N. 2003. Hierarchical scheme for LC-MS identification of chlorogenic acids. J. Agr. Food Chem. 51: 2900–2911.

de Denus, S., Spinler, S.A., Miller, K. and Peterson, A.M. 2004. Statins and liver toxicity: a meta-analysis. Pharmacotherapy 24: 584–591.

De Lima, T.C., Klueger, P.A., Pereira, P.A., Macedo-Neto, W.P., Morato, G.S. and Farias, M.R. 1998. Behavioural effects of crude and semi-purified extracts of Syzygium cumini Linn. Skeels. Phytother. Res. 12: 488–493.

Demonty, I., Ras, R.T., van der Knaap, H.C., Duchateau, G.S., Meijer, L., Zock, P.L., Geleijnse, J.M. and Trautwein, E.A. 2009. Continuous dose-response relationship of the LDL-cholesterol-lowering effect of phytosterol intake. J. Nutr. 139(2): 271–284.

Doi, K., Kojima, T. and Fujimoto, Y. 2000. Mulberry leaf extract inhibits oxidative modification of rabbit and human low-density lipoprotein. Biol. Pharm. Bull. 23: 1066–1071.

Dong, X.F., Gao, W.W., Tong, J.M., Jia, H.Q., Sa, R.N. and Zhang, Q. 2007. Effect of polysavone (alfalfa extract) on abdominal fat deposition and immunity in broiler chickens. Poult. Sci. 86: 1955–1959.

Elekofehinti, O.O., Kamdem, J.P., Kade, I.J., Rocha, J.B.T. and Adanlawo, I.G. 2013. Hypoglycemic, antiperoxidative and antihyperlipidemic effects of saponins from *Solanum anguivi* Lam. fruits in alloxan-induced diabetic rats. S. Afr. J. Bot. 88: 56–61.

Fatima, S.S., Rajasekhar, M.D., Kumar, K.V., Kumar, M.T.S., Babu, K.R. and Rao, C.A. 2010. Antidiabetic and antihyperlipidemic activity of ethyl acetate:isopropanol (1:1) fraction of *Vernonia anthelmintica* seeds in streptozotocin induced diabetic rats. Food Chem. Toxicol. 48: 495–501.

Feng, L.J., Yu, C.H., Ying, K.J., Hua, J. and Dai, X.Y. 2011. Hypolipidemic and antioxidant effects of total flavonoids of *Perilla Frutescens* leaves in hyperlipidemia rats induced by high-fat diet. Food Res. Int. 44: 404–409.

Fogelman, A.M., Shechter, I., Seager, J., Hokom, M., Chird, J.S. and Edwards, P.A. 1980. Malodialdehyde alternation of low density proteins leads to choresteryl ester accumulation in human monocyte-macrophages. Proc. Natl. Acad. Sci. U.S.A. 77: 2214–2218.

Fuhrman, B., Buch, S., Vaya, J., Belinky, P.A., Coleman, R., Hayek, T. and Aviram, M. 1997. Licorice extract and its major polyphenol glabridin protect low-density lipoprotein against lipid peroxidation: *in vitro* and *ex vivo* studies in humans and in atherosclerotic apolipoprotein E-deficient mice. Am. J. Clin. Nutr. 66: 267–275.

Gao, H., Long, Y., Jiang, X., Liu, Z., Wang, D., Zhao, Y., Li and Sun, B.L. 2013. Beneficial effects of Yerba Mate tea (*Ilex paraguariensis*) on hyperlipidemia in high-fat-fed hamsters. Exp. Gerontol. 48: 572–578.

Ghoul, J.E., Smiri, M., Ghrab, S., Boughattas, N.A. and Ben-Attia, M. 2012. Antihyperglycemic, antihyperlipidemic and antioxidant activities of traditional aqueous extract of *Zygophyllum album* in streptozotocin diabetic mice. Pathophysiology 19: 35–42.

Ghule, B.V., Ghante, M.H., Saoji, A.N. and Yeole, P.G. 2009. Antihyperlipidemic effect of the methanolic extract from *Lagenaria siceraria* Stand. fruit in hyperlipidemic rats. J. Ethnopharmacol. 124: 333–337.

Gliozzi, M., Walker, R., Muscoli, S., Vitale, C., Gratteri, S., Carresi, C., Musolino, V., Russo, V., Janda, E., Ragusa, S., Aloe, A., Palma, E., Muscoli, C., Romeo, F. and Mollace, V. 2013. Bergamot polyphenolic fraction enhances rosuvastatin-induced effect on LDL-cholesterol, LOX-1 expression and protein kinase B phosphorylation in patients with hyperlipidemia. Int. J. Cardiol. 170(2): 140–145.

Gonzalez-Paramas, A.M., Esteban-Ruano, S., Santos-Buelga, C., de Pascual-Teresa, S. and Rivas-Gonzalo, J.C. 2004. Flavanol content and antioxidant activity in winery byproducts. J. Agr. Food Chem. 52: 234–238.

Gosain, S., Ircchiaya, R., Sharma, P.C., Thareja, S., Kalra, A., Deep, A. and Bhardwaj, T.R. 2010. Hypolipidemic effect of ethanolic extract from the leaves of *Hibiscus sabdariffa* L. in hyperlipidemic rats. Acta Pol. Pharm. 67(2): 179–184.

Haji Faraji, M. and Haji Tarkhani, A. 1999. The effect of sour tea (*Hibiscus sabdariffa*) on essential hypertension. J. Ethnopharmacol. 65: 231–236.

Handa, M., Murata, T., Kobayashi, K., Selenge, E., Miyase, T., Batkhuu, J. and Yoshizaki, F. 2013. Lipase inhibitory and LDL anti-oxidative triterpenes from *Abies sibirica*. Phytochemistry 86: 168–175.

Heck, C.I. and de Mejia, E.G. 2007. Yerba Mate tea (*Ilex paraguariensis*), a comprehensive review on chemistry, health implications, and technological considerations. J. Food Sci. 72: 138–151.

Hoehne, F.C. 1939. Plantas e substâncias vegetais tóxicas e medicinais, São Paulo, Brasil, Graphics, p. 56.

Imai, K. and Nakachi, K. 1995. Cross sectional study of effects of drinking green tea on cardiovascular and liver diseases. Br. Med. J. 310: 693–696.

Irudayaraj, S.S., Sunil, C., Duraipandiyan, V. and Ignacimuthu, S. 2013. *In vitro* antioxidant and antihyperlipidemic activities of *Toddalia asiatica* (L.) Lam. leaves in Triton WR-1339 and high fat diet induced hyperlipidemic rats. Food Chem. Toxicol. 60: 135–140.

Ishiguro, K., Yahara, S. and Yoshimoto, M. 2007. Changes in polyphenols contents and radical-scavenging activity of sweetpotato (*Ipomoea batatas* L.) during storage at optimal and low temperature. J. Agr. Food Chem. 55: 10773–10778.

Islam, M.S., Yoshimoto, M., Yahara, S., Okuno, S., Ishiguyro, K. and Yamakawa, O. 2002a. Identification and characterization of foliar polyphenolic composition in sweetpotato (*Ipomoea batatas* L.) genotypes. J. Agr. Food Chem. 50: 3718–3722.

Islam, M.S., Yoshimoto, M., Terahara, N. and Yamakawa, O. 2002b. Anthocyanin composition in sweetpotato (*Ipomoea batatas* L.) leaves. Biosci. Biotechnol. Biochem. 66: 2483–2486.

Islam, M.S., Yoshimoto, M., Ishiguro, T., Okino, K. and Yamakawa, O. 2003. Effect of artificial shading and temperature on radical scavenging activity and polyphenolic composition in sweetpotato (*Ipomoea batatas* L.) leaves. J. Am. Soc. Hortic. Sci. 128: 182–187.

Jagetia, G.C. and Baliga, M.S. 2004. Effect of Alstonia scholaris in enhancing the anticancer activity of berberine in the Ehrlich ascites carcinoma-bearing mice. J. Med. Food 7: 235–244.

Javekar, A.R. and Halade, G.V. 2006. Hypoglycemic activity of *Cassia auriculata* in neonatal streptozotocin-induced non-insulin dependent diabetes mellitus in rats. J. Nat. Remedies 1: 14–18.

Jia, X., Chen, Y., Zidichouski, J., Zhang, J., Sun, C. and Wang, Y. 2008. Co-administration of berberine and plant stanols synergistically reduces plasma cholesterol in rats. Atherosclerosis 201: 101–107.

Jones, P.J., Demonty, I., Chan, Y.M., Herzog, Y. and Pelled, D. 2007. Fish-oil esters of plant sterols differ from vegetable-oil sterol esters in triglycerides lowering, carotenoid bioavailability and impact on plasminogen activator inhibitor-1 (PAI-1) concentrations in hypercholesterolemic subjects. Lipids Health Dis. 28: 1–9.

Joshi, S.G. 2000. Textbook of Medicinal Plants. Oxford, IBH Publishing Co., New Delhi, pp. 119.

Kamesh, V. and Sumathi, T. 2012. Antihypercholesterolemic effect of *Bacopa monniera* linn. on high cholesterol diet induced hypercholesterolemia in rats. Asian Pac. J. Trop. Med. 949–955.

Kannur, D.M. 2011. Antidiabetic and antihyperlipidemic activity of Bonducella (*Caesalpinia bonducella*) seeds. pp. 237–244. *In*: Preedy, V.R., Watson, R.R. and Patel, V.B (eds.). Nuts and Seeds in Health and Disease Prevention. Elsevier, London.

Kasetti, R.B., Rajasekhar, M.D., Kondeti, V.K., Fatima, S.S., Kumar, E.G.T., Swapna, S., Ramesh, B. and Rao, C.A. 2010. Antihyperglycemic and antihyperlipidemic activities of methanol:water (4:1) fraction isolated from aqueous extract of *Syzygium alternifolium* seeds in streptozotocin induced diabetic rats. Food Chem. Toxicol. 48: 1078–1084.

Katan, M.B., Grundy, S.M., Jones, P., Law, M., Miettinen, T. and Paoletti, R. 2003. Efficacy and safety of plant stanols and sterols in the management of blood cholesterol levels. Mayo Clin. Proc. 78(8): 965–978.

Katsube, T., Imawaka, N., Kawano, Y., Yamazaki, Y., Shiwaku, K. and Yamane, Y. 2006. Antioxidant flavonol glycosides in mulberry (*Morus alba* L.) leaves isolated based on LDL antioxidant activity. Food Chem. 97: 25–31.

Kirtikar, K.R. and Basu, B.D. 2002. Indian Medicinal Plants, Vol. 1. Lalit Mohan Basu, Allahabad, India.

Kong, W., Wei, J., Abidi, P., Lin, M., Inaba, S., Li, C., Wang, Y., Wang, Z., Si, S., Pan, H., Wang, S., Wu, J., Wang, Y., Li, Z., Liu, J. and Jiang, J.D. 2004. Berberine is a novel cholesterol-lowering drug working through a unique mechanism distinct from statins. Nature Med. 10: 1344–1351.

Kono, S., Shinchi, K., Ikeda, N., Yanai, F. and Imanishi, K. 1992. Green tea consumption and serum lipid profiles: a cross-sectional study in northern Kyushu, Japan. Prev. Med. 21: 526–531.

Kou, M.C., Yen, J.H., Hong, J.T., Wang, C.L., Lin, C.W. and Wu, M.J. 2009. *Cyphomandra betacea* Sendt. phenolics protect LDL from oxidation and PC12 cells from oxidative stress. LWT-Food Sci. Technol. 42: 458–463.

Kumar, B.S.A., Lakshman, K., Jayaveea, K.N., Shekar, D.S., Khan, S., Thippeswamy, B.S. and Veerapur, V.P. 2012a. Antidiabetic, antihyperlipidemic and antioxidant activities of methanolic extract of *Amaranthus viridis* Linn in alloxan induced diabetic rats. Exp. Toxicol. Pathol. 64: 75–79.

Kumar, D., Parcha, V., Maithani, A. and Dhulia, I. 2012b. Effect and evaluation of antihyperlipidemic activity guided isolated fraction from total methanol extract of *Bauhinia variegata* (linn.) in Triton WR-1339 induced hyperlipidemic rats. Asian Pac. J. Trop. Dis. 2(2): 909–913.

Kumar, S., Inamdar, N., Nayeemunnisa and Viswanatha, G.L. 2011. Protective effect of lemongrass oil against dexamethasone induced hyperlipidemia in rats: possible role of decreased lecithin cholesterol acetyl transferase activity. Asian Pac. J. Trop. Med. 4(8): 658–660.

Kumaran, A. and Karunakaran, R.J. 2007. Antioxidant activity of *Cassia auriculata* flowers, Fitoterapia 78: 46–47.

Kuppusamy, U.R. and Das, N.P. 1994. Potential of β adrenoreceptor agonist-mediated lipolysis by quarcetin and fisetin in isolated rat adipocytes. Biochem. Pharmacol. 47: 521–529.

Küpeli Akkol, E., Avcı, G., Küçükkurt, I., Keleş, H., Tamer, U., Ince, S. and Yesilada, E. 2009. Cholesterol-reducer, antioxidant and liver protective effects of *Thymbra spicata* L. var. *spicata*. J. Ethnopharmacol. 126: 314–319.

Lahaye, M. and D. Jegon. 1993. Chemical and physical–chemical characteristics of dietary fibers from *Ulva lactuca* (L.) Thuret and *Enteromorpha compressa* (L.) Grev. J. Appl. Phycol. 5: 195–200.

Latha, R.C.S. and Daisy, P. 2011. Insulin-secretagogue, antihyperlipidemic and other protective effects of gallic acid isolated from *Terminalia bellerica* Roxb. in streptozotocin-induced diabetic rats. Chem. Biol. Interact. 189: 112–118.

Lee, J.S., Bok, S.H., Jeon, S.M., Kim, H.J., Do, K.M., Park, Y.B. and Choi, M.S. 2010. Antihyperlipidemic effects of buckwheat leaf and flower in rats fed a high-fat diet. Food Chem. 119: 235–240.

Li, J., Liu, T., Wang, L., Guo, X., Xu, T., Wu, L., Qin, L. and Sun, W. 2012. Antihyperglycemic and antihyperlipidemic action of cinnamaldehyde in C57blks/j Db/db mice. J. Tradit. Chin. Med. 15; 32(3): 1–2.

Li, T., Li, S., Du, L., Wang, N., Guo, M., Zhang, J., Yan, F. and Zhang, H. 2010. Effects of haw pectic oligosaccharide on lipid metabolism and oxidative stress in experimental hyperlipidemia mice induced by high-fat diet. Food Chem. 121: 1010–1013.

Lii, C.K., Lei, Y.P., Yao, H.T., Hsieh, Y.S., Tsai, C.W., Liu, K.L. and Chen, H.W. 2010. *Chrysanthemum morifolium* Ramat. reduces the oxidized LDL-induced expression of intercellular adhesion molecule-1 and E-selectin in human umbilical vein endothelial cells. J. Ethnopharmacol. 128: 213–220.

Lin, L.Y., Peng, C.C., Liang, Y.J., Yeh, W.T., Wang, H.E., Yu, T.H. and Peng, R.Y. 2008. *Alpinia zerumbet* potentially elevates high-density lipoprotein cholesterol level in hamsters. J. Agr. Food Chem. 56: 4435–4443.

Liu, M.J., Wang, Z., Ju, Y., Zhou, J.B., Wang, Y. and Wong, R.N. 2004. The mitotic-arresting and apoptosis-inducing effects of diosgenyl saponins on human leukemia cell lines. Biol. Pharm. Bull. 27: 1059–1065.

Liu, X., Sun, Z., Zhang, M., Meng, X., Xia, X., Yuan, W., Xue, F. and Liu, C. 2012. Antioxidant and antihyperlipidemic activities of polysaccharides from sea cucumber *Apostichopus japonicus*. Carbohydr. Polym. 90: 1664–1670.

Luo, Q.F., Sun, L., Si, J.Y. and Chen, D.H. 2008. Hypocholesterolemic effect of stilbenes containing extract-fraction from *Cajanus cajan* L. On diet-induced hypercholesterolemia in mice. Phytomedicine 15: 932–939.

Ma, D.Q., Jiang, Z.J., Xu, S.Q., Yu, X., Hu, X.M. and Pan, H.Y. 2012. Effects of flavonoids in *Morus indica* on blood lipids and glucose in hyperlipidemia-diabetic rats. Chin. Herb. Med. 4(4): 314–318.

Magalhaes, M.E. 2005. Mechanisms of rhabdomyolysis with statins. Arq. Bras. Cardiol. 85 (Suppl. 5): 42–44.

Maki, K.C., Lawless, A.L., Reeves, M.S., Kelley, K.M., Dicklin, M.R., Jenks, B.H., Shneyvas, E. and Brooks, J.R. 2013. Lipid effects of a dietary supplement softgel capsule containing plant sterols/stanols in primary hypercholesterolemia. Nutrition 29: 96–100.

Marinangeli, C.P.F., Varady, K.A. and Jones, P.J.H. 2006. Plant sterols combined with exercise for the treatment of hypercholesterolemia: overview of independent and synergistic mechanisms of action. J. Nutr. Biochem. 17: 217–224.

Matsuda, H., Chisaka, T., Kubomura, Y., Yamahara, J., Sawada, T., Fujimura, H. and Kimura, H. 1986. Effects of crude drugs on experimental hypercholesterolemia. I. tea and its active principles. J. Ethnopharmacol. 17: 213–224.

Matsui, Y., Kumagai, H. and Masuda, H. 2006. Antihypercholesterolemic activity of catechin-free saponin-rich extract from green tea leaves. Food Sci. Technol. Res. 12: 50–54.

Matsuzawa, N., Takamura, T., Kurita, S., Misu, H., Ota, T., Ando, H., Yokoyama, M., Honda, M., Zen, Y., Nakanuma, Y., Miyamoto, K. and Kaneko, S. 2007. Lipid-induced oxidative stress causes steatohepatitis in mice fed an atherogenic diet. Hepatology 46: 1392–1403.

Miettinen, T.A., Puska, P., Gylling, H., Vanhanen, H. and Vartiainen, E. 1995. Reduction of serum cholesterol with sitostanol-ester margarine in a mildly hypercholesterolemic population. New Engl. J. Med. 333(20): 1308–1312.

Mizoguchi, K., Yuzurihara, M., Ishige, A., Sasaki, H. and Tabira, T. 2002. *Saiko-ka-ryukotsu-borei-to*, an herbal medicine, prevents chronic stress-induced disruption of glucocorticoid negative feedback in rats. Life Sci. 72: 67–77.

Mollace, V., Sacco, I., Janda, E., Malara, C., Ventrice, D., Colica, C., Visalli, V., Muscoli, S., Ragusa, S., Muscoli, C., Rotiroti, D. and Romeo, F. 2011. Hypolipemic and hypoglycaemic activity of bergamot polyphenols: from animal models to human studies. Fitoterapia 82: 309–316.

Muramatsu, K., Fukuyo, M. and Hara, Y. 1986. Effect of green tea catechins on plasma cholesterol level in cholesterol-fed rats. J. Nutr. Sci. Vitaminol. 32: 613–622.

Murty, D., Rajesh, E., Raghava, D., Raghavan, T.V. and Surulivel, M.K. 2010. Hypolipidemic effect of arboreum plus in experimentally induced hypercholestermic rabbits. Yakugaku Zasshi 130(6): 841–846.

Muzzarelli, R.A.A. and Muzzarelli, C. 2006. Chitosan, a dietary supplement and a food technology commodity. pp. 215–248. *In*: C.G. Biliaderis and M.S. Izydorczyk (eds.). Functional Food Carbohydrates. Francis and Taylor, Orlando, USA.

Nadeem, S., Raj, C. and Raj, N. 2012. The influence of *Alpinia calcarata* extract on the serum lipid and leptin levels of rats with hyperlipidemia induced by high-fat diet. Asian Pac. J. Trop. Biomed. 2(3): 1822–1826.

Nadkarni, A.K. 1976. Dr. K.M. Nadkarni's Indian Materia Medica, Vol. 1. Popular Prakashan, Bombay, India.

Nagaraju, N. and Rao, K.N. 1990. Folk medicine for diabetes from Rayalaseema of Andhra Pradesh. Workshop on Medicinal Plants, Ayurvedic College, Tirupati 9–11th February.

Nomura, S., Hattori, N., Sakakibara, I. and Fukuhara, S. 2001. Effects of *Saiko-ka-ryukotsu-borei-to* in patients with hyperlipidemia. Phytomedicine 8(3): 165–173.

Nwinyi, S.C.O. 1992. Effect of age at short removal on tuber and shoot yields at harvest of five sweet potato (*Ipomoea batatas* (L.) Lam) cultivars. Field Crop. Res. 29: 47–54.

Ochani, P.C. and D'Mello, P. 2009. Antioxidant and antihyperlipidemic activity of *Hibiscus sabdariffa* Linn. leaves and calyces extracts in rats. Indian J. Exp. Biol. 47: 276–282.

Oki, T., Masuda, M., Furuya, S., Nishiba, Y., Terahara, N. and Suda, I. 2002. Involvement of anthocyanins and other phenolic compounds in radical-scavenging activity of purple-fleshed sweet potato cultivars. J. Food Sci. 67: 1752–1756.

Olszanecki, R., Jawien, J., Gajda, M., Mateuszuk, L., Gebska, A., Korabiowska, M., Chłopicki, S. and Korbut, R. 2005. Effect of curcumin on atherosclerosis in apo E/LDLR-double knockout mice. J. Physiol. Pharmacol. 56: 627–635.

Onyenekwe, P.C., Ajani, E.O., Ameh, D.A. and Gamaniel, K.S. 1999. Antihypertensive effect of roselle (*Hibiscus sabdariffa*) calyx infusion in spontaneously hypertensive rats and a comparison of its toxicity with that wistar rats. Cell Biochem. Funct. 17: 199–206.

Ottestad, I., Ose, L., Wennersberg, M.H., Granlund, L., Kirkhus, B. and Retterstøl, K. 2013. Phytosterol capsules and serum cholesterol in hypercholesterolemia: a randomized controlled trial. Atherosclerosis 228: 421–425.

Qi, H., Huang, L., Liu, X., Liu, D., Zhang, Q. and Liu, S. 2012. Antihyperlipidemic activity of high sulfate content derivative of polysaccharide extracted from *Ulva pertusa* (Chlorophyta). Carbohydr. Polym. 87: 1637–1640.

Rajanandh, M.G., Satishkumar, M.N., Elango, K. and Suresh, B. 2012. *Moringa oleifera* Lam. A herbal medicine for hyperlipidemia: A preclinical report. Asian Pac. J. Trop. Dis. 2(2): 790–795.

Rajesh, R., Chitra, K. and Paarakh, P.M. 2012. Anti hyperglycemic and antihyperlipidemic activity of aerial parts of *Aerva lanata* Linn Juss in streptozotocin induced diabetic rats. Asian Pac. J. Trop. Biomed. 2(2): 924–929.

Ramachandran, S., Rajasekaran, A. and Kumar, K.T.M. 2011. Antidiabetic, antihyperlipidemic and antioxidant potential of methanol extract of *Tectona grandis* flowers in streptozotocin induced diabetic rats. Asian Pac. J. Trop. Med. 4(8): 624–631.

Ramachandran, S., Naveen, K.R., Rajinikanth, B., Akbar, M. and Rajasekaran, A. 2012. Antidiabetic, antihyperlipidemic and *in vivo* antioxidant potential of aqueous extract of *Anogeissus latifolia* bark in type 2 diabetic rats. Asian Pac. J. Trop. Dis. 2(2): 596–602.

Ramirez-Tortosa, M.C., Mesa, M.D., Aguilera, M.C., Quiles, J.L., Baró, L., Ramirez-Tortosa, C.L., Martinez-Victoria, E. and Gil, A. 1999. Oral administration of a turmeric extract inhibits LDL oxidation and has hypocholesterolemic effects in rabbits with experimental atherosclerosis. Atherosclerosis 147: 371–378.

Ramjiganesh, T., Roy, S., Nicolosi, R.J., Young, T.L., McIntyre, J.C. and Fernandez, M.L. 2000. Corn husk oil lowers plasma LDL cholesterol concentrations by decreasing cholesterol absorption and altering hepatic cholesterol metabolism in guinea pigs. J. Nutr. Biochem. 11: 358–366.

Ratnawati, H. and Widowati, W. 2011. Anticholesterol activity of Velvet Bean (*Mucuna pruriens* L.) towards hypercholesterolemic rats. Sains Malays. 40(4): 317–321.

Ravi, K., Rajasekaran, S. and Subramanian, S. 2005. Antihyperlipidemic effect of *Eugenia jambolana* seed kernel on streptozotocin-induced diabetes in rats. Food Chem. Toxicol. 43: 1433–1439.

Rechinger, K.H. 1982. Flora Iranica, Vol. 150, Akademische Druck-u, Verlagsanstalt, Graz, pp. 218.

Reckless, J.P.D. and Lawrence, J.M. 2003. Hyperlipidemia (Hyperlipidaemia). pp. 3183–3192. *In*: Caballero, B., Trugo, L. and Finglas, P. (eds.). Encyclopedia of Food Sciences and Nutrition (Second Edition). Academic Press, USA.

Renaud, S. and de Lorgeril, M. 1992. Wine, alcohol, platelets, and the French paradox for coronary heart disease. Lancet 339: 1523–1526.

Rodrigues, H.G., Diniz, Y.S., Faine, L.A., Galhardi, C.M., Burneiko, R.C., Almeida, J.A., Ribas, B.O. and Novelli, E.L. 2005. Antioxidant effect of saponin: Potential action of a soybean flavonoid on glucose tolerance and risk factors for atherosclerosis. Int. J. Food Sci. Nutr. 56: 79–85.

Rosenfeld, M.E., Khoo, J.C. and Miller, E. 1991. Macrophage-derived foam cells freshly isolated from rabbit atheroscleotic lesions degrade modified lipoproteins, prompt oxidation of low-density lipoproteins, and contain oxidation specific lipid-protein adducts. J. Clin. Invest. 87: 90–99.

Ruiz-Larrea, M.B., Mohan, A.R., Paganga, G., Miller, N.J., Bolwell, G.P. and Rice-Evans, C.A. 1997. Antioxidant activity of phytoestrogenic isoflavones. Free Radic. Res. 26: 63–70.

Sachdewa, A., Nigam, R. and Khemani, L.D. 2001. Hypoglycemic effect of *Hibiscus rosa sinensis* L. leaf extract in glucose and streptozotocin induced hyperglycemic rats. Indian J. Exp. Biol. 39: 284–286.

Sajjadi, S.E., Atar, A.M. and Yektaian, A. 1998. Antihyperlipidemic effect of hydroalcoholic extract, and polyphenolic fraction from *Dracocephalum kotschyi* Boiss. Pharm. Acta Helv. 73: 167–170.

Santhosh Kumari, C., Govindasamy, S. and Sukumar, E. 2006. Lipid lowering activity of *Eclipta prostrata* in experimental hyperlipidemia. J. Ethnopharmacol. 105: 332–335.

Sathivel, A., Raghavendran, H.R.B., Srinivasan, P. and Devaki, T. 2008. Anti-peroxidative and anti-hyperlipidemic nature of *Ulva lactuca* crude polysaccharide on d-galactosamine induced hepatitis in rats. Food Chem. Toxicol. 46: 3262–3267.

Schoenfelder, T., Pich, C.T., Geremias, R., Ávila, S., Daminelli, E.N., Pedrosa, R.C. and Bettiol, J. 2008. Antihyperlipidemic effect of *Casearia sylvestris* methanolic extract. Fitoterapia 79: 465–467.

Sedaghat, R., Roghani, M., Ahmadi, M. and Ahmadi, F. 2011. Antihyperglycemic and antihyperlipidemic effect of *Rumex patientia* seed preparation in streptozotocin-diabetic rats. Pathophysiology 18: 111–115.

Shahaboddin, M.E., Pouramir, M., Moghadamnia, A.A., Parsian, H., Lakzaei, M. and Mir, H. 2011. *Pyrus biossieriana* Buhse leaf extract: An antioxidant, antihyperglycaemic and antihyperlipidemic agent. Food Chem. 126: 1730–1733.

Sharma, S.K., Mishra, H.K., Sharma, H., Goel, A., Sreenivas, V., Gulati, V. and Tahir, M. 2008. Obesity, and not obstructive sleep apnea, is responsible for increased serum hs-CRP levels in patients with sleep-disordered breathing in Delhi. Sleep Med. 9: 149–156.

Shin, J.E., Han, M.J., Song, M.C., Baek, N.I. and Kim, D.H. 2004. 5-Hydroxy-7-(4'-hydroxy-30-methoxyphenyl)-1-phenyl-3-heptanone: a pancreatic lipase inhibitor isolated from *Alpinia officinarum*. Biol. Pharm. Bull. 27: 138–140.

Sparow, C.D., Parthasarathy, S. and Steinberg, D.A. 1989. Macrophage receptor that recognizes oxidized LDL but not acetylated LDL. J. Biol. Chem. 264: 2599.

Sridevi, M., Kalaiarasi, P. and Pugalendi, K.V. 2011. Antihyperlipidemic activity of alcoholic leaf extract of *Solanum surattense* in streptozotocin-diabetic rats. Asian Pac. J. Trop. Biomed. 1(2): 276–280.

Srinivasan, S. and Pari, L. 2013. Antihyperlipidemic effect of diosmin: A citrus flavonoid on lipid metabolism in experimental diabetic rats. J. Funct. Foods 5: 484–492.

Stensvold, I., Tverdal, A., Solvoll, K. and Foss, O.P. 1992. Tea consumption-relationship to cholesterol, blood pressure, and coronary and total mortality. Prev. Med. 21: 546–553.

Surana, S.J., Gokhale, S.B., Jadhav, R.B., Sawant, R.L. and Wadekar, J.B. 2008. Antihyperglycemic activity of various fractions of *Cassia auriculata* Linn. in alloxan diabetic rats. Indian J. Pharm. Sci. 70: 227–229.

Taira, J., Ohmi, N. and Uechi, K. 2007. Characteristics of folic acid and polyphenol in Okinawan sweet potato (*Ipomoea batatas* L.) foliage. Nippon Shokuhin Kagaku Kaishi 54: 215–221.

Taira, J., Taira, K., Ohmine, W. and Nagata, J. 2013. Mineral determination and anti-LDL oxidation activity of sweet potato (*Ipomoea batatas* L.) leaves. J. Food Comp. Anal. 29: 117–125.

Tang, L.Q., Wei, W., Chen, L.M. and Liu, S. 2006. Effects of berberine on diabetes induced by alloxan and a high-fat/high-cholesterol diet in rats. J. Ethnopharmacol. 108: 109–115.

Trinick, T.R. and Duly, E.B. 2005. Hyperlipidemia. pp. 479–491. *In*: Caballero, B., Allen, L. and Prentice, A. (eds.). Encyclopedia of Human Nutrition (Second Edition), Academic Press, USA.

Vaya, J., Belinky, P.A. and Aviram, M. 1997. Antioxidant constituents from licorice roots: isolation, structure elucidation and antioxidative capacity toward LDL oxidation. Free Radic. Biol. Med. 23: 302–313.

Vaya, J., Mahmood, S., Goldblum, A., Aviram, M., Volkova, N., Shaalan, A., Musa, R. and Tamir, S. 2003. Inhibition of LDL oxidation by flavonoids in relation to their structure and calculated enthalpy. Phytochemistry 62: 89–99.

Veeramani, C., Aristatle, B., Pushpavalli, G. and Pugalendi, K.V. 2010. Antihypertensive efficacy of *Melothria maderaspatana* leaf extract on sham-operated and uninephrectomized DOCA-salt hypertensive rats. J. Basic Clin. Physiol. Pharmacol. 21(1): 27–41.

Veeramani, C., Al-Numair, K.S., Chandramohan, G., Alsaif, M.A. and Pugalendi, K.V. 2012. Antihyperlipidemic effect of *Melothria maderaspatana* leaf extracts on DOCA-salt induced hypertensive rats. Asian Pac. J. Trop. Med. 5(6): 434–439.

Verma, N., Singh, A.P., Amresh, G., Sahu, P.K. and Rao, V. 2011. Protective effect of ethyl acetate fraction of *Rhododendron arboreum* flowers against carbon tetrachloride-induced hepatotoxicity in experimental models. Indian J. Pharmacol. 43(3): 291–295.

Verma, N., Amresh, G., Sahu, P.K., Rahu, V. and Singh, A.P. 2012a. Antihyperglycemic and antihyperlipidemic activity of ethyl acetate fraction of *Rhododendron arboreum* Smith flowers in streptozotocin induced diabetic rats and its role in regulating carbohydrate metabolism. Asian Pac. J. Trop. Biomed. 2(9): 696–701.

Verma, P.R., Deshpande, S.A., Kamtham, Y.N. and Vaidya, L.B. 2012b. Hypolipidemic and antihyperlipidemic effects from an aqueous extract of *Pachyptera hymenaea* (DC.) leaves in rats. Food Chem. 132: 1251–1257.

Vijayaraj, P., Muthukumar, K., Sabarirajan, J. and Nachiappan, V. 2013. Antihyperlipidemic activity of *Cassia auriculata* flowers in triton WR 1339 induced hyperlipidemic rats. Exp. Toxicol. Pathol. 65: 135–141.

Villareal, R.L., Tsou, S.C., Lo, H.F. and Chiu, S.C. 1982. Sweetpotato tips as vegetables. pp. 313–320. *In*: Proceedings of the First International Symposium, AVRDC, Vol. 59. Shanhua, Taiwan.

Wang, T., Choi, R.C.Y., Li, J., Bi, C.W.C., Ran, W., Chen, X., Dong, T.T.X., Bi, K. and Tsim, K.W.K. 2012. Trillin, a steroidal saponin isolated from the rhizomes of *Dioscorea nipponica*, exerts protective effects against hyperlipidemia and oxidative stress. J. Ethnopharmacol. 139: 214–220.

Wang, Y., Jia, X., Ghanam, K., Beaurepaire, C., Zidichouski, J. and Miller, L. 2010. Berberine and plant stanols synergistically inhibit cholesterol absorption in hamsters. Atherosclerosis 209: 111–117.

Wardah, T. Sapondi and Wurlina. 2007. Identification of active compounds *P. buxifolius* leaf ethanol extract and its effect on preview serology and hematology of broiler chickens infected by the virus of Newcastle. J. Drug Nat. Mater. 6: 88–95.

Wardah, T. Sapondi, Bimo Aksono, E. and Kusriningrum, H. 2012. Reduction of intracellular lipid accumulation, serum leptin and cholesterol levels in broiler fed diet supplemented with powder leaves of *Phyllanthus buxifolius*. Asian J. Agr. Res. 6: 106–117.

Wu, C.Z., Inoue, M. and Ogihara, Y. 1997. Antihyperlipidemic action of a Traditional Chinese Medicine (Kampo Medicine), Ogi-Keishi-Gomotsu-To-Ka-Kojin. Phytomedicine 4(4): 295–300.

Xia, D., Wu, X., Yang, Q., Gong, J. and Zhang, Y. 2010. Anti-obesity and hypolipidemic effects of a functional formula containing *Prumus mume* in mice fed high-fat diet. Afr. J. Biotechnol. 9: 2463–2467.

Yang, H.O., Ko, W.K., Kim, J.Y. and Ro, H.S. 2004. Paeoniflorin: an antihyperlipidemic agent from *Paeonia lactiflora*. Fitoterapia 75: 45–49.

Yang, T.T. and Koo, M.W. 1997. Hypocholesterolemic effects of Chinese tea. Pharmacol. Res. 35: 505–512.

Yang, T.T. and Koo, M.W. 2000. Chinese green tea lowers cholesterol level through an increase in fecal lipid excretion. Life Sci. 66: 411–423.

Yoshimoto, M., Yahara, S., Okuno, S., Islam, M.S., Ishiguro, K. and Yamakawa, O. 2002. Antimutagenecity of mono-, di-, and tricaffeoylquinic acid derivatives isolated from sweetpotato (*Impomoae batatas* L.) leaf. Biosci. Biotechnol. Biochem. 66: 2336–2341.

Yu, P.Z., Li, N., Liu, X.G., Zhou, G.F., Zhang, Q.B. and Li, P.C. 2003. Antihyperlipidemic effects of different molecular weight sulfated polysaccharides from *Ulva pertuse* (Chlorophyta). Pharmacology Research 48: 543–549.

Zargari, A. 1996. Medicinal Plants (6th ed.). Tehran University Publications, Tehran.

Zeng, X.H., Zeng, X.J. and Li, Y.Y. 2003. Efficacy and safety of berberine for congestive heart failure secondary to ischemic or idiopathic dilated cardiomyopathy. Am. J. Cardiol. 92: 173–176.

Zhishen, J., Mengcheng, T. and Jianming, W. 1999. The determination of flavonoid contents in mulberry and their scavenging effects in superoxide radicals. Food Chem. 64: 555–559.

Pharmacological Aspects of *Tribulus terrestris* Linn. (Goksura): Progress and Prospects

Anita Patil* and Bipin D. Lade[a]

Introduction

Natural products from plants are extracted for their medicinal properties and used as biological answers for human diseases. Remedial plants are like a library for pooling several bioactive products that hold curable property. Thus therapeutic plants are constantly explored for isolation of novel compounds that have potential healing properties against various diseases. A large number of plant's species around 35,000 are under surveillance for tracking medicinal values against human disorders. Efforts are being made for isolation of exact compounds of a bioactive nature for specific disorders (Koshy Philip et al. 2011). The *T. terrestris* is one of the ethnomedicinal plants known for its curative values.

Tribulus terrestris also called Puncture Vine is an annual flowering (yellow colour) plant belonging to the family Zygophyllaceae (Abirami and Rajanderan 2011) found in hot places grown in various countries of sub tropical parts from the world such as India, Pakistan (Di Sansebastiano 2013), China, Africa, France, Australia, southern and western Europe (Abirami and Rajanderan 2011, Di Sansebastiano 2013), Bulgaria , Mexico (Mohd et al. 2012) and North America. *T. terrestris* is a common weed plant with divaricate spines, found in sandy soil and waste lands. Since ancient times, *T. terrestris* has been explored as a traditional medicine in China and Ayurvedic medicine and known as Gokhshura in the Sanskrit language in India. (Ukani et al. 1997) (see Table 18.1 for other names of *T. terrestris* in India).

In Turkey, it is used as a folk medicine for blood pressure and cholesterol. However the importance of *T. terrestris* came in 1990's when several athletes used this plant as a testosterone enhancer for increasing their performance in the Olympic Games (Di Sansebastiano 2013, Pokrywka et al. 2014).

It has been reported that *T. terrestris* possess antimicrobial, antihypertension, diuretic, anti-acetylcholine, haemolytic activity, stimulate spermatogenesis and show antitumour activity (Mohd et al. 2012). Plant extracts of *T. terrestris* have been successfully used to control blood pressure and

Department of Biotechnology, Lab No. 106, Plant Secondary Metabolite Lab., Sant Gadge Baba Amravati University, Amravati 444602 (M.S) India.
[a] Email: dbipinlade@gmail.com
* Corresponding author: anitapatil@sgbau.ac.in

Table 18.1. Vernacular name of *T. terrestris* in various Indian languages.

Indian (language)	Other names of *T. terrestris*
Assamese	Gokshura, Gokshurkata
Bengali	Gokhri, Gokshura
English	Caltrops root
Gujarati	Betha gokharu, Nana gokharu, Mithogokharu
Hindi	Gokhru
Kannada	Sannaneggilu, Neggilamullu, Neggilu
Malayali	Nerinjil
Marathi	Sarate, Gokharu
Oriya	Gukhura, Gokhyura
Punjabi	Bhakhra, Gokhru
Tamil	Nerinjil, Nerunjil
Telugu	Palleruveru

cholesterol *in vivo* (Chu et al. 2003). In Iraq, it is used as a folk medicine such as tonic, aphrodisiac, painkiller, astringent, to kill parasitic worms from gastro, antihypertensive, diuretic lithon-triptic and urinary anti-invectives (Abirami and Rajanderan 2011). In southern Europe, the plant parts of *T. terrestris* such as roots, stems and the leaves are used for the formulation of tonic. Besides its many advantages in treatment of dysfunction, the fruits are also used in cooling and diuretic purposes, which help in removing harmful toxins and metabolic waste by increasing urination. Besides this, it is also used as a tonic in calculus affection and urinary disorders.

In China *T. terrestris* has been used for treatment of cutaneous pruritus, edema and inflammation (Abirami and Rajanderan 2011). The extract of *T. terrestris* is used for formulation of cream with antibacterial, anti-inflammatory, antiviral activities (Alexis 2001, Alexis 2005). The combinations of *T. terrestris* extracts with metals are found in several Asian patented literatures. The plant extracts with a combination of metals for preparation of antiviral pharmaceutical compositions have been already worked out (Alexiev 2003a, 2003b). Other patents describe the uses of the extract for skin disorders (Li 2010) and successfully help in dilating the skin permeability and stimulating the generation of melanophore (Zhang 2011) and numerous cosmetic applications (Jing 2009, Ke 2010).

Initial literature search retrieved numerous articles stating *Tribulus terrestris* is an important curable plant famous for its use in Indian and Chinese systems of medicine for several diseases. In India, these are classified under mishrak varga as 'Dashmoola' in Ayurveda and in chemotaxonomy as Saponin glycoside (Vyawahare et al. 2014). It contains various phytochemical such as alkaloids, resins, tannins, sugars, sterols, flavonoids and saponins that are potential candidates for pharmaceutical industries. This plant has been used in dietary supplements, dermatological, cosmetic, hygiene, cough, kidney failure and even as an anticancer agent.

The extracts of *T. terrestris* have characteristic diuretic, astringent, analgesic, aphrodisiac, antiurolithic, immunomodulatory, antidiabetic, cardiotonic, central nervous system, anti-inflammatory, analgesic, antitumour, antibacterial, anthelmintic, larvicidal and anticariogenic activities. A preliminary analytical technique such as TLC is discussed for identification and separation of drug valued compounds.

This chapter discusses the pharmacologic aspects of *Tribulus terrestris* and describes its habitat, botany, minor and major bioactive constituents with their pharmacologically important products. All this valuable information will assist scientists, lecturers, pharmacologists, research students of under-graduates and post-graduates of botany and pharmacy for drafting diverse research work.

Taxonomical position

Kingdom	:	Plantae
Division	:	Phanerogams
Subdivision	:	Angiospermae
Class	:	Dicotyledonae
Subclass	:	Polypetalae
Series	:	Disciflorae
Order	:	Giraniales
Family	:	Zygophyllaceae
Genus	:	*Tribulus*
Species	:	*terrestris* Linn.

Habitat

T. terrestris is usually found in waste land in dry areas of the tropical world. They grow easily in deserts and in poor growth conditions. Its seeds are caltrop like the name puncture vine, devil throne and goat head. Figure 18.1 shows the picture of whole plant, fruit and seed of *T. terrestris*.

Figure 18.1. (A) Whole Plant of *T. terrestris* along with yellow colour flower, (B) Fruit green colour, (C) Dried Fruit, (D) Fruit powder of *T. terrestris*. The picture A of *T. terrestris* was collected from natural habitat from SGB Amravati University campus, Amravati (M.S) India.

Morphology

Tribulus terrestris is a silky herb reproduced by seed and occurs in dry places. It can grow erected and draws support for light and minerals. The stem produced from the crown of plant and branched in several alternate short stems bearing leaves arrived from main stem. Its fruits are spherical in shape, five cornered and light yellow in colour. The roots are brown in, branched slender, fibrous, a cylindrical tap roots system making them resistant to drought conditions. Flowers are yellow sized approximately 1-2.5 cm in diameter and have five petals. Flowering continues from July to September (Vyawahare et al. 2014). Its seeds are oily and are encapsulated in hard stony cells and produce two varieties, one that is sweet and the other bitter (Mohd et al. 2012). The detailed anatomical study of the root, stem and leaf of *Tribulus terrestris* for its ecological adaptation was done by Nikolova and Vassilev (2014).

Uses

T. terrestris is very valued for the broad range of properties such as asthma, cough, splenetic diseases, heart disorders, aching of limbs, striating urinary stones, aphrodisiac, anti inflammatory, anthelmintic, diuretic and used in enemas. The leaves are used in pharmaceutical industry for preparation of herbal tonic, it plays important role in, digestive problems, to increase spermatozoa, to enhance sexual desire and treat urinary tract infection. The fruits are used for coughs, spermatorrhoea, scabies, anemia, opthalmia and haemostatic (Vyawahare et al. 2014). The fruit infusion is exclusively used for kidney disease. The stem is astringent and is used in gonorrhea. Ash of the whole plant is used for rheumatoid arthritis and roots are reported to be aperients, demulcent and used in the preparation of tonics (Mohd et al. 2012). *Tribulus terrestris* therapeutic plants are pioneered for their pharmacological properties and large numbers of pharmaceutical products are available in the national and international market. These ready for use products with desired curative properties are cost effective, making them affordable in lower as well as higher economic countries. *T. terrestris* products are easily prepared from roots, shoots, seeds and flowers which are the main source for raising a profitable standard of the pharmaceutical laboratory.

Phytoconstituents

In order to rearrange phytochemical constituent data several journals such as Journal of Medicinal Plants Research, Brazilian Journal of Nutrition, Journal of Ethnopharmacol, The Journal of Alternative and Complementary Medicine, Pharmacognosy Reviews, Tikrit Journal of Pure Science, African Journal of Biotechnology, International Journal of Chemical Science, Annuals of NewYork Academy of Science, Experimental Biology and Medicine, Quality Standards of Indian Medicinal Plants and others were consulted. All available literature has supported that *T. terrestris* is a pharmacologically important plant with diuretic (Ukani et al. 1997), anti-inflammatory and anticancer property. Phytochemicals such as alkaloids, flavonoid, phenols and saponins (Singh et al. 2013) were reported to be present in *T. terrestris.* The *T. terrestris* extracts (fruit, bark, roots and leaves) contain a resin, fat and mineral matter and showed potential therapeutic properties. The inorganic constituents found in *T. terrestris* are chloride, calcium, sulphate, potassium, magnesium and total alkalinity. The fruits contain (0.001%) of alkaloid, essential oil, resin and nitrates (Vyawahare et al. 2014). Phytoconstituents are just like immunological weapons for medicinal plants that are potentially used to overcome various biotic and abiotc stress conditions. The minor and major phytoconstituents present in *T. terrestris* have been described in Table 18.2.

Minor constituents

Usually phytoconstituents found in *T. terrestris* are alkaloids, resins, tannins, sugars, sterols, flavonoids like rutin, quercetin, cinammic acid, amides, lignanamides (tribulusamides A and B), kaempferol, protodioscin and prototribestin (Evstatieva et al. 2011). Hashim et al. (2014) reported phytosteroids and glycosides. However major phytochemicals produced by this plant are steroidal saponins such as prototribestin, dioscin, protodioscin (Gauthaman et al. 2003), furostanol (De Combarieu et al. 2003),

Table 18.2. Chemical constituents present in different parts of *Tribulus terristris* Linn.

Sr. no.	Phytochemical classes	Plant parts	Chemical constituents	References
1	Saponins	Fruit	TerrestrosinsA-E, terrestrosins F-K, neotigogenin, desgalctotigonin, F-gitonin, ruscogenin, desglucolanatigonin, gitonin, diosgenin, hecogenin, chlorogenin, Tribulosaponin A-E, Isoterrestrosin-B, 25-D spirosta-3,5 diene. Protodioscin and Prototribestin	(Gupta et al. 2011, Tilwari et al. 2011, Ponnusamy et al. 2011, Singh et al. 2013, Gincy et al. 2014)
		Root	Diosgenin, Neotigogenin, Tigogenin, hecogenin gitogenin	(Vyawahare et al. 2014)
		Flower	Diosgenin, ruscogenin, Spirosta-3,5-diene	(Vyawahare et al. 2014)
		Aerial plant	Diosgenin and its acid chlorogenins, dioscin, gitogenin	(Vyawahare et al. 2014)
		Whole plant	Protodioscin	(Vyawahare et al. 2014)
2	Flavonoids	Fruit, leaves	Kaempferol, Kaempferol-3-glucoside, Kaempferol-3-rutinoside, Quercitin, Rutin	(Ponnusamy et al. 2011, Gincy et al. 2014)
		Aerial plant	Four glycoside of Kaempferol, three glycoside of quercitin, 8 glycoside of isorhamnetin	(Vyawahare et al. 2014)
		Leaves	Kaempferol, isorhamnetin	(Vyawahare et al. 2014)
		Flowers	Quercitin, rutin	(Vyawahare et al. 2014)
3	Alkaloids	Seeds	Harmine	(Ponnusamy et al. 2011)
		Herb	Harman	(Vyawahare et al. 2014)
4	Cinnamic amide	Fruits	Feruloytyramine, Terestriamide, Coumaroyltyramine,7-methyldroindanone	(Vyawahare et al. 2014)
5	Phytosterol	Root, flower	Campesterol, stigmasterol, β-Sitosterol	(Vyawahare et al. 2014)
6	Lignanamides	Fruits	Tribulusamide A,B	(Vyawahare et al. 2014)
7	Sugar	Fruits, Flowers	Glucose, rhamnose, reducing sugar	(Vyawahare et al. 2014)
8	Acid	Leaves	Vanillic acid, synergic acid, Melilotic acid, p-coumaric acid	(Vyawahare et al. 2014)

380 *Therapeutic Medicinal Plants: From Lab to the Market*

spirostanol, sitosterol glucoside (Conrad et al. 2004) terrestrosins A-E, terrestrosins F-K, neotigogenin, desgalactotigonin, F-gitonin, tigogenin, tigogenin-3-O-β-D-xylopyranosyl (1 → 2)-[β-D-xylopyranosyl (1 → 3) [β-D-xylopyranosyl (1 → 4)-[α-L rhamnopyranosyl (1 → 2)]-β-D-galactopyranoside (Tandan and Sharma 2010) gitogenin, beta-Sitosterol, spirosta 3,5-diene, stigmasterol, hecogenin, neohecogenin, ruscogenin (Akram et al. 2011), tribulosaponin B, metilprotodiostin, terrestrozin H, prototribestin, gracillin (Kozlova et al. 2011). Errestribisamide, 25R-spirost-4-en-3,12-dione, Tribulusterine, N-p-coumaroyltyramine, vanillin terrestriamide, aurantiamide acetate, xanthosine, fatty acid ester, ferulic acid, p-hydroxybenzoic acid, β-sitosterol, Diosgenin (Tandan and Sharma 2010). All these constituents confirm the potential of these plant extracts for various ailments.

Major constituent is steroidal saponins

Saponins are a diverse group of compounds produced by a combination of triterpene or steroid aglycone and one or more sugar chains produced by the leaves, roots, seeds and fruit (Di Sansebastiano 2013) and used against various diseases, helping in restoration of normal functions of the body. Chemical structures of a typical spirostanol saponin and furastanol saponin are shown in Fig. 18.2A, B, C (Powers and Setzer 2015). They have characteristic (surfactant) properties used in industry for natural product formulation, food, cosmetics and pharmaceutical industry (Francis et al. 2002). Saponins stimulate spermatogenesis by increasing activity of sertoli cells, decreasing urinary oxalate excretion, reducing activity of the liver enzyme GAO (glycolate oxidase) and GAD (glycolate hydrogenase) spirosteroid saponin from plant extracts characterized for the antifungal treatment. Seventeen unique spironosaponins have been identified to the previously reported saponins (Chen et al. 2004, Zhang et al. 2006). The saponins are promising structures for rational drug design of P-gp inhibitors. The study by Ivanova et al. (2009) showed inhibition of MDR of cancer cells by saponins and isomers. They found *Methylprototribestin* as the effective resistance modified and the growth inhibitory dose (ID50) of the compounds ranged from 12.64 to 20.62 µg/ml. Recently, two new steroidal saponins were isolated from the fruits of *Tribulus terrestris* whose structures were determined by using Perkin Elmer 241MC spectropolarimeter at room temperature. NMR analysis was measured on 1H-NMR (600 MHz) and 13C-NMR (150 MHz): Bruker DRX-300 and DRX-600 spectrometer with TMS as internal standard. TOF of micromass spectrometer was used to measure HRESI–MS. TLC was carried out on plates precoated

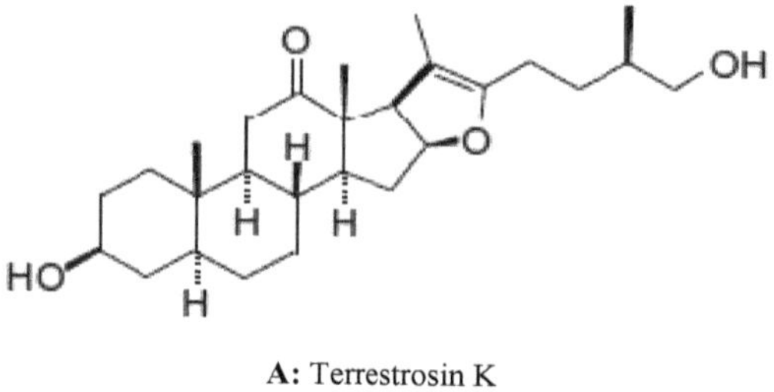

A: Terrestrosin K

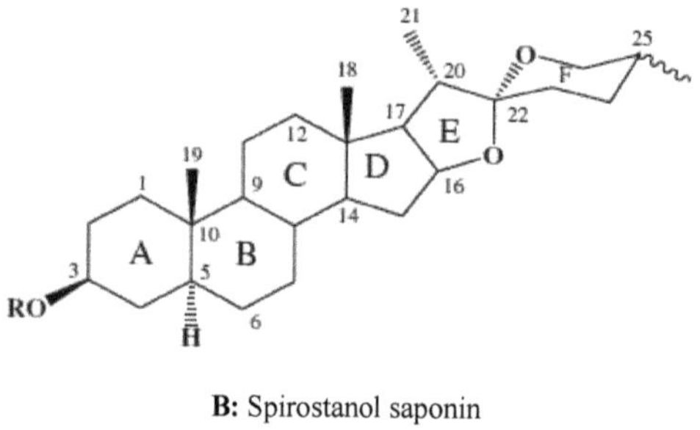

B: Spirostanol saponin

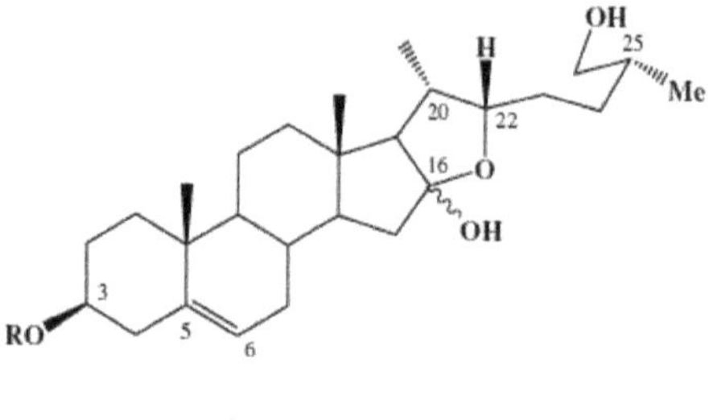

C: Furostanol saponin

Figure 18.2. Chemical structures of spirostanol saponin and furastanol saponin. (A) Terrestrosins A, (B) Spirostanol saponin, (C) Furastanol saponin.

with RP-18 gel (Merck) and silica gel F254 (Qingdao Marine Chemistry Ltd.). Those two structures are (1) 26-O-β-D-glucopyranosyl-(25R)-5α-furostane-12-one-3β, 22α,26-triol-3-O-β-D-glucopyranosyl (1 → 4)-β-D-galactopyranoside. (II) 26-O-β-D-glucopyranosyl-25(R)-5α-furostan-12-one-3β,22α,26-triol-3-O-α-L-rhamnopyranosyl-(1 → 2)-O-[β-D-glucopyranosyl-(1 → 4)]-β-D-galactopyranoside (Chen et al. 2013). The activity of the bioactive compound is analyzed using TLC and HPLC techniques. The compound is basically isolated from the extract with the help of TLC profiling using the suitable solvent system (Lade et al. 2014). The future drug compound is scraped from analytical TLC and used for identification of functional group (FTIR analysis), UV spectrophotometer, GCMS, LCMS, Bioautography for antimicrobial compound (Patil and Paikrao 2012), NMR (structure identification) and mass spectroscopy.

Assay/analytical method

Distinctive colours of bands on the TLC plate represent a definite class compound. As shown in Fig. 18.3, the blue colour band at Rf value is 0.91 corresponds to saponin. Thus a saponin can be detected easily using its Rf value. The positive VSA (vanillin sulphuric acid) test for metabolite identification confirms the presence of saponins in *T. terrestris* as shown in Fig. 18.4. This test is carried out by spraying of solution,

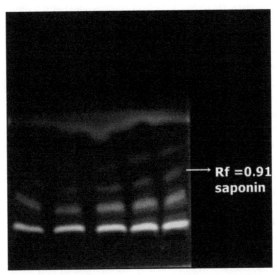

Figure 18.3. TLC of crude extracts of *Tribulus terrestris*, at Rf 0.91 appeared saponin under UV transilluminator at 365 nm.

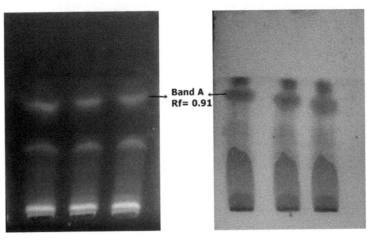

Figure 18.4. Derivatization of saponin from on TLC plates after spraying VSA (vanillin sulphuric acid) reagent.

I (1% ethanolic vanillin) and followed by spraying of solution II (10% ethanolic sulphuric acid). After heating in an oven at 110°C for 5 to 10 minutes, the plate shows a brownish black band confirming the presence of saponins (Tandon and Sharma 2010).

Available market supplements

Tribulus extract is usually found in combination with other sex-enhancing herbs in various libido products of the Australian pharmacy. Herbal medicines prepared from this plant extracts may range from 250 to 750 mg doses. The main contents in various herbal medicines prepared from *T. terrestris* extracts have various percentage of saponin (20%, 40%, 45%, 60%) and 20%, 40% of protodioscin. Even it's sale in powder form consists of 10; 20 and 40% extract powder. Several *T. terrestris* herbal and pharmacological products are available in markets on a large scale, a few are summarized in Table 18.3.

Antiurolithic activity

T. terrestris is one of the well-known herbs for the ailment of kidneys. It has been successfully used for treatment of urinary tract diseases. It increases the flow of urine and helps remove calcareous stones. It was found that biochemical parameters in urine, serums were restored in a dose-dependent manner. Substances that increase the output of urine are called diuretics (Ukani et al. 1997). Diuretic activity was a characteristic feature for the presence of potassium salt in higher concentration, which is confirmed in albino rats and potential extract to inhibit nucleation and growth of the CaOx crystals along with cytoprotective property. Joshi et al (2005) studied the effect of aqueous extract of *T. terrestris* Linn. and *Bergenia ligulata* Linn. against the growth of calcium oxalate monohydrate crystals for their inhibition activity by double diffusion gel growth technique using U-tubes.

The activity of *T. terrestris* was investigated on nucleation and the growth of the calcium oxalate (CaOx) crystals. It was found that the extract showed a concentration dependent inhibition of nucleation and the growth of CaOx crystals. However, it also has a cytoprotective role (Aggarwal et al. 2010). An experiment was conducted where an artificial urolithiasis was induced in an albino rat and orally administrated doses of 25, 50, 100 mg/kg for four months. In this case, urolithiasis was observed, which was consistent with the dose amount. When these artificial induced albino rats were subjected to *T. terrestris* extracts, the formation of oxalate activity was inhibited. Supplements that are derived from *T. terrestris* shows reduce activity in liver enzymes such as GAO, GAD (Mohd et al. 2012). The Gokshuradi churna a *Tribulus terrestris* extracted product was studied and provided scientific proof for Gokshuradi churna as an antiurolithiatic, where calcium oxalate crystallization was induced by the addition of 0.01 M sodium oxalate solutions in synthetic urine and nucleation method (Srinivasa et al. 2013).

The hydroalcoholic extract of fruits of *Tribulus terrestris* (Gokhru, HAEG), at a dose of 50 mg/kg combining with comestible diet, which is the composed of magnesium, potassium, phosphorus and calcium. The antiurolithic activity of the hydroalcoholic extracts of Gokhru at a dose of 50 mg/kg in combination with comestible was evaluated by ethylene glycol (0.75% W/V) induced hyperoxalurea in group II, III, IV and V animals (Chavva et al. 2013). Here, Group I was taken as normal, group II was taken as control, group III as standard group, group IV and V received HAEG and HAEG along with the diet. BUN and creatinine were estimated in the serum. Oxalate, calcium, phosphate, magnesium and uric acid were estimated in the urine. The results were as expected and has prevented the development of urolithiasis in rats treated with ethylene glycol. The results positively conclude the synergetic effect produced when HAEG was used in combination with comestible showing better antiurolithic activity. A novel antilithic protein of ~ 60 kDa showed cytoprotective potency was obtained from *T. terrestris* (Chhatre et al. 2014).

Table 18.3. *Tribulus terrestris* containing pharma products along with its name and main constituents and its price.

Sr. no.	Company Name	Product name	Main constituents	Other ingredients	Price	Country	References/ link
1	Himalaya Global Holding limited	Tribulus	Fruit extract	18% saponins	$14.95	Texas	http://www.himalayahealthcare.com/herbalmonograph/land-caltrops.htm#5
2	Himalaya herbal health care	Tribulus	Organic powdered or aerial parts	1% saponins	$14.95	Texas	http://www.himalayausa.com/products/pureherbs/gokshura.htm
3	Baoji Oasier Bio-Tech Co., Ltd.	*Tribulus terrestris* 90% saponins	90% saponins	-	$40	China	http://www.alibaba.com/trade/search?SearchText=Tribulus+Terrestris+
4	Daxinganling Lingonberry Organic Foodstuffs Co., Ltd.	GMP) High quality *Tribulus Terrestris*	Extract	(Saponins 40%–90%)	$40	China	http://www.alibaba.com/trade/search?SearchText=Tribulus+Terrestris+
5	Himalaya Drug Company	Gokshura	250 mg extract of Gokshura	-	€11,00	India	http://www.ayurveda24h.com/en/ayurvedic-medicine/36-gokshura-himalaya-tribulus-terrestris.html
6	Int'l IMP/EXP Trading ETC Unlimited	I Force nutrition	active saponins and protodioscin	-	$9.44	US	http://www.amazon.com/IForce-Nutrition-IFO1015-670-1501/dp/B004HXNYDG/ref=pd_sbs_hpc_4?ie=UTF8&&refRID=0R3QP26PH3WJJ2541E3C
7	Now Foods	Tribulus 1000 mg 45% Extract	45% saponin	Vegetarian Formula	$20.99	US	http://www.amazon.com/Foods-Tribulus-1000mg-ExtractTablets/dp/B00IDNV56Q/ref=pd_sbs_hpc_6?ie=UTF8&&refRID=0R3QP26PH3WJJ2541E3C
8	Ramdev Baba Products Online Patanjali products	Divya Yograj Guggulu	Total Powder	Many herbal constituents	$9.50	India	http://www.shopayurveda.com/product/divya-patanjali-gokshradi-guggul/
9	Sorvita Health Products	Sorvita *Tribulus terrestris*	*Tribulus terrestris* Extract	• Sorry, this item is not available in • Image not available *Tribulus terrestris* Extract 1000 mg: 95% Steroidal Saponins, 80% Protodioscin. 90 Capsules	$18.95	USA	http://www.amazon.com/Tribulus-Terrestris-Extract-1000mg-Protodioscin/dp/B00EULVH2K

Table 18.3. contd....

Table 18.3. contd.

Sr. no.	Company Name	Product name	Main constituents	Other ingredients	Price	Country	References/ link
10	Sport speed	*Tribulus terrestris* 1000 mg x 90 capsules, 95%	Steroidal Saponins, 80% Protodioscin	maltodextrin, HPMC, magnesium stearate	$17.81	USA	http://www.amazon.com/Tribulus-Terrestris-Extract-1000mgProtodioscin/dp/B00EULVH2K/ref=pd_sim_hpc_3 8?ie=UTF8&refRID=0Z7JYAE2XY5A T7HE91SX
11	True strength optimum nutrients	TRIBULUS 625 CAPS	625 mg blend of *Tribulus terrestris* powder	Tribulus 625 Caps contain sterols, flavonoids.	$10.99	US	http://www.optimumnutrition.com/ products/tribulus-625-caps-p-245.html
12	Ultimate Nutrition	Bulgarian Tribulus, 90 Capsules	100% Bulgarian Tribulus (aerial part): 750 mg (*Tribulus Terrestris*)	gelatin, cellulose, dicalcium phosphate and magnesium stearate	$14.95	America	http://www.a1supplements. com/Bulgarian-Tribulus-90-Capsules-p-2383.html
13	Ultimate super food	Ojio *Tribulus terrestris* extract - 2 oz	45% saponins		$7.9 9	USA	http://ultimatesuperfoods.com/store/ Products/Supplements/ayurveda/ TRIBUL.?t=True
14	Vaidya Atreya Smith	CAP-206 Gokshura	Gokshura fruit powder		€19.95	UK	http://www.atreya.com/ayurveda/ Gokshura-Tribulus-terrestris.html
15	Xian BZ Biotech Co., Ltd.	BZ China *Tribulus terrestris* Extract	Total Powder		$50	China	http://www.alibaba.com/trade/search?Se archText=Tribulus+Terrestris+
16	Xi'an Xuhuang Bio-Tech Co., Ltd.	*Tribulus terrestris* extract	total saponins 98%	20%–98% saponins	$15/ Kilogram	Shanghai/ Tianjin/china	http://www.alibaba.com/product-detail/tribulus-terrestris-extract-total-saponins-98-_1310801869.html?s=p
17	Ramdev Baba Products Online Patanjali Yog Vidyapeeth	Divya Patanjali Gokshradi Guggul	Shoot extract	Gokshura root (*Tribulus terrestris*) Guggulu resin Musta root Amalaki fruit Bibhitaki fruit Haritaki fruit Pippali fruit Ginger root Black Pepper fruit	• 25	India	http://www.shopayurveda.com/product/ divya-patanjali-gokshradi-guggul/

Cardiotonic activity

The (tribulosin) saponins *Tribulus terrestris* is effective at cardiac ischemia/reperfusion injury and the underlying mechanism in rats. The tribulosin reduced the myocardial apoptosis rate and treated rats showed reduced MDA, AST, CK and LDH contents with elevated activity of SOD (Zhang et al. 2010). The major phytochemical saponin is positive in response to dilate the coronary artery and improve circulation in blood vessels. A clinical trial conducted, supported this data, 406 patients with coronary heart disease were treated with *T. terrestris* extract. As a result, total effective rate of remission of angina was 82.3% and efficacious rate of ECG was improved by 52.7% as compared to the control, which was 35.8% (Mohd et al. 2012, Chhatre et al. 2014).

Aphrodisiac activity

It has been reported that a compound from *T. terrestris* extracts called protodioscin, a steridiol saponin is responsible for increasing sexual desire. An experiment conducted by (Gauthaman et al. 2003) confirms that sexual behaviour and intracavernous pressure (ICP) measurements in rats scientifically support that protodioscin (PTN) as an aphrodisiac. They found an increase in ICP, which confirms the proerectile aphrodisiac property responsible for an increase in androgen and release of nitric oxide from the nerve endings of endothelium (Mohd et al. 2012). In order to prove an aphrodisiac effect of a *T. terrestris* the extract effect on relaxation of the Corpus Cavernosum (CC) was investigated (Do et al. 2012). The organ bath was used to study the relaxation effects and mechanism of action of the *T. terrestris* extract on rabbit CC. The intracavernous pressure (ICP) was calculated after oral administration of the extract for one month to evaluate the relaxation response of the CC *in vivo*. Results concluded that the ICP measured after oral administration of the *T. terrestris* extracts for one month was higher than that measured for the control group.

In an *in vivo* study, the *T. terrestris* extract showed a significant concentration-dependent increase in ICP, suggesting its role to improve erectile function. They are also used for treatment of infertility and libido disorders in men and women (Kostova et al. 2005). The effect of *T. territeris* on nicotinamide adenine dinucleotide phosphate diaphorase (NADPH-d) activity and Androgen Receptor (AR) immunoreactivity in rat brain studies were increased in both NADPH-d (67%) and AR immunoreactivity (58%) in TT treated group was observed. The increased in AR and NADPH-d positive neuron is due to the androgen increasing property of TT and suggests its aphrodisiac activity (Fatima et al. 2015).

Central nervous system (CNS) activity

T. terrestris extract effects and show CNS stimulant activity. In a study conducted by (Deole et al. 2011) it was shown that the harmine and β-carboline alkaloid present in extracts are responsible for antidepressant and anxiolytic activity in Swiss albino mice. Harmine is an inhibitor of monoamine oxidase, which helps to increase a level of dopamine within the brain. The experiment conducted on Swiss Albino mice shows antidepressant and anxiolytic activity on the 260 mg/kg dose of tablet containing the major content of a root and fruit extract. Similarly, the *in vivo* central nervous system stimulant or depressant locomotor activity has been studied suggesting that ethanolic leaf extracts in amount of 100 mg/kg shows CNS stimulant activity (Ravichandran et al. 2014).

Immunomodulatory activity

Various solvent systems used for extraction of *T. terrestris* and a major phytochemical saponin isolated from fruit shows increase in phagocytosis (Tilwari et al. 2011). The alcoholic extract enhanced humoral antibody and delayed the type hypersensitivity response which increased specific immune response.

Antidiabetic activity

T. terrestris extract specifically steroidal saponin significantly reduced the level of serum glucose, serum triglyceride and serum cholesterol when administrated to rats. The extract prepared by boiling methods showed inhibition of gluconeogenesis in mice (Li et al. 2001). It is found that *T. terrestris* extract decrease fasting glucose in diabietic rats by declining glycosylated haemoglobin, total cholesterol, triglycerides and LDL cholesterol (Tantawy et al. 2007). The extract causes dilation of the coronary artery and improves the circulation helping in cardiac complication of diabetes. Thus, it is successfully used for diabetes helping in lowering blood glucose and lipid levels (Chhatre et al. 2014).

Anti inflammatory activity

The anti inflammatory compounds could be screened using TM DFS high resolution GC-MS Thermo Fisher Scientific Inc. The extracted compounds run parallel to the standard known compounds and could be identified easily. *T. terrestris* is one of the medicinal plants used for inflammatory disorders. Its methanol extracts shows a dose-dependent inflammatory activity in rats (Baburao et al. 2009). The TT extracts have been used successfully in combination with other herbs for anti inflammatory activity. *In vitro* anti inflammatory activity polyherbal extract (PHE) was prepared and used for evaluation. It was made by mixing appropriate extract volume of viz, *Saraca indica, Symplocos racemosa, Hemidesmus indicus, Aloe vera, Asteracantha longifolia, Erythrina indica* and *Tribulus terrestris* (Deattu et al. 2012). However, TT ethanolic extract potentially inhibits mediators of proinflammatory cytokines, for example. Tumour Necrosis Factor-alpha (TNF-α) and interleukin (IL)-4 in the macrophage cell line effect various inflammatory conditions. The water decoctions of roots for a comparative study of few plants for their anti inflammatory activity administered orally using *in vivo* Carrageenan induced rat paw edema model. Plants at a dose of 1.8 ml/kg b.w. exhibited highest anti inflammatory activity for *Aegle marmelos* (28.20%), *Premna optusifolia* (25.78%), *Oroxylum indicum* (24.15%), *Desmodium gangeticum* (26.74%), *Uraria picta* (21.49%). However, they found less activity of *Tribulus terrestris* compared with their experimental plants (Nagarkar et al. 2013).

In order to hunt for anti inflammatory compounds from TT, the GC-MS analysis was performed and 25 major peaks on the chromatograms were identified by comparing with known anti inflammatory compounds. Those compounds are mono terpene lactone(−)-loliolide (immunosuppressive activity), benzenedicarboxylic acid and octadecadienoic acid derivatives (antitumour activity) and bis phenols shows anti inflammatory activity (Mohammed et al. 2014).

Analgesic activity

The analgesic activities of TT extracted in solvents have been evaluated. It was found that the extract of *T. terrestris* showed pain reducing activity by studying male mice using formalin and tail flick test. It was found that the methanolic extract of 100 mg/ml is sufficient to produce an analgesic effect. The effect of the extract was less than morphine and higher than aspirin (Heidari et al. 2007). The ethanol extracts of the TT leaves at doses of 50, 100 and 200 mg/kg was evaluated against the standard drug indomethacin at a dose of 20 mg/kg, p.o. It showed increased tolerance effect in acetic acid-induced writhing characterized by lowering the number of writhings in rats ($p < 0.01$) and confirming its analgesic activity (Ansari et al. 2012).

Anticancer activity

The aqueous extract of *T. terrestris* has inherent chemopreventive potential. A swiss albino male mouse was induced to papillomagenesis by using synthetic carcinogenic and a dose of 800 mg/kg aqueous extract was found sufficient to reduce tumour formation. The extract from a root show higher chemopreventive potential as compared with fruit extracts when used in the same concentration in papillomagenesis in mice (Kumar et al. 2006). It is reported that aqueous extract block proliferation of HepG2 cells and induces apoptosis thus helping against liver cancer cells. It was shown that aqueous extract has radioprotectant ability and serves to protect against radiation damage (Neychev et al. 2007). Major constituents saponins possess cytotoxic activity on human fibroblasts. As saponins concentration increases the incorporation of [3H] thymidine in DNA decreases, which shows a decline in proliferation rate (Neychev et al. 2007).

Anthelmintic activity

The extract from distinctive origins such as Australia, Bulgaria, China, India, Moldova, Romania, South Africa and Turkey possessed different properties. It is reported that 50% methanolic extract of Indian *T. terrestris* (whole plant) shows anthelmintic activity and it is found that tribulosin and sitosterol glycosides are the main agents responsible for same (Joshi et al. 2008). The extracts of *T. terrestris* show *in vitro* anthelmintic activity against *Caenorhabditis elegans*. The methanolic extract is more effective than petroleum ether, chloroform and water, against the nematode (Chhatre et al. 2014).

Antibacterial activity

All parts such as leaves, stems, fruits and roots of *T. terrestris* have potential antibacterial properties. Some species from Turkey and Iran showed antibacterial activity against *Eterococcus faecalis, Staphylococcus aureus, Escherichia coli* and *Pseudomonas aeruginosa.* The methanol extracts of fruits possessed potent bacteriocidal property against gram-positive (Mohammed 2008). Indian *T. terrestris* fruit and leaves extract are potent bactericidal against *E. coli* and *S. aureus* (Chhatre et al. 2014). The antimicrobial effect of total extracts of *T. terrestris* and its fraction containing benzoxazine derivative (terresoxazine) was studied against 10 Gram positive and negative and candida spp. by cup plate and disc diffusion methods. The total extracts showed an antibacterial effect on *E. coli, P. aeruginosa* and *B. subtilis.* The fraction of benzoxazine had no activity on tested bacteria. It is confirmed that antibacterial effects of the total extract is due to other bioactive compounds present in total extract of TT (Vala et al. 2014).

Larvicidal activity

The leaves of *Tribulus terrestris* have been tested for larvicidal and repellent effect. The leaf extract is extracted in various solvent systems and have been used for its activity. The solvent system such as ethanol, acetone and petroleum ether extracts were found to be the best against 3rd instar larvae and adults of mosquito, *Ae. aegypti* the vector of dengue fever. TT extracts that evoked 100% repellency or biting deterrence was petroleum ether extracts at a dose of 1.5 mg/cm2 compared with 100% repellency for commercial formulation, N, N-diethyl-3-methylbenzamide (DEET) at the same dose was found best (El-Sheikh et al. 2012). The fruit extracts of *Tribulus terrestris* showed larvicidal activity on the larva of *An. stephensi* and *Ae. aegypti, Cx. quinquefasciatus* (Bansal et al. 2013). The repellent or antifeedant, larvicidal, pupicidal has been observed by petroleum ether extract of the leaves of TT (Chhatre et al. 2014).

Side effects

There are no papers suggesting side effects for use of extract of *T. terrestris* on *in vivo* studies. Table 18.4 summarizes several biological properties of *T. terrestris.*

Conclusion and future prospects

Tribulus terrestris extract are traditionally curable medicines used by several nations such as India, China, European nations and America. It is extensively used for severe ailments of cardiovascular, urinary tract diseases and induced spermatogenesis. The complete plant has tremendous medicinal and pharmacological activities such as a diuretic, aphrodisiac, antiurolithic, immunomodulatory, antihypertensive, antidiabetic, anticancer, anthelmintic, antibacterial, analgesic and anti inflammatory. Several potential herbal drugs have been formulated using available knowledge of therapeutic values of *T. terrestris*. The field of genetic engineering and immunology could be used for construction and formulation of engineered pharmaceutical

Table 18.4. Summary of various biological activities shown by extract of *T. terrestris*.

Activity	Note	Reference
Antiurolithic activity	An artificial induced albino rats, subjected to *T. terrestris* extracts, the formation of oxalate activity was inhibited.	(Aggarwal et al. 2010, Mohd et al. 2012, Verma et al. 2013, Vyawahare et al. 2014)
Cardiotonic activity	In an experiment, 406 patients with coronary heart disease treated with *T. terrestris* extract. In result, total effective rate of remission angina was 82.3% and efficacious rate of ECG was improved to 52.7% as compare to control, which was 35.8%. Chinese drug .Xinnao Shutong is prepared from crude saponins of *T. terrestris*, which is used effectively in treatment of coronary disease.	(Mohd et al. 2012, Verma et al. 2013, Vyawahare et al. 2014)
Aphrodisiac activity	"Protodioscin" a steridiol saponin is responsible for increasing sexual desire.	(Gauthaman et al. 2003, Do et al. 2012, Vyawahare et al. 2014, Fatima et al. 2015)
CNS activity	*T. terrestris* extract contains harmine and β-carboline alkaloid responsible for antidepressant and anxiolytic activity in Swiss albino mice.	(Deole et al. 2011) (Ravichandran et al. 2014)
Immunomodulatory activity	An alcoholic extract show increase in humoral antibody and delayed type hypersensitivity response which shows increased specific immune response.	(Tilwari et al. 2011, Fatima et al. 2015)
Antidiabetic activity	*T. terrestris* extract specifically steroidal saponin lowers serum glucose, serum triglyceride and serum cholesterol. The extract cause's dilation of coronary artery and improves the circulation helping in cardiac complication of diabetes.	(Li et al. 2001, Verma et al. 2013, Vyawahare et al. 2014, Fatima et al. 2015)
Anti inflammatory activity	The ethanolic extract inhibits mediators of tumor necrosis factor-alpha (TNF-α) and interleukin (IL)-4 in macrophage cell line that effect on various inflammatory conditions.	(Baburao et al. 2009, Mohammed et al. 2014, Fatima et al. 2015)
Analgesic activity	The methanolic extract of 100 mg/ml is sufficient to produce an analgesic effect.	(Heidari et al. 2007, Fatima et al. 2015)
Anticancer activity	The extract from a root and fruit show chemopreventive potential against Papillomagenesis in mine.	(Kumar et al. 2006, Vyawahare et al. 2014, Fatima et al. 2015)
Anthelmintic activity	The extracts of *Tribulus terrestris* show *in vitro* anthelmintic activity against Caenorhabditis elegans.	(Joshi et al. 2008, Chhatre et al. 2014, Vyawahare et al. 2014)
Antibacterial activity	The extracts of *Tribulus terrestris* show bactericidal property against Enterococcus faecalis, Staphylococcus aureus, Escherichia coli, Pseudomonas aeruginosa and gram-positive. Spirosaponins is found to be responsible for antimicrobial activity.	(Mohammed. 2008, Joshi, et al. 2008, Verma et al. 2013, Vyawahare et al. 2014, Vala et al. 2014, Fatima et al. 2015)
Larvicidal activity	The petroleum ether extract of the leaves of TT show larvicidal activity against the third instar larvae and adults of the mosquito.	(El-Sheikh et al. 2012, Chhatre et al. 2014, Bansal et al. 2013)
Antioxidant activity	The di-pcoumaroylquinic acid derivatives possess potent antioxidant activity.	(Dimitrova et al. 2012, Vyawahare et al. 2014, Fatima et al. 2015)

products with better efficiency and efficacy. Further molecular exploration would be required for understanding the mechanism of its shifting against several diseases and to profile novel drugs.

Acknowledgment

The authors gratefully acknowledge RGSTC, Mumbai for providing financial assistance. Bipin Lade thanks to UGC, New Delhi for BSR Research Fellowship.

References

Abirami, P. and Rajanderan, J. 2011. GCMS analysis of *Tribulus terrestris*. Asian J. Plant Sci. Res. 1(4): 13–16.

Aggarwal, A., Tandon, S., Singla, S.K. and Tandon, C. 2010. Diminution of oxalate induced renal tubular epithelial cell injury and inhibition of calcium oxalate crystallization *in vitro* by aqueous extract of *Tribulus terrestris*. Int. Braz. J. Urol. 36 (4): 480–489.

Akram, M., Asif, H.M., Akhtar, N., Shah, P.A., Uzair, M., Shaheen, G., hamim, T., Ali Shah, S.M. and Ahmad, K. 2011. *Tribulus terrestris* Linn: a review article. J. Med. Plants Res. 5(16): 3601–3605.

Alexiev, B. 2003a. Pharmacological composition based on biologically active substances obtained from *Tribulus terrestris*. Patent Application WO2003/070262.

Alexiev, B. 2003b. Pharmacological substance from *Tribulus terrestris*. Patent Application WO2003/070261.

Alexis, B. 2001. Natural, anti-bacterial, anti-inflammation, anti-virus, anti- herpes cream. Patent Application WO2001/011971.

Alexis, B. 2005. Treatment of vulvovaginitis with spirostanol enriched extract from *Tribulus terrestris*. Patent Application US2005/0112218.

Ansaria, J.A., Sayyed, M. and Amruzzama, Q. 2012. Analgesic and anti-inflammatory effect of ethanol extract of *Tribulus terrestris*. Res. J. Pharma. Bio. Chem. Sci. 3(4): 1365–1372.

Baburao, B., Rajyalakshmi, G., Venkatesham, A., Kiran, G., Shyamsunder, A. and Gangarao, B. 2009. Anti-inflammatory and antimicrobial activities of methanolic extract of *Tribulus terrestris* linn plant. Int. J. Chem. Sci. 7: 1867–1872.

Bansal, S.K., Singh, K.V. and Sharma, S. 2013. Larvicidal potential of (*Cleome viscose*) and Gokhru (*Tribulus terrestris*) against mosquito vectors in the semi arid region of western Rajasthan. J. Env. Bio. 35: 327–332.

Chavva, P., Myreddy, J., Srikanth, S., Chidrawar, V.R. and Rao, U.M. 2013. Estimation of antiurolithic activity of Gokhruand comestible in experimental urolithic rats. Asian J. Pharm. Res. Dev. 1(6): 62–69.

Chen, G., Su, L., Feng, S.G., Lu, X., Wang, H. and Pei, Y.H. 2013. Furostanol saponins from the fruits of *Tribulus terrestris*. Nat. Prod. Res. 27(13): 1186–1190.

Chen, H., Xu, Y. and Jian, Y. 2004. Application of *Tribulus* spirosteroid saponin compound in preparation of antifungal medicine. Patent Application CN1428349.

Chhatre, S., Nesari, T., Somani, G., Kanchan, D. and Sathaye, S. 2014. Phytopharmacological overview of *Tribulus terrestris*. Pharmacogn. Rev. 8(15): 45–51.

Chu, S.D., Qu, W.J., Li, M. and Cao, Q.H. 2003. Research advance on chemical component and pharmacological action of *Tribulus terrestris*. Chinese Wild Pla Res. 22: 4–7.

Conrad, J., Dinchev, D., Klaiber, I., Mika, S., Kostova, I. and Kraus, W. 2004. A novel furostanol saponin from *Tribulus terrestris* of Bulgarian origin. Fitoterapia. 75: 117–122.

Deattu, N., Narayanan, N. and Suseela, L. 2012. Evaluation of anti-inflammatory and antioxidant activities of polyherbal extract by *in vitro* methods. Res. J. Pharm. Biol. Chem. Sci. 3(4): 727–732.

Decombarieu, E., Fuzzati, N., Lovati, M. and Mercalli, E. 2003. Furostanol saponins from *Tribulus terrestris*. Fitoterapia. 74: 583–91.

Deole, Y.S., Chavan, S.S., Ashok, B.K., Ravishankar, B., Thakar, A.B. and Chandola, H.M. 2011. Evaluation of antidepressant and anxiolytic activity of *Rasayana Ghana* tablet (a Compound Ayurvedic formulation) in albino mice. Ayu. 32(3): 375–379.

Dimitrova, D.Z., Obreshkova, D. and Nedialkov, P. 2012. Antioxidant activity of *Tribulus terrestris*—a natural product in infertility therapy. Int. J. Pharm. Pharmc. Sci. 4(4): 508–511.

Do, J., Choi, S., Choi, J. and Hyun, J.S. 2012. Effects and mechanism of action of a *Tribulus terrestris* extract on penile erection. Sexual dysfunction. Korean J. Urol. 54: 183–188.

El-Sheikh, T.M.Y., Al-Fifi, Z.I.A. and Alabboud, M.A. 2012. Larvicidal and repellent effect of some *Tribulus terrestris* L. (Zygophyllaceae) extracts against the dengue fever mosquito, Aedes aegypti (Diptera: Culicidae). Journal of Saudi Chemical Society. http://dx.doi.org/10.1016/j.jscs.2012.05.009.

Evstatieva, L. and Tchorbanov, B. 2011.Complex investigations of *Tribulus terrestris* L. for sustainable use by pharmaceutical industry. Biotechnol. Biotechnol. Equip. 25: 2341–2347.

Fatima, L., Sultana, A., Ahmed, S. and Sultana, S. 2015. Pharmacological activities of *Tribulus terrestris* Linn: a systemic review. Wor. J. Pharm. Pharmac. Sci. 4(02): 136–150.

Francis, G., Kerem, Z., Makkar, H. and Becker, K. 2002. The biological action of saponins in animal systems: a review. Br. J. Nutr. 88: 587–605.

Gauthaman, K., Ganesan, A. and Prasad, R. 2003. Sexual effects of puncture vine (*Tribulus terrestris*) extract (protodioscin): an evaluation using a rat model. J. Altern. Complement Med. 9: 257–265.

Di Sansebastiano, G.P., De Benedictis, M., Carati, D., Lofrumento, D., Durante, M., Montefusco, A., Zuccarello, V., Dalessandro, G. and Piro, G. 2013. Quality and Efficacy of *Tribulus terrestris* as an Ingredient for Dermatological Formulations. Open Dermatology J. 7: 1–7.

Gincy, M.S., Mohan, K. and Indu, S. 2014. Comparative phytochemical analysis of medicinal plants namely *Tribulus terrestris, Ocimum sanctum, Ocimum gratissinum, Plumbago zeylanica.* Eur. J. Biot. Bios. 2(5): 38–40.

Gupta, V.K. and Arya, V. 2011. A review on potential diuretics of Indian medicinal plants. J. Chem. Pharm. Res. 3(1): 613–620.

Hashim, S., Bakht, T., Marwat, K.B. and Jan, A. 2014. Medicinal properties, phytochemistry and pharmacology of *Tribulus terrestris* l. (Zygophyllaceae). Pak J. Bot. 46(1): 399–404.

Heidari, M.R., Mehrabani, M., Pardakhty, A., Khazaeli, P., Zahedi, M.J., Yakhchali, M. and Vahedian, M. 2007. The analgesic effect of *Tribulus terrestris* extract and comparison of gastric ulcerogenicity of the extract with indomethacine in animal experiments. Ann. N Y Acad. Sci. 1095: 418–427.

Ivanova, A., Serly, J., Dinchev, D., Ocsovszki, I., Kostova, I. and Molnar, J. 2009. Screening of some saponins and phenolic components of *Tribulus terrestris* and *Smilax excelsa* as MDR modulators. *In Vivo* 23: 545–550.

Jayanthy, A., Deepak, M. and Remashree, A.B. 2013. Pharmacognostic characterization and comparison of fruits of *Tribulus terrestris* L. and *Pedalium murex* L. Int. J. Herb. Med. 1(4): 29–34.

Jing, H. 2009. Pawpaw coix seed facial mask. Patent Application CN102100658.

Joshi, D.D. and Uniyal, R.C. 2008. Different chemo types of Gokhru (*Tribulus terrestris*): A herb used for improving physique and physical performance. Int. J. Green Pharm. 158–161.

Joshi, V.S., Parekh, B.B. and Joshi, M.J. 2005. Herbal extracts of *Tribulus terrestris* and *Bergenia ligulata* inhibit growth of calcium oxalate monohydrate crystals *in vitro.* J. Crystal Growth 275(1-2): 1403–1408.

Ke, X. 2010. Formula of Chinese medicine capable of allowing skin to be white and tender. Patent Application CN102188631.

Koshy Philip.sim kae shin, abd malek, syarifah NSA. Rahman. 2011. Am. Applied Sci. 8. 1713.

Kostova, I. and Dinchev, D. 2005. Saponins in *Tribulus terrestris*—chemistry and bioactivity. Phyto. Rev. 4: 111–137.

Kozlova, O., Perederiaev, O. and Ramenskaia, G. 2011. Determination by high performance chromatography, steroid saponins in a biologically active food supplements containing the extract of *Tribulus terrestris.* Vopr Pitan. 80: 67–71.

Kumar, M., Soni, A.K., Shukla, S. and Kumar, A. 2006. Chemopreventive potential of *Tribulus terrestris* against 7, 12- dimethylbenz (a) anthracene induced skin papillomagenesis in mice. Asian Pac. J. Cancer Prev. 7: 289–294.

Lade, B.D., Patil, A.S., Paikrao, H.M., Kale, A.S. and Hire, K.K. 2014. A comprehensive working, principles and applications of thin layer chromatography. Res. J. Pharm. Bio. Chem. Sci. 5(4): 486–503.

Li, A. 2010. Chinese herb medicine for treating pruritus skin diseases. Patent Application TW201014611.

Li, M., Qu, W., Chu, S., Wang, H., Tian, C. and Tu, M. 2001. Effect of the decoction of *Tribulus terrestris* on mice gluconeogenesis. Zhong Yao Cai. 24: 586–588.

Mohammed, M.J. 2008. Biological activity of saponins isolated from *Tribulus terrestris* (Fruit) on growth of some bacteria. Tikrit J. Pure Sci. 13.

Mohammed, M.S., Khalid, S.H., Ahmed, W.J., Garelnabi, E.A.E. and Mahmoud, A.M. 2014. Analysis of anti-inflammatory active fractions of *Tribulus terrestris* by high resolution GC-MS. J. Pharmacogn Phytochem. 3(2): 70–74.

Mohd, J., Akhtar, A.J., Abuzer, A., Javed, A., Ali, M. and Ennus, T. 2012. Pharmacological scientific evidence for the promise of *Tribulus terrestris.* Int. Res. J. Pharm. 3(5): 403–406.

Nagarkar, B., Jagtap, S., Nirmal, P., Narkhede, A., Kuvalekar, A., Kulkarni, O. and Harsulkar, A. 2013. Comparative evaluation of anti-inflammatory potential of medicinally important plants. Int. J. Pharm. and Pharmac. Sci. 5(3): 239–243.

Neychev, V.K., Nikolova, E., Zhelev, N. and Mitev, VI. 2007. Saponins from *Tribulus terrestris* L. are less toxic for normal human fibroblasts than for many cancer lines: Influence on apoptosis and proliferation. Exp. Biol. Med. (Maywood). 232: 126–133.

Nikolova, A. and Vassilev, A. 2014. A Study on *Tribulus terrestris* L. anatomy and ecological adaptation. Biotech. Biotechnologi Equi. 25(2): 2369–2372.

Patil, A.S. and Paikrao, H.M. 2012. Bioassay guided phytometabolites extraction for screening of potent antimicrobials in *Passiflora foetida* L. J. Appl. Pharm. Sci. 2(9): 137–142.

Patil, A., Patil, S., Mahure, S. and Kale, A. 2014. UV, FTIR, HPLC confirmation of camptothecin an anticancer metabolite from bark extract of *Nothapodytes nimmoniana* (J. Graham). Ame J. Ethnomedicine 1(3): 174–185.

Pokrywka, A., Obmiński, Z., Lenczowska, J.M., Fijałek, Z., Lepa, E.T. and Grucza, R. 2014. insights into supplements with *Tribulus terrestris* used by Athletes. J. Human Kin. 41: 99–105.

Ponnusamy, S., Ravindran, R., Zinjarde, S., Bhargava, S. and Kuma, A.R. 2011. Evaluation of traditional Indian antidiabetic medicinal plants for human pancreatic amylase inhibitory effect *in vitro.* Evidence-Based Compl. Alt Med. Article ID 515647. 1–10.

Powers, C.N. and Setzer, W.N. 2015. A molecular docking study of phytochemcial estrogen mimics from dietary herbal supplements. *In Silico* Pharmacol. 3: 4. 2–63.

Ravichandran, S., Ajithalekshmi, S., Santhanamari, M., Poomari1, K., Theodore, E.A. and Selvakumar, S. 2014. *In vivo* central nervous system locomotor activity and phytochemical analysis of the *Tribulus terrestris* (Linn) leaf extracts. Asian J. Res. Bio. Pharm. Sci. 2(2): 56–61.

Singh, R., Hussain, S., Verma, R. and Sharma, P. 2013. Anti-mycobacterial screening of five Indian medicinal plants and partial purification of active extracts of *Cassia sophera* and *Urtica dioica.* Asian Paci. J. Trop. Med. 13; 6(5): 366–371.

Srinivasa1, A.P.B., Kuruba, L., Khan, S. and Saran, G.S. 2013. Antiurolithiatic activity of Gokhsuradi churan, an ayurvedic formulation by *in vitro* method. Adv. Pharm. Bull. 3(2): 477–479.

Tantawy, W.H.E.I. and Hassanin, L.A. 2007. Hypoglycemic and hypolipidemic effect of alcoholic extract of *Tribulus alatus* in streptozocin induced diabetic rat: a comparative study with *Tribulus terrestris* (Caltrop). Ind. J. Exp. Bio. 45: 785–790.

Tilwari, A., Shukla, N.P. and Devi, U. 2011. Effect of five medicinal plants used in Indian system of medicines on immune function in wistar rats. Afr. J. Biotechnol. 10(73): 16637–16645.

Ukani, M.D., Nanavati, D. and Mehta, N.K. 1997a. A review on the Ayurvedic herb *Tribulus terrestris* L. Anci. Sci. Life. 17(2): 144–150.

Ukani, M.D., Nanavati, D.D. and Mehta, N.K. 1997b. A review on the Ayurvedic herb *Tribulus terrestris* L. Anci. Sci. Life. 17(2): 1–6.

Vala, M.H., Makhmor, M., Kobarfar, F., Kamalinejad, M., Heidary, M. and Khoshnood, S. 2014. Investigation of Antimicrobial effect of *Tribulus terrestris* against some gram positive and negative and *candida* spp. Nov. in Biomed. 3: 85–89.

Verma, P., Galib, Patgiri, B.J. and Prajapati, P.K. 2013. A *Tribulus terrestris* Linn. phytopharmological review. J. Ayu. and Hol. Med. 1(3): 37–43.

Vyawahare, J.N. and Damle, M.C. 2014. *Tribulus terrestris*: an overview. International J. Pharm. Res. Dev. 6(08): 136–145.

Zhang, J.D., Xu, Z. and Cao, Y.B. 2006. Antifungal activities and action mechanisms of compounds from *Tribulus terrestris* L. J. Ethnopharmacol. 103: 76–84.

Zhang, S., Li, H. and Yang, S. 2010. Tribulosin protects rat hearts from ischemia/reperfusion injury. Acta Pharmacologica Sinica. 31: 671–678.

Zhang, Y. 2011. Formulas of internal medicine and external medicine for treating leucoderma. Patent Application CN102240337.

Vitex negundo: Bioactivities and Products

Gauravi Agarkar, Priti Jogee, Priti Paralikar and Mahendra Rai*

Introduction

In recent times, medicinal plant research has gained a lot of attention in the scientific community. The immense potential of medicinal plants has been already proven by many old traditional therapies such as Ayurveda, Unani and tribal medicines, now these medicinal plants are being identified and screened for their bioactivities by modern scientific techniques creating evidence in literature, and contributing to the field of pharmacology (Tapsell et al. 2006, Triggiani et al. 2006, Ahuja et al. 2015). According to World Health Organisation (WHO), about 80% of the daily health care needs are fulfilled by traditional use of aromatic and medicinal herbs (Chowdhury et al. 2009). Despite numerous developments in the field of medicine and synthetic drug chemistry, medicinal herbs are the precursors of many commonly prescribed drugs and are one of the major sources of raw material for their industrial production.

Several clinical and pharmaceutical explorations have been carried out to isolate and identify the active principles of particular medicinal plants and to study their mode of action for various ailments. These active principles are mostly secondary metabolites synthesized by plants which vary in their contents and activity, both qualitative and quantitative based on the plant species, location, environmental conditions and available resources (Vishwanathan and Basavaraju 2010). This means the market value of such herbal products mainly depend on their active components. One such valuable plant is *Vitex negundo* Linn. found in the Indian continent which is a very important medicinal plant commonly known as Nirgundi (Meena et al. 2010). There is a popular local quote in Bhangalis in the western Himalayan region of India, quoted by Uniyal et al. (2006) translates as, "A man cannot die of disease in an area where *Vitex negundo*, *Adhatoda vasica* and *Acorus calamus* are found". This underlines the significance of these medicinal plants and that the people should be aware of their uses.

V. negundo features innumerable medicinal properties and has been widely used for the treatment of a variety of diseases (Prajapati et al. 2004). It shows various therapeutic activities such as antioxidant, anti-inflammatory, antibacterial, antifungal, analgesics, anticonvulsant, hepatoprotective, bronchial relaxant,

Department of Biotechnology, SGB Amravati University, Amravati-444602, Maharashtra, India.
* Corresponding author: mahendrarai@sgbau.ac.in; mkrai123@rediffmail.com

anti-arthritic, polymenorrhoea, amenorrhea, erratic menstrual cycles, traumatic epilepsy, eating disorders, drug abuse, etc. (Azarnia et al. 2007, Tripathi et al. 2009, Petchi et al. 2011, Khare et al. 2014). It also works well in the treatment of skin infections, superficial bruises, injuries and sores (Azarnia et al. 2007, Tripathi et al. 2009). All plant parts of *V. negundo* are useful, but its roots and leaf extracts have been found to be more active (Azhar-ul-Haq et al. 2006). Roots of *V. negundo* help in treating disorders such as rheumatism, piles, dyspepsia whereas the oil extracted from leaves boost the brain function and enhances hair growth. Besides this, the essential oil of *V. negundo* has a large demand in food, flavor, cosmetics, perfume and pharmaceutical industries (Azarnia et al. 2007, Tripathi et al. 2009). Phytochemical studies of *V. negundo* revealed the variety of compounds including secondary metabolites such as volatile oils, flavonoids, lignans, terpenes (triterpenes, diterpenes, sesquiterpenes) and steroids (Chandramu et al. 2003, Maurya et al. 2007). These compounds are mainly responsible for diverse bioactivities of *V. negundo*. It has also been proved as an efficient bio-control agent against different pests and insects (Vishwanathan and Basavaraju 2010).

Owing to the variety of applications of *V. negundo*, a good range of its products are available in the market. Considering this vast array of applications of *V. negundo* and its increasing demand for commercial products at the industrial level, there is a need for conservation and systematic cultivation of this plant before it may go to the category of endangered plants. Efforts are being taken in this direction with the help of *in vitro* culture techniques which offer viable means of mass multiplication and germplasm conservation of plant species (Bajaj 1988). Vishwanathan and Basavaraju (2010) suggested tissue culture technique as better alternatives for faster propagation and conservation of this economically important plant as it is not possible by conventional methods due to its slow growth rate and poor viability of seeds. Additionally, enhancement of secondary metabolites production by using *in vitro* cell culture techniques thereby decreasing the dependence on fully grown plants in nature can be achieved (Vishwanathan and Basavaraju 2010).

The present chapter summarizes the ethnomedicinal uses of *V. negundo* focusing its importance as a medicinal plant species, its various important pharmacological activities and the updated information about commercially available products in the market.

Plant distribution and morphology

V. negundo is found mostly in tropical environments. It is distributed mainly in Indo-Malaysia region including Afghanistan, India, Pakistan, Bangladesh, Bhutan, China, Malaysia, Myanmar and some other countries. It is ubiquitous throughout India along banks of rivers and coastal areas as it grows best in moist environments. It is most common in grasslands, waste lands, water bodies and mixed open forests. It is also widely cultivated in Europe, North America and West Indies (Rastogi et al. 2010, Ladda and Magdum 2012, Rana and Rana 2014). Its taxonomic position as described by Rana and Rana (2014) is as follows, Kingdom—Plantae, Subkingdom—Tracheobionta, Superdivision—Spermatophyta, Class—Magnoliopsida, Subclass—Asteridae, Order—Lamiales, Family—Verbenaceae, Genus—*Vitex*, Species—*negundo*. Genus *Vitex* includes about 250 species distributed globally mainly in tropical and subtropical regions (Li et al. 2014).

V. negundo grows as a shrub or tree up to 2 to 8 m in height. It is an erect shrub with a reddish brown to gray colour and a somewhat rough bark. Phyllotaxy shows opposite, palmately compound (3–5 leaflets) leaf (Fig. 19.1). Leaflets are lanceolate, about 4 to 10 cm long with serrate margin and pinnate venation. The central leaflet possesses a stalk and the terminal leaflet is quite long. Leaves are hairy on dorsal and pubescent on surface. Inflorescence is panniculate receme, 10 to 20 cm long bearing lavender colored flowers. The hermaphrodite flowers bear fragrance and entomophilous pollination is found in them. Pea sized black berries are found. Fruits are oval in shape, capsulated with four valves (Rastogi et al. 2010, Li et al. 2014, Rana and Rana 2014).

Figure 19.1. *Vitex negundo* plant.

Ethnomedicinal uses

The knowledge of indigenous people about the medicinal properties of plants and techniques for the use of various plant parts for a particular disease is considered as traditional medicines or ethnomedicines. Nature is a warehouse of medicinally important plants and since the last few decades, phytochemicals are gaining much importance in pharmaceutical industries, because of their merits over synthetic medicines. As phytomedicine helps to recover the body to its natural state rather than simply curing a disease, herbal remedies are gaining more popularity (Silver and Bostia 1993, Chhamata 2009). Both countries, India and China are major contributors for the increasing popularity of the ethnomedicines worldwide by applying the remedies described in their traditional therapies like Ayurveda, Unani, Chinese herbal medicines, Siddha, folk medicines, etc. The plant *V. negundo* is used as a medicine mostly in the Indian subcontinent and a lot of study related to its medicinal use was conducted in India.

The leaves of *V. negundo* are most popular for their medicinal values. Leaf extract is used as antifeedant, dried leaves act as a fumigant and when fresh leaves are burnt with grass, their smoke acts as an insecticide and is very powerful as a mosquito repellent (Ladda et al. 2012, Pushpalatha et al. 1995). Chawala et al. (1992) studied its activity against swelling by applying it in different ways. It can be used in an aqueous leaf extract form or can be used as a plaster by applying as a paste of crushed leaves. Ingestion of its seeds with sugar can be effective against swelling. Aqueous extract of roots is very effective against toxins. It can be used to prevent spreading of toxins in the body, in combination with bark extract and proved effective against scorpion stings. Also the antidote activity of roots is very effective for snake venom (Samy et al. 2008).

Roots, leaves and flowers when used in combination showed good results as an astringent, diuretic and tonic (Arora et al. 2011). Leaf decoction is effective to cure wounds, ulcers, allergy, catarrhal fever, gout, whooping cough, flu and to improve eye sight. However, some researchers have studied the efficacy of the leaf and have shown that if leaves are stuffed in a pillow, may provide relief to a patient from headaches, eye diseases, cataract, watery eyes (Kirtikar and Basu 1984, Joshi and Joshi 2000, Khan and Rashid 2006, Au et al. 2008, Ong 2008, Ladda and Magdum 2012). Roots have been proved as effective as leaves, bearing medicinal values. Use of tincture from roots to cure dysentery was studied and when applied in powdered form, proved its medicinal value against dyspepsia and colic diseases. Decoction of root can be used by bronchitic and asthamatic patients (Warrier et al. 2002, Ivan 2005, Basavarju et al. 2009). In addition to this, *V. negundo* has many medicinal properties which are listed in Table 19.1.

Table 19.1. Ethnomedicinal uses of *V. negundo* in various regions of world.

Region	Plant part used	Ethnomedicinal uses to cure	References
Bangladesh	Leaf	Weakness, Headache, Vomiting, Malaria, Black fever	Khan and Rashid 2006
China	Leaf and root	Common cold, Flu and Cough	Au et al. 2008
Nepal	Leaf	Sinusitis, Whooping cough	Joshi and Joshi 2000
Pakistan	Leaf, fruits	Chest pain, Back pain, Allergy, Skin disease, Dentistry	Hamayun 2005, Shah et al. 2006, Zabihullah et al. 2006
Philippines	Stem bearing flowers	Cancer	Graham 2000
Sri Lanka	Leaf	Eye disease, Toothache, Rheumatism, as a tonic, carminative and vermifuge	Kirtikar and Basu 1984
India	Flowers, powder of root and bark, leaf, stem, seeds with sugar cane	Gastrointestinal disorders, Diarrhea, Dysentery, Headache, Asthama, Bronchitis, Dyspepsia, Cholera, Skin diseases, Eczema, Carbuncles, Leprosy, Burns, Jaundice, As diuretic, astringent and tonic, Swelling, Antitoxin, Antifeedant, Fumigant and Insecticide	Avdhoot and Rana 1991, Chawala et al. 1992, Pushpalatha et al. 1995, Ivan 2005, Saikia et al. 2006, Pattanaik et al. 2008, Samy et al. 2008, Basavaraju et al. 2009, Warrier et al. 2009, Arora et al. 2011, Ladda and Magdum 2012
Malaysia	Leaf, shoot, fruits	Headache, Gynecological disorders, Dysfunctional uterin, Gonorrhoea, Dysmenorrhoea, Galactogogue	Ong 2008, Tandon et al. 2008, Tasduq et al. 2008

Phytoconstituents of *V. negundo*

The plant *V. negundo* has an extensive number of phytoconstituents (secondary metabolites) which have immense medicinal value. Most parts of the plant *V. negundo* have this property as they contain complex mixture of secondary metabolites (Chowdhury et al. 2009). Many reports showed that leaves, roots, bark, flowers, fruits and seeds possess a diversity of chemical compounds among them as listed in Table 19.2. Also the structures of some important phytoconstituents are shown in Figs. 19.2, 19.3, 19.4, 19.5, 19.6 and 19.7.

Table 19.2. Diversity of phytoconstituents of different parts of *V. negundo.*

Plant body part	Phytoconstituents	References
Leaves	5-hydroxy-3,6,7,3',4'-pentamethoxyflavone	Banerji et al. (1969)
	6'-p-hydroxybenzoyl mussaenosidic acid; 2'-p-hydroxybenzoyl mussaenosidic acid	Banerji et al. (1969) Sehgal et al. (1982) Sehgal et al. (1983)
	5, 3'-dihydroxy-7,8,4'-trimethoxyflavanone; 5,3'-dihydroxy-6,7,4'-trimethoxyflavanone	Achari et al. (1984)
	Viridiflorol; β-caryophyllene; sabinene; 4-terpineol; gamma-terpinene; caryophyllene oxide; 1-oceten-3-ol; globulol	Singh et al. (1999)
	Betulinic acid [3β-hydroxylup-20-(29)-en-28-oic acid]; ursolic acid [2β-hydroxyurs-12-en-28-oicacid]; *n*-hentriacontanol; β-sitosterol; p-hydroxybenzoic acid	Chandramu et al. (2003)
	protocatechuic acid; oleanolic acid; flavonoids	Surveswaran et al. (2007)
Bark	Flavone glycosides-6-β-glucopyranosyl-7-hydroxy-3',4',5',8-tetramethoxyflavone-5-O-α-Lrhamnopyranoside, 3',7-dihydroxy-4',6,8-trimethoxyflavone-5-O-(6"-O-acetyl-β-Dglucopyranoside), 3,3',4',6,7-pentamethoxyflavone-5-O-(4"-O-β-D-glucopyranosyl)-α-L-rhamnopyranoside, 4',5,7-trihydroxyflavone-8-(2"-caffeoyl-β-D-glucopyranoside), 3',5,5',7-tetrahydroxy-4-methoxyflavone-3-O-(4"-O-β-D-galactopyranosyl) galactopyranoside. Leucoanthocyanidines, leucodelphindin methyl ether, leucocyanidin-7-O-rhamnoglucoside, luteolin, acerosin	Telang et al. (1999) Das et al. (1994)
	terpenes, sterols, phenolic compounds, alkaloids, organic acid, β-sitosterol, glucosides, anthocyanines, and p-hydroxybenzoic acid	Dhakal et al. (2008)
Root	2β, 3α-diacetoxyoleana-5,12-dien-28-oic acid; 2α,3α-dihydroxyoleana-5,12-dien-28-oic acid; 2α,3β-diacetoxy-18-hydroxyoleana-5,12-dien-28-oic acid; vitexin and isovitexin	Srinivas et al. (2001)
	Negundin-A; negundin-B; (+)-diasyringaresinol; (+)-lyoniresinol; vitrofolal-E and vitrofolal-F	Azhar-Ul-Haq et al. (2004)
	acetyl oleanolic acid; sitosterol; 3-formyl-4.5-dimethyl-8- oxo-5H-6,7-dihydronaphtho (2,3-b)furan	Vishnoi et al. (1983)
Flowers	a-selinene, germacren-4-ol, carryophylene epoxide, (E)-nerolidol, p-cymene,and valencene	Khokra et al. (2008)
Fruit	β-selinene, a-cedrene, germacrene D and hexadecanoic acid	Khokra et al. (2008)
Seed	angusid; casticin; vitamin-C; nishindine; gluco-nonitol; p-hydroxybenzoic acid; sitosterol	Khare (2004)
	3β-acetoxyolean-12-en-27-oic acid; 2α,3α-dihydroxyoleana-5,12-dien-28-oic acid; 2β,3α diacetoxyoleana-5,12-dien-28-oic acid; 2α,3β-diacetoxy-18 hydroxyoleana-5,12-dien-28-oic acid	Chawla et al. (1992a, 1992b)
	6-hydroxy-4-(4-hydroxy-3-methoxy-phenyl)-3-hydroxymethyl-7-methoxy-3, 4-dihydro-2-naphthaldehyde	Zheng et al. (2009)
	β-sitosterol; p-hydroxybenzoic acid; 5-oxyisophthalic acid; *n*-tritriacontane, *n*-hentriacontane; *n*-pentatriacontane; *n*-nonacosane	Khare (2004)

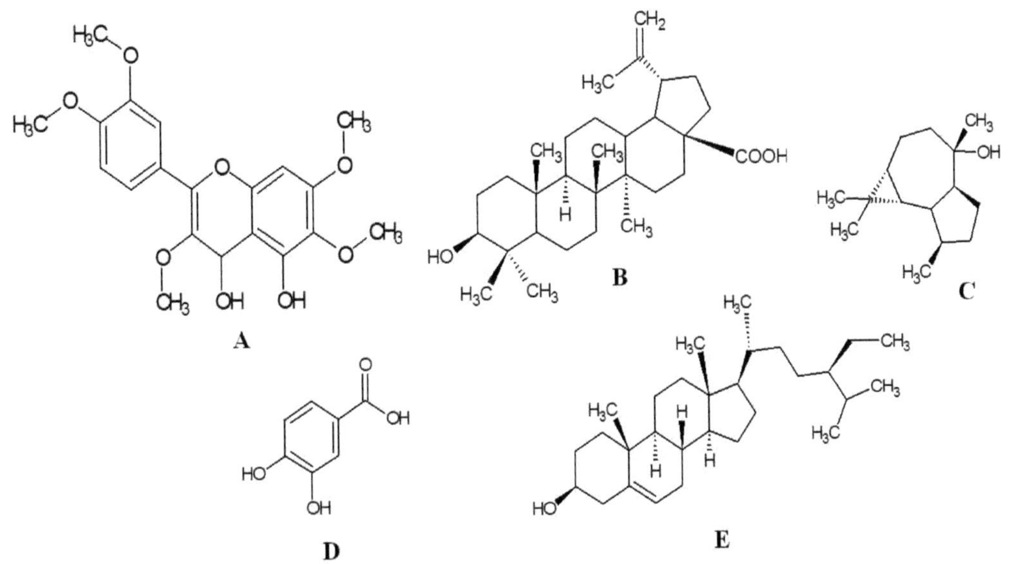

Figure 19.2. Phytoconstituents of Leaves; (A) 5-hydroxy-3,6,7,3',4'-pentamethoxyflavone, (B) Betulinic acid, (C) Viridifloral, (D) Protocatechuic acid, (E)-β-Sitosterol.

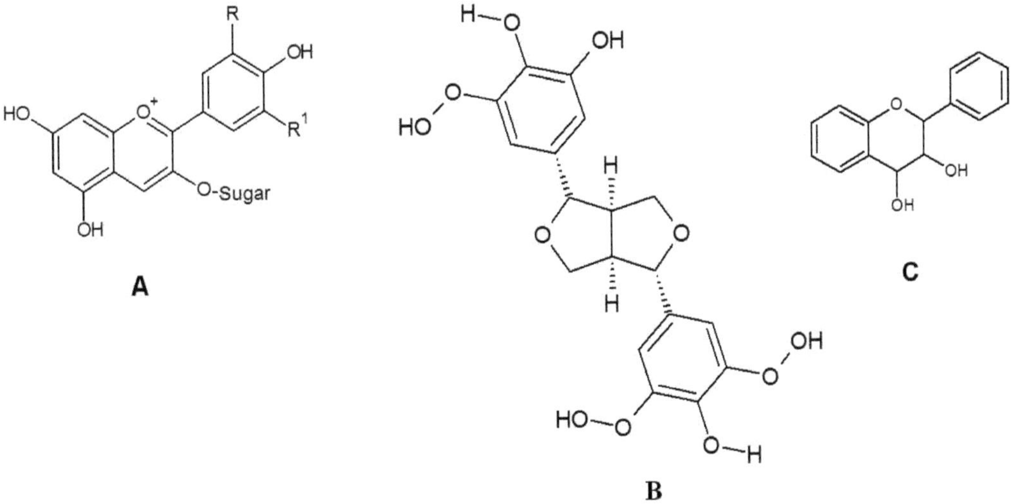

Figure 19.3. Phytoconstituents of Bark; (A) Anthocyanines, (B) Diasyringaresinol, (C) Leucoanthocyanidin.

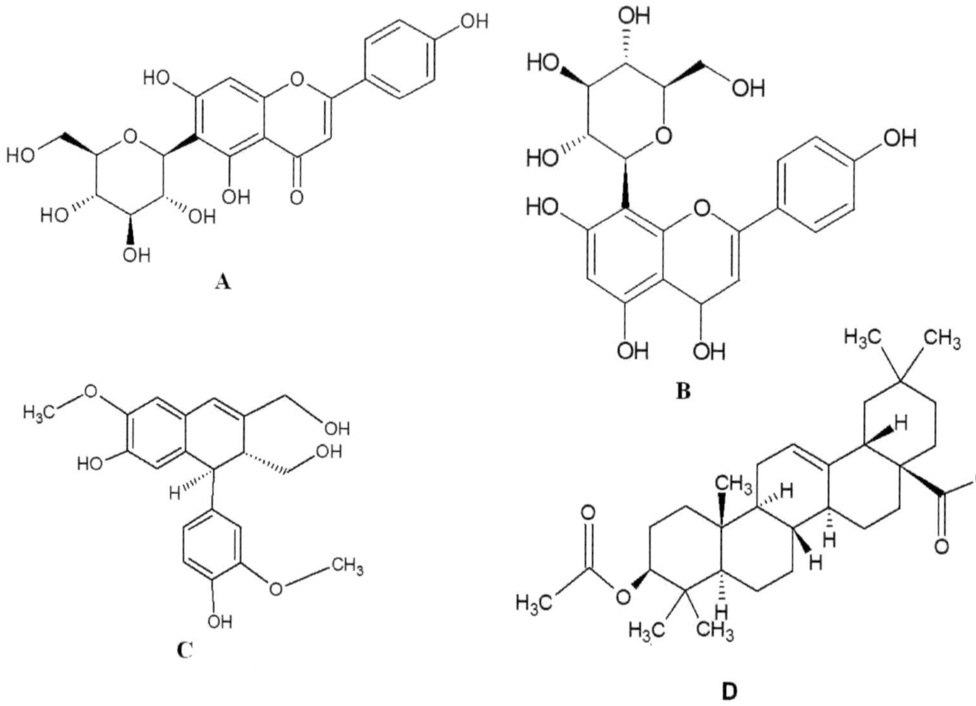

Figure 19.4. Phytoconstituents of Root; (A) Isovitexin, (B) Vitexin, (C) Negundin B, (D) Acetyl oleanoic Acid.

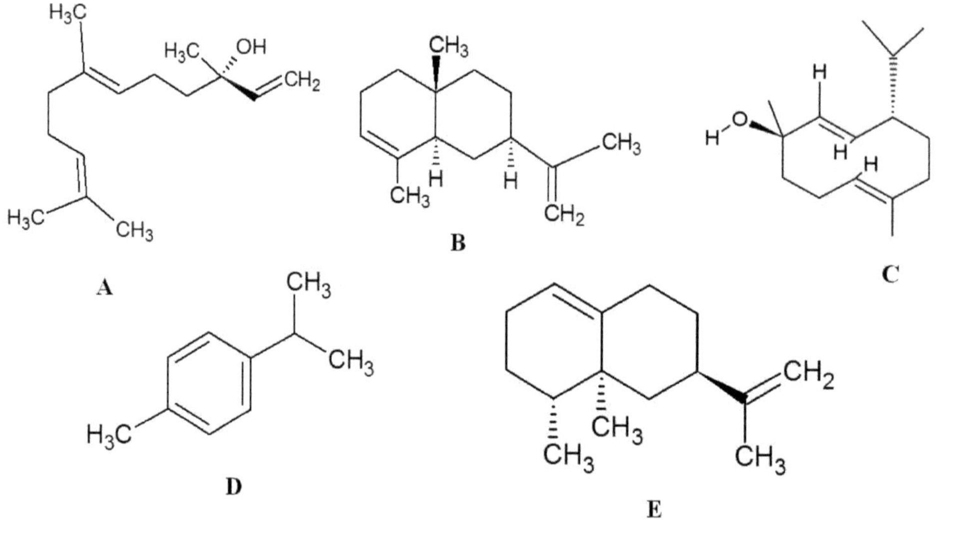

Figure 19.5. Phytoconstituents of Flower; (A) E-Nerolidol, (B) a-Selinene, (C) Germacrene-4-ol, (D) P-Cymene, (E) Valencene.

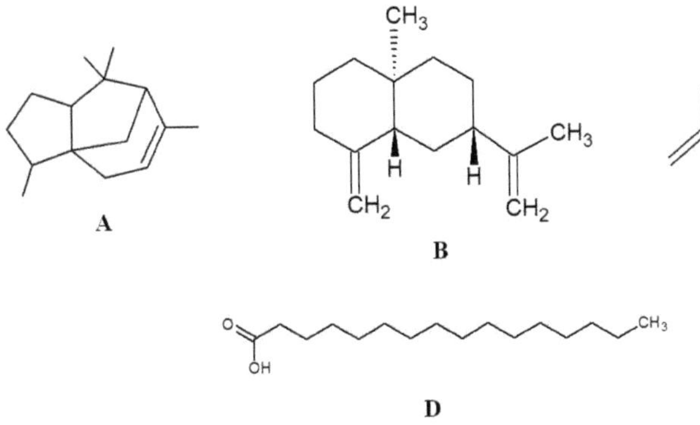

Figure 19.6. Phytoconstituents of Fruit; (A) alpha-Cedrene, (B) β-Selinene, (C) Germacrene D, (D) Hexadecanoic acid.

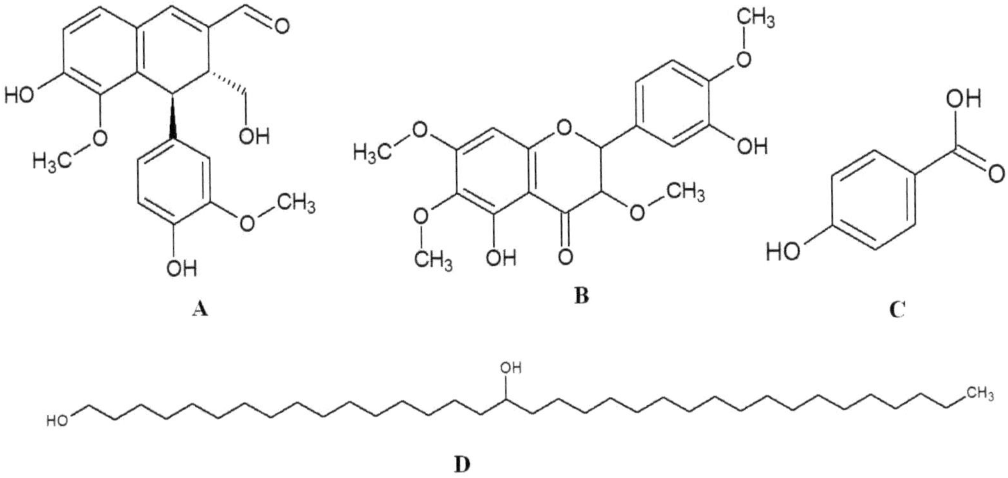

Figure 19.7. Phytoconstituents of Seed; (A) 6-hydroxy-4-(4-hydroxy-3-methoxy-phenyl)-3-hydroxymethyl-7-methoxy-3,4-dihydro-2-naphthaldehyde, (B) Casticin, (C) p-Hydroxybenzoic acid, (D) n-Pentatriacontane.

Important biological activities of *V. negundo*

All parts of plant *V. negundo* contribute to a large range of bioactivities against various microorganisms, insects, pests, antitumor, anti-inflammatory, antioxidant activity, etc. which are mentioned in Table 19.3 and few of them are discussed in detail.

Table 19.3. Various bioactivities of *V. negundo*.

Plant part	Biological activity	Activity against	References
Leaves	Antibacterial activity	*Bacillus cereus, Bacillus megaterium, Staphylococcus aureus, Sarcina lutea, Pseudomonas aeruginosa, Salmonella typhi, Shigella boydii, Shigella dysenteriae Vibrio mimicus, Vibrio parahemolyticus, Klebsiela pneumoniae, Vibrio cholera, Streptococcus mutans, Salmonella paratyphi, Eschericia coli, Proteus mirabilis, Bacillus subtilis, Enterobactor aerogens, Raoultella Planticola, Agrobacterium tumifacians*	Chowdhury et al. 2009, Rose and Cathrine 2011, Khare et al. 2014, Keerti et al. 2012, Padder et al. 2015
	Anti-fungal	*Alternaria alternata, Curvularia lunata, Trichophyton entagrophytes, Cryptococcus neoformans, Aspergillus niger, Candida albicans, Saccharromyces cevevaceae*	Guleria et al. 2006, Sathiamoorthy et al. 2007, Aswar et al. 2009, Chowdhury et al. 2009
	Insecticidal	*Callosobruchus maculates, Phthorimaea operculella, Sitotroga cerealella, Aphis citricola, Aphis gossypii, Myzus persicae*	Paneru et al. 2001, Das 1995, Rajendran et al. 2008, En-shun et al. 2009
	Antifilarial activity	*Setaria cervi* filarial parasite (*in vitro*)	Sahare and Singh 2013
	Larvicidal Antifungal	*Anopheles subpictus, Culex tritaeniorhynchus, Culex quinquefasciatus, Anopheles stephensi, Plutella xylostella, Cnaphalocrocis medinalis, Plasmodium falciparum*	Kamaraj et al. 2008, Karmegam et al. 1997, Kannathasan et al. 2007, Rahuman et al. 2009, Nathan et al. 2006
	Anti-asthmatic activity	Rat, Mice and G. pigs (*in vivo*)	Patel et al. 2009
	Antiepileptic activity	Swiss albino mice and male Albino Wistar rats	Kumar et al. 2011
	Antitumor activity	Swiss albino mice	Kannikaparameswari and Indhumathi 2013
	cytotoxic activity	Brine shrimp nauplii (aquatic crustaceans)	Chowdhury et al. 2009
	Hepatoprotective activity	Rat (*in vivo*)	Tondon et al. 2008
	Mosquito repellent	*Culex tritaeniorhynchus, Aedes aegypti*	Karunamoorthi et al. 2008, Hebbalkar et al. 1992
	Antioxidant activity	B6F10 Melanoma cell lines, WRL68 cell lines	Khare et al. 2014, Kadir et al. 2013
	Anticancer activity	HepG2 and WRL68 cell lines	Kadir et al. 2013
	Acaricidal activity	*Rhipicephalus* (Boophilus) *microplus*	Singh et al. 2014
	Anti-inflammatory activity	Albino rats (*in vivo*) Wistar albino rats of both sexes	Tandon and Gupta 2006, Tomar et al. 2015
	Antinociceptive activity	Mice (*in vivo*)	Gupta and Tandon 2005
	Anti-feedant	*Spodoptera litura, Achoea janata*	Sahayaraj et al. 2008
	Anti-filarial	*Brugia malayi*	Sahare 2008
Roots	Anxiolytic activity	Mice (*in vivo*)	Adnaik et al. 2009
	Anti-snake venom activity	*Vipera russellii, Naja kaouthia*	Alam and Gomes 2003
	Free radical scavenging activity	-	Maheshwari et al. 2015
Seeds	Effect on reproductive potential	Male rats	Das et al. 2004
Bark	Spermicidal and Post-coital Anti-fertility activity	Rats	Vasudeva et al. 2012
	Insecticidal activities	*Tribolium castaneum* (herbst)	Chowdhury et al. 2009

Insecticidal activity

Insecticidal activity of *V. negundo* was extensively studied compared to other cidal activities. Paneru and Shivakoti (2001) tested its insecticidal activity against *Callosobruchus maculatus* (pulse beetle). Oil extracted from *V. negundo* was mixed with lentil grains at various concentrations and the insects were exposed to these treated grains. The mortality rate of insects was recorded and it showed great potential as an insect repellent. The same substrate, i.e., essential oil of *V. negundo* was tested against two strains of moth *Corcyra cephalonica* and *Sitotroga cerealella*. The study showed that oil in combination with ethyl formate or CO_2 have excellent cidal activity in the adult moth than in eggs (Rajendran and Shriranjini 2008). A comparative study of Jiang et al. (2009) for three species of aphids showed that dichloromethane extract of seeds of this plant had higher toxicity for all three test organisms. Not only this, when its cytotoxicity was compared with commercially available insecticide imidacloprid, synergistic effect was found. The study revealed that insecticidal activity of imidacloprid was enhanced in combination with the seed extract.

Antifilarial activity

Antifilarial activity of *V. negundo* was studied by Sahare et al. (2008). A comparative study was conducted for three herbs, one of them was *V. negundo*. Extract of roots and leaves were studied and the results exhibited that the root extract of this plant at a concentration of 100 ng/ml, have total inhibition activity against microfilariae, causative agent of filaria.

Mosquito repellent activity

Apart from all other activities, leaves of *V. negundo* are also important as a mosquito repellent. While much work has not been done on this aspect, some investigations are interesting. Oil extracted from leaves of this plant contains polar compounds, found after fractionation of oils by column chromatography. These polar compounds were proved to possess mosquito repellence ability when studied against strain *Aedes aegypti*. This can protect against a mosquito bite for one–three hours in a single time application (Hebbalkar et al. 1992). Karunamoorthi et al. (2008) have studied the cidal activity of leaves against larvae of *Culex tritaeniorhynchus*. Petroleum ether extract of leaves was tested against this larvae and the experiment was carried out by the use of *V. negundo* leaf extract made from strips. These experimental strips were given to volunteers to wear and observed for their efficacy against mosquito bites. The complete protection was observed at the concentration 2 mg/cm^2 for eight hours. Similarly, the larvicidal activity of *V. negundo* was demonstrated against larvae of *Culex tritaeniorhynchus* by Adnaik and Pai (2008) and *Culex quinquefasciatus* and *Anopheles stephensi* by Pushpalatha and Muthukrishnan (1995).

Anti-inflammatory activity

Anti-inflammation activity refers to the property of a chemical compound or drug that helps to reduce swellings or inflammation in the target organ. *V. negundo* owns this property. A study of Murugesan et al. (2014) revealed the same. According to them, methanolic extract of *V. negundo* leaves have the ability as inhibitors of cyclooxygenase-2 (COX-2) enzyme which synthesizes prostanoids, the cause of the swelling. While performing experiments, they used aspirin and ibuprofen, commercially available drugs as control, and compared their docking score and gliding energy with the phytoconstituents of *V. negundo* plant. The results depict that, bioactive compound of the plant binds to the active site of the target molecule, i.e., 6COX_A protein molecule and had shown good results. All parts of plant *V. negundo* have a lage application in therapeutics. Seeds are also a good source of anti-inflammatory active compounds. Seed extract of this plant is a rich source of phenylnapthalene type lignin, it decreases the inflammatory factors present in serum as well as increases the anti-inflammatory cytokines that down regulates the COX-2 enzyme and helps to reduce pain in arthritic patients (Zheng et al. 2014). Many reports are available on

Table 19.4. Commercially available products of *V. negundo*.

Manufacturer	Name of product	Used in
Baidyanath Ayurved bhawan Pvt. Ltd., Kolkata, India	Nirgundi Tail	Pain relief oil
Phlemex forte	*Vitex negundo* L. Lagundi leaf (Syrup)	Cough, Asthma
Himalaya Drug Co., Bangalore	Acne-n-Pimple Cream	Acne and skin eruptions
	JointCare B cream	Rheumatic disorders
	Muscle and Joint Rub	Muscle strains, musculoskeletal disorders
	Pilex tablet and cream	Hemorrhoids (Piles)
	Rumalaya gel and tablets	Inflammatory musculoskeletal disorders
	V-Gel	Vaginitis, Cervicitis
	Liv-52	Liver detoxifier and healer
NOW food	Vitex Women's Health	Supports hormonal balance in women
eVaidyaji (natural herbal product)	Nirgundi powder	Improve immunity of body (richest source of vitamin C)
Hamdard Laboratories, New Delhi, India	Jigrine	Liver ailments
Dey's Medical, Kolkata, India	Itone Eye Drop	Eye ailments
Ambica Research & Development Pvt. Ltd., New Delhi, India	Amgesic Arthritis Tablet	Arthritis
Surya Herbal Ltd., Noida, India	Relieef Cream	Joint and Muscle pain, Stiff back
	Rheumanaad Tablet and Cream	Rheumatic Pain, Sprain
	Ostranil Gel	Osteoarthritis, Lumbago
IndSwift Ltd., Chandigarh, India	Arthrill Capsules and Massage oil	Arthritis, Joint pain, Frozen shoulder, Gout, Cervical spondylitis
VHCA Herbals, Haryana, India	Nigundi Ghrita	Traumatic weakness
Axiom Ayurveda, Ambala, Haryana, India	Nirgundi Panchang Swarasa	Useful in fever, acidity, migraines, headaches
Vindhya Herbals, Bhopal, India	Nirgundi Churna	Kidney pain, removes calculus in urinary system
Bellan Pharmaceuticals, Vadodara, Gujrat, India	Nirgundi Ghan Vati	Rheumatoid arthritis
Pascual Laboratories, Inc., Philippines	ASCOF Lagundi tablet and syrup	Cough and Asthma

the study of anti-inflammatory activity of *V. negundo*. Tandon (2005) has reviewed this aspect and reported that leaves of *V. negundo* have an antihistaminic activity, membrane stabilization and antioxidant activity. All these activities are due to its action mode similar to non steroidal anti inflammatory drugs (NSAID) (Telang et al. 1999).

Anti-bacterial activity

Several reports have demonstrated the significant activity of *V. negundo* against a large diversity of bacteria. Panda et al. (2009) found that the ethanol and methanol extracts of *V. negundo* leaves were inhibitory against both gram-positive and gram-negative bacteria while petroleum ether and chloroform extracts of bark were more active against gram-positive bacteria. Similarly, Rastogi et al. (2009) displayed significant antibacterial activity of ethanolic extract comparable to standard antibiotic tetracycline. Bacterial pathogens such as *Salmonella paratyphi*, *Vibrio cholerae*, *Klebsiella pneumoniae*, *Streptococcus mutans*, and *E. coli* were susceptible to the ethanolic leaf extract of *V. negundo* (Rose and Cathrine 2011).

Anti-fungal activity

In vitro antifungal activity of *V. negundo* fruits was evaluated against *Candida albicans*, *C. glabrata*, *Microsporum canis*, *Aspergillus flavus* and *Fusarium solani*. Ethanolic extract of fruit seeds exhibited significant activity against *F. solani* and moderate against *M. canis* while no effect on *C. albicans* (Mahmud et al. 2009). Gautam and Kumar (2012) showed that bound flavonoids of flowers of *V. negundo* were most potent against *A. flavus*, *A. niger*, *C. albicans*, *Trichophyton mentegrophytes*. Bioactivity guided fractionation of ethanolic leaf extract of *V. negundo* resulted in the isolation of new flavone glycoside along with five known compounds. The new flavone glycoside had significant antifungal activity against *T. mentagrophytes* and *Cryptococcus neoformans* with MIC 6.25 µg/ml (Sathiamoorthy et al. 2007). Several other authors also provided references for the antifungal activities of *V. negundo* extracts (Guleria and Kumar 2006, Aswar et al. 2007, Shaukat et al. 2009).

Commercial products of *V. negundo*

The pharmacological potential of *V. negundo* has been effectively utilized in the formulation of commercial products by different companies dealing with Ayurvedic medicines for various ailments. All such commercial products of *V. negundo* have been displayed in Table 19.4.

Conclusion

The plant *V. negundo* has great medicinal values which are being exploited to some extent with the commercial products presently available. Still, there is a lot of scope for research and exploration of new bioactivities. In the present scenario, to make the best use of its therapeutic potential, systematic experimental studies must be carried out. The validation of traditional knowledge using modern scientific approaches contribute to the pharmacological research leading to the formulation of new drugs and commercial therapeutic products. While commercializing medicinal products quality consciousness, legal requirements, ethical considerations and economical benefits are important and to achieve this, traditional knowledge must be revived and reaffirmed. Detailed studies on the phytochemistry and the mode of action of bioactive metabolites are necessary for further extending their applications in therapeutic advancement.

Considering the medicinal and commercial importance of *V. negundo*, care should be taken, not to disturb the natural biodiversity of such a vital herb, so that it may not be overexploited. Initiatives should be taken to conserve these plant species *in situ* as well as *ex situ*. The pharmaceutical companies should be obliged to efficiently cultivate these plants as per their requirements for manufacturing of the product, thus preserving its natural biodiversity.

References

Achari, B., Chowdhuri, U.S., Dutta, P.K. and Pakrashi, S.C. 1984. Two isomeric flavones from *Vitex negundo*. Phytochem. 23: 703–704.

Adnaik, R.S. and Pai, P.T. 2008. Laxative activity of *Vitex negundo* linn. Leaves. Asian J. Exp. Sci. 22(1): 159–160.

Adnaik, R.S., Pai, P.T., Sapakal, V.D., Naikwade, N.S. and Magdum, C.S. 2009. Anxiolytic activity of *Vitex negundo* linn. in experimental models of anxiety in mice. Int. J. Green Pharm. 3: 243–247.

Ahuja, S.C., Ahuja, S. and Ahuja, U. 2015. Nirgundi (*Vitex negundo*)—Nature's Gift to Mankind. Asian Agri-History 19(1): 5–32.

Alam, M.I. and Gomes, A. 2003. Snake venom neutralization by Indian medicinal plants (*Vitex negundo* and *Emblica officinalis*) root extracts. J. Ethnopharmacol. 86: 75–80.

Arora, V., Lohar, V., Singhal, S. and Bhandari, A. 2011. *Vitex negundo*—A Chinese Chaste Tree. Int. J. Pharm. Inn. 1(5): 9–20.

Aswar, P.B., Khadabadi, S.S., Kuchekar, B.S., Rajurkar, R.M., Saboo, S.S. and Javarkar, R.D. 2009. *In vitro* evaluation of anti-bacterial and anti-fungal activity of *Vitex negundo* (Verbenaceae). Ethnobotanical Leaflets 13: 962–967.

Au, D.T., Wu, J., Jiang, Z., Chen, H., Lu, G. and Zhao, Z. 2008. Ethnobotanical study of medicinal plants used by Hakka in Guangdong. China. J. Ethnopharmacol. 117: 41–50.

Avadhoot, Y. and Rana, A.C. 1991. Hepatoprotective effect of *Vitex negundo* against carbon tetrachloride induced liver damage. Arch. Pharm. Res. 14(1): 96–98.

Azarnia, M., Ejtemaei-Mehr, S., Shakoor, A. and Ansari, A. 2007. Effects of *Vitex agnus castus* on mice fetus development. Acta Medica Iranica. 45(4): 263–270.

Azhar-Ul-Haq, A. Malik, Anis, I., Khan, S.B., Ahmed, E., Ahmed, Z., Nawaz, S.A. and Choudhary, M.I. 2004. Enzyme inhibiting lignans from *Vitex negundo*. Chem. Pharm. Bull. 52: 1269–1272.

Azhar-ul-Haq, A. Malik, Khan, M.T.H., Anwar-ul-Haq, Khan, S.B., Ahmad, A. and Choudhary, M.I. 2006. Tyrosinase inhibitory lignans from the methanol extract of the roots of *Vitex negundo* Linn. and their structure-activity. Phytomed. 13: 255.

Bajaj, Y. 1988. Biotechnology in Agriculture and Forestry. Springer Verlag, Berlin.

Banerji, A., Chadha, M.S. and Malshet, V.G. 1969. Isolation of 5-hydroxy-3,6,7,3',4' pentamethoxyflavone from *Vitex negundo*. Phytochem. 8: 511–512.

Chandramu, C., Manohar, R.D., Krupadanam, D.G.L. and Dashavantha, R.V. 2003. Isolation characterization and biological activity of betulinic acid and ursolic acid from *Vitex negundo* L. Phytother. Res. 17(2): 129–134.

Chawla, A.S., Sharma, A.K., Handa, S.S. and Dhar, K.L. 1992a. Chemical investigation and anti-inflammatory activity of *Vitex negundo* seeds. J. Nat. Prod. 55: 163–167.

Chawla, A.S., Sharma, A.K., Handa, S.S. and Dhar, K.L. 1992b. A lignan from *Vitex negundo* seeds. Phytochem. 31: 4378–4379.

Chowdhury, J.A., Islam, M.S., Asifuzzaman, S.K. and Islam, M.K. 2009. Antibacterial and cytotoxic activity screening of leaf extracts of *Vitex negundo* (Fam: *Verbenaceae*). J. Pharm. Sci. Res. 1(4): 103–108.

Das, B. and Das, R. (nee Chakrabarti). 1994. Medicinal properties and chemical constituents of *Vitex negundo* Linn. Indian Drugs 31(9): 431–435.

Das, G.P. 1995. Plants used in controlling the potato tuber moth, *Phthorimaea operculella* (Zeller). Crop Prot. 14: 631–636.

Das, S., Parveen, S., Kundra, C.P. and Pereira, B.M.J. 2004. Reproduction in male rats is vulnerable to treatment with the flavonoid-rich seed extracts of *Vitex negundo*. Phytother. Res. 18: 8–13.

Dhakal, R.C., Rajbhandari, M., Kalauni, S.K., Awale, S. and Gewali, M.B. 2009. Phytochemical constituents of the bark of *Vitex negundo*. J. Nepal Chem. Soc. 23: 89–92.

En-shun, J., Ming, X., Yu-qing, L. and Yu-feng, W. 2009. Toxicity of *Vitex negundo* extract to aphids and its co-toxicity with imidacloprid. Chinese J. Appl. Ecol. 20: 686–690.

Gautam, K. and Kumar, P. 2012. Extraction and pharmacological evaluation of some extracts of *Vitex negundo* linn. Int. J. Pharm. Sci. 4(2): 132–137.

Graham, J.G., Quinn, M.L., Fabricant, D.S. and Farnsworth, N.R. 2000. Plants used against cancer–an extension of the work of Jonathan Hartwell. J. Ethnopharmacol. 73: 347–377.

Guleria, S. and Kumar, A. 2006. Antifungal activity of some Himalayan medicinal plants using direct bioautography. J. Cell Mol. Biol. 5: 95–98.

Gupta, R.K. and Tandon, V.R. 2005. Antinociceptive activity of *Vitex-negundo* Linn. leaf extract. Ind. J. Physiol. Pharmacol. 49(2): 163–70.

Hamayun, M. 2005. Ethnobotanical studies of some useful shrubs and trees of district Buner, NWFP, Pakistan. Ethnobotanical Leaflets 9.

Hebbalkar, D.S., Hebbalkar, G.D., Sharma, R.N., Joshi, V.S. and Bhat, V.S. 1992. Mosquito repellant activity of oils from *Vitex negundo* Linn. leaves. Ind. J. Med. Res. 95: 200–203.

Ivan, A.R. 2005. Medicinal Plants of the World: Chemical Constituents, Traditional and Modern Medicinal Uses. Humana Press, New Jersey, 510 p.

Jabeen, A., Khan, M., Ahmad, M., Zafar, M. and Ahmad, F. 2009. Indigenous uses of economically important flora of Margallah Hills National Park, Islamabad, Pakistan. Afr. J. Biotech. 8: 763–784.

Jayasree, T., Arpitha, T., Kavitha, R. and Kishan, P.V. 2012. Evaluation of anticonvulsant activity of ethanolic extract of *Vitex negundo* in Swiss albino rats. Int. J. Pharm. Phytopharmacol. Res. 1(4): 161–165.

Jiang, E.S., Xue, M., Liu, Y.Q. and Wang, Y.F. 2009. Toxicity of *Vitex negundo* extract to Aphids and its co-toxicity with Imidacloprid. Chinese J. App. Eco. 20: 686–690.

Joshi, A.R. and Joshi, K. 2000. Indigenous knowledge and uses of medicinal plants by local communities of the Kali Gandaki watershed area, Nepal. J. Ethnopharmacol. 73: 175–183.

Kamaraj, C., Rahuman, A. and Bagavan, A. 2008. Antifeedant and larvicidal effects of plant extracts against *Spodoptera litura* (F.), *Aedes aegypti* L. and *Culex quinquefasciatus* Say. Parasitol. Res. 103: 325–331.

Kannathasan, K., Senthilkumar, A., Chandrasekaran, M. and Venkatesalu, V. 2007. Differential larvicidal efficacy of four species of *Vitex* against *Culex quinquefasciatus* larvae. Parasitol. Res. 101: 1721–1723.

Kannikaparameswari, N. and Indhumathi, T. 2013. Haematological and cytotoxic effect of the ethanolic extract of *Vitex negundo*. Int. J. LifeSci. Bt & Pharm. Res. 2(1): 247–253.

Karmegam, N., Sakthivadivel, M., Anuradha, V. and Daniel, T. 1997. Indigenous plant extracts as larvicidal agents against *Culex quinquefasciatus* Say. Bioresour. Technol. 59: 137–140.

Karunamoorthi, K., Ramanujam, S. and Rathinasamy, R. 2008. Evaluation of leaf extracts of *Vitex negundo* L. (Family: Verbenaceae) against larvae of *Culex tritaeniorhynchus* and repellent activity on adult vector mosquitoes. Parasitol. Res. 103: 545–550.

Khan, N. and Rashid, A. 2006. A study on the indigenous medicinal plants and healing practices in Chittagong Hill tracts Bangladesh. African J. Trad. Compl. Alt. Med. 3: 37–47.

Khare, C.P. 2004. Encyclopedia of Indian Medicinal Plants. Springer, Berlin.

Khare, P., Kumar, N. and Kumari, T. 2014. Evaluation of antibacterial and antioxidant activity of phytochemical constituents obtained from leaves of *Vitex negundo*. IJIRAE. 1(10): 189–195.

Khokra, S.L., Prakash, O., Jain, S., Aneja, K.R. and Dhingra, Y. 2008. Essential oil composition and antibacterial studies of *Vitex negundo* linn. Extracts Ind. J. Pharm. Sci. 70(4): 522–526.

Kirtikar, K.R. and Basu, B.D. 1984. Indian Medicinal Plants. Bishen Singh Mahendra Pal Singh, Dehradun, India 3: 697.

Kumar, K.P., Vidyasagar, G., Ramakrishna, D., Reddy, I.M. and Gupta Atyam, V.S.S.S. 2011. Antiepileptic activity for methanolic extract of *Vitex negundo* leaf against different animal models. J. Chem. Pharm. Res. 3(4):159–165.

Ladda, P.L. and Magdum, C.S. 2012. *Vitex negundo* Linn.—ethnobotany, phytochemistry and pharmacology—A review. Int. J. Adv. Phy. Bio. Chem. 1(1): 111–120.

Maurya, R., Shukla, P.K. and Ashok, K. 2007. New antifungal flavonoid glycoside from *Vitex negundo*. Bioorg. Med. Chem. Lett. 17(1): 239–242.

Meena, A.K., Singh, U., Yadav, A.K., Singh, B. and Rao, M.M. 2010. Pharmacological and phytochemical evidences for the extracts from plants of the genus *Vitex*—A review. Int. J. Pharm. Clin. Res. 2(1): 01–09.

Merlin Rose, C. and Cathrine, L. 2011. Preliminary phytochemical screening and antibacterial activity on *Vitex negundo*. Int. J. Curr. Pharm. Res. 3(2): 99–101.

Murugesan, D., Ponnusamy, R.D. and Gopalan, D.K. 2014. Molecular docking study of active phytocompounds from the methanolic leaf extract of *Vitex negundo* against cyclooxygenase-2. Bangladesh J. Pharmacol. 9: 146–153.

Nathan, S.S., Kalaivani, K. and Murugan, K. 2006. Behavioural responses and changes in biology of rice leaffolder following treatment with a combination of bacterial toxins and botanical insecticides. Chemosphere 64: 1650–1658.

Ong, H.C. 2008. Tumbuhan liar-khasiat ubtan and kegunaan lain. Kuala Lumpur: Utusan publication& distributors Sdn Bhd. 121.

Panda, S.K., Thatoi, H.N. and Dutta, S.K. 2009. Antibacterial activity and phytochemical screening of leaf and bark extracts of *Vitex negundo* L. from similipal biosphere reserve, Orissa. J. Med. Plants Res. 3(4): 294–300.

Paneru, R.B. and Shivakoti, G.P. 2001. Use of botanicals for the management of pulse beetle (*Callosobruchus maculatus* F.) in lentil. Nepal Agri. Res. J. 4(5): 27–30.

Patel, J., Shah, S., Deshpande, S. and Shah, G. 2009. Evaluation of the antiasthmatic activity of leaves of *Vitex negundo*, Asian J. Pharma. Clin. Res. 2: 81–86.

Petchi, R.R., Vijaya, C., Parasuraman, S., Alagu Natchiappan and Devika, G.S. 2011. Anti-arthritic effect of ethanolic extract of leaves *Vitex negundo* Linn. (Verbenaceae) in male albino Wistar rats. Int. J. Res. Pharm. Sci. 2(2): 213–218.

Prajapati, D.S., Purohit, S.S., Sharma, A.K. and Kumar, T. 2004. A Handbook of Medicinal Plants. Agrobios India, Jodhpur.

Pushpalatha, E. and Muthukrishnan, J. 1995. Larvicidal activity of a new plant extracts against *Culex quinquefasciatus* and *Anopheles stephensi*. Ind. J. Malariol. 32(1): 14–23.

Rahuman, A., Bagavan, A., Kamaraj, C., Vadivelu, M., Zahir, A., Elango, G. and Pandiyan, G. 2009. Evaluation of indigenous plant extracts against larvae of *Culex quinquefasciatus* Say (Diptera: Culicidae). Parasitol. Res. 104: 637–643.

Rajendran, S. and Sriranjini, V. 2008. Plant products as fumigants for stored-product insect control. J. Stor. Prod. Res. 44: 126–135.

Rana, S. and Rana, K.K. 2014. Review on medicinal usefulness of *Vitex negundo* Linn. Open Access Lib. J. 1: 508.

Rastogi, T., Ghorpade, D.S., Deokate, U.A. and Khadabad, S.S. 2009. Studies on antimicrobial activity of *Boswellia serrata*, *Moringa oleifera* and *Vitex negundo*: A comparison. Res. J. Pharmacog. Phytochem. 1(1): 75–77.

Sahare, K.N. and Singh, V. 2013. Antifilarial activity of ethyl acetate extract of *Vitex negundo* leaves *in vitro*. Asian Pac. J. Trop. med. 689–692.

Sahare, K.N., Anandhraman, V., Meshram, V.G., Meshram, S.U., Reddy, M.V., Tumane, P.M. and Goswami, K. 2008. Anti-microfilarial activity of methanoloic extract of *Vitex negundo* and *Aegle marmelos* and their phytochemical analysis. Ind. J. Expt. Bio. 46: 128–131.

Sahayaraj, K. and Ravi, C. 2008. Preliminary phytochemistry of *Ipomea carnea* Jacq and *Vitex negundo* leaves. Int. J. Chem. Soc. 6(1): 1–6.

Samy, R.P., Thwin, M.M., Gopalakrishnakone, P. and Ignacimuthu, S. 2008. Ethnobotanical survey of folk plants for the treatment of snakebites in Southern part of Tamil Nadu, India. J. Ethnopharm. 115: 302–312.

Sathiamoorthy, B., Gupta, P., Kumar, M., Chaturvedi, A.K., Shukla, P.K. and Maurya, R. 2007. New antifungal flavonoid glycoside from *Vitex negundo*. Bioorg. Med. Chem. Lett. 17(1): 239–42.

Sehgal, C.K., Taneja, S.C., Dhar, K.L. and Atal, C.K. 1982. 2'-p-Hydroxybenzoyl Mussaenosidic acid, a new Iridoid Glucoside from *Vitex negundo*. Phytochem. 21: 363–366.

Sehgal, C.K., Taneja, S.C., Dhar, K.L. and Atal, C.K. 1983. 6'-p-Hydroxybenzoyl Mussaenosidic acid, an Iridoid Glucoside from *Vitex negundo*. Phytochem. 22: 1036–1038.

Shah, G.M. and Khan, M.A. 2006. Common medicinal folk recipes of Siran Valley, Mansehra, Pakistan. Ethnobotanical Leaflets 10: 49–62.

Shaukat, M., Huma, S., Umbreen, F., Arfa, K. and Ghazala, H.R. 2009. Antifungal activities of *Vitex negundo* L. Pak. J. Bot. 41(4): 1941–1943.

Singh, V., Dayal, R. and Bartley, J. 1999. Volatile constituents of *Vitex negundo* leaves. Planta Medica. 65: 580.

Srinivas, K.K., Rao, S.S., Rao, M.E.B. and Raju, M.B.V. 2001. Chemical constituents of the roots of *Vitex negundo*. Ind. J. Pharm. Sci. 63: 422–424.

Surveswaran, S., Cai, Y., Corke, H. and Sun, M. 2007. Systematic evaluation of natural phenolic antioxidants from 133 Indian medicinal plants. Food Chem. 102: 938–953.

Tandon, V.R. and Gupta, R.K. 2005. An experimental evaluation of anticonvulsant activity of *Vitex negundo*. Ind. J. Physiol. Pharmacol. 49: 199–205.

Tandon, V.R. and Gupta, R.K. 2006. *Vitex negundo* Linn. (VN) leaf extract as an adjuvant therapy to standard anti-inflammatory drugs. Ind. J. Med. Res. 124(4): 447–50.

Tapsell, L.C., Hemphill, I., Cobiac, L., Patch, C.S., Sullivan, D.R., Fenech, M., Enrys, S., Keogh, J.B., Clifton, P.M., Williams, P.G., Fazio, V.A. and Inge, K.E. 2006. Health benefits of herbs and spices: the past, the present, the future. Med. J. Australia 185: S4–S24.

Telang, R.S., Chatterjee, S. and Varshneya, C. 1999. Study on analgesic and anti-inflammatory activities of *Vitex negundo* Linn. Ind. J. Pharmacol. 31: 363–366.

Triggiani, V., Resta, F., Guastamacchia, E., Sabba, C., Licchelli, B. and Ghiyasaldin, S. 2006. Role of antioxidants, essential fatty acids, carnitine, vitamins, phytochemicals and trace elements in the treatment of diabetes mellitus and its chronic complications. Endocr. Metab. Immune Disord. Drug Targets 6: 77–93.

Tripathi, Y.B., Tiwari, O.P., Nagwani, S. and Mishra, B. 2009. Pharmacokinetic-interaction of *Vitex negundo* Linn. & paracetamol. Ind. J. Med. Res. 130: 479.

Uniyal, S., Singh, K., Jamwal, P. and Lal, B. 2006. Traditional use of medicinal plants among the tribal communities of Chhota Bhangal, Western Himalaya. J. Ethnobio. Ethnomed. 2: 14–21.

Vasudeva, N., Sharma, S.K. and Mor, A. 2012. Spermicidal and post-coital antifertility activity of *Vitex negundo* stem bark. J. Herbs Spices Med. Plants 18(4): 287–303.

Vishnoi, S.P., Shoeb, A., Kapil, R.S. and Popli, S.P. 1983. A furanoeremophilane from *Vitex negundo*. Phytochem. 22: 597–598.

Vishwanathan, A.S. and Basavaraju, R. 2010. A Review on *Vitex negundo* L.—A medicinally important plant. EJBS. 3(1): 30–42.

Warrier, P.K. and Nambiar, V.P.K. 2002. Indian medicinal plants: A compendium of 500 species. Chennai: Orient Longman Private limited 5: 387.

Zabihullah, Q., Rashid, A. and Akhtar, N. 2006. Ethnobotanical survey of Kot Manzary Baba valley, Malakand Agency, Pakistan. Pak. J. Pla. Sci. 12: 115–121.

Zheng, C.J., Tang, W.Z., Huang, B.K., Han, T., Zhang, Q.Y., Zhang, H. and Qin, L.P. 2009. Bioactivity-guided fractionation for analgesic properties and constituents of *Vitex negundo* L. seeds. Phytomed. 16: 560–567.

Zheng, C.J., Zhao, X.X., Ai, H.W., Lin, B., Han, T., Jiang, Y.P., Xing, X. and Qin, L.P. 2014. Therapeutic effects of standardized *Vitex negundo* seeds extract on complete Freund's adjuvant induced arthritis in rats. Phytomed. 21(6): 838–46.

Index

Milton Keynes UK
Ingram Content Group UK Ltd.
UKHW050455071024
449327UK00015B/388